The Quest for Food

Harald Brüssow

The Quest for Food

A Natural History of Eating

 Springer

Dr. Harald Brüssow
Chemin de la Chaumény 13
CH-1814 La Tour de Peilz
Switzerland
e-mail: haraldbruessow@yahoo.com
harald.bruessow@rdls.nestle.com

Library of Congress Control Number: 2006932833

ISBN-10: 0-387-30334-0   e-ISBN-10: 0-387-45461-6
ISBN-13: 978-0387-30334-5   e-ISBN-13: 978-0-387-45461-0

Printed on acid-free paper.

9 8 7 6 5 4 3 2

springer.com

To my mother Lydia, who gave me her curiosity for knowledge,
To my father Ernst, who added his interest for books,
To my wife Margret and my daughter Friederike, who shared
my fascination for biology.

"The wheel of Life" bronze from K. Franke (art piece in the possession of the author)

Preface

When you go into a scientific library or look through the catalogues of scientific publishers, you will quickly find books from food scientists, food technologists, food chemists, food microbiologists, and food toxicologists. Agronomists, nutritionists, and physicians have written on food, and last but not least cooks. What I missed was a book on food written from the perspective of a biologist. When Susan Safren, the food science editor from Springer Science + Business Media, LLC, invited me to write a book, I decided that I would write this book on food biology.

What I had in mind was a survey on eating through space and time in a very fundamental way, but not in the format of a systematic textbook. The present book is more of an ordered collection of scientific essays.

Contents. In *Chapter 1*, I start with a prehistoric Venus to explore the relationship between sex and food. Then I use another lady—Europe—to investigate the strong links between food and culture. I then ask what is eating in a very basic but simple physicochemical sense. In *Chapters 2 and 3*, I embark on a biochemistry-oriented travel following the path of a food molecule through the central carbon pathway until it is decomposed into CO_2 and H_2O and a lot of ATP. My account does not intend to teach biochemistry, but to use recent research articles from major scientific journals to look behind food biochemistry. In *Chapter 4*, we explore the evolution of eating systems over time starting with the primordial soup, going into the RNA world, and then into the fascinating eating world of cells. I follow here the historical time line and you should not be too surprised that most of these chapters is dedicated to the prokaryotic cells and its nutritional biochemistry. Don't blame me for a microbiological bias. For the larger part of the biological evolution on Earth, the living world was represented mostly, if not exclusively, by microbes. In *Chapter 5*, I give "higher" organisms their full rights. I selected animal-oriented research papers under an ecology perspective. The actors come first from the ocean, then its borders. To unite plants and animals and to put land-based biology on center stage, I choose herbivory as a read thread for the second part of this chapter.

In *Chapter 6*, we investigate food stories from a behavioral viewpoint, first with animals and then with humans, where our march through time reaches from animals to early hominids into human history, even politics. If we would end here, we would miss a major point, namely that human eating stories cannot be seen in isolation. To dismantle again our anthropocentric view of the world, *Chapter 7* will show us as food for many predators. Microbes as the invisible rulers of the world make again their strong appearance on the scene. I end my chapter with an outlook on a few selected chapters from agronomy and the problems to feed a growing world population with the help of science and technology. It is also a story about pessimism and optimism in life.

Scope. I am deliberately speaking of eating systems because my definition of eating is very inclusive. It also covers gases (CO_2, O_2, H_2, and CH_4 take central places in the arguments of this book), electrons, and protons. Prokaryotic systems, "eating" very strange compounds and using exotic energy sources, are prominently treated. You will see another aspect in this book on food: The arena where the story is unfolding is planet Earth. Several of the chapters stress the link between the biosphere and the atmosphere and touch questions of climate change.

Writing Style and Readership. The reader will find two types of chapters. One type tells a linear story in an easy essay style. These chapters should be accessible to any science student, perhaps even an educated layman or laywoman. The other type organizes a story around recent research articles in major scientific journals. Here you might get the impression of a somewhat heterogeneous patchwork when I try to integrate a couple of research papers into a story. As the level of arguments in research journals is frequently very complex, the second type of chapters is more difficult to read. I tried to work out the essence of the papers without simplifying them too much. These parts are better suited to advanced students.

When it comes to the technique of writing, I tried to animate the flow of arguments with historical remarks, anecdotes, or personal reflections, which are not conventional parts of a scientific textbook, but are essential for a scientific essay since they are necessary to provoke thought. To avoid false scientific claims for my personal opinions, these passages are printed in italics. As these passages often interrupt the flow of the main arguments, I have chosen frequent subheadings to structure a chapter.

Reading Recommendation. Some chapters on bioenergetics, photosynthesis, and bacterial metabolism in Chapters 2 and 3 require some background knowledge. If you need a recall of your biochemistry knowledge, you will find the required backup information with a mouse click on the virtual bookshelf of the National Center for Biotechnological Information. Please go to www.ncbi.nih.gov/ and on the opening page you find on the first row of the header "Books." Clicking on it, you can search numerous science books or you can choose a specific one. Personally I recommend *Biochemistry* by Berg or *Molecular Biology of the Cell* by Alberts when you need backups for the present book. If you need help for more classical subjects of biology, you will find a first orientation on the Internet with Wikipedia. Useful web sites are provided at the end of the book after the reference section. The second part of the book is easier to follow and may in part even please a larger public. Even when I have written the book as a logical flow of arguments, I do not think that it must be read in a linear fashion. In fact I strongly encourage you to selective reading. I was also a selective author. Start with the chapter where the heading arouses your interest.

Illustrations. Most figures in this book go back to illustrations from landmark science publications from the late nineteenth century Germany. I used them for several reasons. One is esthetical—I like these old figures because they are so artful and I owe my thanks to the Brockhaus Verlag, who allowed me to present these figures from the "Brehm" and the "Kerner," multivolume treatise of zoology and botany, respectively, and "Meyers Konservationslexikon" to an international readership. The "Brehm" illustrations were not only used by Charles Darwin in his book *The Descent of Man*, but became a standard in the educated German-speaking household. Where the Brehm fails in the world of the microbes, my colleagues from our microscopy group (M. L. Dillmann, M. Rouvet) helped me with modern pictures. Apart from this historical aspect, the old figures were also meant as a contrast to the quoted research papers, which were mostly published over the last few years to provide a topical review. However, this mania for actuality neglects the fact that science also relies on tradition of knowledge. We can lose insight when we are not looking back. The idea for illustrating the Dramatis Personae, the actors of the play of life, came from a reviewer (Ted Farmer, Uni Lausanne), who suggested that classical zoology and botany might not be that present in the modern generation of molecularly trained biologists. The figures are also meant as moments of relaxation and fun when reading through a sometimes demanding text. I hope that you will enjoy this survey of eating through space and time. Personally I learned a lot when writing this book.

6 June, 2006   Harald Brüssow

Acknowledgments

I have to thank many people who helped me to write this book. First and foremost my wife Margret—without her support at home, her helpful comments in the early writing phase, and her willingness to forgo her husband for many evenings and weekends, the book could never have been written. For critical comments on the early chapters and help with the references, I thank my yearlong collaborator Anne Bruttin. I sent out chapters of the book manuscript to a number of scientists at different institutions for critical reading. I was overwhelmed by the time investment and dedication of these busy scientists. I received indepth comments for larger passages of the book from Uwe Sauer (ETH Zurich), Jan Roelof van der Meer (Uni Lausanne), Edward Farmer (Uni Lausanne), and Michel Goldschmidt-Clermont (Uni Geneva). For smaller sections of the book, I received helpful comments from Martin Loessner (ETH Zurich), Barbara Stecher (ETH Zurich), Otto Hagenbüchle (ISREC Epalinges), and Laurent Keller (Uni Lausanne). I also have to thank my colleagues Nicolas Page, Marcel Juillerat, Heribert Watzke, and Bruce German, all working at the Nestlé Research Center in Lausanne, for discussion and helpful comments. When I lacked Swiss colleagues for critical reading, I sent several passages of the book to foreign scientists. Here I owe my thanks for their critical reading to Stephan Beck (Sanger Center Cambridge), Roger Glass (CDC Atlanta), Todd Klaenhammer (UNC Raleigh), and Robert Haselkorn (Uni Chicago).

As the book deals with many subdisciplines of biology, I had to rely on the expert knowledge of textbook authors. I want to quote here my main sources of insight, and I will directly recommend these books for the interested reader who wants to know more about the subject. In zoology this was R. Brusca's *Invertebrates* (Sinauer 2002); in plant biology B. Buchanan's *Biochemistry & Molecular Biology of Plants* (Am. Soc. Plant Physiol. 2000); in Microbiology J. Lengeler's *Biology of the Prokaryotes* (Thieme 1999) and L. Prescott's *Microbiology* (McGraw Hill 2002); in biochemistry D. Nelson's *Principles of Biochemistry* (Freeman 2005) and L. Stryer's *Biochemistry* (Freeman 1995); in genetics M. Jobling's *Human Evolutionary Genetics* (Garland 2004); in evolution M. Ridley's *Evolution* (Blackwell 2004), S. Gould's *The Book of Life* (Norton 2001), and S. Jones' *Human Evolution* (Cambridge 1992); in ecology M. Begon's

Ecology (Blackwell 1996); and in ethology C. Barnard's *Animal Behaviour* (Pearson 2004). The legends of the old figures and the species names were checked with the *Encyclopedia Britannica* for actuality.

The vast majority of the figures come from three publications that appeared in the late 19th and early 20th century in Germany, namely 1) Brehms Tierleben-Allgemeine Kunde des Tierreiches in 13 Bänden. Vierte Auflage, Herausgeber: Otto zur Strassen. Bibliographisches Institut Leipzig und Wien 1911–1918, 2) Pflanzenleben von A. Kerner von Marilaun, in drei Bänden, dritte Auflage, neubearbeitet von A. Hansen, Bibliographisches Institut Leipzig und Wien 1913, 3) Meyers Konversations-Lexikon, ein Nachschlagwerk des allgemeinen Wissens, fünfte Auflage, in 17 Bänden, Bibliographisches Institut Leipzig und Wien 1894–1897, in the given order with respect to numbers. I thank the Brockhaus Verlag for permission to reproduce these figures.

Five line drawings around photosynthesis come from J. Lengeler's "Biology of the Prokaryotes". I thank the Thieme Verlag for permission to reproduce them.

The electron microscopy pictures come from our microscopy group at the Nestlé Research Center, some of them were published and I owe my thanks to the journals (Journal of Bacteriology, Microbiology and Molecular Biology Reviews from ASM and Current Opinion in Microbiology from Elsevier) for reproducing them here again. The line drawing of the Willendorf Venus is from my son Felix Brüssow.

Finally, I have been working for a quarter century in the research department of a major food industry (Nestlé in Switzerland). However, the book was written in my free time, and I express here my personal ideas and no company opinions. I cannot and I do not want to claim the authority of a company for my personal views. This means that the responsibility for the factual mistakes and misunderstandings and perhaps sometimes controversial interpretations is entirely mine.

6 June, 2006, *La Tour de Peilz, Switzerland*   Harald Brüssow

Contents

1
A Nutritional *Conditio Humana*

A Few Glimpses on Biological Anthropology

An Early Venus and Breastfeeding: The Quest for Food and Sex as Driving Forces in Biology

Basic Forces in Biology

The genetically defined Darwinian evolution theory is for many scientists the great unifying theory of biology. In contrast to theories in physics, its outline is not formulated in the language of mathematics, but in simple semantic headlines like "the survival of the fittest," "eat or be eaten," or more recently "the selfish gene." While there is undeniably truth in these captions, they need illustrations not to remain just slogans. If you could ask animals about their daily preoccupations, they would name food, sex, and avoidance of predation. Due to a long cultural process, our interests might be more varied. However, the quest for food remained one of the underlying determinants of human life and is probably only dominated by our interest in sex. Both interests are eminent biological forces. Actually, both motives are intertwined as they both deal with survival, the only goal in biological evolution. We need food for our personal survival, and we need sex for the survival of our genes in our children. One might even argue that the quest for food is the primordial drive in biological evolution since sex was invented relatively late in evolution. Prokaryotes (organisms without a cell nucleus—eubacteria and archaea) propagate essentially without sex, and they were over long periods the only players in the evolutionary game.

Humans understood the complex relationship between these two basic biological forces long before scientists thought on these problems. I will illustrate this by a piece of prehistoric art: The Venus of Willendorf (Austria), sculptured more than 20,000 years before the present (Figure 1.1). This is perhaps the earliest, still-preserved sex symbol of human history. What you see is a naked woman with her sex exposed. Even more eye-catching is the breast of the woman and her sheer proportions (the size of the original sculpture, enshrined in a Vienna museum, is actually small). Her hair is treated with much care. Apparently, this woman

FIGURE 1.1. *Homo sapiens*, represented by the Venus of Willendorf, a 20,000-year-old idol of womanhood from Austria. The sculpture illustrates the intersection of the quest for sex and food.

was perceived as beautiful, a fertility symbol, an idealization of womanhood or perhaps even the Great Mother idol before the father Gods took over. The biological importance of nakedness and the exposed sex needs no comment. But the sculpture promises you more than sex; it signals the survival of your genes by good food. This is an important consideration since finding a partner willing to mix his or her genes with yours and create a new being sharing half of your genes is only part of your genetic success. Therefore, a modern biologist and the Ice Age hunter perceive biological underpinnings in the other sexual attributes of this early Eve: The broad pelvis of this Venus means likely survival of mother and child during delivery. Here humans have a serious problem. The tremendous evolution of the human brain already in utero was not accompanied by a corresponding evolution of the human birth channel. Possibly, this was the price we had to pay for walking upright. Biologists speak here of a trade-off between the need to walk upright necessitating a slender pelvis, and the need to have a sufficiently large opening in the middle of the pelvis for giving birth. Apparently, evolution cannot easily serve two masters. In apes there is a comfortable difference between the inner diameter of the pelvis and the head diameter of the baby, and giving birth is literally a child's play. In contrast, delivery in humans is a risky exercise, and the head has to pursue a complicated movement to get out. Before the development of the caesarian, men had good reasons to be interested in the hips of the women with whom they wanted to have sex.

Breastfeeding

The enormous breast of the Venus conveys the same message of survival to the biologist and the early hunter: Big breast might be a sign of good health, which increased chances of survival for the child in a time when no replacement for breast milk was around. The dominance of the breast in the Venus of Willendorf demonstrates the importance of the food argument for the early hunters during the Ice Age. Many contemporary men are still hunters in that respect because in psychological tests many men look first for the breast in a casual encounter with a nice woman. Then they look for the pelvis and surrounding, or they do it in the opposite order. *The advantage of being a biologist is that one need not be ashamed of instincts because you know why it is so and why it is good the way it is.* In fact, we admire in women two of the major inventions of our family characteristics. For zoologists we are Chordata, then vertebrates (both traits refer more or less to our backbone), but then we are vertebrates with jaws, our final family attribution is Mammalia (the breast-feeders), Theria (we are viviparous, but this is found in other animals, too), and our last evolutionary invention refers to the placenta (the embryo grows in the womb of the mother, where it is nourished by the mother via a special feeding organ). If you look at our zoological attributions, you see easily three major evolutionary inventions that deal directly with the quest for food.

Competition between the quest for food and sex leads to fascinating interplays not only in this female fertility symbol. Throughout almost the entire period of human evolution, infant survival has critically depended on breastfeeding.

Human milk contains about 7 g of the milk sugar lactose per 100 g of milk. Therefore, lactose accounts for about 30% of the caloric value of whole milk. Commonly, a period of exclusive breastfeeding is followed by the introduction of solid food, partial breastfeeding, and weaning. In early human tribes, the period of breastfeeding was probably much longer than today since there were no readily available alternative food sources. Now let's take a deeper look into breast milk.

Apart from being an important sexual display organ, the breast fulfills a major secretory function and thus becomes the most important source of nutrition for the mammalian young after delivery. The fundamental functional unit is the alveolus, which is surrounded by contractile myoepithelial cells and adipose tissue. If you follow it from the outside, you have first the nipple surrounded by the areola. This is the area of oily secretion that lubricates the suckling process and the place where many fine jets of milk emerge. The milk comes from the underlying lactiferous ducts, which widen to an ampulla just under the areola as a kind of milk reservoir. The ducts split then in an elaborate duct system leading to the secretory lobules. They contain the functional units of the lactating breast, the alveolus. It has a typical glandular structure. The central cavity contains the secreted milk and is lined by a secretory epithelium built by the alveolar cells. They are underlied by a basal lamina in which you find myoepithelial cells and, further down, capillaries. The alveolar cell shows a rich endoplasmic reticulum, indicating substantial protein synthesis capacity, and an extensive Golgi apparatus, from which many types of vesicles emerge. Some vesicles contain proteins (final concentration in the milk: 0.9 g/100 ml; lactalbumin, casein), others contain sugars (milk content of lactose: 7.1 g/100 ml) and salts (milk Ca^{2+}: 33 mg/100 ml), and intracellular milk lipid droplets (only small chain fatty acids are synthesized in the breast) acquire a membrane envelope when they pinch off the cell surface (milk fat: 4.5 g/100 ml). Not all milk proteins are synthesized in the alveolar cell; for example, immunoglobulins come from the circulation and experience a transcellular endocytosis/exocytosis transport. Water and salts take a paracellular or a transcellular transport way.

Lactose Synthesis

Lactose is synthesized in the alveolar cell by lactose synthetase. This enzyme consists of two subunits. Subunit A is a galactosyl transferase. Its normal function is heteroglycan synthesis: It couples galactose to N-acetyl-glucosamine. Its affinity for glucose is too low (i.e., its K_m is high) to allow lactose synthesis. To achieve this, it must associate with subunit B (alias α-lactalbumin), and then subunit A prefers glucose as substrate. During gestation, the peptide hormone prolactin stimulates the growth of the alveoli and induces the synthesis of subunit A. Simultaneously the steroid hormone progesterone suppresses the synthesis of subunit B. Immediately after delivery, the progesterone levels fall and free the way for the synthesis of subunit B and hence lactose. Lactose synthesis is a costly exercise. The recipe is the following: You take galactose from the diet and activate it by the enzyme galactokinase. You split ATP, your first investment, to create galactose 1-phosphate. Then you activate the galactose 1-phosphate by coupling it to UTP, the preferred activated carrier of sugar moieties for biosynthetic pathways. UTP is a nucleotide, actually the same as that used in RNA synthesis. The enzyme, which does this job, uses UDP-glucose as a cosubstrate mediating in an energy-free step a hexose sugar exchange at the UDP moiety. If I say energy neutral, you have to realize that the initial synthesis of UDP led to the splitting of an energy-rich pyrophosphate bond. This cost must also be counted in your balance sheet. In the final act, lactose synthetase transfers the galactose part of UDP-galactose on glucose leading to the disaccharide lactose.

Why did nature choose lactose for this job in mammals? I know of only a few marine mammals that do not contain lactose in their milk (e.g., the *Sirenia*), while dolphins and whales have lactose in their much more lipid-rich milk (here the young needs quickly a fat layer for thermal insulation). Why produce a molecule that you have to synthesize with extra metabolic input (the synthesis of enzymes is also costly)? In addition, the human gut cannot directly absorb disaccharides. Lactose must be digested by a tailor-made enzyme, lactase. It digests lactose to the constituting monosaccharides, which can be absorbed. In fact, despite its composition of two ubiquitous monosaccharides, lactose is in nature a very exotic compound found only in low amounts in forsythia flowers and some tropical shrubs. This fact is not even well known to biologists because they all hear during their university studies about the lactose operon, which is one of the great paradigms of molecular biology. For the food scientist, another group of bacteria, the lactic acid bacteria, are a workhorse of industrial microbiology because they can transform lactose into lactic acid. They are the starter organisms for a variety of industrial milk and food fermentation processes. However, lactose is an evolutionarily new substrate even for these bacteria. This is still evident from the careless way in which some of the most popular dairy bacteria deal with their lactose-digesting enzyme, the β-galactosidase. They encode it on plasmids, mobile extrachromosomal DNA elements that are easily gained, but also easily lost. Stable genes are better integrated into the genome. The ancestors of *Escherichia coli* and lactic acid bacteria have not seen lactose before the arrival of mammals on the planet in the late Triassic, around 210 My (million years years) ago. Only then *E. coli* discovered the mammalian and somewhat later the avian gut as an ecological niche; actually the split of *E. coli* from *Salmonella typhimurium* by molecular means is also dated to about 200 My ago. The association of lactic acid bacteria with milk is even much younger and dates probably to the domestication of animals in the Neolithic Revolution, some 10,000 years ago. Only from that time on, milk became an ecological niche. Before this event, milk was a fast food, taking the shortest way from the producer (breast, udder) to the end consumer (baby, calf).

To come back to our question, why invent a new molecule that needs, in contrast to glucose, the central sugar fuel in biology, extra energy and proteins for its synthesis, and new enzymes for its absorption in the gut of the mammalian young?

Intermezzo

Here comes now the importance of philosophy in biology. If you accept the idea of a sometimes Byzantine decoration in biology (a "l'art pour l'art" argument), you shrug your shoulders and go to more urgent questions. In fact, this decoration aspect of nature is not farfetched. This is the overwhelming first impression a biologist and a layman get when they look at nature. However, Darwin has told us differently and taught us to see all organisms as the product of natural selection in the struggle for life, as the title of his 1859 book reads. On the basis of this theory, we can anticipate a purpose behind many biological phenomena. In fact, a lot of what appears as pure decoration in biology reflects hard selection pressures. However, not all biological phenomena are useful from an engineering point of view. Darwin has clearly seen this in his 1871 book *The Descent of Man and Selection in Relation to Sex*. The peacock's tail is the classical illustration for an apparently nonadaptive structure, which hinders more than it helps, but the hens have chosen this as a marker for good mate selection. Males that were interested in offspring were obliged to follow this trend, sometimes even into extinction as we will see with the Irish elk in a later chapter. In biochemistry biologists have learned another lesson when they first encountered what was then called futile cycle, biochemical pathways that were seemingly only wasting metabolic energy. However, it turned out that much of this seemingly wasteful cycling was the price for finer regulation of metabolic pathways.

Why Lactose?

Therefore, one might suspect a regulatory function behind using lactose as a major energy carrier in lactation. In the following, I will explore this hypothesis, which nicely serves as an illustration for the intimate links between food and sex, two major forces in biology. My working hypothesis is that lactose was designed by evolution as a compromise signal to serve both fundamental instincts. I do not pretend that this is the correct explanation. We probably do not know enough about this process to already have an overview of the entire puzzle, with only a few pieces lacking. However, hypothesis building is an essential activity in scientific research. To get from the level of idle armchair speculation to a useful working hypothesis, the biologist has to design an experiment that will verify or falsify the predictions of the hypothesis. In biology the experiment is the ultimate arbiter on hypotheses; we have not yet reached the level of understanding in biology that would allow a theoretical biology, as it is the case in physics. So for the pleasure of arguing let's follow the hypothesis, but the following paragraphs will not provide you an answer, at best only a meaningful question. Yet, this is the fun of science as a game of unending questions. With progress of time and knowledge, we succeed in asking better questions. The technologist and the physician are definitively interested in the answers provided by biological research to develop new tools, procedures, or medical treatments, while the biological researcher is already looking for the next question after the last experimental answer.

Lactase

Now I will provide some background data underlying the hypothesis. Lactase is a tightly regulated enzyme. Lactase-phlorizin hydrolase (LPH), the way it is called in full by biochemists, is an intestinal microvillus membrane enzyme. It is a large protein with a length of 1,900 amino acids, consisting of four homologous parts suggesting its origin from two successive gene duplication events. It is posttranslationally modified by heavy glycosylation and cleavage into two parts. The C-terminal part contains the two active sites, while the N-terminal half functions in the correct folding. The C-terminal parts form a homodimer and insert into the top of the microvilli membranes. The *LCT* gene encoding LPH is located on chromosome 2q21, covers more than 50 kb, and shows 17 exons resulting in a 6-kb-long mRNA. It is relatively heavy with its 220-kDa weight. It is not only a critical gut enzyme for neonatal nutrition but also shows a very distinct pattern of developmental expression that differs from that of other digestive enzymes (for a recent review and literature references, see Grand et al. 2003). In the second half of gestation, the intestinal lactase activity begins to increase in the fetus. High levels are reached in the third trimester and remain high in the first year of life. In the second year of life, the lactase level starts to decline. In most human populations, children reach low levels of lactase, characteristic for adults, at about 5 years of age. All the investigated mammals show a conspicuous decrease in intestinal lactase activity at the time of weaning.

Animal experiments showed that the control of the lactase gene expression is at the level of transcription (Lee et al. 2002). A 1.2-kb-long DNA sequence upstream of the rat lactase gene is sufficient to confer in transgenic mice a spatiotemporal pattern of gene expression to a fluorescent reporter gene that mirrors that of the lactase gene. The reporter gene was not only expressed in the appropriate locations of the small intestine but was also switched on with birth and switched off with weaning. Apparently, this small piece of rat DNA contains a navigator and a timer signal for lactase expression. Withholding lactose in the feeding of animals had no effect on the lactase expression level. Also under this condition, the rat transgene in the mouse context was switched off at about 3–4 weeks.

Interpretations

Life, reproduction, and weaning times differ greatly between mammals and in correlation with this the lactase expression seems to be regulated. In humans, lactose malabsorption is very rare in infants, but in the second year of life children start to show this condition. For example, at the age of 18 months 60% of Bangladeshi children show lactose malabsorption; this prevalence is greater than 80% after the third year of life. *There is a simple argument for the understanding of this observation. At about 18 months, children are weaned. Since lactose is not encountered in food (except in dairy products), it would be a waste of energy to maintain the synthesis of an enzyme that is no longer needed. Suppressing lactase synthesis could thus be a form of metabolic economy. This is exactly what bacteria are doing; when the lactose supply dwindles, E. coli stops producing the enzyme β-galactosidase. However, this argument might be too simple for a complex multicellular organism such as a mammal where many futile biochemical cycles are maintained for regulatory purposes. In fact, adding lactose after the weaning did not induce resumed lactase expression.*

One might, therefore, suspect an evolutionary reason to suppress lactase synthesis. An interesting alternative explanation is at hand. Perhaps, breast-feeding has another function besides the nutrition of the baby—it suppresses the ovary function. And here we come to an interesting interplay between eating and sex. The pathway is somewhat complicated, but relatively efficient, although not a recommended contraception method. The stimulus of the suckling travels from the breast to the spinal cord via sensory nerves; neurons from the spinal cord then inhibit neurons in the arcuate nucleus and the preoptic area in the hypothalamus. This inhibition decreases the production of the gonadotropin-releasing hormone by these centers. These released factors normally induce the secretion of the follicle-stimulating hormone and luteinizing hormone from the anterior pituitary. The end result of this neuroendocrine circuit is that breast-feeding delays ovulation and the normal menstrual cycle. Suckling intensity and frequency determine the duration of anovulation and amenorrhoea, i.e., inter-rupted fertility, in well-nourished women. There is a lot of biological sense in this regulatory system. In hunter-gatherer societies, the survival chances of mother and child are not increased when the mother would have to provide extra calories both for the feeding of the baby via breastfeeding and for the growth of the new embryo in the uterus via its blood supply to the placenta. Concentrating on one child before conceiving the next is a better strategy. However, if breastfeeding is maintained for an extended period, the conception of the next baby is delayed. The spacing between pregnancies has to be optimized to obtain the maximal number of surviving offspring. This process is most likely controlled by a strong selection pressure. Too long anovulatory periods means less children and this is to the detriment of the population in times of high childhood mortality, which probably was the case during early human history. The extinction of the tribe might be the consequence. From this evolutionary reasoning, one could argue that the suppression of lactase synthesis in the intestine of the child is the trigger for the onset of a next round of ovulation and the next pregnancy. The timing of 18–24 months for the disappearance of the lactase corresponds possibly to what natural selection has calculated as optimal spacing of pregnancy in humans to achieve maximal reproductive success. It also fits to the spacing which most European parents find ideal even in face of dwindling absolute numbers of pregnancies. (Despite its demographic impact on European societies, I have not read about convincing evolutionary reasons for this phenomenon.)

The mechanism linking lactose to a new pregnancy could be very simple: without lactase, lactose is no longer split and thus not absorbed from the intestine. If the baby continues to consume milk, the undigested lactose reaches the colon, where it is digested by gut bacteria. However, they ferment lactose into gas as end product. The accumulation of gas leads to bloating of the gut, discomfort, and flatulence. In addition, the osmotic drag of the nonabsorbed lactose in the gut leads to water inflow and diarrhea-like symptoms. Breastfeeding becomes quickly uncomfortable for the baby and the mother will start to search for a weaning food. These are the very symptoms many adults from Asia experience after the ingestion of a quart of milk.

While this scenario is plausible and has some explicative power, it is not said that it is the correct explanation. Predictions of the hypothesis must be tested. One series of tests would be to check whether the average weaning time is correlated with the shutdown time of lactase synthesis. This would be correlation-type evidence. Another prediction could be that a transgenic mouse, which produces lactase from a constitutive promoter, has significantly prolonged intergestation periods. Such a result would suggest a causal relationship. Whatever the mechanism, we see here a general principle in biology that there are trade-offs between competing interests.

Beauty and Fitness

Returning to the Venus of Willendorf, you might find the biological interpretation of the figure with the food–sex conflict farfetched. You might not even find this Venus sexually attractive. There are good reasons to be skeptical. Who does guarantee the mother quality of this fertility symbol? The enormous breast might actually contain more adipose tissue than milk-producing alveolar tissue. The broad hips could suffer from the same deception, hiding enormous fat layers and a small pelvis. For the Ice Age hunters, fat might have had a different meaning than for us. Fat represented a good thermal insulator and fat was a good caloric reserve in times of food shortage. The rapid release of insulin after a high carbohydrate meal was perhaps a good thing in the early human evolution. It allowed the buildup of fat stores. In the words of a paper written 40 years ago: diabetes mellitus was a thrifty genotype rendered detrimental by progress (Neel 1962). When you can get plenty of food as part of the world population now, obesity becomes a problem and is now a risk factor for numerous diseases. Perhaps this was unknown in early human history, where overeating was probably not a widespread phenomenon. Later societies might have realized the health risks associated with obesity and the beauty ideal changed. Consequently, women idols became slimmer (there remained, however, an interesting controversy on this point between different cultures and different art periods, even between the advertised women ideal and the societal reality). Yet prominent breast and pelvis were kept high in esteem as can for example be seen in Mediterranean fertility goddesses from early Crete. This beauty ideal survived the next 4,000 years into our days where actresses are characterized by circumferences of their breast, waist, and hip. I once read the nice evolutionary interpretation that the large breast and basin and small waist circumferences (wasp-waist) are in fact a simple empirical algorithm for optimization of survival factors corrected for fatness as a confounding factor.

You might smile about these examples linking aesthetic ideals to our concern about food. But at the end, we are in zoological terms Mammalia (mammals) defined by breasts and breastfeeding. The obvious interest of males (and women) with the female breast is thus only a biological reverence to one of the two reproductive novelties introduced by these animals (the other one being in utero carriage of the embryo; I will speak about this later). Milk was actually even projected to the heaven. In Greek mythology, our galaxy, the "milky way," was created when Hera spilled her milk after pulling off Heracles from her breast. Interestingly, Heracles was not her son, but an extramarital son of her husband Zeus. Hera was able to breastfeed without having recently delivered a baby. I come back to Zeus in the next section.

Lady Europe's Liaison with a Bull: The Spread of Agriculture and Dairy Cultures

Mythos and Collective Memory

Personally, I believe that a lot of collective memory, which goes back to the early eating habits of the human race, can still be retrieved from mythology and fairy tales. The ancient Greeks had a lot of stories which tell you why a given region got its name. I found one story especially perplexing. My home continent Europe got its name from an Asian lady, Europe, who is identified in this myth as living in the Levant part of the Fertile Crescent. The Greek God Zeus abducts the Asian beauty disguised as a beautiful bull when mixing in the cattle herd belonging to the father of the bride. Zeus brings her to Crete, where she conceives Minos, a legendary king of the Minoic culture preceding the classical Greek culture. Minos had a son (extraconjugal I guess) called Minotaurus, a hybrid human dominated by a bull's head (small wonder—the grandfather was a bull-god). Young people from Athens were sacrificed to the Minotaurus living in a special labyrinth. I will dare to attract the lightning bolts of Zeus or classical philologists in telling you a biologist's interpretation of this story. The labyrinth is the city of Knossos, the people from geographical Greece (at that time neither Greeks nor Athens existed) were subjugated by a bull-revering culture coming from the Fertile Crescent, where Crete was an outpost ready to go into the Agean world (the culture of Mycenae was the continental complement to Crete). The Asian princess is riding the bull; she comes as a colonizer, not as a victim of a rape. Her sons are associated with an archeologically proven bull cult in Crete. We see here a conflict between a female and a male element. Zeus is the god of the new world; he is keen to associate himself with the new cult of the bull. However, he is constantly in fear of his wife Hera when he is looking for extramarital love affairs. In my opinion, we see here a transition period from an older matriarchic (Mesolithic?) into a newer patriarchic (Neolithic?) human community. In fact the ancient Greeks still kept the memory that several waves of invasion came into Greece. Why do I tell you this story? The Western civilization is still in the tradition of the myth I recalled. Do you know why the stock exchange in Frankfurt has a bull sculpture at its entrance? The answer is quite simple; they deal there with goods and money and the English word

FIGURE 1.2. *Bos primigenius*, the Ur or auerochs, which became extinct with the shooting of the last animal in 1627. The figure is probably the only authentic picture, discovered in 1827 in Augsburg/Germany, but later lost again.

"pecuniary" still recalls the Latin word for cattle or sheep (Figure 1.2). The first expression of wealth was the possession of domesticated bigger ruminants. For Americans there is an even clearer link, the grand-grand-child of Europe's marriage with the steer-god Zeus is the American Minotaurus, the "cowboy." The economic reality is that Northern and Central Europe and areas colonized by them (North America and Australia) became to a large extent dairy cultures where cows have a high position in folklore (e.g., Switzerland) or became modern myths in movies (cowboy films). There is even a fine detail in the cowboy: Unlike his grandmother Europe, he is not riding a bull, but a horse. This is an important detail because the cowboy thus symbolizes the merger of two distinct traditions, that of the horse-riding Kurgan pastoralists and that of the cattle-domesticating farmers from Anatolia, the two cultures discussed as the cradle of European agriculture. I will now try to bridge mythology and biology when retracing the origin of agriculture in Europe.

The Dairy Culture Comes with a Language

Cattle were domesticated in the Near East. The domestication of wheat can even be located more precisely in eastern Anatolia. There is now much archeological evidence as to how the invention of early farming traveled from east to west Anatolia and from there to Greece and the Balkans, then splitting into different waves before reaching perhaps 4,000 years later the western-most outposts of

Europe. Agriculture was such a success story in Neolithic Europe that it transformed the continent. *The Europeans apparently kept the memory of their origin. In reverence for the place of invention of this new food production technology, they adopted the name of this Asian princess for the newly colonized continent. Actually, these tribes brought a lot more than just the seeds of wheat and their cattle. The spread of agriculture brought with it the spread of the Indo-European languages. Traditionally, in university courses natural and historical sciences had few common points. Archeology is now a hybrid between both branches, and its success and popularity is a good sign that both branches will have more points in common in the future. In fact, there is only one reality in the world, and our categorization of it into different branches of knowledge reflects more the structure of our brain capacities than the structure of the world surrounding us.*

I will illustrate this new marriage with a fascinating research report (Gray and Atkinson 2003). Its title is—the authors will excuse me—barbarian for true philologists (*Language-Tree Divergence Times Support the Anatolian Theory of Indo-European Origin*). The report is written in the scientific language of our days, which is frequently not even accessible to fellow scientists of other scientific specializations. Some major scientific journals account for these language problems and have easier-to-read comments on major scientific breakthroughs. The comment for this article carries the more digestible title "Trees of Life and Language" (Searls 2003). In the later passages of this book, I have tried to retrace major biological discussions surrounding the quest-for-food issue (admittedly defined relatively generously) with recent research work. I have deliberately quoted as references both types of articles, the original research and the comments. You can easily distinguish them in the reference list by the complexity of the language. Now to the languages as an object of research: Every language has a core vocabulary of 100–200 words that represents most fundamental concepts expressed in any language, which are also relatively resistant to change. This set of words is called, after its discoverer, the Swadesh list. He looked then for corresponding words in related languages that derive from a common ancestor word, a "cognate." Formally, this concept corresponds to the notion of "orthologs" in the language of genes. As genes and languages are two different means of information transfer, there might even be a deeper reason to treat both with the same statistical methods (Solan et al. 2005). In fact, the Oxford zoologist and neo-Darwinist R. Dawkins treated in his book *The Selfish Gene*, genes and memes at a comparable level. Bioinformaticians pointed out that many techniques used in bioinformatics are grounded in linguistics (Searls 2002). The close formal relatedness is also revealed in book titles like W. Bodmer's accounts of the human genome project (*The Book of Man*). It is intuitively evident that languages that share more cognates are closely related and have thus split from a common ancestral language at a more recent time period. With these principles in mind, the authors of the above-mentioned article used new and powerful statistical methods to derive a tree for the Indo-European languages. This tree does not hold surprises for professional linguists, but it is worth retracing the major nodes in view of our princess Europe myth. At the base of the tree is Hittite, the language spoken by an empire located in Anatolia that dealt with the Egyptian pharaohs on an equal base. One of the next nodes leads to a branch from which in later times the Greek and Armenian developed. The next node points to the Albanian in the west and to Persian, Indian, and central Asian languages in the east. The next nodes are filling the European space in two different major branches. One branch covers the south, west, and north of Europe with the Romanic, Germanic, and Celtic languages, respectively. A relatively separate branch covers Eastern Europe with Slavic and Baltic languages. In fact, the geographical distribution fits roughly with the dispersal of the new agronomical technologies revealed by archeologists. However, the authors went a step further and also back. They used Swadesh's idea of glottochronology, which corresponds to the problem of a molecular clock in phylogenetic trees applied to languages. Swadesh determined the average rate of word substitutions over time for recent language histories and then he calculated backward. Gray and Atkinson did just this with their language tree and calculated the Anatolian node to about 8700 BP. This dating was robust against many statistical considerations that could blur the data. Taken at face value, this tree addresses a hotly discussed issue of the dispersal of the Indo-European languages. Was this dispersal horse-driven by pastoralists from the Kurgan area in what is today Ukraine or cattle-driven by farmers from Anatolia (Balter 2004). I will again deal with both hypotheses when reviewing domestication. The dating of the language tree fits better to the Anatolian scenario than the Kurgan scenario, which should yield a younger date of about 6,000 years ago.

Cultural Adaptation Versus Gene Flow

It is wrong to anticipate that all scientific disputes follow the rules of the impassionate observer *sine ira et studio*. In fact, there is a trend that the more heated the disputes are the more difficult it is to obtain experimental data. The Kurgan–Anatolian discussion is still harmless in comparison with a discussion as to how agriculture actually spread over Europe. One camp is the proponent of the demic diffusion model, also known as the wave of advance. In this model, made popular by the work of L. Cavalli-Sforza in the 1980s, spread is stimulated by population growth due to increased availability of food and local migratory activity. In the opposite model called the cultural diffusion model, the farmers did not move, it was the technology that moved by imitation. Hunter-gatherer groups copied the more successful food production methods as soon as they came in contact with the agriculturists. The speed of the spread is thus not in dispute between both camps. For example, as soon as the new technology reached Britain even populations living at the border of the sea abandoned fishing and mussel collection for the new agrarian food as demonstrated by carbon isotope distributions in ancient bones of "Proto-Britons." The evidence derives from the marked preference of the photosynthetic CO_2 fixation process for the lighter ^{12}C over the heavier ^{13}C isotope (Richards et al. 2003). In the purely demic model, the farmers replaced the hunter-gatherers genetically. This does not necessarily mean aggressive inter-action; they might have outnumbered the ancient population, which was anyway relatively sparse. The dispute goes about the question whether the geographical distribution of genes in the current European population allows to differentiate the genetic heritage of the Paleolithic *Homo sapiens* populations living as hunter-gatherers from the Neolithic *H. sapiens* populations coming as farmers. Overlaid on these old gene fluxes are genetic changes due to later invasion events (e.g., Cimberns, Huns, and Mongolians) and distortions caused by epidemic diseases (e.g., the AD 1348 pestilence, see likewise B. Tuchman's *The Distant Mirror*). Clarity was expected first from classical genetic data, which supported the demic model (Sokal et al. 1991). Mitochondrial data suggested that we Europeans are still largely Paleolithic in our genetic outfit. Y chromosome analyses were equivocal. A paper cosigned by a major demic protagonist (Cavalli-Sforza) came surprisingly to the acculturation conclusion (Semino et al. 2000). However, this didn't settle the dispute. The same dataset was reinvestigated by another group using more sophisticated statistical methodology, and they came to the support of the demic diffusion model, criticizing the previous report for statistical flaws (Chikhi et al. 2002). These contradictory interpretations of the same data seem to be diagnostic of the current discussion in that field that has perhaps generated more heat than light. It must, however, be admitted that this is an extremely demanding area of inquiry. The methods of biology are perhaps also very difficult to adapt to intrinsically historical events that are not repeatable in the laboratory. The entire field seems to be in a flux as recently demonstrated by a provocative publication that dated the ancestor of all living humans to a time period much more recent than the Neolithic Revolution (Rohde et al. 2004).

Linear Pottery Culture

Another recent paper addresses the question "culture-or-gene sweep" at its source, namely with DNA analysis from skeletal remains from an archaeolog-ically well-defined culture, the "Linearbandkeramiker." This term is not only German but directly an example of the composed nouns possible in German. But this is the least problem here, and we decompose the word into the more handy English term "linear pottery culture." This 7,500-year-old culture is widely distributed across Germany and Hungary and marks the origin of farming practices in temperate Europe. Within a mere 500 years, this culture has reached the Ukraine in the east and Paris in the west. Mitochondrial DNA could be amplified from 24 individuals. The striking result was that six of them belonged to a rare branch of mitochondrial DNA, which is today only found in 0.2% of the Europeans. The authors concluded that modern Europeans did not descend from these early farmers (Haak et al. 2005). In their scenario, the farming culture spread without the people originally carrying these ideas. This contradicts, of course, the conclusion of the Chikhi (2002) paper stating that less than 50% of the genes of modern Europeans can be traced to indigenous hunter-gatherers. The contradiction between both data sets is not so dramatic as it might appear. The former study investigated Y chromosome sequences, which trace only the male line. To be precise, the authors concluded that the early farmers left no maternal descendants in modern Europe—mitochondrial DNA is only inherited from the maternal line. The reason now became clear: Mitochondrial DNA from the sperm is actively broken down in the egg only hours after fertilization. It is thus possible that indigenous hunter-gatherer females intermarried with male early farmers in polygynic relationships, thereby diluting out the early farmer mitochondrial type. This conjecture would fit with the typical longhouse of these early farmers showing three or four hearths (Balter 2005). Truth might actually be a mixture from the concepts of both schools of thought, some cultural adaptation and some demic diffusion.

The Lactase Mutation

Despite this slippery slope, I will expand on a famous mutation that is possibly directly involved in the success of dairy cultures in Europe and many regions colonized by Europeans overseas. I resume here the discussion of the previous section. Many medical textbooks deal with lactose intolerance as a disease entity. In medical terminology, it is called adult hypolactasia. This "disease" is an interesting test case for several issues. One is the apparent "Eurocentrism" of the disease entity. Physicians have defined here the human wild type as a disease simply on the basis that in many European countries mutant subjects were in the majority. In fact, as we have seen in the last section, loosing the expression of lactase after weaning is the physiological norm in mammals and persistence of the lactase activity in adulthood is a clear mutation. Europeans are, in this respect, mutants. This case illustrates the dilemma for organizations like the WHO to define health and disease in objective forms. Any disease definition is arbitrary because the healthy gold reference standard is an abstract idea. If you look on it worldwide you see foci of lactase persistence into adulthood outside of Europe in four African populations, the Peul (Fulani), the Tuareg, the Tutsi, and the Beja, all are pastoralists, some with exclusive milk and no meat use (Fulani), all with a dairy culture. Nearby agricultural populations with whom they partially mixed (e.g., Hutus for the Tutsis) do not show this high lactase persistence level. In Europe you see a clear gradient with highest prevalence rates in classical dairy countries. When looking naively on the prevalence maps you might be tempted to anticipate a founder mutation that became more and more fixed with the spread of agriculture further west (France, UK) and north (Germany toward Netherlands, Denmark, and then Scandinavia). Here I will not stipulate a demic diffusion process with lactase persistence as a genetic marker. Living now for many years in Switzerland known for its coexistence of many languages and traditions at a crossroad of Europe, I have adopted their compromising character. The spread of agriculture over Europe probably did not come exclusively from Anatolia, but also from the Kurgan area. From what we know of human nature, demic diffusion and acculturation were probably both involved in this process. I will only raise the point that what today are classical dairy societies are also populations where a high percentage of lactase persistence is found. This suggests an adaptive advantage for this mutation.

No clear data are available to say what is the cause and what is the effect. The ability to digest lactose into adulthood is a clear nutritional advantage for a dairy culture: It allows adults to drink the milk they produced. More specific hypotheses have been formulated; one of the most plausible is the effect of milk consumption on improved calcium absorption (Flatz and Rotthauwe 1973), but this view was not shared by other researchers. Calcium can help prevent rickets, probably by reducing the breakdown of vitamin D in the liver as demonstrated in Nigerian children (Thacher et al. 1999). That the consumption of milk improves bone health is not farfetched and was actually suggested by a study with 103 healthy Italian subjects working in hospitals; 55 proved to be lactose malabsorbers in H_2 breath tests, while 29 of them experienced in addition intolerance symptoms like diarrhea, abdominal pain, bloating, and flatulence. Bone mineral density was significantly lower in the lactose intolerant subgroup when compared with the lactose tolerant subjects (Di Stefano et al. 2002). It remains to be explored whether the increased height achieved by some Northern European populations over the last decades and centuries is linked to these dairy food habits. Notable are the milk distribution programs in schools.

Gene-Culture Coevolution

Biologists are, with few exceptions, reluctant to reach out of the area of biological research. There are a few exceptions like Cavalli-Sforza, who started to study gene-culture coevolution in human history. In fact, humans inherit two types of information from their ancestors: one is strictly vertical (by genes), another is both vertical and horizontal ("oblique," by education and cultural transmission). The interaction of both factors makes what we call human culture. Do we have with the lactase persistence phenotype such a showcase for a gene dairy culture coevolution? Feldman and Cavalli-Sforza applied this approach to the question whether selection pressures following the adoption of dairy farming led to the spread of the lactase persistence. This analysis showed that the persistence allele will increase to the observed high frequency within about 300 generations (this is the time period that elapsed since the Neolithic Revolution) only if there is a strong cultural transmission of milk consumption (quoted from *Jobling: Human Evolutionary Genetics*). Only when both factors work together such a coevolution can be observed. If the lactase allele traveled on its own, unrealistically high selective advantages must be postulated to achieve the observed prominence in Scandinavian countries. While the lactase persistence gene is still a good candidate for such a "cultural" gene and a holistic human genetics, what is actually known about this mutation? The likely mutation of the lactase gene was studied in a large Finnish pedigree with lactase nonpersistence (Enattah et al. 2002). Sequencing of the coding regions and of the promoter of *LCT* showed no variations that correlated with the persistence phenotype. Haplotype mapping identified a C/T dimorphism located exactly 13,910 bp ahead of the *LCT* gene that completely associated with the phenotype. A second nucleotide G/A dimorphism was located 22,018 bp ahead of the gene, which correlated with most, but not all cases. The mutations are already in introns of the next upstream gene. The first mutation disrupts a consensus-binding site for the transcription factor AP-2, and this may explain the change in the developmental regulation of lactase transcription (Kuokkanen et al. 2003). US geneticists have now extended the evidence to a formal population genetics level and typed 100 single-nucleotide polymorphisms covering 3 Mb around the lactase gene. In northern European populations, two alleles associated with lactase persistence mark a common haplotype that extends over 1 Mb. By using new statistical tests, they demonstrated that this haplotype arose rapidly due to recent selection over the past 5,000–10,000 years. The signal of selection is one of the strongest seen for any human gene in the genome (Bersaglieri et al. 2004). Also the Finnish researchers have now extended their studies to 37 populations on four continents. They searched for those populations having the greatest number of DNA sequence variability around the lactase gene mutations, arguing that these are those in whom the lactase persistence arose first. They identified Ural farmers as the likely source for the spread of this mutation. The trait developed sometime in the period 4,800 to 6,600 years ago—clearly a point for the camp of the Kurgan hypothesis and also one for the demic diffusion (Kaiser 2004, in a recent congress report). Veterinary geneticists have investigated the genes encoding the six most important milk proteins in 70 European cattle breeds and found a striking geographical coincidence between high diversity of these genes and the locations of European Neolithic cattle farming sites and present-day lactose tolerance in Europeans. The plot is thus thickening for one of the strongest cases of gene–culture coevolution between cattle and humans identifying cow's milk as one of the most important factors shaping human nutrition (Beja-Pereira et al. 2003).

These are only the first steps in the endeavor to disentangle the complex interaction between genes and cultures. However, it is interesting to note that the gene in question affects a food consumption capacity that makes an agricultural and food production system possible that shaped much of Europe and areas colonized by them. In fact, in Europe still other traditions became associated with this dairy culture, for example the Protestant work ethics. One of the founding fathers of modern sociology has shown the intimate links between religious beliefs and economy (Max Weber in his fascinating book The Protestant Ethics and the Spirit of Capitalism). There is perhaps more than one reason why a bull decorates the entrance to the Frankfurt stockmarket.

The history of lactose consumption has still other ramifications. Lactic acid bacteria ferment lactose into lactic acid. The acidification of milk leads to dairy products like yogurt and cheese that resist the growth of bacterial spoilage organisms and transform milk into relatively stable food. At the same time, fermented dairy products solve the lactose problem in two different ways. Ripened cheese has only trace amounts of lactose. Yogurt, the result of a much shorter fermentation process, contains still 60–70% of the original milk lactose concentration. However, in yogurt lactose coexists with high concentrations of lactic acid bacteria that produce β-galactosidase. This enzymatic activity could be demonstrated in the intestine of adults eating yogurt. In this way, dairy products became a stable and digestible form of milk even to adults lacking lactase activity.

Basic Concepts on Eating

Raw Food for Thought

Energy

Before embarking on our journey into the quest for food, let's get a quantitative idea of the average food intake by a typical middle-age male living in central Europe. He consumes daily 84 g of protein, 101 g of fat, 263 g of carbohydrates (alimentary fibers add only further 20 g), and 21 g of alcohol. This intake corresponds to a total energy content of 10,360 kJ (or 2,476 kcal). If we calculate the daily consumption of O_2, we get the appreciable quantity of 360 l oxygen daily consumed by a single adult. This means every individual voids daily nearly 2 m^3 of oxygen from air. The energy needs of a human being are composed of three different layers: at the bottom is the resting metabolic rate. As a rule of thumb, it is given as 1 kcal/kg body weight and hour. It covers the maintenance of the body temperature and the vital functions of the resting cardiovascular, pulmonary, kidney, and brain functions. For a 75-kg human this adds up to an astonishing 1,800 kcal/day.

You spend a further 10% of your energy balance for the thermic effect of feeding. If you follow a volunteer over the day by indirect calorimetry or in a respiration chamber, he will consume about 1 kcal/min during sleep; this value will increase to 1.6 kcal for several hours after the three meals. Before we can extract energy from food, we first have to invest in its chewing, digestion, and absorption. The investment costs are highest with proteins, where 20% of the food energy is spent in its assimilation, while only 5 and 3% of the energy content from food carbohydrates and fats is spent on its adsorption. The most variable component of the daily energy expenditure is the thermic effect of physical activity. In a sedentary individual, this part might be as low as 100 kcal/day, whereas in a highly active individual who is either performing strenuous work or demanding sports the energy requirement of activity can be as high as an additional 3,000 kcal/day.

Macronutrients

From a catabolic viewpoint, the three major classes of macronutrients can replace each other. This is biochemically understandable since all their degradation pathways lead into the citric acid cycle. Nutritional epidemiologists have documented a vast variety in the composition of the diet in different human cultures, which concurs with this biochemical conclusion. However, hormonal regulation directs carbohydrates (mainly absorbed as glucose) into energy generation directly after meals, whereas fatty acid oxidation covers 70% of the energy production between the meals. A lipid-free diet leads to deficiencies since polyunsaturated fatty acids cannot be synthesized by humans; fat-soluble vitamins will not be absorbed. Likewise, a protein-free diet will lead to deficiencies since nine amino acids are essential, i.e., they cannot be synthesized by humans. In addition, amino acids are the carrier of nitrogen and sulfur for the human body. Amino acids provided by proteins serve three main functions: they are required for protein synthesis on ribosomes; they are the precursors in the biosynthesis of purines, pyrimidines, and porphyrins; and finally they are the starting material for gluconeogenesis. Nitrogen is obligately lost at a daily rate of about 50 mg/kg body weight; 24 g is thus the minimal amount of protein needed by an adult. Finally, about 120 g of glucose is needed daily by an adult mainly to cover the nutritional needs of the brain. An essential function of glucose in the diet cannot, however, be claimed since glucose can be resynthesized via gluconeogenesis from the glycerol moiety of many lipids and glucogenic amino acids. A carbohydrate-free diet leads to an increase of fatty acids and ketone bodies in the serum, with the risk of a ketoacidosis.

Water

We rely on a daily water intake to compensate for water loss via urine, feces, sweat, and exhaled air of about 1,500, 100, 550, and 350 ml, respectively. Life is defined as an open system that can only function when matter and energy flow through the system. This is another application of Heraclit's "*panta rhei,*" everything flows. Water is actually the vehicle, which organisms use for this flow. Water has to be lost as urine, where it carries away all chemical waste that accumulates in the metabolism that the body cannot use any longer. We loose water with sweat because the evaporation of water on the skin cools us down. Actually, when we are sometimes dubbed naked apes, this indicates no imperfection. The loss of the hairs allows us sustained physical activity under high ambient temperature without killing us by overheating. The heat loss in a cold environment then becomes a problem, but here we have "borrowed" the fur from animals to warm us.

Gas

When we have already done the step to perceive H_2O and O_2 as foodstuff, one should be consequent and include still other reduced carbon sources (e.g., hydrocarbons) and further O_2/CO_2 gas exchanges into our energy balance. If you use a car to get to work, heat or cool your home, and prepare your food in an oven or store it in a refrigerator, you need energy, a lot of energy in fact. A substantial part of this energy comes from fossil fuel. In your car, you are burning plant material that grew in the forests of a distant past and was transformed into oil by geological processes. Chemically, this energy consumption strikingly resembles eating in biological systems. In both the biological and the technical systems, energy-rich, reduced organic carbon compounds are burnt with oxygen to CO_2 and H_2O; the starting material is or was foodstuff (some bacteria still make a living from oil or natural gas). The end products of both types of burning are identical whether they come from a technical combustion chamber or from mitochondria. It is thus more than a metaphor when we speak of the "energy-hungry" man of the industrialized countries. It reminds us that we should carefully calculate our per capita energy needs and gas exchanges in an overall energy and matter balance sheet. Human civilization is burning relatively large amounts of fossil fuel, resulting in a measurable increase of the amount of CO_2 in the atmosphere. Since CO_2 is a greenhouse gas, these increases have the potential to increase the global temperature with all associated climate changes. An increased production of CO_2 remains without consequence if it could be balanced by an increased fixation of CO_2. The only chemical process that could adsorb these huge amounts of released CO_2 is the Calvin cycle run by microbes and plants. In our extended definition of food consumption, we will discuss trials to increase the global CO_2-fixing process by spurring the Calvin cycle.

Thermodynamics Made Simple

What does thermodynamics mean for our understanding of eating? To paraphrase, eating is our answer to the challenges of the laws of thermodynamics. If you are not a scientist you might wonder about the wicked force that bows all organisms under the yoke of this bitter law of eating. As a scientist, you are not searching for a magical *vis vitalis* organizing the biological theater surrounding you. The "eat" is the first commandment of the laws of thermodynamics. The first law of thermodynamics stated in popular terms is that energy cannot be created new, it can only change form. The second law is about molecular disorder, entropy: There is a tendency in nature toward ever-greater disorder in the universe. Organisms behave superficially as if these laws do not apply to them. What contrast to entropy when from millions to billions of individual nucleotides a handful of giant DNA molecules are synthesized that contain the precise base sequence coding for the construction of an organism. Do organisms violate the laws of thermodynamics? The solution to this paradox is that organisms are never in equilibrium. Order cannot be conserved in nature statically. Organisms have to be constructed as open dynamic systems, which exchange both matter and energy with the surrounding. What appears in equilibrium is in fact a dynamic steady state. If you could look into the organisms, you would see a continuous breaking and creating of chemical bonds. Even if you are not any longer growing, cells are renewed at an incredible rate. The crucial difference between life and death is that dead organisms do not any longer consume energy and thus become victims of the laws of thermodynamics. Except for some latent forms of life (e.g., spores or seeds), stopping to eat means relentless decay to ever increasing disorder until organisms cannot any longer be distinguished from the piece of earth on which they died. In fact, even the dead organism, be it a fallen tree, a leaf, or the carrion of a mouse, is still far from the equilibrium with it's surrounding. It is still an energy-rich resource for many forms of life. The chemical energy stored in their decaying bodies can still power the life processes of other organisms. Thus, most organisms will not decay due to physical and chemical weathering, but will literally be eaten up. We can thus conclude that organisms need constant input of energy to maintain their astonishing order. The magic power to keep this order in a steady state is derived—at least for us animals—from food. When we eat, we take an object from our surrounding that contains stored chemical energy usable for us. If the food happens to be a banana, for example, this is starch, which we digest to glucose, which we burn to CO_2 and H_2O by using oxygen, which we also extract from the surrounding atmosphere. The banana was synthesized by a plant that transformed the light energy of the sun in the process of photo-synthesis into sugars like sucrose and starch. We energize our life processes by exploiting the chemical energy put down in molecules created by other organisms. Actually, when we eat, we can cope with the laws of thermodynamics. For example, when we eat a banana, we digest a giant starch molecule into its constituting glucose monomers; the glucose is further oxidized to even smaller units according to the equation $C_6H_{12}O_6 + 6O_2 \rightarrow 6CO_2 + 6H_2O$. Oxidation is the basis of our energetics; we live from the flow of electrons "downhill" from higher to lower electrochemical potential. Reduced organic compounds are oxidized by molecular oxygen which itself becomes reduced to water. Oxidations and reductions are always coupled reactions. The flow of the electrons from glucose to oxygen underlies our energy metabolism, as discussed later. If you count the molecules during the eating process, you will also see that the number of molecules has increased: First, from a single large starch molecule to many hundred glucose molecules; then in the transformation of glucose into carbon dioxide and water. At the left-hand side of the above chemical equation, we have seven molecules and at the right-hand side we end up with 12 molecules that are randomly dispersed. The entropy of the system, the degree of disorder, has clearly increased. We "buy" the order of our body by the creation of disorder during the digestion of the food. We transform the energy contained in the reduced organic food components into ATP that powers our anabolism and physical activity. By eating, we fulfill the laws of thermodynamics and at the same time act seemingly against it when creating order from disorder.

Different Ways of Life

Heterotrophs Versus Autotrophs

Humans like all animals need preformed, reduced organic molecules commonly called food for sustaining life. However, animal food sources are always other organisms. This property explains their scientific name with respect to their nutrition. They are called heterotrophs; this Greek word means literally nutrition (*trophe*) from other (*heteros*) organisms. However, this leads to a circular argument. A living world cannot be constructed with animals alone since life would literally be eaten up if there were not other organisms that have learned to feed themselves. These organisms are called autotrophs, Greek for "self-feeders." Of course thermodynamic arguments do not allow such self-sufficient organisms.

Energy cannot be created de novo, therefore these organisms had to learn to use the energy stored in inorganic chemical compounds; these organisms are consequently called chemoautotrophs and they occur exclusively in prokaryotes. Other organisms learned to use the energy contained in the sunlight that bathes the planet earth; these organisms are logically called photoautotrophs from the Greek word for light. The most visible form of photoautotrophs is the terrestrial plant, but they are also found in eukaryotic algae and in prokaryotes. Naively, one would expect heterotrophs and autotrophs to use fundamentally different forms of biochemistry. To a certain extent, this is true and one can illustrate this by two fundamental reactions restricted to autotrophs: photosynthesis and nitrogen fixation. These two reactions lead to the biological fixation of atmospheric CO_2 and N_2 in chemical forms, usable to heterotrophs.

Metabolic Complexity Does Not Need Morphological Complexity

The biochemical endowment of organisms, which master these fundamental reactions, allows the existence of the complex world of heterotrophic animals. It is surprising that these complex reactions were mastered by genetically very humble organisms which took center stage in the evolution of life: cyanobacteria. This is a strong argument against a trend toward a crown of creation and increasing complexity anticipated in many popular thoughts on evolution. Cyanobacteria populated the earth quite early, long before animals of the simplest form appeared on our planet, and this is only logical because cyanobacteria created the conditions that allowed animal life, namely an oxygen-containing atmosphere. On a second look, however, the biochemistry of heterotrophs and autotrophs is not that different. Photosynthesis and respiration share many characteristics down to the molecular details. At a more fundamental level, photosynthesis and respiration are to a certain extent reverse reactions. In fact, according to the endosymbiont theory, cyanobacteria became the chloroplasts in algae and plants and proteobacteria, that contained the respiratory chain, became the mitochondria found in most eukaryotes, including plants and animals. The fundamental similarity between the photosynthetic and the respiratory chain probably reflects the fact that both electron membrane transport systems derived from a common ancestor bacterium. Perhaps the most fundamental similarity between both systems is that they handle (in first approximation) oxygen.

Bacterial Feeding Modes

When dealing with bacterial nutrition, the feeding modes and consequently the terminology become complicated. Bacteria like all living organisms need a carbon source because carbon is the backbone of all organic molecules and thus a building block of all biological bodies. In addition to the need for building blocks, biological systems need energy. There are only two known sources of energy available to organisms: Phototrophs use light as their energy source; chemotrophs obtain energy from the oxidation of chemical compounds (either organic or inorganic). Finally, organisms need "electrons" for life.

Lithotrophs ("rock eaters") use reduced inorganic substances as source for electrons, whereas organotrophs extract electrons from organic compounds. The bacterial nutritional requirements can come in many combinations. Hence cyanobacteria are "photolitho-autotrophs." This tongue-twisting word means that cyanobacteria fix CO_2 from the atmosphere, power their metabolism with light energy, and extract electrons from an inorganic compound (in cyanobacteria H_2O). Anoxic photosynthetic bacteria from the Green nonsulfur bacteria belong to "photoorgano-heterotrophs." This means that light provides the energy, but organic compounds provide the carbon source and the electrons. Most bacterial pathogens are "chemoorgano-heterotrophs," that is, organic compounds are needed as carbon, energy, and electron source (we as animals would fall into this microbial category). The fourth type is "chemolitho-autotrophs." These organisms can use CO_2 as carbon source, and inorganic chemicals as energy and electron source.

The Central Metabolic Pathway

Hans Krebs, the discoverer of the citric acid cycle, distinguished three stages in the extraction of energy from foodstuff. In the first step, large moderately reduced molecules of the food are broken down into smaller subunits. No energy is extracted from food during this preparatory stage; in fact, energy has first to be invested. In the second stage, the building blocks of the major food ingredients are further degraded to still fewer simple subunits. In fact, most of them end up as pyruvate in probably one of the oldest energy-yielding pathways—glycolysis. The antiquity of this pathway is deduced from its presence in all three domains of life. Its enzyme set and their regulation has experienced major modifications to account for the special nutritional needs of the very different organisms, while the chemical intermediates of this reaction pathway have remained essentially unchanged. This uniformity is testimony to the common origin of all extant organisms, irrespective of being prokaryotes or eukaryotes. The chemical structure of several compounds involved in energy extraction from food is indeed very revealing with respect to the time period when the precursor to the current central metabolic pathway developed. ATP, NADH, FAD, and Coenzyme A all contain ribonucleotides, more specifically ADP (Figure 1.3). The wide distribution of ribonucleotides not only as information molecules as in mRNA but also as central biochemical compounds and the uniform conservation of these central molecules of life over all forms of extant life can hardly be explained by chance. In fact, many biologists interpreted the very use of ribonucleotides as evidence that these central biochemical compounds are billions of years old and their roots are still in the RNA world preceding the current DNA world.

To recover the energy retained in pyruvate, many organisms go to the third step of the energy extraction process, which consists of the citric acid cycle and oxidative phosphorylation (respiration).

However, the central metabolic pathway is not only involved in energy extraction, it deals as much with the creation of biological matter in anabolic

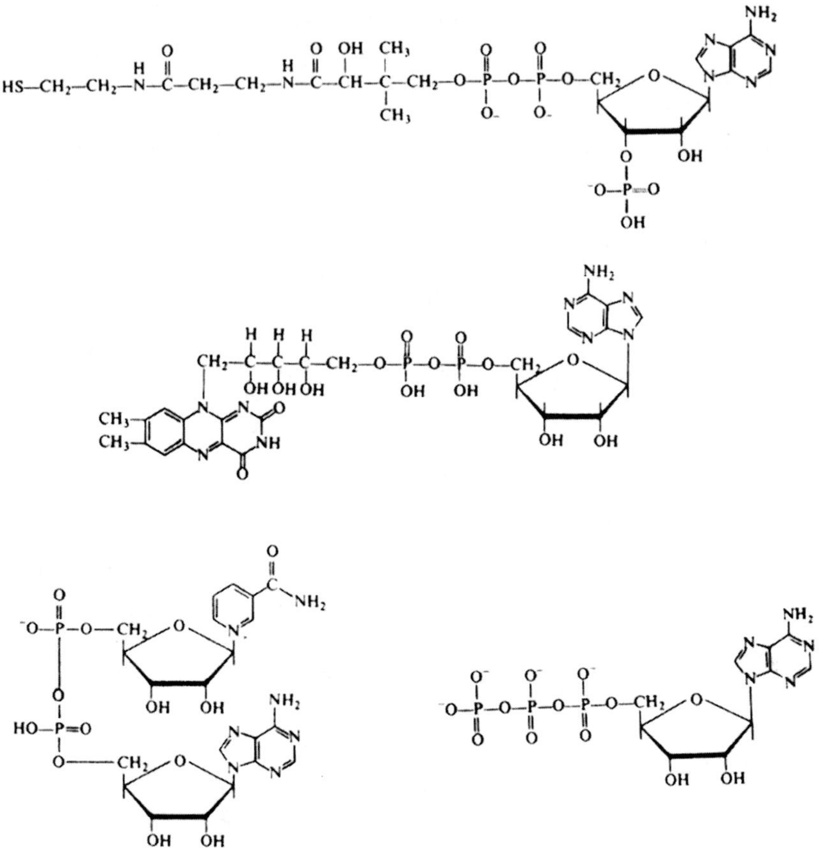

FIGURE 1.3. The most basic biomolecules are the cofactors ATP (bottom right), NAD (bottom left), FAD (center), and CoA (top), which are central to our metabolism. They are chemically closely related relics of the RNA world.

reactions. In humans food is also the source of biological compounds that we cannot synthesize ourselves. For example, we humans lack the enzymes to synthesize nine out of the 20 amino acids needed for protein synthesis. These so-called essential amino acids must be provided with the food. If only one of these essential amino acids is lacking, a negative nitrogen balance results. More protein is degraded than synthesized and if the organism does not find a food source that contains an essential food ingredient, it is doomed. It starves and eventually dies in the presence of ample food calories. Each organism therefore depends on a balanced food composition to meet its metabolic needs. How this is regulated via selective food intake is not known in detail, but that it is finely regulated is a physiological fact. In a sense, the central metabolic pathway is the stock exchange market for the energy and matter needs of an organism and the interplay of offer and demand applies as well to market economies as the regulation processes in biochemistry. The fine-tuning of the metabolic control in organisms would not be possible if the body would trade in too many chemicals. It is thus an ultimate question of molecular economy to reduce the incredible complexity of organisms to a small number of basic goods, i.e., the central precursors of the biochemical building blocks. *Ex pluribus unum* turned upside down. This need for molecular economy explains the modular design of biochemistry. We find recurring sets of activated carriers in many biochemical reactions and unifying motifs in biochemical reaction mechanisms. Some central metabolites take the role of relative universal trading goods for the bartering of matter in the organisms while ATP plays the role of universal money on the biological energy market.

A Few Words About ATP

The parallel with market economy is more than just superficial. Many biochemical reactions are controlled by the energy charge of the cell. This term is defined by a simple arithmetic formula $[ATP]+\frac{1}{2}[ADP]/[ATP]+[ADP]+[AMP]$, which describes the energy status of the cell. Theoretically, it can take values between 0 and 1. Cells are designed to buffer chemical reactions against wild fluctuations. As they maintain the pH and critical ions within precise physiological limits, the energy charge also is kept within narrow limits (0.80–0.95). As in market economies, ATP-generating (catabolic) pathways are stimulated by low energy charge and inhibited by a high-energy charge. The opposite is true for ATP-utilizing (anabolic) processes.

Why did ATP become the universal energy currency? At a simplistic level, the answer is relatively clear. ATP has a large standard free energy of hydrolysis ($\Delta G^{\circ\prime} = -7.3 \, \text{kcal/mol}$), i.e., ATP has a strong tendency to transfer its terminal or γ phosphate group to water. However, there are other phosphorylated biochemical compounds with even greater phosphoryl-group transfer potential like phosphoenolpyruvate or creatine phosphate, but there are also compounds with lower phosphoryl potential like glucose 6-phosphate. The advantage of ATP is its intermediate phophoryl potential. It can easily transfer a phosphate group to numerous compounds and thereby activate them chemically. On the contrary, it can still accept phosphate groups from other physiological phosphate carriers. There is another reason for using ATP. Without a catalyst, ATP does not transfer its phosphate group. Despite its large negative $\Delta G^{\circ\prime}$, ATP hydrolyzes spontaneously only on the timescale of hours or days. This kinetic stabilization of ATP makes it a very handy compound for the control of its hydrolysis with enzymes.

One of the big inventions of organisms is the coupling of thermodynamically unfavorable reactions (with positive $\Delta G^{\circ\prime}$) that would not occur spontaneously to ATP hydrolysis. Because the overall free-energy change for a coupled series of chemical reactions is equal to the sum of the free-energy changes of the individual steps, ATP hydrolysis with its $\Delta G^{\circ\prime} = -7.3 \, \text{kcal/mol}$ can pull such an unfavorable reaction. For example, if the transformation of compound 1 into compound 2 has a $\Delta G^{\circ\prime} = +4 \, \text{kcal/mol}$, you have at equilibrium a 1,000-fold more compound 1 than compound 2. If you couple it to ATP hydrolysis, the high phosphorylation potential [ATP]/[ADP] [Pi] maintained in cells at about 500 will now result in 100,000-fold more compound 2 over compound 1. If the unfavorable reaction has a $\Delta G^{o\prime}$ value that exceeds $+7\,\mathrm{kcal/mol}$, it cannot any longer be energetized by ATP hydrolysis. However, this is not a major problem, you have to invest a second or a third ATP molecule to power this reaction.

With few exceptions, ATP hydrolysis as such cannot accomplish work; it simply creates heat. And heat is a difficult form of energy with respect to its conversion into other forms of energy. When a biochemical reaction shows ATP hydrolysis in the chemical equation, it usually indicates a two-step process. Mostly a phosphoryl group is transferred to a nondescribed cofactor or to a side chain of an amino acid of the enzyme molecule to which it is linked via a covalent bond. Only in a second step, the group transferred from ATP is displaced, yielding the product molecules. Another error concerns the notion of "high-energy phosphate bond"—there is no energy in the P–O bond of ATP per se. In fact, it needs an input of energy to break this as any other chemical bond. The free energy released by the hydrolysis of ATP does not come from the specific bond that is broken, but from the fact that the products of the reaction have a lower free-energy content than the reactants.

Now why is ATP and not another common ribonucleotide the universal energy currency? The answer is not known. In the citric acid cycle, another high-energy compound is created, namely GTP. GTP is also the phosphoryl donor in protein synthesis. Does this mean that GTP was the earlier energy currency? Probably not: The citric acid cycle in plants creates ATP and not GTP. Both nucleotides seem to be quite equivalent as basic energy currency. In fact, the enzyme nucleoside diphosphokinase keeps both nucleotides in equilibrium: GTP + ADP ↔ GDP + ATP. Why is UDP the activated carrier of glucose and CDP the activated carrier of phosphatidate in metabolism? Could ADP or GDP not fill this ticket? It is certainly important to use different ribonucleotides for different biochemical tasks to regulate them separately. Why ADP became the preferred phosphoryl-carrier is chemically not obvious and it might simply reflect an initial chance event that later became fixed when the central metabolism was elaborated in evolution.

ATP is the immediate donor of free energy in our body and not a storage form of energy; the latter role is taken by other molecules (direct: creatine phosphate, indirect: glycogen, fat). This role of ATP predicts a high turnover rate for ATP. In fact, the average ATP molecule is consumed within a minute after its synthesis. If you calculate the total amount of ATP synthesized by a resting human being per day you get the incredible number of 40 kg of ATP. The need for ATP synthesis gets even higher when we perform heavy work. Under these conditions, ATP is used at the rate of 0.5 kg/min.

2
Some Aspects of Nutritional Biochemistry

The Central Carbon Pathway

Why is Glucose the Central Fuel Molecule?

Nature has invented several ways to synthesize ATP. One solution is a series of coupled chemical reactions catalyzed by soluble enzymes in the cytoplasm that result in substrate-level phosphorylation. The most prevalent starting substrate for such an energy-yielding pathway is glucose, which is decomposed in all higher cells by a carefully orchestrated and evolutionarily fixed way called glycolysis.

There is apparently something special about glucose since it is so widely used by organisms of all kind. In our diet, glucose comes in fruit juice, as starch and glycogen (the polymeric storage form of glucose in plants and animals, respectively), and in the disaccharides saccharose (table sugar) and lactose. For herbivores glucose comes also as cellulose, a glucose polymer from plant cell walls. The latter is arguably the quantitatively most abundant biological molecule on earth. Hemicellulose, which consists mostly of xylose and arabinose, represents 15–30% of plant material and is thus of great importance to herbivores. Even if they are part of our food, we cannot deal with cellulose and hemicellulose. Only few other hexoses are found in our diet. These are galactose (from lactose), mannose (mainly from glycoproteins), fucose (in milk oligosaccharides), and fructose (in fruit juice, saccharose). In the series of pentoses, ribose and deoxyribose dominate since they constitute the backbone of all nucleic acids that we eat with our food.

An organism that does not know to handle glucose deprives itself of the most important organic carbon source on our planet. The prominent role of glucose is also reflected by the fact that nearly all other ingested sugars are transformed in the body into glucose before they can be used for energy production (fructose is an exception). Conversely, glucose is the starting material for all monosaccharide synthesis in our body, and also the carbon skeleton of glucose provides the starting material for the synthesis of amino acids and lipids. *In fact, the first cells were probably more concerned to make glucose than to degrade it. Although it looks different today, the initial function of glycolysis was gluconeogenesis. We get into the realm of evolutionary speculation when we ask why glucose became the universal cellular carbon currency, but a few arguments can be given. The following argument for glucose is of course circular but nevertheless true: Glucose is used by so many organisms because it is so prevalent in the organic world. Any newly evolving biological system finds itself in the tradition of life that existed before and with it. Even historical accidents thus become a biochemical necessity.* However, there are also a few chemical arguments for the prominence of glucose in biology. Take glucose in the chair conformation. Glucose has the smallest axial substituents imaginable for hexoses—hydrogen atoms. This gives it an energetically privileged position within hexoses. This stable chemical structure of glucose makes it attractive for metabolism; compared with other hexoses, it has a lower tendency to react nonenzymatically with proteins. That this concern is not a mute point is demonstrated by the problems of diabetic patients, who do not manage to keep their blood glucose levels at a carefully regulated level. Excess blood glucose results in the glucosylation of hemoglobin and of proteins from the vascular tissue, which is at the basis of the medical problems in diabetic patients. Finally, glucose can be formed from formaldehyde under simulated prebiotic conditions. *Glucose was thus simply around when Nature started to tinker with primitive metabolism.*

Glycolysis

Glycolysis is the central energy-providing process in an astonishing diversity of organisms. In many organisms, it is also the sole source of energy. This also applies to several tissues of our own body. The Greek word "glycolysis" can be translated as "sweet-splitting," and this is a very precise description of its chemistry. Glycolysis is the controlled degradation of the six-carbon sugar glucose into the central intermediate of metabolism, pyruvate. The conversion is linked to the gain of two molecules of ATP and two of NADH, but glycolysis is more than just energy gaining. Recall from your biochemistry courses that many intermediates of glucose degradation are also the starting points for several biosynthetic pathways.

Its Origin

The design of the central carbon pathway thus evolved under dual constraints. The conservation of this pathway in so many organisms could suggest that there is only one possible chemical solution to the central metabolic pathway. Alternatively, it could reflect a "frozen accident" of evolution. In this second scenario, the pathway developed over a long time period, where it probably showed many variations, but when the chemical reactions fitted neatly together, it was fixed by selective pressure. No organism could tinker with alternative pathways when living on carbohydrates. They could only refine its regulation to their peculiar needs (what most organisms actually did), but they were not allowed to redesign its chemical path because of selective constraints. As all living organisms are evolutionarily linked (as expressed in the concept of the universal phylogenetic tree), protoglycolysis was then inherited from the ancestor cell, which first successfully fixed this invention, by all its descendents. It is probably safer to restrict this argument to eukaryotes. Many, if not most, microbes use the Entner–Doudoroff pathway for "glycolysis" and the glycolytic pathway for gluconeogenesis. The very fact that glycolysis is found in all kingdoms of life (archaea, eubacteria, and eukaryotes) speaks in favor of its antiquity. We should therefore treat this reaction sequence with much respect as we might look through it deep into the biological past. Fitting with the above arguments, the chemical intermediates are exactly the same over all organisms; only the cofactors and enzymes show variations. For example, some bacteria, protists and perhaps all plants have a phosphofructokinase that uses pyrophosphate instead of ATP in glycolysis. The very fact that the Entner–Doudoroff pathway replaces the Embden–Meyerhof pathway (glycolysis) in so many microbes (Fuhrer et al. 2005) is a frequently used argument for its even older origin.

The Reactions

So what happens in glycolysis? Ten enzymes transform glucose into pyruvate. The six-carbon sugar glucose is symmetrically split at about halfway of the reaction pathway into two three-carbon compounds. This splitting reaction separates the preparation phase of glycolysis from the energy-yielding phase. The motto of this first part is again the old dictum "there is no free meal in biology." Before you can extract the chemical energy stored in the glucose molecule, you have to invest energy in the form of two phosphorylation steps. There is chemical reason in these two steps, but the motivation differs for the two steps. If a cell wants to keep glucose for its own use, it has to mark it as its own. The cell labels it with a tag that prevents the flow of glucose across the plasma membrane following the concentration gradient of glucose. Cells found a very simple, although costly measure: phosphoryl transfer to glucose by hexokinase. This adds chemical charges to the otherwise electrically neutral glucose molecule. Glucose needs a transport protein to get across the membrane. The argument is different for the next ATP energy consumer step in glycolysis. The enzyme phosphofructokinase adds a second phosphate group at the opposing end of the six-carbon sugar. The electrostatic repulsion of the two negative charges puts the ring structure under strain. This strain is exploited in the following reaction where aldolase catalyzes the splitting of the C6 sugar into two C3 sugars.

With glyceraldehyde 3-phosphate (G3P) starts the second phase of glycolysis, which you can call the payoff phase. Chemically something has already happened. The entropy has increased: One molecule became two molecules, but the glyceraldehyde is still at the approximately same oxidation level as the starting compound glucose. However, if you look at the end product of the glycolytic pathway, pyruvate, you see a definitively more oxidized compound. Pyruvate shows a carboxylate and a keto-group. At these positions, G3P has an aldehyde and a hydroxyl group. Two separate oxidation steps (only one uses NADH) have thus occurred, and the cells have learned to harness the energy released by each oxidation step in a chemically usable form. A crucial tool of the enzyme G3P dehydrogenase is a critical cysteine residue. Its thiol group establishes a covalent linkage to the aldehyde carbon of G3P. An adjacent histidine residue helps here as a base catalyst: It accepts the hydrogen from the thiol and stabilizes the resolution of the double bond between the C and O atoms in G3P, which allows the covalent binding of the substrate to the enzyme. Now comes the enzyme cofactor NAD^+ into play. It abstracts a hydride (a hydrogen atom that took away the two binding electrons) from the substrate. To fill the hole in the electronic shell of the terminal carbon of G3P, an intramolecular electron transfer has to occur, which reestablishes the C=O double bond. You have now a thioester intermediate. This thioester bond has a very high standard free energy of hydrolysis, and the enzyme keeps this high-energy bond for energy fixation. It achieves this goal by excluding water from the reactive center, and it allows only a phosphate group to detach the substrate from the enzyme surface. Instead of hydrolysis a phosphorolysis occurs and creates 1,3-bisphosphoglycerate. This is an anhydride between two acids, namely the carboxylate and the phosphoric acid. Its standard energy of hydrolysis is $-49.3\,kJ/mol$, much higher than that of ATP hydrolysis ($-30.5\,kJ/mol$). The next enzyme in the glycolytic pathway then does the logical step: It transfers the phosphate group from 1,3-bisphosphoglycerate to ADP creating ATP (although it is not sure whether it is physically the same P_i).

Energetics

Now let's look at the stoichiometry of glycolysis and the energy sheet: glucose + $2NAD^+ \rightarrow 2$ pyruvate $+ 2NADH + 2H^+$; $\Delta G'^\circ = -146\,kJ/mol$. This free-energy change is conserved in two ATP molecules, which need 61 kJ/mol for their synthesis. Under the actual concentrations of the reactants in the cell, the efficiency of the process is more than 60%. However, if you stop the reaction here, only a small percentage of the total chemical energy contained in glucose is recovered (recall the complete burning of glucose with oxygen to H_2O and CO_2 yields $-2,840\,kJ/mol$). However, not all cells have this option because either they do not have access to oxygen or they lack a respiratory chain. Furthermore, early in the history of life, molecular oxygen was not present in the atmosphere precluding glucose degradation using oxygen as electron acceptor.

Variations on a Theme

Johann Sebastian Bach has written variations on a single musical theme that should especially appeal to scientists by their near mathematical logic. Mother Nature is an equally creative composer and has tried a lot of variations around the theme of glycolysis. Let's first take a small variation where only a few side notes are altered.

Pyrococcus

The first theme is played by an extreme character: *Pyrococcus abyssi*. This is quite an exotic prokaryote, what biologists actually call an extremophile. It belongs to Archaea, the third kingdom of life. However, it is not just its phylogenetic affinity that makes this organism interesting. It is a hyperthermophile; its optimal growth temperature is 96 °C, the minimum and maximum temperatures of growth are 67 and 102 °C, respectively. You might raise your eyebrows since this is above the boiling point of water. This is, however, not an issue for *P. abyssi*. As the name suggests, ("the fire globule of the deepest depth") its habitat is a hot area of the seafloor. This location solves the paradox: The hydrostatic pressure at the seafloor exceeds 600 atm, and at this pressure, 102 °C hot water cannot boil. Pyrococci are motile by flagella and reduce elemental sulfur to sulfide under strictly anaerobic conditions. All this would sound to our forefathers as hell and like a proof for the most dreadful believes of the Dark Ages. Yet, some like it hot, and life might actually have started under such conditions. Even under hellish conditions, you can make your living comfortable if you are adapted to it. In pyrococci all metabolites of glycolysis are identical to those of humans; the differences concern only the cofactors (Sapra et al. 2003). *Pyrococcus* uses ADP as phosporyl donor in the first two phosphorylation steps of glycolysis, and instead of NAD^+ it uses ferrodoxin and tungsten. Pyruvate is decarboxylated to acetyl-CoA (here again pyrococci use a ferredoxin oxidoreductase). The free energy of the CO_2 release reaction is stored in the thioester bond to coenzyme A and is used for the synthesis of a further ATP with the concomitant release of acetate. The wide phylogenetic and ecological distribution and conservation of glycolysis speaks for one of the oldest sugar degradation pathway invented in biological systems on earth.

Entner–Doudoroff Pathway

Now comes another variation on the glycolysis scheme, but this time, new elements are introduced. This second variation is called the Entner–Doudoroff pathway, but remember that many microbiologists believe that this pathway preceded the gylcolytic pathway in evolution. This series of reactions starts quite similar to the entrance reaction of glucose phosphorylation by ATP like in glycolysis. Since the enzyme phosphofructokinase is missing in many bacteria, an alternative path has to be taken. The C1 glucose position is oxidized from the aldehyde to the carboxylate oxidation level. Then follows a dehydration step with a familiar keto–enol tautomerization in a six-carbon sugar acid. In the next step, the keto group of the substrate forms a Schiff's base with a lysine from the enzyme, which catalyzes the splitting of the compound into pyruvate and G3P. With the latter, we are again at the midway of the glycolysis, and the next steps follow as usual. However, there is a difference. The other half of the molecule is pyruvate and thus already at the end of the glycolytic pathway. You gain therefore only two ATP from the transformation of G3P to pyruvate. Since you had to invest one ATP in the initial phosphorylation of glucose, your net gain is only one ATP from this pathway, which is a meager half of the exploitation of glucose in glycolysis.

Due to this energetic limitation, this pathway is only used in aerobic bacteria that can use the NADH produced in the oxidation of glucose to gluconate and in the oxidation of G3P in a respiratory chain. The only fermentative bacterium using this pathway is *Zymomonas*. This bacterium is adapted to environments with high sugar concentrations permitting such a wasteful metabolism (Seo et al. 2005). This variation also contains a lesson. The Entner–Doudoroff pathway demonstrates clearly that the glycolytic pathway is not the only way of sugar degradation and in fact two further alternatives exist in microbes (e.g., the phosphoketolase and the *Bifidobacterium bifidum* pathways).

The Pentose Phosphate Pathway

I have on purpose chosen the Entner–Doudoroff pathway because it is actually a hybrid. The early steps of this path up to 6-phosphogluconate are practically identical to the pentose phosphate pathway; the later steps are identical to glycolysis. In biochemistry the same elements are frequently used in new combinations. The pentose phosphate cycle is itself a new variation on the glucose degradation scheme. Such variations are important to cells since they allow them to maintain several pathways with distinct and even competing metabolic goals in parallel. Nature must only introduce regulated enzymes that are responding to metabolic signals that allow an appropriate channeling of the substrate flows. If this is not possible, higher cells also have the option of locating competing pathways into different cellular compartments. Actually, the function of the pentose phosphate pathway is not energy metabolism (catabolism). Its goal is to provide precursors (ribose 5-phosphate) and cofactors (NADPH) for biosynthetic pathways (anabolism). The destiny of ribose is clear: It is the precursor to nucleotides that make RNA, DNA, and a number of coenzyme nucleotides. NADPH is the cellular currency of readily available reducing power. In biochemistry textbooks, it is frequently stated that catabolic reactions are generally oxidative, while anabolic reactions are generally reductive. Catabolic enzymes like glucose 6-phosphate dehydrogenase from the Entner-Doudoroff pathway use NAD^+, while the enzyme catalyzing the same reaction in the pentose phosphate pathway uses $NADP^+$. This is, however, often not true. In *E. coli* and many other bacteria, it is the same enzyme that does both jobs with one cofactor. In animals the intracellular ratio $NAD^+/NADH$ is high (actually 700 in a well-fed rat), which favors hydride transfer from the food substrate to NAD^+ and thus channels electrons into the respiratory chain. In contrast, the ratio $NADP^+/NADPH$ is low (0.014 in the same rat), which favors the hydride transfer from NADPH into biosynthetic pathway.

The biochemical details of the pentose phosphate pathway are actually quite complicated despite the fact that it consists basically of a combination of a few basic chemical reaction types. If you see the written partition of this pathway (the interested reader is encouraged to consult standard biochemistry books), it is all too apparent that it shares one major biochemical motif, namely the interconnected *trans*-ketolase and *trans*-aldolase reactions, with the Calvin cycle in the dark reaction of photosynthesis for CO_2 fixation. Actually, the Calvin cycle is not an invention of photosynthesis since it also represents the major, although not the only, CO_2 fixation pathway in nonphotosynthetic prokaryotes. *It can thus also claim substantial antiquity.*

Gluconeogenesis

Nature tried something, which is also very popular in music, namely a reversal of the glycolytic theme. This movement also has a biochemical name and is called gluconeogenesis. On the biochemical partition, this pathway reads like a reversal of glycolysis at first glance. But this is of course thermodynamically not possible.

In reverting from phosphoenolpyruvate (PEP) to glucose, gluconeogenesis uses the same intermediates as glycolysis, but two enzymes differ. These are two phosphatases, which reverse the phosphorylations done by phosphofructokinase and hexokinase in glycolysis. The hydrolysis of glucose 6-phosphate to glucose appears to be a chemical child's play. In reality it is a surprisingly complex reaction. Glucose 6-phosphate is transported into still another compartment (the lumen of the endoplasmic reticulum), where a complex of five proteins catalyze the reaction.

A popular start point for gluconeogenesis in the liver is lactate coming from the exercising muscle and erythrocytes. Lactate is first converted into pyruvate by lactate dehydrogenase. This pyruvate becomes the starting material for glucose resynthesis. Liver glucose travels back to the muscle where glycolysis powers the movement and generates lactate. This physiologically important metabolic highway between muscle and liver is the Cori cycle.

Glycolysis takes place in the cytoplasm, which is not so for gluconeogenesis, which calls different cellular compartments into action. Cytoplasmic pyruvate is first transported into the mitochondrion where it is carboxylated to oxaloacetate. This is an important biochemical reaction since oxaloacetate is not only a stoichiometric intermediate in gluconeogenesis, but also the carbon skeleton for the *trans*-amination reaction creating the amino acid aspartate and a catalytic intermediate of the citric acid cycle. Oxaloacetate leaves the mitochondrial matrix after reduction by NADH to malate. In the cytoplasm, malate is reoxidized to oxaloacetate. Oxaloacetate is then simultaneously decarboxylated and phosphorylated to yield PEP. To drive this reaction, one GTP has to be spent. If not enough GTP is available, oxaloacetate cannot be used for gluconeogenesis and goes into the citric acid cycle, which creates GTP. Thus the energy status of the cell decides in what pathway oxaloacetate is actually used.

Evolution of Metabolism

All the above-mentioned pathways are either chemically related or occur even in the same cell. Apparently, a relatively small set of basic chemical reactions provides the very fabric of life as we know it on our planet. These basic reactions come in many variations and are also variously reassorted into new biochemical themes. Nature is very modular in its construction and always reuses old clothes to make new suits. It is actually not correct to say that nature uses the same chemicals; in fact, nature uses a rather limited set of enzymes that creates these chemical intermediates. This is a subtle, but important difference. The question is thus not whether these compounds are special and cannot be replaced by other chemical compounds. In the following section we will see two variant glycolytic intermediates that do not figure in the main chemical pathway, but are important chemical regulators of intermediary metabolism. The crucial contribution was the invention of the protein enzymes that learned to handle a minimal set of chemical reactions. The cell does not play biochemistry like a student who learns pathway by pathway when advancing from chapter to chapter in the biochemistry textbook. If you take the simplest case of a unicellular prokaryote, you have actually a bag filled with a viscous solution containing a high protein concentration and a large number of small chemical compounds. Metabolism means that a food chemical is introduced into the system, turned over by the protein enzymes, and useless end products of the cellular chemical exercises are excreted from the cell. Some enzymes come in complexes and the substrate is actually reached from one enzyme to the next. However, many other enzymes from metabolic pathways that are so neatly assembled in the biochemistry textbook occur without much architectural order somewhere in the cytoplasm. They are exposed to all substrates and product chemicals at the same time and must do their job. Diffusion is thus the limiting factor for chemical communication in the living cell. Diffusion is not such a barrier for small substrate molecules that reach every point in the relatively large mammalian cell within a tenth of a second. Proteins have a much lower diffusion coefficient and are frequently retained by protein–protein interactions. This means that a pathway is in fact an event that is only reconstructed by the human mind.

Most enzymes do not know much about the other enzymes in the same pathway and do not know whom to deliver what. The only thing they have learned in evolution is to do the more or less specific catalytic reaction and to interpret the chemical environment from the small molecules it meets. If every enzyme would only execute its function every time it meets its substrate, only a rather primitive metabolism could be built. This was certainly the way primitive protein enzymes reacted early in the biochemical evolution. At this level, protein enzymes function like a perception. This term comes from a device that portrays the basic reaction of synapses, which process an input signal into an output signal. In the case of enzymes the input is the substrate, the output the product chemical. Already this system has some flexibility since the input–output relationship can take a linear, hyperbolic, or sigmoidal form according to the construction of the enzyme. A cross talk between proteins is possible only if the product of one enzyme becomes the substrate of another enzyme, and this relationship must hold for many proteins. It is evident that this restriction necessitates that early metabolism could only be built with enzymes that spoke a common chemical language. One can surmise that the chemicals of the central intermediary pathway must belong to this Esperanto chemical vocabulary understood by all living systems. To get a primitive metabolism the enzymes had to travel together through organisms if the communication between them should not be interrupted resulting in chemical deadlocks in the cell with the accumulation of a product that could not any longer be processed. This need for the chemical Esperanto is probably also the reason why the variant ways of glucose handling share so many common chemical intermediates. A network can only be created if sufficient numbers of common nodes are shared. Only stepwise could variations be introduced which had to use the common elements. To remain in the language picture, in later steps of evolution when the organisms reached already some maturity, they could differentiate and develop in addition to the Esperanto their local language, which is only understood in their corner of the biological world.

Evolution of Complexity

Nature apparently soon discovered that it could use proteins as computational elements in the living cell (Bray 1995). The individual protein elements became stepwise more complicated. The next step in the development of metabolic networks was taken when proteins learned to read more than one chemical signal as in allosteric enzymes. They did not automatically process the substrate into the product; if, for example, the end product of the biosynthetic pathway accumulated (thus indicating no need for further synthesis), it was sensed by the allosteric enzyme and the catalytic activity dropped. Aspartate transcarbamoylase from the de novo pyrimidine nucleotide synthesis pathway fits into this category. It "reads" with its six catalytic subunits the two substrates aspartate and carbamoyl phosphate and transforms them into the product N-carbamoylaspartate. Concomitantly it "reads" with its six regulatory subunits the end product of this pathway CTP. Binding of CTP shifts the K_M for aspartate to a higher concentration. The next step in the evolution of computational devices is proteins that function as molecular switches; an example is CaM kinase. It binds Ca^{2+} complexed with calmodulin. This binding activates the kinase activity and CaM kinase phosphorylates many different other target proteins. Another variation of the scheme is represented by glycogen synthase, the enzyme making the storage form of glucose in animals. It is the target of six protein kinases and several protein phosphatases that add or take away a phosphate group from the enzyme affecting its enzymatic activity. The kinases and phosphatases acting on glycogen synthase themselves come under the control of other signal chemicals, including hormones, such that different organs can now speak with the regulated enzyme. The chemical cross talk possibilities could still be increased when nucleotide-binding proteins were included into the network or protein phosphorylation cascades were invented. Now much more sophisticated switches than simple ON/OFF decisions could be constructed; AND and OR or NOT gates could be built into the metabolic network. It appears that the picture of the blind watchmaker applies also to the design of metabolism. Over long evolutionary periods, the coordination of metabolism was optimized by changing rate and binding constants of enzymes and then their chemical cross talk, all step by step in an interactive random way until the system as a whole performed in a selectively advantageous way. Redesigning the basic rules became from a given degree of complexity impossible. There was only a single way forward: you could only overlay new layers of complexity on the old layers. Interestingly, the increasing complexity in biological systems was actually not achieved by substantially increasing the number of enzymes. In fact, we differ from our gut bacterium *E. coli* only by a factor of 10 with respect to gene number. Apparently, there are constraints on how many different molecular nodes you can introduce into a metabolic network. The control circuits increased substantially in complexity, but less so the number of chemical reactions.

Glycolysis from the side of the chemical intermediates is very similar in our gut bacterium *E. coli* as in our own body. However, *E. coli* has only to integrate this pathway into the needs of a single cell, which is already a great job. In contrast, glycolysis in humans needs the coordination of something like 10^1 organs, 10^2 tissues and about 10^{13} cells (to speak only of orders of magnitudes). It is apparent that the same set of chemical reactions came here under the control of enzymes that have to integrate a far more complex set of signals. The mathematical description of glucose flow in *E. coli* is a showcase of systems biology (more on it further down), while glucose handling in human diabetes is still beyond a detailed understanding despite the large number of biomedical researchers working in this high priority field of contemporary medicine. Before presenting more data on metabolic networks as revealed by systems biology approaches, I want to review some more "traditional" issues around glycolysis.

Variant Glycolytic Intermediates

2,3-Bisphosphoglycerate

First, I want to illustrate the role of variant glycolytic intermediates with two examples. The first one is 2,3-bisphosphoglycerate (BPG) from erythrocytes. The fact that erythrocytes, the dedicated oxygen transporters of our body, derive their energy from glycolysis might sound paradoxical. Apparently, nature found that respiration and dedicated oxygen transport could create conflicts of interest and preferred to muzzle erythrocytes energetically. However, this is not a major handicap because erythrocytes swim in a carefully controlled glucose solution—blood. Glucose is taken up and phosphorylated to glucose 6-phosphate. Then comes an interesting metabolic split into competing pathways within the same cell. Five to ten percent of the glucose is used in the pentose phosphate pathway for NADPH production. The reason is clear: NADPH is needed to cope with the oxidation stress imposed by its transport functions. The remaining 90% of the glucose goes into glycolysis where erythrocytes show an interesting sideway. Part of 1,3-BPG is not transformed in an ATP-generating step to 3-phosphoglycerate. Instead a mutase transforms part of it into 2,3-BPG. *Quite substantial amounts of this compound can accumulate in erythrocytes. A specific phosphatase transforms it back into the glycolytic inter-mediate 3-phosphoglycerate, but this means a missed opportunity with respect to energy gain. The strong selective forces acting on all living systems will assure economical solutions to organisms. If this principle is violated (it is in fact frequently violated because organisms do not search necessarily the cheapest solution, they also have to search adaptable solutions), you can generally make the bet that nature had a hindsight. Against common wisdom, wasteful solutions can even be imposed by selection. If you look at this question at the whole animal level, you can think of the tail of the peacock or the antlers of the extinct Irish elk, which is a metabolically wasteful, even harmful development with respect to predation or nutrition but nevertheless imposed by selection forces (this time by the sexual preferences of females for a big male sexual display organ).* Mean and lean is thus not necessarily the optimal solution in biology. In fact BPG is an important regulator of oxygen transport. It binds to the hemoglobin and lowers hemoglobin's affinity for oxygen by stabilizing the T state. The physiological level of 5 mM BPG in the blood (this is a substantial amount and equals the steady-state concentration of blood glucose) assures that 38% of the oxygen cargo is delivered in the peripheral tissue. If you now climb without adaptation to high altitudes, you have a problem. At 4,500 m above sea level, the partial oxygen pressure is only 7 kPa, less oxygen is bound in the lungs, and the hemoglobin would only release 30% of its cargo in the tissue. As a quick fix solution, the blood BPG level will have risen after a few hours to 8 mM. This rise shifts the hemoglobin oxygen binding, assuring again 37% oxygen release. After returning to the sea level, the BPG levels decrease again. A sideway of glycolysis thus becomes an important regulator of respiration.

Fructose 2,6-Bisphosphate

Another example is fructose 2,6-bisphosphate. The enzyme, which carries the impossible name of 6-phosphofructo-2-kinase/fructose 2,6-bisphosphatase, leads to the synthesis of this compound that resembles another glycolytic intermediate. As even biochemists cannot pronounce easily this tongue twister, this enzyme is shortly called PFK-2. This enzyme is really a maverick: if you read its name it does one thing (kinase: a phosphate transfer) and its opposite (phosphatase: hydrolysis of a phosphate group). Actually, if you look at its domain structure, these are two fused enzymes. Not enough with that oddity, it synthesizes fructose 2,6-bisphosphate, which closely resembles the glycolytic intermediate fructose-1,6 bisphosphate.

Phosphofructokinase, also called PFK-1, is the committing step into glycolysis and the most important regulator of the flux of glycolysis. The enzyme is inhibited by ATP; quite logically, glycolysis decreases when the energy charge of the cell is high. PFK-1 is also inhibited by H^+. This control loop prevents lactate dehydrogenase excessively reducing pyruvate to lactate. Phosphofructokinase is further inhibited by citrate. High citrate concentrations signal ample supply of biosynthetic precursors to the citrate cycle. Additional glucose is thus not needed to fill up these pools. However, the most powerful activator of PFK-1, which overrides the other signals, is fructose 2,6- bisphosphate synthesized by PFK-2. This compound increases the affinity of PFK-1 for its substrate from a sigmoidal to a hyperbolic velocity–concentration curve and offsets the inhibitory effect of ATP on PFK-1. This is now another case where a close relative of a glycolytic intermediate fulfills a regulatory role. *Are these relics from the try-and-error phase of early glycolysis when many compounds were explored for their suitability to construct glycolysis, the major metabolic highway of cellular life? Did one compound actually made it to a pathway intermediate and the other got the consolation price to be an important regulator? Or did Nature try on purpose the related compounds for regulation that are sufficiently similar to existing intermediates that they can be made and used by modifications of the available enzymes, while at the same time being sufficiently different to allow separate control?*

Lactate and Ethanol Fermentation: A Bit of Biotechnology

Pyruvate is a very versatile intermediate. In the textbooks of biochemistry, it is the end product of glycolysis, but due to the need for reoxidation of NADH, the carbon metabolism cannot stop here. A one-step reaction fulfilling this requirement is the reduction of pyruvate to lactic acid by lactate dehydrogenase. In our body, only L-lactate is synthesized because our lactate dehydrogenase strongly prefers hydrogen transfer from the A site of NADH. This is not the case in all organisms. In lactic acid bacteria, some lactate dehydrogenases prefer the A site, others the B site of NADH; the latter lead to the production of D-lactate. Some bacteria produce mixtures of D,L-lactic acid.

Lactic Acid Bacteria

Streptococcus thermophilus, which ferments yogurt in symbiosis with *Lactobacillus bulgaricus*, is a homolactic starter bacterium (Figure 2.1). This means it produces primarily lactate from lactose. It excretes this lactate as a waste product together with protons despite the fact that it still contains a lot of chemical energy. This excretion of protons is driven by the membrane-bound bacterial ATPase. Lactate excretion also leads to an acidification of the fermented milk, leading to a precipitation of milk proteins and the buildup of a semisolid food matrix. Due to their peculiar metabolism, lactic acid bacteria develop an acid resistance (down to a pH of 3.5) that is greater than that of most other bacteria. In fact, during the spontaneous fermentation of cabbage to sauerkraut, which relies on the bacteria naturally associated with the vegetables, one sees a succession of different populations of lactic acid bacteria (*Leuconostoc* sp. followed by *Lactobacillus plantarum*, which is likely driven by a phage that kills *Leuconostoc*). The succession is also dictated by their different degrees of acid resistance. The final products, sauerkraut or yogurt, are relatively stable and can be stored much longer than fresh lettuce or milk, respectively. However, the exclusive production of lactate by streptococci is observed only when the cells grow in the excess of substrate like lactose in milk. When the substrate is offered in growth-limiting amounts, some streptococci change to further exploitation of pyruvate as far as

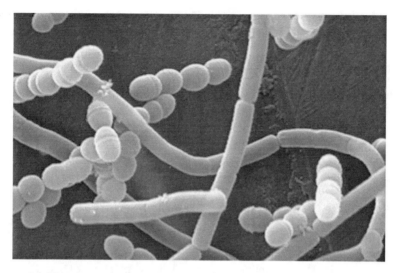

FIGURE 2.1. The picture shows the bacterial consortium (starter bacteria) which achieves yogurt fermentation. The culture consists of *Streptococcus thermophilus*, the short chains of globular cocci in the picture, and *Lactobacillus bulgaricus*, the chains of elongated cells. Despite their different morphology, both bacteria are phylogenetically closely related low GC content Gram-positive bacteria. The scanning electron microscope reveals cell division, but nothing from the interior of the bacterial cell.

their catalytic activities reach. Pyruvate is, for example, decarboxylated, and the energy of the reaction is stored in the thioester bond of acetyl-CoA. However, streptococci do not possess a Krebs cycle. They had to invent another way to exploit the energy contained in acetyl-CoA. They achieve this by the transfer of the acetyl group on a phosphate. The acetylphosphate created in this reaction can then transfer the phosphate on ADP, creating another molecule of ATP and acetate. Many other chemical end products of fermentation exist that give the specific fermented food its flavor and attraction to the human consumer. Some are also more decorative like in Swiss cheese production where the end products are propionic acid, giving this cheese its particular flavor, and CO_2, which gives decorative holes in the Swiss cheese. Lactic acid bacteria can even produce ethanol by two reduction steps of acetyl-CoA to acetaldehyde and then to ethanol.

Zymomonas and Ethanol

In fact, the dairy industry is second in the food industries and follows the largest branch, producing alcoholic beverages. Yeasts play the dominant role in industrial alcohol fermentation, but the bacterium *Zymomonas* might play an increasing role in the future. Both organisms reach the same maximum alcohol concentration of 12% by fermentation. However, under batch fermentation and continuous fermentation conditions, *Zymomonas* produces about 10-fold higher amounts of ethanol per biomass. In other words, less biomass has to be disposed of for the same amount of ethanol produced, which is of technical interest. *Zymomonas* uses the Entner–Doudoroff pathway for sugar degradation, and it yields only half of the ATP than glycolysis provides to yeast, hence the produced biomass is much smaller.

The brewing of beer is a complicated process, some would say even an art. The yeast *Saccharomyces cerevisiae* can ferment glucose to ethanol. Under anaerobic conditions, this is the only mode of energy production. In the presence of oxygen, respiration occurs. Since glucose respiration yields much more ATP than glucose fermentation, the yeast cell must compensate this yield difference by a higher glucose consumption under anaerobic condition when compared to aerobic condition ("Pasteur effect"). However, alcoholic fermentation may set in even under aerobic conditions if the glucose concentration surpasses a critical threshold value ("Crabtree effect"). This metabolic flexibility makes yeast a difficult organism for the food industry. Mastering its ethanol production capacities by a timely change from oxygenic to anoxygenic metabolism in yeast makes beer making an art. Industrially yeast is thus not an easy beast to tame.

Genetic Engineering

Zymomonas has an exclusively anaerobic metabolism and lends itself to a more straightforward industrial processing. *Zymomonas* has traditionally been used for the production of alcoholic beverages. For example, the popular Mexican drink pulque is made with *Zymomonas* from the sap of the agave plant. It has thus some industrial potential, but it can only use a relatively limited number of sugars as carbon substrates for alcohol production: glucose, fructose, and sucrose. These are valuable sugars also needed in food and feed, but what about less valuable carbohydrate sources like xylose. The latter is a waste material produced as a by-product of industrial pulp and papermaking. What is the prospect of metabolic engineering? Will it be possible to introduce genes into *Zymomonas* that allow its growth on waste containing xylose? Genetic engineers constructed a shuttle vector that can travel between *E. coli* and *Zymomonas*, which carries two xylose assimilation genes and a *trans*-ketolase and an aldolase that can funnel the pentose sugar via the pentose pathway into the Entner–Doudoroff pathway (Zhang et al. 1995). The xylose genes were placed under the control of a strong constitutive *Zymomonas* promoter, namely that of the glyceraldehyde dehydrogenase gene. The recombinant bacterium did what the researchers hoped—it could grow on glucose or on xylose, and it converted xylose to ethanol at high yield. This was not a trivial result. Not only does it allow gaining a valued compound from a waste product (perhaps not in your beer, but as bio-fuel in heating or in a car), it has even important theoretical implications. The experiment tells that you can add a few genes, which open new nutritional possibilities to a microbe. By acquiring a few crucial genes, it can conquer a new environment. The metabolism of this organism is not in such a poised equilibrium that any new metabolic trafficking would upset the cell. This result is also of substantial theoretical interest for the metabolic network discussion, which we will touch in a different section. Encouraged by these results, the genetic engineer tried to open other food sources for *Zymomonas*. Xylose is mainly found in hardwood, but there are also valuable energy crops like switchgrass that contain large amounts of arabinose. The researchers introduced into the same shuttle plasmid three arabinose-degrading genes. The strategy was the same, and it paid off: *Zymomonas* could also grow on arabinose. Vice versa, Lonnie Ingram has tried to introduce *Zymomonas* genes into *E. coli* to produce ethanol in this workhorse of the biotechnology industry (Ohta et al. 1991; Tao et al. 2001). Since *E. coli* has many pathways starting from pyruvate, the construction of such strains also necessitated the inactivation of *E. coli* genes to channel the metabolism into the desired direction. *Klebsiella oxytoca*, a cousin of *E. coli*, thus became an ethanol-producing bacterium by genetic engineering.

A Short Running Exercise

Energy Stores for Muscles

Animal life in contrast to plant life is defined by a fundamental property: locomotion. To move around you need muscles. The movement of muscles is powered by molecular motors, which consist of thin and thick filaments that slide past each other during contraction. The molecular interplay of myosin and actin is powered by ATP. We run differently when we anticipate a short distance sprint intended for catching a prey or escaping a predator (or nowadays for a sports event) or when we envision a medium or long distance run. We dose the running speed differently for the simple reason that we cannot sustain the sprinter's speed for very long. The reason for this behavioral adaptation is our empirical knowledge of our physiology. Yet this empirical knowledge can be rationalized in the light of our biochemical knowledge. The stores of ATP in the muscles are low. They were calculated to be 200 mmol for a 70-kg person with a total muscle mass of 28 kg. This keeps you running at full speed, perhaps for 2 s. Therefore for a 100-m catch or escape sprint you need to tap another energy resource. This is creatine phosphate; our reference person contains of it the equivalent of 400 mmol ATP. It is biochemically more inert than ATP; hence, the cell exploits it as a transient energy buffer. As creatine phosphate can directly transfer its high-energy phosphate to ADP, ATP is quickly regenerated. However, this store does not bring you to the 100-m mark. Therefore you need to exploit the next energy store: glycogen, a polymeric form of glucose residues. Muscles have by far the greatest glycogen store of the human body followed by the liver. According to the energetics of the glycolytic and respiratory metabolism, the equivalent of the same amount of muscle glycogen of our test person is worth either 7,000 or 80,000 mmol ATP. However, there is a difference between both modes of glycogen use. Your running speed will decide about life or death: you get the prey animal you want to eat, which assures your nutritional survival over the next days or weeks, or you don't get it with potential dire nutritional consequences. Or stated more dramatically, you fall victim to a predator or you happily escape from this attack, assuring your physical survival. Running speed is a direct function of the maximal *rate* of ATP production, and it differs between glycolysis and respiration. If you use glycolysis, you get ATP relatively quickly from substrate-level phosphorylation. If you bet on the respiratory chain, you get more for your money, but it takes longer to get to the superior level of ATP. The reason is simple: Metabolizing glucose to CO_2 needs more time because there are more chemical reactions and more cellular compartments involved. The glycolytic rate is about twofold higher than the oxidative rate of ATP production. A human sprinter will thus opt for glycolytic use of glycogen, and this brings him over the 100-m distance. The muscle recalls this running spree: After the run, its ATP level is down from 5.2 to 3.7 mM, its creatine phosphate has diminished from 9.1 to 2.6 mM, while the blood lactate level is up to 8.3 from initially 1.6 mM. During intensive exercise, the blood is flooded with lactate from the muscle. It is the job of the liver (and partly the kidney) to take care of lactate. Both organs synthesize glucose from lactate via gluconeogenesis. They deliver this newly synthesized glucose into the blood stream. The muscle can move on in using glucose to sustain its strenuous exercise.

Liaison Dangereuse: Lactate, Cancer, and the Warburg Effect

Imaging Techniques

Modern imaging techniques in medicine have revolutionized the biochemical investigation that can now also be made with human beings. Here I will discuss only one technique, positron-emission tomography (PET). This technology uses short-lived isotopes of carbon (^{11}C), nitrogen (^{13}N), oxygen (^{15}O), and fluorine (^{18}F), which are produced in a cyclotron. Their half-life ranges from 2 (^{15}O) to 110 min (^{18}F) and represents therefore only a low-radiation burden to the subject. These instable isotopes loose a positron and fall back to a stable nuclear level. The positron carries the mass of an electron, but in contrast to the negatively charged electron, the positron carries a positive charge. After emission from the instable isotope, the positron travels a few millimeters in the tissue until it meets an electron. This leads to a matter–antimatter collision, mass annihilation, and emission of gamma rays. A positron camera, which is arranged cylindrically around the subject, measures this emission. A computer calculates the position of the collision and constructs virtual cuts of the patient, which are called tomographs. The instable isotopes allow the labeling of many biochemical substrates. Take glucose: Glucose is transformed into 2-deoxyglucose (a hydrogen atom replaces the hydroxyl group) and then into 2-^{18}F-2-deoxyglucose, or FDG (the instable fluorine replaces this hydrogen). Fluorine takes about the size of the hydroxyl group and can thus fool the cells. It is taken up like glucose and is phosphorylated by hexokinase like glucose. However, the next enzyme of glycolysis is cleverer, it denies further metabolism. In tissues with low glucose 6-phosphatase, the glucose analogue is now confined to the cell (metabolic trapping). Tissues with higher glucose turnover can now be localized. The turnover can even be quantified and expressed as μmol glucose/100 g tissue × min. Since the brain gets more than 99% of its energy from glucose and since the technique has a substantial anatomical resolution when combined with nuclear spin tomography, PET can now localize brain areas in the active subject performing different tasks (calculations, speaking, reading, etc.). FDG-PET has another important clinical application: It detects malignant tumors with a specificity and sensitivity near to 90% if the cancer is greater than $0.8 \, cm^3$.

Warburg Effect

The biochemical basis for this diagnostic method goes back to one of the founding fathers of biochemistry. Nearly 70 years ago, Otto Warburg observed that in cancer cells the energy metabolism deviates substantially from that of the normal surrounding tissues. The activity of a number of glycolytic enzymes, like hexokinase, phosphofructokinase, and pyruvate kinase, and the glucose transporter are consistently and significantly increased. Glycolysis proceeds about 10-times faster in most solid tumors than in neighboring healthy cells. The Warburg effect is in fact the negation of the Pasteur effect: While normal cells change their energy metabolism from glycolysis to aerobic respiration as oxygen becomes available, the accelerated glycolysis of the cancer cell is maintained in the presence of oxygen. It was hypothesized that this phenotype is a consequence of early tumor growth, when clusters of cells grow without any vascularization. The absence of blood vessels means no oxygen, and the precancerous cells get an energy problem that they try to fix by an increased glycolysis. In fact, the center of these cancerous lesions frequently shows a necrotic zone. It was therefore logical to implicate an adaptation to hypoxia (low oxygen pressure) as the key to this phenomenon. Indeed, hypoxia-inducible factor 1 (HIF-1) turned out to be a key transcription factor. It upregulates a series of genes involved in glycolysis, angiogenesis, and cell survival. HIF-1 is a heterodimer of two constitutively transcribed subunits HIF-1α and HIF-1β. The complex is regulated via instability of HIF-1α. Under normal oxygenic conditions, HIF-1α undergoes ubiquination, i.e., it gets a chemical death certificate fixed at it such that it is proteolytically degraded in a proteasome—a destruction complex that dutifully destroys all proteins with this ubiquitin signal. As one would expect for such a destructive signal, its allocation is under careful control. Before the death tag can be fixed, a critical proline residue must first be hydroxylated by one enzyme, and this enzyme is active only in the presence of oxygen and iron. After this first chemical modification, a tumor suppressor protein binds, and ubiquination can occur (Lu et al. 2002). So far, so good. HIF-1 plays an important role in normal cellular metabolism when the oxygen level falls as in the exercising muscle where it is only one tenth of that in the nearby capillary blood. Here it makes sense when HIF-1 increases the flux through glycolysis. Mice, which got their HIF-1 gene knocked out, do not show this increased glycolysis; they do not accumulate lactate while muscle ATP remains at the same level, thanks to an increased activity of enzymes in the mitochondria. Mutant mice show first a better endurance under exercise conditions. However, under repeated exercise, the muscles of these mutants showed extensive histological damage. HIF-1 thus achieves a self-protection of the muscle (Mason et al. 2004). However, in tumors HIF-1 is not inactivated, even when oxygen is present. Tumor cells do not change for oxygenic respiration. Glycolysis remains high, and lactate levels elevated. It was even argued that lactate became a chemical club for the cancer cell, not unlike how lactic acid bacteria eliminate competing bacteria by lactate production. The malignant cell developed during carcinogenesis a marked acid resistance, which is not found in the healthy surrounding cells. The cancer cell can thus poison the tissue and erode its way through the tissue into the next blood vessel to build metastases (Gatenby and Gillies 2004). We see here how a metabolic end product gets, perhaps, a selective function in the fight for resources, here for a malignant cell clone against its parent organism. We will later see another comparable case in cyanobacteria, which evolved oxygen in their new metabolism, which became a poison for competing, but oxygen-sensitive bacteria. Otto Warburg predicted that we would understand malignant transformation when we understand the cause of increased glycolysis in cancer cells. We understand now some parts of the control cycle from hypoxia over HIF-1 to increased glycolytic flux. One of the latest news at the time of the writing was the binding of HIF-1 to a DNA consensus sequence called hypoxia response element (HRE) in the promoter region of PFK-2. Binding of HIF-1 increases the transcription of this important regulator of glycolysis.

Glucokinase at the Crossroad of Cellular Life and Death

Apoptosis

Cancer cells not only have an increased metabolic rate but are also less likely to commit suicide when damaged. Cell suicide is an everyday event in multicellular organisms and occurs by a carefully orchestrated process called apoptosis. That energy metabolism and cell death are intimately linked processes was already indicated by the fact that mitochondria play a key role in apoptosis. Many apoptotic signals converge on mitochondria where they cause the release of cytochrome c from mitochondria. Cytochrome c in the cytoplasm in turn triggers the activation of a group of proteins called caspases, which lead to an ordered cell destruction from within. Recently glucose and a key enzyme of glycolysis became implicated into another pathway leading to apoptosis, further strengthening the links between energy metabolism and programmed cell death. The proapoptotic player is designated with the acronym BAD. The activity of BAD is regulated by phosphorylation in response to growth and survival factors (Downward 2003). In liver mitochondria, BAD exists in a large protein complex consisting of two proteins that decide on the phosphorylation status of BAD. One is the protein kinase A, which still needs an anchoring protein to get into this complex, and the other is protein phosphatase 1. The fifth partner was, however, a surprise (Danial et al. 2003). It was also a kinase; specifically it was glucokinase, a member of the hexokinase family. This enzyme catalyzes the first step of glycolysis, it phosphorylates glucose to glucose 6-phosphate, and is also involved in glycogen synthesis in the liver cell. Addition of glucose to purified mitochondria induced the phosphorylation of BAD. Liver cells deprived of glucose go into apoptosis unless BAD is also missing. Glucokinase activity was markedly blunted when the three phosphorylation sites on BAD were mutated. Insulin stimulates in the liver cell not only glucose transport, but recruits hexokinase to the mitochondria and stimulates glycolysis and ATP production. Conversely, withdrawal of growth factors decreases the glycolysis rate, O_2 consumption, and closes the VDAC channel, which mediates the ADP for ATP exchange over the mitochondrial membrane. This channel closure is also linked to cytochrome c release from the mitochondria. In this way, the glucose metabolism is crucially linked to cell survival at least in liver cells.

Metabolic Networks

Quantitative Biology

To understand the complexity of biological systems, we must analyze it. This means that we have to dissect it either literally with a lancet or intellectually by teasing apart the constituting elements. In this dissected way, we learn biology in the university textbooks. Individual pathways are presented chapter by chapter; only in later chapters of the textbooks, some integration of metabolism is offered. So far as learning is concerned, this is a fine method. However, humans tend to mix the representation of a thing with the real thing. In a word of Zen philosophy, the finger, which points to the moon, is not the moon. The analytical method needs the synthetic method as a necessary complementation. This is still relatively easy in chemistry where analytic and synthetic chemistry are classical and complementary branches. In view of the awful complexity of biological systems in chemical terms, biologists tend to be highly satisfied when they have pulled the molecules apart. However, in simple, but highly investigated systems like the gut bacterium *E. coli*, engineers and physicists started to synthesize the whole picture from its constituents by using computational methods. The mathematical treatment of the metabolism has a tradition that dates back into the 1970s. To predict cellular behavior, each individual step in a biochemical network must be described with a rate equation. As these data cannot be read from genome sequences, and since empirical data are lacking, Edwards and Palsson (2000) tried to model metabolic flux distributions under a steady-state assumption. Based on stoichiometric and capacity constraints, the *in silico* analysis was able to qualitatively predict the growth potential of mutant *E. coli* strains in 86% of the cases examined. In follow-up studies, *in silico* predictions of *E. coli* metabolic capabilities for optimal growth on alternative carbon substrates like acetate were confirmed by experimental data (Edwards et al. 2001).

Optimal Growth Rate Planes

Annotated genome sequences can be used to construct whole cell metabolic pathways. In particularly well-investigated systems like *E. coli* and *S. cerevisiae*, limiting constraints can be imposed on the *in silico* calculated model based on mass conservation, thermodynamics, biochemical capacity, and nutritional environment such that optimal growth rates can be calculated for common carbon substrates. For *E. coli* strain K-12, optimal growth rate planes (phenotype phase plane analysis) were calculated for substrate and oxygen uptake at different temperatures, which avoided both futile cycling and excessive acidic waste excretion. When they did this calculation with different substrates, they found growth rates on the line of optimality for a number of substrates (Ibarra et al. 2002). The absolute growth rate decreased successively when going from glucose to malate, succinate, and acetate. When the cells were serially subcultured on a given substrate, a small but significant increase in growth rate of about 20% was observed. This means that the cells were overall more or less well poised to make their living from these carbon sources, but small increases could still be obtained by selection procedures. However, the predicted optimal growth of *E. coli* on the carbon source glycerol was not achieved by the cells. Was the model wrong? Interestingly, after 700 generations of *E. coli* growth on glycerol the growth rate nearly doubled and came close to the optimal growth rate predicted by the *in silico* model. Apparently, the starting organism was never adapted to growth on glycerol, but could achieve the adaptation by a trial-and-error process demonstrating substantial flexibility in the metabolic network of *E. coli* toward changes in its food source.

Fiat Flux

Other scientists used the central metabolism of *E. coli* for elementary flux mode. This major pathway contains 89 substances and 110 reactions connected in 43,000 elementary flux modes (glucose is involved in 27,000 of these modes). Corresponding to biological intuition, glucose is a much more versatile substrate than acetate: Glucose, in this model, can be used in more than 45 different ways than acetate (Stelling et al. 2002). Glucose was also the only substrate that could be used anaerobically without additional terminal electron acceptors. Their calculations showed a remarkable robustness of the central metabolism. Mutants with significantly reduced metabolic flexibility (important nodes were removed affecting the flow along certain lines) still showed a growth yield similar to wild type. This calculation fits well with gene knockout experiments in *E. coli*: fewer than 300 out of the 4,000 genes are essential in the sense that the deletion of one of them prevented growth in a rich medium (Csete and Doyle 2002). Also, the network diameter did not change substantially. The calculations showed that the cell has to search a trade-off between two contradictory challenges. On one side is flexibility, i.e., the capacity to realize a maximal flux distribution via redundant node connections, and on the other side is efficiency, i.e., an optimal outcome like cell growth with a minimum of constitutive elements like genes and proteins. This trade-off control should correlate with messenger RNA (mRNA) levels. The metabolic network structure should help to understand the large amount of mRNA expression data, nowadays provided by microarray analysis. *In silico* prediction fitted relatively well with mRNA expression data from *E. coli* growing on glucose, glycerol, and acetate. An editorial on this report noted that biochemical progress has been driven in the past mainly by new observations rather than theories. Stoichiometric analysis will not only push biology more into the direction of the "exact" sciences like chemistry, but also boost hypothesis-driven experimental research in biology (Cornish and Cárdenas 2002). Recent data underline that this might not be a vain hope. Bioengineers developed the first integrated genome-scale computational model of a transcriptional regulatory and metabolic network for *E. coli*. This model accounts for 1,010 genes of *E. coli* including 104 regulatory genes that control the expression of 479 genes (Covert et al. 2004). The model was tested against a data set of 13,000 growth conditions. The growth matrix tested a large number of knockout mutants for growth on media that varied for carbon and nitrogen food sources. Remarkably, the experimental and computational data agreed in 10,800 of the cases. The discrepancies can be used to update the *in silico* model by successive iterations, and it can also generate hypotheses that lead to uncharacterized enzymes or noncanonical pathways. The updated *E. coli* model can now predict a much higher percentage of the observed mRNA expression changes and highlights a new paradox like the reduction of mRNA levels for a regulatory gene when the protein was in fact activated. System biologists can now start to guide the experimentalist at least in well-investigated prokaryotic systems such as *E. coli*.

Highways ...

In silico predictions of *E. coli* metabolic activities must account for 537 metabolites and 739 chemical reactions (*E. coli* has a 4 Mb genome). Their interconnection results in a complex web of molecular interactions that defies the classical metabolic maps commonly pinned to the walls of research laboratories. The group around Barabasi used the Edwards and Palsson model on *E. coli* but used a less detailed analysis (topology versus stoichiometry). They ran the explicit calculations with two out of the 89 potential input substrates of *E. coli*, namely on a glutamate-rich or a succinate-rich substrate. The biochemical activity of the metabolism is dominated by several "hot" reactions, which resemble metabolic superhighways (Almaas et al. 2004). These highways are embedded in a network of mostly small-flux reactions. For example, the succinyl-coenzyme A synthetase reaction exceeds the flux of the aspartate oxidase reaction by four orders of magnitude. The results are somewhat disappointing because they are obvious and confirm what we knew from the older metabolic wall charts.

... And Small Worlds

In a follow-up study, the same group worked with the genome sequences from 43 organisms representing all three domains of life and compared their metabolic networks (Jeong et al. 2000). On the basis of the gene annotations, pathways were predicted and data from the biochemical literature were integrated into the model. They based the analysis on a mathematical theory developed in 1960, the classical random network theory. The biochemical network was built up of nodes, the substrates, which are connected to one another through links, which are the actual metabolic reactions. Their first question concerned the structure of the network. Was it exponential or scale free? In biological parlance, this reads, Do most nodes have the same number of links, which follow a Poisson distribution, or does this distribution follow a power-law? That is, most nodes have only a few links while a few, called hubs, have a very large number of links. The answer— according to this study—was clear-cut. In all three domains of life, the power-law described the network structure. This means that a handful of substrates link the metabolic ensemble. Then the physicists investigated the biochemical pathway lengths. Many complex networks show a small-world character, i.e., any two nodes can be connected by relative short paths along existing links. The physicists calculated in their graph-theoretic analysis an average path length of about 3. However, their computer program considered enzymatic reactions as simple links between metabolites. When another group used a more detailed approach considering the pattern of structural changes of metabolites (from the traditional biochemical perspective a much better alternative), the average path length became 8. They summarized the difference in the title of their paper: "The Metabolic World of *Escherichia coli* is Not Small" (Arita 2004). The discussions become somewhat difficult to follow for biologists having just a biochemical education level, but no training in mathematics. However, problems even with sophisticated computer programs become evident for simple-minded biologists. For example, you can grow *E. coli* in the laboratory using just D-glucose as sole carbon source. Thus in reality, all carbon atoms in any metabolite must be reachable from D-glucose. In the best current calculations, this applies only for 450 out of the 900 *E. coli* metabolites. The small-world character of the programs might thus reflect more of their incapacity to reach the other half of the metabolites from glucose than their real short pathway length. Arita (2004) suspected that many carbon atoms in metabolites become reachable from D-glucose via cyclic metabolic pathways. Especially the TCA cycle might be the maelstrom that equalizes the carbon atoms for all cellular metabolites pointing just to another central function of this central turning wheel of metabolism.

Yeast: Many Nonessential Genes

To avoid a too theoretical section, I will mention two biological problems on which metabolic network analysis was applied in a way that is directly understandable for mainstream biologists. To set out the problem, let's start with a basic statement. Gene disruption is a fundamental tool of molecular genetics, where the phenotypic consequence of the loss of the gene function can be determined. We are so much impressed by the effect of some mutations especially in the field of medical genetics that one might a priori expect that most genes are needed for survival of a given organism in its habitat. For organisms with facile genetic methods and known genome sequence, a systematic experimental approach to this problem can be done. For example, 96% of the 6,000 genes of the baker's yeast *Saccharomyces cerevisiae* (Figure 2.2) were neatly replaced by a deletion cassette, and the resulting mixture of mutants was tested in a rich medium for growth (Giaever et al. 2002). This procedure allowed to assess growth defects as small as a 12% decreased growth rate compared with the parental wild type. The surprise was great: Under laboratory conditions, 80% of yeast genes seem not to be essential for viability. As the rich medium might hide the need for genes, the researchers conducting this genetic tour de force used growth conditions that asked adaptive responses from the yeast cells. The yeasts were grown with restrictions in amino acid availability and changes in sugar carbon source, osmolarity/salinity, and alkaline pH. These more selective conditions added only few further genes to the list of essential functions; this was most marked in the alkali selection protocol, which revealed 128 alkali-sensitive mutants. Another surprise was the observation that genes that are upregulated in expression studies during these selective conditions were mostly dispensable for the survival of the cell.

Yeast: Genome Duplication

There are different possible explanations for these surprising results. One was actually proposed by the scientists of this study: They ascribed the low number of essential genes to the highly duplicated nature of the yeast genome. Indirect

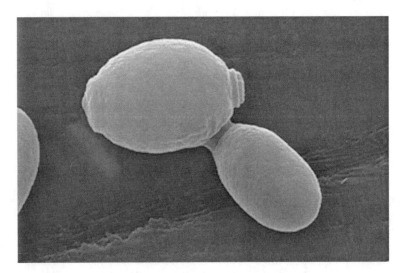

FIGURE 2.2. *Saccharomyces cerevisiae*, a yeast used in wine, beer, and bread production, rivals the role of lactic acid bacteria for importance in food processing. This yeast has been dubbed the *E. coli* of the eukaryotes because it is the best-investigated "higher" organism. The picture shows how daughter cells bud from a yeast mother cell. Note that *S. cerevisiae* is not much longer than many bacteria.

molecular evidence that the entire yeast genome has been duplicated was already around since 1997 (Wolfe and Shields 1997). An elegant confirmation of this hypothesis was provided by scientists from Boston, who sequenced another yeast, *Kluyveromyces waltii*, and compared it to the genome of *S. cerevisiae* (Kellis et al. 2004). The trick was that *Kluyveromyces* had diverged from the ancestor line of *Saccharomyces* before this duplication event. By a comparative genome analysis, it became apparent that the two genomes were related by a 1:2 mapping, with each region of *Kluyveromyces* corresponding to two regions of *Saccharomyces*, as expected for whole genome duplication. The analysis demonstrated that *S. cerevisiae* arose from complete duplication of eight ancestral chromosomes and subsequently returned to functionally normal ploidy by massive loss of nearly 90% of the duplicated genes. The *S. cerevisiae* genome is today only 13% greater than the *K. waltii* genome. The loss occurred in small steps involving at the average two genes. The losses were balanced and complementary in paired regions, preserving at least one copy and virtually each gene in the ancestral gene set. When they looked more carefully into the 450 duplicate gene pairs, they observed frequently accelerated evolution in one member of the gene pair. Apparently, one copy tended to preserve the original function, while the other copy would be free to diverge. The derived gene tends to be specialized in function, in their cellular localization or in their temporal expression and sometimes developed a new function. This interpretation fits nicely with genetic analysis conducted 1 year earlier by other yeast geneticists (Gu et al. 2003b). They distinguished singleton and duplicate genes in the yeast genome. When they created null mutants for these genes, they observed that duplicate genes had a significantly lower proportion of genes with a lethal effect of deletion than singleton genes (12 vs. 29%). They found, with lesser statistical support, that duplicate genes showed a lesser degree of mutual functional compensation when their degree of genetic divergence had increased. Duplicate genes were also more flexible with respect to singletons when tested through different growth conditions.

Yeast Metabolic Network

However, the authors of this study admitted that the high frequency of genes that have weak or no fitness effects of deletion call for alternative interpretations like compensation through alternative pathways or network branching. And with this hypothesis, we are back to metabolic network analysis. L. Hurst and colleagues analyzed this alternative explanation by using an *in silico* flux model of the yeast metabolic network (Papp et al. 2004). The model included 809 metabolites as nodes, which are connected by 851 different biochemical reactions when transport processes of external metabolites were included. Their analysis indicated that seemingly dispensable genes might be important, but only under conditions not yet examined in the laboratory. This was their dominant explanation for apparent dispensability accounting for a maximum of 68% of the dispensable genes. Compensation by duplicate genes accounted for at most 28% of these events, while buffering by metabolic network flux reorganization was the least important process. In fact, they observed that the yeast metabolic network had difficulties in tolerating large flux reorganizations. Isoenzymes were selected in order to enhance gene dosage, which then provides higher enzymatic flux instead of maintaining alternative pathways. These data have important implications for our perception of genome organization. Apparently, despite its ancient duplication event, yeast is not a peculiar case. Other model organisms like *E. coli* (Gerdes et al. 2003), *Bacillus subtilis* (Kobayashi et al. 2003), and *Caenorhabditis elegans* (Kamath et al. 2003) yielded only 7–19% of essential genes for laboratory growth. This contrasts well with the high percentage of the essential genes in the *Mycoplasma genitalium* genome of up to 73% (Hutchison et al. 1999). In this series, *Mycoplasma* is clearly the odd man out. Its genome is apparently the result of secondary genome reduction occurring in an intracellular parasite with strict host and tissue specificity. This parasite has given up most of its "dispensable" genes simply because it has given up the idea to conquer alternative habitats beyond its super-specialized niche.

Lateral Transfers into Networks

The situation is different for *E. coli*. Against popular belief, its niche is not the genetics and molecular biology laboratory, but the intestine of mammals and birds. This is a complicated and highly competitive microbial community in a dynamic physiological context where the host takes precautions to expel the gut commensals. Once out of this niche, *E. coli* must survive in the environment before it reaches the next gut; recent data suggest that it does so in soil. *E. coli* became a successful pathogen in mammals and birds, small wonder that different *E. coli* strains differ by 0.5–1 Mb in their genome content. Most of these gene acquisitions were by horizontal gene transfer. The question is now how metabolic networks incorporate new genes acquired during adaptive evolution in bacteria. Gene duplication, the main source of evolutionary novelty in eukaryotes, plays only a minor role in *E. coli. E. coli* K-12 contains few duplicated enzymes in its metabolic network, and all but one seem to be ancient (Pal et al. 2005). Most changes to the metabolic network of *E. coli*, which occurred in the past 100 million years since its divergence from *Salmonella*, are due to horizontal gene transfer (Figure 2.3). Only 7% of the genes that are horizontally transferred into the metabolic network of *E. coli* are essential under nutrient-rich laboratory growth conditions. Genes that contributed most to the evolution of the metabolic network and thus to the evolution of proteobacteria were generally environment specific. The flux balance analysis of these authors also explains why most of these horizontally transferred genes are not expressed under laboratory conditions. They concluded that the evolution of networks is largely driven by adaptation to new environments and not by optimization in fixed environments. Proteins contributing to peripheral reactions (nutrient uptake, first metabolic step) were more likely to be transferred than enzymes catalyzing intermediate steps and biomass production.

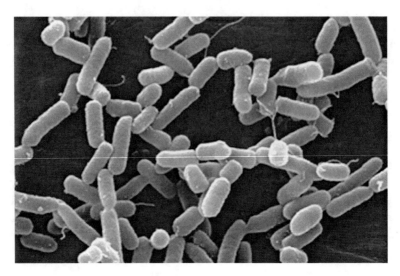

FIGURE 2.3. *Salmonella typhimurium*, a close relative of the gut bacterium *Escherichia coli*, is the prototype of Gram-negative bacteria belonging to the Proteobacteria group. They resemble each other morphologically, but differ in their pathogenic potential.

Salmonella in Mice

It is a laboratory abstraction to analyze the metabolic network of a microorganism growing as an isolated culture in a broth medium. In nature most microorganisms occur in association where metabolites are exchanged between different bacterial species. In the case of commensals or pathogens, bacteria also acquire metabolites from the eukaryotic host. Metabolic network reconstruction for an isolated microbe thus gives only an incomplete picture. This was recently realized by German scientists who tried to identify targets for new antimicrobials derived from a model of in vivo *Salmonella* metabolism (Becker et al. 2006). They used two mouse *Salmonella* infection types, the typhoid fever and the enteritis model. Proteome analysis identified 228 and 539 metabolic enzymes in the spleen and the cecum from the typhoid and enteritis model, respectively. The *Salmonella* genome probably encodes 2,200 proteins with putative metabolic functions. The proteome analysis underestimated the true number of expressed proteins due to the technical challenge to differentiate minor bacterial proteins against a background of high host proteins. However, a large discrepancy remains between both figure sets, which is explained by the authors with two interpretations. First, the metabolic network of *Salmonella* shows extensive metabolic redundancies reducing the number of truly essential bacterial genes. Second, *Salmonella* has apparently an access to a surprisingly diverse pool of host nutrients. Auxotrophic mutants of *Salmonella*, which could not grow without supplements in broth culture, retained substantial virulence in the typhoid fever model. Apparently *Salmonella*, which resides in the supposed nutrient-poor murine macrophage phagosome, still finds access to various amino acids, purine and pyrimidine nucleosides, glycerol, sialic acid, hexoses, pentoses, vitamines, and electron acceptors for both aerobic and anaerobic catabolism. With this section, we get to the next level of complexity where metabolic networks of interacting organisms must be described to achieve a realistic model.

De Revolutionibus Orbium Metabolicorum

Revolutionary Histories

The introduction of the heliocentric system into astronomy is hailed as the Copernican revolution and the section heading is paraphrasing the title of the famous book of this Polish canonicus. Actually our understanding of the word "revolution" has changed over time. For people like Nicolaus Copernicus, Latin was their scientific language as scientists are nowadays speaking English. In Latin, "revolution" has a direct physical meaning, "revolvere," from which it is derived, means simply to rotate. However, the idea that the earth is revolving around the sun was such a rotating and revolutionary idea at his time that Nicolaus Copernicus preferred—against all instincts of modern scientists—to have his work only published on his deathbed. The concepts underlying oxidative phosphorylation—the subject of our next section—are also revolutionary for modern biology in the double sense. They taught us how energy is mechanistically extracted from foodstuff, and this discovery revolutionized biochemistry. In addition, two rotating biochemical devices stand at the beginning and at the end of the process. On the start, this is the Krebs or citric acid cycle revolving intermediates that are organized into a cyclic pathway. At the end is a small molecular motor that mechanically rotates when fuelled by a proton gradient and which thereby synthesizes ATP. This process of oxygenic respiration is true molecular biology, i.e., the description of fundamental biological processes at a molecular level. If one considers the electron transport chain in the inner mitochondrial membrane, part of the description is even at the subatomic level. Basic life processes can now be understood in chemical detail: We understand now where the oxygen we breathe is used in our metabolism, where CO_2 we exhale is created, and where and how the 40 kg of ATP is created that our metabolism has to produce day by day to power our life processes. The site of all the above-mentioned activities is the mitochondrion. We are dead within minutes when the blood circulation or lung respiration stops, reflecting the fact that more than 90% of our ATP derived from foodstuff is gained from oxidative phosphorylation. Much of the anatomy and physiology of animals is explained by the biochemical needs of mitochondria. We are thus leaving the cytoplasm, the site of the glycolytic enzymes. This change of place of action from the cytoplasm to a cell organelle means also the crossing of a watershed.

Mitochondria as Bacterial Endosymbionts

Endosymbiont Hypothesis

Mitochondria resemble prokaryotes in numerous important properties. Like the bacterium *E. coli*, mitochondria are enveloped by two membranes. The inner membrane, which is much larger in area than the outer membrane, folds into the matrix building the so-called cristae. The matrix contains soluble enzymes, including those of the citric acid cycle, the β-oxidation of fatty acids, prokaryotic type ribosomes, tRNAs, and a small circle of DNA. Nowadays only a smaller part of the mitochondrial proteins are encoded on this mitochondrial DNA chromosome. The inner membrane contains complexes I–IV of the electron transport chain and the ATP synthase also called complex V. The mitochondrion is wrapped by a second membrane, creating an intermembrane space corresponding to the periplasmic space of bacteria. Like bacteria, mitochondria reproduce by fission. All these similarities make the conclusion inescapable that mitochondria were once eubacteria that invaded the ancestor of the modern cells.

Proof for this endosymbiotic origin of mitochondria came from the genome of mitochondria. At first glance, this seems not to be an obvious source of information: The mitochondrial genomes are linear or circular and range in size from <6kb (kilobase, 1,000 bp) in the malaria parasite to 360 kb in the model plant *Arabidopsis*, the water cress. The latter encodes despite its large size only 32 proteins; >80% of its DNA is actually noncoding. Protozoa at the basis of the eukaryotic tree were considered as the most interesting study objects for the origin of mitochondria since some of them lack mitochondria (more on that subject in a later section). Therefore many mitochondrial genomes from protozoa were sequenced. The breakthrough was achieved in 1997 when a *Nature* report entitled "An Ancestral Mitochondrial DNA Resembling a Eubacterial Genome in Miniature" was published. This genome belonged to a free-living freshwater protist called *Reclinomonas americana* (Lang et al. 1997). Its 69 kb genome encodes 97 genes. Protein-coding genes contributed to complexes I–IV of the respiratory chain, the ATP synthase, the mitochondrial-specific translation apparatus (ribosomal proteins), and most spectacularly a eubacterial RNA polymerase. The mitochondria from higher eukaryotes use strangely a phage-like RNA polymerase of totally unknown origin. The informational content of this mitochondrial genome is greater than that of the sum of all other sequenced mitochondrial genomes (Gray et al. 1999). Thus it is not only the most bacteria-like genome, but also the closest relative to the ancestral protomitochondrial genome.

Rickettsia

One year later, another genome sequence was published in *Nature* and this is now the most mitochondria-like genome within eubacteria (Andersson et al. 1998; Gray 1998). If one considers that the evolution of animals could not have occurred without the capture of this powerhouse of the cellular energy metabolism, the closest bacterial relative of mitochondria is an unlikely character. It is *Rickettsia prowazekii*, the causative agent of epidemic, louse-borne typhus. *This is a dreadful human disease, which wrote many chapters in human military history. Pericles died from it in Athens at 430 BC during the Peloponnesian War; it caused the retreat of Napoleon's troops from Russia; and infected 30 millions and killed three million people mainly in the Soviet Union in the wake of the First World War. If you are dissatisfied with the tongue-twisting nature of many species names in biology, you should do justice to that of this microbe since it honors the names of a US pathologist and a Czech microbiologist, who both died when investigating this pathogen.*

Rickettsia's life cycle belongs to the annals of the history of eating, so close is its relationship with the subject of this book. Rickettsia are transmitted by lice. When lice feed on the blood of a human infected with rickettsias, the bacteria infect the gut of the insect, where they multiply and are excreted in the feces a week later. If the insect then takes a blood meal on an uninfected person, it defecates on the skin. This irritation causes the person to scratch and thereby the bite wound is contaminated by rickettsias. The bacterium spreads with the bloodstream and then infects the endothelial cells of the blood vessel, causing all the clinical sequels, the rash and the high mortality of 50% if untreated. This does not sound like a close relative of the energy-producing machine of the cell. Yet its genome sequence reveals the parallels. Rickettsia enters the host cell, escapes the phagosome and multiplies in the cytoplasm of the eukaryotic cell. It is an intracellular bacterium, and this lifestyle caused a dramatic reshuffling of its genome and of its metabolism. The amino acid metabolism has practically disappeared, so did the nucleotide biosynthesis pathway. Small wonder since these metabolites and ATP are amply provided by the host cell. With its 1,100 kb size, the genome of rickettsia is small, yet other intracellular bacteria went even further with their genome reduction as demonstrated by *Mycoplasma* with only 470 kb large genomes. Selection could not maintain nonessential genes, and the process of gene loss is apparently still ongoing as demonstrated by the unusual 24% of noncoding DNA and frequent point mutations in the rickettsia genome. However, rickettsia maintained their capacity to produce energy. Later in the infection process, when the cells get depleted of cytosolic ATP, the intracellular pathogen switches to its own energy system. Like in mitochondria, pyruvate is imported into the bacterium, converted by pyruvate dehydrogenase (PDH) into acetyl-CoA. The TCA cycle enzymes are present in rickettsia as well as the respiratory chain and the ATP synthesizing complex. However, the glycolytic pathway is conspicuously absent. Suddenly, although in pathological disguise, we have here a metabolic parallel to mitochondria. Phylogenetic analysis of the cytochrome *c* oxidase gene indicated that the respiratory systems of rickettsia and mitochondria diverged about 1.5–2 Gy (giga years, one billion years) ago, shortly after the amount of oxygen in the atmosphere began to rise. This is the most appropriate moment for the evolution of oxygen-based respiratory systems. However, mitochondria are not derived from rickettsia. Instead both genomes derive from an α-Proteobacterium ancestor and both lines followed an independent path of genome reduction. Mitochondria went much farther down this way. In fact, many mitochondrial proteins are now encoded by genes located in the chromosomes of the nucleus. The current endosymbiont theory envisions a successive transfer of genes from the protomitochondrial genome to the host genome. This transfer stabilizes the endosymbiontic relationship since mitochondria that lost essential parts of their gene content can no longer escape from this close metabolic relationship between the ex-bacterium and its eukaryotic host. *Actually, mitochondria became slaves of the modern cells and the successive loss of genetic autonomy can be read from the decreasing size and information content of mitochondrial DNA in animals. However, when we explore the energy metabolism of mitochondria in the following chapters, we should not forget that we look into a basic blueprint invented by α-Proteobacteria that was only perfected in the extant mitochondria to suit the needs of the eukaryotic master.*

Pyruvate Dehydrogenase: The Linker Between Pathways

Haeckel's Principle

Glycolysis is confined to the cytosol, at least as far as animals are concerned. The Krebs or citric acid cycle in contrast takes place in the matrix of the mitochondrion. One might suspect that this location in different cellular compartments might be a practical solution to separate pathways, intermediates, and metabolic fluxes allowing their separate regulation. This is certainly true; however, we should not forget that biochemistry is very conservative, and we might see here still the metabolic endowments of the different cells that made up the eukaryotic cell. Glycolysis was the energy-providing process of the ancestor cell providing the cytoplasm of the eukaryotic cell, whereas the citric acid cycle was the heritage of the α-Proteobacteria leading to mitochondria. This conservative nature of biochemistry becomes even more evident when two pathways leading to a comparable end product are maintained in two cellular locations. Take the synthesis of membrane lipids. In plants, chloroplasts, the home of the photosynthetic apparatus, synthesize membrane lipids in a typical prokaryotic pathway. Apparently, chloroplasts in Ernst Haeckel's words recall phylogeny and remember that they were derived from bacteria, in their case cyanobacteria. In parallel, plant cells synthesize membrane lipids along a typical eukaryotic pathway in the endoplasmic reticulum. Membrane lipid synthesis in plants still reflects the distinct origin of the cellular compartments. This separation is maintained even if it obliges modern plants to a collaboration between different cellular compartments. We should thus be cautious when arguing from teleological principles in biology, history might be an equally strong argument.

Pyruvate

Back to pyruvate: If the cell wants the full pay for its glucose food, it has to change the compartment. Pyruvate has to cross the two membranes of the mitochondrion to reach the place of its complete oxidation to CO_2 and H_2O. The outer mitochondrial membrane is not a major barrier to its transport, but the inner membrane is selectively permeable—and for good reason as we will hear soon. Gases and water move rapidly across this membrane; small uncharged molecules like protonated acetic acid achieve this too. However, pyruvate is charged under physiological conditions and needs therefore a carrier to cross the membrane. Pyruvate gets into the mitochondrial matrix by an antiporter: It takes pyruvate in and OH^- out. It is thus an electroneutral transport because no net charges are moved across the membrane. The transport is driven by the pH gradient across the membrane created during electron transport in the respiratory chain. In the matrix, pyruvate is received by PDH. This is a remarkable enzyme for several reasons: It is so large that its structure can be visualized by cryoelectron microscopy (Gu et al. 2003a). In fact, it is five times larger than ribosomes, the cellular protein synthesis machines. The inner core consists of 60 molecules of enzyme E2. This enzyme contains a long flexible linker molecule, made from lipoic acid fixed to a lysine side chain. The active components are two thiol groups that can undergo reversible oxidation to a disulfide bond, leading to a five-membered heterocyclic ring. The reduced linker swings to the E3 enzyme, which is also located in the core of the multienzyme complex with 12 copies. The task of this enzyme is the reoxidation of the reduced linker dihydrolipoyl with FAD, hence its name dihydrolipoyl dehydrogenase. However, NAD^+ is the ultimate electron acceptor in this oxidation reaction, the reduced NADH is fed into the electron transport chain. Then the reoxidized disulfide linker swings back from E3 and touches the outer rim of the complex consisting of a shell of many E1 enzymes. The lipoate linker assures substrate channeling: the five-reaction sequence on the enzyme complex never releases the intermediates. This measure prevents other pathways keen on this central metabolite from stealing it from PDH.

Pyruvate Dehydrogenase

The outer shell made of E1 actually takes care of the pyruvate. E1 is called a PDH, but the chemical reaction it mediates is actually an oxidative decarboxylation. The carboxyl group of pyruvate leaves as CO_2; this is the first carbon from glucose leaving our body as a gas in our breath. We should think on this brave enzyme when we exchange oxygen from the air surrounding us and trade it against CO_2 in our expiration. The task is actually not simple: E1 needs the cofactor thiamine pyrophosphate for this task. As many cofactors of enzymes it has a somewhat complicated structure and our body does not take the pain to synthesize it. However, this cellular economy comes at a price, we have to take it up from our food where it is better known as vitamin B1. The active group in this reaction is the C-2 carbon from the central thiazolium ring with a relative acidic proton. If this proton dissociates, a carbanion (a negatively charged carbon) is created, which is a powerful nucleophil, a chemical substance searching for a positively charged compound. It finds its partner in the partially positive carbonyl group of pyruvate. The positively charged nitrogen atom of the thiazolium ring acts now as an electron sink. The traveling of the electrons leads to the leaving of the carboxyl group as CO_2. In pyruvate decarboxylation, we now get a hydroxyethylgroup at thiamine, which leaves as acetaldehyde. However, in PDH, a different path is followed: The hydroxyethyl group is oxidized to the level of a carboxylic acid creating acetate. The two electrons, which are extracted from the substrate, are taken up by the disulfide group of the flexible E2 linker. The acetyl group created by this oxidation–reduction reaction is first esterified to the thiol group of the linker and then transferred to the thiol group of coenzyme A. Now we have the acetyl-CoA, the fuel of the citric acid cycle. The acetyl group is linked in a high-energy thioester group, which actually conserves the energy of the oxidation reaction. We understand now also why E2 is called a transacetylase, more specifically a dihydrolipoyl transacetylase.

PDH is also an important site of regulation: The enzyme complex is inhibited by ATP, acetyl-CoA, NADH, and fatty acids. This is a logical design, all these compounds indicate that fuel supply is high and the energy charge of the cell is also high. There is thus no need to feed further pyruvate into the catabolic pathway. Equally logical is the activation of the PDH by AMP, CoA, NAD^+, and Ca^{2+}. These compounds signal low energy and food charge. Calcium is released from contracting muscles: Energy is thus urgently needed and PDH must channel pyruvate into the citric acid cycle.

Deficiency of Cofactor and Enzyme

PDH is a remarkable enzyme not only for its size but also because it needs five different coenzymes. Four cannot be synthesized by our body and are thus vitamins: thiamine, riboflavin (in FAD), niacin (in NAD), and pantothenate (in CoA). Pantothenate deficiency is very rare in humans and has only been observed in prisoners of war suffering from severe malnutrition. The symptoms were a strange numbness in the feet. More severe and more prevalent is actually thiamine deficiency. Polished rice lacks thiamine, which is mainly found in the removed hulls of the rice. In populations that live mainly from rice, thiamine deficiency is known as beriberi. There are different forms of it, a cardiac form in infants, and dry beriberi in adults with peripheral neuropathy. Also alcoholics show a form of thiamine deficiency known as Wernicke–Korsakoff syndrome, which is linked to neurological symptoms. The problem with alcohol is that it represents what is called in nutrition "empty calories." Food comes here in a relatively pure chemical form of ethanol and is only minimally "contaminated" by vitamins as in conventional food items.

I now want to mention a fortunately very rare human condition, PDH deficiency. Human diseases that affect tissues with high-energy requirements are often caused by defects in the mitochondrial functions. Mitochondrial dysfunctions are also discussed as causes of type-2 diabetes (Lowell and Shulman 2005).

PDH deficiency affects mainly the central nervous system: Key symptoms are developmental delay, feeding difficulties, lethargy, ataxia, and blindness. Early death is the inevitable outcome. The chemical signs can directly be understood from the biochemistry. The decreased production of acetyl-CoA results in reduced energy production. Pyruvate and lactate accumulate in the body and lead to metabolic acidosis. With the block in the link from glycolysis to the citric acid cycle, why is the energy production down? There are in fact other feeder pathways into the TCA cycle. A major alternative is the β-oxidation of fatty acids. Fatty acids linked via their carboxyl group to CoA in the cytosol can be *trans*-esterified to carnitine in the outer mitochondrial membrane (or the intermembrane space, the biochemical details have not yet been settled), and they get there via a carnitine transporter in the inner mitochondrial membrane into the matrix. Here a second carnitine acyltransferase transfers the fatty acid back to CoA. The next steps are straightforward: Four stereotype successive enzyme reactions release electrons and acetyl-CoA in multiple rounds from the imported fatty acids. In the first step, a dehydrogenase abstracts electrons from the fatty acid and introduces a double bond between the α- and β-carbon atom, hence the name β-oxidation. In the second step, a hydratase adds water to the double bond. The resulting hydroxyl group at the β-carbon position is oxidized by another dehydrogenase to a keto-group. In the last step, a free CoA group attacks the bond between the α and β carbon and splits it by a thiolysis. This means that at every turn of the cycle, again, an activated, but shortened acyl-CoA group is created. The reduced $FADH_2$ and NADH feed their electrons into the respiratory chain and the split acetyl-CoA enters the TCA cycle. Why should then energy production be a problem? High energy demanding tissues like the heart are actually covering their energy needs nearly exclusively by β-oxidation of fatty acids followed by oxidative respiration. In fact, infants with PDH deficiency do not show cardiac insufficiency; they suffer mainly from neurological symptoms. That neurons have a high-energy demand is not a sufficient explanation. However, in contrast to the heart, the brain lives nearly exclusively from glucose as carbon source, fatty acids can thus not replace the glucose in the brain mitochondria.

On the Value of Mutants

Pediatricians have then searched solutions to this problem in nutritional biochemistry literature. Actually, during starvation, glucose becomes limiting because the body can only store a limited amount of glucose as glycogen. As the brain is an absolutely vital function for survival, we can of course not give up this organ simply because it can only use glucose as carbon fuel. In fact, nature knows about this problem and after a few days of starving, the brain learns to use ketone bodies as an alternative fuel. Ketone bodies are produced in the liver essentially by the condensation of two acetyl-CoA molecules leading to compounds like aceton, acetoacetate, or hydroxybutyrate. However, ketone diets have not yielded the remedy expected from textbook biochemistry knowledge (Wexler et al. 1997). *What you do when your biochemistry model fails is to develop an animal model of the disease and study what went wrong. This was actually done in the case of PDH deficiency. The investigated mutant is called noa. The names of mutants are left to the discretion of the researchers. Some of them are fanciful like "bobbed" (a fly with short bristles linked to a female short hair dress in the 1920s), others descriptive as "krüppel" or "fushi tarazu" but only understandable for parts of the scientific community due to language barriers.* "Noa" follows a tradition in microbiology where the mutant names tend to be acronyms: "noa" in full reads <u>n</u>o <u>o</u>ptokinetic response <u>a</u>. This mutant is in the zebrafish, a popular pet animal of geneticists. The major advantage of this animal is the fact that biologists can study the living mutant animal under the microscope. However, the observation of *noa* does not necessitate a microscope. The zebrafish shows expanded melanophores (pigmented cells), no feeding behavior, lethargy and premature death. Noa has a defective E2 subunit of PDH (Taylor et al. 2004). The phenotype is relatively easy to understand: Like children, the fish larvae show elevated levels of pyruvate and lactate. Interestingly, the ATP/ADP ratio is normal despite the lower energy production (the TCA cycle is blocked). The ratio can only be maintained because the ATP consumption is decreased. A similar regulation is done by fish in anoxic water: It reduces swimming activity and visual function. In fact, photoreceptors belong to the most energy-demanding cells of the vertebrate body. As the receptors do not see light under these conditions, the brain gets the information of darkness and the fish adapts to this misperceived night by a darkening of its skin via the expansion of the melanine deposits. Now comes the ketogenic diet and with it an arousal of the lethargic animals: They start to swim and eat (their aquarium delicacy are paramecia, a protist known to all children getting their first light microscope). The retina resumes its activity; the fish larvae get brighter again, survive, and resume growth. However, they show increasing growth retardation and mimic thus the inefficiency of the ketogenic diet already observed in infants. This is now an important message: Our understanding of the vertebrate metabolism is not yet so developed that we could easily do nutritional engineering in humans. However, the stakes are set, and there are many pathological situations where nutritional interventions would be a highly wanted addition to the medical toolbox.

Why is the Citric Acid Cycle so Complicated?

Principles

In colloquial physiological speech, we speak of burning the food when we extract energy out of it. If you burn sucrose chemically, you get a lot of energy because this is a strongly exergonic reaction, but you get it as heat. Heat is a relatively worthless form of energy because you cannot convert heat into other forms of energy. And organisms have multiple tasks to perform, and all need to be powered by an energy input. The pervading principle is to do small, but controlled steps downhill the overall exergonic reaction pathway of food burning and to conserve the redox energy in energy forms usable for the cell. The principle of the small chemical steps of energy extraction from food is realized in the citric acid cycle. Actually the most striking aspect of this pathway is that it is not linear like glycolysis, starting with one compound and ending up with another. It is a truly cyclic process.

The Chemical Steps

A four-carbon compound, oxaloacetate, condenses with the two-carbon compound acetyl-CoA, produced by PDH to give a six-carbon compound, citrate. After a dehydration step, a hydration step follows, and citrate is nearly reconstituted in isocitrate. *Here you should protest: If you were taught that nature uses the most economical of all possible solutions, then you would not expect that the oxidative degradation of acetyl-CoA takes the detour to first lengthen the molecule to a six-carbon compound and then doing the illogical steps of first a dehydration and then its inverse, namely a hydration.* The reaction is catalyzed by the enzyme aconitase. Notably, an aconitase is also found in the cytoplasm, but here its function is not in the citric acid cycle (there is none in the cytoplasm), but here it is an iron-responsive element. It is a bifunctional enzyme: It can catalyze the citrate to isocitrate reaction, but it can also bind specific mRNAs and interfere with their translation. As iron metabolism was most likely an earlier activity than oxygenic respiration, we might suspect that aconitase is a late recruit to the citric cycle, but this is a pure speculation. Back to the cycle: Next follow two successive decarboxylation steps. Decarboxylations are very popular steps in biochemistry; they often drive reactions that would otherwise be highly endergonic. Two molecules of CO_2 leave the cycle and counterbalance the addition of the two carbon atoms introduced in the acetyl group of acetyl-CoA. Notably, the two carbon atoms that leave the cycle are not those that have just entered the cycle with acetyl-CoA. It is of central importance to the following steps of energy conservation that a hydroxyl group in isocitrate is oxidized to the ketone oxidation level (α-ketoglutarate). The abstracted electrons are recovered by higher organisms in one molecule of NADH; in microbes, the acceptor is always NADPH. Then follows a nearly exact copy of the reaction catalyzed by the PDH complex, the E3 subunit is in fact identical between both enzymes. In contrast, the E1 and E2 subunits differ because they must display a distinct binding specificity, but the cofactors and the reaction mechanisms are absolutely identical. Here we see again the modular organization of the metabolism. PDH and α-ketoglutarate dehdrogenase certainly derive from a common ancestor (and related enzymes are also found in amino acid degradation pathways). The energy gained by oxidation is conserved in the energy-rich thioester bond (succinyl-CoA). Then comes a difference to acetyl-CoA. The thioester bond in acetyl-CoA is used to drive the condensation of oxaloacetate with acetyl-CoA to citrate, while the energy in the succinyl-CoA bond is recovered in the anhydride bond of GTP. The remaining reactions of the citric acid cycle follow two goals: First, the extraction of further electrons from the food molecules, and second, the reconstruction of oxaloacetate, the starting compound of the cycle. The central C–C bond in succinate is oxidized to a C=C bond in fumarate, the abstracted electrons create $FADH_2$. Water is added to the double bond yielding malate. The hydroxyl group in malate is oxidized to a keto-group, which creates another NADH and most importantly leads again to oxaloacetate. The cycle is closed and oxaloacetate is ready to fuse again with acetyl-CoA to restart the cycle. In principle only catalytic amounts of intermediates of the citric acid cycle are needed, and the steady-state concentrations of oxaloacetate are definitively extremely low ($<10^{-6}$ M). Here we are back again.

Alternatives?

As a critical reader you might ask whether this cycle is not unduely complicated. Is it not possible to oxidize pyruvate directly? The answer is yes, of course. Chemically, these are two successive decarboxylations: first to formate (HCOOH) and then to molecular hydrogen H_2. On paper this reaction looks quite simple, but the enzyme catalyzing this reaction, pyruvate-formate lyase is strictly anaerobic, the reaction doesn't work in the presence of oxygen. This is an important restriction because it prevents cells like *E. coli* from using this pathway in the presence of oxygen where it can metabolize glucose via glycolysis and the citric acid cycle. This gives obviously the most energy from glucose. If oxygen is lacking, but an alternative electron acceptor like nitrate is present, *E. coli* uses first nitrate respiration. In the absence of an electron acceptor, *E. coli* changes to fermentation pathways. In fact, *E. coli* has different options, which testify its metabolic versatility. In one it transforms pyruvate into acetyl-CoA, which is then degraded to ethanol by two reduction steps and, not oxidation steps. Actually you sacrifice reducing equivalents (NADH), and the cell has to excrete an energy-rich two-carbon compound, ethanol. This is, by the way, the reason that absolute abstinent people still have a detectable blood alcohol level produced in the gut by *E. coli*.

Beside other options (e.g., excreting acetate), *E. coli* has the above-mentioned oxygen-sensitive pyruvate-formate lyase that catalyzes the first decarboxylation. However, this enzymatic reaction is not simple at all: this enzyme is under a complicated network of transcriptional control (Fnr, NarL, ArcA repressors/activators) and posttranscriptionally regulated by an activase (Act) and a deactivase (AdhE). The explication of this control web would lead too far, but its essence is the control for the most efficient glucose use under different metabolic conditions. Then it needs a partner: the decarboxylation of formate, i.e., its splitting into CO_2 and H_2, is mediated by formate hydrogen lyase. Again this is a complicated enzyme. Formate-hydrogen lyase is a multicomponent membrane-associated complex, and at least 12 genes contributed by two operons (*hpc* and *hyp*) are involved. The enzyme complex requires a molybdenum cofactor, Ni and Fe. It requires in addition that an internal stop codon be read by an unusual tRNA that recognizes the stop codon as a signal for the insertion of a selenocysteine. The expression of the enzyme is also under transcriptional control. It depends on the alternative sigma factor σ^{54} that changes the promoter recognition of the RNA polymerase.

So a possible answer is this: The citric acid cycle is in fact not so complicated at all; the seemingly simpler chemical pathway of two successive decarboxylations is with respect to the enzymes probably even more complicated. At the end, the citric acid cycle uses only nine enzymes. In addition, strictly anaerobic enzymes lost a lot of importance when the atmosphere of the earth got increasingly rich in oxygen. Furthermore, the TCA cycle took over a number of biosynthetic service function and became thus the hub of the central metabolism of higher organisms.

The Horseshoe TCA Pathway

E. coli's Problem

Like glycolysis, the TCA cycle comes in many variants. What at first glance appears as a complicated cycle is in fact a simple and malleable device. I will present an interesting variant of the TCA cycle. It describes how it turns in the absence of oxygen as electron acceptor. If higher organisms use the TCA cycle in the catabolic mode, it is linked to energy production in the respiratory chain using molecular oxygen as electron acceptor. In fact, even if *E. coli* prefers oxygen for growth, under physiological conditions, i.e., in its gut ecological niche, *E. coli* does not see much oxygen. Except for a few niches in tropical ecosystems offering high nutrient concentrations and high temperature, *E. coli* is not known to grow in freshwater of temperate ecosystems (Winfield and Groisman 2003). This inability of *E. coli* to grow in freshwater is actually the underlying logic to use *E. coli* as an indicator of fecal contamination in recreational water. This poses a dilemma for *E. coli* when it gets expulsed from the gut by peristalsis and defecation and has to find a new host before it starves to death. *Some freshwater microbiologists seem to suggest that a residual growth of E. coli is also observed in freshwater from temperate ecosystems, but I have not found published reports supporting this claim. In fact, many crucial questions in the ecology of E. coli are still seriously under-investigated.*

Problems with the Reductionist Principle

There are two reasons for this ignorance. One is a research principle called reductionism. This working principle states that a biological process should be studied under the simplest and most standardized conditions and many laboratories should work with the same standardized organisms to create maximal synergism. This principle was introduced into biology by physicists that went in the 1940s into biology to tackle the question of Erwin Schrödinger, "What is life?" by using the approaches of physics. Schrödinger raised only the question, but the next generation of physicists like Max Delbrück actually worked with E. coli and its phages to address this question with concrete experiments. Their work was a scientific bombshell and led to a new discipline, molecular biology, which made biology for the first time an exact science. The success of this approach was dramatic. In the nonspecialized research journals, the biological sciences overshadowed over the last decades all other natural science branches combined. But also triumphs have their price. The extremely successful reductionist approach with E. coli discouraged researchers to ask questions like, What is the metabolism of E. coli in the gut? Where does it grow actually in the gut? E. coli is found in the gut lumen, but here it is apparently starving and has a very long generation time.

E. coli's Solution

If *E. coli* does not grow outside of the intestine (pathological conditions set aside), why is then the TCA cycle maintained which needs a respiratory chain? We know that *E. coli*'s growing fraction in the gut is not found in the gut lumen, but as small microcolonies within the mucus overlaying the gut epithelia. *It is possible that the mucus-associated E. coli microcolonies still capture enough oxygen from the blood vessels of the gut mucosa. However, the metabolic fluxes in these E. coli microcolonies have not yet been investigated.* Researchers have studied the metabolism of *E. coli* grown in vitro under anoxic conditions with intestinal mucins as sole carbon source. The microarrays showed that under these conditions the enzymes of the citric acid cycle are not active, while the glycolytic pathway is fully activated (Chang et al. 2004). However, the TCA cycle is not totally down under anoxic conditions when tested under in vitro conditions: Most enzymes work at 5–20% of their aerobic (aerobic and oxic is largely synonymous as is anoxic and anaerobic) activity level. There is only one enzyme of the cycle, which is really nonfunctional: This is the ketoglutarate dehydrogenase complex, which catalyzes the reaction from ketoglutarate to succinyl-CoA. This enzyme is the cousin of the PDH complex discussed above. Notably, the two decarboxylation reactions in the TCA cycle are essentially irreversible reactions, while the rest of the cycle except for the citrate synthase reaction are perfectly reversible reactions. This allows now an important reorganization of the cycle. Our chemical wizard *E. coli* does not give up the cycle, but it splits it into two arms. One branch is oxidative and leads via the normal pathway of the TCA cycle until ketoglutarate. As the reactions stop here, ketoglutarate gets a new mission. The keto-group becomes the acceptor of amino groups leading to the amino acid glutamate and thus into anabolic pathways. As the glutamate cannot take up the entire metabolic flux from glucose, the PDH activity is downregulated, too (small wonder because it is so similar to the ketoglutarate dehydrogenase (KDH) complex in its reaction mechanism) and the pyruvate-formate lyase mentioned above takes over and channels part of the glycolytic flow into various excreted products. However, since the TCA pathway is not cycling, an anaplerotic (fill-up) reaction is needed to keep the two separate arms under substrate flow. This is achieved in *E. coli* by carboxylation of PEP to oxaloacetate. The latter can now lead into the oxidative part via condensation to citrate or can feed into the reductive branch of the TCA cycle, which runs now in reverse direction to fumarate or succinyl-CoA. Recall that all these TCA reactions are fully reversible. The two metabolites play different roles: Fumarate with its double C=C bond is a suitable electron acceptor for *E. coli* yielding succinate with the reduced C–C single bond under conditions when neither nitrate or oxygen are available as electron acceptors. Succinyl-CoA is also an important precursor to anabolic pathways, like heme synthesis.

The Horseshoe Cycle

Other bacteria have in fact permanently lost the KDH and run the TCA cycle like *E. coli* under anoxic conditions. This is the so-called horseshoe TCA pathway operated by obligate chemolithoautotrophs. The name "horseshoe" is meant in the literal sense since the open cycle printed on a paper resembles now a horseshoe. These bacteria use the TCA enzymes only for biosynthetical purposes and not for energy generation. These bacteria cannot grow on common carbon sources because they have lost the transporters for sugar import. They must fix inorganic CO_2 via the Calvin cycle and gain energy from membrane-bound cytochromes. Small surprise that they are only competitive in environments that are especially poor in organic nutrients.

Split TCA cycles are in fact common among bacteria. Out of 17 microbial genomes surveyed *in silico*, only four appeared to encode all the genes necessary for a complete, canonical TCA cycle. Lack of KDH often accompanies anaerobic or microaerophilic metabolism. In some bacteria, the TCA half-cycles are joined by alternative enzymes. In *Mycobacterium tuberculosis*, which adapted to persistence in human macrophages, the lack of KDH is apparently imposed by the need to synthesize high amounts of glutamate from ketoglutarate as a "compatible solute" and as osmoprotectant to withstand the high osmotic pressure an intracellular bacterium experiences in its host cell. The two branches of the TCA cycle are joined by succinic semialdehyde, which is synthesized by α-ketoglutarate decarboxylase (Tian et al. 2005b). As this bypass enzyme does not exist in humans, it represents an excellent target for chemotherapy of tuberculosis.

This modified TCA pathway in *E. coli* and these chemolithotrophs provide another answer to the complicated chemical design of the TCA cycle. Even in organisms that live most of their time under anoxic conditions, the TCA cycle is important since it has crucial anabolic functions. This is also the case for animals like us, which run the cycle under double mission. The TCA cycle is called amphibolic, from Greek αμφι, meaning "on both sides." It serves catabolic and anabolic processes. If you draw precursors from the TCA cycle, you drain the cycle by diminishing the flow of matter through it. It becomes therefore of crucial importance to replenish the intermediates. The lowest steady-state concentrations are measured for oxaloacetate. Therefore nature has logically decided for three different replenishing reactions for oxaloacetate leaving from the glycolytic intermediates pyruvate or PEP. These reactions are called anaplerotic, from Greek ανα-πληρόω, meaning "to fill up." A fourth pathway practiced by plants, some invertebrates, and microorganisms (including *E. coli* and yeast) is still another way of creating oxaloacetate by condensing two acetyl-CoA into oxaloacetate via the glyoxylate cycle.

History Might Matter: An Argument on Chance and Necessity

Determinism?

In our metabolism, the TCA cycle runs exclusively clockwise in the conventional biochemical representations even if it is used both for energy-delivering (catabolic) and for substrate-providing (anabolic) reactions. This puts of course substantial constraints on the chemical design of the intermediates from such a pathway and it seems not unreasonable to anticipate that there might not be many chemical compounds that could fill in the ticket. Are there only one or a few chemical solutions to construct such a cycle that has to fulfill so many chemical constraints or in other words: Is the chemistry of the cycle deterministic?

Caveats

Before embarking on this question, I must perhaps play down our concept of the TCA cycle as the only pathway mediating the complete oxidation of carbohydrates to CO_2. New pathways are even described for *E. coli* when you investigate it under nutritional conditions, which come close to the natural situation. Microbes typically subsist under conditions of starvation (absence of nutrients) or hunger (suboptimal supply of nutrients) in their natural environment. In laboratory media (conditions of feast with excess of glucose), *E. coli* experiences catabolite repression, and its metabolism cannot be compared to the hunger situation. The hungry *E. coli* shows a hitherto unknown pathway, called the phosphoenolpyruvate-glyoxylate cycle (Fischer and Sauer 2003). This pathway combines the glyoxylate shunt with PEP carboxykinase to oxidize PEP completely to CO_2. Thus, under different physiological conditions, different types of metabolisms are observed with the same organism. I recall that the metabolism of *E. coli* in the gut has to my knowledge not yet been explored. There is another caveat. Metabolic flux analysis has emerged as key technology to quantify the in vivo distribution of molecular fluxes through the metabolism of model microbes with industrial relevance. When a wider range of bacteria was investigated, it turned out that the generally held view that the Embden–Meyerhof–Parnas pathway ("glycolysis") is the major route of glucose catabolism may be a misconception (Fuhrer et al. 2005). In this study, the Entner–Doudoroff pathway was the almost exclusive route of glucose catabolism, whereas the EMP pathway was mostly absent and the pentose phosphate pathway served exclusive biosynthetic functions. If we consider, in the following, the TCA cycle as written in stone, we should keep these caveats about "textbook biochemistry" in mind.

We have seen several times that history eminently matters in biology. As I am writing a natural history of eating, the question when the TCA cycle originated is of some importance and it might also provide arguments for the historical chance versus necessity debate. Some biochemists argue that respiration is much more complicated than glycolysis and was possible only after the rise of oxygen in the atmosphere. It should therefore be a much later invention. Personally,

I do not really buy these arguments. Oxygen is not the only electron acceptor in respiration as demonstrated by many prokaryotes. There is thus no reason to wait on the arrival of oxygen, which might mean several 100 millions years after the evolution of cyanobacteria. It is furthermore not said that the TCA was invented for its current use. We have seen at several occasions that nature never discards old inventions, but uses and reuses them in a new context. That might well have been the case also for the TCA cycle.

Chlorobium

The case is made by the green sulfur bacterium *Chlorobium*. This is not an exotic bacterium; it is widely distributed and a prominent member in the cycling of sulfur in the biosphere. This organism is a photoautotrophic organism with a very peculiar organization of the photosynthetic apparatus. In many respects, this apparatus looks quite primitive and is considered by some microbiologists as a possible precursor of photosystem I (PSI). The characteristic feature of *Chlorobium* is vesicles attached to the inside of the cytoplasmic membrane. The attachment surface of the vesicle shows a crystalline baseplate structure (Figure 2.4). Apposed to the baseplate membrane of the vesicle is a type I photosynthetic reaction center associated with a bacteriochlorophyll *a* containing light-harvesting protein. The light energy is stored in a proton gradient, and this gradient is used to drive ATP synthesis in an F_1F_0-type ATP synthase as in many other photosynthetic prokaryotes. The vesicle is also called chlorosome because it is filled with rod-like structures that consist of stacked aggregates of bacteriochlorophyll *c*, obviously not in a protein-bound form. Bacteriochlorophyll *c* molecules collect the light and channel it via bacteriochlorophyll *a*, embedded

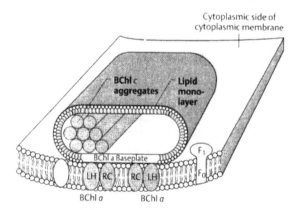

FIGURE 2.4. *Chlorobium*: chlorosome organization. Chlorosomes are attached by a proteinaceous baseplate to the cytoplasmic side of the cytoplasmic membrane. They absorb light via the linearly arrayed bacteriochlorophyll *c*, *d*, or *e*, which are arranged as rod-shaped elements in the chlorosome. The baseplate harbors bacteriochlorophyll *a* containing light-harvesting complexes LH. Below the baseplate are the reaction centers RC. F_1F_0 is the ATP synthase. (courtesy of Thieme Publisher).

in the baseplate, and then to the reaction centers located in the membrane. About 1,000 bacteriochlorophylls in the chlorosome serve a single reaction center. Later we discuss about purple bacteria that have a photosynthesis apparatus resembling photosystem II (PSII). Cyanobacteria have in contrast an already evolved photosynthesis apparatus consisting of a PSII/cytochrome $b_6 f$/PSI like in modern plants, which probably represents a combination of the more ancient photosystems evolved in purple and green sulfur bacteria. It is thus relatively safe to conclude that the photosynthesis system in cyanobacteria evolved later than the photosynthesis system we see in *Chlorobium*.

The Reductive TCA Cycle

What does this indirect conclusion mean for our argument about the age of the TCA cycle? In fact, a lot: *Chlorobium* has a TCA cycle that runs with exactly the same intermediates like the TCA cycle in many other forms of life. As this bacterium obtains energy from its photosystem, it can use its TCA cycle for other purposes. You might suspect that it is used in this organism as a hub of the intermediary metabolism to provide precursors for anabolic pathways. In fact, despite the fact of using the same intermediates, it shows a dramatic difference with modern TCA cycles: It turns counterclockwise. We have seen that the TCA cycle releases the CO_2 from the molecules of our food that we recover as CO_2 gas in our breath. If the cycle turns in the opposite sense, it must also do the opposite with CO_2, namely CO_2 fixation. This so-called reductive citric acid cycle is nothing exotic in microbiology. Indeed, it is one of the four basic mechanisms of CO_2 fixation in prokaryotes. Identical reductive TCA pathways were identified in sulfate-reducing (*Desulfobacter*) and Knallgas bacteria (*Hydrogenobacter*) as well as archaea (*Thermoproteus*). There are several indirect reasons that speak in favor of the antiquity of this pathway and that the TCA cycle as we know it now is only an adaptation of this old invention to the rise of oxygen in the atmosphere. This would again be an illustration that Mother Nature never rejects an old invention, but reuses them in a way like the Greek god Proteus, who was constantly changing his form with each emergence from the sea.

Let's look somewhat into the reductive TCA cycle. Despite using the same intermediates, some enzymes must be different. In order to reverse the cycle, three irreversible reactions of the oxidative TCA cycle must be catalyzed by alternative enzymes. The use of these alternative enzymes, which work with NAD(P)H and ferredoxin as reductant for the reductive carboxylations, has a remarkable consequence: The entire cycle is now fully reversible. This means that this cyclic pathway can serve both purposes in the same organism. It can mediate CO_2 fixation when powered by an independent ATP supply (the reductive TCA cycle is actually less energy devouring than the more widely distributed Calvin cycle, using only five instead of nine ATP per triose phosphate synthesized). The outlet of the reductive cycle is of course also different. ATP citrate lyase catalyzes the reversible reaction: citrate $+$ CoASH $+$ ATP $\leftrightarrow$ oxaloacetate $+$ acetyl-CoA $+$ ADP $+ P_i$. The first triose is created by a third carboxylation reaction: acetyl-CoA $+ CO_2 +$ ferredoxin red $+ 2 H^+ \leftrightarrow$ pyruvate $+$

CoASH + ferredoxin ox. Alternatively, it can serve as a pathway for the end oxidation of acetyl-CoA. The decision of the direction is made depending on the substrate supply and the energy charge of the cell.

Questions

Some biochemistry books come with a chart of the metabolic pathways. The very fact that the central metabolic pathway is formed by the connected glycolysis and TCA cycle for so many extant organisms speaks in favor of the existence of this pathway in the universal ancestor of life. Some biochemists argued that we can treat it as a virtual fossil. It could not be changed after its invention due to the high selective advantage conferred by this invention and the high interconnectivity of the central metabolism. However, its initial design—so goes the argument—was a chance event, and its conservation reflects only a historical accident. The alternative is the hypothesis that it represents an optimally successful chemical solution to designing biochemical networks and if life would be recreated under comparable environmental conditions as on the young earth, it would end up with a rather similar solution. Harold Morowitz and colleagues (2000) argued in that sense. They imagined a shell structure for the metabolism of autotrophs. In their view, the core and hence the oldest biochemical fossil is the reductive TCA cycle. The first outer shell is the synthesis of amino acids derived from amination of the ketoacids generated in the core metabolism. The second shell of the metabolic chart contains the reactions that incorporate sulfur into amino acids. The third shell reactions deal with the synthesis of dinitrogen heterocycles leading to the invention of bases and from there to nucleic acids. This onion-type view of the metabolism is in itself nothing very heretical and could even reflect the temporal order of the evolution of biochemistry on earth. The interesting argument is that in their view the inner core of the metabolism might be necessary and deterministic. Their suggestion is that any aqueous carbon-based life anywhere in our universe will resemble the intermediates of the TCA cycle. They did a bold chemical approach by applying a simple set of a priori rules. For example, they based the core chemistry on $C_xH_yO_z$ compounds with certain permitted indices, favoring for feasibility reasons small molecules. They preferred water-soluble compounds and those having low heats of combustion. While making some other selection rules based on chemical plausibility, they run the rules with the online Beilstein, an enormous encyclopedia on organic chemistry where the print Encyclopedia Britannica version is a pocket edition in comparison. From the 3.5 million entries in the Beilstein emerged 153 molecules, and hold and believe, all 11 members of the reductive citric acid cycle were in. Leslie Orgel (2000), an eminence in biochemical evolution, was not impressed by the result. He doubted that some of the rules were a priori not necessary and unknowingly selected for the intermediates they wanted to identify. In fact, is it chemically compulsory to base the selection procedure on the reactivity on the carbonyl function in $C_xH_yO_z$ compounds? Can alternative worlds not be built on boron or silicon chemistry? Evolutionary analysis with computer-generated *virtual organisms is now a subject in the biological research literature. Why should physical organisms not be based on silicon bonds?*

Metabolic Crossroads in Ancient Landscapes: NAD or NADP—That's the Question

Logic of Life?

Is there a logic of life in the biological observations or just an enormous variation of unique solutions somewhat tamed by frozen accidents? Many philosophical answers to the organization of the living might tell you more about the state of mind and temperament of the thinker than about the state of nature. Most biologists will probably lack a clear global answer to the Lucretian "De natura rerum," but in their zeal for details they might search an answer in a particular biological phenomenon to which they can easily and without regret dedicate decades of their professional life. Psychologically, this concentration on details of a single organism chosen from millions of species makes sense only if you start from the underlying hypothesis that your observations will provide you data that are also of relevance to other researchers or even the philosophical orientation of a thoughtful layman. Practically, you have in the biological research literature a large number of specialty journals and only a handful of nonspecialized journals. However, one might suspect that this differentiation does not split the biological observations into those that apply to a single organism and those that are of wide applicability. This differentiation will reflect more the depths of the experimental approach and the brilliance of the research group, which determines the degree of generalization that you derive from your observations. In fact, the evolution theory puts at the same time large limits (at the phenotypical level where an enormous space of ecological possibilities remains to be exploited) and narrow limits (by the very meaning of descent, all organisms remain linked and thus related). The choice of model organisms like E. coli, the yeast, or the cress Arabidopsis is a logical choice only if the solution found with these individual organisms are of heuristic value for millions of other species. As biologists are generally not too keen on theoretical discussions, let's take a few examples from model organisms and you decide yourself from the experimental details what perception of nature you derive from it.

Isocitrate Dehydrogenase

To stick to the theme of the previous sections, I have chosen isocitrate dehydrogenase, the enzyme in the TCA cycle that catalyzes the oxidative decarboxylation of isocitrate to form α-ketoglutarate. This enzyme is one of the contributors of the CO_2 in your breath. However, before the enzyme removes CO_2 from the oxalosuccinate reaction intermediate, it abstracts a hydride, which it transfers to NAD^+ or $NADP^+$. Normally, enzymes do not like this ambiguity, either you work as a decent catabolic enzyme and you use then the nonphosphorylated form of the coenzyme or you are dedicated to anabolism, then you use the phosphorylated coenzyme. We have discussed that the TCA cycle can turn in both directions, and even in the standard clockwise representation, it can fulfill catabolic and anabolic functions. Isocitrate dehydrogenase belongs to a large, ubiquitous and very ancient family of enzymes. Most family members use NAD^+ to oxidize their substrate, but not all. *E. coli,* for example, uses $NADP^+$. None is able to use both. The phylogenetic family tree demonstrates that NAD^+ use is the ancient trait. $NADP^+$ use evolved independently several times, but this "later" use is still very old since it probably dates to the oxygenization of the atmosphere (Zhu et al. 2005). Interestingly, there is some convergent evolution in these independent events since the same suite of amino acids changes occurred in both eubacterial and archaeal lineages to adapt to the binding of $NADP^+$. The geometry of the binding pocket for both NAD^+ and $NADP^+$ is known from high-resolution crystallographic structures, which allowed US scientists to engineer the binding specificity of the *E. coli* enzyme from $NADP^+$ to NAD^+. What does this switch mean to the *E. coli* cell? To answer this question, the scientists grew the cell with the NAD^+ and the NADP enzyme, and the cell with the altered enzyme turned out to be fitter when the cells were grown on glucose. This is a rather surprising result because it would mean that the change of coenzyme specificity was maladaptive. As such a result is a violation of the most fundamental laws of Darwinism, the researchers searched further. To the reassurance of the traditional picture, the adaptive value of the NADP-enzyme form became clear when the cells were grown on acetate on which the $NADP^+$-enzyme outcompeted the NAD^+-enzyme in less than 10 generations.

Glucose Versus Acetate Food

Does this result make sense with our knowledge of biochemistry? Stated otherwise, is there logic in this design? Let's compare glucose and acetate. Glucose is a highly reduced and thus an energy-rich compound. NADPH, the reducing power for anabolic reactions, is obtained by diverting energy-rich carbon from glycolysis into the oxidative branch of the pentose phosphate pathway. Acetate, a highly oxidized and thus an energy-poor compound, is, for example, derived from pyruvate, the end product of glycolysis. NADPH can thus not be produced from acetate via the pentose phosphate pathway. Alternative enzymes must supply this compound for anabolism. Metabolic flux analysis in *E. coli* showed that isocitrate dehydrogenase provides 90% of the NADPH for anabolism.

E. coli's Problem

Taking this observation at face value it would mean that the ancestor of *E. coli* has not seen much glucose and had to live on less energetic food like acetate. This is probably also true for *E. coli* living today, and on theoretical reasons this must be so because any organisms must be adapted to the conditions it encounters today or to be precise, it had encountered yesterday. Since *E. coli*'s only known ecological niche is the intestine of mammals and birds, where it populates the large intestine with modest titers, it will not see much glucose. The vertebrate body has already absorbed glucose from food into the bloodstream. The digestion of complex carbohydrate, which remains in the gut, is not the strong side of *E. coli*—it must leave this job to better carbohydrate digesters in the colon like *Bacteroides* and *Bifidobacterium*. The half-oxidized waste of other bacteria now becomes the food for *E. coli*. I must stress here that this sketch is only a probable scenario, not the result of experimental measurements. We know surprisingly little about the metabolism of *E. coli* in its natural niche, the gut. To give you a taste for the beauty of biochemistry—or the logic of life, but here I do not want to anticipate your judgment—let's go back to the Zhu paper (Zhu et al. 2005).

Glyoxylate Cycle

These authors searched the genomes of prokaryotes and found a strong correlation between strains possessing an $NADP^+$-dependent isocitrate dehydrogenase and those showing an isocitrate lyase. The latter is the enzyme that diverts isocitrate from the TCA cycle into the glyoxylate cycle. A lyase is by definition an enzyme that catalyzes cleavage (or in reverse direction, additions) in which electronic arrangements occur. Isocitrate lyase cleaves the six-carbon compound isocitrate into the four-carbon compound succinate and the two-carbon compound glyoxylate. To keep the cycle running, glyoxylate condenses in two reactions with acetyl-CoA to reconstitute isocitrate in a shortcut of the TCA cycle which avoids two CO_2-releasing steps of the TCA cycle. It now becomes understandable why cells that want to grow on acetate need isocitrate lyase to provide carbon for biosynthesis and $NADP^+$-dependent isocitrate dehydrogenase for the supply of reducing equivalents for anabolism. However, we now have a problem: Isocitrate can take two ways—one releasing CO_2 from organic acids in catabolism, the other sparing carbons for anabolic use. *E. coli* must regulate both ways to respond to its actual needs, but it cannot clamp down either pathway entirely when growing on acetate. The solution is as simple as appealing. The cell encodes in the *aceK* gene a kinase/phosphatase that phosphorylates the isocitrate dehydrogenase and inactivates it. During growth on acetate, about 75%, but not 100%, of the enzyme is inhibited. The kinase/phosphatase activity of the regulating enzyme is regulated by intermediates of glycolytic and TCA pathways and the energy level of the cell such that *E. coli* can now channel its metabolic flow according to its needs.

When Zhu et al. looked through the genomes of prokaryotes, the tight association between these two isocitrate-handling enzymes was independent of taxonomical group (archaea, firmicutes, proteobacteria), metabolic lifestyle, and habitat (you find there auto-, hetero-, chemo-, and lithotrophs). Proudly, the authors declared at the end of their research article that it is apparently possible to reconstruct not only what occurred, but also how it occurred and why it occurred. An ancient adaptive event that occurred billions of years ago can thus be reconstructed by a combination of genomics, protein engineering, and chemostat experiments.

From NAD to NADP

In a follow-up study, the lab of Antony Dean worked with another enzyme involved in the biosynthesis of leucine, which uses NAD^+. They engineered six amino acids in the protein, which progressively transformed the enzyme into an $NADP^+$-binding instead of an NAD^+-binding enzyme, and they investigated the adaptive value of the mutant enzymes (Lunzer et al. 2005). The genotype—phenotype fitness map showed that NAD use is the global optimum for this enzyme. The reason can also be understood with basic knowledge in biochemistry. To perform optimally both for catabolic and anabolic pathways, cells keep NADPH concentrations higher than those of $NADP^+$, while NAD^+ concentrations are maintained higher than those of NADH. The enzymatic reaction leading to leucine involves the familiar scheme of an oxidative decarboxylation, creating NADH and CO_2 as side-products. If the modified enzyme uses $NADP^+$ as coenzyme, it will experience product inhibition by the high cellular concentration of NADPH, which will not occur from NADH because it is kept at a low level in the cell.

The Logic and Adaptive Value of Metabolic Cycles

Plant Cycles

Life has evolved on earth. This statement sounds pretty trivial. Nevertheless, one should not overlook that the development of life on a planet that revolves around itself with a periodicity, which we call a day, puts strong constraints on life. Biological clocks have therefore evolved so that clock outputs are in phase with the Earth's rotation. Circadian clocks—by definition—synchronize biological events with the day–night cycles and they have evolved at least four times in organisms. Nowhere is the adaptive value of this clock more apparent than in photosynthetic organisms that "feed" on light. Already cyanobacteria and even more plants coordinate their metabolism with the light–dark cycle. In higher plants, there are circadian rhythms in transcript abundance of genes associated with chlorophyll synthesis and the light-harvesting apparatus. One should expect that this synchronization between light offer and metabolic demand improves the competitiveness of plants. The proof that this synchronization improves photosynthetic effectiveness was provided recently (Dodd et al. 2005). The researchers varied the light–dark periods in their experimental setting from 10–10 to 12–12 to 14–14 h, yielding artificial "days" of 20-, 24-, and 28-h duration, respectively. In this light "feeding" regime they tested wild-type *Arabidopsis*, long- and short-period clock mutants, and an arrhythmic plant overexpressing the oscillator component. The answers were clear. Leaves contained more chlorophyll when the oscillator period matched that of the environment, but did this improve photosynthesis? The answer was a clear yes: net carbon fixation increased with the matching of the rhythm and the rhythmic stomatal opening and closure played a key role. The fixed carbon was actually used for increasing the aerial biomass of the plants; the leaf area was visibly greater and the plants looked clearly different. The enhanced growth and survival became more pronounced when the plant mutants competed with each other. Importantly, no mutant was favored under all light regimes, but only when a match between the endogenous and exogenous cycle was achieved. Circadian clock function was apparently selected during plant evolution.

Yeast Oscillations

Not all biological cycles are imposed by the 24-h rhythm of the Earth's rotation. The budding yeast *Saccharomyces* shows cycles of alternating glycolytic and respiratory activity that was described already 40 years ago. After growth to high concentration, followed by a starvation period, and then continuous supply of low-level glucose, the cells started to cycle with a rhythm of about 5 h. Biologists followed the events with expression studies. Microarray analysis was done every 25 min and revealed a periodic expression of about half of the yeast genome (3,500 genes), with a periodicity of 5 h (Tu et al. 2005). The oxidative phase, characterized by an intense burst of respiration, shows increased expression of genes involved in amino acid synthesis, RNA metabolism, and protein translation. All these processes have high energy demands that are best covered by ATP produced in abundance in the respiratory chain. The oxidative phase is followed by a reductive phase, which comes in two forms. The first is when the cell begins to cease its oxygen consumption. Genes involved in mitochondria biogenesis, DNA replication, and cell division show now peak expression levels. The second phase is characterized by nonrespiratory modes of metabolism. Protein degradation, autophagy, peroxisome, and vacuolar functions are now transcribed. Notably, cell division is confined to the nonrespiratory phase, which would allow minimizing oxidative damage to DNA. The observed fluctuations really reflect a metabolic cycle where respiration is followed by a glycolytic fermentative metabolism resulting in an accumulation of ethanol and acetate in the medium. The acetate is then charged on acetyl-CoA and prepares the cell for the next respiratory burst, consuming the metabolites accumulated in the previous period—the researchers distinguished therefore a building and a charging reductive phase. In contrast to prokaryotic cells, the eukaryotic yeast cell contains organelles. Some of them like the vacuoles experience even a visible change during the yeast metabolic cycle, the mitochondria cycle in biochemical terms. However, the yeast cell has evolved with the metabolic cycling a temporal compartmentalization of cellular processes, which are mutually exclusive. Tu and colleagues speculated that this peculiar temporal separation of cellular processes was a means of coordinating incompatible biochemical activities contributed by the two-cell types that probably fused during the birth of the eukaryotic cell, i.e., the reductive nonrespiratory cell and the oxidative respiratory cell. With respect to the logic-of-life argument, it is interesting to note that temporal compartmentalization of metabolic function might also take place during the circadian cycle of flies and mice. In fact, useful inventions tend to be made independently in nature. The cyanobacteria have learned relatively early in the evolution two lessons in biochemistry that are crucial for life on earth: oxigenic photosynthesis and nitrogen fixation. The first develops oxygen, and the second must occur under strictly reductive conditions. *Synechococcus elongatus* has found a solution that is later reused in yeast and higher plants. These two biochemically incompatible pathways are executed at temporally distinct phases of the circadian cycle: photosynthesis in the light, nitrogen fixation in the dark (Nagoshi et al. 2004).

3
Bioenergetics

Oxygen

The Origin of the Electrons and Biochemical Cycles: Anatomy of Complex II

The Biological Meaning of Oxidation

Now it is time to look into the underlying mechanism that creates biological usable energy from the oxidation of foodstuff. How is this feat achieved? The key to this process is the very meaning of the word "oxidation." In the simplest chemical sense it means the addition of oxygen, which is actually achieved in the central pathway: carbon is linked in glucose, on the average, to a hydroxyl group, a hydrogen, and two other carbon atoms. These carbons end up as atoms linked to two oxygen atoms: CO_2. Reduction by this same simple definition is, in contrast, the addition of hydrogen atoms. One can also formulate it indirectly: Oxidation is the abstraction of hydrogen atoms. Let's investigate this basic process with a reaction of the TCA cycle. Succinate is a four-carbon dicarboxylic acid with a central H_2C–CH_2 bond. Fumarate, the next intermediate in the TCA cycle, is an identical four-carbon dicarboxylic acid, except that it contains a central HC=CH bond. Fumarate is the oxidation product of succinate: Two hydrogens were abstracted from succinate by succinate dehydrogenase (SDH) and transferred to the enzyme cofactor FAD, creating $FADH_2$. Succinate was thus oxidized to fumarate, which follows its fate in the TCA cycle. We will, however, discuss the two abstracted hydrogen. This pursuit is facilitated by the solution of the structure from this enzyme of the TCA cycle (Hederstedt 2003; Yankovskaya et al. 2003). This enzyme is unusual because it is the only membrane-bound enzyme of the TCA cycle and thus already belongs to the electron transport chain. In this respiratory disguise, the TCA cycle enzyme is also called complex II. Like all other respiratory chain complexes, it consists of multiple subunits. Subunits C and D pass the membrane with α-helices several times and hold two cofactors in place: ubiquinone and heme b. Subunits B and A are stacked on the top of subunits C and D and pile up on the cytoplasmic side (in *E. coli*; in mitochondria, this side corresponds to the matrix side). The roughly globular top A subunit contains the succinate binding site and directly next to it lies the FAD coenzyme. As you suspect correctly, the coenzyme cofactors are doing the job. But what is FAD? In full it reads flavin adenine dinucleotide. The somewhat complicated name hides that this is basically a molecule very familiar to you. Like NAD^+, the major electron acceptor, it consists of an AMP subunit linked to another nucleotide flavin mononucleotide (FMN). Like the nucleotides that are part of our genetic material, FMN consists of a phosphate group esterified to a five-carbon sugar linked to a base by an N-glycosidic bond. In FMN the sugar and the base deviate from the standard sugars and bases used in nucleic acids. The reactive part of the cofactor is the base, a nitrogen containing a three-ring system (in chemical jargon the isoalloxazine ring of FMN). This ring system accepts two hydrogen atoms (two electrons and two protons) from the nearby substrate succinate in a two-step reaction. The first step creates a semiquinone and the next a quinone. The semiquinone is a stable free radical form of FAD, and FAD thus has a greater flexibility than does NAD^+ for electron transfer reactions as it can couple one or two electron transfer reactions. Next in the line are three Fe–S centers arranged as convenient stepping stones for electron hopping from FAD via the B subunit to the ubiquinone, located near the matrix side within the lipid membrane. In this way, none of the individual electron transfer reactions have to bridge distances greater than 11 Å, allowing for rapid transfer reactions.

Iron–Sulfur Centers

Here we need a chemical stop. Iron–sulfur clusters are very popular in biochemistry and come in different configurations. In the simplest form, an iron atom is tetrahedrally coordinated to four sulfur atoms provided by cysteine residues of a protein. The next levels of complexity are two irons and two inorganic sulfides held in position by four cysteine groups. This form is denoted as [2Fe–2S]. The next layer is called a [4Fe–4S] cluster and involves, as the name indicates, four iron atoms and four inorganic sulfides, once again held together by four cysteine residues yielding a beautiful cage structure. *These cofactors or variants of it are found in complex II and even more complicated inorganic cages are found in other prominent enzymes, which mediate other basic life processes, like the nitrogenase involved in nitrogen fixation. We can take these factors as molecular fossils. We are probably dealing with vestiges of early biochemical reactions mediated by inorganic molecules that were in existence before the invention of enzymes, which are a relatively late invention in evolution. The iron–sulfur complexes found in modern enzymes are probably the molecular link to an iron–sulfur world that formed the basis of life processes coming out from the prebiotic era. This was probably followed by an RNA world that developed the genetic code and the early enzymes. The molecular vestiges of this world are also very evident in modern enzymes and are visible in the many nucleotide cofactors used at a crucial step in the enzymatic reaction, in complex II represented by FAD. The iron atom participates in the electron shuttle by changing back and forth between two ionic forms of iron, namely Fe^{2+} and Fe^{3+}.*

Quinones

The final destination, at least for the moment, is ubiquinone, also known as coenzyme Q. This molecule is an electron shuttle. Its name is aptly chosen and represents a contraction of two words: the Latin ubique, for everywhere, this part describes the wide distribution in all biological systems; and quinone, a phenol derivative containing two oxygens (=O) in *para* position. Quinone is transformed into a quinol (−OH) by a two-step electron transfer (plus two proton additions), yielding a phenol derivative with two hydroxyl groups. As in FAD reduction, a relatively stable radical is formed as an intermediate. In FAD it is a nitrogen radical, whereas in ubiquinone it is an oxygen radical. A radical is chemically characterized by an unpaired electron in the valence shell of the atom, endowing the compound with a high chemical reactivity. This two-step electron transfer chemistry has definitive advantages in electron chains as it allows flexibility and the coupling of one and two electron transfer processes. However, a quinone alone will not mix into the lipid phase of biological membranes. To do just that, ubiquinone is equipped with a long isoprenoid side chain. It commonly contains something like 10 isoprene units, but the exact length depends on the species. This long hydrocarbon chain gives ubiquinone the lipid solubility, which it needs to shuttle electrons in the inner mitochondrial membrane (or the cytoplasmic membrane of bacteria).

Cyclic Processes in Biology

Here we need another reminder for the antiquity of the basic biochemical reactions. Take photosynthesis, a process that also relies heavily on electron transport in membranes. Photosynthesis uses another membrane-soluble electron transporter called plastoquinone. The first half of this word refers to its location in chloroplasts, the organelles mastering photosynthesis. Quinone indicates similarity with ubiquinone, and if you compare both structures, you see that they only differ in the smaller substitutions of the quinone ring system. *It is thus a safe bet that both derive from a common quinone precursor and are likely to represent one of the earliest inventions in biochemistry. In fact, if you compare the organization of the respiratory chain and the photosynthesis apparatus, you will see other striking similarities in both systems, despite their very different tasks. Here we get to a very important argument for the early history of eating. Respiration leads finally to the reduction of molecular oxygen to water; photosynthesis is, at the end, the splitting of water and the release of molecular oxygen. Respiration is thus in a way the reverse of photosynthesis. Similar situations are met in the pair of glycolysis and gluconeogenesis or in the pair of oxidative and reductive citric acid cycle. Cyclic processes are necessary on earth: One process builds a compound, another process decomposes this compound. The cyclic nature is necessary because life has to use and reuse the same atoms over and over. The amount of atoms of each kind is fairly defined on earth, only relatively few escape into the space and only negligible quantities* *of matter arrive from the cosmos. Therefore a basic rule for life was set early on: You use the atoms you find on the surface of the planet for your play of life. This is a 10^x piece Lego play (you know these famous construction stones that children play with?). Admittedly, it is a giant Lego, but the number of atoms is counted. Life is confronted with the same dilemma as that confronted by children: If you use all your Lego stones in one big construction, then you have a nice toy, but it is static. If you want to continue your game, you have to destruct it and then you can start with a new construction idea. Life probably went through this dilemma of using up the Lego stones early on in the development of life in the primordial soup, and it learned from this lesson to build life on the basis of two opposing processes. This was necessary both at the level of individual organisms and at the level of the cycling of matter on the global planetary scale. The very fact that the three above-mentioned basic processes use partially the same chemical intermediates seems to suggest that they derive from a shared ancestral process. The reversible reductive TCA cycle found in some prokaryotes is probably still very near to the common ancestor process. Glycolysis and gluconeogenesis are relatively separated processes (spatially or regulation-wise) in an extant organism, and the fully reversible forward–backward process does not have a direct biochemical complement any longer. Respiration and photosynthesis are only distantly related and share few intermediates, but many basic principles. It is tempting to speculate that the increasing distance separating these processes also provides a relative timescale for the evolution of these basic opposing pathways. According to this argument, TCA might represent the youngest, glycolysis–gluconeogenesis the older process, and respiration/photosynthesis the oldest invention. This temporal order also fits logical constraints. Only photosynthesis (or its complement in chemoautotrophs) could provide the basic organic material, which could then yield the foodstuff for the earliest heterotrophs in protoglycolysis. When evolution established more complex forms of life, there was also an increasing need to incorporate compounds into the early forms of life still further, hence the need for the TCA cycle serving the outer and thus later shells of biochemistry in the Morowitz model (Morowitz et al. 2000).*

You see that I have on purpose chosen complex II for the illustration of biological oxidation. You can literally see how we are eating electrons. With this enzyme, we are at the intersection of the central metabolic pathway into the respiratory chain, at the transition from soluble enzyme to membrane-bound enzymes. If you see the arrangement of cofactors in SDH, you can see how an ancient cofactor- and metal-mediated process could have happened, primitive and disordered, with many side reactions. The proteins came into play to order what they already found in the earliest forms of biochemical life (do we really dare to call it life?). If you look at the structure of complex II, you get the impression that the protein does not participate too much in the electron transport process. Its task seems to be to fix the different cofactors in the optimal relative position in order to facilitate the wanted reaction and to suppress unwanted side reactions.

Fumarate Reductase: The Dangers with Oxygen

On Electron Transfers

The need for the evolution of protein enzymes such as oxidoreductases was dictated by chemical constraints. Outside catalytic sites, organic substrates are resistant to oxidation/reduction because single electron transfers to form a radical intermediate are rarely followed by a rapid second electron transfer to form new stable redox states. A natural engineering principle for biological oxidation–reductions calls therefore for a two-electron donor or acceptor in close proximity (<6 Å) to the substrate (Page et al. 1999). This is elegantly solved in complex II with FAD in 4.6-Å distance to succinate (Yankovskaya et al. 2003). Chemical modeling showed that a proximity of the centers <14 Å provides electron transfer chains with a robustness to substantial variance in ΔG (Page et al. 1999). What does this mean? First, the protein structure itself does not participate too much in the electron transfer and can thus be ruggedly constructed; its only real constraint is the appropriate holding of the redox centers. Second, electrons spontaneously flow downhill along the reduction potential: from half-reactions with more negative standard electrical potential to those with more positive electrical potential. A higher (more positive) potential expresses a greater electron avidity of the chemical reaction. However, the detailed analysis of many oxidoreductases showed that many enzymes with multiple electron transfer reactions contain also very endergonic reactions. Oxidoreductase holds the reaction centers in close proximity to allow energetic tunneling, i.e., uphill electron flow. In complex II, none of the centers are separated by more than 16 Å, very close to the predicted limit for high-efficiency electron transfers. The transfer process must be efficient for the speed of the process. It cannot allow side reactions: In complex II, $FADH_2$ sends single electrons down the electron transport chain. Only ubiquinone collects again the two initial electrons in a two-step reaction. If one of the electrons misses its next signpost, probability is high that a very electron-avid substance is around, oxygen. Capture of electrons by oxygen is not good for a biological system; it creates what is called ROS (reactive oxygen species). When O_2 receives electrons, superoxide radical (O_2^-), hydrogen peroxide (H_2O_2), and hydroxyl radical (OH·) may be formed. These oxygen derivatives can harm many structures in the cell. If you look into the structure of complex II, you see a redox center off the normal electron pathway, a heme b ring. The heme consists of a porphyrin ring and the central iron ion is bound by the four nitrogen atoms of the porphyrin. A number of hemes are distinguished (a, b, c, d, o), which differ by their substitutions at three positions of the porphyrin ring. The central iron ion can shuttle between Fe^{3+} and Fe^{2+} and thus catch an erring electron that missed its way to ubiquinone. The porphyrin ring is a very popular biochemical structure; it is used not only in proteins from the respiratory chain (cytochromes) but also in oxygen transport in the organism (hemoglobin). A derivative with a magnesium ion complexed in the porphyrin ring is chlorophyll, the pigment central to photosynthesis. Color is a characteristic property of these compounds that is equally important for their physiological function as for their chemical differentiation.

The high redox potentials of the last Fe–S (+65 mV) and the offside heme b (+35 mV) pull the electrons from FAD by creating a real electron sink. In this way, 98% of FAD stays oxidized at all times, preventing electrons from slipping to molecular oxygen, which has access to the electrons because FAD is exposed to the cytoplasm. Without heme b, electrons could build up on FAD (Yankovskaya et al. 2003).

Fumarate Reductase

This redox situation is nicely illustrated in fumarate reductase (Iverson et al. 1999). We have already discussed this enzyme when speaking of the TCA cycle in *E. coli* under anaerobic conditions. The TCA cycle is reshuffled to produce fumarate, which becomes the electron acceptor in the absence of O_2 and nitrate as preferred electron acceptors. Under these conditions, *E. coli* uses menaquinone as an electron donor. Menaquinone (derived from *methyl naphthoquinone*) is chemically very similar to ubiquinone (it differs by the addition of a methyl group and a further benzene ring to the quinone). In comparison to complex II, the electron flow is reversed in fumarate reductase. Electrons pass from the menaquinone through three Fe–S centers to FAD, as in complex II. FAD maintains a comparable redox potential in both enzymes, but the Fe–S centers become substantially more negative such that FAD is now the most positive partner despite its −50 mV. This means that FAD is mainly reduced in fumarate reductase. This is needed for its functioning: As demonstrated in another fumarate reductase (Lancaster 1999), a hydride from the N5 position and a proton from the N1 position of the FAD base pass to fumarate, leading to the reduction of the double bond and thus to succinate. Why then is the reduced FAD not dangerous in fumarate reductase, while Mother Nature built in a safety valve in SDH? The answer is very simple: *E. coli* uses fumarate reductase only when there is no oxygen around. Under oxygenic conditions, it uses SDH. In principle the redox reaction between fumarate and succinate is thermodynamically reversible and both reactions could occur in *E. coli* by using the same enzyme. Indeed, *E. coli* can grow aerobically only with fumarate reductase in the complete absence of SDH. The electrons thus run in the opposite direction. However, under these conditions *E. coli* produces substantially more hydrogen peroxide and superoxide (Messner and Imlay 2002). Here you clearly see the advantage of the extra heme design in SDH. Actually, those prokaryotes that use the TCA cycle in both the reductive and the oxidative mode work with two distinct enzymes for the reversible fumarate–succinate reaction.

An Instructive Nematode Mutant

If you are not really convinced about the danger of living with oxygen, I can prolong the evidence list. Let's speak about a small nematode, which eats *E. coli* on Petri dishes in the laboratory with pleasure. The natural habitat of this small roundworm is the soil where it also lives of soil bacteria. Its popularity is not only smallness but also its precise cell number (just for fun: females contain 959 and males 1,031 somatic nuclei). Actually, a complete cell division fate map, a sequenced genome, and an extensive collection of mutants make this worm attractive to molecular biologists. The mutant worm called *mev-1* fits our context. This is really an interesting mutant: Unlike the wild type, its life span decreases dramatically as oxygen concentrations are raised from 1% to 60% (Ishii et al. 1998). These poor worms show molecular signs of prematured aging like the accumulation of carbonyl groups in proteins. This points to oxygen damage, and this hint attracted researchers to map the mutation. And they struck gold: The defect is explained by a point mutation changing a glycine to a glutamic acid residue in the SDH. If you look into the location of this change, the connection to straying electrons becomes even more convincing. The negatively charged glutamic acid is only two amino acids removed from the positively charged histidine, which holds the extra heme ligand in place.

The Handling of Molecular Oxygen

Complexity

Even if the basic fabric of life—once investigated in sufficient detail—is quite logical, biology tends to be disturbingly complex. The underlying reason is that life's main characteristic is diversity. Organisms come in many forms and Mother Nature shows a nearly exuberant joy to entertain herself with the ever-new forms of life as exemplified by the 800,000 species of beetles. Diversity comes also within a single organism. Our body, for example, contains perhaps 100 different tissues. Now we are confronted with a conundrum: In comparison with E. coli, which builds no tissues and consists only of a single cell, we construct these many tissues and integrate their functions with perhaps just 10 or 20 times more genes than this gut bacterium. We normally explain this difference by our superior regulatory network and alternative splicing. If one looks into the regulatory mechanisms in E. coli controlling oxygenic metabolism, one might actually wonder whether we are really smarter with our regulation processes. It is thus intuitively clear that a number of genes in our body must have been conscripted to more than one function.

Succinate Dehydrogenase as Oxygen Sensor

This point can again be convincingly demonstrated with SDH. As we saw, this protein is an enzyme of the TCA cycle and a component of the respiratory chain at the same time. Now we will see a third function: It is possibly also an oxygen sensor for the body. The story starts again with a mutation: a hereditary paraganglioma in humans characterized by the development of benign, vascularized tumors in the head and neck. The most common tumor site is the carotid body, a chemoreceptive organ in the bifurcation of the Arteria carotis that senses oxygen levels in the blood. This is a very special sensory organ. It is extremely small (its weight is a mere 2 mg), but for that it receives an extraordinary blood flow, exactly 40-fold higher than the next big consumer, the brain, if expressed per unit weight. It also shows a threefold higher metabolic rate than does the brain. So what is this small ganglion doing to justify this blood supply? Actually, its major task is the detection of hypoxia (low oxygen partial pressure) in the arterial blood. In parallel, it also responds to high blood CO_2 concentrations (hypercapnia) and a too high acid load (acidosis). The three signals are independently sensed, but the common output is the same. K^+ channels are inhibited, leading to the stimulation of an afferent nerve, which tells the respiration center in the Medulla oblongata to increase the ventilation rate. This measure will solve all three problems: It transports more O_2 into the tissue and takes away the CO_2 and acidity. Sensors in the aortic arc regulate our hunger for oxygen. Physiologists suspected for a while that the actual O_2 sensor is a hemoprotein. The mapping of the hereditary paraganglioma mutation was revealing: it was located in 11q23 (Baysal et al. 2000). In human genetics, this reads chromosome 11, long chromosome arm (q), band 23. This is in fact the location of SDH. Once mapped, the geneticists surveyed the region using sequence analysis. And it was nearly too good to be true: One family showed a missense mutation in histidine 102, which in the *E. coli* enzyme is the side chain binding heme b. As one would suspect, the story is more complicated (e.g., the mutant was only genetically active when inherited from the father, a phenomenon explained by genetic imprinting—the gene "recalls" by a chemical modification whether it came from the father or the mother), but all this did not change the fact that all mutants mapped into the SDH gene. Pathologists provided a further hint. They had observed that the carotid body tumor occurred about 10 times more frequently in persons living in high altitude than those living at sea level. This observation suggested that the tumor actually represents a hyperplastic response to the prolonged sensing of hypoxia by the carotid body. From this observation, it is only one step to the following hypothesis: Heme b in SDH is the actual oxygen sensor; if inactivated it relays a hypoxia message, irrespective of what is the real arterial oxygen level. This induces HIF-1 locally (recall our discussion of the Warburg effect in cancer cells), which leads to proliferation and vascularization as observed in the mutant patients (Baysal et al. 2000). Indeed, SDH deletions are common in many carcinomas.

Insect Respiration

At the same time, life is dictated by the fear of oxygen. If oxygen is such a double-sided sword, one should expect that animals search for and flee oxygen at the same time. There is a nice illustration for this predicted behavior in insects. Here we first need a short excursion into insect anatomy and physiology because their respiratory system differs fundamentally from ours. However, we should refrain from looking at it as a primitive system. If bacteria are the true rulers of the world, we might dismiss them because of their tiny size. This we cannot do with insects because they are visible to the eye, extremely numerous, and represented by a breathtaking diversity of species. *There are good reasons for seeing them as the crown of terrestrial life. We appreciate great animals more than small animals and our moviemakers try to instill horror in us by increasing mutant flies to the size of elephants. Is this a real fear?* In fact, the largest insect lived 280 million years ago and was the famous dragonfly *Meganeuron* with an impressive wingspan of 70 cm. This size might frighten some fainthearted individuals, but it is not really terrifying. *It is a widespread misconception even among biologists that the insect tracheal system, which relies exclusively on gas diffusion, limits the size of insects to the Meganeuron range. There were probably other constraints on size evolution in insects—the respiration system is not the limiting factor.*

The tracheal system is charged to bring oxygen to the mitochondria of insects where food molecules are oxidized to CO_2. In mammals like us, two systems take part in this process. At the periphery, there is the respiratory system, which consists of the mouth and nose taking up ambient air, a tubing system of decreasing caliber (trachea, bronchi, bronchioli) carrying the air into the lungs. In the lung alveoles, air and blood come in close contact and oxygen leaves the air and goes into the bloodstream. This causes problems of solubility and diffusion and therefore we need specialized vehicles for oxygen transport: At the protein level, this is the oxygen-transport protein hemoglobin; at the cellular level, this is the erythrocyte, the red blood cell, an extremely simplified cell packed with hemoglobin. To get oxygen to the end consumer, i.e., the mitochondria in the various body tissues, blood must be provided to these tissues by the pumping activity of the heart. Under great pressure (and energy expense), this oxygenated blood is then pressed into a sophisticated arborification system of vessels of decreasing size. The smallest, the capillaries, deliver the oxygen to the tissues in need. With 21% oxygen content in the atmosphere, the partial pressure of oxygen at sea level is 20.4 kPa. In the lung, it is already reduced to about 13 kPa; at the capillary/tissue interface, it is further reduced to 5–4 kPa. In accordance with the calculations of A. Krogh done as early as in 1919, oxygen concentrations sufficient for mitochondrial respiration are limited to a distance of 150 μm from the capillary. Oxygen pressure in the cell of this zone is 0.4 kPa. Why do I tell you these well-known facts? Of course, I need the figures for comparison with the respiratory system in insects. Insects do not work with this double transport system: air/lung and water/blood circulation. The heart of insects is a weak pump circulating the hemocoelic fluid. Insect blood has only nutritive functions and does not carry oxygen. Many types of hemocytes are reported in the insect "blood," some function in familiar processes such as wound healing and clotting, but none in oxygen transport or storage. The body surface of insects, the cuticle, is impermeable to water and gas. To allow gas exchange to occur, insects have up to 10 pairs of spiracles (air openings) along the thorax and abdomen. Some insects have a flow-through system with air entering in some anterior spiracles and leaving via posterior spiracles. This order can change with the transition to activities necessitating higher oxygen consumption like flying or in different larval stages. These spiracles have typically an outer hair filter that prevents dust, debris, and parasites from entering. Then follows a mouthpart, commonly a muscular valve under the control of the internal partial pressure of O_2 and CO_2. Notably, in resting insects the spiracles are generally closed.

The outer part of the tracheal system is sclerotized and gas filled. The caliber of these air channels gets finer and finer. Trachea split in many tracheoles that are kept open by ring-like stabilizers called taenidia. They allow substantial extensions of the tracheal system without collapse of the air channels when the insect body changes length in pumping movements. This muscular activity also drives the airflow. However, diffusion of oxygen in air is 200,000 times faster than in any aqueous system. Thus oxygen reaches the end consumer efficiently. A particularity of the tracheal system is that it gets so fine at the end that its final ramifications enter the flight muscle cells and deliver the oxygen directly to the mitochondria. Let's have a look at the oxygen partial pressures. When the spiracles are fully open, the oxygen partial pressure at the tip of the tracheoles should be about 19 kPa, much more than that achieved by the vertebrate system. However, this is a mixed blessing. If oxygen is really so dangerous, you should keep away from too much of oxygen. In resting insects, this is achieved by closure of the spiracles. During this closure period, which lasts 30 min, the pO_2 decreases to about 3–4 kPa. This closure period is followed by a flutter period of short openings of the spiracle, which leads to short spikes of CO_2 release (Hetz and Bradley 2005). In the resting insect, the next opening occurs only in 90-min intervals leading to short but massive CO_2 release. When the physiologists changed the oxygen concentration from 6 kPa (hypoxia) to 50 kPa (hyperoxia), the intratracheal pO_2 was kept constant at about 4 kPa, the apparent set value for the insect, which is remarkably close to that in the vertebrate capillary bed. At hypoxia the spiracles open frequently to assure oxygen supply, the opening frequency decreases proportionally with the ambient oxygen pressure. At high pressure, the openings are only dictated by the need to release the accumulated CO_2. The insect respiratory system is thus overdesigned for the massive oxygen needs during periods of high oxygen demand, like in flying when oxygen is quickly reduced to water.

Aging Hypotheses

The free-radical hypothesis of aging postulates that senescence is due to accumulation of molecular oxidative damage, caused largely by oxidants that are produced as by-products of normal metabolic processes, namely in cellular respiration. A logical prediction based on this hypothesis is that the elevation of antioxidative defenses should delay aging and extend the life span, hence the popularity of antioxidants in human nutrition. The hypothesis got a boost when data were published that demonstrated an extension of the life span, delayed loss in physical performance, and lesser oxidative damage of cellular proteins in transgenic flies overexpressing superoxide dismutase and catalase

(Orr and Sohal 1994). However, in later publications, the same group realized that the 14–30% increase in life span with respect to control flies was only observed in short-lived flies, whereas no or even slightly decreased life spans were observed in transgenic flies from more long-lived *Drosophila* stocks (Orr and Sohal 1994). In another insect system, namely ants, life spans vary by two orders of magnitude within genetically identical individuals belonging to different casts: Queens live >28 years, workers 1–2 years, and males only a few weeks. In striking contrast to predictions of the oxygen hypothesis of aging, male ants showed the highest expression levels of superoxide dismutase measured by two independent methods (Parker et al. 2004).

Although the superoxide dismutase hypothesis in insects is no longer tenable, the oxygen hypothesis of aging is not yet dead. Synthetic antioxidants extended the life span of the nematode *Caenorhabditis elegans* (Melov et al. 2000). In addition, feeding experiments with *E. coli* lacking ubiquinone (coenzyme Q) showed extended adult life span in *C. elegans*. The authors hypothesized that the bacterial coenzyme acquired with the diet is also used in the respiratory chain of the worm. They proposed a model in which the reduced level of ubiquinone also reduced the production of ROS, which in turn positively influenced the life span. The reduction in dietary coenzyme Q level even caused a life span increase in worms that already have a genetically determined longer lifetime (*daf* mutants). In recent years, another metabolic hypothesis of aging got increasing support: caloric restriction (for a recent review see Bordone and Guarente 2005). An appealing early hypothesis in this field held that decreasing calorie intake increased longevity simply by decreasing metabolism and thus the production of ROS in the mitochondria. Convincing as it sounded, the idea got no experimental support. The metabolic rate in calorie-restricted mice did not decline when normalized to their smaller body weight. Over their extended lifetime, their metabolic output was even greater than that of ad libitum-fed mice (Masoro et al. 1982). Respiration actually increased in *C. elegans* and yeasts (Lin et al. 2002). Despite this clear evidence, the catalase camp has not yet admitted defeat. After controversial and contradictory findings in invertebrate models of aging, researchers turned to mammals. In transgenic mice they targeted human catalase, which is normally expressed in the peroxisome, to mitochondria (Schriner et al. 2005). Indeed, they obtained increased catalase expression in heart and skeletal muscle. In addition, median and maximum life span was extended by 5 months in the transgenic mice. This was, however, less than that achieved by caloric restriction in mice (Longo and Finch 2003). All hallmarks of the aging process were attenuated in the catalase-expressing mice: Cardiac pathology was less severe in old mice, they showed less H_2O_2-induced inactivation of the mitochondrial enzyme aconitase, and PCR showed less mitochondrial DNA deletions compared with the nontransgenic controls. Mitochondrial DNA mutations are important for the aging process in mice: When a proofreading deficient form of the mitochondrial DNA polymerase γ was expressed in mice, the animals showed more mutations and the signs of progeria (accelerated aging; Kujoth et al. 2005). The accumulation of the mitochondrial mutations was not associated with increased markers of oxidative stress but with the induction of apoptotic markers. Recent results suggest that mitochondrial ROS increase markedly *after* proapoptotic signals. They are a transmitter, but not the inducer of controlled cell death. The mediator is the mitochondrial protein p66SHc, which utilizes reducing equivalents of the mitochondrial electron transfer chain to create a proapoptotic ROS in response to specific stress signals (Giorgio et al. 2005).

What we saw before with SDH also applies to mitochondria: They are multifunctional—they are the main source of cellular ATP; they oxidize fatty acids; they synthesize phospholipids; they are involved in calcium signaling; and via ROS generation and apoptosis, they decide on the life and death of the entire cell.

Social Feeding in Worms Explained by Oxygen Avoidance

Nematode Biology

The nematode *C. elegans* became a model organism in molecular biology (Figure 3.1). As it has a fixed quota of 302 neurons, it became possible to decipher even the neurobiological basis of relative complex behavior reactions. One of these success stories of recent neurobiological research is the social eating in stressful situations (Sokolowski 2002). However, I promised you further

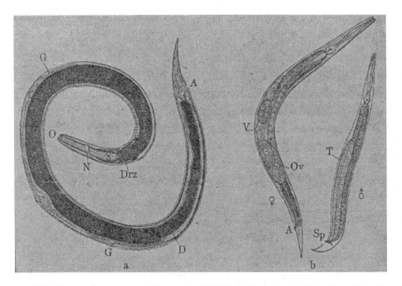

FIGURE 3.1. Nematode worms, *left*: a parasitic Rhabdonema, *right*: a female and male free-living Rhabditis form, morphologically closely related to *Caenorhabditis elegans*, which is in this book a main actor in oxygen avoidance, aging, toxic bacterial food avoidance, and plant root attack. It is a predator of the rootworm and a host to sulfur-oxidizing bacteria. The mouth, secretory gland, gut, and anus are marked in the figure by the letters O, Drz, D, and A, respectively.

evidence of how animals avoid too high oxygen concentrations for fear of reactive oxygen damage. Therefore I must tell this story in a different order. First, some background information on nematodes. The name of this "worm" (one of the least defined terms in zoology) derives from the Greek "nema," meaning thread. In fact, the threadlike appearance is a major characteristic of this group. Most of the nematodes are microscopic in size; *Caenorhabditis* with its 1-mm length is already a large specimen. Zoologists have counted up to 90,000 nematodes in a single rotten apple; some soils contain 3 million nematodes per square meter. Nematodes are abundant in many environments (marine from shore to abyss, freshwater, soils; some are important gut parasites of mammals, others punch plant roots with stylets and suck the cell sap). A few are carnivores feeding on other animals, whereas *Caenorhabditis* eats bacteria. Its digestive system is rather simple: It consists of a mouthpart leading into a buccal cavity. The cuticule-lined esophagus (also called pharynx) is elongated and separated in different chambers, representing more muscular regions for grinding and a glandular part for external digestion. The muscle pumps the preyed bacteria into the intestine, which has absorption epithelia not unlike our own intestine with microvilli. In this part of the gut also, internal digestion is observed. Undigested waste is expelled via an anus located at the posterior end of the worm. If you look at a cross-section of the worm, you distinguish an external cuticle, underlaid by a ring of muscles interspersed with some nerves and a few excretory cells. The inner part of the body cavity (also called coelom) is filled up with the intestine and to an even larger extent by reproductive organs. You search in vain for a heart, a circulatory system, lungs, and a respiratory system. Gas exchange is accomplished by diffusion across the cuticle and by the movement of the body cavity fluids. Parasitic nematodes sport both an aerobic and, within the gut of the host, an anaerobic metabolism. If you do not have a respiratory system, your possibilities to control the oxygen level are rather limited. How does *Caenorhabditis* assure minimal oxygen concentrations to maintain oxidative phosphorylation, but flees from too much of oxygen to avoid oxygen damage?

Feeding Control

Behavioral control of locomotion is the answer, and genetic research led to its elucidation. However, at the outset, there was a behavior difference. The common laboratory *C. elegans* strain N2 is a solitary feeder: It shows reduced locomotion and high dispersal activity when encountering food, for example, a lawn of bacteria. This behavior is what would be predicted by the optimal foraging theory. However, there is another type in the population—the social feeder. It continues to move rapidly toward food and eventually aggregates on the border of the lawn (Sokolowski 2002). As the genetics of this worm is well developed, the difference in the behavior could be traced to a single amino acid change in the *neuro*peptide *r*eceptor protein *1* (NPR-1). This protein is related to neuropeptide Y, and this is a significant link. Neuropeptide Y is the major orexigenic hypothalamic factor that controls appetite in mammals. Orexigenic is better known from its opposite anorexia and this means that it stimulates feeding behavior. This was an interesting link for worm geneticists who unleashed their powerful tools on this question. They introduced the green fluorescent protein gene into the *npr-1* gene. This fusion protein identifies its location in the animal when observed under a fluorescence microscope. The protein was found in sensory neurons (as all 302 neurons are known in *Caenorhabditis*, the geneticist could be more precise and could identify the individual cells by their code names, e.g., the URX or the AUA cell; Coates and de Bono 2002). They knew that AUA is the principal synaptic target of URX, the observations therefore made sense, but they could not yet tell what sense. Expression of the "solitary" *npr-1* gene in just these few neurons of the "social" feeder was sufficient to establish solitary feeding, and this behavioral transition occurred within 1 h after they artificially switched on the gene (this is commonly done by a heat-regulated promoter). This experiment identified the most important inhibitor of social feeding. The importance of the sensory neurons ASH and ASL for social feeding was also demonstrated by microsurgical experiments. The scientists destroyed just these individual cells with an extremely focused laser beam, resulting in the loss of the social feeding (de Bono et al. 2002). Now the researchers had another piece of the puzzle in their hands: These two sensory neurons were implicated in aversive responses to noxious stimuli. Social feeding in worms might not have the same function as in humans, where it fosters group coherence. It might be a response to a repulsive cue. But what was it? Was it the food bacterium? This does not sound terribly logical, but this was the interpretation explaining most of the results. To get to a more satisfying interpretation by integrating their results into other elements of current knowledge, the researchers set out to explore the signaling pathway leading to this behavior. They knew that cyclic nucleotide-gated ion channels are important transducers of sensory signals in both vertebrates and invertebrates. The genome of *Caenorhabditis* has been sequenced and many mutants facilitated the search. To make a long story short, by this line of research the *tax-2*/*tax-4* genes coding for subunits of a cGMP-gated cation channel were placed in the signaling pathway to the identified sensory neurons, which mediated this conspicuous bordering and aggregation behavior. This was a satisfying interpretation: A cGMP-dependent protein kinase had an important effect on the foraging behavior of other invertebrates. In the fruit fly *Drosophila* the foraging gene *for* encodes a kinase PKG stimulated by cGMP. The gene *for* comes in two forms *sitter* and *rover*, fly geneticists like funny names, but here the names are clearly describing the behavior of larva on food lawns. The *sitter* larvae are rather immobile on food, whereas *rover* hover around the lawn (Osborne et al. 1997). Likewise, in honeybees a cGMP-dependent PKG is upregulated when bees switch from nursing in the hive to foraging for nectar and pollen in the environment (Ben Shahar et al. 2002).

Oxygen Control

However, with all this detective work, you have not yet seen the link to oxygen. The same researchers constructed a chamber for *Caenorhabditis*, which allowed applying a gas-phase gradient over the lawn. Here the worm showed its preference for about 8% oxygen, lower oxygen levels as well as higher oxygen levels corresponding to those in our atmosphere were avoided (Gray et al. 2004). The previously identified mutant worms in the *tax-2/tax-4* ion channel had lost this behavior of searching for the optimal oxygen concentration. Now the researchers had a smoking gun: They could search for upstream signals for this ion channel. As the channel is cGMP gated, it was logical to look for soluble guanylate cyclase homologs in the genome. GCY-35 was the right candidate for the upstream regulator: It was expressed in the appropriate sensory neurons, and its knockout led to the loss of the behavior of searching for an optimal oxygen concentration. And sometimes researchers are also lucky: GCY-35 has a heme iron that binds molecular oxygen; the protein is therefore the likely molecular sensor of the worm for oxygen. This oxygen sensor is not without precedence: Aerotaxis in Archaea and Bacteria is transduced by myoglobin-like proteins (Hou et al. 2000). Now the pieces of the puzzle came together (for the moment at least, as science is a story without an end). Oxygen diffuses rapidly through small animals like this worm, and it does not become a limiting factor for respiration until it falls below 4% in the air surrounding the animal. Higher concentrations do not contribute to respiration but increase only the risk for oxidative damage. Low oxygen in the soil may actually signal the presence of food in the form of actively growing bacteria that consume oxygen more quickly than the time in which oxygen can diffuse through soil.

Electrons

The Chemiosmotic Hypothesis

From Lavoisier to Mitchell

After all these cautionary notes on oxygen, let's follow how Nature is playing with fire in molecular respiration to get energy out of foodstuff. It was Antoine Lavoisier who first formulated the modern concept of bioenergetics when he described respiration in animals as nothing but a slow combustion of carbon and hydrogen. He likened this internal fire in us to the fire stolen from the heavens by the ingenious Prometheus, making life possible in the earthlings. However, it is a silent fire; evolution has taken all its diligence to subdivide this combustion into smaller steps to recover the energy in a convertible form that can be used to sustain all energy-demanding life processes. Lavoisier made this truly revolutionary discovery very appropriately in the first year of the Great French Revolution, only to be decapitated 5 years later as a counterrevolutionary. Actually, it took a relatively long time for biochemists to get further new insights into how Nature is running this slow combustion process. Only in 1948 was oxidative phosphorylation associated with mitochondria by Eugene Kennedy and Albert Lehninger. The breakthrough came with a seminal Nature paper published in 1961 by Peter Mitchell (1961). He pointed out that the orthodox view that oxidative phosphorylation is a variant of substrate-level phosphorylation could not explain a number of observations available to biochemists at the time. The observations that were at odds with the classical interpretation were the lack of an energy-rich chemical intermediate, the uncoupling of redox reactions from ATP-producing processes by compounds lacking a common denominator, and strange swelling and shrinkage processes occurring in mitochondria. He came up with a simple hypothesis, which radically changed the ideas about how biological systems conserve energy.

A centerpiece of his chemiosmotic hypothesis was the role of the charge-impermeable cell membrane. Hydrogen ions are generated on one side of the membrane (later called the P side, P for positive because of the H^+charge) and consumed on the other side, leading to an excess of OH^-over H^+ ions (consequently this side was later termed N side for the negative charge of the hydroxyl ion). The three simple line drawings of his paper already had all elements familiar to modern biochemists: a reversible ATP synthase; electron transport in the membrane; cytochromes, quinones, and vectorial transport of protons across the membrane.

Unifying Principles in Biology

Another important suggestion was that essentially the same process applies to energy generation in mitochondria and in chloroplasts. This is quite a remarkable suggestion because in other respects these two processes look very different. However, this energetic link between two different organelles fitted into the 1960s, which saw also other unifying principles in biology such as the universality of the genetic code. This unifying concept allowed integrating two lines of research. In 1954 Daniel Arnon discovered that spinach chloroplasts produced ATP when illuminated. The next logical step after the chemiosmotic hypothesis was to prove that the establishment of a proton gradient across the membrane could drive ATP synthesis. This was actually achieved by André Jagendorf in 1966. Chloroplasts were incubated in the dark in a pH 4 buffer for a longer time period to allow an acidification of the thylakoid lumen of the chloroplasts and then suddenly transferred to a pH 8 buffer. As predicted by Peter Mitchell, the pH gradient could be used for ATP synthesis. Indeed, the elusive high-energy state of oxidative and photosynthetic phosphorylation was the proton gradient across the membrane and not a chemical intermediate.

Seeing is Believing

The major advances in structural biology in the last 15 years revealed the crystal structures of nearly all elements of the electron transport chains in the mitochondrial and chloroplast membranes. Part of the predictions of the chemiosmotic hypothesis can now be followed at atomic resolution. Also, the other partner of this energy conversion process, ATP synthase, yielded to the analysis of biochemists, and we now have an understanding of this smallest electromotor of the world at the resolution of crystal structures. As these processes are the ultimate basis of the quest for food, I will select three topics from this area for more detailed discussion. The first is an electron and proton transducer shared by oxidative and photosynthetic phosphorylation; the second is the oxygen gas eater and major proton pumper of mitochondria; and the third is the revolutionary (because revolving) device of ATP synthesis with which I will close this section.

Anatomy of the Respiratory Chain

A Road Map

The basic design of the respiratory chain is described briefly. You have the inner mitochondrial membrane as the basic construction device. Owing to its lipid character, this membrane is practically impermeable to protons; this characteristic allows the maintenance of a proton gradient across the membrane. This proton gradient conserves the energy of the oxidoreduction processes that occurred in the preceding catabolic pathways (mainly glycolysis, β-oxidation followed by TCA cycle). The reducing equivalents from these catabolic oxidation processes are fed into the electron transport chain via NADH or $FADH_2$. The feeding pathways use different entrance portals into the membrane, but all converge on ubiquinone. Despite this convergence, not all entrance portals are equal energetically. NADH created in the TCA cycle enters via complex I into the mitochondrial electron transport chain. Complex I transports electrons, and protons are also coupled to this process. This complex thus contributes actively to the energetization of the membrane. In contrast, succinate oxidation feeds electrons into the chain via $FADH_2$, but it uses complex II, as we have seen. There was no pedagogical omission when I did not mention coupled electron transport processes for this complex. No protons are pumped by this complex. This difference is not just a difference between $FADH_2$ and NADH as reductants. Also, NADH from glycolysis uses a low-energy entrance port into the mitochondrial electron transport chain. Of course, the glycolytic NADH cannot feed complex I because its electron-hungry mouthpart is at the matrix side of the inner mitochondrial membrane. Complex III does not use ubiquinone as an electron transporter any longer; it charges a small soluble protein at the outer mitochondrial side (P site) with an electron, and this cytochrome c protein carries the electron to the end consumer in this chain, complex IV. This membrane complex carries the donated electrons to the final electron acceptor, oxygen, reducing it to water.

The Principles

The essence of this electron transport process is the transport of protons from one side of the membrane to the other, creating the proton gradient. The electrons leave the chain with the water molecules opening the pathway for the next electrons fed into the respiratory chain. Of course, the proton gradient does not build up beyond a certain steady-state level and is constantly used to drive a rotation process in ATP synthase (also called complex V), which creates ATP from ADP and P_i. Thus the proton gradient is diminished, and the energy of proton translocation is conserved in the ATP molecule. Finally, the ATP has to be transported out of the mitochondrial matrix to power all energy-requiring processes of the cell. Overall, this summary sounds rather trivial and does not highlight half a century of painstaking efforts of biochemists, biophysicists, structural biologists, and cell biologists to unravel the process. In fact, compared with the chemical reactions occurring in the cytoplasm, the membrane-bound processes in respiration occur at a different level of complexity that are only rivaled by the process of photosynthesis.

Comparison of Respiration with Photosynthesis

As can be gleaned from the comparison of the schemes for the photo-synthetic and the respiratory electron transport chain, both processes share fundamental similarities:

- Four large complexes are found in both membranes.
- Complex I in the inner mitochondrial membrane takes electrons from NADH coming from the matrix side and feeds them into the electron transport chain. PSI in the chloroplast thylakoid membrane takes electrons from the photo-synthetic electron transport chain and feeds them into reactions, leading to NADPH production.
- Complex IV takes electrons from the transport chain to reduce oxygen to water by picking up protons. PSII splits water into oxygen and protons and delivers electrons into the transport chain.
- In both systems the electron transport chain uses quinones as an electron carrier (ubiquinone in mitochondria, plastoquinone in chloroplasts).
- Likewise, both electron transport chains use membrane-associated small soluble metalloproteins that can be reduced by a cytochrome complex and carry only one electron (plastocyanin in chloroplasts, cytochrome c in mitochondria).
- The chloroplast ATP synthase is so similar to the mitochondrial ATP synthase that biochemists make the tacit assumption that biochemical details described in the better-investigated mitochondrial complex apply also to the chloroplast protein complex until something contrary is proven.
- Finally, complex III from mitochondria and the cytochrome $b_6 f$ complex in chloroplasts (Kuhlbrandt 2003; Stroebel et al. 2003) share so many character-istics in architecture and function that their evolution from a common ancestor is very likely. We will look into these proteins in the next chapter.

Overall, there are too many similarities between both major energy-yielding processes to explain them by chance or evolutionary convergence. In fact, the most likely explanation is that both processes were derived from a common prokaryotic heritage. Photosynthesis is without doubt the more fundamental process: Oxygen did not exist in the early atmosphere and was literally created by photosynthesis. Light, the driving force of photosynthesis, was present from the beginning of time on the planet earth, whereas reduced organic compounds that are catabolized in mitochondria first had to be synthesized in autotrophic organisms, including photoautotrophs. It thus seems logical to anticipate that oxygenic respiration was initially copied from photosynthesis, when the products of photosynthesis had accumulated to levels making this process energetically possible. As we have discussed in preceding sections, oxidative respiration seems to be the reversal of oxygenic photosynthesis (this specification is necessary because both nonoxygenic respiration and nonoxigenic photosynthesis exist in extant organisms).

The Oxygen Cycle

Ecologists know a number of nutrient cycles in the biosphere at a global level: The most prominent are the carbon, sulfur, nitrogen, iron, and manganese cycles. Minor cycles deal with additional metals and phosphorus. The oxygen cycle is, however, rarely mentioned. The splitting of water, the most abundant substrate/electron donor on earth, into oxygen, protons, and electrons can only be reversed by oxygenic respiration and combustion processes. Perhaps the shuttling of oxygen atoms between molecular oxygen and water is not considered as a cycle because it is not in equilibrium. If we later speak about problems in the carbon cycle in the context of the increasing CO_2 levels, we should not forget that these increases—important as they are for our current climate development—are small in comparison with the past CO_2 levels life has experienced on earth and the spectacular rise of atmospheric oxygen. In fact, the absolute CO_2 levels in our current atmosphere are 700-fold smaller than the oxygen level (0.03% vs. 21%). Life on earth has successfully coped with a concentration that is much higher than the present CO_2 concentration (Mojzsis 2003; Kaufman and Xiao 2003; it has also faced much hotter climates than that in the present, temperatures that would probably melt away human civilization). However, oxygen made a prodigious progress to 21% of the current atmosphere. Oxygen is—as we have seen in the previous section—a necessary but harmful compound for animals. When we are concerned about the future of our civilization and when we look with much justified concern to the increase in atmospheric CO_2, we should as biologists be equally concerned about the increase in atmospheric oxygen concentration.

Where are we with respect to the attainable oxygen level created by photosynthesis? Biophysicists recently conducted an interesting experiment that perhaps provides an answer. They were interested in identifying an intermediate in the water-splitting reaction occurring at the Mn_4Ca cluster of PSII. As the process is described by a simple chemical equation: $2H_2O \leftrightarrow O_2 + 4H^+ + 4e^-$, they reasoned that they would detect an intermediate of this enigmatic reaction when increasing the oxygen pressure and slowing the forward reaction. Simple as their reasoning was, it worked out (Clausen and Junge 2004). An unexpected low pressure of 2.3 bar of oxygen at a pH of 6.7 half-suppressed the oxygen evolution. Two arguments indicated that the oxygen pressure that will slow water splitting might even be lower than 2.3 bar. The actual pH at the luminal side of the thylakoid membrane is actually more acidic than the pH 6.7 used by the biophysicists in their experiments. Furthermore, the scientists calculated that the PSII operates at the outmost redox span that can be driven by the absorption of light quanta. Their conclusion was that the present atmospheric pressure of oxygen is close to the expected obtainable maximum. So life will not face hyperoxic conditions.

NADH Oxidation in Complex I

It would be logical to start our journey through the respiratory chain with the first step, namely the feeding of the electrons extracted from the food molecules and carried by NADH into the mitochondrial electron transport chain. A detailed description of what happens during this stage was, however, not possible until quite recently. The complicated structure of complex I, which in mitochondria consists of 46 different polypeptide chains, has not been resolved by protein crystallographic analysis. First hints about the structure were provided by high-resolution electron microscopy analysis showing an overall L-shaped protein complex consisting of a hydrophobic arm in the membrane and a hydrophilic arm extending into the matrix of the mitochondrion or the cytoplasm of the bacterium. British scientists have now solved the structure of the hydrophilic arm from the complex of *Thermus thermophilus* (Sazanov and Hinchliffe 2006). A protein complex from a thermophilic prokaryote has the advantage of greater stability and lesser complexity: the entire complex I consists only of 14 subunits. The available structure covers eight of them and half of the 550-kDa-sized complex. The structure suggests a path for the electron from the cytoplasmic NADH to the vicinity of the quinone, the membrane-soluble carrier of electrons to complex III. Subunit 1 contains FMN, the primary electron acceptor of complex I. The FMN cavity can accommodate one NADH molecule comfortably; the adenine ring of NADH is fixed between two phenylalanine rings by aromatic stacking interaction, allowing a facile hydride transfer to the isoalloxazine ring of FMN. From there it is only a small step for the electron to the first and from there to six further iron–sulfur redox centers of the complex, mostly in the 4Fe–4S cluster form with cubane geometry. The electron transfer is favored electrically: The two-electron midpoint potential of FMN is at neutrality $-340\,mV$, the iron–sulfur centers are isopotential at about $-250\,mV$, while the quinone, the ultimate electron acceptor, is at $-80\,mV$. The *Thermus* complex I uses menaquinone and not ubiquinone as mitochondria do, the latter exerts a midpoint potential of $+110\,mV$, an even more forceful electric pull. The iron–sulfur redox centers are separated by less than $14\,Å$, the reach for electron transfer in biomolecules (Page et al. 1999). The crystallographers were puzzled as to why complex I uses so many redox centers. They came up with two answers.

The first answer is functionality, namely that of a safety valve. From NADH to FMN you have a two-electron transfer, whereas the iron–sulfur centers can only accept a single electron. This creates the risk of a reactive FMN radical at the surface of the molecule, which must wait until the quinone has taken up the electron from the redox chain, which is the rate-limiting step. Nature has found an elegant solution: subunit 2 carries another iron–sulfur center in the reach of FMN. FMN thus transfers its two electrons quickly, one on the iron–sulfur center in the redox chain and the other on this temporary storage hidden from the surface. This safety valve minimizes ROS production during turnover of complex I, the prime danger when playing with the electronic fire of oxygen chemistry.

The second answer is suggested by a fourth iron–sulfur center in subunit 3. Three are in the redox chain, one is at 24-Å distance, which is too far away for electron transfer. The authors interpret it as an evolutionary remnant. In fact, the different subunits from complex I resemble proteins such as ferredoxin, different hydrogenases, and even frataxin structurally. The latter is a mitochondrial iron storage protein, where a gene defect leads to a severe neurodegenerative disease, Friedreich's ataxia. Interestingly, frataxin is an iron donor to aconitase. Remember that this TCA enzyme has a second function in converting inactive 3Fe–4S to an active 4Fe–4S cluster. The surprising structural affinities of the complex I subunits suggested to the crystallographers that the complex assembled in evolution from many smaller building blocks, which had their own biochemical functions. This process can still be seen when considering the mitochondrial complex I. The central subunits are functionally related to the 14 subunits known from the prokaryotic complex I, forming a central catalytic core. This attribution is supported by the observation that seven of these core subunits are still encoded by the mitochondrial genome. About 30 further accessory subunits decorate the periphery of the mitochondrial complex I. They were probably recruited during the evolution of the eukaryotic respiratory complex from independent sources. It is painful to realize that we know literally nothing about the functions of these eukaryotic-specific complex I subunits. This statement illustrates major gaps in our biochemical knowledge concerning the most basic aspects of biochemistry, namely how organisms extract energy from the food they eat. Transporting electrons is only one aspect of complex I, the other is the generation of a proton gradient across the membrane, which is then used for driving ATP synthesis. The overall stoichiometry of complex I can be described as $NADH + 5H^+(N) + Q \rightarrow NAD^+ + QH_2 + 4H^+$ (P). P stands for the positive side of the inner membrane (intermembrane side of the mitochondrion), N for the negative side of the membrane (the matrix side). The structure of the hydrophilic arm from complex I provided no answer to the question of how electron transport is coupled to the mechanism of proton pumping, which probably occurs in the hydrophobic membrane arm of complex I.

Cytochromes bc_1 and b_6f: The Linkers in and between Respiration and Photosynthesis

The Art of Crystallography

In the current form of scientific research, it is not uncommon for notable scientific discoveries to be made simultaneously and independently by two or more groups of researchers. This was also the case for the mitochondrial complex III alias cytochrome bc_1 where three groups reported the structure of this protein from bovine heart mitochondria (Xia et al. 1997; Iwata et al. 1998; Zhang et al. 1998). This observation was made again when two independent groups reported the corresponding structure of the chloroplast protein complex (Stroebel et al. 2003; Kurisu et al. 2003). The structures differed only in crystallographic details with respect to the resolution of minor subunits or details. The resolution is mainly determined by the quality of the crystals. Unfortunately, the first step in this process—the growing of the crystal—is still more an art than a science, which explains why it took both labs more than 10 years to resolve the structure of cytochrome b_6f. The instability of the protein is frequently the major problem. The two groups used different tricks to overcome this problem. The group that worked with the protein complex from the unicellular alga *Chlamydomonas* engineered a six-histidine chemical tag at the C terminus of cytochrome f to allow a fast protein purification on a nickel column. The group working with the cyanobacterial protein capitalized on the inherently greater stability of proteins from thermophilic prokaryotes. However, they had to add back some membrane lipids to get good crystals. Overall, they spent more than 10 years to get the structures (Kuhlbrandt 2003). Yet, the results justified the substantial efforts.

Cytochrome b_6f Anatomy

To begin with, the cytochrome complex from the algae and the cyanobacterium have essentially the same structure. This is a remarkable observation that under-lines again the endosymbiont hypothesis. In this case, it means that the chloroplast of green algae like *Chlamydomonas* and green land plants are derived from an ancestor of cyanobacteria. This crystallographic observation is but a piece in the puzzle of the endosymbiont hypothesis. To appreciate the conservative nature of evolution, I should mention that *Chlamydomonas* chloroplasts and cyanobacteria are separated by about 1 billion years from their common ancestor.

As in the case of the mitochondrial cytochrome bc_1, the cytochrome b_6f crystal structure shows a dimer. The complex is composed of two major cytochromes: Cytochrome b_6 spans the inner mitochondrial membrane with many α-helices and has only small extensions beyond the level of the membrane. Cytochrome f shows more β-pleated sheet structure and looks like a "q." The head part is mainly localized in the lumen of the thylakoid membrane system (this is the P side of the chloroplast photosynthetic membrane; it corresponds to the space in the mitochondrion between the inner and the outer membranes). This part of cytochrome f contacts the small soluble one-electron carrier plastocyanin, which carries the electron to PSI. The handle of the q crosses the membrane and intermingles with one long helix into the helices of cytochrome b_6. A very similar q structure was also observed for the third major partner of this complex: the Rieske Fe–S protein. Its head is located at the luminal side between the heads of the two cytochromes f. The head is linked via a hinge to the membrane-spanning handle.

The Importance of Proximity

The hinge has an important physiological function, as already demonstrated before for the corresponding Rieske Fe–S protein in the mitochondrial complex III. The hinge allows a nodding of the head. This nodding brings the iron–sulfur redox center alternatively nearer to the reduced plastoquinone (or ubiquinone) binding site or the heme group in cytochrome f (or cytochrome c_1 in mitochondria). The nodding of this protein is thus an essential part in decreasing the distance between the different redox centers. As we have seen before, the distance is a decisive factor for rapid electron transfer reactions (including tunneling) between redox centers. The major job, like in other oxidoreductases, is done by a series of iron ions that shuttle between Fe^{2+} and Fe^{3+} states. The irons come as heme complexes or as iron–sulfur clusters. Now comes a really remarkable observation: Despite the long period of time that separated the current mitochondria and chloroplasts and their prokaryotic ancestors from a common ancestor, you can still superpose the relative crystallographic position from the low- and high-energy heme b (b_L and b_H, respectively), the iron–sulfur cluster, and, to a somewhat lesser extent, heme c. You can also superpose the transmembrane helices from both protein complexes at the level of the b_L heme.

Differences

However, both complexes also differ: The b_6f complex contains a further heme and a chlorophyll ring as well as carotenoids. These are clear tributes to the chloroplast attribution and might be involved in cyclic electron transport during photosynthesis. Another marked difference is seen at the N side of the two enzymes. This is the matrix side for the mitochondrion (where the TCA cycle is localized); here cytochrome bc_1 shows many further subunits that reach $80\,\text{Å}$ into the matrix (compared with only $30\,\text{Å}$ at the P side). The N side of the chloroplast is called stroma (this is the "cytoplasm" of the chloroplast, where the CO_2-fixing dark phase of photosynthesis occurs and Rubisco is localized). Only small loops of cytochrome b_6f extend into the stroma.

The Q Cycle

Now we have the framework for electron transport and the creation of the proton gradient—the essence of the energy conservation from food oxidation or the photosynthetic light reaction. How is this achieved? Once again, the lead was provided by Peter Mitchell, who designed the Q cycle in a theoretical paper published in 1971. The details of this process can now be projected on the crystal structures, as the binding sites of inhibitors that interfere with different steps of the Q cycle were localized. The position of the inhibitor stigmatellin, which blocks the oxidation of hydroquinone, is called Q_o site ("o" stands for out; this site is also called Q_p because it is close to the membrane face pointing to the intermembrane space in mitochondria and the thylakoid lumen in chloroplasts). The Q_o site is situated between the b_L heme and the Fe–S cluster. What happens at that side? The fully reduced hydroplastoquinol arrives at this position. It is stripped off its two electrons and two protons in two steps. The oxidized plastoquinone diffuses away and mixes with the quinone pool in the membrane. The electrons/protons released from the plastoquinol take two different paths: One e^-/H^+ pair heads toward the Fe–S cluster. The proton leaves for the P space via the Fe–S protein. The electron travels from the Fe–S redox center to the heme f in the cytochrome f; from here the electron travels to the Cu redox center in the soluble plastocyanin protein. The other electron/proton pair from plastoquinone takes a totally different way: The proton leaves for the P space via the cytochrome f protein. The electron travels to the b_L heme and from there to the b_H heme. Near this heme, the Q_i site, the binding site for an oxidized quinone, is located. In cytochrome bc_1, the Q_i site was identified by the binding of the inhibitor antimycin. This Q_i site is near the N face of the membrane. The oxidized plastoquinone or ubiquinone from the membrane pool is bound here, and in a two-step process receives a single electron from b_H each time. Importantly, to create the fully reduced quinol, two protons have to be taken up from the N space. The loss of protons results in an excess of hydroxyl ions in the N space, which makes it more negative (hence the N for negative) and more basic. The reduced quinol mixes into the membrane and then docks at the Q_o site, and the cycle is closed. At first glance, this cycling of electrons seems futile, but if you look at the stoichiometry of the overall reaction you realize that the release of two protons at the P site and the uptake of two protons at the N site correspond to a net build-up of a gradient of four protons across the membrane. The food or light energy carried in the electrons is now stored in the protons.

Interchangeability in Cyanobacteria

You should realize that I have deliberately mixed the chloroplast and mitochondrial complex in the above discussion. This is entirely justified because so many aspects are so similar that one could argue that essentially the same protein complex is used as an intermediate in both respiration and photosynthesis. That this is not a vain claim is demonstrated by cyanobacteria, whose ancestors became the chloroplasts of the modern photosynthetic cells. Cyanobacteria are metabolically very versatile cells. One aspect is that they are literally cells for all seasons. During daytime or in the upper sunlit layers of the ocean, they generate ATP by photosynthesis using PSII and PSI in series, like in modern algae and plants. At night or in lower and thus darker ocean layers, they obtain ATP by oxidative phosphorylation using protein complexes corresponding to complexes I and IV of modern mitochondria. These two processes occur in the same cell and across the same membrane, which can be used for energy storage: the involuted cytoplasmic membrane of the bacterial cell. As we know, both processes need a linker, the photosynthetic process requires something like cytochrome b_6f and the oxidative process something like cytochrome bc_1. Now comes the surprise: Both processes use the same molecular intermediates. Both respiration and photosynthesis feed electrons into a plastoquinone pool, which reduces the same b_6f/bc_1 complex, which reduces the same shuttle protein cytochrome c_6; c_6 thus plays the role of plastocyanin when reducing PSI and cytochrome c when reducing complex IV. We can really see Mother Nature playing around with biochemical modules in the quest for new inventions. This observation in cyanobacteria points strongly to the common origin of both electron transport processes. Probably part of the versatility of the intermediate complex is the fact that the electron transfer processes occur close to equilibrium and are thus fully reversible (Osyczka et al. 2004).

Protons and ATP

Proton Pumping and O_2 Reduction

Cytochrome c Oxidase

We should also discuss complex IV of mitochondria for two reasons. Cytochrome c oxidase is the protein for which we breathe, have lungs, and have hemoglobin in our blood. This protein allows us a life in high gear because we can extract more energy out of the same foodstuff than in the absence of oxygen. Actually, we humans depend so much on it that we die in the absence of oxygen. Our consciousness and then our brain is quickly blotted out when we are prevented from breathing. Fishes are not so delicate because they might meet anoxic zones during swimming. Even if they survive with the help of glycolysis, their locomotion in anoxic zones resembles a slow-motion movie. The other reason for distinguishing cytochrome c oxidase is that it actually tunnels protons from one side of the membrane to the other. The proton gradient is thus not just created by selective release and uptake of protons at both membrane faces, as in complex III.

Structure–Function Relationships

As for the other studies in bioenergetics, the resolution of the 3-D structure of complex IV contributed greatly to the understanding of its function. The first structure was obtained for the cytochrome c oxidase from the soil bacterium *Paracoccus denitrificans* (Iwata et al. 1995). The bacterial enzyme has the advantage that it consists only of four subunits, in contrast to bovine heart enzyme, which musters 13 subunits (Yoshikawa et al. 1998). As the bacterial enzyme fulfills the same biochemical functions with its four subunits, it was the first target for structure–function studies. In fact, the role of the supplementary subunits in the mammalian enzyme is not very clear at the moment. The core structure of both enzymes consists of subunits I–IV, which are essentially identical despite the wide evolutionary distance that separates a soil bacterium from a mammalian. The protein structure is not very conspicuous and is described briefly. Subunits I–IV provide an array of membrane-crossing α-helices. The only exception is part of subunit II, which shows a C-terminal globular domain containing a 10-stranded β-barrel. This domain sticks out of the inner mitochondrial membrane and reaches out into the P site, the space between both mitochondrial membranes. It is the docking site for cytochrome c, which carries one electron from complex III to complex IV. *It is revealing how the researchers who solved the first cytochrome c oxidase structure in 1995 depicted the enzyme in a 2002 paper dealing with questions of electron and proton translocations (Ruitenberg et al. 2002). The enzyme is shown as an empty eggshell, the authors depicted only two hemes, several metal ions, and a few amino acid side chains inside. Of course, they wanted to present a simplified scheme of the enzyme that does not distract the reader from the essential message of this report. However, this concentration on the cofactors of the enzyme illustrates also an important evolutionary principle. Crucial for the action of this enzyme are the metal ions; we are again in the realm of bioinorganic chemistry and the protein backbone is reduced to a mere placeholder.* This is not to be taken negatively: If the electron transfer reactions should proceed in an efficient way, the redox centers must be optimally oriented. There are two principal arguments for this: The first is speed. The individual cytochrome c oxidase enzyme handles more than 500 electron transfers per second. The other is efficiency. Side reactions away from the beaten tract and electric short circuits for the electron are not allowed. The enzyme deals with an extremely dangerous reaction: the transfer of electrons to oxygen. Very reactive compounds like peroxy species are created as reaction intermediates during its enzymatic cycle. Peroxy species are ROS. If they are not quickly processed into harmless end products like water, biological damage is assured as we have seen in the previous section.

Electron Transfers

As electron transfers are one of the most basic processes in the design of life on earth, one can deduce from the omnipresence of metal ions and the Corrin ring in its various forms (heme, chlorophyll) that life had actually a quite metallic start. The first redox center near the docking site of cytochrome c is two copper ions in subunit II, also called Cu_A center. It is also called a binuclear center and should not be mixed with another binuclear center further down in the electron path of this enzyme. The two copper ions from the Cu_A center are complexed with the thiol groups from two cysteine residues of subunit II. Cu_A thus resembles the 2Fe–2S center of iron–sulfur proteins. Next in the line of the electron transfer is heme a, followed at nearly a right angle by a second heme a_3. The latter faces a second copper ion, called Cu_B. As mentioned above, this heme

a_3/Cu_B complex is also called a binuclear center. A Fourier map revealed a residual density between the iron and the copper ion of this binuclear center. The three-atom group was interpreted as a peroxide group, an intermediate in the oxygen reduction sequence at this center. The bovine enzyme was also investigated in the presence of two potent inhibitors of cytochrome oxidases and respiratory poisons, CO and N^{3-} (Yoshikawa et al. 1998). The Fe from heme a_3 and the Cu_B are bridged by carbon monoxide or by azide (Yoshikawa et al. 1998), explaining the striking toxicity of both compounds. Under physiological conditions, molecular dioxygen (O_2) is bound at this site. The Fe–Cu_B center then becomes an important catalytic site. As in other redox systems, the iron ion shuttles between the Fe^{2+} and Fe^{3+} states, Cu_B between Cu^{1+} and Cu^{2+} states. In a sequential action, four electrons received via the above-described electron transport chain are channeled into the oxygen binding site and unloaded onto the molecular oxygen.

Protons Get Involved

In addition, four protons are added to the active site. A hypothetical reaction scheme was developed in the early 1990s (Babcock and Wikstrom 1992). The Fe^{3+}–Cu^{2+} center is first reduced by two successive electron transfers to a Fe^{2+}–Cu^{1+} center, which then binds O_2 to Fe^{2+}. Oxygen is then stabilized as a peroxy species (O_2^{2-}) by internal electron transfers from both Fe and Cu. The third electron reduces again the Cu_B. Then come two protons, which are added to the peroxy species, followed by the oxygen–oxygen bond scission. One oxygen is now bound to Cu_B and the other to Fe as a ferryl group ($Fe^{4+} = O$). Subsequently, two protonations release two water molecules and the cycle can start again. As a bottom line, four electrons end their travel through the respiratory chain in two molecules of water. Each electron transferred from cytochrome c to the O_2 reduction site is accompanied by the uptake of one proton from the matrix side. The two water molecules released from the O_2 reduction site carry away four electrons from the respiratory electron transport chain and four protons from the matrix side. The uptake of these four protons adds to the proton gradient across the membrane.

Proton Pumping

During the transport of the four electrons, four protons are pumped across the membrane. The kinetic order and the pathway of this proton pumping was a subject of substantial debate. The initial argument was based on the dependence of the redox equilibria on the membrane potential and Marten Wikström (1989) argued that the steps reducing the "peroxy" and "oxyferryl" intermediates, abbreviated as states P and F, respectively, are each coupled to the translocation of two H^+ across the membrane. However, on theoretical grounds, Hartmut Michel (1998) challenged the idea that protons are transported only during the oxidative half-cycle of this reaction. This argument had to be taken seriously as it came from the lead scientist who unraveled the first cytochrome c oxidase structure. Therefore, Marten Wikström's lab turned back to the bench. They inserted the enzyme into a liposome (a membrane vesicle) and measured the proton movements directly by a pH electrode (Verkhovsky et al. 1999). They used some smart tricks: They poisoned the enzyme with CO, then broke the CO bond to the enzyme by photolysis, which allowed the enzyme preparation to go into a single oxygen cycle because of the excess of competing CO. The measurements were clear: 3.5 and 4.5 charges were translocated during the oxidative and the reductive phase, respectively. The model that all protons were pumped during the oxidative phase of the oxygen cycle was thus invalidated. After finishing the chemical reaction, the four metal centers of the enzyme were fully oxidized and in some way the energy driving the delayed charge separation must have been stored in strained chemical bonds of the O form of the enzyme. If the reduction phase follows immediately, two further protons are pumped; if this does not occur, the potential energy is lost as heat (Rousseau 1999). Hartmut Michel's group then reinvestigated the case by accumulating an enzyme state in the reductive phase. Then they followed the membrane potential of liposomes containing this enzyme. They used another trick: an enzyme with a single amino acid replacement (D124N, aspartic acid replaced by asparagine; Ruitenberg et al. 2002). When they fed a single electron into the enzyme either in the reductive or in the oxidative phase of the cycle, they observed a characteristic electric difference: The slow photopotential change was not observed with the mutant protein. This identified D124 as part of the proton translocation pathway in both phases of the enzyme cycle (D-pathway). The pathway of the proton could be tracked until the protein vicinity of heme a. The protons used to create water from oxygen take another path, which is called the K-pathway. This path got its name from the lysine (K) residue K354, which was of critical importance for this reaction to occur. However, despite its importance in the energy metabolism of all higher life forms and especially animals, the molecular mechanisms of proton pumping driven by electron transfer in cytochrome c oxidase have not been determined in any biological system. In a recent study, a Swedish group used a variant enzyme that differed from the wild type by the replacement of a single amino acid. This mutation slowed the electron transfer from the ferryl iron to the oxygen by 200-fold such that the absorbance change to an indicator dye, which was associated with proton release, could be followed spectroscopically in real time. Their observation suggests a scenario where protons are pumped and are not coupled in time to any electron movement within the enzyme (Faxen et al. 2005).

M. Wikström and colleagues, who worked on this problem with cytochrome oxidase for more than 30 years, came forward with a different interpretation. They combined sensitive electrometric measurements with time-resolved optical spectroscopy. This allowed them to simultaneously monitor the translocation of electric charge equivalents and electron transfer within complex IV, both in real time (Belevich et al. 2006). They developed a three-stage model for these processes. In the first step, an electron transfer occurs from redox center heme a to the binuclear site. This step is coupled to the transfer of a proton from the carboxylic E278 residue at the end of the D-pathway to an unidentified protonable site at the opposite side of the heme a-binucleate plane. In the next step, E278 is reprotonated from the N side of the membrane via the D-pathway. In the following step, a substrate proton is transferred from E278 to the binuclear site, mediating the stepwise reduction of the bound O_2 to two water molecules. Thereafter, the proton above the heme centers is ejected toward the P side and at the same time E278 is reprotonated. Overall, four protons are pumped across the membrane for the four electrons guided to the binucleate center where the oxygen reduction to water occurs. In addition, four substrate protons are taken up by the binucleate center to form the product, water. So far this is the latest interpretation of the electron-coupled proton transport in cytochrome c oxidase. However, the best-characterized proton pump for energy conservation purpose is currently found in bacteriorhodopsin, and we will discuss this membrane protein in one of the next chapters.

Why Oxygen?

Before changing the subject, I will add a chemical reflection of G. Babcock and M. Wilkström (1992) as to why Nature has chosen molecular oxygen for the conservation of energy in cell respiration. The fitness of oxygen for biology derives from two aspects of its structure and chemistry. On one side, there is a strong thermodynamic driving force for its reduction to water by biological reductants derived from foodstuff. On the other side, oxygen is kinetically stabilized. In the ground electronic state, O_2 is in a triplet state, having two unpaired electrons imposing spin restrictions on its reaction with two-electron donors. One-electron transfer events are also slow because of the very large difference between the O–O bond energies in O_2 and the super-oxide radical O_2^-. This kinetic inertia of oxygen prevents the spontaneous burning of organic material and allows a controlled handling of this potentially dangerous molecule by the active sites of enzymes. As we have seen in a previous section, there is still substantial danger lurking in the reactivity of oxygen for cellular systems. *Life was probably easier in the absence of oxygen and would probably never have evolved in its presence. This was, however, not a risk because only later was oxygen created by a biological process, which represents one of the pivotal inventions in the history of life: photo-synthesis.*

Purposeful Wastefulness

Bacteria Have Lower P/O Ratios than Do Eukaryotes

I mentioned in several sections the universality of biochemistry and in our survey I deliberately changed from respiration to photosynthesis, from mitochondria to chloroplast, from eukaryotes to prokaryotes. In the field of bioenergetics, this principle is certainly valid because the eukaryotes only developed further what was already invented by prokaryotes. This observation is vividly illustrated when comparing the crystal structures of the corresponding respiratory protein complexes, which increased with respect to the complexity of the associated protein subunits while maintaining the basic design of the reactive centers. However, this principle misses an important point. Prokaryotic respiratory chains are not only simpler in design—they also function differently with regard to the energetics of the reactions. Biochemistry textbooks frequently present the human situation with four membrane complexes, which produce 38 ATP per glucose oxidized, yielding a P/O ratio of 3. In this respect, most if not all bacteria are different. Their respiratory chain misses at least one complex, and the best they can get is a P/O ratio of 2, which is the case with *E. coli*; in other prokaryotic systems, this ratio is only 1. If microbes are the unseen majority and the true masters of the world, why is their metabolism not optimally organized? Here we nearly get to a philosophical question, which has very important biological implications. If philosophy is at stake, we should carefully weigh our words. In fact, what vexes us is not the question that bacteria are not optimally organized. They probably are—otherwise they would already have been replaced by other and better-adapted organisms in their respective niches. In reality, what we mean by our question is why bacteria are not maximizing the use of food by complete oxidation of sugars to CO_2, like we are doing. In fact, this statement is frankly wrong: We saw a case in our running exercise where we use fermentation instead of respiration to get away quickly. As we saw in the example, we trade under this special condition rate versus yield of ATP synthesis because our priority at this moment is rate; we deliberately accept a wasteful metabolic situation. Before we get deeper into the topic, let me present a bacterium known for its wasteful metabolism.

Gluconobacter

Gluconobacter oxydans exhibits so-called oxidative fermentation, which was considered by microbiologists as an unusual metabolic feature of energetic waste-fulness (McNeil and Harvey 2005). Rather than fully oxidizing a wide variety of substrates, it oxidizes them only to the level of organic acids, aldehydes, and ketones, which it excretes into the medium. For example, glucose from the medium enters the periplasm (the space between the outer and the inner cell membranes) where it is processed by a glucose dehydrogenase inserted in the outer leaflet of the cytoplasmic membrane into gluconic acid. This organic acid leaves the periplasm via a porin in the outer membrane to accumulate in the extracellular medium. We like this organism for its apparent wasteful metabolism and exploit it industrially. If you offer it the sugar alcohol D-sorbitol, it gets transformed into L-sorbose by a similar pathway. The L-sorbose is excreted into the extracellular space in large amounts where it can further be converted to ascorbic acid in industrial vitamin C synthesis. The genome sequence confirmed that *Gluconobacter* does not possess a complete TCA cycle and has only a simple respiratory chain (Prust et al. 2005). It lacks the proton-translocating complexes I and IV, and its ability to translocate protons in the course of redox reactions is thus rather limited. The consequences are clear: The low energy-transducing efficiency results in very low growth yields and biomass for *Gluconobacter*. However, this organism does not lack oxidoreductases, it contains 75 ORFs coding for such enzymes that are actively expressed. Many sugar derivatives can be transported into the cytoplasm where they are degraded by the oxidative pentose phosphate pathway. Does all this make sense? Stated differently: Is this clearly nonmaximal food exploitation an optimal evolutionary strategy? Giving the matter a second thought the answer is yes. *G. oxydans* thrives best in environments where sugars and alcohols are abundant. Its natural habitats are flowers and fruits, and it is industrially also known as a spoilage organism of alcoholic beverages and soft drinks. Substrates are thus not limiting. *Gluconobacter* worked out a clever strategy: By taking up sugars and quickly excreting sugar acids, it lowers the pH of the medium and thus inhibits the growth of competitors. When these competitors are eliminated, it takes up the previously excreted organic acids for further processing. Maximal exploitation is thus not necessarily optimal food exploitation. What is optimal is apparently context-dependent.

Evolution of ATP-Producing Pathways

A fascinating theoretical paper provides deeper insights into this context dependence (Pfeiffer et al. 2001). Its basic arguments are a mixture of thermodynamic and evolutionary game theory. The latter describes evolution toward the inefficient use of a common resource. The thermodynamics is straightforward: Heterotrophs degrade organic substrates into products with lower free energy. The energy difference is in part conserved by production of ATP. If the entire free-energy difference is conserved as ATP, a maximum ATP yield is obtained. The reaction is in thermodynamic equilibrium and the rate of ATP production is low. If, in contrast, part of the energy difference is used to drive the reaction, the rate of ATP production increases, but the ATP yields decrease. There is thus a trade-off between the two strategies. Sugar fermentation stands for high ATP rate, and sugar respiration for high ATP yield. Many organisms are playing this "stone–paper–scissors" game. For example, cell populations using a pathway with low yield but high rate can invade and replace a population using a high-yield/low-rate pathway. Thus evolution will select inefficient pathways when cells are in competition for a shared resource. This is the case for the exploitation of external resources like in the early phases of organic material decomposition by *Saccharomyces*, which uses fermentation even in the presence of oxygen and an intact respiratory chain. Now the reasoning of the paper gets on its way. Organisms that ingest food items remove that material from the environment, and competition is no longer an issue. Now the priority changes: Maximal yield is selected and this could explain why multicellular organisms capitalize on the high ATP yields afforded by respiration. This sounds logical, perhaps trivial, but the authors push their hypothesis further. They argue that respiration is only an option in multicellular organisms because the cells moved from competition to cooperation. Now if this treaty is broken, the traitor will use fermentation to steal the shared resource. This sounds theoretical, but put a tumor cell for traitor and the theory comes with a wonderful evolutionary explication for the Warburg effect, which we discussed in one of the earlier sections. Here you might argue that this match was a chance event. The authors counter with another example. Some dimorphic fungi show a yeast-like unicellular stage and a mycelium-like multicellular stage. Notably, their sugar metabolism differs and the difference again concurs with the theoretical prediction: The unicellular stage uses fermentation and the multicellular stage uses respiration. It seems that we are definitively getting closer to a theoretical biology.

This line of thought got popular and led to an interesting research article entitled "*Resource Competition and Social Conflict in Experimental Populations of Yeast*" (MacLean and Gudelj 2006). In fact, the authors investigated an old problem called "The tragedy of the commons," namely how can a group of individuals that cooperate resist invasion by cheaters that selfishly use common resources? They used for their research isogenic yeast cells that differed in glucose metabolism: cooperative respirers against selfish respiro-fermenters. The only material basis for the conflict was a higher maximum rate of glycolysis in the "cheaters." Cheaters had the advantage of a higher rate of ATP production but suffered from the accumulation of high concentrations of the incompletely oxidized toxic intermediates ethanol or acetate. As these intermediates are washed out in the chemostat conditions, coexistence between both cell types was not possible and cheaters took the upper hand. Spatial structure in the incubator, however, promoted the evolution of cooperation that could efficiently exploit local resource patches. The authors concluded that cooperation can persist in well-mixed populations in the absence of kin recognition, policing, or rational behavior.

Fiat Lux

Science and Religion

This section starts with a famous wording from the book of Genesis. The politic columns of some scientific weeklies are filled over the recent years with the quarrel of creationists in the USA with the lessons of biology in high school courses. The creationists claim a truth for the first chapter of the Genesis over the evolutionary theory. As a biologist, I cannot quite understand the heat of this debate. My personal patron is St Thomas, the disciple of Jesus who always wanted to see the evidence before he believed a statement. Jesus accepted this disciple and simply put St Thomas' hand into his wound to convince him about his identity. Thus doubt is an accepted attitude by the founder of Christianity. Furthermore, the report of the Genesis coincides in several important messages with current scientific hypotheses. Autotrophs precede heterotrophs in this report. Life was created in the oceans before it was founded on land. Humans are latecomers of creation following animals. Another deep insight of the Judeo-Christian cosmology is the primate of light in the world. "Fiat lux" (...) were the first words of God in this cosmology. Light is also at the beginning of biology as seen by biologists. The book of Genesis and the book of Nature differ thus not so much in details of the evolution of life (who says that the days in the life of God do not fill millions of years on the human timescale?), but more in the forces setting this process in motion. Scientists start with the working hypothesis that known physical and chemical laws underlie the processes in biology. The very success of biology in the second half of the twentieth century testifies the strong predictive power of this working hypothesis. Many simple life processes are now understood in molecular and sometimes even in atomic detail. The "fiat lux" in biological disguise as described in the following paragraphs is a wonderful illustration of this principle. Does the fact that we now understand in atomic detail a proton pump in a salt-loving prokaryote tell you that God exists or does not exist? Already the formulation of this question reveals the pure nonsense of this discussion. Science and religion argue on two different levels. Biology describes how life is organized on earth, whereas true religion describes how human life should be organized on earth. The picture painted by biology on the conditio humana is quite often not very flattering for us, but we have to live with it as a foundation. If our reason or ethical consciousness tells us that we should react differently than Mother Nature has told us, then no biologist will quarrel with you as a truly religious person. If you accept the hypothesis that medicine is applied biology, biology has already started to offset basic tenets of evolution by ethical considerations, which are relatively alien to Nature, at least until the development of consciousness in biological organisms. In the book of life, nowhere is it written that you should respect life. In fact, organisms unscrupulously compete with each other for a place under the sun not only between species but also as fiercely within the confines of a species. Medicine has already offset the basic rules of the survival of the fittest by defining the conservation of any form of human life whether being fit or unfit for life. This life-sustaining activity creates new opportunities for mutations to travel through populations and to impact on the genetic structure of populations. All the major religions define basic rules of compassion with fellow organisms (some restrict this to human life, others include animal life too). This is in fact anti-Darwinism in the sense that a new ethically motivated set of rules is formulated for life on earth. This is—I guess—fine with most biologists. Some might warn that if you do this consequently, humans will become self-domesticated and evolution will go on; even if you set new rules, evolution will only change direction. This is probably true, and it needs the insights of the best minds to foresee the application of ethical rules on the biological evolution of life and whether under these new rules life remains sustainable on earth. Probably, biologists will claim that the exclusive preoccupation of religious thought with humans (indirectly as subjects to God or directly as fellow to other humans) is misplaced. God might feel a wider responsibility, which goes beyond humans and will probably cover the whole creation anywhere in the cosmos not just here on our spaceship earth. In my view, it is time that the major illuminated religions make their peace with science: In fact, Rome has officially buried its quarrel with Galilei. However, currently I do not see much interest from either side to get into a dialog.

The "fiat lux" was uttered a second time, namely by the philosophers of the Enlightenment. Science, as it is conducted today, is founded in the works and philosophy of Copernicus and Galilei. The professional philosophers of the Enlightenment provided in a sense the justification for the mathematical and empirical approach of the first modern scientists. Here I will use the double sense of bringing light into a discussion by using perhaps one of the oldest devices of light-to-energy conversion that was elucidated by using electromagnetic waves ("light") of very short wavelength for its structural analysis.

Halobacterium

What I also wanted to express with this motto is that a light-driven proton pump is still our best insight into the functioning of a membrane protein, storing energy in a proton gradient. The process is worthy of being recounted here because it was not only the topic of more than a thousand publications and entire conferences, but it turned out to be a paradigm for one of fundamental processes in biology. At first glance, the place of action is one of the least where you would expect insight into basic life processes: the Dead Sea. This lake is called dead because of its hostility to life, but it does not exclude all life forms. As you might expect, we will find there again extremophiles, and this time in the form of extreme halophiles, salt-lovers. *Halobacterium halobium* has optimal growth rates at about 3–4 M NaCl but will also grow near salt saturation at 36%. This fact explains why many marine salterns are intensively red colored—due to a pigment found in halobacteria. This organism has adapted its metabolism to high internal salt concentration of 4–7 M KCl. Under normal conditions, such high intracellular salt concentration would inactivate enzymes and burst membranes. Not so in *Halobacterium*: Its enzymes, ribosomes, plasma membrane, and cell wall are actually stabilized by high salt. If ever the external NaCl concentration falls below 1.5 M NaCl, its cell wall literally disintegrates. Otherwise it is metabolically a relatively normal character. Halobacteria are aerobic heterotrophs that require complex nutrients such as proteins or amino acids for food. However, there is a problem: Oxygen is not very soluble in concentrated salt solutions and the environment of halobacteria becomes temporarily anoxic. Under this condition, *Halobacterium* relies on a second power source: a light-sensitive proton pump in its membrane. Under low oxygen tension, crystalline patches of bacteriorhodopsin are inserted into the cytoplasmic membrane, which then appears purple in color. This purple membrane is quite peculiar because it consists only of 25% lipid, while 75% of it is the inserted protein. The protein adsorbs light, and the light energy mediates the pumping of protons from the cytoplasm into the periplasm. The proton gradient powers the synthesis of ATP by an ATP synthase, like in respiration or in photosynthesis. Indeed, we have here a primitive form of photosynthesis. Interestingly, true photosynthesis was never invented in Archaea, and thus it is only consequent that this light reaction does not depend on chlorophyll as light harvester. However, it is not an exotic pigment that absorbs light: Halobacteria rely on a pigment that is also used in the vision of human beings: retinal (vitamin A). Also, the protein part of the prokaryotic bacteriorhodopsin and the human opsin show a remarkably similar structure. Both belong to the large class of seven-membrane proteins, one of the most popular transducers in our body when coupled to G proteins. This versatility of the rhodopsin becomes already apparent in halobacteria: They use light to transport chloride ions into the cell via a halorhodopsin to maintain high intracellular KCl concentrations. Furthermore, two rhodopsins are used as real photoreceptors, one for red and another for blue light. The protons pumped by rhodopsin drive also the flagellar motor; with the help of this motility system, halobacteria can search the optimal light position in the water column. However, halobacteria cannot grow for prolonged time periods under anoxic conditions because oxygen is required in the reaction that splits carotenes into the aldehyde retinal.

Bacteriorhodopsin Structure

A high-resolution X-ray structure of an early intermediate in the bacteriorhodopsin photocycle was determined in 1999 by a multi-national effort (Edman et al. 1999). The protein structure is quickly described. It consists of seven transmembrane helices termed A–G. The helices are arranged in a roughly cylindrical structure. Retinal is linked via its aldehyde to the terminal amino group of lysine 216 from helix G as Schiff's base. The other end is wedged deep in the protein. During its photocycle, bacteriorhodopsin passes through a series of structural intermediates with well-defined lifetimes and spectral properties. If the crystal is cooled to low temperatures, early intermediates can be excited by illumination with light of specific wavelength. However, due to the low thermal energies at low temperatures, the protein cannot transverse the energy barriers to the following cycle intermediate. The compound is thus effectively frozen in a given state. A year later, not less than three other frozen intermediates of this photocycle were analyzed by X-ray crystallography (Royant et al. 2000; Sass et al. 2000; Subramaniam and Henderson 2000). A reviewer even spoke from a true movie reconstructed from the different time frames of this photochemical process (Kuhlbrandt 2000). The primary process is the photochemical isomerization of retinal (Figure 3.2). The ground state of retinal is the all-*trans* configuration: All carbons from this poly-isoprene are in *trans* configuration. The conjugated double bonds then absorb green light; this energy absorption causes an isomerization around the C13–C14 double bond from the all-*trans* to the 13-*cis* configuration. Even minute details of this process have been investigated. For example, with the help of ultrashort laser light pulses of less than 5-fs duration (fs = 10^{-15} s), an intermediate tumbling state between the two isomers could be identified by real-time spectroscopy (Kobayashi et al. 2001). Likewise, the ultrafast evolution of the electric field within bacteriorhodopsin after retinal excitation was measured by monitoring the absorption changes of one of four tryptophan residues that sandwich the retinal (Schenkl et al. 2005).

The light-induced isomerization causes a transfer of the proton from the Schiff's base to the Asp 85 residue in helix C. This transfer is aided by a slight central transitory kink in helix C. The deprotonated retinal now straightens and

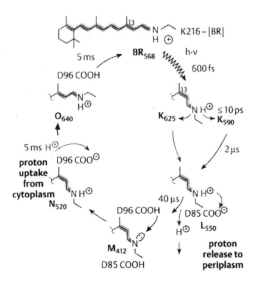

FIGURE 3.2. Bacteriorhodopsin in the purple membrane of *Halobacterium halobium*: the photocycle of the pigment retinal. Light-dependent *cis–trans* isomerization of retinal and transient deprotonation of the Schiff's base causes a proton translocation across the membrane. Ground states and intermediates are indicated with letters K, L, M, N, O where the subscripts denote the absorbance maxima in nanometers. Lysine K216 is the residue forming the Schiff's base, aspartic acid residues D85 and D96 are essential for proton translocation. (courtesy of Thieme Publisher).

pushes away the upper half of helix F. This movement brings the deprotonated retinal into the vicinity of Asp 95 from helix C, from which it abstracts a proton. The tilting of helixes F and G has now created a transient opening at the cytoplasmic side of the bacteriorhodopsin. Asp 96 can now be reprotonated by a proton from the cytoplasmic side of the membrane. In the next cycle intermediate, the proton at Asp 85 travels via a network of hydrogen bonds and water molecules past Arg 82 from helix C into the outside medium. In this way, the absorption of a light quantum has caused the transfer of a proton from the inside to the outside of the cell against a 10,000-fold concentration gradient ready to drive the ATP synthase. Thirty years of ingenious experimentation thus yielded a complete motion picture of bacteriorhodopsins in action.

Proteorhodopsin

A new type of phototrophy was detected in the sea through the cloning and sequencing of large genomic DNA sequences from an uncultivated microorganism: bacterial rhodopsin (Beja et al. 2000). Proteorhodopsin, as it was called, is a retinal-containing integral membrane protein that functions as a light-driven proton pump. The protein resembled the bacteriorhodopsin from the halophilic *Halobacterium* that we have discussed in the previous paragraph. Should marine microbiologists have overlooked an unrecognized phototrophic pathway in the ocean's photic zone? To answer this question researchers from the Monterey Bay Aquarium designed physiological experiments with membrane preparations from bacterioplankton from their aquarium (Beja et al. 2001). Like in bacteriorhodopsin, they observed a photochemical reaction cycle of 15 ms. Furthermore, after optical bleaching of the pigment, they could restore the cycle by adding all-*trans* retinal. Flash photolysis experiments provided an estimate of 2×10^4 proteorhodopsin molecules per cell, which carried the code name SAR86. This is the same order as bacteriorhodopsin in *Halobacterium*, where the protein occupies a substantial part of the cell surface forming tightly packed crystalline arrays. This amount of light-driven proton pumps should make a significant contribution to the energy metabolism of this cell. In their aquarium, the oceanographers found many closely related variants of this protein, whereas samples from Antarctic marine plankton were less related and also showed a different spectral property—a 37-nm blue shift away from the Monterey prote-orhodopsin maximum. They obtained water from different depths at Hawaii and observed the same blue shift when changing from surface water to 75-m depth samples. This blue shift is apparently an adaptation to the environmental light conditions because a similar blue shift was observed in totally unrelated organisms. Rod and cone visual pigment rhodopsin from closely related fish species in Lake Baikal showed a similar blue shift with increasing depth of their habitat.

The physiological role of proteorhodopsin cannot be evaluated well when working with an uncultivated microorganism. Therefore, a metabolic analysis became much easier when a proteorhodopsin was identified in *Pelagibacter ubique* (Giovannoni 2005). This is a prominent cell, but I keep this fascinating story for a later section ("the most abundant cells on earth are on a small diet"). Here I will only provide some data on its proteorhodopsin, which has the structural and kinetic features of a rhodopsin proton pump (Giovannoni, Tripp et al. 2005). Its photocycle rate is 13 ms for light-grown and 34 ms for dark-grown cells. Its genome encodes the enzymatic pathway for β-carotene biosynthesis and a *blh* gene, which mediates the cleavage of β-carotene to retinal. The proteorhodopsin gene is associated with a downstream ferredoxin gene, a genetic constellation, which is found in half of the Sargasso Sea DNA sequences (Venter et al. 2004). However, no genes for carbon fixation were found in *P. ubique*—the cell is clearly a heterotroph. This is understandable, as the proteorhodopsin membrane protein provides a transmembrane electro-chemical potential to drive ATP synthesis, but unlike in photosynthesis, no reduced nucleotide cofactors are produced in this form of phototrophy. However, its physiological role remains still enigmatic: *P. ubique* shows the same growth rate when grown under diurnal light or in darkness. Proteorhodopsin occupies about 20% of the cell surface, which is then no longer available for nutrient transporters; the cell is confronted with a metabolic trade-off between energy provision via phototrophy and heterotrophy. The fun in science today is that you can wait for the answer coming with the postman who brings your favorite scientific journals to your home.

Prokaryotic rhodopsins were first described in extremely halophilic Archaea, then came proteorhodopsin detection in Eubacteria and recently the cycle was closed when proteorhodopsin was also detected in Thermoplasmatales—Archaea from the Euryarchaeota branch (Frigaard et al. 2006). In fact, about 10% of the Euryarchaeota in the photic zone of the ocean contained the proteorhodopsin gene adjacent to their small subunit ribosomal RNA gene. The presence of a bacterial-like proteorhodopsin gene in Archaea was unexpected and probably reflects the result of lateral gene transfer between marine planktonic Bacteria and Archaea. However, such a transfer should not surprise. The proteorhodopsin-based phototrophy is extremely suited for lateral gene transfer. The transfer of just two genes, namely the membrane pump and the carotene-cleavage enzyme, can confer phototrophy and thus a significant light-dependent fitness contribution. That this contribution is linked to the presence of light was underlined by the finding that archaeal-like proteorhodopsin was abundant in the upper 130 m of the ocean, became scarce at 200 m, and was absent at 500-m depth.

The Smallest Motor of the World

We had heard about proton gradients, before we dealt with ATP as universal energy currency. How do we get the ATP, which powers our metabolism, from the food molecule? Now we need a machine that closes the loop, a machine that exploits the H^+ energy and converts it into ATP. This molecular motor is the ATP synthase. The unraveling of the structure and function of the ATP synthase is just another success story of modern biology. As this single enzyme complex has the gigantic daily task to synthesize in each of us 40 kg ATP as chemical energy to power our life, it really merits our attention. The ATP synthase is found not only in mitochondria but also in chloroplasts and in prokaryotes.

The Boyer Model

I will start to retrace the inquiry into this enzyme with the elucidation of the 3-D structure of the F_1 part of this enzyme complex, achieved in the lab of John Walker in Cambridge. This is not to say that the story really starts here. In fact, a major part of the work that earned Paul Boyer a Nobel Prize was already done at that time. I will only quote two of his major contributions: The first was the observation that the synthesis of ATP on the enzyme does not require much energy. On the β-subunit of the enzyme, ATP and ADP plus P_i are in equilibrium (Zhou and Boyer 1993). The fact that an exergonic irreversible chemical reaction in solution should be a readily reversible reaction on the enzyme surface is really surprising. On the basis of this observation, Paul Boyer proposed a rotational catalysis mechanism, the "binding change mechanism" (Boyer 1993). In his model, three catalytic sites are distinguished on the F_1 protein complex: O, for *o*pen site with very low affinity for nucleotide and phosphate ligands and catalytically inactive; L, for *l*oose site, loosely binding ligands and also catalytically inactive; and T, for *t*ight site, tightly binding ligands and catalytically active. The three sites interact and interconvert. As the synthesis of ATP occurs spontaneously on the enzyme, energy input is only required to release the newly synthesized ATP from the F_1 enzyme complex by transforming the T into an O site. This is achieved by a cyclic three-step rotation within the enzyme. In the following paragraphs, I will quote a few key papers published in the last 10 years that put flesh on the bones of this model.

The F_1 Structure

As in other cases, structural biologists took the lead, and the opening chapter of recent ATP synthase research was written with the publication of the structure from the F_1-ATPase from bovine heart mitochondria at 2.8-Å resolution (Abrahams et al. 1994). F_1 is that part of the ATP synthase that reaches into the mitochondrial matrix, the stroma of the chloroplast, or the cytosol of bacteria, which is the N side of the membrane conserving the proton gradient. This part of the enzyme cannot synthesize ATP, but it can catalyze the reverse reaction: the hydrolysis of ATP. To drive the condensing reaction, it needs to combine with the membrane-located F_0 part of ATP synthase. Together, the $F_1 F_0$ enzyme complex can synthesize ATP when powered by the proton gradient across the membrane. The structure of the F_1 complex was already revealing and confirmed previous observations reached by a combination of biochemical and biophysical experiments. The knob-like portion of F_1 is a flattened sphere 10 nm in its longer axis, consisting of alternating α- and β-subunits arranged like the sections of an orange. Three α- and β-subunit pairs were observed, conforming with the Boyer model. The structure of the α- and β-subunits were quite similar: An N-terminal six-stranded β-barrel was observed, followed by a central domain consisting of both α-helices and β-pleated sheets and finally a C-terminal bundle of α-helices. The nucleotide binding sites could be located at the interface between the α- and β-subunits, with the major contributions made by the β-subunits. Three different nucleotide-binding sites could be readily identified, but their association with the O, L, and T sites of the model was initially not so clear because the researchers used an ADP inhibitor and not ADP for the crystallization process. The crystal structure revealed a tight binding of the ATP and thus provided an answer to the question why F_1 binds ATP with a much higher affinity ($K_d = 10^{-12}$ M) than it binds ADP ($K_d = 10^{-5}$ M). The center of this F_1 "orange" contains some space, filled with the two α-helices from the γ-subunit. This axis of the F_1 wheel comes in a coiled conformation and extends toward the membrane, assuring the contact to the membrane-bound F_0 protein part.

A Mechanistic Model

Four years later, a group from Berkeley developed with these structural and a number of biophysical data a mechanistic model for the principal features of the F_1 motor. According to their calculations, the principal motion of the β-subunits consists of a hinge motion that bends the top and bottom segments toward one another by about 30° (Wang and Oster 1998). This bending stress is converted in the F_1 complex into a rotatory torque on the central γ-subunit. The central axis then presses in a sequential pattern on the three catalytic sites changing their microenvironment and thus mediating the transitions from empty to ADP-bound, to ADP and P_i-bound, to ATP-bound, and again to an empty site. In this model, the β-subunits contain a passive and an active spring function. This conclusion was based on the observation that the ATP synthase cannot function as a heat machine because it works with a mechanical efficiency approaching 100%, an impossible requirement for a heat engine. F_1 must conserve the free energy of nucleotide binding in elastic strain energy.

The F_0 Structure

However, F_1 was only half of the story because it only hydrolyzes ATP—an important part was still missing: This was the F_0 part. The next move came again from the Cambridge crystallographers, who solved the structure of the F_1F_0 complex from yeast in 1999 (Stock et al. 1999). In fact, not all subunits were contained in the final crystal (the external handle subunits a and b were lost) and the δ- and ε-subunits were not well resolved. Otherwise, the structure was very clear: Each c subunit consists of two α-helices that run back and forth across the membrane. Ten c subunits are closely packed together, forming a ring. An end-on view of the c ring (parallel to the plane of the membrane) shows two circles of helices. The crystallographers were first surprised by the observation that the ring did not consist of 12 subunits as widely anticipated. This means that the 3-fold symmetry of the F_1 part does not match the 10-fold symmetry of the c ring. However, the principle of symmetry mismatch was also found in other molecular rotary machines like the DNA packaging enzymes filling the genome into the head of bacteriophages or the flagellar motor of bacteria.

Higher-resolution data were recently obtained for two Na^+-ATPases. One was a bacterial F-type ATPase, which synthesizes ATP at the expense of ion-motive force by the transport of a sodium cation. The c ring of F_0 reveals a concave barrel with a pronounced waist and an inner septum, the lipid membrane. Eleven subunits make up the c ring. A central Glu residue in the hairpin helices binds the Na^+ (Meier et al. 2005). The rotor from the V-type Na^+-ATPase solved for another bacterium shows a highly similar overall structure for the K ring of the V_0 subunit, except that it consists of 10 subunits (Murata et al. 2005). V stands for vacuolar and indicates that the function of this ATPase is not ATP production but the transport of ions across the membrane against a concentration gradient at the expense of ATP hydrolysis to ADP. In this way, highly acidic vacuoles are created, like in citrus fruits. The investigated bacterial ATPase translocates sodium ions across the membrane.

The Next Model

In parallel the theoreticians from Berkeley around George Oster were again active. They now developed a model for energy transduction in ATP synthase

(Elston et al. 1998). They distinguished a rotor and a stator part like in technical motors. The rotor part consists of the γ-subunit from F_1 and the c ring from F_0, which are connected via the ε-subunit at the surface of the membrane. The stator part consists of a single a-subunit in the membrane, the b-subunit handle, connected via the δ-subunit to the F_1 sphere consisting of the three α- and β-subunits. Protons flow through the channels at the subunit a–c interface. This flow generates a torque by driving the rotor against the stator. This motor must produce a sufficient torque to generate three ATPs per revolution of the motor. The model was even more explicit. It postulates two aqueous half channels that connect the P and N sides of the membrane. Asp 61 from the c ring subunits is the central proton carrier. The positively charged Arg 210 from the single a-subunit interacts with the negatively charged Asp 61, according to Coulomb's law. If a proton gradient exists across the membrane, protons enter from the P side into the lower half channel of the a-subunit, board the rotor, rotate with the c ring to the right by a full turn, and encounter the exit half channel in subunit-a from where they leave to the N side. The rotation of the c ring driven by the flow of the proton down the gradient is transmitted to the connected γ-subunit, and its movement with respect to the α–β subunits in F_1 achieves the ATP release. As the process is fully reversible, ATP hydrolysis can also drive the rotor and can in the reverse mode build up a proton gradient. The theoreticians predicted that under these two modes the rotor revolves in the opposite direction.

The newer data are still compatible with the previously proposed two-channel and two-ring Brownian ratchet for torque generation in these tiny electromotors (Junge and Nelson 2005). The essential Glu must be negatively charged when facing the positive stator. It must be neutralized by Na^+ binding when facing the hydrophobic membrane lipids. Two access channels for the ions are on either side of the stator. The ion enters through the lower access channel, binds to the Glu, and allows a counterclockwise move after relieving the electrostatic repulsion.

Biophysicists Observe the Rotation

Now the task to visualize these predicted movements of the rotor was in the hand of biophysicists and engineers. Several Japanese laboratories set out to analyze the mechanical details of these rotation processes using experiments, which pushed analytical methods to their extremes.

When Paul Boyer first formulated his rotational catalysis model for the ATP synthase, few people believed this idea. Then came the structure of F_1, which provided circumstantial evidence for the Boyer model. The final visual evidence was provided by the Tokyo Institute of Technology (Noji et al. 1997; Block 1997). The researchers used some tricks: They engineered a 10-histidine tag to the β-subunits of F_1. This tag binds nickel ions avidly, and this affinity allowed binding the complex to a glass coverslip coated with nickel. In this configuration, the central γ-subunit sticks into the airspace. At the tip of the γ-subunit, they introduced a cysteine residue. Using a common chemical sticker (biotin–streptavidine), they fixed a long fluorescence-labeled actin filament to the central rotor. They added ATP to the construct and observed the coverslip in an epifluorescence microscope, and what they saw was really breathtaking. The long actin filament turned like a propeller and those that were not entangled turned anticlockwise. However, the revolution was not occurring in 120° steps as predicted by the Boyer model, but rather was smooth and slow. The observed maximal speed was only four revolutions per second, suggesting a heavy viscous drag on the motor when it swings the actin through the medium.

In the next study, the researchers put the manipulated enzyme on the top of a bead to allow an unhindered movement and more detailed observations (Yasuda et al. 1998). Not surprisingly, the rate of rotation decreased with the length of the actin filament and the concentration of ATP. When the Japanese physicists from Keio University calculated the torque, they got an astonishing 100% efficiency of ATP hydrolysis into mechanical work when using the Boyer model based on three ATPs per revolution. But does the enzyme really go ahead in three steps of 120°? When the researchers worked at very low ATP concentrations, they could actually visualize this three-step mode of revolution and even determine the dwelling times of stops between the individual steps. An even sharper resolution of the events at the F_1 ATPase complex was obtained when the Japanese researchers decreased the viscous drag on the small motor. To achieve this, they replaced the fluorescent actin filament by a colloidal gold bead of 40 nm. This is still four times larger than the sphere of the 10-nm F_1 complex, but only a fraction of the 1- to 2-μm-long actin filament. The viscous friction is thus only 10^{-3} of that for actin, which allowed high-speed rotation of the engineered complex. Under these experimental conditions, each 120° step could even be resolved into two substeps: A 90° substep was associated with the ATP binding reaction and a 30° substep was linked to the product release of ADP and P_i. A careful analysis of the reaction kinetics proposed two 1-ms dwell times between both substeps, which they interpreted as the time needed to expel the first and then the second hydrolysis product of ATP (Yasuda et al. 2001; Schnitzer 2001).

Biophysicists Turn the Motor

In the Oster model, the rotor consists of a γ-subunit connected to the c ring. ATP hydrolysis should thus also drive an actin filament fixed to the c ring. This prediction was experimentally demonstrated by another Japanese group working with the F_1F_0 complex (Sambongi et al. 1999). Until now the biophysicists have used ATP to drive the motor. Theoretically, it should also be possible to synthesize ATP by driving the rotor. This is actually the way ATP is synthesized by F_1F_0 ATP synthase in the cell. This was actually demonstrated in two papers published in the past year. The technical obstacles turned out to be formidable. The gold beads were replaced by magnetic beads; the beads were then rotated with circularly arranged magnets in a medium containing ADP and P_i as substrates. In addition, the system contained a sensitive detector (luciferin–luciferase system) that emits a photon when it captures and hydrolyzes ATP. Surprisingly, the biggest problem was not an engineering problem, but the residual contamination of ADP with ATP, which obliged the researchers to work with microvolumes in the reaction chamber and consequently with small signals. However, the proof of principle was provided: The photon counts were significantly greater in the ATP synthesis than in the ATP hydrolysis rotation direction or at no rotation (Itoh et al. 2004; Cross 2004). Fittingly, the latest data were provided by those researchers around Hiroyuki Noji who started this type of experimentation in 1997. In their most recent experiment, they used F_1 molecules linked to magnetic beads enclosed in femtoliter (10^{-15} l) reaction chambers and rotated in a clockwise (ATP synthesizing) direction by magnetic tweezers. Then they stopped the reaction. The synthesized ATP now powered the anticlockwise rotation of the manipulated F_1 linked beads. The speed of the anticlockwise rotation was proportional to the amount of synthesized ATP. Interestingly, the mechanochemical coupling was low for the $\alpha_3\beta_3\gamma$ complex, but it reached 77% efficiency when the complex was reconstituted with the ε-subunit (Rondelez et al. 2005). If perfect mechanical coupling is assumed, the enzyme should produce three ATPs per clockwise turn. The complex containing the ε-subunit showed, upon forced rotation, an average of 2.3 ATPs produced per revolution, and the best traces of the recordings reached the postulated theoretical value of 3. The F_1 complex lacking the ε-subunit showed a coupling ratio of only 0.5 ATPs per turn. However, in the opposite, spontaneous, and anticlockwise rotation, which occurred in the presence of ATP, both complexes consumed three ATPs per revolution. This work showed an unexpected importance of the ε-subunit in the synthesis of ATP, but its exact function and location was not elucidated at the time of writing.

We are here at the end of our travel through that part of bioenergetics that describes the extraction of usable chemical energy from foodstuff. I promised you revolutions and you got perhaps more than you wanted and your head is swirling with biochemical cycles and the rotations of the smallest but most widely distributed motor on our planet. You can summarize the essence of respiration in a nutshell: Three forms of energy follow sequentially: first a redox potential (NADH), then a proton-motive force (H^+), and finally a chemical energy (ATP). It might be important to repeat that it is not the $\sim$P bond in ATP that stores the energy. ATP works by keeping the ATP/ADP ratio far away from its thermodynamic equilibrium, as nicely explained in F. M. Harold's book *The Vital Force*. There is thus nothing metaphysical in the "energy-rich bond" of ATP as still stated in some biochemistry books.

4
The Evolution of Eating Systems

The Beginning of Biochemistry

LUCA

In contrast to chemistry and physics where many processes can be analyzed outside of a specific geological time frame, it makes no sense to consider biological systems without knowledge about their development in time (Nisbet and Sleep 2001). Or in the famous word of the biologist Theodosius Dobzhansky: "Nothing in biology makes sense except in the light of evolution." All forms of cellular life—as far as we know them from Earth—are linked to LUCA, the *l*ast *u*niversal *c*ommon *a*ncestor in science slang. All extant organisms, including viruses, synthesize their proteins by the same synthetic machinery (ribosomes); use nucleic acids for information transfer, and decode this message according to a (practically) universal code. We are used to the yardstick of ribosomal RNA sequence, which allows constructing the universal tree of life. Some organisms are still changing their place on this tree or have not yet found a comfortable place. Even the tree analogy has recently been challenged and weblike phylogenetic relationships have been proposed for bacteria to account for the prominent role of lateral gene transfer in prokaryotes. However, the common origin of all extant life on Earth has not been questioned seriously. The only nagging questions are presented by viral genomes. We will touch this issue in one of the subsequent sections.

Traces

Genes are not the only witnesses of the common origin of life; we also find many traces for the shared origin of life in noninformational biological molecules that relate to the creation of biochemical energy and the synthesis of biological molecules. For example, the central precursor metabolites are virtually identical in essentially all organisms. This uniformity is testimony to the common origin of all extant organisms irrespective of if they are prokaryotes or eukaryotes. The wide distribution of ribonucleotides not only as information molecules but also as central biochemical compounds and the universal energy currency can hardly be explained by chance and were quoted as evidence that these central metabolic reactions are billions of years old and are rooted in the RNA world preceding the current DNA world. There are even more distant worlds of the early life that left traces in the biochemistry of extant organisms (Lazcano and Miller 1996). The many bioinorganic compounds in contemporary cells might be derived from an earlier iron–sulfur world. Not surprisingly, chemists took the lead in exploring this world of early eaters. We will now go back to the abiotic creation of critical organic compounds, the invention of metabolic cycles, the birth of the cell, and the appearance of the earliest animals in the Precambrian fossils before we get into the more familiar world of extant eaters.

A Soup as a Starter? The Origin of Biochemical Cycles

Life

The origin of life has fascinated humans from the beginning of human thinking. All major religious beliefs offered their version on the moment and the circumstances of creation. In this tradition, we also start our natural history on eating with this beginning of all biology because it provides a nice illustration of thermodynamic principles and a lively controversy about the most fundamental divide in biology, namely that between heterotrophic and autotrophic ways of life. At the dawn of life there were no other organisms that you could eat. For logical reasons, one would therefore suspect that life started with autotrophic systems. However, the earliest experiments in the origin of life research gave heterotrophy a head start. Here one should remember that life did not start with cells in the modern sense but with entities that showed some form of primitive metabolism and replication that satisfied the simplest definitions of living material. Some of the biochemical activities are so basic that one might wonder whether a contemporary biologist would detect this form of life when being on a hypothetical mission on planet Mars or the Jupiter moon Europe. Therefore you do not require complicated biochemistry from the beginning.

The Miller Soup

The crucial experiments in the early "origin of life" research were conducted by Stanley Miller in the 1950s when he was a graduate student at Chicago in the Laboratory of Harold Urey, a physicist who discovered the heavy hydrogen isotope deuterium. An electric discharge, simulating lightning, was passed in their laboratory through a mixture of methane (CH_4), ammonia (NH_3), water (H_2O), and hydrogen (H_2), what was then thought to represent the primitive atmosphere of the early Earth. Very strikingly, these experiments yielded many organic compounds, including amino acids and other substances fundamental to biochemistry. The yields of the amino acids glycine and alanine were a startling 2% of the added methane. Shortly thereafter it was demonstrated that hydrogen cyanide (HCN), also considered a likely component of the early atmosphere, and ammonia condense on exposure to heat or light to produce adenine, one of the four nucleotide bases. These experiments provided the foundations for the "primeval soup" hypothesis. According to this thinking, these organic compounds accumulated in some shallow parts of the early oceans to critical levels that allowed the first metabolic cycles to organize. Some form of molecular self-organization led to vesicle building, competition between vesicles for the first "food" supply, and thus the start of Darwinian evolution. Later followed the invention of informational molecules: The proposals ranged from clay to precursors of modern RNA allowing some primitive forms of self-replication. These first living systems were fed by the primeval soup and when their number increased and the soup was exhausted, these living systems had already entered into the RNA world and were already clever enough to invent autotrophic forms of life. Part of the idea went back to the "coacervate" ideas from Oparin, the father of the modern origin of life research in the 1920s. More recent theories claimed that membranes and vesicles came late in this scenario.

Meteorites

Strong support for this abiotic origin of the first "food" molecules came from astrophysicists. They knew from spectroscopic techniques that the primary biogenic building blocks were also the products of interstellar chemistry. An impressive confirmation of the Urey–Miller experiments was provided by the analysis of the Murchison meteorite, a carbonaceous chondrite recovered from Australia. The meteorite contained both in quality and in relative quantity the same amino acids as produced in the Urey–Miller experiments. The hypothesis was thus chemically rather plausible. As the Earth suffered intense bombardment from space during its early history (Tiedemann 1997), a substantial amount of biogenic building blocks were thus introduced from outer space (Orgel 1998). In one hypothesis, which once had strong advocates in astrophysicists, cells might even have been introduced into the early Earth from space. This is the old panspermia hypothesis that got new headlines with the purportedly bacteria-like microfossils in a Martian meteorite found in the Antarctica.

The Atmosphere

However, some scientists had doubts about the primordial soup recipe (Maden 1995). The soup of biogenic molecules created on Earth by the Urey–Miller process or imported from space was in their view too dilute to serve as a starting material for a primitive heterotrophic metabolism. Then came doubts about the composition of the primitive atmosphere. Our planet might not have been able to hold H_2 in its gravity field. If hydrogen was lost, CO_2 rather than CH_4 was the main carbon source, and the atmosphere was no longer strongly reducing ($CH_4 + N_2$, $NH_3 + H_2O$, or $CO_2 + H_2 + N_2$), but was perhaps neutral ($CO_2 + N_2 + H_2O$). Under such conditions, the yield of amino acids by electric discharge would be drastically reduced to unrealistically low levels. Fifty years after his first experiment, Stanley Miller conducted trials in a weakly reducing CO–CO_2–N_2–H_2O atmosphere (Miyakawa et al. 2002). This experiment also gave a variety of biogenic compounds with yields comparable to those obtained from strongly reducing atmosphere. The necessary amount of CO in his postulated atmosphere had to be supplied by the impact of comets and asteroids. This not implausible corollary supported at least partially the primeval soup hypothesis.

Yet the last word was not yet spoken in this discussion. Physicists have recently reexamined the theory of the diffusion-limited escape of hydrogen into space and found that it was by far not as rapid as previously assumed. Their calculation of hydrogen escape and volcanic outgassing made an atmospheric hydrogen mixing of more than 30% possible. Because the CO_2 concentration was likely to be less, a H_2/CO_2 ratio of greater than 1 would result in the early atmosphere. Under these conditions, electric discharge would produce 10^7 kg amino acids per year (Tian, Toon et al. 2005). In such a hydrogen-rich early Earth atmosphere, the Miller soup scenario again gains credibility.

Banded Iron

Reassuringly, there are alternative sources of reducing equivalents even when H_2 was lacking in the atmosphere. The discovery of banded iron formations (BIFs) in Precambrian strata around the world offered a possible source for reducing equivalents. UV light can energize the transfer of an electron from ferrous (Fe^{2+}) ions to protons with the generation of molecular hydrogen from hydrogen ions. Nascent hydrogen can be used as a reducing agent for other chemical reactions whereas the ferric (Fe^{3+}) iron produced precipitates as insoluble mixed iron oxide (magnetite), explaining the BIF.

Iron–Sulfur Worlds

The heterotrophic hypothesis was challenged by Günter Wächtershäuser, a patent attorney from Munich without a laboratory but trained as a chemist and soaked with the philosophy of science from Karl Popper (Hagmann 2002). Wächtershäuser (1990) proposed an autotrophic theory in which pyrite formation is the earliest energy source of life. In fact, the ferrous sulfide/hydrogen sulfide system creating pyrite and hydrogen ($FeS + H_2S \rightarrow FeS_2 + H_2$) is a powerful reducing agent yielding more than adequate free energy ($\Delta G° = -9.23$ kcal/mol, $E° = -620$ mV) to reduce CO_2 to any desired organic compound. Furthermore, Wächtershäuser invoked surface metabolism before enzymes and templates were developed. According to him, metabolism commenced with 2-D systems, i.e., "life" in a very extended definition, characterized by surface bonding of anionic organic ligands to the cationic mineral surface of pyrite. The primordial carbon fixation occurred in this theory via an archaic, reductive citric acid cycle pulled by pyrite formation, and the di- and tricarboxylates were surface bonded in the nascent state to the pyrite surface. Finally, he elaborated an autotrophic origin for the whole of the central metabolism based on an iron–sulfur world. He used a principle, which he called retrodiction, i.e., the use of knowledge from the biochemistry of extant organisms to make "pre"dictions about the chemistry in the primeval biological world. As these predictions are about events in the past, he consequently calls it "retro"dictions. This is quite an interesting approach, which is based on the conviction that, first, all forms of life can be traced back to a single last common ancestor and, second, evolution has substantially developed its inventions, but never discarded its successful early solutions. Wächtershäuser is explicit and predicts enzymes and nucleic acids as products and not as precursors to this archaic metabolism. As one would expect, this autotrophy theory sparked interest and controversy at the same time and the early exchange of arguments between the two camps in *Two-Dimensional Life?* (de Duve and Miller 1991) and *Life in a Ligand Sphere* (Wächtershäuser 1994) makes fascinating reading. De Duve and Miller concede that Wächtershäuser's theory is imaginative and original, but they argue that it is not plausible in the framework of aqueous solution chemistry. The two sources of free-energy-sulfide oxidation and anionic bonding are not enough to drive metabolism. In addition, the introduction of a surface chemistry is not enough to escape from random solution chemistry. Notably, they question whether the conformity of the theory with the science philosophy of Karl Popper has any explicative value, arguing that philosophical conformity does not constitute a scientific reason.

The Earliest Pathways

Glycolysis mediates the anaerobic exploitation of the energy stored in sugar molecules, but the energy gain is modest. Energetically, it was a great progress in evolution when organisms learned to channel pyruvate into the citric acid cycle, where many reducing equivalents are abstracted from the metabolites to power oxidative phosphorylation. However, this cannot be the primary function of this cycle as oxygen was not available during the early periods of biological evolution. In fact, three groups of eubacteria and at least one archaeon drive the citric acid cycle in the opposite direction. Wächershäuser's proposal of a reductive, energy-consuming, primitive TCA precursor cycle for creating complex organic molecules is therefore not implausible. In addition, the use of iron–sulfur centers in several enzymes from the glycolytic or TCA cycle recalls somewhat the pyrite chemistry invoked by Wächtershäuser. It should not surprise us that biochemistry keeps memories of its evolutionary past, somewhat like E. Haeckel's famous words that ontogeny repeats phylogeny. However, the problem with Wächershäuser's hypothesis is that the experimental conditions described in his publications using ferrous sulfide/hydrogen sulfide failed to reduce CO_2 (Keefe et al. 1995). A prominent origin of life researcher also questioned the plausibility of the self-organizing potential of biochemical cycles in the absence of genetic polymers (Orgel 2000).

Peptide Synthesis

No theoretical biology exists that could be compared to theoretical physics. Basic new concepts can be developed theoretically in biology, but the complexity of biological processes and the limited applicability of mathematics in biology prevent the calculation of whether a given hypothesis works out. The crucial test in biology is the experiment or field observation. Therefore Wächtershäuser had to team up with microbiologists and then organic chemists to provide evidence for his theory. One experiment suggested that the earliest organisms fed on CO and CO_2 at volcanic sites or hydrothermal vents. In the presence of NiS and FeS found at these sites, a mixture of CO and methylmercaptane (CH_3SH) were converted into an activated thioester, which hydrolyzed to acetic acid (Huber and Wächtershäuser 1997, 1998). A C2 compound was created from two C1 compounds. Interestingly, the reductive acetyl-CoA pathway, a CO_2-fixing series of reactions in sulfate-reducing bacteria and methanogens, also uses nickel in the binding reaction of one CO_2 molecule. This enzyme again illustrates the conservation of primordial conditions in an extant enzyme. Subsequently the formation of peptide bonds between amino acids was achieved under the postulated primordial conditions (Keller 1994; Huber and Wächtershäuser 1998). Geologists from the Carnegie Institution demonstrated the production of such a central metabolite as pyruvic acid wherever reduced hydrothermal fluids pass through iron sulfide-containing crust (Cody et al. 2000). Recently, not only the formation but also the degradation of peptide bonds were shown in the presence of freshly coprecipitated colloidal (Fe, Ni)S (Huber et al. 2003). This opens the possibility for a primordial peptide cycle where peptide bonds are continuously formed and degraded on metal surfaces forming a dynamic chemical library that scans the space of structural possibilities for small peptides.

On Timescales in Biology

Life is Old

Religious beliefs commonly quote living things when alluding to short-lasting events on Earth. Humans are likened to grass in the field, which perishes quickly. In the longing for duration, many religions fix their hopes on sky-rising mountains. Solitude in a mountainous area surrounded by a barren desert is the closest you can get on Earth to feel eternity. The common denominator in these religious perceptions is the absence of other life forms that would recall the weakness of life. I think many biologists would disagree with this view. In fact, mammals are older than the Alps. The only change you have to do to appreciate this viewpoint is to see life as a connected chain of generations and not as an individual life span. Even if you raise your eyes to the sky, you will see many stars that cannot compete with the collective lifetime of the mammalian evolutionary line. Our sun and its associated planetary system belong to relatively older objects in the cosmos. Sure, the iron core of our home planet testifies that the solar system was collected from the debris of an earlier star that used up its fuel in sequential nuclear fusion events. Iron is an end point of the fusion because it has an especially stable nuclear configuration that does not allow further fusion events to occur.

Timescales

As the essence of science is about giving precise numbers, I will try to express major events in the history of life in years. The beginning of time in the interpretation of science comes with the explosion of the cosmic fireball, dubbed big bang. This process occurred perhaps 15 billion years ago. The oldest stars in our galaxy formed about 10 billion years ago. Our sun was born about 5 billion years BP. The formation of the Earth with its present mass goes back to 4.5 billion years. The primordial oceans began to form 4.2 billion years ago. The prebiotic chemistry described in the preceding section is commonly dated to 200 million years, following the creation of this ocean. This prebiotic world, creating the molecules that would later make up living systems, was followed by a 200 million-long pre-RNA world, where the first information storage molecules were invented. Two hundred million years later, at about 3.8 billion years BP, scientists located a full-blown RNA world, where energy-yielding processes, genetic information storage, and enzymatic catalysis were all based on RNA. In the current scenario (Joyce 2002), the first DNA- and protein-based life-forms developed a mere 200 million years later to be followed by a spectacular diversification of life ignited at 3.6 billion years ago. The diversification of life manifested itself as a rapid ramification of the tree of life as visualized today by the tree based on the sequence of the RNA making up the small subunit of the ribosomes and an impressive metabolic diversification of the prokaryotic cells. In view of the rapid development of life on Earth, it is rather surprising how long it took before eukaryotes, multicellular life, and animals evolved—the fossil record yields 2.1 billion years BP for the eukaryotic cell and 580 million years for the first body fossils of animals. Before dealing with this problem, we need to address some problems linked to the genesis of metabolism in the earliest life-forms.

Early Autotrophy?

When the first oceans formed at 4.2 billion years BP, life was not yet in a safe cradle. The early Earth was still in its roaring twenties. Actually, a lot of the roaring came from outside in form of a cosmic bombardment. The scars of these events are still visible to the naked eye when looking at the cratered face of the moon. As the moon lacks an atmosphere and a hydrosphere, erosions could not destroy these signs as happened on our planet. Geological evidence pointed to asteroid impacts that might have sterilized the planet as recently as 3.8 billion years ago (Maher and Stevenson 1988; Sleep et al. 1989). Calculations suggested that the greater of these impacts boiled away all but the deepest layers of the oceans. Water was temporarily found only as water vapor in the atmosphere. However, these events are already fully in the RNA world. The scene becomes even more disturbing when considering the evidence from the oldest rocks on Earth found in the Itsaq Gneiss Complex from Greenland (Hayes 1996; Mojzsis et al. 1996). Itsaq is Greenlandic for "ancient thing," and indeed we have here the oldest remains of sediments in the geological record. The sediment is metamorphosed; it is thus futile to expect fossil evidence. The scientists looked into the rocks with an ion-microprobe mass spectrometer, which allowed the differentiation of the ^{12}C and ^{13}C carbon isotope content in features as small as $20\,\mu m$ associated with apatite, a phosphate mineral known for its biological association (e.g., in our bones). The carbon content was also measured in acid-insoluble carbonaceaous residues (kerogen) of the Isua belt of BIF. It was isotopically light. No known geological process could explain this enrichment for the ^{12}C isotope, whereas biochemical processes such as photoautotrophic carbon fixation are known to prefer the lighter carbon atom and would thus easily explain the observation. However, this interpretation leads to a paradox as it leaves almost no time between the end of the so-called late heavy bombardment and the first appearance of life. Of course, not knowing a purely geological process that could lead to a fractionation of carbon isotopes is perhaps not enough to quote this as evidence for life processes as an alternative. However, you should not forget that "soon" afterward, the first microfossils were reported in 3.5 billion-year-old rocks (see below). Biochemical processes 3.8 billion years BP are perhaps not so exotic at all. Furthermore, life might be a quick development in any system that offers right conditions and thus a cosmic imperative as stated in a book of the cell biologist Christian de Duve (1995). In the end, we might be excited or disappointed when realizing in future, from expeditions back from Mars, that life existed there in forms comparable to life on Earth, as already suggested in a highly contested paper on a Martian meteorite (McKay et al. 1996).

The RNA World

Reactivity

The RNA world (Gilbert 1986) is commonly considered as the period when biological information was first laid down into nucleic acids. At first sight, it might seem illogical to store genetic information in RNA and not in DNA. The supplementary C-2 hydroxyl group makes RNA more susceptible to degradation and also causes more problems in nonenzymatic polymerization reactions. The condensation of activated nucelotides could result in 5'-5', 5'-3', and 5'-2' links and thus rather complex macromolecules and not just the 5'-3' links of modern information-carrying nucleic acids. However, the greater reactivity of RNA over DNA was perhaps the very reason why RNA was favored as a chemical. In fact, in the early RNA world, protein enzymes were not yet developed, and RNA had also to fulfill enzymatic functions. In this context, the greater chemical and structural flexibility of RNA is a definitive advantage.

Peptide Nucleic Acids

To get started, the Watson–Crick base pairing turned out to be of central help in information transfer as it could sort out the fitting of unsuitable base pairs. Base pairing was even a strong organization principle for polymers that were potential precursors of RNA in a pre-RNA world. One candidate polymer is a peptide nucleic acid consisting of a peptide-like backbone contributed by the amino acid glycine. In this polymer the base is linked via a methylencarbonyl group to the peptide. Even these exotic compounds can form base pairs (Egholm et al. 1993). Before RNA catalyst could take the alignment into their hands, the mineral surfaces might have catalyzed the synthesis of long prebiotic polymers (Ferris et al. 1996).

The Central Role of RNA

In the late 1960s, Woese, Crick, and Orgel proposed the general idea that evolution based on RNA replication preceded the appearance of protein synthesis (Crick 1968; Orgel 1968). It was suggested that RNA was also the sole catalyst of biochemical reactions at that time. The major involvement of three classes of RNA molecules (ribosomal, transfer, and messenger RNA) in protein translation was seen as a clear evidence for the importance of RNA in this central process of synthesis of biological structure. Other lines of thought came from biochemistry. We mentioned already the observation that ribonucleotides, especially AMP, provided the chemical skeleton for a number of central coenzymes (ATP, NADH, FAD, coenzyme A) still in use in modern metabolism. Coenzymes were thus interpreted as molecular fossils (White 1976).

Molecular Fossils in Biochemistry

Steven Benner and colleagues (1989) developed the idea further and stated that the modern metabolism is a palimpsest of the RNA world. A palimpsest is a term borrowed from the work of historians. They characterize with it a parchment that has been inscribed two or more times. In fact, a parchment was too precious a support for new writing to throw the old written manuscript away. The ancient text was erased and replaced by the new one. However, the erasing of the old text was frequently imperfect and the text remained still partially legible. Benner argued in that analogy that remnants of the ancient metabolism can still be observed in extant organisms. We already encountered this conservative nature of evolution several times.

The argument became famous with the zoological eye argument in evolutionary discussions. How could such a perfect image-building organ like the eye evolve from very imperfect image-building precursors that hardly offered any selective advantage or that showed no image-forming capacity at all? This discussion has actually created more heat than light. If you look into the molecular basis of light perception, you will see a seven-membrane protein (opsin) associated with retinal that shuttles between an all-trans and a 11-cis conformation. This should immediately remind you that you have seen this motive long before in evolution with the light-induced proton pump from the archaeon of the Dead Sea. Now a very similar opsin called melanopsin was identified in the eyes of mammals that had no image-producing functions. Yet it detects light and can adjust our body clocks to the circadian rhythm (Foster 2005; Melyan et al. 2005; Qiu et al. 2005). Many molecules would have first evolved for other functions and would have taken over new functions when new opportunities opened up.

tRNA as an Example

I want to illustrate this subject with the tRNA molecule in its common cloverleaf representation. At one end of the molecule is the anticodon arm, and at the other end is the amino acid acceptor arm. In this constellation, you see perfectly the adaptor molecule: The anticodon loop reads the appropriate codon from the mRNA and brings in the corresponding amino acid for the peptidyl transfer center on the ribosome. How could this complex machinery evolve when there was not yet a translation machine or proteins in place? A possible answer to this conundrum came with a hypothesis formulated by Weiner and Maizel (1987). They started with the observation that X-ray crystallography revealed an L-like structure for tRNA. The top horizontal half contains the TψC arm and the amino acid acceptor stem ending with CCA in all tRNAs irrespective of what amino acid is bound. The bottom vertical half is built by the D arm and the anticodon arm. Their reasoning is that tRNA did not evolve for serving in protein synthesis. The top half was in their genomic tag hypothesis the signal that marked single-stranded RNA for replication in the RNA world. There are sound reasons for the specific use of a CCA end for this task as they respond best to the problem of getting started for RNA synthesis when lacking a primer without loosing the end parts of the genomes (the telomer problem). The 3'-terminal A would be the minimal telomeric sequence and untemplated addition of this A residue would be required to regenerate the genome before each new round of replication. The G:C base pairs are bound by more hydrogen bonds and are further stabilized by strong base stacking to compensate for the absence of a primer. This scenario is not only chemically plausible, but CCA ends and tRNA as primers are found in diverse elements such as the bacterial virus Qβ, the plant cauliflower mosaic virus, or the animal retroviruses.

Modular Evolution

Thus, tRNA did not evolve for protein synthesis function but for RNA replication function. The lower half of the tRNA is likely a later acquisition, when new functions were recruited. *However, new functions in evolution were frequently acquired without loosing old functions. This puts selection forces under constraints. Many molecules are not free to evolve new functions and maximal efficiency; they must also maintain the old functions. Also as molecular partners evolved, a complicated set of coevolving pathways was set in motion. Very early in the evolution of biological molecules, the necessity of coevolution resulted in the freezing of old solutions, and newer solutions were only overlaid on an essentially conserved core function. Molecules were thus frozen in time and became molecular fossils that can be dug out in extant organisms. Here the famous words of Lucretius from his poem "De natura rerum" on natural history, namely "Natura non saltus fecit" (Nature makes no leaps), gets a modern meaning.*

Split tRNA Genes

A recent observation by Dieter Söll's lab provides an exciting illustration of this principle (Randau et al. 2005). His lab has long been interested in the origin of the tRNA molecule. They examined Nanoarchaeota, minute microbes that are located at the root of the Archaea lineage, before the Archaea split into the Euryarchaeota and Crenarchaeota branches. They were attracted to this organism because it lacked genes for four tRNA species. After some searching they found them all. Significantly, they all presented as two half genes: The conserved T loops and 3'-acceptor stem on one side and the D-stem plus 5'-acceptor stem in another unit. The primary transcripts of these tRNA halves included the intervening complementary sequences at the position of separation, which could build a duplex RNA. Did this primitive Archaeon conserve traces of the precursor RNA molecules from the prebiotic world before they were permanently joined for their new job? The tRNA preserves still other traces of the extensive rework for its new task. For a tRNA to participate in protein synthesis, it must carry a 3'-terminal CCA sequence to which the amino acid is esterified. In *Nanoarchaeum* this CCA sequence is not encoded in the tRNA half gene, but is added posttranscriptionally by an enzyme (Xiong et al. 2003). It thus seems that the evolution of protein translation needs has recruited older genetic elements to perfect the new task and that the primitive extant organisms have still preserved traces of this process.

The Ribosome is a Ribozyme

Ribosome Structure

Protein synthesis is one of the crucial processes for creating biomass, the ultimate aim in the quest for food. Until this day, this important enzymatic task is fulfilled by RNA. There are actually only a few, short peptides that are synthesized by protein enzymes (e.g., the tripeptide glutathione). And the ribosome? The role of proteins in this process became doubtful, when it was shown that the peptidyl transferase activity of ribosomes was unusually resistant to protein extraction procedures from this ribonucleoprotein complex (Noller et al. 1992). Scientists knew that the ribosomal proteins were pretty small when compared with the central RNA molecule of each subunit. The molecular weights of the ribosomal proteins range from 6,000 to 75,000 in *E. coli*, while the 23S rRNA in

E. coli musters a molecular weight of about one million. Despite the numerical abundance of the different protein species, 65% of the total ribosomal mass is represented in RNA. Nevertheless, the crystal structures of the ribosome and its subunits came as a shocking and exciting surprise (Ban et al. 2000; Yusupov et al. 2001). The ribosome is essentially a complicated tangle of RNA strands and coils that form the shape of the ribosome. And what about the proteins? They bravely decorate some bits of the ribosome surface here and there, leaving most of the surface unoccupied. A few protein helices sneak into the ribosome, but one gets the impression that they serve more a structural stabilization function to the RNA than actually playing a functional role.

No Proteins at the Active Site

In a companion paper, Steitz and collaborators examined the details of the peptidyl transfer center which they located with two substrate analogs (Nissen et al. 2000). Both analogs contained puromycin as basic structure. Puromycin resembles the terminal aminoacyl-adenosine portion (the above-mentioned CCA) of the acceptor stem from an amino acid-charged tRNA. It can thus bind the A site of the ribosome. Puromycin contains a free amino group, which is linked by the ribosome to the carbonyl group of the growing peptide chain bound at the P site. This leads to a chain termination reaction, and the peptidyl puromycin derivative dissociates from the ribosome. One compound was the Yarus inhibitor, which mimics the tetrahedral carbon intermediate produced during peptide bond formation (Moore and Steitz 2002). Now comes the surprise: There is no ribosomal protein in the vicinity of the puromycin–ribosomal small subunit cocrystal. The nearest are four globular proteins, but they do not reach the peptidyl transfer center with their extension by some 20 Å—too far away to expect a catalytic action from them. If there is only RNA around the catalytic site, then the ribosome must be a ribozyme, an RNA enzyme.

Catalysis by RNA?

Ribozymes have been described in another context, for example, in self-splicing introns (Doudna and Cech 2002). However, a ribozyme in one of the most crucial and abundant biosynthetic functions around us—this was not really expected. Remember that a single *E. coli* cell contains something like 15,000 ribosomes. The labs of T. Steitz and, independently, S. Strobel went to some length to investigate the mechanism of this RNA-catalyzed reaction (Nissen et al. 2000; Muth et al. 2000). Both labs deduced that adenosine 2451 (the number refers to the position of the nucleotide in the 23S rRNA sequence from *E. coli*) played a crucial role as a general acid–base catalyst. The proposed reaction mechanism resembles the function of the catalytic histidine residue in the hydrolytic cleavage of peptide bonds by chymotrypsin. Both groups knew that the pK value from the N3 position of the adenine is far too low to fulfill this function. Steitz et al. deduced that the stereochemical configuration of nucleotides in the vicinity of the critical A2451 residue changes the pK value by an appropriate hydrogen bonding to the neighboring guanosines. Strobel and colleagues actually investigated the pK at the peptidyl transferase center using chemical probes and found a single adenosine, namely A2451, with a neutral pK. They pointed out that this nucleotide A2451 is conserved in every living organism in all the three biological kingdoms. When they mutated A2451 to any of the other three nucleotides, the expression of this mutated ribosome led to a dominant lethal phenotype. The circle of arguments in favor of a ribosome ribozyme seems to be closed. Science is a dynamic process. Hardly a year after these breakthrough discoveries that changed our view of the persistence of the RNA world until our day came an important modification of the reaction mechanism. Researchers constructed nucleotide substitutions into a cloned 23S rRNA from *Thermus* and targeted all critical nucleotides deduced by the combined efforts of Steitz and Strobel. They expressed these RNAs in vitro and probed them by a common in vitro peptidyl transfer assay: the linkage of formyl-[^{35}S] methionine tRNA, the usual starter tRNA, to puromycin. With most base changes in A2451 and the associated guanosines, a residual transfer activity was observed (Polacek et al. 2001). They deduced that the transpeptidation reaction on the ribosome does not occur through chemical catalysis but by properly positioning the substrates of protein synthesis.

Demise of the RNA World

These structural and functional data have important consequences for our view of the RNA world. Proteins are not involved in the catalytic activity of the ribosome; these functions are fulfilled by the proper scaffold of RNA. Some proteins sneak with extensions into the ribosomal interior. However, here their task is mainly to fill the void between the RNA helices and to neutralize the phosphate backbone charges. Even if the 23S rRNA might not be catalytically active in the stricter sense of the word, RNA is a very flexible molecule. Despite its limited chemical functionality when compared to proteins, RNA is capable of performing all the reactions of protein synthesis. In vitro evolution has been used to develop ribozymes that catalyze peptide bond formation with a mere 190-nucleotide-long RNA molecule at a reasonable rate (Zhang and Cech 1997). It could be speculated that the pure RNA precursor of the modern ribosomes has invented peptide synthesis to stabilize its polyanionic structure in a more efficient way than with simple counterions. This invention became one of the turning points in the evolution of life because it led to the evolution of proteins. The much wider chemical functionality of the 20 amino acids in proteins offered more chemical possibilities if you learned to store this information. Evolution could become less constrained when RNA did not have to play the role of genetic storage and enzymes with the same molecule. RNA invented two helpers: DNA as a more passive storage form of genetic inventions and proteins as more versatile functional arms of RNA. RNA maintained the crucial mediator position between the two new key inventions that led to the biological world that we know today. But this burden sharing also led to the demise of the RNA world; RNA lost its exclusive right as the crucial biological polymer. However, RNA is not out. Molecular biologists frequently use aptamers made from RNA by in vitro evolution to catalyze specific enzyme reactions. The process is much more rapid than conducting the same process with protein enzyme modification. Increasingly, RNA worlds are still discovered in the DNA/protein world. One new area is the increasing role of riboswitches in metabolic regulation, and the other is the long-known world of RNA viruses. We will shortly discuss both as they touch our survey of the quest for food.

Metabolic Control by Riboswitches

How were metabolic controls handled in the RNA world where the only agents around were ribonucleotides? This seems at first glance an impossible historical question about a period 3.8 billion years before our time. One might suspect that an answer was lost in the depth of time. However, remember that the book of life was likened to an overwritten parchment. If you read carefully between the lines or behind the lines, you can still find strong evidence for such ancient processes in living organisms. In fact, you do not have to search an answer in exotic organisms, which survived the times retreated in an impossible ecological niche. No, the answer came straight from two mainstream bacteria: *E. coli* and *Bacillus subtilis*, the best-investigated Gram-negative and Gram-positive organisms, respectively, of bacteriology. Key discoveries were made as recently as 2002.

Riboswitches in *Bacillus*

Riboflavin synthesis needed for the coenzymes FMN and FAD were intensively studied in *B. subtilis*, but the mechanisms of its regulation remained undefined. As in other biosynthetic pathways in bacteria, a repressor that binds the end product of the biosynthetic pathway was suspected and the activated repressor then binds to the operator, preventing wasteful synthesis of enzymes that are not needed. However, intensive searches did not identify this hypothetical repressor. Instead, an untranslated leader region was found upstream of the first gene of the operon. A closer inspection revealed an evolutionarily conserved sequence that could fold into a characteristic RNA structure called a riboflavin box (Mironov et al. 2002). A similar thiamin box was identified ahead of the thiamin synthesis operon (Winkler et al. 2002). These RNA structures can exist in two alternative states: One causes premature transcription termination, and the other prevents it. The trick is that FMN modulates transcriptional termination by binding to the nascent RNA. As FMN is a fluorescent compound, its binding to the RNA riboflavin box could be directly followed by the extinction of the fluorescence after RNA binding. RNase treatment led to reappearance of the fluorescence. The thiamine system was investigated in some detail in *E. coli*. Mutations in the stem-loop structure identified a thiamin pyrophosphate (TPP) binding site, which discriminated the substrate against closely related forms like thiamin phosphate by 1,000-fold. TPP binding changes the conformation of the leader RNA, which reduces the access of ribosomes to the mRNA. The translation of the enzymes is thus prevented when the end product thiamin is available. Some mRNAs still carry natural aptamer domains that bind specific metabolites directly to the RNA, leading to modifications of gene expression. The change in gene expression is mediated via an expression platform in some riboswitches. This riboswitch operates as an allosteric sensor of its target compound. Notably, the sensed molecules are TPP, coenzyme B12, and FMN, which emerged as biological cofactors during the RNA world. Apparently, these modern riboswitches in "modern" bacteria are molecular fossils from the RNA world.

The Guanosine Riboswitch

In the meantime, a couple of other riboswitches were identified in the biosynthethic pathways of *B. subtilis*, leading to adenine, guanosine, lysine, and S-adenosyl methionine. A striking discriminative power of RNA aptamers for guanosine and the structural basis for this specific binding were defined in the guanosine riboswitch. Apparently, guanosine binding activates a catalytical function in the riboswitch, leading to an increased spontaneous cleavage of the leader RNA and thereby a decreased protein expression (Mandal et al. 2003). A riboswitch formed by the combination of a metabolite binding domain and a self-cleavage domain was clearly demonstrated in the RNA sequence upstream of the *glmS* gene of *B. subtilis*. This enzyme uses fructose 6-phosphate and glutamine to synthesize glucosamine-6-phosphate, an initiating step in cell wall biosynthesis. In the absence of glucosamine-6-phosphate, the half-life of the mRNA is 4 h. The saturation of the ribozyme with the ligand glucosamine-6-phosphate apparently activates the self-cleavage capacity of the ribozyme, and the half-life of the mRNA falls to less than 15 s (Winkler et al. 2004). These examples show clearly that quite sophisticated metabolic control mechanisms could be built purely with RNA elements. Until quite recently biologists thought that most regulatory functions in modern organisms have been taken over by protein–DNA interactions.

The presented data demonstrate that some of the oldest regulatory systems developed in the evolution of life have still survived until our days. At least 2% of the *B. subtilis* genes are still regulated by riboswitches—they are apparently efficient enough to survive in the presence of protein systems controlling gene expression. *When the progress in science is mentioned, it is frequently said that we are standing on the shoulders of giants. As biological organisms, we should acknowledge that we are also standing on the shoulders of our ancestors. In fact, as demonstrated by the universal tree of life depicted with the sequences from ribosomal RNA, all extant organisms on Earth can trace their origin to the same ancestor, LUCA. Really all? Are there no traces of a biological world before LUCA? I will summarize the arguments in the next two sections.*

Let Others do the Job: Viral Relics of the RNA Worlds

Viruses

If translation is such an energy-demanding step, it would only be logical to outsource this activity. The best would be receiving this service without paying for it. In fact, this strategy is extremely common in biology: All viruses do it. Biologists still do not know whether to regard viruses as living things. Actually, virus means poison in Latin, and this is quite a good definition of viruses because they are a nuisance to any form of cellular life on Earth. Some biologists avoided the contentious issue of whether viruses are living, and they defined viruses operationally as "obligate parasites of the cell's translational system." This is a very pertinent definition; any cellular life-form on Earth, however reduced it might have become, still maintains its own translational machinery, including degenerate bacteria like the subcellular organelles mitochondria and chloroplast, which became captured slaves of the eukaryotic cell quite early in the history of life. Viruses are known to have genomes that exceed 1 Mb in size and thus are larger than the genomes from intracellular parasitic bacteria, but no virus that encodes ribosomal genes has ever been found. *This possession and nonpossession of ribosomes is such a watershed in biology that tempts one to speculate that viruses are heirs of life-forms that evolved before the invention of ribosomes.*

Messengers of the RNA World?

Can such a hypothesis of the preribosomal origin of viruses be defended? In alternative views, viruses are escaped genetic elements of cells that have attained some autonomy to the detriment of the cell from which they originated. Virologists believe that RNA viruses are geologically very young and very dynamic. The mutation rate of RNA viruses is so high that it is difficult to imagine how viruses could be living fossils from a geological period before cellular life-forms were invented. Nevertheless, a handful of viruses look pretty much as living fossils of this time period.

Viroids

A group of plant pathogens are called viroids; they cause striking, although not well-known, diseases. One is cadang-cadang, the ominous killer of more than 30 million coconut palms in the Philippines. The disease started in the 1930s; the palms lost their leaves and the barren trunks remained as a devastated scene. Despite this large impact, the etiological agent is extremely simple: It is a 246- or 247-nt-long RNA. This RNA does not code for a protein. We are so used to proteins in our current biological world that even biologists asked how such a short noncoding RNA could have such an effect. The situation is even more perplexing: Overall we know about 50 such viroids, without exception plant pathogens. None is larger than 400 nt of RNA. Normally, I would have said "contains an RNA genome of larger than 400 nt," but this is semantically wrong. The RNA is not the genome of a pathogen—the RNA *is* the pathogen and the entire infectious agent. When looking into this outlandish RNA, we get a glimpse into an RNA world justifying the suspicion that we deal here with living fossils from the RNA world. The viroids exist in two major forms: the majority belongs to the Pospiviroidae. The name is a contraction of the strain type and stands for *po*tato *spi*ndle tuber *viroid*. Others are called apple scar skin or citrus bent leaf or grapevine yellow speckle. The striking observation is now that these small agents not only are infectious but also cause distinctive pathologies. How can this be possible if you have less than 400 nt in your tool box? All known viroids are single-stranded circular RNAs. However, this RNA assumes a highly base-paired rod-like conformation. These rods can be further divided into different regions: two terminal regions flank a pathogenicity, a central conserved region, and a variable region. The pathogenicity region is about 30 bp long. A second group of viroids, called Avsunviroidae for *Av*ocado *sun*blotch *viroids*, has a somewhat more complicated RNA, not in size but in secondary structure. Their RNAs adopt a branched conformation at the center of the molecule, where Pospiviroidae have the central conserved region. Viroid replication occurs through RNA-based rolling circle mechanisms in which the infecting monomeric (+) circular RNA is transcribed within the infected plant cell by a cellular RNA polymerase into head-to-tail (−) multimers that serve as templates for a second RNA–RNA transcription step. The resulting head-to-tail (+) multimers are cleaved into unit-length strands and subsequently ligated into the final progeny monomeric (+) circular RNA by an RNase and an RNA ligase from the cell. *Arabidopsis* has the enzymatic machinery for replicating viroids of the Pospiviroidae family, but it lacks the elements that mediate the spread of the infectious agent along the plant (Daros and Flores 2004). One could therefore dub viroids as "obligating parasites of the cell's transcriptional machinery."

The Hammerhead Ribozyme in Viroids

However, viroids are not devoid of all enzymatic activity. Avsunviroidae, which replicate in the chloroplast and not in the nucleus as the Pospiviroidae, do not code for a protein but contain a hammerhead ribozyme that cleaves a specific nucleotide position in the multimeric viroid RNA. Autocatalytic ligation of viroid RNA was also proposed for one member since atypical 2′-5′ phosphodiester bonds were observed (Cote et al. 2001). The hammerhead ribozyme leads us directly back into the RNA world (Doudna and Cech 2002). It consists of three short helices connected at a conserved sequence junction where the handle touches the hammer. This is also the place of its own site-specific autocatalytic cleavage. With only 40 nt, the hammerhead is the smallest ribozyme and was also the first to be solved in 3-D structure—it resembles more a broad Y structure than a hammer (Pley et al. 1994). Since then, additional crystal structures have provided snapshots of the RNA at several steps along the catalytic reaction pathway (Scott et al. 1996). A likely reaction mechanism could be deduced: The 2′-ribose hydroxyl adjacent to the scissile phosphate linking the two nucleotides is activated by abstraction of its proton. The resulting nucleophile attacks the phosphorus atom, which bridges the 2′-3′hydroxyls of the top ribose and the 3′-5′

hydroxyls of the adjacent riboses in the intermediate state. Then a proton is given to the 5' hydroxyl to stabilize the developing negative charge on the leaving group oxygen of the bottom nucleotide. Whether the viroids are really direct descendants of the RNA world—and thus living fossils—can be questioned on the basis of in vitro evolution experiments. The hammerhead ribozyme is the most efficient self-cleaving sequence that can be isolated from randomized pools of RNA (Salehi-Ashtiani and Szostak 2001). Hammerhead RNAs might have arisen multiple times during the evolution of functional RNA molecules.

Pathogenesis of Viroids

Recently, an original model was proposed to explain the pathogenic potential of viroids (Wang 2004). The authors made two basic experimental observations. Symptoms of viroids could be duplicated when just a hairpin RNA from a Pospiviroid was genetically expressed in a plant. Symptoms could be greatly reduced when the plants expressed an RNA silencing suppressor. Gene silencing is a conserved biological response to double-stranded RNA (dsRNA). As dsRNA is a replication intermediate of many pathogens, silencing mediates resistance to pathogenic nucleic acids. RNA interference, as it is also called, was first discovered in the nematode worm, where introduction of double-stranded RNA by injection or by feeding bacteria expressing this dsRNA (Timmons and Fire 1998) resulted in sequence-specific gene silencing (Fire et al. 1998). Viroid replication led to dsRNA replication intermediates. The cellular protein complex Dicer breaks this dsRNA down to about 22-bp fragments (Hannon 2002). These fragments are taken up by the RNA-induced silencing complex (RISC). RISC unwinds the dsRNA and uses the now-single-stranded RNA to guide the complex to a target mRNA substrate. If the viroid RNA contains exactly 22-nt-long copies of plant gene mRNA then the RISC would bind this plant mRNA and degrades it. The lack of gene expression then leads to symptoms. In this way, small noncoding RNA could easily lead to a variety of complex symptoms in infected plants.

The Smallest Human Virus

Hepatitis delta virus (HDV) represents the next step in the complexity ladder of viruses. With a 1.7-kb-long single-stranded RNA molecule, it is the smallest human virus. As the Greek letters of the hepatitis viruses denote the order of their discovery, it was the fourth ($\delta = D$) and thus relatively recent addition to the human hepatitis viruses. For a human virus, it has an unusual character. Viroid RNA can be placed with satellite RNA on the same phlyogenetic tree, which also included HDV (Elena et al. 1991). Satellites are a heterologous collection of subviral agents, which comprise nucleic acid molecules that depend for their replication on a helper virus. HDV's helper virus is hepatitis B virus, HBV. HDV thus exploits another virus to get its genome packaged and to reach the human liver cell for transcription and protein synthesis. Currently, HDV is the only viroid-like agent outside of the plant kingdom. It earns its name "virus" by the fact that it encodes a protein. In fact, the protein coding sequence is found on the copy of the infecting RNA, the antigenome, produced by the cellular RNA polymerase II in the liver cell. HDV is thus a negative strand RNA virus. The liver cell contains about 300,000 copies of the genome, 50,000 copies of the antigenome, and 600 mRNAs from the antigenome. From the single gene HDV encodes two proteins: The first is the δAb-S, a 200-aa protein consisting of a coiled coil dimerization domain, a nuclear localization signal, and RNA binding motifs. The HDV antigenome is edited at position 1012 by a cellular dsRNA-activated adenosine deaminase, which replaces an adenosine with a guanosine (Polson et al. 1996). This posttranscriptional editing eliminates a stop codon and allows the translation of a longer δ Ab-L protein (the second protein). Despite its only 19-aa C-terminal extension, the two proteins have quite different functions. δAb-S is involved in genome replication, while δAb-L leads the HDV rodlike RNA to the cytoplasm for assembly of the virus particle. The outer protein shell of the HDV particle is made exclusively from the surface antigen of the helper hepatitis B virus. It is thus not surprising that HDV has the same tropism as HBV—the liver is the exclusive target—and this is biologically meaningful since HDV can only replicate in patients that are chronically infected with HBV. Clinically, HDV infections are more severe than other human hepatitis virus infections. Worldwide about 300 million people are chronically infected with HBV and more than 20% of them are coinfected by HDV in parts of South America and Asia. Feeding on a latecomer of evolution, namely humans, can thus be a quite successful strategy. However, with its roots, HDV goes still into the RNA world. The self-cleaving ribozyme of HDV assists in genome replication. A 72-nucleotide segment was crystallized, and its structure revealed a compact core consisting of five helical segments connected as an intricate nested knot (Ferre-D'Amare et al. 1998). The self-scission reactions are buried deep within an active-site cleft produced by juxtaposition of the helices. A cytidine is positioned to activate the 2′ hydroxyl of the ribose for nucleophile attack on the phosphorus. The cytidine apparently acts as a general base in the reaction mechanism. The analysis of 10 crystal structures of the HDV ribozyme revealed that a conformational switch controls the catalysis; thus, the ribozyme remarkably resembles protein enzymes like ribonucleases for which conformational dynamics are an integral part of their biological activity (Ke et al. 2004).

Perhaps some elements of the evolution of biological complexity are still found in contemporary viruses, either as living fossils or as modern reinventions of ancient motifs.

Messengers from a Precellular DNA World?

The Outer Reaches of Viral Complexity

The greater stability of DNA allowed the evolution of substantially larger genomes. The largest known RNA genome is only 62-kb long. Notably it is found in a reovirus, which possesses dsRNA, and mimics thus a basic property of the double-stranded DNA (dsDNA), the genome material of all cellular life. When it comes to dsDNA, viral genomes are not necessarily smaller than the genomes of cellular life. The recently defined Mimivirus, infecting an amoeba, showed a 1.2-Mb-large genome (La Scola et al. 2003). There is thus no genome size gap that separates the worlds of viruses and cells. The minimal genetic need of a viable cell has been estimated for *B. subtilis* to be 562 kb, close to the actual size of the *Mycoplasma genitalium* genome with its 580 kb. Transposon mutagenesis studies suggested that up to 350 of the 480 protein-coding genes from *M. genitalium* are essential under laboratory growth conditions (Hutchison et al. 1999).

The major difference which distinguishes the world of viruses from the world of cells is in nutrition: All modern viruses need cells for their propagation, no virus can develop in a growth broth, and they all depend on ribosomes for protein synthesis and lack an energy metabolism of their own. One should observe some caution with categorical statements in biology: They are correct until they are disproven. The recent report of an extracellular development of two long tails in a lemon-shaped virus infecting a thermophilic archaeon shows for the first time a viral morphogenetic activity outside of the host cell (Häring et al. 2005).

Mimivirus

The surprises are still limited: The giant genome of the Mimivirus did not encode ribosomal genes (Raoult et al. 2004). Like all viruses, it is a molecular parasite that depends on a cell for its protein synthesis capacity. Mimivirus has the coding capacity for complementing a few metabolic pathways. It takes part in glutamine metabolism, and it encodes a number of glycosyltransferases for the biosynthesis of di-, oligo-, and polysaccharides, three lipid-manipulating enzymes, and a few enzymes that synthesize nucleotide triphosphates. However, Mimivirus does not actively take part in any energy generation from food substrates. However, in contrast to all the previously described viral genomes, Mimivirus takes the pain to encode three proteins involved in protein synthesis (three tRNA synthetases), two proteins implicated in transcription (two subunits of RNA polymerase II), and two enzymes involved in DNA synthesis. All seven proteins belong to a set of proteins, which are universally found in all forms of cellular life. This gave scientists the first chance to probe the position of a virus on the universal tree of life by using the seven concatenated universally shared proteins from Mimivirus. On this tree, Mimivirus branches out near the origin of the Eukaryota domain but is equidistant from the four main eukaryotic kingdoms: Protista, Animalia, Plantae, and Fungi. The affinity to prokaryotic sequences is more distant. This places Mimivirus somewhere at the root of the Eukaryota. This conclusion is still somewhat vague, and the authors propose that further large virus genomes should be sequenced to get more clarity.

Viruses and the Universal Tree

In fact, none of the other viruses from the database ever found a place on the universal tree. There is one straightforward interpretation for this observation. It sees viruses as members of a biological world that does not belong to the tradition of all extant cellular life. Viruses are simply not derived from the last universal common ancestor of cellular life. This evolutionary line is defined by the possession of ribosomes. Are viruses derived from life-forms that originated before the creation of LUCA? In this interpretation, viruses never possessed a protein translation machinery. They are derived from a variety of "genomes" that predated LUCA. Perhaps due to the evolution of a protein synthesis apparatus, LUCA developed a superior competitiveness that threatened to push aside all previous life designs. The previous life-forms had missed the train because they could not simply acquire the protein synthesis apparatus. In fact ribosomal genes are generally very resistant to lateral gene transfer, hybrid ribosomes have been constructed in the laboratory, but the organisms lost fitness. Only recently was a case for an rRNA gene transfer between distantly related bacteria documented (Miller, Augustine et al. 2005).

One might speculate that these previous life-forms had only one option when they wanted to compete with LUCA and its progenies. They had to become parasites of LUCA. Only by this way, they could share the new invention and perpetuate their genomes through time. This is of course pure speculation at the moment. However, enough viral genome sequences are out to allow constraining at least the wildest speculations.

Viral Genomics and Phage Lambda

First, viruses are, with respect to the DNA sequence space, a separate world to that of cellular life. It will thus not be very obvious to derive viruses from a patchwork of sequences that escaped from the cellular DNA sequence space. Second, the viral sequence space has perhaps been dramatically underestimated and might equal or even surpass that of cellular life. Third, viruses do not have a common denominator. In fact, viruses are surely polyphyletic—no common origin can be postulated for them as for the extant forms of cellular life. There are at least a dozen of fundamentally different genomic lineages in viruses. One line and probably one of the oldest is represented by the tailed phages (Figure 4.1). One of the best known is phage lambda. Lambda-like phages do not evolve along linear lines of descent, but by a different, modular mode of evolution. The λ genome can conceptually be divided into a dozen or so modules, i.e., gene sets of related function (head genes, tail genes, DNA replication genes, lysis genes, etc.). Each module is represented by a number of alleles (i.e., sequence-unrelated genes that fulfill the same genetic function, e.g., head capsid morphogenesis). The different modules are free to recombine to give new phages. A striking observation was that this gene map is conserved between phages that have lost DNA sequence or protein sequence similarity. The evolutionary reach of this conserved gene map goes very far since it is observed in Gram-negative (*E. coli*) and Gram-positive bacteria of high (*Streptomyces*) and low GC-content (*Bacillus*) and even one branch of Archaea (*Euryarchaeota*). The generalized gene map of lambdoid phages has a further unexplained but thought-provoking property that was already noted more than 30 years ago by R. Hendrix and S. Casjens. *The order of the genes in the structural modules corresponds to their location on the*

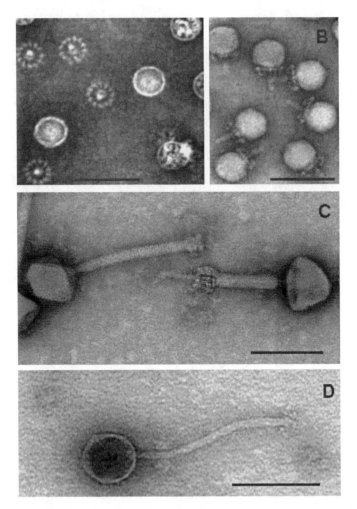

FIGURE 4.1. Tailed phages represent the vast majority of the viruses infecting bacteria. The figure represents the three major forms of tailed phages with examples from low GC-content Gram-positive bacteria. *Top*: Podoviridae are phages with short tails. At the *left*, phages are seen from the bottom (the axis of the wheel is the tail and the rim the baseplate) or the top (the circular structures). At *top right*, the phages are seen from the side. *Middle*: Myoviridae are phages with long contractile tails; at the *left* is a phage with an extended tail, at the *right*, one with a contracted tail. *Bottom*: Siphoviridae are phages with long, but noncontractile tails. Morphologically similar phages are found in many classes of bacteria. Some biologists believe that tailed phages are the oldest forms of viruses.

phage particle in the head-to-tail orientation, recalling the conservation of hox gene order with the site of action along the body axis of animals from the head to the tail. Likewise in phages, first comes the head genes, followed by the neck genes, then the tail genes, the baseplate, and then the tail fiber genes.

These observations put lambdoid phages in an interesting context: They are perceived as a superorganism that can draw modules from a large gene pool shared by many similarly organized phages (Hendrix et al. 1999).

The Progenote

This property of lambda sounds very similar to the pre-LUCA life as conceived by Carl Woese (1998) with his progenote concept. The progenote has already a genome but not yet an individualized metabolism. It represents somewhat a community concept, the addition of many different genomes associated with many different metabolisms (like in modern bacterial consortia). These traits are frequently exchanged precluding the definition of an individual entity on which Darwinian selection could easily work. Selection pushed only the entire progenote communities. C. Woese is explicit in his progenote model; a protein synthesis machinery had not yet evolved. Translation became fixed in what he called a genote. He distinguished a hot phase of extensive lateral gene transfer in the progenote world from a cold phase of evolution in the genote world, where lateral gene transfer definitively cooled down. *Aren't lambdoid phages looking somewhat like a progenote, while LUCA is a genote? At least this progenote character of phages explains why viruses generally lack ribosomal genes and a fixed metabolism, irrespective of their genome size. It raises the possibility that they evolved from life-forms that preceded LUCA, the genote. The difficulty in the transfer of ribosomal genes between cells explains why viruses could not participate in the protein synthesis machinery otherwise than becoming a parasite.*

Structure Conservation in Viruses Speaks for Antiquity

When the *Sulfolobus solfatericus* virus STIV was investigated by cryoelectron microscopy, it revealed at a 27-Å resolution a virion morphology that was reminiscent of known viruses, despite the lack of any sequence similarity of its proteins (Rice et al. 2004). Apparently the structure of viruses still maintained a certain degree of conservation when all sequence similarity was erased by long evolutionary periods of separation. For example, the same double β-barrel "jelly roll" motif was previously identified in the *E. coli* bacteriophage PRD1, in the mammalian adenovirus (Benson et al. 1999), and more recently in PBCV-1—a virus infecting the green algae *Chlorella* (Nandhagopal et al. 2002). Several virologists concluded from these observations that the structural similarities imply a common ancestry, revealing a viral lineage spanning all three domains of life (Benson et al. 2004). The conserved structure corresponds to a conserved "self" of the virions in contrast to the "nonself" parts of the genome, which represents the adaptations of each virus to the lifestyle imposed by the host and its ecological particularities. Similar claims for broad range relationships were made for the virion structure of tailed phages from Bacteria and Archaea, herpesvirus infecting vertebrates (Newcomb et al. 2001) or reovirus, and phage φ6 (Butcher et al. 1997). The simplest interpretation of these observations is that there were already viruses resembling modern adenoviruses, herpesviruses, and reoviruses before the divergence of cellular life into the contemporary domains of Bacteria, Archaea, and Eukarya, about three billion years ago (Hendrix 2004).

The Importance of Being Lipid Enveloped

The Darwinian Threshold

Biologists tend to fix a time mark of 3.6 billion years BP for the start of the DNA/protein world. In C. Woese's thinking, the roots of the universal tree still predate the invention of the cell. We have here a communal metabolism where the community knows a lot of metabolism, but the individual did not yet exist. In his words, the beginning of the universal tree is still in the "chaos of a universal gene-exchange pool" (Woese 2002). He does not believe the dictum that the complex emerges from the simple applies to the early phases of life. In his view, the complex metabolism of the community was in an event, which he calls a "Darwinian threshold," cut down into smaller units having only a simple, abridged or even partial metabolic capacity of the whole. *Different chapters of the early textbook of biochemistry were distributed among the early cells, and they were asked to succeed on their own with their limited capacities and competing with each other instead of the communism-like cooperation that prevailed before. This has a bit the taste of a biological theory of the original sin, and selection from that time onward had a wide field of activity to choose the fittest cell. There is, however, good evidence that cells came after the first branching of the rRNA phylogenetic tree. The evidence refers to the chemical nature of the membrane lipids, the material, which is supposed to achieve the compartmentalization of the early cells. To end the phase of biological communisms and to start the hard world of biological competitiveness, it became important to distinguish "mine and yours." You need a cell membrane that separates your chemical space with all its resources you obtained in your quest for food, the metabolites that you formed, the proteins and their encoding genes from the outside world. What was speculated on the origin of the cell?*

The Age of Confinement

In Oparin's hypothesis on the origin of life, the formation of coacervates—a type of self-organizing protocell—played a prominent role. M. Eigen, in contrast, believed that organization into cells was postponed as long as possible because constructing boundaries like lipid membranes posed more problems than it solved. Liposomes are essentially impermeable, not only to macromolecules but also to most small hydrophilic molecules and ions. However, confined systems are necessary to start the Darwinian competition between macromolecular assemblies, which leads to steady improvements of the design of early forms of life. G. Blobel designed a way around this dilemma by postulating empty vesicles that absorbed macromolecules to their surface, early enzymes, and also macromolecular complexes like the precursors of ribosomes and chromosomes. If these vesicles started folding and nearly closing into small droplets surrounded by the double membrane of the original vesicle, something like a Gram-negative bacterium with its cytoplasmic and periplasmic membrane could have resulted. To address the problem of transport of molecules across the membrane, de Duve proposed that hydrophobic peptides instead of phospholipids confined the earliest protocells. Wolfram Zillig proposed that the bacterial and the archaeal lipid membrane emerged independently by replacing such an ancestral nonlipid membrane made of proteins. In fact extant organisms still sport something like this: The membrane of *Halobacterium* consists of 75% protein, namely the tightly packed bacteriorhodopsin, which works as a light-driven proton pump and only 25% lipid. This system might come close to ancestral membrane systems.

The Tale of Two Lipids

The independent origin of the membrane lipids can still be read from modern biochemistry. Membranes in all three domains of life are basically built on triose phosphate derivatized with long hydrocarbon chains at two positions. However, here stops the similarity also. In Archaea it is the C1 position of glycerol, which carries the phosphate group, while in Bacteria and Eukarya the phosphate is attached to the C3 group. In Archaea a hydrocarbon chain is linked via an ether linkage to the C2 and C3 position of glycerol. In Bacteria and Eukarya, the hydrocarbon chain is linked via an ester linkage to the C1 and C2 position of glycerol. Also the chemical nature of the hydrocarbon chain differs between the domains of life: In Bacteria and Eukarya, it is a straight long-chain fatty acyl group, while in Archaea it is highly methyl-branched, mostly saturated isopranyl chains (isoprenyl without double bonds). When inspecting the membrane lipids from Archaea, one gets the impression that they were still trying out different solutions when confining their Archaean protocells. Some have an internal hydroxyl group at the C3-linked hydrocarbon residue, whereas others have a covalent bond between the ends of the hydrocarbon chains, forming a large cyclic molecule. Still other fuse two standard Archaean phospholipids by head-to-head condensation of the C20 hydrocarbon chains leading to a giant 84 C atoms containing heterocyclic compound with four O atoms. In bacteria, much less variability of the basic chemical design of the membrane lipids is found, and only in the lowest branches of Bacteria, variants are found to contain long-chain-C30 dicarboxylic acids instead of the common fatty acyl groups. On the basis of this chemical evidence, the conclusion that the lipid membranes of Archaea and Bacteria evolved independently seems inescapable. Some microbiologists have pointed out that ether lipids might have advantages over fatty ester lipids especially at high temperatures, which is interesting with regard to the hypothetical hyperthermophilic origin of the cell. However, this high temperature start is not shared by all biologists (Wuarin and Nurse 1996), and fatty acid ester lipids can easily be adapted to diverse and changeable environments by modifying the chemical identity of the fatty acids when playing with C=C double bonds and *cis–trans* isomerizations. These tales of two lipids (Wächtershäuser 2003) are difficult to reconcile with other observations. The basic paradox being that Bacteria and Eukarya share an essentially identical membrane lipid chemistry and a closer metabolism, while Archaea and Eukarya share much more related information processing machineries. The easiest explanation is that the lipid membranes and thus the cells, in the modern sense, were invented after the first differentiation of the protein translation apparatus and also the transcription and DNA replication apparatus.

Sorting Out a Metabolism

How should we imagine the precells? Otto Kandler was quite explicit in his prediction. He defined precells as "metabolizing self-producing entities exhibiting most of the basic properties of the cell but unable to limit the frequent mutational exchange of genetic information." Even if they were somehow spatially confined, their promiscuous exchange of genetic material prevented the development of individuality. Precells were genetically still one coherent population that was "multiphenotypical" and distributed over a variety of habitats, each harboring a different subpopulation. Some populations may have been autotrophic and others heterotrophic, some anaerobic and others microaerophilic, and some H_2 producers and others H_2 consumers. Perhaps the relatively late invention of membrane lipids put a lid on this boiling incubation vessel of gene exchange and made physically an end to all easy exchange of the highly charged nucleic acids informational molecules across the hydrophobic double-layered lipid membrane. In the words of C. Woese, life "cooled down" from the roaring twenties into the better-behaved world of Darwinian selection. *Perhaps the invention of the membrane lipids was the physical correlate of his Darwinian threshold in the evolution of life. If everything was actually in place, it could also explain why life needed so short time to develop from the invention of the DNA/protein world into cells within a mere 100 million years.*

Physical Models for Protocells

Other recent proposals for the origin of the cell include cell-like holes in mineral surfaces resembling weathered feldspar surfaces. Orthogonal honeycomb networks of $0.5\,\mu m$ width might have served as a mold for the first cells. The mineral surface would provide the catalytic surface for the start of biochemical evolution; phosphorus and transition metals important to energize life would be readily available. The mineral would protect against UV radiation, and the construction of a lipid lid would separate the contents of the tubes in the mineral from the outside world. The tubes of weathered feldspar are still inhabited by extant soil bacteria (Parsons et al. 1998). Other proposals are bolder, but chemically not less plausible. Aerosols are formed by wind-driven wave action followed by bubble bursting at the ocean surface. A surfactant partial monolayer covers the sea surface; evaporation of water in the airborne particle creates a complete monolayer. Organic molecules comprise about 50% of the mass of the upper tropospheric aerosol particle with much higher concentrations of carbon and nitrogen and trace elements than in the seawater. The mobility of cell-size aerosol particles through a wide range of temperature and radiation fields makes them ideal chemical reactors. Upon their downward movement and reentry into the ocean, they could acquire a bilayer structure when passing through the sea surface surfactant layer (Dobson et al. 2000). Life in the gaseous phase is not as farfetched as it might appear on the first look. *Stramenopila* protists (diatoms and consorts) have been found living in clouds, in the atmosphere. Meteorologists showed that cellular and protein particles injected directly from the biosphere constitute a major portion of atmospheric aerosols, which need to be included into improved climate models (Jaenicke 2005). These observations challenge our concept to see life only in the context of liquid water; we might have to include gas planets with an extended atmosphere into scenarios for extraterrestrial life.

The Beginning of Competition

Jan Szostak and his group have tried to address the precell problem from first physicochemical principles. The clay montmorillonite is a prominent ingredient of Cairns' hypothesis on the origin of life because it can catalyze the polymerization of RNA from activated ribonucleotides. It also accelerated the spontaneous conversion of fatty acid micelles into vesicles. These vesicles can divide without dilution of their content when extruded through a small pore (Hanczyc et al. 2003). In a recent contribution, his work showed competitive behavior between model protocells. RNA complexed with counterions gets encapsulated into vesicles and exerts an osmotic pressure on the vesicle membrane. This pressure drives the uptake of fatty acids from adjacent vesicles that were relaxed due to the absence of osmotically active RNA. The RNA-containing vesicles grow in size, while the empty vesicles shrink (Chen et al. 2004). A primitive Darwinian competition for "food" and structure molecules has started.

Early Eaters

What is at the Root?

Aquifex and the Tree

The ribosomal RNA gene sequence analysis attributed the deepest branch position of the Bacteria domain to hyperthermophiles such as *Aquifex aeolicus* suggesting that such organisms were the earliest living bacteria. This led to the idea that life started in a hot environment. The hypothesis that the rRNA tree truly reflects the evolutionary history of living bacteria was shaken when the extent of horizontal gene transfer (HGT) in bacteria became apparent. If genome fragments are displaced easily and to a large extent between different bacterial species, then a phylogenetic positioning based on a single marker gene could lead to erroneous conclusions. To account for the ambiguity created by various contributions of HGT, evolutionary biologists used concatenated sets of protein sequences conserved across a wide range of prokaryotes to create a type of average tree rather than one based only on the 16S rRNA gene. Even when 32 protein sequences are used for the timing of the tree, *Aquifex* remains the deepest branch in bacteria. Whether *Aquifex* really resembles a basic bacterial lifestyle or whether its position reflects a tree-constructing artifact has not yet been settled. Nowadays *Aquifex* sp. occupy very rare niches, which may only be a distant mirror of habitats it populated billions of years ago. As the genome of one of its members has been sequenced, we can get a tentative idea about how a self-sufficient bacterial life could have looked like before the invention of photosynthesis. *Aquifex aeolicus* is thermophilic. It lives in hot springs of the Yellowstone. With a growth maximum at 95 °C, it is at the temperature limit of bacterial life.

Aquifex Metabolism

Karl Stetter succeeded in the cultivation of *A. aeolicus* using only inorganic components and a reducing atmosphere of $H_2/CO_2/O_2$ in a volume ratio of about 80:20:1. It is an obligate chemolithoautotroph; obligate because it does not grow on organic substrates such as sugars and amino acids, but uses inorganic compounds such as CO_2, and H_2 to gain energy and to make biomass. Its genome is a mere 1.5 Mb in size, which is relatively small for bacterial standards (Deckert et al. 1998).

Apparently it does not need a large genome to be autotroph. You do what you have to do to power and to feed your metabolism, and you can live without genes involved in metabolic regulation, transport of substrates into the cell, and various other degradative pathways. The other side of the coin is that you are not terribly flexible. As an autotroph, *Aquifex* obtains all necessary carbon by fixing CO_2 from the environment. The reductive TCA cycle fixes two molecules of CO_2 to form acetyl-CoA. The TCA cycle also provides the precursors for pentose and hexose monosaccharides biosynthesis, probably via gluconeogenesis using pyruvate. Glycolysis, pentose phosphate pathway, and glycogen synthesis and catabolism are present. *Aquifex* can conduct oxygen respiration with enzymes that can use oxygen concentrations as low as 8 ppm.

Knallgas Metabolism

Aquifex is motile by monopolar peritrichous flagella and uses an as-yet undefined chemotaxis system, which may respond to different gases. *Aquifex* gains energy by hydrogen oxidation: $2H_2 + O_2 \rightarrow 2H_2O$. This is chemically a very vigorous reaction, and the bacteria are therefore called "Knallgas" bacteria from the German word for this reaction. Of course this reaction does not occur as such but is carefully hidden in complex enzyme reactions. We do not know the enzymes for H_2 oxidation in *Aquifex* very well. They have been studied in some detail in a different bacterium, *Alcaligenes eutrophus*. It contains two hydrogenases. A membrane-bound nickel–iron hydrogenase takes up molecular hydrogen, which it decomposes into $2H^+$ and $2e^-$. The complex transfers the electrons to a cytochrome *b* and then to an electron transport chain ending with a cytochrome *c* oxidase, which reduces oxygen to water. Now where does the oxygen come from in an anoxic world? Perhaps *Aquifex* has lived in the oxygen oases postulated for the reduced Archaean oceans. The bacterium does not need a lot of oxygen to grow: already 8 ppm is sufficient, suggesting that it was adapted to conditions prevailing before the rise of the atmospheric oxygen levels by oxygenic photosynthesis.

Aquifex contains in its genome two further oxidoreductases: one of which plays a role in nitrate reduction, and the other might function in sulfur respiration. This suggests that the ancestor of *Aquifex* could have used nitrate or sulfate as alternatives to oxygen.

Thermotoga

While *Aquifex* thus potentially fulfills the properties expected for an autotrophic lifestyle before the advent of photosynthesis, this may not apply to all thermophilic organisms near the base of the phylogenetic tree. For example *Thermotoga* has an optimal growth temperature of 80 °C and was isolated from a geothermally heated marine sediment. Its physiology and genome are that of a typical heterotroph: It metabolizes many simple and complex sugars, including glucose, sucrose, starch, but remarkably also cellulose without relying on complicated cellulose degrading enzyme complexes. It dedicates an elevated percentage of its genome on substrate transport and sugar metabolism (Nelson et al. 1999). Nowadays *Thermotoga* lives in environments rich in organic material that was created by photosynthesis. This suggests that the metabolism of *Thermotoga* is evolutionarily more recent than its branching position on the bacterial tree suggests. However, *Thermotoga* is not only capable of fermentation but can also reduce Fe(III) to Fe(II) in the presence of H_2 as an electron donor (Vargas et al. 1998). This seemingly trivial physiological experiment has substantial evolutionary implications because it suggests that perhaps the more ancient lifestyle of *Thermotoga* was hydrogen oxidation with Fe(III) and the fermentation metabolism is a more recently acquired trait. Geochemical evidence pointed out that Fe(III) probably was a very early terminal electron acceptor. In conclusion many arguments speak in favor of the last common ancestor being a metabolically sophisticated respiratory organism. More primitive microorganisms might have even started earlier with the capacity to transfer electrons from H_2 to extracellular Fe(III) as the widespread Fe(III) reduction capacity in hyperthermophilic organisms seems to indicate. Perhaps biology started with an iron age and remained in it for a while.

Hydrogen and Bioenergetics

Chemosynthesis

Most of the Earth's biomass is considered to be the product of photosynthesis. However, here we have a problem for the early eaters. Photosynthetic bacteria are not found at the root of the phylogenetic tree, neither fossil nor isotope evidence traces photosynthesis back to the earliest eaters, and temperatures above 70 °C are considered to be incompatible with photosynthesis. All these arguments exclude photosynthesis as the energy-generating process at the hot root of the tree of life. Primary productivity must have derived from chemosynthesis based on the oxidation of reduced inorganic or organic sources. To avoid circular conclusions, one would look for organisms that do not depend on other organism's metabolism, i.e., autotrophs. If one looks for inspiration with respect to the energy basis of early cellular life, one might be well advised to investigate the bioenergetics of organisms, which like it hot in order to account for the thermophilic root argument. Geothermal ecosystems like the Yellowstone springs, which harbor a hyperthermophilic microbial community, might be a good starting point. However, the power basis of this ecosystem was not clear until very recently (Nealson 2005). To microbiologists, the smelly hydrogen sulfide emanating from these springs suggested a sulfur-driven ecosystem. Alternatives were methane, short chain hydrocarbons, or reduced metals such as Fe(II) and Mn(II). The problem was that all these resources were highly variable in the different springs. Only H_2 at concentrations appropriate for energy metabolism was ubiquitous (Spear et al. 2005). This chemical observation concurred with a biological observation: Community DNA analysis via PCR-amplified rRNA genes pointed to a dominance of *Aquificales* and *Hydrogenobacter* in these microbial communities—and these were just the abovementioned root organisms. Collectively, 90% of the sequences belonged to organisms that rely on H_2 as an energy source. Thermodynamic modeling showed that H_2 oxidation is the favored process under oxygen-limited conditions. This multidisciplinary approach showed that these boiling sulfurous ponds were in fact driven by hydrogen of geochemical origin. The source of this geochemical hydrogen is not well understood, most likely the formation of hydrogen-rich fluids is the result of reaction of water with ultramafic rocks at moderate temperatures and pressures (Sleep et al. 2004).

A Deep, Hot Biosphere

These new data excited microbiologists and earth scientists alike since they demonstrated that ecosystems that exist are entirely uncoupled from the energy of the Sun. These are geologically powered dark ecosystems as postulated by T. Gold (1992) in his theoretical article "The Deep, Hot Biosphere". The description of such ecosystems has far-reaching consequences. It means that all parts of our planet that remain in the physicochemical range of microbial life (e.g., $<110\,°C$ or for more daring characters $<150\,°C$) were and still are "infected" with life. Life is thus not limited to the surface of the planet, like continents or oceans. This means that the potential range of life would be substantially extended and potentially includes suitable subsurface ecosystems on planetary bodies outside of the Earth–Sun radius like Mars or Jovian satellites. Light is no longer the sole possible motor of life. The experiment-guided thinking on a prephotosynthetic Earth is now possible. The Yellowstone study was not the first claim for a lithoautotrophic microbial ecosystem based on H_2 as energy source. Deep basalt aquifers were reported to contain SLiME, an acronym for active, anaerobic subsurface lithoautotrophic microbial ecosystem

(Stevens and McKinley 1995). In most cultured samples from these aquifers, autotrophic microbes outnumbered the heterotrophic organisms, and they could grow with H_2 as the sole electron donor. The H_2 concentrations were in many deep aquifer samples in the 1 mM range and thus 100-fold higher than would be expected from microbial fermentation of organic matter. However, this interpretation was contested. Another group reported that hydrogen is not produced from basalt at an environmentally relevant, alkaline pH (Anderson et al. 1998). They claimed that the more likely energy source for microorganisms in the basalt aquifer is organic matter fermentation of dissolved organic carbon (DOC) found in groundwater. However, even this opposing group stated that they did not exclude that such subsurface communities can subsist from reduced gases emanating from deeper layers in the Earth. In fact they themselves reported such an ecosystem in Idaho as we will see at the end of the next section. Another group reported a hydrogen-driven microbial community near a hydrothermal vent in an oceanic ridge. This community produced methane and was dominated by hydrogen-utilizing Methanococcales as primary producers and *Thermococcus* as fermenters (Takai et al. 2004). As in the case of the Yellowstone study, their conclusions were backed by a multidisciplinary approach.

Methanogenesis

Methane and the Faint Young Sun

Methane is a source of perplexity for scientists as indicated in titles such as *Deciphering Methane's Fingerprints* (Weissert 2000) or *Resolving a Methane Mystery* (DeLong 2000). These editorials suggest a criminal case based on indirect evidence. This is as such nothing unusual in science because most of the scientific evidence is indirect and only linked to a coherent picture by logical conclusions that can be overturned by each new discovery. However, the case for methanogenesis is especially troublesome because huge amounts of methane are created, but they never reach the atmosphere. At the same time, methane is discussed as a greenhouse gas that rivals the role of CO_2 in our current atmosphere and even more in the atmosphere of the early Earth. To begin with the beginning, we have the case of the faint young Sun: For the first 3.5 billion years of Earth's history, the Sun burned only about 70–90% as bright as today. From the radiance budget, the Earth should have been entirely frozen, yet there is evidence for the persistence of liquid oceans through this time period and it is difficult to imagine how life could develop under an ice shield. The solution to this paradox is greenhouse gases. Their presence in the atmosphere retained much of the irradiance received from the Sun and prevented a cooling of our planet into a global snowball. This conclusion is so far undisputed; however, which gas actually played the decisive greenhouse role is a contentious issue. Until quite recently, methane was the favored agent after it had replaced the previously preferred CO_2 from this role in the scientific discussion. Now the pendulum seems to swing back to CO_2 after some inconsistencies with the CO_2 hypothesis have found a possible explanation (Ohmoto et al. 2004). In a comment on this latest turn of the greenhouse discussion, a geologist comments that "a universal theme in studies of the early Earth is that big stories are told with little data and lots of speculation" (Lyons 2004). The closer we get with the events to the present, the less speculative the stories become. Yet, they still remain rather indirect.

Methane Inclusions in the Pilbara Craton

Japanese geologists achieved a technical tour de force that pushed back the time horizon for the origin of methanogens by 700 millions years over previous estimates (Ueno et al. 2006). They extracted methane-containing fluids from 3.5 billion-year-old hydrothermal precipitates from the Pilbara Craton in Australia and analyzed the carbon isotope composition of the methane. The origin of the methane was distinguished by their analysis as microbial (emitted by metabolic activity from methanogens), thermogenic (generated by thermal decomposition of organic matter), or abiotic (produced by inorganic chemical reactions between carbon dioxide and molecular hydrogen). The primary fluid fill, which was entrapped during mineral growth, showed that the methane was significantly depleted for the heavier carbon isotope relative to the coexisting carbon dioxide, suggesting a biological origin. Also the lack of higher hydrocarbons in the sample argued against a thermogenic origin. Taken at face value, it suggests the presence of methanogenic microbes in a rock sample that dates just 300 million years after the estimated origin of life on our planet. This diagnosis could be backed by other observations. The Apex chert from this geological formation also contains [34]S-depleted pyrites, which were possibly produced by sulfate-reducing microbes. Furthermore the Apex chert is famous for its claim to the oldest microfossils described in J.W. Schopf's book *The Cradle of Life*. However, this claim for fame is not accepted by all geologists. In addition, methanogenesis occurs only in one branch of the Archaea, the Euryarchaeota and is thus not a primitive character of all Archaea. Nevertheless the peculiar chemistry of the cofactors used in methanogenesis speaks for a very old process.

Oceanic Anoxic Events

A more recent methane story is based on carbon isotope measurements in fossil wood, which suggested to the authors of this study (Hesselbo et al. 2000) that the so-called Early Toarcian oceanic anoxic event (183 Ma ago) was produced by voluminous and extremely rapid release of methane from gas hydrate contained in marine sediments. This event is characterized by high rates of organic carbon burial, high paleotemperatures, and significant mass extinction. The proposed scenario links increased CO_2 release to greenhouse warming, which warms in turn the deep water and causes a massive release of methane via destabilization of gas chimneys by a temperature rise of only 6°C (Pecher 2002). Methane reacts with oxygen leading to its disappearance in the ocean (hence the anoxia and mass extinction) and via the creation of further CO_2 ($CH_4 + 2O_2 \rightarrow CO_2 + 2H_2O$) to an amplification of the greenhouse effect and hence further temperature increase. There is even a speculation that the Permian/Triassic extinction event (250 Ma) was also caused by methane release, but this time perhaps by an extraterrestrial object hitting the Earth (Weissert 2000). But is there enough methane in subsurface reservoirs to affect the global climate? The answer is: yes—below a water column of 300 m, the high pressure and low temperature cause methane to form ice-like crystals of methane hydrate that can be localized by seismic imaging (Wood et al. 2002). Geologists calculated that about 10,000 giga-tons of carbon is amassed in these reservoirs, which equals the amount of reduced carbon found in all other fossil fuels combined. Methane of biological origin is most abundant in deep marine sediments but economically important accumulations of methane have also been demonstrated in shallow organic-rich shale from Michigan at a depth of less than 600 m (Martini et al. 1996). However, except for the abovementioned catastrophic events (but recall the recent concern on the permafrost region of Siberia responding to global warming), the deep sediment methane does not reach the atmosphere, the greenhouse-relevant amounts of methane come from sources linked to human activities such as rice cultivation, livestock, biomass burning, and landfills to quote the most important (Hogan et al. 1991). Why do the apparently huge amounts of methane produced in the subsurface not reach the atmosphere? Part of the answer is that methylotrophic bacteria, located at the borderline of oxic and anoxic zones of sediments and wet soils, oxidize CH_4 to CO_2. The reaction proceeds by four two-electron oxidation steps with methanol, formaldehyde, and formate as intermediates. Methane is chemically very inert, and the splitting of the stable C–H bond (dissociation energy 435 kJ/mol) needs molecular oxygen in the chemical attack. If oxygen is not present, methylotrophs cannot handle methane. However, when looking at depth profiles in marine sediments, methane is reoxidized well below the oxygenized sediment layer. The clarification of how this is achieved is one of the recent, big success stories of microbial ecology. However, to understand this fact, we need first to consider methane synthesis.

Cofactor Chemistry of Methanogenesis

Methanogenesis is restricted to Archaea, eubacteria have actually never learned this job in evolution. A number of particularities surround this metabolic pathway. Methanogens (organisms able to synthesize methane) are found only in the Euryarchaeota domain of Archaea, but in 17 different genera. The secluded character of this pathway is also documented by the biochemistry of this reaction sequence catalyzed by seven enzymes. I will describe one pathway in somewhat more detail. Reducing CO_2 to CH_4 does not seem a very complicated business under the condition that the reaction occurs in the absence of oxygen. Yet, this pathway uses eight different cofactors, and seven of these coenzymes are unique to Archaea. First, CO_2 is added to an amino group of a methanofuran (furan is a five-membered heterocyclic ring containing an oxygen atom) and concomitantly reduced to a formyl group. The formyl group is then transferred to the next coenzyme, tetrahydromethanopterin, which resembles tetrahydrofolate not only in structure but also in the reaction mechanism: The formyl group is integrated into a five-membered ring structure and then reduced to a methyl group. However, the similarity with tetrahydrofolate is limited to the pteridine ring and the adjacent phenyl ring; the remainder of the two coenzymes is entirely different. The reduction of the formyl group is done by coenzyme F420. This is a strange hybrid between the two universal hydrogen carriers $FADH_2$ and NADH, which are the ubiquitous redox carriers in other organisms. In fact, F420 shows a structure comparable to FAD. As we interpreted before, many coenzymes give insights into the distant past of biochemistry, and F420 behaves as if Nature was still experimenting with coenzymes before it settled for FAD and NAD in the majority of the organisms. Methanogenesis is known to be an old process, and it might not be farfetched to interpret its somewhat exotic cofactors as molecular fossils. In the next step, the methyl group is transferred from tetrahydromethanopterin to the thiol group of coenzyme M (CoM), an ethane derivative containing at one end a thiol and at the other end a sulfonate $(-SO_3^-)$ group.

Methyl-CoM Reductase

The final step of methanogenesis is done by methyl-CoM reductase, a 300 kD protein organized as a hexamer of three different protein subunits $(\alpha_2\beta_2\gamma_2)$, which forms two identical active sites (Ermler et al. 1997). The active site contains yet another coenzyme F430, a cyclic tetrapyrrole, which differs from related compounds (heme, chlorophyll, corrinoids) by having the smallest system of conjugated double bonds and a complexed central nickel atom. The ring system is noncovalently bound to the active site and is connected via a 30-Å-long narrow channel to the protein surface. The nickel is in the plane of the ring and coordinated to each of the pyrrole rings; below the plane, it is coordinated to a glutamine residue of the protein and above the plane to the methyl group of CoM. CoM is almost parallel to the plane of F430. In the channel, there is a third coenzyme CoB. CoB consists of a thiol group, a heptane chain, followed by a threonine and a phosphate group. CoB locks the channel even to water such that the enzyme reaction takes place in the hydrophobic protein environment. The crystal structure of two different oxidation states of the methyl-CoM reductase revealed the reaction mechanism. By changing its oxidation state, the nickel atom forms a metal-organic compound with the methyl group of CoM. The thiol group of CoM is reconstituted probably from the neighboring thiol proton of CoB. Nickel abstracts a further electron from the reconstituted CoM thiol, which completes the reduction of the methyl group to the methane and induces the disulfide bond formation between CoM–S–S–CoB. This causes a 4 Å shift in the position of CoM, and the negatively charged sulfonate group is now positioned next to the positive nickel atom, which pushes the linear heterodimer of the two coenzymes out of the channel. The reduction of this oxidized coenzyme heterodimer is achieved by a membrane-bound enzyme complex that uses H_2 as a reductant. The enzyme complex contains several iron–sulfur clusters and FAD and acts as an electrogenic proton pump. The energy released by methanogenesis is conserved as a proton gradient and not via substrate level phosphorylation.

Four Pathways ...

Methanogenic Archaea are a diverse group of anaerobic organisms that obtain energy for growth by converting a limited number of substrates to methane. Biochemists have identified four pathways of methanogenesis.

- The CO_2 reduction pathway involves the reduction of CO_2 to CH_4 with H_2 as the electron donor.
- The methyl reduction pathway also uses H_2 as electron donor but reduces methanol to methane after transfer of the methyl group to CoM.
- The acetoclastic pathway (see below).
- The methylotrophic pathway uses the disproportionation of C1-compounds such as methanol and methylamine, to CO_2 and CH_4.

These four principle reactions are probably not the only ways of methanogenesis. Recently a fifth pathway was described that transforms acetate into CO_2 and formic acid and couples this reaction to the reduction of methanol to CH_4 (Welander and Metcalf 2005).

... And Some Need of Cooperation

Specifically the acetoclastic reaction is the predominant methane-forming reaction in nature, catalyzed by *Methanosarcina* and *Methanothrix*. This is a disproportionation reaction, where the carbon in acetate, of medium oxidation state, is split into carbon at +4 oxidation state (CO_2) and −4 oxidation state (CH_4). In this reaction, acetate is first phosphorylated, and then transferred to HS–CoA yielding CH_3–CO–S–CoA. The acetyl rest is cleaved in two parts; the terminal methyl is transferred to tetrahydromethanopterin to follow the above-mentioned methanogenesis pathway. The central CO moiety is oxidized with H_2O to CO_2, and the resulting H_2 is used for the reduction of CoM–S–S–CoB. Methanogens thus need cooperation with primary and secondary fermenting bacteria that provide the necessary acetate for methanogenesis. However, the acetate concentration in the subsurface sediment is surprisingly low with only $12\,\mu M$. Is this sufficient to drive the huge methane production in the sediment? The answer is yes: If ocean sediment is heated, it releases substantial amounts of acetate of up to $24\,mM$. The maximum release was observed at $40\,°C$ (within the range of 10–60°C), speaking in favor of a biological production of acetate and not a chemical process. The acetate concentration increased with depth from 450 m and paralleled the distribution of methanogenesis (Wellsbury et al. 1997). The heating apparently activated fermentative bacteria; methanogenesis is thus the effort of metabolic cooperation between different prokaryotes.

Ecology

Actually the earliest investigation of marsh fire gas, as methane called thence, goes back to the chemist John Dalton, who collected in the 1800s the biogas CH_4 from anoxic ponds (Parkes 1999). Two hundred years later, German marine microbiologists collected methane from enrichment cultures taken from anoxic ditch sediments. In an anoxic mineral medium containing hexadecane—a C16-alkane—as sole organic carbon source, methane could be recovered in the headspace. No such gas was formed when hexadecane was omitted from the medium. However, the experiment needed a lot of patience: It took nearly a year before the first significant amounts of methane were released from the culture (Zengler et al. 1999). Alkanes contain only apolar sigma bonds and are thus very recalcitrant to biological degradation in the absence of enzymes that can use molecular oxygen for the chemical attack. The long delay indicates the difficulty of the chemical task. Sequencing of 16S rRNA demonstrated the members of this enrichment community. It consisted of bacteria belonging to the δ subclass of Proteobacteria, which mediated the degradation of hexadecane into acetate according to the equation: $4C_{16}H_{34} + 64H_2O \rightarrow 32CH_3COO^- + 32H^+ + 68H_2$. This reaction has a high negative ΔG of $-929\,kJ/mol$. The reaction is clearly exergonic and thus thermodynamically feasible, but the difficult nature of cracking of the alkanes makes it a kinetically very slow reaction. Several strains mediating this reaction belong to the *Syntrophus* cluster, a very appropriate name since this Greek name means "eating together." In fact the reaction products prepare now the next wave of food degradation done by Archaea belonging to the *Methanosaeta* cluster. These are acetoclastic methanogens, meaning they split acetate according to the equation: $32CH_3COO^- + 32H^+ \rightarrow 32CH_4 + 32CO_2$. When looking at these two equations, you remark that $68H_2$ from the first and $32CO_2$ from the second reaction were left untouched. They are now taken over again by another group of methanogens belonging to the *Methanospirillum* cluster, which synthesize methane according to the following equation: $68H_2 + 17CO_2 \rightarrow 17CH_4 + 34H_2O$. The measured fluxes in the enrichment culture fitted closely these equations, if a partial incorporation of hexadecane into the biomass of the cells and a small use of hydrogen by sulfate reducing *Desulfovibrio* bacteria was accounted for.

History

In the abovepresented scenario, we saw methanogens at the last step of the food chain that wring the last free chemical energy from organic compounds that are nearly exhausted. This is a quite impressive show with respect to our theme of the quest for food. Not the least crumbles are left on the table of Mother Nature. However, it does not explain why methanogenesis is considered a very old, perhaps one of the oldest energy delivering processes in biochemistry (as suggested by tree-building and the apparent antiquity of its coenzymes). To be considered as candidates for the root of the biochemistry of cellular life, methanogens should be able to do their job without fermentative bacteria (which then might not have existed), without an obvious organic carbon source for the reducing equivalents (because at the beginning, these compounds might have been scarce). Since one type of methanogenesis can run on CO_2 and H_2, geothermal hydrogen and CO_2 of volcanic origin could represent the primary energy source of methanogens instead of organic carbon. Such a community has recently been described in a hot spring from Idaho (Chapelle et al. 2002). The underlying rocks are of volcanic origin and are devoid of organic carbon, and there are sources of geologically produced hydrogen. The subsurface water was consequently hot (nearly 60°C), anoxic, devoid of organic carbon, but contained H_2. Diagnostic PCR tests based on rRNA sequences demonstrated a 99% preponderance of Archaea; sequencing defined two groups of methanogens. We have here a plausible case for prokaryotes that can make a living from the most basic food molecules H_2 and CO_2. These data might also be important for the model building and the search of extra-terrestrial life.

Plant Methane Sources

Since methane absorbs solar radiation strongly at infrared wavelengths, it is the most important greenhouse gas after CO_2. This role of methane for global warming motivated an increased interest into the sources of methane. Currently about 530 million tons (Mt) of methane are released per year. Top biological sources are wetlands (145 Mt), ruminants by eructations (90), rice agriculture (60), and termites (20). Human activity contributes considerably with biomass burning (50), energy generation (95), and landfills (50), while marine sources are negligible (15) (Lowe 2006).

It was, for example, estimated that microorganisms living in anoxic rice soils contribute 10–25% of global methane emission (Lu and Conrad 2005). The scenario is approximately as follows: Between 30 and 60% of the carbon fixed by photosynthesis in the leaves is allocated to the roots. A major part of this fixed carbon (estimates range from 40 to 90%) enters the soil in the form of root exudates, lost cells, and decaying roots. In wetlands and rice paddies, this carbon flow feeds the methane production by microbes. However, then, the picture gets somewhat less clear because the soil microbiota is considered in most ecological studies as a black box. This black box is not an arbitrary simplification; it reflects the fundamental lack of knowledge in soil microbiology. From the practical importance of the soil for agronomy to climate change, this ignorance cannot be excused except by the bewildering complexity of this system. Relatively crude first insights into the rice rhizosphere were recently published (Lu and Conrad 2005). The researchers exposed plants to a pulse of $^{13}CO_2$; this heavy carbon isotope was taken up by the leaves within 1 hour. The rapid ^{13}C labeling of CH_4 in the soil pore water demonstrated that the methanogenesis in the rice rhizosphere was highly active and tightly coupled to plant photosynthesis. The heavy ^{13}C was also incorporated into RNA from the soil microbiota. One group of methanogens became specifically labeled. This rice cluster I Archaea from the soil plays a key role in CH_4 production from plant-derived photosynthate.

This no-name group reflects perfectly our extremely restricted knowledge in the field of soil microbiology. However, microbiological ignorance is only one limitation in our balance sheet of methane emissions. Until quite recently, biologists thought that biological methane derives from processes in anoxic environments—hence the importance of wetland, rice paddies and the guts of ruminants, and termites for this emission. A recent paper challenges this view by demonstrating that methane is readily produced in situ by terrestrial plants under oxic conditions (Keppler et al. 2006). The new process shows a number of surprising characteristics. The experiments were conducted under 20% oxygen atmosphere—anaerobic acetate fermentation and CO_2 reduction thus cannot contribute the methane. Microbial origin was made unlikely because methane production was measured in leaves sterilized with γ-radiation and in plants, which were not grown in soil. The doubling of the methane production with steps of 10 °C temperature increases—and this up to 70 °C—suggests a chemical, but not an enzyme-catalyzed reaction. The authors suspect that methoxyl groups from the plant pectin and lignin are precursor to the plant-derived methane via an unknown chemical reaction. The shock goes even further when looking at the magnitude of the process. It contributes an estimated 150 Mt to the yearly methane budget with 100 Mt attributed to tropical forests and grassland. This observation could neatly explain the plume of methane observed over the tropical forests, seen from satellite observation. As this report has political dimensions in the context of the Kyoto Protocol, it will unleash a lively debate among researchers and politicians alike.

Methanotrophs

Where Remains the Produced Methane?

Large amounts of methane are produced in marine sediments, constituting perhaps twice the amount of all known fossil-fuel stores, and lie buried beneath the sea floor. If it remains untouched, it is stable and can be stored. However, if it escapes into the water column, methane is apparently consumed before contacting aerobic waters or the atmosphere. If the C–H sigma bond is so hard to crack, who is then reversing methanogenesis? Actually, with the exception of the very last step catalyzed by methyl-CoM reductase, all other steps of methanogenesis are in principle reversible. Therefore microbiologists searched organisms in sediments that could conduct the reversal of methanogenesis under the specific environmental conditions met. The task was now to find them in this environment (otherwise methylotrophs are very well-known). A possible tracer is methane itself since it is renowned for containing less ^{13}C than virtually any other product on Earth. Therefore an organism living on methane should have carbon compounds that are likewise characterized by unusual low ^{13}C composition. Ecologists searched 500-m deep sediments known to decompose methane hydrate for such marker molecules. In lipid extracts from this methane seep, they found two ether lipids ("archaeol") characteristic of Archaea that were so extremely depleted for ^{13}C that they could not be produced by methanogens that use CO_2 as their carbon source, but must have been derived from methanotrophs that use CH_4 as carbon source (Hinrichs et al. 1999). The investigation of the 16S rRNA sequences revealed Archaea that formed their own new branch distinct from the previously known methanogens. Its nearest neighbors were the *Methanosarcinales*. Another important hint was provided by the association of these new Archaea with sulfate-reducing bacteria.

Reverse Methanogenesis

This association confirms an observation of geologists who observed that sulfate becomes depleted downward through the sediment. Conversely methane describes an upward depletion curve, and the minima of both depletion curves intersect in the sediment (DeLong 2000). This is a strong hint that both processes are mechanistically linked. How can this be imagined? Reverse methanogenesis can be written as $CH_4 + 2H_2O \rightarrow CO_2 + 4H_2$. This would be energetically favorable if the H_2 end product is rapidly removed. This could actually be done by sulfate-reducing bacteria according to the equation $H^+ + 4H_2 + SO_4^{2-} \rightarrow HS^- + 4H_2O$. The exchange good would be H_2 traveling from the "reverse methanogen" to the sulfate reducer; this type of metabolic interaction is termed syntrophy. In the following section, we will see that this scenario is also central to an original hypothesis on the origin of the eukaryotic cell. German microbiologists working off the coast of Oregon got striking evidence for this syntrophy (Boetius et al. 2000). Methane ascends there along a fault; the crest of this fault is populated by large clams of the genus *Calyptogena* and thick bacterial mats of *Beggiatoa*, a strong hint to gas seeping, which provides HS^-, which these organisms need for their living. Depth profiles showed decreasing sulfate and increasing sulfide concentrations crossing at 3 cm sediment depth. These high sulfate reduction zones are restricted to areas of methane seeping. At the same depth, the researcher found a peak with nearly 10^8 cells/cm^3 sediment forming a remarkable prokaryotic consortium. The average consortium cluster consisted of about 100 coccoid archaeal cells; they were surrounded by about 200 sulfate-reducing bacteria. The consortia apparently matured over time because consortia consisting of less than 10 cells and consortia containing 10,000 cells were detected. The surrounding cells were characterized as *Desulfococcus*—the hypothesis of reverse methanogenesis seemed to work out. The next piece of the puzzle was netted in an expedition to the Black Sea. Here the microbiologists had spotted a thick microbial mat over a cold anoxic methane seep. The mat provided so much biomass that they could do protein chemistry directly with the recovered material without the need for further cultivation. What they found fitted again well with the reverse methanogenesis hypothesis: It was an abundant protein, which they called nickel I protein. It contained nickel in a variant of the F430 coenzyme. They purified the protein and obtained its sequence. It showed a high degree of similarity with methyl-CoM reductase from various methanogenic Archaea, but differed in the active site of the

α-subunit which binds the F430 coenzyme. The next step came from US oceanographers working off the coast of California. They applied environmental genome sequencing techniques and obtained 4.6 Mb of DNA sequence for such a methane-oxidizing community. They found DNA sequences from sulfate-reducing bacteria and those of Archaea related to the *Methanosarcinales* lineage. In fact they identified all genes from the methanogenesis pathway, with the possible exception of the enzyme catalyzing the reduction of the methylene to the methyl group on the tetrahydromethanopterin coenzyme (Hallam et al. 2004). The reverse methanogenesis plot in the sediments thickens, and we start to understand why so little methane from the sediment escapes into the water column.

The Rice Field

Although this is reassuring from a climate change viewpoint, these results should also stimulate research into environments that definitively release methane into the atmosphere such as rice paddies, which contribute about 30% of the annual emission of methane into the atmosphere. It has been calculated that the estimated increase of the human population over the next three decades can only be fed when the rice production increases by about 60%. This will only be possible when more nitrogenous fertilizers are used in rice cultivation. The methane production by methanogens in a rice paddy is kept at least partially in check by other methanotrophic bacteria, which belong to the α- and γ-subgroup of proteobacteria. These microbes have an obligate oxygenic metabolism and need oxygen for attacking the C–H bond in methane. In flooded rice fields, they do not find enough oxygen and associate therefore with the roots of the rice plants, which provide them with the necessary oxygen to oxidize methane, which diffuses from the anoxic bulk soil to the rhizosphere. NH_4^+ fertilization has now three partially compensating effects. First, at the plant/ecosystem level nitrogen fertilization increases plant growth. More organic carbon sinks and feeds the methanogen in the soil. The result is: CH_4 emission goes up. Second, at the microbial community level, nitrogen fertilization stimulates the growth of methane-oxidizing bacteria; their enzymatic activity increases, and the CH_4 emission goes down. Third, at the biochemical level ammonium salts inhibit the monooxygenase of the methanotrophs, the critical enzyme for breaking the first C–H bond in methane. Again the CH_4 emission goes up (Schimel 2000). Thus, the net result of fertilization cannot be predicted easily and must be determined in mesocosm experiments (Bodelier et al. 2000). The results were reassuring for the climate question: Nitrogen addition stimulated rather than inhibited the methane-oxidizing activity in the root zone of rice and resulted in an increased incorporation of label from methane into the fatty acids typical of methanotrophic bacteria of both subgroups. This interplay of competing mechanisms is typical of many ecological situations and makes predictions of the effect of anthropogenic interventions on the global climate a very tricky task.

The Peat Bog

The largest part of methane formed in wetland ecosystems is recycled and does not reach the atmosphere. How this actually happens was not clear until recently. Dutch ecologists took a deeper look into peat bogs and demonstrated that submerged *Sphagnum* mosses—the dominant plants in many of these habitats—consume methane, which is then incorporated into plant sterols (Raghoebarsing et al. 2005). Balance analysis showed that methane acts as a significant carbon source for this moss contributing up to 15% of the fixed carbon. This observation explains the riddle why peat lands show low primary productivity and nevertheless high carbon burial. Peat bogs are composed of lawns and pools, in the latter of which *Sphagnum* (Figure 4.2) grows below the water table. Methane oxidation was more prominent in the submerged than in the top parts of the moss. Did they find another surprise, after methane producing

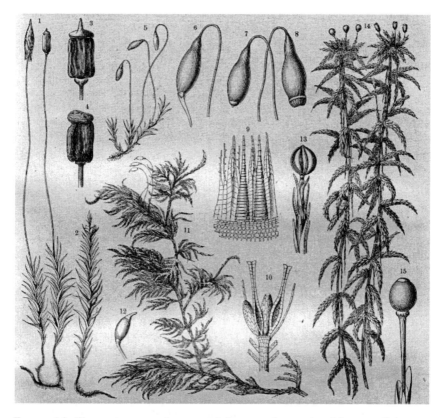

FIGURE 4.2. The peat moss or bog moss belongs to the species-rich genus *Sphagnum* of the class Musci, Bryophyta (mosses). At the *right* side is *Sphagnum acutifolium*. In the figure you also find two representatives of another group of Musci, the Bryidae, here depicted with *Hylocomium splendens* (*center*) and *Polytrichum commune* (*left*). The corresponding sporangia of these mosses are also shown.

plants, now methane consuming plants? Yes and no. Yes because the methane consuming activity was definitively localized into the outer cortex of the moss stem structure, but no: the biochemical activity was contributed by methanotrophic α-Proteobacteria. Microscopic examination showed that these bacterial symbionts lived in the hyaline cells. These are dead, water-filled cells specialized in solute transport. Coccoid cells built cubic clusters, which differ only by lack of intracytoplasmic membranes from known acidophilic methanotrophic *Methylocella palustris*. When the plant was exposed to ^{13}C-labeled methane, the label became incorporated into the plant components via a two-step reaction: Methane oxidation to CO_2 was followed by CO_2 fixation via the Calvin cycle. In this way, photosynthate lost to methanogens is recycled via symbiontic methanotrophs into the plant. The global role of peat bogs for the carbon cycle should not be underestimated: In the Northern hemisphere, it comprises an estimated one third of the carbon stored in soil.

A Canal Receiving Agricultural Runoff

The Netherlands is a country gained from the sea—small wonder that canals dissect many regions. The country is also intensively used for agriculture. The fertilizer nitrate gets thus commonly into the canal water. This creates an interesting interface where upward fluxes of methane generated by anaerobic decomposition of organic material, mainly cellulose, meet downward flows of nitrate or nitrite, creating a sharp oxic/anoxic interface. Global biogeochemical cycles are maintained by prokaryotes using C1 compounds such as methane or carbon dioxide as the most extreme-reduced and extreme-oxidized form of carbon. Each step in such a cycle is catalyzed by a different group of microbes that associate into an ecological guild to achieve a reaction cycle. Microbiologists believed that they knew the major reactions in the carbon cycle, but they were unaware of anaerobic oxidation of methane. They knew that sulfate can act as oxygen donor but could also nitrate or nitrite play this role? Thermodynamically such a reaction is possible, even energy-yielding (-928 kcal/mol methane): $3CH_4 + 8NO_2^- + 8H^+ \rightarrow 3CO_2 + 4N_2 + 10H_2O$. The oxidation of methane must be coupled to denitrification (nitrate reduction to nitrogen gas). An enrichment culture showed the chemical changes consistent with the above equation. It preferred nitrite to nitrate; when nitrite was consumed, it could change to nitrate as a substrate. You must be patient as microbial ecologist, but the Dutch scientists knew that it could take some time to see the chemical reaction. So they incubated the culture for 16 months. The labeled methane substrate ended up in bacterial and archaeal cells. They could finally visualize the members of the consortium. The Archaea, distant relatives of marine methanotrophs, formed a central cluster inside a matrix of bacterial cells, belonging to a new division of bacteria lacking a cultivated member. The consortium showed a 1:8 cell ratio. A further step in the carbon cycle was described, which was previously overlooked, and, as a bonus, two new and geochemically important prokaryotes were discovered (Raghoebarsing et al. 2006).

Sulfur Worlds

A Deep-Sea Beginning of Bacterial Life?

In Wächtershäuser's ideas on the origin of metabolism, iron and sulfur take center stage. Many microbes live still today in an iron–sulfur world as we will see in the next section. There is also some indirect fossil evidence for this link. Pyritic filaments (FeS_2), perhaps the fossil remains of threadlike microorganisms, were found in 3,200 My-old deep-sea volcanogenic massive sulfide deposits from the Pilbara Craton in Australia (Rasmussen 2000). These fossils suggest that iron- or sulfur-oxidizing microbes were prevalent on the early Earth nearby deep-sea volcanoes or hydrothermal vents. A substantial amount of reduced sulfur enters at continental fracture zones. In regions of seafloor spreading, lava comes in contact with cold ocean water, contracts on cooling, and allows seawater to enter several kilometers deep into the earth crust. This water stream gets charged with metals, hydrogen sulfide, and hydrogen and returns to the seafloor either at low speed and low temperature or more spectacularly at high speed and high temperatures in so-called black smokers. The surrounding of the black smokers teems with microbes despite the high pressure of 2,000–3,000 m water depth and temperatures exceeding 100 °C. Notably the sulfide-oxidizing bacterium *Thiomicrospira* dominates the microbial community.

Sulfur Reduction

Sulfur is recycled in the living environment through the action of many different organisms. In assimilatory sulfur reduction, sulfate is assimilated for biosynthetic purposes into organic sulfur compounds. Bacteria and plants synthesize their amino acid cysteine from sulfate and serine. Some bacteria transform sulfate to sulfide in dissimulatory sulfur reduction to gain energy. This reaction is conducted by a large group of bacteria starting their genera name with *Desulfo*, for example *Desulfovibrio*. Also a few Archaea manage this reaction (e.g., *Archaeoglobus*). Still other bacteria mediate the mineralization of organic sulfur compounds into H_2S (the smell of rotten eggs belongs to their legacy). The basis for the energy yielding reactions in dissimulatory sulfur reduction is still enigmatic. The reactions start with endergonic reactions: $ATP + SO_4^{2-} \rightarrow AMP-S + PP$, followed by: $AMP-S + 2H \rightarrow AMP + HSO_3^- + H^+$. These are endergonic reactions catalyzed by soluble enzymes. To pull these reactions, an exergonic reduction to sulfide must follow. These reductions liberate enough free energy ($\Delta G° = -152 \, kJ/mol$ for $SO_4^{2-} + H^+ + 4H_2 \rightarrow HS^- + 4H_2O$) to allow comfortable growth. In fact sulfate-reducing bacteria compete quite well with other bacteria in marine sediments and seawater, where the sulfate concentration is as high as 28 mM.

Sulfur Oxidation

To maintain the cycling of matter, other bacteria must achieve the oxidation of H_2S back to elemental $S°$, sulfite SO_3^{2-}, and then sulfate SO_4^{2-}. As you gain energy from the reduction of sulfate to sulfide by molecular hydrogen, how can you gain energy from the reverse reaction without violating the laws of thermodynamics? The answer is straightforward: You change the reaction partners of sulfur. The dissimilatory sulfur reducers work under anoxic conditions. In the absence of oxygen, these bacteria can use molecular hydrogen as a reductant. If you transfer the end product of this exergonic reaction, i.e., H_2S, into oxygenic conditions, you get again an energy-yielding process, actually even a strongly exergonic reaction. *Beggiatoa*, a gliding bacterium of the order *Thiotrichales* (literally, sulfur filaments, a γ-Proteobacterium), gains energy from this reaction: $HS^- + O_2 \rightarrow SO_4^{2-} + H^+$; $\Delta G° = -798\,kJ/mol$. *Thiobacillus thiooxidans* belongs to the so-called colorless sulfur bacteria and is a β-Proteobacterium; it exploits the reaction: $2S° + 3O_2 + 2H_2O \rightarrow 2SO_4^{2-} + 4H^+$; $\Delta G° = -588\,kJ/mol$. The front-runner with respect to energetics is the following reaction: $5S_2O_3^{2-} + 8NO_3^- + H_2O \rightarrow 10SO_4^2 + 2H^+ + 4N_2$; $\Delta G° = -3,925\,kJ/mol$, conducted by the colorless sulfur bacterium *Thiomicrospira denitrificans*. *Why should you bother to exploit something so complicated and so stingy with respect to energetics as dissimulatory sulfur reduction with a meager $\Delta G° = -152\,kJ/mol$, when the menu card offers such highly exergonic reactions? As usual there is no free meal in biology, you must work hard or do an ingenious invention to get your food and you have to defend it to other contenders.* Just when looking to the above chemical reactions of sulfur oxidation, problems are immediately apparent. I will mention five, and how bacteria got around them.

Beggiatoa

Let's start with *Beggiatoa*—it cannot be cultivated with its two substrates $HS^- + O_2$ in a homogeneous culture because the two components would spontaneously react with each other and leave only bits of the meal for the bacterium. If you want to grow *Beggiatoa* in the laboratory, you need a special tube with a mineral medium containing the sulfide at the bottom, a second mineral medium containing bicarbonate, and a headspace with air. If you leave the tube for several days, HS^- diffuses up and oxygen diffuses down into the solid medium, and they overlap in a relatively broad zone in the agar medium. If you added *Beggiatoa* to the same tube, HS^- and O_2 show a much sharper gradient and only a very narrow meeting zone. Of course this is also the zone where you find *Beggiatoa*. How did it get there? *Beggiatoa* is a glider, and this motility directs it to the transition zone. Since the two chemical reactants are just touching each other at very low concentrations, the bacterium can now compete well with the spontaneous chemical reaction. Logically the natural habitat of *Beggiatoa* is the oxygen/sulfide interface of sediments under seawater layers of more than 100 m. This depth excludes photosynthetic sulfur-oxidizing competitors. Using their capacity to glide on solid surfaces, *Beggiatoa* can follow the interface as it moves during diurnal and tidal cycles. With this simple trick and its capacity to fix CO_2 via the Calvin cycle, *Beggiatoa* grows well when looking at its size;

the smallest are 10 μm and the largest are 100 μm in diameter. Its productivity is also high—it covers many marine sediments with 1-mm thick white layers.

Thiomicrospira

If you are not a good glider, you can compensate for this defect—for example, by teaming up with a glider, which must not necessarily be another prokaryote. The nematode *Catanema* is covered on its skin with a very regular array of sulfur-oxidizing bacteria. *Catanema* lives in shallow marine sediments and moves up and down thus transporting the bacteria either to oxygen- or sulfide-containing sediment layers. The 16S rRNA sequence data suggest that the bacterium on its skin is related to *Thiomicrospira*. Apparently the bacteria profit from being exposed alternatively to chemical environments tailored to its metabolic needs, while the nematode is protected from sulfide toxicity by the metabolic activity of the associated bacteria. Metabolic labeling with a heavy carbon isotope demonstrated that *Catanema* derives nearly all of its carbon from the sulfide-oxidizing bacterium.

Thioploca

Even more prominent are the several centimeter-thick colonies of *Thioploca* cells found off the coast from Chile over thousands of square miles. *Thioploca* cells are 30 μm in diameter and many centimeters long. Several cells live in bundles covered in a sheath of slimy material, which gave them the nickname "spaghetti bacterium." *Thioploca* actively creeps out of the sheath and accumulates nitrate in intracellular vacuoles. Then they glide back into their sheath where a high sulfide concentration is found. An individual cell can glide 10 cm deep into the sulfide-rich sediment. Nitrate is the electron acceptor for the oxidation of sulfide to sulfate and replaces oxygen that is not found in this anoxic environment.

Thiomargerita

Actually, the front-runner in size is the bacterium *Thiomargerita namibiensis*, its cells are so big that they are almost visible to the naked eye. This "sulfur pearl" (so its translated scientific name) is found off the coast of Namibia and measures 100–300 μm in diameter with a maximal size of 750 μm. How does this bacterium get sulfide and nitrate together? Nitrate is found in the overlaying seawater, and sulfide in the bottom mud of the coast. *Thiomargerita* waits simply for the next storm, which mixes both layers. During this mixing period, the bacterium takes up the nitrate and stores it in a large central vacuole, which takes practically all of the cell volume. Actually only a small rim of cytoplasm surrounds this vacuole, which stores up to 800 mM nitrate. *Thiomargerita* thus resembles more than superficially a human fat cell. The cytoplasm contains elemental sulfur granules, the other partner for the energy reaction. The sheer size of the latter two bacteria demonstrates that they can make with their adaptation a comfortable living. However, they also challenge our concepts of the diffusion-imposed smallness of prokaryotes.

Thiobacillus

Now to the last example: *Thiobacillus*. This bacterium produces sulfuric acid in substantial amounts while yielding energy. To make a living from sulfur oxidation, it must resist acid. An extreme case is *Thiobacillus ferrooxidans*, which survives pH values below 1. I mentioned this organism in the previous section as an illustration that bacteria do not occur in isolation in nature and that biochemical pathways of bacteria should best be studied as community-connected pathways of cohabiting organisms.

Metagenomics and the Strange Appetite of Bacteria

Even a casual look on the universal phylogenetic tree will teach you that the genetic diversity on our globe is not dominated by the so-called higher life-forms (eukaryotes), but by prokaryotes. However, it is not so much the outer appearance of bacteria that is diverse, but their nutritional appetite. Nature sports bacteria that derive metabolically useful energy from the oxidation of—in our view—unpalatable inorganic compounds such as hydrogen, carbon monoxide, reduced sulfur and nitrogen compounds, iron or manganese ions. These ways of living might appear exotic to us, but they are essential for the cycling of matter on our planet. In addition, the strange appetite of bacteria is used industrially in such diverse areas as wastewater treatment, bioremediation, or mining.

Acid Mine Drainage Ecosystem

Let us consider mining in somewhat more detail, and you will realize how prokaryotes make a living in the iron–sulfur world. The microbial oxidation of metal sulfides to sulfuric acid and dissolved metal ions is exploited as an inexpensive method of leaching low-grade metal ores. FeS_2 (pyrite) contained in coal is oxidized by *T. ferrooxidans* leading to the formation of Fe^{3+} and sulfuric acid (acid mine drainage), which kills all life coming in contact with it except those microbes making a living from it. A biofilm, growing in this extreme environment (pH of 0.8!), was recently investigated by DNA cloning and sequencing. This work is of substantial importance for several reasons. Microbial genome sequencing has fundamentally changed our perception of the microbial world. Until recently, only cultivated microbes were sequenced. However, microbiologists do not know how to cultivate the majority of the bacteria present in the environment as individual strains. Therefore, they would like to be able to study microbial ecosystems as a whole.

Community Sequencing

One of the ideas to do this is to sequence the DNA of all the microbial members in a community at once in a so-called metagenomic analysis. The sequencing of this biofilm from acid mine drainage represents for the first time the genetic blueprint for an—admittedly simple—entire ecosystem (Tyson et al. 2004). The sequencing and the specific microscopic detection of individual bacteria revealed the presence of two groups of *Leptospirillum* bacteria constituting 85% of all cells and *Ferroplasma* archaea representing 10% of the cells in the biofilm. This extremely acidophilic biofilm is a self-sustaining community that grows in the subsurface of the Earth and receives no energy input from light and no significant input of fixed organic carbon or nitrogen from external sources. According to the genome sequences, the *Leptospirillum* bacteria possess all the genes to fix CO_2 via the Calvin cycle. In view of the large number of sugar and amino acid transporters present in the archaeon *Ferroplasma*, this cell has apparently opted for a heterotrophic lifestyle. Surprisingly, the majority group of *Leptospirillum* bacteria, which represent 75% of all cells in this biofilm, has no complete nitrogen pathway. They gain energy from iron–sulfur oxidation. These bacteria constitute the chemilithoautotrophs in the community. The nitrogen input into the community seems to stem from a minority group of *Leptospirillum* bacteria, which can fix N from N_2 in the air. Despite their only 10% representation, these bacteria represent the keystone species in this ecosystem.

After the metagenomics analysis, the US research consortium went a step ahead and performed a community proteomics analysis (Ram et al. 2005). They separated the proteins recovered from the biofilm by 2-D chromatography, followed by mass spectrometry and then identified peptides that corresponded to 6,000 of the 12,000 predicted proteins. As an illustration of our ignorance, the biofilm library was dominated by novel proteins that were either annotated as hypothetical (because they lacked homology to proteins with functional assignments) or unique (they had no match in the database). Apparently, these proteins fulfill functions that are outside of the imagination of microbiologists. If one considers that 42% of the genes in the acid mine econiche were annotated as "hypothetical," then the constituents of the biofilm may harbor still a lot of surprises. The category of hypothetical genes is underrepresented in the proteome analysis because this group of genes may contain many inactive or nonfunctional genes. The comparison between the genomics possibility and the proteomics reality paints a clearer picture of the biofilm. For example proteomics suggests that carbon fixation occurs via the acetyl-CoA pathway. Nitrogen fixation seems to be a sporadic activity, which was low at the time of sampling. Abundant is an extracellular cytochrome, the primary iron oxidant in this system, which couples biology and geochemistry in a metal-rich (Fe in near molar concentration) acidic environment. Proteins involved in protein refolding and response to oxidative stress are highly expressed demonstrating the difficulty of survival in such extreme environments.

These studies illustrate an important observation in microbial ecology: Even a single very specialized niche can offer a living to at least four different bacteria. The bacteria in this acid mine biofilm can coexist in the same environment because of the nutritional interactions between them. Specifically important is often the complementation of metabolic pathways; the waste product from one microbe becomes the food for another. This nutritional complementation allows the cycling of matter in an ecosystem and only thereby the stable maintenance of metabolic activity over long time periods without exhausting the raw material for their respective metabolisms. This mine ecosystem is exceptional. Most communities are shaped by gradients of external input of nutrients as observed in sediments.

Problems of Iron Oxidation

The major, if not sole source of energy in acid mine drainage ecosystem is the oxidation of ferrous iron (Fe^{2+}). At neutral pH, ferrous iron is rather insoluble and bacteria have to compete with atmospheric oxygen for Fe^{2+} oxidation. At low pH, Fe^{2+} has a higher solubility and becomes stable in the presence of oxygen. Understandably many iron-oxidizing bacteria are therefore also acidophilic. *Leptospirillum* oxidizes iron from the ferrous (Fe^{2+}) to ferric (Fe^{3+}) state, but the ferric iron produced in this reaction is then used as a chemical oxidant for pyrite according to the equation: $FeS_2 + 14Fe^{3+} + 8H_2O \rightarrow 15Fe^{2+} + 2SO_4^{2-} + 16H^+$. *Leptospirillum* is thus also a sulfur-oxidizing bacterium, and the equation explains the abundant acidity produced by this bacterium. The large pH difference between the outside (pH 0.8) and the inside of the cell (pH 6) maintains a permanent proton gradient across the cell membrane, which can drive the ATP production by the ATP synthase. However, this gradient potentially leads to acidification of the cell's interior, and several requirements must be met. First, the cytoplasmic inflow of protons must be neutralized ($^1/_2O_2 + 2H^+ + 2e^- \rightarrow H_2O$). This is done by the transfer of electrons extracted from the oxidation of Fe^{2+} in the periplasm ($2Fe^{2+} \rightarrow 2Fe^{3+} + 2e^-$) and the concomitant build up of water from molecular oxygen. The electrons are transported across the membrane via a cytochrome and Cu protein-containing electron transport chain. Second, the huge pH gradient must be balanced by an inverted electric potential to keep the proton motive force in a manageable range. Finally, the cell must protect itself against the external acidity. Interestingly *Leptospirillum* contains a cellulose synthesis operon. Cellulose is with few exceptions (*Acetobacter xylinum*) only produced by plants. The cellulose produced by *Leptospirillum* may coat the biofilm, causing it to float and protecting it against chemical and physical harm.

Metagenomics

As demonstrated by this example, we now dispose of a new approach to address the metabolic interaction of entire ecosystems by the genome sequencing of all microbes constituting the ecosystem. Importantly it is not necessary to cultivate the constituting members of this community. In fact microbiologists suspect our ignorance of microbial nutrition is frequently the cause for viable, but noncultivable microbes in the environment. If you take the situation of syntrophism (i.e., cooperation in which both partners depend entirely on each other to perform the metabolic activity performed), it will be very difficult to cultivate the participating microbes as pure cultures. Cultivation methods will thus systematically underrepresent or suppress microbes involved in complicated metabolic interactions. This problem is circumvented if the entire DNA found in a given habitat is extracted, cloned, and sequenced. The genomes of the constituting microbes are post hoc reconstructed by computer analysis, the metabolic potential of the genomes is deduced by bioinformatic analysis, and by inference the metabolic interaction between the microbes in the given environment is deciphered. Of course, at the current level of sequence analysis at a community level, this is a research tool limited to very special communities that are constituted in its majority by a few dominant microbial strains like in the ecosystem described above.

Soil—The New Frontier

Most "real" microbial communities are extremely complex, and you will not be able to assemble genomes with the current sequencing capacities. A group of bioinformaticians argued that this is not even necessary (Tringe et al. 2005). Prokaryotic genomes show a high gene density of 1 gene per 1 kb of DNA. The current read length in high-throughput shotgun sequencing projects of environmental samples is 700 bp; you get in this way a protein catalogue of the entire community. This catalogue can then be developed into a fingerprint of the particular environment. The researchers compared prokaryotic shotgun projects from nutrient-rich soil (food is plant material), nutrient-rich deep-sea whale fall (food is lipid-rich bones from the whale, see a next chapter for details), and nutrient-poor surface ocean samples from the Sargasso Sea. The different samples showed clear-cut differences: 100 Mb of sequenced prokaryotic soil DNA encoded 73 cellobiose phosphorylases involved in the degradation of plant material, 700 Mb DNA from the sea yielded none. In contrast, the Sargasso Sea library gave 466 light-driven proton pumps of rhodopsin type, while none was found in the soil. The sea sample yielded many sodium ion exporters, while the soil gave many genes involved in active potassium channeling, reflecting the distinct abundance of these ions in the marine and terrestrial environment. The researchers were nevertheless interested to get at least an estimate for the diversity of prokaryotic life in these ecological niches. When they used ribosomal RNA libraries, they estimated from the accumulation of new sequences with increasing sequencing effort ("rarefaction curves") that the soil contained more than 3,000 different ribotypes, while the whale fall showed less than 150 (Tringe et al. 2005). Statisticians from Los Alamos came to much higher estimates for soil bacterial diversity. Likewise Scandinavian microbiologists have applied an old technique which was developed to estimate the genetic complexity of eukaryotic genomes. The method is based on DNA reassociation kinetics and allows measuring single copy genes (Britten and Kohne 1968). By applying this method to the genetic complexity of a microbial community in a given niche, they calculated 10,000 bacterial species per gram soil as a rule of the thumb (Torsvik et al. 1990). The Los Alamos statisticians pointed out that these estimations are erroneous because they assume that all bacterial species in the sample are equally abundant like the single copy genes in a genome. This is, however, an unrealistic assumption for bacterial diversity. They estimated the species number with different abundance models and arrived to a nearly 100-fold higher figure: 800,000 bacterial species per gram of pristine soil, containing a billion bacterial cells. In soils containing high metal pollution levels, the overall bacterial count remained unchanged, but the number of different bacterial species fell to 10,000. While some bacteria can thrive on heavily metal polluted soils, the genetic diversity was drastically reduced by the contamination. Soil genomics is the next challenge of the DNA sequencing community (Gewin 2006) as soil is a new frontier in biology.

Bacterial Species Diversity

Formal taxonomic recognition of a new bacterial species requires its deposition into a collection, which is only possible if it can be cultivated. This explains why microbiologists have currently only about 8,000 described bacterial species. However, based on molecular techniques not involving cultivating, the number of different bacterial species is already higher than 100,000. Yet this figure is still lower than the 800,000 described species of beetles. There are two potential answers to this paradox. One proposal is that of the population geneticist Haldane who said "the Creator had an inordinate fondness of beetles." Many observations now suggest that almost all insects harbor different endosymbiontic bacteria. I suspect therefore that the Creator had an even greater fondness of bacteria, and the conundrum of the low bacterial species number is a taxonomical problem with organisms having few traits observable to the eyes. Now, with several hundreds of bacteria sequenced, we do not yet reach a saturation of the DNA sequence space for bacteria. This observation suggests a bewildering genetic diversity in this domain of life. In fact what we call a bacterial species might correspond in eukaryotic taxonomic terms to genera. It is not unusual that different isolates from the same bacterial species differ by 5–15% in their gene content. In comparison we (H, sapiens) differ from chimpanzee (Pan troglodytes) by less than 2% at the DNA sequence level. This could mean that each bacterial species is already differentiated in many different lineages representing distinct ecotypes. There are even good arguments from evolutionary biologists why clonally organisms like bacteria have to split in different ecotypes. According to this line of thinking, bacteria have to occupy different niches to survive periodic sweeping selection events. A broader spectrum of metabolic types is probably also needed to exploit an environment. This principle was already seen when studying different metabolic types of sulfur-oxidizing bacteria in freshwater environments during energy-limiting growth conditions. Autotrophic, mixotrophic, and heterotrophic types coexisted, and their relative frequency shifted with the relative input of reduced inorganic sulfur and organic substrates availability.

Nutritional Interactions

Photosynthesis Combined with Sulfur Oxidation

Nutritional interactions are common in nature and many turn around sulfur oxidation. Some situations seem to violate the laws of thermodynamics. For example, there are bacteria that oxidize H_2S under anoxic conditions, i.e., in the absence of oxygen. Of course, no laws of physics are broken. In such cases, scientists postulate and actively search for unknown reactions that would reconcile the observations with such fundamental laws as those of thermodynamics. In this case, scientists did not have to search far. Bacteria that can oxidize H_2S under anoxic conditions are phototrophic bacteria. To gain energy, they must first "invest," which is done with the help of light energy. The "colored" sulfur oxidizers come in two principal groups: purple sulfur bacteria-like *Chromatium* and green sulfur bacteria-like *Chlorobium*. They are found in eutrophic lakes, where they are located at the oxic/anoxic interface that just receives enough light. The upper oxygen layer of the lakes is not suitable for their metabolism since sulfide would be spontaneously oxidized by oxygen. Hydrogen sulfide can thus only accumulate in the lower layers of the lakes.

The light exploitation of these bacteria is quite remarkable: Their light harvesting system is tuned to those wavelengths that penetrate relatively deep into the water layer and which are not absorbed by the green algae and cyanobacteria in the top water layers. They achieve that by using carotenoids for light adsorption and much more antenna bacteriochlorophyll around the reaction center (RC) than bacteria in the upper layer. In that respect, they can live with light intensities as low as 5–0.1% of that found in surface water. The green sulfur bacteria (*Chlorobium*) are the more efficient light users and more tolerant to higher sulfide concentrations (up to 4 mM), explaining why they find their niche in the water column below the purple sulfur bacteria. A "green sulfur" *Chlorobium* species was even found at 80-m depth in the Black sea, where it lived from phototrophic sulfide oxidation. The end product of their sulfide oxidation is not always sulfate; frequently they oxidize sulfide only to elemental sulfur, which is then deposited as extracellular (*Chlorobium*) or intracellular sulfur globules (*Chromatium*). Phototrophic sulfur bacteria also show other conspicuous intracellular structures: gas vesicles that allow the bacteria to float to the optimal position in the opposing light/sulfide gradients. The regulation of the buoyancy is achieved by a size control of the vesicles and reversible aggregation of the cells, which changes their sedimentation velocity. Some purple sulfur bacteria also possess flagella, which endow them with active movement.

Bacterial Consortia

Why does *Chlorobium* store elemental sulfur extracellularly, while *Chromatium* keeps it intracellularly? The reason for this difference is at first glance not obvious. If elemental sulfur is still a resource for energy gain by further sulfur oxidation, then why does *Chlorobium* throw this source of potential energy away? If elemental sulfur is only a waste for these bacteria, why does *Chromatium* keep it inside the cell? When microbiologists tried to culture *Chlorobium*, they observed an associated bacterium, *Desulfuromonas acetoxidans*. As the name indicates, this is a heterotrophic bacterium that oxidizes an organic substrate (here acetate) to CO_2. The electron acceptor for this bacterium is elemental sulfur, which is reduced by *Desulfuromonas* to H_2S. The extracellular sulfur globules thus act as electron carrier between both bacteria restoring the electron needs of the phototrophic *Chlorobium*. S and H_2S thus serve as redox shuttles. In this case, both bacteria can still be grown separately. However, cocultivation of both bacteria leads to a dramatic increase in biomass production over the growth in isolated culture. The extracellular sulfur globules have a biological sense, since they make the cooperation between different bacteria possible. Other such consortia exist in which both members cannot be cultivated separately. For example, many phototrophic consortia show a large central chemotrophic bacterium associated with small phototrophic bacteria that cover the internal cell. The large cell is flagellated and confers motility to the consortium. The flagellum of the chemotroph seems to be used for a phototactic response that places the phototrophs nearer to a light source. Furthermore, it is thought that the central cell reduces sulfate to H_2S, which can again be used by the phototrophs, whereas the phototroph might provide organic carbon to the internal cell.

Animal/Bacterial Consortia

A rich and exotic fauna surrounds the black smokers. Amphipods and mussels graze the bacteria that create organic matter by autotrophic carbon dioxide fixation. Some of them grow to unusual sizes: *Calyptogena* is a nearly 1-kg-heavy mussel. They reach their spectacular size with the help of sulfide-oxidizing bacteria that live as endosymbionts in the gill cells of this mussel. Another exotic beast is the 2-m-long tube worm (Figure 4.3) *Riftia* sporting beautifully red-colored gills. These gills adsorb O_2, CO_2, and H_2S from the seawater and supply it via a primitive heart into a blood circulation system. H_2S binds to the hemoglobin of the worm giving the bright red color to the gill plumes. Remarkably, this worm does not possess a gut. Instead of filtering food, it supplies this H_2S- and CO_2-charged blood to a special organ, called trophosome, where sulfur-oxidizing bacteria live inside the trophosome cells and produce organic matter using the reducing power of H_2S to fix CO_2 via the Calvin cycle. CO_2 is transported freely, dissolved in the blood, or bound to hemoglobin. It could not yet be established whether the symbiotic bacteria feed the host via excretion of organic matter or are eaten up by the worm as entire cells. Physiology experiments could not be conducted with the worms since they do not survive the transfer to the low pressure of our terrestrial laboratories. The transfer of a 2-m-long worm in compression chambers is of course no easy task.

The transmission of the bacterial symbiont is of central importance to the host when both are linked up by obligate symbiotic relationships as in the case of the tube worm. Yet the worm disperses by larvae that show a well-developed digestive tract with a ventral mouth; a buccal cavity; a foregut, midgut, and hindgut; and a terminal anus. The larval gut is even functional since it contained bacteria, but they were undergoing degradation due to the digestion process. No symbionts were detected in the larval gut. Marine biologists investigated *Riftia* worms of different age by thin-sectioning to understand how the symbiont colonizes its host (Nussbaumer et al. 2006). Interestingly

FIGURE 4.3. Tube-dwelling polychaete worms from the Sabellida order, phylum Annelida, also called feather-duster worms. The mouth (peristomium) bears a crown of branched, feathered tentacles that project from the tube made from calcareous material. The tentacles function in gas exchange and ciliary suspension feeding. The depicted species is *Spirographis spallanzani*.

the bacterium–worm interaction starts as an infection process. The bacterium penetrates the worm via the skin. The bacterium disperses in the cytoplasm of the worm's mesentery. Then the zoologists observed apoptosis of the symbiont-infected tissues, which resulted in the formation of the trophosome. In this nutritive organ of the worm, the symbiont becomes confined into a vacuole. When this nutritive relationship has been established, the worm abandons its digestive tract.

Hydrothermal Vents as a Cradle of Life?

Nutritional Oasis

Today thermal vents are considered as nutritional oases in the huge stable desert of the ocean. The productive zone of the oceans extends to 100 m depth at maximum. In the open ocean, the organic constituents of sedimenting detritus are nearly completely oxidized on their way down to the bottom, mostly within the first 1,000 m. Deep-sea ocean sediment receives less than 1% of the primary production created at the ocean surface layer. The microbial activity at deep-sea basins (3,000–6,000 m, representing 77% of the ocean-depth profile) is restricted to the digestion of these sedimenting particles ("marine snow"). However, the fracture zones that crisscross the oceans, where tectonic plates meet and interact, turned out to be sites of a surprising range of microbes and animals. The energy basis for these unanticipated forms of life is provided by chemicals washed out from the Earth's crust and not by light, and may be similar to the situation on the early Earth. This rich fauna around hydrothermal vents is a proof for the nutritional richness of an environment lacking light. Although vents at the time of the early Earth did not contain animal life, microbes seem to have thrived in the ample food sources delivered by the vents.

Shelter

Life around the vents may have offered other important advantages than just nutrition over life at the ocean surface. These advantages might have been crucial during a period where life was threatened by the collision with great impactors that would have repetitively boiled away the superficial layers of the Archaean ocean. Life at the surface of the ocean was also pretty dangerous for other reasons: As oxygen was not yet emitted by oxygenic photosynthesis into the atmosphere, a protective ozone layer was still lacking. Strong ultraviolet light was thus impinging on the surface of the Earth and the ocean causing damage to the genetic material. The Earth's mean surface temperature may well have been below the freezing point of water around 2 Ga (giga anni/years: a billion years) ago (the "snowball Earth" scenario). However, there is independent geological evidence for the presence of liquid water on Earth in its early history. This was probably the effect of greenhouse gases in the early atmosphere that could compensate for the dimmer Sun in the heat budget of the Earth. However, we do not know how comfortable the surface temperatures actually were and therefore a more constant temperature near hydrothermal vents may have been more favorable for the evolution and sustenance of thermophilic bacteria.

Light

Infrared light is radiated from the black smokers and some invertebrates sport even functional eyes in this visible darkness, which allows them to come close to the source of microbe-generated food without getting too close to the heat of the smoker and thus risking to get cooked. This is a nice physiological adaptation, but other observations on light use near hydrothermal vents are even more challenging. Take the statement: Light energy from the Sun drives photosynthesis to provide the primary source of nearly all the organic carbon that supports life on Earth. This claim seems pretty unassailable since it took the necessary precaution not to exclude some contribution of chemosynthesis to the organic carbon. A fascinating paper challenges this tenet on a quite unexpected ground. The researchers made a cruise trip to a deep-sea black smoker, and they cultivated a green sulfur bacterium from the plume directly above the orifice of the smoker. The surprise was that it was classified as *Chlorobium* and showed the characteristic light-harvesting structures only found in green sulfur phototrophic bacteria, the chlorosome. The light-harvesting pigments showed adaptations to capture the low photon flux emitted from the smoker. The light intensity of this geothermal light is greatest at wavelengths in excess of 700 nm. While faint, it was not weaker than the sunlight flow captured by green sulfur bacteria living at 80-m depth in the Black Sea (Beatty et al. 2005). In contrast to the anaerobe chlorobia isolated from the anoxic lower layers of the Black Sea, this deep-sea isolate was relatively resistant to exposure to oxygen. The bacterium was thus well adapted to live in the otherwise dark and oxygenated depth of the ocean.

A Photosynthetic Beginning of Cellular Life?

The Origins

Ten years ago, a prominent geologist published a daring hypothesis that expressively traces the origin of the photosynthetic light capturing system to the detection of infrared light from submarine vents (Nisbet et al. 1995). The authors of the *Chlorobium* paper do not describe their isolate as a direct descendent of a line of photosynthetic organisms that have continuously occupied this deep-sea hydrothermal vent. Their major argument is the time horizon for these events. The appearance of anoxygenic photosynthesis on Earth is currently dated to >3 Ga ago (De Marais 2000), well before the evolution of oxygenic photosynthesis that led to the evolution of oxygen that started somewhat earlier than about 2 Ga ago (Kasting and Siefert 2002). Compared to these timescales, hydrothermal vents are ephemeral phenomena at the individual level, and the authors expected many vent-to-shore and vent-to-vent exchanges of green sulfur bacteria in a more recent past. However, these data raise the issue of the age of photosynthesis.

Cyanobacteria Fossils

For biologists, the development of the protocell is as thorny an issue as the development of the first biochemical cycles. In addition all protocell models postulate still a long evolutionary way to the simplest bacterial cell. Therefore it came as a surprise when paleontological evidence was presented for early cellular life at about 3.5 Ga ago. It would mean that life developed into cyanobacteria in less than 500 My. The sediments in the Apex cherts from the Australian Warrawoona formation dated to this period contained suggestive evidence of cyanobacteria-like microfossils. This work was the culmination of painstaking efforts in microscopic paleontology. It essentially involved finding the oldest untransformed sedimentary rocks, cutting them into thin light-transmitting slices, and searching them under the microscope for structures resembling modern bacteria. This work was built on the efforts of a generation of geologists (Tyler, Barghoorn, Cloud, Glaessner, and Timofeev) who hunted for this evidence, vividly described in a book of W. Schopf (*The Cradle of Life* 1999). The morphological evidence was unequivocal for the microfossils from the Bitter Spring Formation (1 Ga ago) and quite convincing for the Gunflint chert (about 2 Ga ago). However, Schopf's oldest finding raised eyebrows because planet Earth is thought, without direct evidence, to have remained molten for several hundred million years after its formation 4.6 Ga ago. This was the Hadean period of the history of the Earth. The name is derived from the Greek god of the inferno, and this name is appropriately chosen. It was the period of heavy bombardment, which was dated from 4.5 to 3.8 Ga ago. Before 3.8 Ga, the uppermost layers of the early ocean would probably have been repeatedly vaporized by large impacts, sterilizing the Earth perhaps with the exception of prokaryotes living in sediments or in anyway hot environments at the bottom of the oceans near hydrothermal vents (Kasting 1993). How could life develop so quickly? Are we back to panspermia? Was the Earth seeded from space?

Not surprisingly, the biological nature of these oldest fossils was challenged. It was reinterpreted as secondary artifacts formed from amorphous graphite within metal-bearing veins of hydrothermal vents and volcanic glass (Brasier et al. 2002). Schopf later used laser Raman imagery to trace the isotopic signature of carbon and to back with this evidence the biological (photoautotrophic) carbon fixation in this 3,500-My-old sediment (Schopf et al. 2002). The jury is still out on this issue, but at a recent meeting W. Schopf conceded that the microfossils are not cyanobacteria after all, while maintaining their bacterial origin (Dalton 2002). This dispute is significant since cyanobacteria are the inventors of oxygenic photosynthesis—a biological process that changed the geochemistry of the Earth and its atmosphere and with that the evolution of life and eating forever.

Chemical Evidence for Cyanobacteria

The most recent geological evidence showed that the rise of the atmospheric oxygen concentration must have occurred 2.2 Ga ago (Bekker et al. 2004). The evidence is of course indirect but can be quite well deduced from the analysis of banded iron formation (BIF), paleosols (ancient soils), and red beds (oxidized subaerial deposits), which all suggest that the transition from a reducing to a stable oxygenic atmosphere occurred only between 2.3 and 1.8 Ga ago. The disputed cyanobacterial fossils predate an oxygenic atmosphere by nearly a billion years. In addition molecular biomarkers of cyanobacteria were only recovered in 2,500-My-old organic-rich sediments and not before (Summons et al. 1999). These biomarkers were hopanoids, a five-ring hydrocarbon compound that resembles the structure of steroids. These compounds are the bacterial equivalent of cholesterol in the eukaryotic membrane. A specific form, 2-methyl-hopanoids, is diagnostic for extant cyanobacteria. It is a very recalcitrant carbon skeleton that survives the anoxic burial of organic carbon. It could thus enter into the kerogen matrix, which makes up petroleum and oil and could thus be found in sediments.

Origin of Respiration

Taken together, these data seem to indicate that cyanobacteria developed not before 2.5 Ga ago. This leads us to a dilemma. Without photosynthesis you have no oxygen and thus no terminal electron acceptor for aerobic metabolism. Likewise without photosynthesis you have no renewable source of reduced organic carbon as food for glycolysis or respiration in heterotrophs. How could respiration evolve before photosynthesis? Evolutionary arguments can perhaps help to reconstruct a sequence of events. Aerobic respiration occurs in all three domains of life: Bacteria, Archaea, and Eukarya. In the latter, oxygen respiration in mitochondria is a bacterial heritage. It is thus likely that the last universal ancestor of current life possessed already a respiratory chain. For example, the cytochrome oxidase subunits I and II, the cytochrome b, the Rieske iron–sulfur proteins, the blue copper proteins, the 2Fe–2S and 4Fe–4S ferredoxins, and the iron–sulfur subunit of succinate dehydrogenase are all found both in Bacteria and Archaea. All aerobic organisms contain oxidases of the cytochrome oxidase superfamily (Castresana and Saraste 1995). Aerobic respiration has thus probably a single phylogenetic origin, and the last universal ancestor had probably a quite elaborate respiratory chain. However, this likely antiquity of the respiratory system does not mean that respiration was oxygenic. Already, pure logic imposes this constraint since the early atmosphere contained only trace amounts of oxygen from the photolysis of water and perhaps localized oxygen oases in biologically highly productive regions of the surface ocean (Kasting 1993). Whether these oases could support the evolution of such a widespread respiratory system based on oxygen as terminal electron acceptor seems doubtful. This dilemma can, however, be solved by a simple hypothesis namely that the respiratory system was initially not using molecular oxygen, but another electron acceptor such as the gaseous NO, which was easily generated in the early atmosphere. According to this hypothesis, the change to oxygen as electron acceptor came much later with the rise of the oxygen levels induced by oxygenic photosynthesis. Organisms had to cope with this dangerous molecule and had to adapt to this chemical club of cyanobacteria if they wanted to stay on the scene. At the same time, evolution recognized that oxygen is a super-fuel for their metabolism when they managed to handle this reactive compound.

Origin of Photosynthesis

Now what does phylogenetic analysis tell us about the origin of photosynthesis? First, let's stick to logic. Even if oxygenic photosynthesis was not invented before 2.5 Ga ago, this does not mean that photosynthesis did not occur before that time. In fact microbiologists know that photosynthesis also comes in an anoxygenic form, which uses electron sources other than water and which therefore does not evolve oxygen. If this form of photosynthesis is older than that of cyanobacteria, the dilemma of respiration preceding photosynthesis is not necessarily solved, but softened. So where do we find photosynthesis? Oxygenic photosynthesis is only found in cyanobacteria (e.g., *Synechocystis*) and the eukaryotic algae and green plants that derived their chloroplasts from cyanobacteria in an endosymbiont catch after the acquisition of mitochondria. Anoxygenic photosynthesis has a wider distribution: It is found in purple bacteria (e.g., *Rhodobacter*), green filamentous bacteria (e.g., *Chloroflexus*), green sulfur bacteria (e.g., *Chlorobium*), and heliobacteria (e.g., *Heliobacillus*). Two observations are striking in this distribution. First, these are all prokaryotes belonging to a single domain, namely Bacteria. No Archaea have ever evolved photosynthesis based on magnesium–tetrapyrrole photosystems. All archaeal exercises in using light energy for life are restricted to a few species using retinal-based photosystems. This observation suggests that tetrapyrrole-based photosystems were invented in Bacteria. Therefore they cannot be traced back to the last universal ancestor. Life must, for a while, have persisted with respiration in the absence of photosynthesis. Second, the five photosynthetic branches of bacteria have nothing in common, except that they all belong to the domain Bacteria. On a phylogenetic tree constructed with the 16S ribosomal RNA gene, they are attributed to distinctly different branches.

Horizontal Gene Transfer

Theoretically, one could argue for selective losses of the photosynthetic capacity in Bacteria and only five, now quite unrelated bacteria maintained this metabolic trait. However, most evolutionary biologists explain this strange distribution by horizontal gene transfer (HGT). The universal tree of life is determined with a DNA sequence that reflects the evolutionary history of the protein synthesis apparatus. This choice was necessary since the ribosome is—as we have heard—a universally shared heritage of cellular life on Earth, which makes it so suitable for reconstructing the history of life in its grand design. However, the history of the ribosomal genes does not necessarily reflect the history of other genome segments in a given bacterium. In Eukarya one is with a few notable exceptions used to the vertical mode of gene transfer, i.e., genes are inherited from an assorted gene set of a mother and a father cell. In prokaryotes there is nothing like a father and a mother since cells divide by a fission-like process where one mother cell yields two daughter cells. At first glance, this would mean even more uniform genetic relationships between bacteria since they should belong to clonal lineages. However, there is a big if, namely if genetic material is not acquired by neighbors belonging to other clonal lineages. In fact, an increasing number of genes in prokaryotic genomes is now suspected to have been acquired by HGT, i.e., they did not come from the parental cell, but from another cell, not necessarily from the same species. This gives prokaryotic genomes a definitive patchwork appearance and makes the reconstruction of evolutionary histories sometimes a difficult business. HGT has apparently occurred with the genes encoding the photosynthetic apparatus in bacteria.

Bacteriochlorophyll Genes

Despite the difficulty in analyzing phylogenetic relationships with genes that suffered extensive HGT, some light on the origin of photosynthesis could be shed with sequence analysis. Carl Bauer's lab based the analysis of the early evolution of photosynthesis on bacteriochlorophyll biosynthesis genes (Xiong et al. 2000). The results were quite clear: Cyanobacteria and *Heliobacillus*, which use chlorophyll a and bacteriochlorophyll g, respectively, form one cluster; the green sulfur and green filamentous bacteria, which both use bacteriochlorophyll c, form another cluster; the purple bacteria, which use bacteriochlorophyll a, were the most basic group when the tree was rooted with an outgroup sequence. This result came in some way as a surprise because the prevailing Granick hypothesis formulated in the mid-1960s stated that chlorophyll biosynthesis evolutionary preceded that of bacteriochlorophyll because it required fewer biosynthetic steps in its production. The newer interpretation groups the light-harvesting pigments on a scale of increasing energy absorption by the tetrapyrrole ring system and fits thus with an anoxygenic-to-oxygenic transition in photosynthesis. Only chlorophylls but not bacteriochlorophylls have the capacity to absorb the energy amounts required to power the oxidation of water. Taken at face value, these data mean that we can deduce an antiquity of anoxygenic photosynthesis in purple bacteria over oxygenic photosynthesis in cyanobacteria. The evolution of photosynthesis thus predates the "great oxidation event" starting 2.2 Ga ago. This line of thought narrows the time gap between the evolution of respiration and photosynthesis by postulating an antiquity of anoxic photosynthesis. HGT of photosynthesis genes clearly complicates the issue, but the major conclusions were confirmed when extending the analysis to a larger gene set. Robert Blankenships' lab took a more comprehensive approach by executing a whole-genome analysis of photosynthetic prokaryotes (Raymond et al. 2002). They defined a photosynthesis-specific gene set (genes present in all photosynthetic bacteria and absent in all nonphotosynthetic bacteria) and confirmed the basic conclusions of Bauer's group: They saw cyanobacteria, *Heliobacillus* and *Chloroflexus* as a favored clustering in their analysis, while *Chlorobium* grouped with *Rhodobacter*. According to that analysis, *Chloroflexus* has acquired photosynthetic genes mainly by HGT. The structural analysis of the reaction centers (RCs) of the different bacterial phototrophs largely concurs with these sequence and genomics analyses (Schubert et al. 1998). The strongest argument for the derived character of cyanobacteria is actually that its dual photosystems PSI and PSII look like a combination of photosystems that had developed independently in the two groups with anoxic photosynthesis.

Photosynthesis

One Cell for all Seasons: The Nutritional Flexibility of Purple Nonsulfur Bacteria

Rhodopseudomonas Biology

Purple bacteria can cause large blooms that stain bogs and lagoons with intensive color, and hence their name. In our survey of eating, we have already met one branch of these bacteria, the purple sulfur bacterium *Chromatium*. Here we will make acquaintance with its cousins, the purple nonsulfur bacteria. Nonsulfur means that they do not make their living from sulfur oxidation, although they know to handle it. In fact, there are few types of food these bacteria do not know to handle since they belong to the most flexible eaters. The rod-shaped bacterium *Rhodopseudomonas palustris* is a member of this metabolically versatile group. Flexible as they are with respect to their biochemistry, so diverse are they with their shape. Some are spirals (*Rhodospirillum*), others are half circles or full circles (*Rhodocyclus*), and still others form thin extensions (prosthecae) and buds (*Rhodomicrobium*). Its name *R. palustris* means in Latin swamp and in fact its favorite habitats are swine waste lagoons, earthworm droppings, coastal sediments, and pond waters. All this does not sound very posh and, not surprisingly, *R. palustris* is found in sewage plants.

Metabolism

Its physiology is quite impressive and reads like a table of contents in a bacterial nutrition textbook. It grows by anyone of the four modes of metabolism that supports life. At times it is photoautotrophic (i.e., it derives energy from light and carbon from carbon dioxide); it can be photoheterotrophic (energy comes again from light, but carbon from organic compounds); in the dark and in the presence of oxygen, it can switch to chemoautotrophic life (i.e., energy comes from inorganic compounds and carbon from CO_2) or directly to chemoheterotrophic life (where energy and carbon are both derived from organic compounds). It degrades plant biomass including lignin monomers, chlorinated pollutants, and it creates hydrogen during nitrogen fixation.

One has nearly the impression that one still looks into the hypothetical communal metabolism of the precell world before patches with defined metabolic capacities were packaged into cells. All the major biochemical pathways—respiration and photosynthesis, carbon and nitrogen fixation, TCA cycle, glycolysis, pentose phosphate cycle, glyoxylate shunt—are realized in this single cell endowed with a circular chromosome of only 5.5 Mb in size (Larimer et al. 2004). This is on the moderately larger side of bacterial genomes but still a small genome in comparison with that of animals. However, metabolically spoken, we are dwarfs with respect to this bacterium. When looking somewhat deeper into the bacterium's genome, our astonishment is still likely to increase.

Photosystem

Like many photosynthetic bacteria of this group *R. palustris* has two light antenna complexes, LH1 and LH2, which funnel photons into its photosynthetic RC. However, the bacterium sports not less than four different sets of LH2 genes (LH stands for Light Harvesting). This redundancy allows harvesting of light of differing qualities and intensities. Two harvesters are flanked by phytochrome systems. Only when the latter are activated by far-red light, they provide the signal for expression of the photosynthesis apparatus (Giraud et al. 2002). Via this feed-forward system, the costly photosynthetic apparatus is synthesized only when light is present, thus promising a return on the synthetic investment. In addition, *R. palustris* contains two types of CO_2-fixing enzymes (Rubisco) and three different nitrogenases. The researchers who sequenced its genome marveled as to how this complex web of potential metabolic reactions operates within the space of a single cell, and how it reweaves itself in response to changes in light, carbon, oxygen, nitrogen, and electron sources.

In the following, I will discuss only one aspect in some more detail—the photosynthesis apparatus of *R. palustris*. This is a rewarding exercise because of the beauty of this basic form of photosynthesis in this metabolic artist. Photosynthesis in *R. palustris* is anoxygenic, meaning that it does not evolve oxygen like the photosynthesis in cyanobacteria, algae, and higher plants. It is likely a more ancestral form of photosynthesis. Judged from their oxygen sensitivity, Rubisco and nitrogenases were born in an anoxic world—purple bacteria fulfill this requirement.

The Genesis of the Photosystems

The photosynthetic system of purple nonsulfur bacteria is clearly the precursor to PSII of cyanobacteria and higher plants, while they lack any complement to PSI. This observation fits well with the idea that photosynthesis, as we see it in cyanobacteria and in chloroplasts, is derived from purple nonsulfur bacteria. The derived character of the photosynthetic apparatus in cyanobacteria becomes even clearer when considering the indications that PSI seems to be derived from the photosynthetic apparatus of another more "primitive" bacterium resembling the green sulfur bacterium *Chlorobium*. Here we see definitively the familiar pattern of evolution from the simple to the complex: The photosynthesis as "invented" by the ancestors of purple nonsulfur and green sulfur bacteria is fused in cyanobacteria to a single two-component system. The cyanobacteria model is maintained in chloroplasts (there are good reasons to believe that they were derived from the ancestors of the extant cyanobacteria), yet the basic scheme experienced some perfection by increasing the number of the constituents that make up the system. *However, we do not always see this trend from the simple to the complex in evolution. With respect to the metabolic complexity, we do not see this trend when comparing purple bacteria with cyanobacteria and higher plants.* R. palustris *is not less developed with respect to its metabolism than "higher" organisms. In fact, it seems that these so-called higher organisms have streamlined their metabolism to exploit and perfect just one of the four basic metabolic modes that* R. palustris *sports. As hypothesized by C. Woese, we see here actually a trend from the complex to the simple in the progress of evolution.*

The Importance of Photosynthesis

Introduction in papers on photosynthesis usually starts with some hype (Figure 4.4). For example, something like this: "Photosynthesis is one of the most important biological reactions on Earth. It provides all of the oxygen we

FIGURE 4.4. Photosynthesis—the quest for light. Plant leaves organized in optimal orientation for light collection, counterclockwise: *Geranium pyrenaicum* (*1*), *Saxifraga aizoon* (*2*), *Campanula pusilla* (*3*), *Thuja* (*4*).

breathe and, ultimately, all the food we eat." Even if as a scientist you try to keep away from emotional excitement, this hype is absolutely justified. However, purple nonsulfur bacteria are only the prerunner, not yet the savior. Their modest upbringing does not yet allow them to play the role of photosynthesis invented by cyanobacteria, which shaped life on the planet forever. For that watershed, you will have to wait until cyanobacteria appear on the scene. The reason is simple: Purple nonsulfur bacteria work with bacteriochlorophyll in their RC—and this pigment is not yet such a strong oxidant as the chlorophyll of cyanobacteria. It is unable to achieve the splitting of water. However, the purple bacteria learned already big lessons from the book of life. The quest for food is a running for energy and the atoms of life. Purple bacteria achieved three major goals. They learned to harness the energy contained in the sunlight that bathed the planet Earth, and they managed to fix carbon and nitrogen from gas in the atmosphere. We have seen already one more modest trial with photorhodopsin in Archaea. Strikingly, this is the only trial to use the energy of light in this otherwise versatile domain of prokaryotes. But what is the photoisomerization of retinal in salt-loving archaea in comparison with the sophisticated and aesthetically appealing sunlight collectors of purple nonsulfur bacteria (Figures 4.5 and 4.6).

Photon and Electron Flow

The gross anatomy of their photosynthetic membrane is quickly described. In the middle sits the RC. It is surrounded by a first girdle of light antenna, LH1, in full light harvester 1. Adjacent to LH1 comes LH2, the other antenna. Photon energy from the sun is first captured by the bacteriochlorophyll pigments

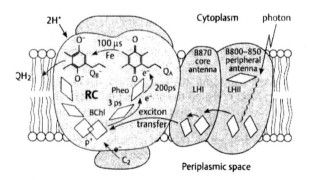

FIGURE 4.5. The primary process in anoxygenic photosynthesis. Light energy is absorbed by the peripheral antenna system LH2, where it is transformed into excitation energy, which migrates within the antenna system to LH1 and then to RC. Here the primary photochemistry starts. The special pair of bacteriochlorophyll P is excited to P* and one electron is transferred within 3 ps via a BChl monomer to bacteriopheophytin (Pheo) and then within 200 ps to ubiquinone A (QA). By formation of P^+Q^- a charge separation is achieved across the membrane and a redox potential difference is formed. (courtesy of Thieme Publisher).

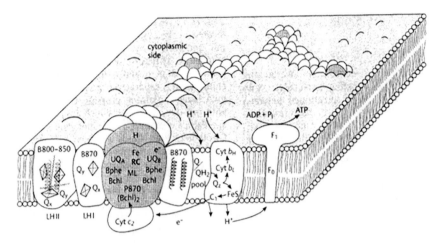

FIGURE 4.6. The anoxygenic photosynthesis center of the purple bacterium *Rhodobacter capsulatus* in a schematic overview. The photochemical reaction center RC consists of the protein subunits M and L (ML, dark gray), capped by protein H on the cytoplasmic side (*top*) and contacted from the periplasmic space by cytochrome c_2 (Cyt c_2), which shuttles electrons between the bc_1 complex, depicted to the *right*, near the ATP synthase complex. In RC you find the special pair of bacteriochlorophylls $(BChl)_2$, two monomeric bacteriochlorophylls, two bacteriopheophytins, the ubiquinones A and B and one Fe atom. LH1 depicts the light harvesting complex with the core antenna B870. Under low light intensities, the peripheral antenna B800-850 is the dominating light harvesting complex (LH2). The pool of quinones in the membrane (Q/QH_2) connects the RC with the cytochrome bc_1 complex (C1). The proton:ATP synthase uses the proton gradient across the membrane to generate ATP from ADP and P_i. (courtesy of Thieme Publisher).

B800 and B850 in LH2 (the number indicates the approximate wavelength in nanometer of maximal absorption; recall that the smaller the wavelength, the higher the energy of the photon). They pass the photon energy to the LH1 bacteriochlorophyll B870, which also acts as a light harvester on its own. The next step in photon transfer is to a bacteriochlorophyll pair in RC. In the RC, photon flux is changed into electron flow across the photosynthetic membrane, the change from photon to electron transport occurs during charge separation in the special bacteriochlorophyll pair. The electrons pass from bacteriochlorophyll via an intermediate to quinones, they mix with the ubiquinone pool in the membrane, which provide the electrons to the cytochrome bc_1 protein complex and from there to a periplasmic protein, cytochrome c. Here at the latest you should have a déjà vu experience: ubiquinone → cytochrome bc_1 → cytochrome c is part of the respiratory electron pathway. And, as in respiration, photosynthetic membrane electron transport is coupled to pumping of protons, which builds up a proton gradient. This concentration gradient powers ATP synthesis via an ATP synthase as in respiration. In light-exposed purple bacteria, cytochrome c shuttles the electron back to the RC and maintains what is called a cyclic electron transport. In fact, this cytochrome c-delivered electron enters again the electron flow through RC, ubiquinone, and cytochrome bc_1 and c, hence the name. Alternatively, electron abstracted from food molecules (mostly organic molecules, but also reduced sulfur compounds) can be fed into the electron flow through the RC.

The Reaction Center

The structure of the photosynthetic RC was solved in 1985 at the Max Planck Institute in Munich (Deisenhofer et al. 1985). This was the first structure for an integral membrane protein, which earned the researchers the Nobel Prize in chemistry. The *Rhodopseudomonas* RC has in its core 11 α-helical regions that span the membrane region. Five helices are each contributed by the L and M protein subunits of RC. Except for one longer loop in M, both subunits L and M can be superimposed and are probably the result of gene duplication. The core region has thus a twofold symmetry axis. The 11th helix is contributed by subunit H, which otherwise builds a globular domain on the cytoplasmic side of the membrane. The periplasmic side of the RC is capped in the crystal structure by the cytochrome c protein. The proteins are mainly structural elements to hold the cofactors in appropriate spacing. The cofactors belong to the oldest inventions of the biochemical evolution: the Corrin rings. Specifically, there are four hemes in cytochrome c, tilted against each other but in a spacing allowing easy electron transfer. Then comes the critical crossing point where photons and electrons meet at the special pair of bacteriochlorophyll, D_L and D_M (P870). The photon transferred from LH1 onto the special pair leads to the charge separation process. The released electron passes within 3 ps (pico $= 10^{-12}$) to the next station, a bacteriopheophytin (a chlorophyll derivative lacking the central Mg^{2+} ion). Then comes the electron transfer to the first ubiquinone Q_A, an isolated iron ion in the symmetry axis, and a second ubiquinone Q_B. Californian scientists provided 10 years later the next exciting X-ray structures. They cooled the RC down to cryogenic temperatures either in the dark or under illumination. They observed one major light-induced structural change that directly clarified the mechanism of the electron–proton transfer (Stowell et al. 1997). The half-reduced ubiquinone makes a 4.5 Å movement toward the cytoplasmic membrane accompanied by a 180° propeller twist around its isopren tail. The high resolution of the structure also visualized two paths of well-ordered water molecules from the cytoplasm toward Q_B^-. These are the water channels that guide the two protons to Q_B. The RC binds the ubiquinone with decreasing affinity in the order $Q_B^- > Q_B > Q_BH_2$. This assures that the half-reduced quinone radical will not leave the RC since this would be biologically dangerous. If the ubiquinone is fully reduced after the second electron transfer, the binding affinity drops, the reduced species diffuses away into the membrane and is replaced by the fully oxidized ubiquinone species, and the cycle can start again. However, how can the reduced ubiquinone diffuse away when we heard before that RC is surrounded by an LH1 ring? The LH1 ring consists of a large $(\alpha\beta)_{16}$ cyclic structure that surrounds the RC located in the central "hole." The core antenna consists of two small proteins α and β, arranged in 16 units, and associated with about 30 bacteriochlorophylls B870 stacked between the α- and β-helices. First, it was thought that the structure is symmetric (Karrasch et al. 1995). However, refinement of the structure revealed that one helix pair is displaced such that a gate is created in the circle. Small wonder: The gate is adjacent to Q_B^- the diffusion pathway for Q_B exchange with the membrane pool of ubiquinone is open (Roszak et al. 2003).

A Second Light Antenna

There is a second more variable antenna complex, LH2, which interconnects the RC–LH1 core units. Its structure has been resolved, and it is of breath taking beauty and comparable to that of LH1, but quite distinct from that of plant or cyanobacteria antenna systems (McDermott et al. 1995). It consists again of only two proteins, α and β, of mere 53 and 41 aa length, respectively. The transmembrane helices of nine α proteins are packed side-by-side to form a hollow cylinder of 18 Å radius. The nine β proteins are arranged, somewhat tilted to the first cylinder, as a second wider cycle of 34 Å radius. A ring of 18 B850 bacteriochlorophyll molecules is located between the α- and β-cylinders. These chlorophylls act like an electronic storage ring, which can delocalize the excited state rapidly over a large area. The bacteriochlorophylls are closely associated, perpendicular to the membrane plane, and only separated by 9 Å Mg–Mg distances (the central magnesium ions complexed in the two chlorophyll rings). A second ring of nine B800 bacteriochlorophylls is located between the individual subunits of the β-cylinder. These chlorophylls are parallel to the plane of the membrane. The elongated carotenoid molecules connect both chlorophyll rings creating a graceful crown structure. When a bacteriochlorophyll molecule is excited by light, its first excited singlet state lasts only for a few nanoseconds. The light harvesting system must therefore be able to transfer the absorbed energy to the RC in a shorter time than that. The dense packaging of the two LH systems with bacteriochlorophylls and their carefully ordered orientation assures this transfer. The energy transfer within the LH2 occurs from B800 to B850 in the incredible time of 0.7 ps and from LH2 to LH1 in 10 ps. The entire system is organized like a funnel, a periphery that collects light at the lowest wavelength, namely carotenoids and B800, followed by transfers to B850 → B870 → RC (Herek et al. 2002). Recently, photosynthetic membranes from purple bacteria were investigated by atomic force microscopy, a technique which reveals the native structure of these membranes since it can work with buffered solutions at room temperature and under normal pressure (Bahatyrova et al. 2004). The pictures showed a slightly chaotic landscape of RC–LH1 rings organized as dimers, which are then further organized as linear arrays of dimers. A protruding central H subunit topped the RC. Smaller, centrally empty rings of LH2 connected the dimer rows as a type of matrix. Apparently, a high degree of order is not essential for transmitting the light energy; the basic requirement is only multiple contacts.

Mutational Outlook

The RC in *Rhodobacter sphaeroides* is functionally similar to the PSII in cyanobacteria as demonstrated by mutated RCs. By changing the protein environment of the special bacteriochlorophyll pair, a higher redox potential was achieved for the bacteriochlorophyll, and tyrosyl radicals were observed in the purple membrane RC analogous to the intermediates of the water-splitting reaction in PSII (Kálmán et al. 1999).

Regulation of Photosynthesis

Photosynthesis is optional in purple bacteria of the *Rhodobacter* group. During strict aerobic growth, the RC and LH systems are not synthesized. The redox components of the membrane cytochrome system are, however, constitutively maintained. The reason is simple: They are used both by photosynthesis and respiration. When the oxygen pressure falls and respiration becomes impossible, characteristic morphological changes occur in this facultative phototrophic bacterium. The cytoplasmic membrane invaginates and builds intracytoplasmic membranes. Under low or no oxygen pressure, the molecular architecture of the antenna system is determined by the light intensity. Under high light, the ratio of bacteriochlorophyll in LH/RC is small; the peripheral LH2 is not well developed. However, under low light conditions, the LH2 system is maximally built up to allow the most efficient light capture to assure the survival of the cell. The synthesis of the different elements (bacteriochlorophyll, LH, and RC proteins) is coordinated. Notable is the clustering of nearly all genes in a 46-kb genome region of *Rhodobacter capsulatus*. Only the proteins from the LH2 complex are encoded in a separate cluster. A two-component sensor–effector system RegA and RegB senses the oxygen level and regulates the promoter use in this gene cluster by changing the length of the transcripts and their abundance by up to a factor of 100. The mRNA stability is posttranscriptionally regulated—the mRNAs encoding more abundant proteins have a longer half-life.

Cyanobacteria and the Invention of Oxygenic Photosynthesis

The Combination of Older Inventions Creates Revolutionary Novelty

Cyanobacteria (Figure 4.7) made one major invention in the quest for food, and this will merit them a place in the history of life: Cyanobacteria invented the water splitting in photosynthesis (Figure 4.8). This process was of such fundamental importance for the further development of life that it changed not only life on Earth, but also the chemistry of the oceans and the atmosphere alike. The importance of this chemical reaction is quickly told. Other forms of photosynthesis need an electron donor, a food molecule, which the bacteria had

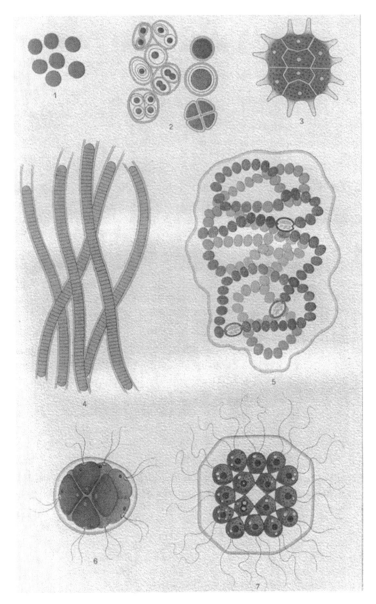

FIGURE 4.7. Cyanobacteria from freshwater and damp soil, which build colonies (CB). *Gloeocapsa* (*2*)—after division the cells remain within multiple gelatinous layers; *Chroococcus* (*3*)—forming a Coenobium colony; *Oscillatoria* (*4*)—diving into long linear threads; *Nostoc* (*5*)—growing like pearls on a string in a gelatinous layer. Note the heterocysts with thicker cell wall. They are prokaryotes and for comparison are shown three true eukaryotes, algae of the Chlorophytes class, belonging to the order Volvocales; *Pandorina* (*6*); *Gonium* (*7*) and the order Chaetophorales; *Pleurococcus* (*1*). Cyanobacteria are next to humans the main player in this book. Their marine ancestors are the "inventor" of oxygenic photosynthesis.

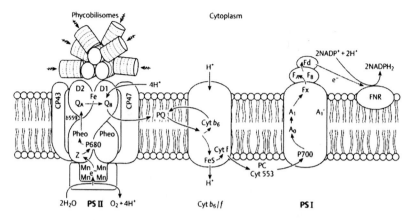

FIGURE 4.8. The oxygenic photosynthesis center of cyanobacteria. Photosystem II (PSII) obtains light energy via phycobilisomes, overlaying PSII at the cytoplasmic side, and the chlorophyll *a*-containing antennae CP43 and CP47. The water-splitting manganese site is at the luminal side. Plastoquinone PQ connects PSII with the cytochrome $b_6 f$ complex, which functions as a proton pump. Plastocyanin (PC) or cytochrome 533 (cyt 533) connects the latter with photosystem I, a plastocyanin: ferredoxin oxidoreductase. PSI is similar to the RC of green sulfur bacteria and *Heliobacterium*. FNR stands for ferredoxin:NADP oxidoreductase. (courtesy of Thieme Publisher).

to search in the biosphere, and its niche was limited to spaces on the planet where this substrate was found. The electron donor in oxygenic photosynthesis is water. This compound is the most abundant molecule on the surface of the Earth since 70% of the planet is covered by oceans. With this invention, cyanobacteria became independent from any special food source: They could thrive everywhere provided that water and light was present. Not surprisingly, cyanobacteria became one of the major primary producers in the ocean. However, the physiology of cyanobacteria is more complex than that because phototrophy is only an option for cyanobacteria, not a necessity. *The release of molecular oxygen was initially probably only a disturbing side effect since means had to be invented by cyanobacteria to handle this potentially dangerous chemical. Probably, relatively early cyanobacteria discovered that molecular oxygen was a smart weapon in the fight for light—it could be used to chase other competing microbes from their surrounding into deeper and thus less sun-exposed areas. Cyanobacteria got a competitor-free place under the sun until the competitors found ways to deal with oxygen.* We find a number of other remarkable biochemical capacities in cyanobacteria: A photosynthetic apparatus consisting of two photosystems which allowed electron transport that could be used both for energy production in the form of ATP synthesis and for the production of reducing equivalents in the form of NADPH. This double exploitation allowed a more efficient

CO_2 fixation via the Calvin cycle. The ample availability of both energy and reducing equivalents made demanding biochemical reactions like N_2 fixation possible. Cyanobacteria became a major contributor of fixed nitrogen in the upper layers of the ocean. If we discuss the different biochemical features of cyanobacteria in this section, we should not forget that it were not cyanobacteria that invented these processes. The CO_2 fixation via the Calvin cycle and the N_2 fixation were already mastered by purple bacteria, for example. It was only the special design of the photosynthetic apparatus in cyanobacteria that allowed an upscaling of these processes to a global level, which was not possible in purple bacteria. Cyanobacteria only intelligently combined what other bacteria invented before them.

Evolutionary Mechanisms for Novelty

Without the power of HGT, such combinations of unique properties in photosynthesis would have taken much longer times to develop. In fact, they would have had difficulties to develop at all because one underlying principle in evolution is that any invention must have had a selective advantage for the inventor however imperfect it is at the actual stage. *It is thus much easier to screen the invention space of the prokaryotic world for usable bits and parts, to combine them and to try them. In such a way, systems that already fulfilled a physiological function in other prokaryotes could be tested for synergistic effects. All this sounds terribly theoretic: How can an organism screen the DNA sequence space for usable parts? From where does the engineering insight come to construct new biochemical solutions? Of course Nature does not follow teleological principles: Evolution has no design that it tries to realize. Nevertheless, the physical processes exist that underlie this depicted scheme. Teleology comes from a combination of two undirected processes: One is the constant shuttling of DNA elements between prokaryotes in the pervasive HGT events. Cells actively take up DNA from the environment in the pursuit of new genes that might be of advantage to the cell. DNA is picked up by transformation either as DNA fragments or as plasmids. Other DNA comes without invitation; phages bombard constantly the bacterial cells and try to get their DNA into the cell. Certain types of phages get their DNA integrated into the bacterial genome and then sometimes contribute functions that are of selective advantage to the cell. And here you have the other player: Overlaid on this random process of DNA trafficking between prokaryotic genetic systems comes another process that blindly tests all organisms for their relative fitness. Fitter solutions get amplified, less fit solutions decrease in frequency and are finally wiped out. A combination of two undirected processes is at work: first, random DNA transfers and DNA mutations (of course, prokaryotic evolution is not governed by HGT, but by the accumulation of successive point mutations in vertical lines of inheritance), and, second, selection, which together create a process that looks directed in the hindsight. R. Dawkins proposed for this process the picture of the "blind watchmaker."*

Cyanobacteria and their Phages

All these processes can be nicely illustrated by cyanobacteria: first, because they are special and a key player in the evolution of life and, second, because they are so intensively investigated. For example, few bacterial systems are so carefully documented for the ecological role of bacteriophages in the natural environment. The mortality of cyanobacteria is determined by phage lysis, but cyanobacteria apparently evolve to escape from this onslaught. The genome analysis of *Synechococcus* shows "variability islands" that resembled "pathogenicity islands" in pathogenic bacteria in representing horizontally acquired genes of selective advantage. In this category, *Synechococcus* showed many genes involved in carbohydrate modification of the cell envelope (Palenik et al. 2003), possibly an adaptation to escape from phage predation and to evade metazoan grazers. A characteristic observation is the coexistence of cyanophages and cyanobacteria in the same environment. Frequently, the titers of strain-specific phage are low ($<10^3$ per ml) in face of high total cyanobacteria abundance ($>10^5$ per ml). Not all phages are strain specific, some have a much broader host range infecting different clades of *Prochlorococcus* and *Synechococcus* (Sullivan et al. 2003). *This might reflect different basic strategies between morphologically distinct phages or the consequence of an arms race between bacteria developing resistance and phages that undermine this defense by a counterdefense. Ironically, this arms race is at the end not necessarily detrimental to the prey. The selection pressure from phages favors the biodiversity in their prey. Phages are essentially nucleic acids wrapped with proteins. Therefore, cells can easily exploit phages as gene transfer particles (Canchaya et al. 2003a).* A striking case is provided by a cyanophage distantly related to T4, which carries the RC proteins D1 and D2 from PSII on the viral genome (Mann et al. 2003). It is now known that most T4-like cyanophages carry photosynthesis genes on the phage genome. *This might be a ruse of war from the phage: To be successful, it needs a metabolically active host. The Achilles heel of photosynthetic energy production is exactly protein D1.* As one will see in the following section, D1 holds the cofactors for the photosynthetic electron transfer chain in place. This is a dangerous business because electrons can spill chemical damage to the carrier protein. Cyanobacteria and to an even greater extent plants have developed a repair system for D1. However, if the repair system is overstretched, D1 is not any longer repaired resulting in photoinhibition. *As this is linked to a decreased energy charge, the phage will have a problem during its replication. This might be the reason why this phage comes with photosynthetic genes as a purely selfish help. As we will see soon, this observation fits neatly to the evolutionary analysis of bacterial photosynthesis: The combination of two photosynthesis systems is apparently the result of extensive lateral gene transfer.*

Photosynthesis genes of cyanobacterial origin have now been described on phages that infect *Prochlorococcus* (Sullivan et al. 2005; Lindell et al. 2004) and *Synechococcus* (Mann et al. 2005; Millard et al. 2004), the numerically dominant phototrophs in ocean ecosystems. These genes include *psbA*, which encodes the PSII core RC protein D1, mentioned above, and high-light-inducible genes *hli*. The latter protect the photosynthetic apparatus from photodamage by dissipating excess light energy. Scientists from Boston designed experiments to test the role of phage photosynthesis genes in *Prochlorococcus* cells infected with a T7-like podovirus (Lindell et al. 2005). The dependence of the phage on ongoing photosynthesis was demonstrated by a decreased phage yield when the infection was done in cells held in the dark or in cells in the light where photosynthesis was poisoned. Despite the effect of phage on host transcription and translation, photosynthesis was sustained during infection. Host *psbA* and *hli* transcription declined substantially, but was nearly compensated by the transcription of the corresponding phage-encoded genes. Some phage biologists had predicted that nonphage genes integrated into the viral genome come as independent transcription units ("morons") that are expressed separately from the remainder of the phage genome (Hendrix et al. 2000). However, this was not the case for this *Prochlorococcus* phage: Its photosynthesis genes were found directly upstream of the major capsid gene, and they were cotranscribed with the phage-head genes, suggesting that they became integral part of the phage genome. Two important conclusions can be drawn from these experiments. First, on a global scale, phage genes play an important role in the conversion of light into chemical energy. Second, phages are important vectors for lateral transfer of photosynthesis genes between bacteria.

In the following, we will take time to investigate some of the arguments with cyanobacteria, not only because they are so important for the cycling of food in the biosphere but also because they nicely illustrate general principles.

Getting Closer to the Water-Splitting Center: Photosystem II

The *Synechococcus–Rhodopseudomas* Crystal Comparison

The central innovation of cyanobacteria is the oxygen-evolving center (OEC), and it is thus logical to start the description with this system. OEC is part of PSII. All bacterial and plant photosythetic systems identified to date belong to two distinct groups, conveniently classified by the terminal electron acceptor of the RC electron transfer chain. In type I systems this is a high potential $Fe_4–S_4$ cluster and in type II systems a quinone-type compound. According to this definition, purple bacteria have a type II system. As already discussed before, cyanobacteria have probably developed their type II system from purple bacteria despite the fundamental difference with respect to the input of electrons into the RC. There is in fact some sequence homology between the major structural proteins from both RCs (Schubert et al. 1998), but it is not so impressive to allow being definitive with respect to a common origin. However, proteins that conserved only weak sequence similarity sometimes keep an impressively conserved 3-D structure, which allows tracing their origin from a common ancestor. The resolution of the crystal structure of PSII from the cyanobacterium *Synechococcus* was in that respect revealing (Zouni et al. 2001). At first glance, PSII from *Synechococcus* looks more complicated than the RC from purple bacteria: It is composed of 17 subunits, it is a homodimer, it contains the amazing number of 36 chlorophylls and seven carotenoids per monomer, and it has a substantially more complicated luminal structure, where the RC from purple bacteria just shows the association with cytochrome *c*. In contrast, at the extraluminal side, *Synechococcus* shows only a flat surface where the photosynthetic complex in *Rhodopseudomonas* is capped by protein subunit H. However, when the inner core of the RC is dissected in *Synechococcus*, similarities to the *Rhodopseudomonas* system become evident: Two groups of five transmembrane helices are arranged in two semicircles, which are interlocked in a handshake motif. The names of the central RC protein players of both photosystems differ, but the arrangement of their α-helices is so similar that it becomes clear that the cyanobacterial photosynthetic proteins D1 and D2 correspond to the subunits L and M in the RC from purple bacteria. The RC in *Synechococcus* is flanked by two antenna proteins—CP43 and CP47. Both consist of six transmembrane helices arranged as trimers of dimers, which coordinate 12 and 14 antenna chlorophylls, respectively. Further, transmembrane helices could be attributed to a cytochrome *b* and numerous smaller proteins, several of which are involved in the stabilization of the dimer interface. If there are still some doubts about the common origin of the RC of both bacteria, they are alleviated when the organization of the cofactors are studied. One finds the same special pair of chlorophyll near the luminal (P side), an accessory chlorophyll, a pheophytin, a fixed Q_A quinone, a nonheme iron, and a free Q_B quinone. The symmetrical arrangement of the cofactor chain is identical, only the relative distances and relative orientations of the planes of the tetrapyrolle rings differ somewhat. The main differences between the *Synechococcus* and the *Rhodopseudomonas* RCs concern those proteins enabling the use of water as electron donor in cyanobacteria. The refinement of the crystal structure to 3.5 Å resolution revealed much more details at the luminal side (Ferreira et al. 2004). The resolution was so fine that it allowed a look into the water-splitting center. This site is the holy grail of oxygenic photosynthesis, and its invention changed the further course of biological evolution on our planet fundamentally.

Bioinorganic Chemistry

Comparisons of cyanobacteria, green algae, and land plants reveal that the same machinery is found at the active site of all O_2-producing organisms that have been studied to date. Since its invention about two to three billion years ago, Nature has not come up with alternative solutions to the water-splitting process performed in the OEC of PSII. In view of the variety of enzymatic sites developed in enzymes for many biochemical reactions, it is nearly unbelievable that only one solution arose. As the oxidation of water involves a complex four-electron, four-proton oxidation reaction, we are here perhaps confronted with one of the thermodynamically most challenging multielectron reactions in biology. Is there really only one chemical solution to this problem? Notably, as in other fundamental reactions in biochemistry, the PSII–OEC possesses an *inorganic RC. These relics of inorganic chemistry in major enzyme systems represent perhaps the takeover of catalysts that developed in what one might call the iron–sulfur world when the first life processes were still mediated by metal centers.*

S-Cycle

The first model for the OEC that integrated many physiological and physico-chemical data was developed in the 1970s and introduced the so-called S-state cycle. In the S0-state, water is bound to a manganese cluster of the OEC. In the model of Hoganson and Babcock (Hoganson and Babcock 1997), each transition to S1, S2, S3, and S4 was linked to the release of one e^- and H^+. The final S4 to S0 transition was linked to the O_2 release and binding of two new water molecules.

Chlorophyll Photochemistry

A basic experimental observation was that oxygen was developed only with every fourth of a sequence of short light pulses. The light pulses activated the special pair chlorophyll P680. To understand this process, we need a bit of photochem-istry. Sunlight comes as a continuum of electromagnetic radiation. However, the light-harvesting apparatus of green plants can only use those wavelengths, which the chlorophylls absorb. Chlorophyll a absorbs in the violet and orange range, while green light is not absorbed giving plants the characteristic green color. A pigment molecule like chlorophyll becomes excited when absorption of light energy causes one of its electrons to shift from a lower-energy molecular orbital to either one of two more-distant, higher energy orbitals. Absorption of red light lifts the electron to the first excited singlet state, while absorption of blue light allows the shift to the second excited singlet state. The singlet state contains electrons with opposite (antiparallel) spins and is relatively short-lived. Once excited the electron can return to the more stable ground state, which it can do in several ways. In relaxation, energy is released as heat. In fluores-cence, the electron returns to the ground state by emitting light at a somewhat longer wavelength. Alternatively, the excited pigment molecule can transfer its excess energy to another molecule in its vicinity. This energy transfer is an important vehicle for the movement of absorbed light energy through an array of pigment molecules in antenna structures. In phototrophic bacteria and plants, many bacteriochlorophyll or chlorophyll molecules are associated with each RC. The transfer of excitation energy from one pigment molecule to another does not involve emission and readsorption of photons but is mediated by a resonance process called Förster energy transfer. Over short length (pigments are separated by only 1.5 Å), the transfer occurs with high speed (1 ps) and high efficiency (99%). It is here where important differences occur between purple bacteria and cyanobacteria; differences, which explain why only the latter are capable of splitting water. In purple bacteria, light is absorbed by a special pair of bacterio-chlorphylls called P865 (because the maximal absorption occurs at 865 nm), in contrast cyanobacteria have P680 chlorophylls (as you suppose correctly they absorb light maximally at 680 nm, the "special pair" is actually a chlorophyll tetramer). Light of shorter wavelength also contains more energy. Due to this lower energetization, the excited bacteriochlorophyll from purple bacterium is not a sufficiently strong oxidant to oxidize water.

PSII Electron Transfers

The crucial process of photosynthesis is the charge separation event where the activated chlorophyll P680* loses its excited electron to an electron-acceptor molecule. This process is the dividing line from photophysics to photochemistry that converts light energy transfer reactions into the electron transport chain in the photosynthetic membrane. The first electron acceptor is a pheophytin (a chlorophyll without a coordinated magnesium ion in the middle of the tetrapyrrole ring system of chlorophylls). This process is ultrarapid (3 ps). The next electron transfer processes are slow by this standard and reach a quinone called Q_A in 200 ps, then a second quinone called Q_B in 100 ps (this slow process still involves an iron atom as electron carrier) resulting in a plastosemiquinone Q_A^- and Q_B^-, respectively. Most of these electron transfers occur in the D1 protein. D1 is therefore most susceptible to photochemical damage. Repair consists of continuous replacement of protein D1. Then a second electron is transferred along this pathway and produces a fully reduced Q_B^{2-} molecule. This molecule takes up two protons from the stromal side of the chloroplast yielding a plastoquinol $Q_B H_2$. It dissociates from the RC and diffuses away in the lipid membrane. PSII feeds thus electrons into the electron transport chain.

By vectorial consumption of stromal protons, it contributes directly to the proton gradient across the thylakoid membrane. However, these are not the only protons created by PSII. Further protons are created in the water-splitting reaction. How does this happen?

The Classical Model for the Manganese Center

After the light-induced ejection of an electron, chlorophyll P680 forms a cationic radical that has a very high oxidizing potential of +1.4 V. This high oxidation power of the P680 radical allows the oxidation of a redox-active tyrosine residue in the D1 subunit to the TyrZ neutral radical. To sustain a cyclic process, the tyrosine side chain must get back what it lost to P680. The electrons are provided by the manganese center, which abstracts them from water. Water is actually consumed by this process. It is decomposed into electrons for the transport chain, protons for the proton gradient across the membrane and what remains is chemical waste for the photosystem and is consequently discarded. This "waste" is actually di-oxygen. Since this waste should become the super-fuel for the evolution of all higher life-forms, its generation merits a few comments.

Initially, chemists imagined a chain of four manganese ions linked via oxygens in a C-shaped cluster. In the S0 state, the water molecules are bound to the manganese ions at the open ends of the C-chain. Enter the neutral tyrosyl radical

TyrZ which abstracts the first e^- and H^+ from one end (S1). This process is repeated at the other end of the manganese cluster leading to two hydroxyl intermediates (S2), which prepares the formation of the O–O bond (S4). This Mn–O–O–Mn chain then collapses and releases subsequently di-oxygen. The manganese ions shuttle during this operations between the 2+ and 4+ oxidation states. This model is chemically very appealing as it shows clear analogies to amino acid radical functions in other enzymatic reactions.

The Clausen–Junge Model

However, none of the postulated intermediate oxidation products of water have been detected so far. Recently, biophysicists from Osnabrück in Germany used a simple trick to stabilize an intermediate. Since the basic reaction of water splitting is simple, namely $2\,H_2O \leftrightarrow O_2 + 4\,H^+ + 4e^-$, they reasoned that if they would rise the concentration of a product of the equation, they could push the equilibrium toward the left side of the equation. The easiest way was to raise the oxygen concentration. And indeed, at an oxygen pressure of 2.3 bar, which corresponds to the 10-fold ambient oxygen pressure, they succeeded to suppress the progression to oxygen by 50%. At saturating oxygen pressure of 30 bar, they could stabilize an intermediate and monitor it by near UV spectroscopy (Clausen and Junge 2004). By doing so, they demonstrated that the reaction at the manganese center can be subdivided into at least two electron transfer events. They suspected that a peroxide intermediate had escaped detection because of its short lifetime under normal pressure conditions. Their experiment demonstrated the backpressure of oxygen on the photosynthetic manganese center. This observation has a notable consequence. Since the start of oxygenic photosynthesis, the oxygen concentration in the atmosphere has been creeping upward from negligible quantities to about 20% of the current air mass. The inhibitory effect by oxygen, which Clausen and Junge observed in Osnabrück, might act as a brake on further oxygen rises in the atmosphere. Photosynthesis, therefore, might be near to its culmination point with respect to oxygen release.

The S4' State

Instead of attempting to trap intermediates by inhibiting dioxygen formation, physicists from Berlin monitored the redox processes at the fully functional manganese complex in real time by X-ray absorption spectroscopy (Haumann et al. 2005). They succeeded to identify the S4 state formed after the absorption of the third of the four photons driving the S-cycle. Their data suggest that the S4 intermediate is not formed by electron transfer from the manganese complex to the tyrosine Y_Z radical as proposed previously, but by a deprotonation reaction probably at an arginine residue of the CP43 protein or directly from the water molecule. The proton moves then in a proton path to the luminal surface. Only then an electron transfer occurs at S4 and leads to a new hypothetical intermediate called S4' in which four electrons have been extracted from the manganese ligands and from the two substrate water molecules in the manganese complex. The manganese is reduced and dioxygen formation occurs with the longest halftime of 1 ms in the S-cycle compared to halftimes ranging from 30 to 200 μs for the other S-state transitions.

The Mn Cube in the Crystal Structure

Is there anything to be added to this chemical scheme for the manganese center from the structural point of view? The crystallographers observed a cube-like cluster built from three Mn and one Ca ions with an outside fourth Mn ion. The Ca ion was coordinated with TyrZ. They proposed that the O=O bond formation occurs by a nucleophilic attack of the water molecule bound to the Ca ion on the water molecule bound by the outside Mn ion. Two nearby amino acid residues from protein D1 stabilize the intermediate. This interpretation fits data where their mutagenesis totally inhibited the evolution of oxygen.

A proton channel emanates from the manganese center. These four protons from water splitting make one side of the membrane more acidic. Together with the four "chemical" protons consumed by the reduction of the plastoquinone Q_B at same side of the membrane, this adds up to a net difference of $8\,H^+$ over the membrane. This electrochemical gradient is part of the proton motive force that energizes the photosynthetic ATPase for ATP synthesis.

Evolutionary Patchwork: Photosystem I

The Z-Scheme

What role is played by PSI in cyanobacteria and plants? Remember that the electrons provided by PSII have to be transferred to $NADP^+$. This process involves first an electron transport chain and a second light-induced charge separation event. Spectroscopic experiments conducted with chloroplasts in the 1940s revealed the existence of chlorophylls that absorb light at 680 and 700 nm. Playing with different excitation wavelengths led to the Z-scheme of chloroplast electron transfer. This scheme remained the leading model for noncyclic photosynthetic electron transfer and applies also to cyanobacteria. The Z analogy results from the vertical placement of each electron carrier on a scale of the redox potential, the zigzag is caused by the redox potential increases caused by light absorption and subsequent electron flow toward more electropositive acceptors.

The Cytochrome b_6f Mediator

A wealth of biophysical, biochemical, and recently also structural data can be summarized as follows. Three membrane complexes compose the chloroplast/cyanobacteria electron transport chain. The first is PSII described in the preceding section. The second complex is the cytochrome b_6f complex: It transfers electrons from the reduced plastoquinone to the oxidized plastocyanin. The third is PSI, discussed further below. The cytochrome b_6f complex from the cyanobacterium *Mastigocladus* (Kurisu et al. 2003) is structurally and functionally similar to the cytochrome b_6f complex from the green algae *Chlamydomonas* (Stroebel et al. 2003) and the cytochrome bc_1 complex (complex III) from mitochondria. All these protein complexes receive electrons from the membrane-soluble quinones (plastoquinone and ubiquinone) and pass the electrons to small soluble proteins (cytochrome c_6/plastocyanin and cytochrome c). The cytochrome f corresponds to the cytochrome c_1 subunit in the mitochondrial complex, likewise the chloroplast cytochrome b_6 resembles the mitochondrial cytochrome b. Both membrane complexes contain a Rieske-type Fe–S protein, and they translocate electrons via a so-called Q cycle. It is obvious that this membrane complex in both electron transport chains derives from a common ancestor. The cytochrome b_6f complex performs the rate-limiting step for the electron transport in the chloroplast, due to the "slow" oxidation of plastoquinol taking 10 ms.

Cytochrome c_6–Plastocyanin Replacements

The small proteins serving as electron carriers between the cytochrome b_6f complex and PSI are worth a further digression. The same function is served with two rather different proteins: cytochrome c_6 and plastocyanin. The former is a heme system wrapped by several protein helices, and the latter contains a copper ion covered by several pleated protein sheets. Despite different electron carrier cofactors and protein surrounding, they fulfill the same function. Actually, they can replace each other even across quite substantial evolutionary distances—first between the algae *Chlamydomonas* then to a somewhat lesser extent between the watercress *Arabidopsis*, the model in most of plant molecular biology, and *Synechococcus*, the model cyanobacterium, despite the fact that the aa-sequence identity is no more than 20% (Gupta et al. 2002). Both proteins are found in *Arabidopsis*, each can be inactivated individually without phenotypical effect to the plant, whereas a double inactivation was lethal. Some data indicate that the plant cytochrome c_6 not only changed the surface charge but also adapted to a new function (Molina-Heredia et al. 2003). However, in cyanobacteria, both proteins can be used alternatively and seem to be expressed under different nutrient limiting conditions. If iron is in limited supply, the copper protein plastocyanin will serve the electron transfer business, whereas under copper limitation the iron protein cytochrome c_6 takes over. Despite the genetic economy imposed by a small genome size, cyanobacteria thus have some flexibility on how to cope with metal deficiencies, which might be one of the major nutritional limitations for their growth in the ocean.

Structural Analysis of Photosystem I

Now back to photosynthesis. Next in the chain is PSI. PSI, illuminated with light of 700 nm, generates a strong reductant (P700*) that is capable of reducing $NADP^+$. In order to do so, it must have a redox potential more negative than -1.14 V. Illumination also creates a weak oxidant that can take the electrons from the reduced cytochrome c_6. The structure of the cyanobacterial PSI was resolved at 2.5 Å resolution (Jordan et al. 2001) and allowed to identify the path of the electron from the P700 chlorophyll dimer to a nearby accessory chlorophyll, then to chlorophyll A_0. Each step crosses about 8 Å of distance. Then follows a nearby quinone, and thereafter the electrons are translocated over greater steps of about 15 Å distance to three different Fe–S clusters, F_X, F_A, and F_B. In this way, the electron travels from cytochrome c_6, the soluble protein at the luminal side, over a 30-Å distance of the membrane to another soluble protein—ferredoxin—located at the opposite stromal side. Ferredoxin does not transfer its electrons directly to $NADP^+$, but via a flavin-containing membrane-bound NADP reductase. The reduction of $NADP^+$ to NADPH consumes a further proton at the stromal side, which adds further to the overall proton gradient, about 12 protons move across the membrane for each set of four electrons created by the water splitting, yielding one O_2 molecule. The absorbed light energy is stored as a proton gradient.

Structural biologists revealed for PSI from *Synechococcus* a rather complex structure consisting of a trimer in cloverleaf configuration. Each monomer contains not fewer than 12 protein subunits and 127 cofactors, including 96 chlorophylls, 2 quinones, 22 carotenoids, and 3 Fe_4–S_4 clusters. The structure showed a dense and bewildering array of transmembrane α-helices. Most are contributed by the major subunits PsaA and PsaB, which are also associated with the majority of the chlorophylls. In contrast to the RC from purple bacteria, which collects light energy through a separate light-harvesting complex, PSI from cyanobacteria has the antenna chlorophylls built into the RC. The chlorophylls come as a ring, which surrounds the central electron transfer path and two peripheral antenna that align their pigments near both membrane faces. All these chlorophylls keep a respectful distance to the central electron pathway and leave it to two chlorophylls to focus the collected light energy on the central pathway.

Evolutionary Patchwork

What does all this complicated structural biology tell us with respect to the origin of the photosystem? First, PsaA and PsaB share similarities in protein sequence and structure. They are probably the result of an ancient gene duplication event. Second, the N-terminal parts of PsaA and PsaB resemble in their chlorophyll organization closely the antenna proteins CP43 and CP47 of PSII. Third, the central C-terminal domains from PsaA and PsaB show a structure like the subunits D1 and D2 from PSII. If these complex data are integrated, the PsaA and PsaB proteins from PSI are not only the result of gene duplication, but each Psa protein from PSI is also the fusion of an antenna and a RC protein from PSII. PSI and PSII are apparently derived from a common ancestor that diversified by both gene duplication and gene splitting and fusion (Rhee et al. 1998).

This similarity with photosystems described in other bacteria can still be extended. The operon encoding the RC proteins from the green sulfur bacterium *Chlorobium* was cloned, and its sequence showed significant sequence similarity with PSI from plants (Buttner et al. 1992). Likewise, the RC protein from *Heliobacillus* could be tracked by sequence similarity to the PSI (Xiong et al. 1998). Summarizing available data, Xiong and Bauer concluded that the various photosynthetic components have distinctively different evolutionary histories and were recruited and then modified from several other preexisting pathways in a mosaic process that appears to have no single well-defined origin. However, one issue has been solved with overwhelming support: Anoxygenic photosynthesis evolved prior to oxygenic photosynthesis.

Speculations on the Origin of Photosynthesis

Science is an enterprise of never-ending questions. Therefore, it is absolutely legitimate to ask the next question: Where did anoxygenic photosynthesis come from? How could such an intricate and complex system evolve, which is functional only when the different elements are correctly assembled? In fact, in the tradition of Charles Darwin, biologists would postulate that primitive prerunners of current photosynthesis already conferred some selective advantage to the organism that it possessed. Darwin dealt already with the argument how the complex eye could have developed, when he argued that the most primitive eyes would have increased the survival of its owner. A geologist and one of the leading characters in the early life discussion Euan Nisbet, from London, came up with an imaginative proposal (Nisbet et al. 1995). He started with the idea that life began near hydrothermal vents. This environment offers sharp frontiers of redox disparities, which allows to power life processes as we have seen in the sections dealing with the early eaters. However, vents are not only a rich environment from the food perspective, due to violence of the underlying physical processes they are also a very dangerous environment. Centimeters decide on survival: Organisms that loose touch with such vents risk starvation, whereas those that come too close risk cooking or poisoning. As the hot turbulent plume of the vent is quite unpredictable, a possibility to locate the suitable border of the plume would be of definitive advantage. Experimentalist provided then a cue: The purple bacterium *Rhodospirillum* showed phototactic behavior—it moves toward infrared light and flees visible light (Ragatz et al. 1994). Such a behavior is what you need when you want to approach a vent without getting cooked. Nisbet and colleagues calculated the thermal radiation of a body at 400 °C under the optical conditions of the deep sea. They found two bands, which fitted nicely that of bacteriochlorophyll. They hypothesized that bacteriochlorophyll evolved as a supplement to chemotrophy. Note that we meet here a second time a link between phototrophy and seeing. The other case is rhodopsin and retinal, which was first used for energy production and then became the chemical basis for vision in animals. Actually, the Nisbet's hypothesis for the origin of photosynthesis is not far-fetched. The shrimp *Rimicaris exoculata*, which lives in the night of the deep sea near hydrothermal vents has highly developed eyes (Van Dover et al. 1989). They function probably as photosensors for the infrared radiation of the hot plumes and assure thus access to the food sources near the vent and at the same time prevent the shrimp from being cooked. Nisbet's hypothesis is even more explicit: If these phototactic bacteria came into shallow water, exposure to UV light became a problem, which could have led to the evolution of other pigments to shield the bacteria from these damaging radiation. In this way, the pigment part (carotenoids) of the photosynthetic apparatus could have evolved without any connection to energy conservation mechanisms.

The Impact of Oxygen on the Evolution of Metabolisms on Earth

Life on earth depends on nonequilibrium cycles of electron transfers, involving only a handful of elements: hydrogen, carbon, nitrogen, sulfur, and oxygen. Before oxygen rose to prominence due to its liberation from water by oxygenic photosynthesis, the first microbial "traders"—to use the terminology of a thoughtful editorial (Falkowski 2006)—consumed electrons from H_2, H_2S, and CH_4 and "sold" the electrons at relatively low price to acceptors such as CO_2 or SO_4^{2-}. Over long evolutionary time periods, metabolic pathways were established that created a planetary "electron market" where reductants and oxidants were traded across the globe. Anaerobic microbes managed one or at most a few redox reactions, but they were interconnected in such a way that a sophisticated set of biogeochemical cycles developed that powered the process of life. This metabolic design, itself more the result of a frozen metabolic accident that evolved only once on earth than of a thermodynamic necessity, run the risk to become obsolete with the introduction of a new player into the electron game, namely molecular oxygen in the atmosphere. Organisms had only three options, they could go into hiding from oxygen, or go extinct, or—and this was the most creative solution—they could accommodate the new player into a redesigned metabolism. In fact, all three options were used. The introduction of O_2 represented a cataclysm in the history of life. Many organisms went probably extinct. Some used anoxic hiding successfully and survived until our days in specialized but sometimes vast ecological niches. However, many organisms used the new energetic possibilities offered by the water–water cycle to go into a high-gear metabolism. An ingenious bioinformatic approach using the KEGG (Kyoto Encyclopedia of Genes and Genomes) database, encompassing nearly 7,000 biochemical reactions across currently 70 genomes, reconstructed some aspects of this metabolic redesign (Raymond and Segre 2006). This analysis of the metabolic networks converges on just four discrete groups of increasing size and connectivity. The networks in the smaller groups are largely nested within those of the larger groups. The networks working in the presence of oxygen have about 1,000 reactions more than the largest networks achieved in the absence of O_2. Novel and augmented pathways occurred largely at the periphery of the network, but this metabolic expansion was not simply the result of the invention of O_2-dependent enzymes, but more the evolution of new reactions and pathways. Today, O_2 is among the most-utilized compounds outcompeting so basic metabolic compounds as ATP or NADH. As evolution is intrinsically conservative even in revolutionary events, selection salvaged and remodeled parts of the old anaerobic machinery and in addition had to develop safety valves against the biohazard of the new super-fuel of cellular metabolism.

The impact of oxygen concentration on the evolution of organisms is not restricted to this most important event in the history of life about two billion years ago. Even on shorter timescales, oxygen concentrations apparently played a driving force for animal evolution. Geologists observed over the last 200 million years the doubling of the oxygen concentration from about 10% in the Early Jurassic to 21% at present time (Falkowski et al. 2005). In the Jurassic, substantial oxygen concentration changes were observed, but a sustained rise of oxygen was observed over the last 50 million years in the Tertiary. As O_2 is created by water splitting in photosynthesis and then reduced back to water by respiration of organic compounds and as both processes are finely equilibrated, net oxygen enrichment of the atmosphere can occur only when organic matter (or pyrite) is buried. This was the case in the Mesozoic and Early Cenozoic with the burial of the biomass produced by eukaryotic phytoplankton, especially the large and nonmotile coccolithophorids and diatoms. The continental margins became the storehouse for organic matter; most of the world's petroleum sources are the result of this burial. The diversification of the placental mammals and birds occurred over the last 50 million years. The geologists argued that this oxygen rise was necessary to drive this evolutionary trend since birds and mammals have as homeothermic animals about three to sixfold higher metabolic demand per unit weight as reptiles from which they are derived. They argued specifically that high ambient oxygen concentrations are needed to allow the development of a placenta. The reason is that the fetus gets its oxygen from two gas-exchange processes conducted in series: the first in the alveole of the lungs from the mother and the second between the maternal and the fetal blood vessels in the placenta. The geologists also noted a parallel trend between oxygen rise and an upward surge in body size of the mammals (Falkowski 2006). In the same line, although with opposite trend, they noted that the relatively rapid decline in oxygen during the end-Permian and Early Triassic might explain part of the extinction of terrestrial animals mainly represented by reptiles at that time (Huey and Ward 2005).

Another spectacular event linked to oxygen increase in the atmosphere occurred in the Late Precambrian. This oxygen increase in the Ediacaran period probably led to the evolution of the earliest animals. Unless these early animals had circulation systems that outcompeted those of present animals, which is unlikely, these animals could evolve only with the push of a substantial oxygen increase. US geologists proposed that this increase was induced by a change in the terrestrial weathering regime (Kennedy et al. 2006). Pedogenic clay—the result of the expansion of a primitive land biota on the soil—has a much higher capacity to adsorb organic carbon material than detrital clay—the mechanically and not biologically ground rocks (Derry 2006). As the time horizon for the early colonization of land by fungi and plants was by molecular methods pushed back to 1,000 and 700 million years ago respectively (Heckman et al. 2001), fungi in association with cyanobacteria (as lichen; Figure 4.9) or with early land plants could fill the ticket for this weathering event preceding the Cambrian Revolution of animal life-forms.

FIGURE 4.9. Lichens growing on rocks of the Central Alps. The brightly colored lichen in the foreground is *Imbricaria caperata*.

The Acquisition of the Atoms of Life

From the previous sections, we got some insight into metabolically versatile cells that manage biochemical pathways that animals have never learned. Purple bacteria are not prominent cells on Earth, neither in numbers nor in ecological distribution. However, their biochemical flexibility should make us think about the basic aspects of the quest for food. One of these necessary thoughts is that organisms need food not only as a source for energy but also as a source for the molecules of life. Animals lost the capacity to synthesize the building blocks that make biological bodies from the inorganic scratch. Many prokaryotes still manage many of these basic biochemical reactions, which introduce the small number of major biological atoms (C, H, O, N, P, and S) into the food chain. These atoms make up the biomass of organisms in our planet. Many prokaryotes learned how to transform inorganic constituents into organic form. Previous sections dealt already with the incorporation of hydrogen from H_2 into biomolecules (hydrogen and bioenergetics, methanogenesis) or S into the cell biomass (sulfur worlds). Oxygen and phosphorus are relatively easy acquisitions and are only shortly mentioned. The focus of the next sections will be the difficult transformation of gaseous CO_2 and N_2 from the atmosphere into biomass.

The Easy Acquisitions: HOP

Oxygen

Virtually, all cellular oxygen is derived from water like, virtually, all oxygen in our atmosphere. In fact, only very few biochemical reactions use O_2 to introduce oxygen into biomolecules. Examples for these exceptions are hydroxylation reactions in a few obligate aerobic bacteria or the introduction of double bonds into saturated compounds. This is the case in Gram-positive aerobic bacteria that use a desaturase, which introduces with the help of molecular oxygen a double bond in the middle of a saturated C 16 (palmityl) acyl bound to the ACP carrier protein. The formation of polyunsaturated fatty acids in cyanobacteria and chloroplasts always requires this O_2-dependent mechanism. In contrast, the synthesis of unsaturated fatty acids is done in anaerobic (and many aerobic) bacteria by a dehydrase that does not depend on oxygen in the so-called anaerobic pathway of unsaturated fatty acids synthesis. The reliance of most bacteria on water for dehydration and oxygen introduction into biomolecules has a very simple reason. Oxygen was only available after the evolution of oxygenic photosynthesis by cyanobacteria. Many biochemical reactions, however, evolved already before this event and therefore had to use the only abundant source of oxygen available to them, namely water. The reaction sequence of oxygen introduction into biomolecules is the same in many modern biochemical reactions: A simple C−C bond in an organic molecule is oxidized to a double C=C bond. Water is then added across this double C=C bond, yielding a hydroxyl group linked via a single bond to the carbon atom.

Hydrogen

Reduction steps in biosyntheses are normally NADPH dependent. This molecule is the activated electron and hydrogen carrier of the cells. In heterotrophs, the reductant NADPH pool is loaded by a number of oxidation reactions of food molecules. One major group is represented by the oxidation of 3-hydroxy acids to the corresponding 3-oxo-acids with the concomitant reduction of $NADP^+$ to NADPH. The other major group of NADPH delivering reactions is the oxidation of the aldehyde group of an aldose to a lactone (an intramolecular ester). This is a prominent reaction in the pentose phosphate cycle, the major metabolic function of which is the provision of reducing equivalents for biosynthesis next to the delivery of pentose sugars for the synthesis of nucleic acid precursors (hence the name). Another source of NADPH is via a transhydrogenase that transfers hydrogen from NADH to $NADP^+$. However, in all these cases, the source of the hydrogen is an organic food molecule, which must first be introduced into the cycling of biological matter. This is actually done in the reaction of PSI described in the previous section. Ferredoxin-$NADP^+$ reductase transfers two electrons from two reduced ferredoxins to $NADP^+$, which catches a proton to create NADPH. Water is thus twice the source of H in biological molecules; the electrons are derived from the water-splitting reaction and the proton from the dissociation of water. That NADPH and not H_2 is the activated carrier of electrons/hydrogen is not surprising. Molecular hydrogen is too small, too volatile, and too reduced and thus too rare in the current oxygen-rich atmosphere to be a suitable source of atomic hydrogen for biomolecules. Since H and O are derived from water, neither atom is limiting for the growth of organisms that stay in contact with water. In some anaerobic bacteria, the donor of H in organic molecules is molecular hydrogen and not water. We will see such a reaction when discussing the CO_2-fixing pathway in acetogenic bacteria.

Phosphorus

In contrast to the previous elements, phosphorus is under many natural conditions a growth-limiting nutrient, not so much because it is rare (0.3% of the earth crust is phosphate), but because of the very low solubility of its Al, Fe, and Ca salts. Consequently, about 90% of the total phosphate mined in the US goes into fertilizers. However, in the open ocean at distance from continents, phosphorus can be limiting because of short supply. Despite these problems for phytoplankton or plant roots with phosphate, from the viewpoint of pure nutritional biochemistry, phosphorus is an easy element. Phosphorus is quickly oxidized by air to phosphate and is also transported as phosphate across the bacterial cell membrane. One gram of bacterial cell contains about 30 mg of phosphorus. ATP is not only the universal energy currency, but also the P carrier inside the cell. Phosphate can also be hoarded as an energy or chemical reserve under the form of polyphosphate granules. The cell handles phosphorus at the same oxidation level as it finds it in its environment, namely as a phosphate.

Thus, phosphate does not need to be reduced as other biological elements derived from the environment.

Trichodesmium and Phosphonate Acquisition

Phosphorus is a vital nutrient in marine ecosystems as expressed in the famous Redfield ratio for marine planktonic organism giving a ratio of C/N/P of 106/16/1. C and N enter the biogeochemical cycle via biological fixation of atmospheric gases, CO_2, and N_2. Gifted organisms like the photosynthetic cyanobacterium *Trichodesmium* can use both gas sources. They fix carbon via the Calvin cycle and nitrogen with their nitrogenase. Due to their N_2 fixation capacity (for which they are called diazotrophs), they are cornerstone organisms for the introduction of N into the oceanic surface water. However, no such atmospheric source exists for phosphorus and the P_i, which is readily available to most organisms, shows with 10^{-9} mol/kg very low concentrations in the tropical Atlantic. Despite that dearth of phosphorus, *Trichodesmium* thrives in that environment. Is nitrogen fixation by *Trichodesmium* not limited by phosphorus concentration in the surface water? Actually, US oceanographers, working in the central Atlantic, demonstrated that nitrogen fixation by *Trichodesmium* is correlated with the phosphorus content and irradiance level (N fixation is costly and needs a lot of photosynthetically produced ATP), but not with iron levels. The latter observation is surprising since N fixation needs an estimated 10-fold higher iron supply than non-N-fixing microorganisms due to the high iron needs of the nitrogenase complex (Sanudo-Wilhelmy et al. 2001). In contrast, German oceanographers studying nitrogen fixation by *Trichodesmium* in the eastern tropical Atlantic diagnosed a colimitation by both iron and phosphorus (Mills et al. 2004). Phosphorus seems to be supplied primarily from deep water containing large amounts of dissolved organic phosphate (DOP), which is released by viral lysis, protozoan grazing, and decomposition of dead cells. Earlier data had already shown that DOP is preferentially remineralized from dissolved organic matter (DOM; Clark et al. 1998). These authors concluded that the selective removal of phosphorus from DOM reflected the nutrient demand of marine microorganisms for this nutrient. When they analyzed the chemical bonds of phosphorus in DOP, they observed two types: phosphorus esters, which many enzymes can split, and phosphonates, a group of compounds containing a C–P bond. The latter represents 25% of DOP, but its stable C–P bond was considered refractory to the attack of marine microbes. Does *Trichodesmium* possess phosphonatases, which would give it a competitive edge in severely phosphorus-limited environments? US oceanographers followed this link by screening the nine sequenced genomes of cyanobacteria and indeed only *Trichodesmium* contained the C–P lyase pathway for phosphonate utilization (Dyhrman et al. 2006). Notably, the gene organization was the same as in *E. coli* and it shared even closer sequence relatedness with the same gene cluster from *Thiobacillus*, a sulfur-oxidizing organism—both are not close relatives of cyanobacteria. The authors concluded that the gene cluster was acquired by HGT. Field data showed that the genes were regulated and expressed only under conditions of phosphate starvation. The dominance of *Trichodesmium* in some parts of the world's oceans becomes now understandable as unique niche adaptations via acquisition of crucial biological atoms of life.

Phosphorus in Plants

The availability of P is one of the major limiting factors for plant growth in terrestrial ecosystems. The problem with P is its low solubility and its high adsorption to soil, which makes it less available to plant roots (Figure 4.10). Roots have developed several strategies to overcome these difficulties. One is an active high affinity uptake system. Root cells transport phosphate at external concentrations of $1\,\mu M$ against internal phosphate concentrations in the millimolar range and against a negative inside potential of the root cell. P influx is coupled to a transient depolarization of the root cell induced by the net cotransport of two to four protons with each P. This leads also to an alkalinization of the soil in the rhizosphere, which decreases the solubility of phosphorus further. The physiological answer in white lupine is clear-cut: It develops proteoid (cluster) roots by proliferation of tertiary lateral roots. This increases the absorption surface. Not enough with that, these proteoid roots release organic acids, mainly citrate and malate, which chelates Al and Fe in the soil thereby releasing P_i from the insoluble $Al–P_i$ and $Fe–P_i$ complexes. Lupines sacrifice a quarter of their total fixed C in photosynthesis to get to soil P_i via excreted organic acids. Another response to P stress in plants is the release of phosphatases. Bacteria and plants cannot absorb organic phosphate; the phosphate must be cleaved either in the periplasm or in the extracellular space to transport phosphate into the cell. A further adaptation to P acquisition is mycorrhizal associations (Figure 4.11).

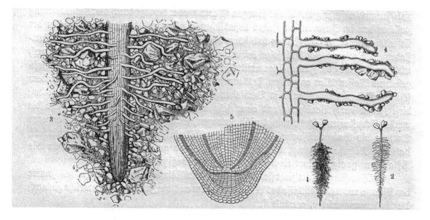

FIGURE 4.10. The lateral root hairs of plants show a close association with soil (before (*1, 3*); after washing (*2,4*)) to favor mineral acquisition. Root apex with meristem (*5*).

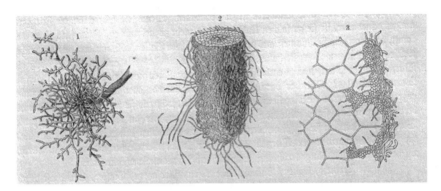

FIGURE 4.11. Mycorrhizal association between roots from Populus (*1*) and Fagus (*2*) and soil fungi. (*3*) shows how the fungal mycelium penetrates the external cell layer of the plant root. We speak about this symbiosis when discussing phosphorus acquisition by plants.

Fungal hyphae associate with plant roots building either a sheath around the root (ectomycorrhiza) or they penetrate into the root cortical tissue (endomycorrhiza). The wide hyphal network in the soil facilitates the absorption of P_i from the soil and its diffusion to the plant root. In this association with fungi, the plant actually trades P_i against fixed carbon. The high affinity phosphate transporters of plants and mycorrhizal fungi show comparable structures consisting of six N-terminal transmembrane regions separated by a cytoplasmic loop from six C-terminal transmembrane regions. The quantity of fixed carbon provided by the plant to the symbiotic fungus is substantial, underlining the importance that the plant attributes to P_i acquisition.

A Demanding Step: Photosynthetic CO_2 Fixation

The Step from Inorganic to Organic Chemistry

Photosynthesis creates a lot of ATP and NADPH. Autotrophs do not sit idle on their accumulated asset, they use it in a very energy-demanding chemical exercise, namely CO_2 fixation (the defining characteristics of autotrophs). This is an important reaction. First with respect to a major dividing line in chemistry: CO_2 is transformed into phosphoglycerate. With that step carbon changes from the realm of inorganic into organic chemistry. Early chemists believed that a "vis vitalis" is necessary to transgress this frontier between the two kingdoms of chemistry. This belief was shattered when in 1828 the chemist Wöhler synthesized urea (a typical organic compound) from ammonium cyanate (a typical inorganic compound). However, deep respect remained for the capacity of bacteria, algae, and plants to achieve this transgression routinely. Perhaps this is the jealousy of the heterotrophic have-nots: We are simply suffocated by CO_2 but not fed.

In contrast, we can make a living with the later products of the photosynthetic CO_2 fixation. This respect for this life-giving reaction is justified: Without exaggeration it can be said that an enzyme feeds the world. If one looks into the biochemical details of photosynthetic CO_2 fixation, it becomes evident that this is not an easy task. Despite all green exuberance surrounding us in many areas of the world, the enzymatic reactions still testify the fundamental difficulties to achieve this transition of carbon from the inorganic into the organic world, i.e., the reduction of CO_2.

Calvin Cycle

The enzyme that catalyzes this crossing of the inorganic to organic carbon is called Rubisco, which reads prosaically as ribulose 1,5-bisphosphate carboxylase/oxygenase. As we will see, "*nomen est omen*": Its systematic name puts already the major problem of this enzyme into the limelight. Ribulose, a 5-carbon keto sugar, esterified to phosphate groups at both its terminal sugar hydroxyl groups, is the starting compound of the famous CO_2-fixing pathway shared by bacteria, algae, and green plants: the Calvin cycle. It is named after the Australian biochemist Melvin Calvin, who fed the easy-to-grow green algae *Chlorella* with radioactive $^{14}CO_2$ and in a time series dropped it into alcohol and separated the cell sap by 2-D paper chromatography. He and his colleagues were startled by the complexity of autoradiographic spots they obtained after 60 s of illumination. However, after a mere 5 s, the pattern consisted of a single spot that was identified as 3-phosphoglycerate. At first glance, this does not seem a major feat to obtain a 3-carbon from a 5-carbon sugar, but when realizing that the reaction yields two of these C3 compounds the stoichiometry of CO_2 fixation is given: $C5 + C1 \rightarrow 2 \times C3$. The business is actually quite tricky, but the underlying chemistry has been elucidated, and the protein context of the active site has been characterized to the structural level (Taylor and Andersson 1997). Rubisco as synthesized by the ribosome binds its substrate ribulose 1,5-bisphosphate, but then the reaction is stalled. It needs another enzyme, Rubisco activase, that expels the bound substrate by using one ATP. This energy-requiring step (it will not be the last energy input into CO_2 fixation) frees the active site and the lysine 201 residue. This amino acid is nonenzymatically carbamoylated, i.e., a CO_2 is covalently bound to the side chain amino group of lysine. The two oxygen atoms of this bound CO_2 and two further acidic side chains from Rubisco now bind a Mg^{2+} ion and the enzyme is ready for its task. Ribulose 1,5-bisphosphate is bound to the active site. The carbamoylated lysine abstracts an H atom from ribulose 1,5-bisphosphate leading to a C=C double bond. Then CO_2 is bound by Mg^{2+} and polarized, i.e., the central carbon atom gets a partial positive charge, preparing a nucleophilic attack by the partially negative charge of the C=C double bond. This leads to an unstable, branched 6-carbon sugar. This intermediate, while still bound to the active site, undergoes an aldol cleavage, which liberates the first 3-phosphoglycerate. The aldol cleavage also creates a negatively charged C atom (a carbanion) that remains complexed to the Mg^{2+} ion. This intermediate is stabilized by the capture of a proton provided by the amino group of lysine 175 from Rubisco. The second molecule of 3-phosphoglycerate can now leave the active site, and the cycle can start again.

Rubisco is Slow

CO_2 fixation is an energy devouring exercise and needs to be inhibited in the night when photosynthesis does not provide fresh ATP and NADPH. This is done in some plants in an ingenious way: They synthesize a transition-state inhibitor—you have to look twice on its structural formula to differentiate it from the true intermediate. In a way, it is a joke that plants have to inhibit Rubisco since it is already naturally an extremely slow enzyme. At 25 °C Rubisco fixes just three CO_2 molecules per second—an exceptionally slow turnover rate for an enzyme. If you realize that a substantial part of the organic carbon of our biosphere goes through this enzymatic bottleneck, you can imagine the stress, which plants, algae, and photosynthetic bacteria must have with this enzyme. In fact, to achieve a reasonable throughput, chloroplasts are literally packed with Rubisco, nearly 50% of the soluble chloroplast proteins are Rubisco. Hence, there is good reason to believe that Rubisco is the most abundant enzyme on Earth.

Rubisco has a High Error Rate

Since Rubisco is such a fascinating enzyme, allow me a philosophical digression. Philosophers of the period of Enlightenment were fascinated by the idea that everything in nature has a purpose. Leibniz marveled at the best of all possible worlds and was ridiculed for that by Voltaire in his "Candide." This eighteenth-century idea of a perfect adaptation of organisms to their environment still holds in popular beliefs on evolution. However, evolution seen by biologists is a less perfect engineer. Instead of coming up with perfect solutions, evolution prefers to tinker with old models even if one could argue that they are out of fashion. A reason for this conservative character of evolution is the fact that nature has not created a free space for new solutions, a selection-free workshop. All new solutions are directly tried on stage and must immediately be competitive. This need of direct functionality prevents redesigning entire pathways from scratch since these entirely new designs are initially very likely to be less fit than older solutions. Therefore, even clearly imperfect solutions are maintained when they work sufficiently well. In some way, therefore, evolution has resulted in a junkyard: historical achievements that were good solutions in the past lost a substantial part of their attraction in the present. Rubisco is an excellent illustration for these principles. As this enzyme controls the entry point of CO_2 from the atmosphere into the biosphere, Rubisco is central to life on Earth. Following Leibniz one would thus expect a perfect enzyme. However, Rubisco is not only slow but also notoriously inefficient as an entry port for atmospheric CO_2 into the biological carbon world. The terminal O in Rubisco stands for *oxygenase. Actually for every three or four turnovers, Rubisco does not work as a carboxylase, but as an oxygenase. What does this mean?*

As an oxygenase, Rubisco adds O_2 to ribulose 1,5-bisphosphate with a given error frequency, specific for each enzyme. The unstable hydroperoxide intermediate splits into a molecule of 3-phosphoglycerate and a C2 compound, 2-phosphoglycolate. Thus in this mode, actually a C5 sugar transformed into a C3 and C2 compound, and nothing was gained with respect to C fixation. Even worse, 2-phosphoglycolate is a rather useless metabolite, and plants had to design a new pathway to salvage these two carbon atoms for their metabolism. This pathway involves the cooperation of three different organelles: chloroplasts, mitochondria, and peroxisomes. The rescue operation shows some elegance that Rubisco lacks. Yet, we should not condemn Rubisco, but consider its environment when it was created.

Rubisco's Structure

More than a thousand sequences have been determined for Rubisco, and a phylogenetic analysis should thus be a straightforward task to clarify the origin of this remarkable enzyme. However, this exercise turned out to be complicated. Until quite recently, two major forms of Rubisco were known: Forms I and II. The structure of Form II Rubisco, the enzyme from purple bacteria like *Rhodobacter* and *Rhodospirillum*, was solved relatively early on (Andersson et al. 1989). It is a homodimer consisting only of L (large) subunits. Five dimers form the enzyme complex $(L_2)_5$. Each L_2 dimer is inclined to the next such that a toroid-shaped decamer is formed (Lundqvist and Schneider 1991). The central structural motif in the L protein is an eight-stranded parallel α/β-barrel, which forms a scaffold harboring the active site. In particular, loop 2 (located between strand 2 and helix 2) contributes three ligands (residues Lys201, Asp203, and Glu204) to a magnesium ion intimately involved in catalysis (Taylor and Andersson 1997). The protein lacking the Mg ion and the carbamoylation of Lys 201 is enzymatically inactive. The structure of Form I Rubisco from plants was also solved in the late 1980s. This enzyme is built up from eight large and eight small subunits forming a large L_8S_8 complex. When viewed from the top, the plant Rubisco is a torus-like enzyme consisting of four lobes each consisting of a large protein subunit, separated by the small subunit. The large subunit is encoded by the chloroplast genome, and the small subunit by the cell nucleus. When looking from the side, one realizes that the enzyme complex is made up of two stacks of the basic torus structure. In the side view, the small subunits occupy the top and bottom parts. The structure shows a prominent central channel. Despite this overall structural difference, the eight-stranded parallel α/β-barrel structure is also identified in the Form I Rubisco. The similarity of the active site is also still reflected by a significant amino acid sequence identity between the L proteins from the Form I and II enzymes, which, however, does not exceed 30%. This similarity is enough to make a derivation from a common ancestor likely.

Evolution of Rubisco

It has been suggested that the Form II enzyme is ancestral over Form I. The arguments are, however, rather indirect: Form II is associated with purple bacteria and Form I with cyanobacteria. As we have argued before, the photosynthesis of purple bacteria is also evolutionarily more ancient with respect to the photosystems. Another argument for the antiquity of Form II is its very low specificity factor S_r in purple bacteria. This value refers to a central enzymatic property of Rubisco. The specificity factor S_r refers to the relative preference of the carboxylase over the oxygenase activity in the enzyme. In *Rhodospirillum* S_r is rather low. Such an enzyme is unable to sustain growth in an aerobic environment. This is not a problem for bacteria like *Rhodospirillum* and *Rhodobacter*, which live in anaerobic niches. This was anyway not a problem when Rubisco was invented more than three billion years ago since at that time the CO_2 concentrations in the atmosphere were much higher than today (geologists place it between 0.1 and 1 bar for pCO_2), while oxygen concentrations were very low (Hessler et al. 2004). However, no clear pattern came out from the phylogenetic analysis: Form II Rubisco is not limited to the α-Proteobacteria like *Rhodospirillum*, but was also identified in β- and γ-Proteobacteria and even in dinoflagellates and unicellular photosynthetic algae (Morse et al. 1995). A further twist to the dinoflagellate story is that this L protein Rubisco is nucleus encoded and not plastid encoded as usual in eukaryotes. The dinoflagellates are mostly marine, photosynthetic algae that inhabit aerobic environments. The possession of a Form II enzyme does not seem to bother this eukaryotic aerobic organism, in fact it grows pretty well. They grow to such high titers in the sea that some forms (e.g., *Noctiluca*) are responsible for much of the luminescence seen in ocean water at night. Other dinoflagellates grow out to what is called a red tide. This sounds like a Biblical scourge. However, mostly it is simply a streak of discolored ocean water of a pinkish orange caused by a sudden massive outgrowth of dinoflagellates to concentrations of 10 to 100 million cells per liter of seawater. However, they merit this Biblical name: Occasionally, dinoflagellate blooms cause toxin release that results in massive fish and, not rarely, even human killing. Unfortunately, not enough is known about the CO_2-fixing mechanisms of dinoflagellates to understand why they use Form II Rubisco.

The Carboxysome

A basic distinction between prokaryotes and eukaryotes is the lack of a membrane-bound nucleus and organelles in prokaryotes. However, this does not mean that bacteria lack organelles entirely. In fact, some cyanobacteria possess organelles involved in CO_2 fixation, but they were overlooked. Electron microscopists knew them for a while, but misinterpreted them as viral particles. The carboxysome, as it is called, shares with viruses the characteristics to form a protein shell not to shelter a viral genome, but to sequester the carbon fixation reaction. This 200-nm-large polyhedral structure is filled with the Rubisco enzyme associated with carbonic anhydrase. The latter mediates the localized conversion of bicarbonate to CO_2, the substrate of Rubisco. Ultrastructural analysis of the carboxysome shows the trick (Kerfeld et al. 2005). The protein shell is built by small proteins, which are arranged in hexameric units, leaving a 7-Å-small, central hole surrounded by positive charges. These hexamers fit to each other forming sheets leaving 6-Å gaps between the units. The charged pores and the gaps probably act by regulating the metabolic flux. The substrates bicarbonate, ribulose bisphosphate, and phosphoglycerate pass, while CO_2 and O_2 would not be attracted. The authors argue that we have here a primitive organelle since it provides a permeability barrier, which is a fundamental property of subcellular organelles. Only in this type of organelle, the protein shell plays the role of the lipid membrane from classical organelles.

Recycling in Biochemistry: Rubisco and the Calvin Cycle

Further Splitting in Type I Rubisco

Back to the phylogeny of Rubisco: Also Form I is not a homogeneous cluster. There you find two major branches: the Green-like and the Red-like cluster. Each cluster splits again into two subclusters. The Green-like cluster separates into a bacterial group and a cyanobacterial/plant group. The Red-likes split into a bacterial group and one of eukaryotic, nongreen algae. The pertinent picture is that the Rubisco phylogenetic tree is not congruent with the rRNA tree—a strong indication that the current distribution of Rubisco genes was shaped by multiple HGT events. We observed this strong trend for lateral inheritance also for the light reactions of photosynthesis. Alternative explanations like ancient gene duplication followed by selective loss were not yet formally excluded, but coexistence of different Rubisco systems in the same organism, which would be evidence for gene duplication, was not yet observed.

Selection for Increased Efficiency

A recurring subject of our survey is the connection between the quest for food and climate change. The relentless increase of atmospheric CO_2 concentrations over the last century has raised concerns about the global heating through this greenhouse gas. One hope is that the increased offer of CO_2 will also increase the productivity of the terrestrial and aquatic photosynthetic organisms. Part of the burnt fossil fuel could thus be bound at least temporarily in the biomass. This hope is tempered by the notorious inefficiency of Rubisco. Its Achilles' heel is its basic design coming from a time when CO_2 versus O_2 distinction was not a selection issue. Perhaps four times in Earth's history (about 2.3 Ga, 0.6 Ga, 300 Ma and 30 Ma) global changes might have selected for more efficient CO_2 fixation. However, apparently it was "easier" for evolution to work on the concentration of CO_2 at the cellular site of Rubisco by CO_2 concentration mechanisms than over amelioration of the specificity factor. This is not to say that Nature has not tried to work on a better specificity of the enzyme.

Rhodospirillum showed only an S_r value of 15, whereas cyanobacteria increased this value to 45, algae to 70, and spinach even to 93. If this series suggests a linear increase with later evolution of the organisms, there are also outliers. For example, the frontrunner is unexpectedly the red alga *Galdieria* with the remarkable S_r value of 238. Since biologists and biotechnologists alike are eager to increase the efficiency of Rubisco, it should be no surprise that the X-ray structure of this enzyme was solved (Sugawara et al. 1999). The outcome was disappointing: The structure of this enzyme and that of spinach could be nearly perfectly superimposed. Both proteins differed only in minute details, and the only major change was in the small subunit of the red alga enzyme. It showed a 30-aa extension at the C terminus, which formed a hairpin-loop that locked neighboring subunits and nearly closed the central channel of the protein complex. The problem is that we do not know much about the function of S subunit, and some data even indicate that it is not absolutely essential for the carboxylase activity. The hope is now with the genetic engineers (Spreitzer and Salvucci 2002), but up to now the experience has shown that single aa mutations have not ameliorated the enzyme. Apparently, change at one site also necessitates compensating changes at one or two other sites of the enzyme if a better enzyme is targeted. *Personally, I would say the prospect is bleak since Nature has probably also tinkered with Rubisco to keep it efficient with the changes of the atmosphere. She has apparently not achieved major breakthroughs and has put her imagination into peripheral ameliorations, like CO_2 transporters, she has hired carbonic anhydrase and other CO_2-fixing enzymes as helpers for Rubisco, she has designed new morphologies in vascular plants (C4 and CAM, i.e. crassulacean acid metabolism plants) and rescue systems for the lost carbon, which show signs of creativity like in fat mobilization in plant seedlings. Rubisco, the enzyme that literally feeds the world, stays there as a powerful reminder of the imperfectness of Nature.*

Form III Rubisco

But where might be the actual roots of this fascinating enzyme? An interesting hint came from the genome sequence of *Methanococcus jannashii*. This organism was practically unknown before it was sequenced (Bult et al. 1996). However, its genome analysis catapulted this Archaeon and methanogen nearly overnight to prominence in the scientific discussion. One of the surprises was a gene that encodes a protein, which shared 41% aa identity with Form I and 34% identity with Form II Rubisco. Since Archaea do not know photosynthesis, the result was at first glance surprising. However, one might anticipate that aa sequence similarity implies related function, namely CO_2 fixation. This was indeed the case; the *Methanococcus* enzyme could incorporate radioactive CO_2 into organic carbon (Finn and Tabita 2003). The CO_2 fixation was not due to alternative biochemical activities. Strikingly, the *rbcL* gene from *Methanococcus* could complement L protein mutants from *Rhodobacter*, which started to fix CO_2. This protein is now referred to as Form III Rubisco.

Rubisco-Like Enzyme and Sulfur Salvage

Then an even more distant relative of Rubisco was unearthed in several genome projects. A Rubisco-like protein, RLP or Form IV Rubisco, was detected in the genome sequence of *Archeoglobus*, another archaeal methanogen, in the sulfur bacterium *Chlorobium* and the Firmicute bacterium *B. subtilis*. Since the latter is the paradigm of Gram-positive bacteria, researchers studied Rubisco's function in this well-investigated bacterium. RLP is encoded by the gene *ykrW*, which belongs to a larger operon whose genes are involved in a methionine salvage pathway. Cells are economical and do not want to loose valuable organic sulfur. Therefore, several bacteria have designed a pathway that transforms by-products of pathways dealing with sulfur-containing metabolites back to methionine, hence the name of this pathway. Methylthioribose is such a by-product of the synthesis of spermidine. Spermidine is a polyamine used in DNA packaging. It is synthesized from ornithine and methionine. In this pathway, methionine is activated by ATP to *S*-adenosylmethionine. Methylthioribose is released when spermidine is condensed from both molecules.

The individual genes of the methionine salvage pathway were expressed, and the metabolites were chemically characterized when the enzymes were added one by one. RLP catalyzed the keto–enol reaction in this pathway (Ashida et al. 2003). The intermediate has chemical similarity to the ribulose 1,5-bisphosphate. And now it became even more fascinating: Rubisco from *Rhodospirillum* could rescue the methionine salvage pathway in the *ykrW*-mutant of *B. subtilis*. The take-home message was clear to the Japanese biochemists: RLP and Rubisco are derived from a common ancestor protein. They argued that RLP in the methionine salvage pathway and in Archaea appeared first on earth—long before the Calvin cycle developed in photosynthetic bacteria, which according to them then recruited this enzyme for the new purpose. *We should not complain about the notorious inefficacy of this enzyme; it is a gift from other pathways in nonphotosynthetic organisms. It is thus rather amazing that Nature used a second hand enzyme for one of its key enzymatic reactions.* From the purely chemical side, it will be interesting to investigate what CO_2 acceptor molecule is used by Form III Rubisco in *Methanococcus*.

Reversing Glycolysis

What is true for Rubisco is possibly also true for the entire Calvin pathway. Nature has not really taken the pain to invent something new for CO_2 fixation. It apparently recycled in the literal sense other probably more ancient pathways. This will become apparent if we take a look at the Calvin cycle, where we will also see for what purpose photosynthesis has created so much ATP and NADPH. ATP is first used to phosphorylate 3-phosphoglycerate, the product of Rubisco's action, to 1,3-bisphosphoglycerate. NADPH is used to reduce it to glyceraldehyde 3-phosphate. This is the reverse of the glycolytic pathway. Notably, the stroma of the chloroplast where the Calvin cycle takes place contains a nearly complete complement of the glycolytic pathway (only phosphoglycerate mutase lacks).

However, the stromal and cytosolic glycolytic enzymes in plants are isoenzymes, i.e., they are products of different genes. The reversal of glycolysis is still followed by the next few steps: Triose phosphate isomerase transforms glyceraldehyde 3-phosphate into dihydroxyacetone phosphate. Aldolase condenses glyceraldehyde 3-phosphate and dihydroxyacetone phosphate into fructose 1,6-bisphosphate. Fructose 1,6-bisphosphatase creates fructose 6-phosphate, here ends the recycling of glycolysis in the reverse mode for the Calvin cycle.

Copying the Pentose Phosphate Pathway

Nature has not searched far to continue with the Calvin cycle. The task is how to get from a C-6 sugar to a C-5 sugar and sparing a C3 (the actual gain from photosynthetic CO_2 fixation). The task is very similar to that of the nonoxidative phase of the PPP (how to create C6, glucose 6-phosphate, from C5—the pentose phosphates—in an efficient way) if only executed in opposite direction. In PPP, the cells run the show in three steps, where sugars with different numbers of C-atoms are reshuffled (*C5 + C5 ↔ C3 + C7* then C7 + C3 ↔ C4 + C6 and *C5 + C4 ↔ C3 + C6*, for details see a biochemistry textbook). If you summarize, the left terms and the right terms, you get: 3 C5 ↔ 2 C6 + C3. For this exercise, you need two enzymes. Transketolase shifts a two-carbon C2 group, transaldolase a three-carbon C3 group between these compounds. The net result of this pathway is the transformation of three pentoses into two hexoses and G3P. These two enzymes thus assure a seamless integration of the PPP with glycolysis. The cell can now meet different metabolic needs by integrating enzymes from glycolysis, the PPP and gluconeogenesis.

If you write down the Calvin cycle, it gives a more complicated set of reactions but, despite that difference, similarities are striking. Similar enzymes (*trans*-aldolases and *trans*-ketolases) and substrates are used. The two underlined reactions are actually copied in both pathways. Differences concern the addition and abstraction of phosphate groups to confer directionality to the process. *It seems apparent that both pathways derive from a common ancestor that juggled with sugars of different length to fulfill its now forgotten biochemical tasks. Actually, the current purpose of a given biochemical cycle might not well reflect its original task, as we had seen before when discussing the TCA cycle.*

Gain of the Calvin Cycle

However, the Calvin cycle does not turn for its own sake. Every third round frees a triose from the cycle. If the photosynthetic organism is a eukaryote, the triose will leave the chloroplast via a phosphate–triose phosphate antiporter into the cytoplasm, where it can enter glycolysis for local energy production. After six turns of the Calvin cycle two supernumerary trioses are formed. They can condense to a fructose and lead to starch synthesis in the chloroplast. Alternatively, both triose phosphates are exported into the cytoplasm where they condense to fructose and then to sucrose. Sucrose and starch synthesis are the major pathways by which excess triose phosphates from photosynthesis are harvested for the plant body. As one can expect, this harvest has not escaped the attention of herbivorous animals, but for the moment photosynthetic life still had some time before evolution showed up with animals. The overall balanced chemical equation of the Calvin cycle shows that 12 NADPH and 18 ATP molecules are consumed to transform six CO_2 molecules into a hexose sugar. Remarkably, NADPH and ATP are produced in the light-dependent reactions of the photosynthesis in the same ratio (2:3) as they are used in the Calvin cycle.

Alternative CO_2 Pathways in Autotrophic Prokaryotes

The Calvin cycle is the dominant mechanism of CO_2 fixation in aerobic eubacteria; it was not yet seen in Archaea. Alternative pathways of CO_2 fixation were described. In a previous section, we have already discussed the reductive TCA cycle. This pathway occurs in a number of bacteria, which we have encountered in our survey: the phototrophic green sulfur bacterium *Chlorobium*, the sulfate-reducing bacterium *Desulfobacter*, the thermophilic Knallgas bacterium *Hydrogenobacter*, and the sulfur-dependent anaerobic archaeon *Thermoproteus* belong to this category.

In addition, there are two further alternative autotrophic carbon-fixing pathways. Less well-investigated is the cyclic 3-hydroxypropionate cycle where two carboxylation reactions release after one turn one molecule of glyoxylate for supplying cellular carbon in the green nonsulfur phototrophic bacterium *Chloroflexus* and possibly in aerobic archaea. In a fourth pathway, two molecules of CO_2 yield one molecule of acetate.

Reductive Acetyl-CoA Pathway

This is the autotrophic CO_2-fixation pathway in strictly anaerobe methanogens (Archaea) and acetogenic eubacteria. In contrast to the two aforementioned fixation pathways, this reaction is noncyclic. Here I will describe the pathway for acetogenic eubacteria. In fact, it is not one pathway, but two parallel lines that fuse in one enzyme. In one line, CO_2 is reduced to formate, and then further reduced to formaldehyde. Since the latter is very toxic for the cell, the reduction is done after transfer of formate to the coenzyme tetrahydrofolate. The coenzyme-bound formyl-group is then stepwise reduced to the methyl group. The ultimate hydrogen donor for these reductions is molecular hydrogen. A hydrogenase catalyzes the reaction $H_2 \rightarrow 2H^+ + 2 e^-$, and keeps the NAD(P)H and an unknown electron donor reduced that participate as cosubstrates in the reduction steps of the coenzyme-bound formyl group. When the first CO_2 is reduced to $-CH_3$, it is transferred to a very complex enzyme with many functions. The methyl group is first complexed to a cobalamine and then to a nickel ion. This is the activity of the methyl-transferase in the enzyme complex. Then comes the carbon monoxide dehydrogenase function: It takes care of a second CO_2, which is reduced to CO. The CO group is bound to an iron–sulfur center in the vicinity of the Ni-bound $-CH_3$. Finally, the acetyl-CoA synthase activity catalyzes the joining of both groups to the enzyme-bound CH_3–CO–acetyl group, which is then liberated as an acetyl-CoA. This enzyme is not only complex, but also versatile. It not only mediates autotrophic carbon fixation but also participates likewise in the total synthesis of acetate in acetogenic bacteria and in the reverse reaction in acetoclastic methanogens. The introduction of acetyl-CoA into the central metabolic pathway necessitates still another new enzyme, a ferredoxin oxidoreductases, which introduces a third CO_2-yielding pyruvate.

A Few Numbers on the History of CO_2 Concentrations

The present concentration of CO_2 in the atmosphere is about $350\,\mu mol\,mol^{-1}$ (ppm; value for the year 1995). With 0.03 vol.% this is low in comparison with the main constituents of our atmosphere (N_2 78.08 vol.%, O_2 20.9 vol.%), however, this low concentration belies its biological and biogeochemical role. The CO_2 concentration is quite a dynamic parameter. There are diurnal changes in CO_2 fluxes: During daytime, there is a downward flux of CO_2 over the vegetation cover due to photosynthetic CO_2 fixation. This trend reverses during the night when there is an upward flux from the soil and the vegetation into the atmosphere due to respiratory CO_2 production. On a typical summer day, peak midnight values might reach above 400 compared to a trough of $300\,\mu mol/mol$ at 1 p.m. However, these values are subject to other influences. One factor is temperature: In the wintertime, no diurnal variation in CO_2 concentration is observed in temperate climate zones since photosynthetic and respiratory activities are substantially reduced. Another factor that influences the CO_2 concentration is the height above the ground. Top CO_2 concentrations are found in the summer directly above the ground (5 cm) due to the rapid decomposition of organic litter, which creates CO_2 as end product of respiration. At the level of the canopy of a forest, the CO_2 concentrations of the atmosphere are not reached due to the intensive CO_2 fixation by photosynthesis.

CO_2 concentrations varied also over a secular and a geological timescale. In 1750 the CO_2 was at 280 ppm; it has steadily risen to 350 ppm in 1995 and still continues to rise. If this process goes unabated, values like 700 ppm are expected in 2100. This should result in a rise of temperature. The underlying physical process is straightforward. During the day, the solar radiation heats up the Earth's surface, and during the night the adsorbed heat is released as infrared radiation that is lost again into space. However, molecules like CO_2, and also nitrous oxide, methane, ozone, and fluorocarbons absorb infrared radiation and thus keep heat in the atmosphere, resulting in a temperature rise like in a green-house effect. While this is undisputed, the rise in temperature, which it might cause globally, is a matter of equally heated discussions. Some models predict a warming by 4 °C for a doubling of the CO_2 concentration with consequent melting of polar ice caps, rise of sea level, and major climatic changes. The cause of the current CO_2 increase is the Industrial Revolution. The energy used by industry and household stems to a large part from fossil fuels like coal, gas, and oil. The reduced carbon compounds stem from photosynthetic processes in past geological time periods. The plant material was not decomposed by aerobic processes, but became stored away under the surface. The large amount of $5-7.5 \times 10^9$ tons of carbon is released annually into the atmosphere by human activity. Hydrocarbons are burned with oxygen to CO_2 and water. This amount might seem small in comparison with the estimated release of 10^{11} tons of carbon by respiration of the world's biota. However, here comes the point of geochemical cycles. The amount of CO_2 released by respiration corresponds approximately to the amount of CO_2 fixed by photosynthesis. In this way, despite an enormous amount of carbon cycling between photosynthesis and respiration, the net concentration of free CO_2 at any time point remained fairly constant over historical times. Fossil fuels did not participate in this cycle until they were literally unearthed in the Industrial Revolution. This extra amount of photosynthetic products from ancient forests now adds to the CO_2 balance. Of course, the nature of geochemical cycles suggests that if you add more CO_2 to the atmosphere you might after a transient increase also increase the fixation of the extra CO_2 by photosynthesis at higher temperature levels expected from the greenhouse effect. It is therefore important to better understand the biochemistry of CO_2 fixation by photosynthetic plants and how this process is influenced by temperature rises. As this process is also crucial to plant nutrition, this discussion affects both our subject of a history of eating and global climatic changes.

When oxic photosynthesis evolved in bacteria about 3 billion years ago, the atmosphere contained 100-fold higher CO_2 levels, but little or no oxygen. According to geochemical mass balance models, a large decline in CO_2 levels occurred during the Carboniferous Period (Figure 4.12) about 300 million years ago accompanied by significant atmospheric oxygen level increases. Was this the consequence of large amounts of plant material being taken out of the carbon cycle when building the coal deposits? CO_2 levels were again low during the late Tertiary Period some 65 million years ago, which has an impact on the evolution of alternative photosynthetic pathways as we will hear soon.

A Tricky Business: N_2 Fixation

Stoichiometry

In many habitats, no suitable nitrogen source is available. If life wants to conquer these environments, it must use an abundant, but chemically inert nitrogen source: dinitrogen of the atmosphere. This molecule is in high supply since it represents 78% of the atmospheric volume. However, the two nitrogen atoms are bound by a triple bond having a bond energy of 225 kcal/mol and is thus highly resistant to chemical attack. On paper the chemical reaction is simple: $N_2 + 3H_2 \rightarrow 2NH_3$. To break the triple bond, quite drastic conditions are needed such as in the industrial Haber–Bosch procedure used in fertilizer factories. N_2 is carried over a metal catalyst at 500 °C and under a pressure of 300 atm.

About 1% of the world's total annual energy supply is consumed by this single industrial process, which provides nitrogen fertilizers for agriculture. It is thus remarkable that many bacteria can reduce dinitrogen to ammonia at ambient temperature and pressure. To achieve this task, bacteria use a remarkable catalyst for this reaction: nitrogenase. However, also for bacteria dinitrogen fixation is an extremely costly exercise when looking at the following chemical equation: $N_2 + 8e^- + 16ATP + 16H_2O \rightarrow 2NH_3 + H_2 + 16ADP + 16P_i + 8H^+$. A lot of reducing power and energy-rich bonds are used for this reaction. But the investment allows growth under conditions that would otherwise not support life. In fact, as we will soon see, bacteria powered the growth of many plants and thus became a precious biological fertilizer in agriculture. Before investigating this economically important bacterial-plant symbiosis, let's explore a bit the structure of this remarkable bacterial enzyme.

The Enzyme

The enzyme responsible for the biological nitrogen fixation is the nitrogenase. The structure of this multiprotein enzyme complex was solved by X-ray crystallography for *Azotobacter*, a free-living, nonsymbiotic, nitrogen-fixing bacterium. This bacterium derives its microbiological fame from the isolation by pioneers of microbiology, namely Winogradsky, Beijerinck, and van Delden, who isolated nitrogen-fixing bacteria between 1895 and 1902. *Azotobacter* is a relative of *Pseudomonas* and belongs in the γ-Proteobacteria group. The enzyme complex is an elongated structure consisting of a central tetramer (in fact a back-to-back duplicated αβ dimer), the MoFe protein. At both ends, the MoFe tetramer is capped by a Fe-protein (each side is again a dimer, but this time a γ_2 homodimer; Schindelin et al. 1997). The Fe-protein, encoded by the *nifH* gene, is actually a dinitrogenase reductase. This means it provides the electrons to the MoFe protein, the proper dinitrogenase, which reduces N_2 to NH_3 as described above. The Fe-protein receives the electrons from soluble electron carriers that vary from organism to organism. This can be ferrodoxin or flavodoxin or other redox-active species. The actual transmitter of electrons is an [4Fe–4S] cluster sitting at the bottom of the γ_2 homodimer near the interface to the MoFe protein. There is a break in symmetry: Although there are two identical γ-subunits, the two proteins hold a single [4Fe–4S] cluster in place, just between their bottom interface that contacts also the MoFe protein. This is a good place for its function because it has to transmit the electron further down into the MoFe protein via an intermediate Fe–S cluster of an even more complicated [8Fe–7S] geometry (the P clusters),

---

FIGURE 4.12. An artistic impression of the Carboniferous plant life. At the left side of the creek, you see from *left to right*: the ferns *Caulopteris*, *Pecopteris*, the climbing fern *Sphenopteris*, and the lying trunk is from the fern *Megaphyton*. The large tree is *Lepidodendron*. At the right side of the creek you see from *left to right*: the trees *Calamites ramosus*, *Cordaites*, *Syringodendron* (lying trunk), the *rightmost* tree is *Sigillaria* and at the *right* side is the climbing fern *Mariopteris*.

which then transmits the electrons to the enzyme's active center. The P cluster resembles in the reduced state two distorted [4Fe–4S] cubes that share one sulfur atom at the corner. In the oxidized state, one of the cubes opens and two iron atoms loose their bond to the sulfur corner. You will probably not be too much surprised to learn that the active site is again borrowed from inorganic chemistry. It is a complex composed of another Fe–S cluster associated with a molybdenum atom, hence the name molybdenum iron cofactor or, in short, FeMoCo. Actually, this compound is at the interface between inorganic chemistry and biochemistry since it is at one side bound to a cysteine and at the other side to a histidine from the α-protein subunit of the MoFe protein and in addition to an organic acid, homocitrate. As if nature wants to show all its flexibility in playing with iron and sulfur atoms for redox processes, this complex can be best described as one [4Fe–3S] cluster bridged by three inorganic sulfur atoms to another [4Fe–3S] cluster. However, most commonly one Fe atom is replaced by a molybdenum (Mo) or less frequently by a vanadium (V) atom. Apparently, we take here a deep look back into time and the beginning of biology in the iron–sulfur world that survived in this enzyme until our days. This is by far not an isolated view into this iron–sulfur world offered in biochemistry. Other redox proteins offer an independent view like the respiratory complex I in the bacterium *Thermus thermophilus* sporting no less than nine iron–sulfur clusters, seven of them forming an electron transport chain separated by less than a critical 14-Å distance (Hinchliffe and Sazanov 2005).

Electron Flow

Now back to the electron flow in the nitrogenase complex: The iron–sulfur centers are aligned, but there is still a problem. Electron tunneling in redox-proteins is an efficient process as long as the relay stations are not separated by more than 14 Å (Page et al. 1999). The [4Fe–4S] cluster in the Fe protein and the P cluster in the MoFe protein are too far away: 18 Å. Now ATP comes into play. When ATP binds to the nitrogenase complex, it changes its conformation. In fact, ATP binds to the external part of the γ-subunits upon which the Fe protein makes a transition from a U-shaped into a Δ-shaped protein. The contact movement of the two γ-subunits induced by ATP binding at the top pushes the [4Fe–4S] cluster at the bottom further 4 Å into the MoFe protein. The distance from the [4Fe–4S] cluster to the P cluster now gets down to 14 Å and efficient electron transfer becomes possible. In the next step, the ATP is hydrolyzed, the phosphate is released, the intersubunit stabilization is decreased, and the two γ-subunits move apart and the proteins dissociate to be ready for the next round of interaction when reduced ferrodoxin and ATP is provided. Now you understand the stoichiometry of the nitrogen fixation, why two ATP are hydrolyzed for each transferred electron. Since the reduction of dinitrogen to ammonia is a six-electron process, you understand why multiple cycles of protein complex formation, ATP hydrolysis, and electron transfers are required for substrate reduction. Both proteins depend critically on each other: ATP is hydrolyzed by the Fe protein only when complexed by the MoFe protein and the MoFe protein accepts electrons only from the Fe protein. Cocrystallization studies with the Fe and MoFe proteins in the absence of ATP, in the presence of ADP, and in the presence of an ATP analogue revealed that essentially only the Fe protein undergoes conformational changes and occupies three different docking sites on the MoFe protein (Tezcan et al. 2005). The Fe protein behaves like a nucleotide switch protein also known from proteins involved in signal transduction pathways.

The Active Site

There is excellent evidence that the FeMo-cofactor cluster is the enzyme's N_2 binding and N_2 reduction site. The electrons are transferred from the [4Fe–4S] cubes of the reductase to the P site and then to the FeMo cofactor of the nitrogenase. During these electron transfers, the iron atoms cycle between Fe^{2+} and Fe^{3+} oxidation states. Mechanistically, it was proposed that three electrons, one at a time, have to be transferred from the Fe to the FeMo protein before N_2 can bind. These three electrons reduce the nitrogen to the level of nitride (N^{3-}) before it can be liberated as ammonia. With the lower resolutions of the crystal structure of the nitrogenase, the nitrogen could not be visualized. With the most recent resolution at $1.16\,\text{Å}$, a central ligand was detected in the inner cage of the FeMo cofactor (Einsle et al. 2002; Smith 2002). The electron density profile is compatible with a nitrogen atom, but not with dinitrogen or even larger species. At the previous lower resolutions, the light atom at the center was over shadowed by the signals of the heavy iron atoms in its vicinity. The newer data also resolve the dilemma of the threefold coordinated six iron atoms of the complex (Kim and Rees 1992). Since each is linked to the central light atom, the fourfold coordination of iron is again respected. Thus, step-by-step the chemical secrets of one of the fundamental reactions in biochemistry are revealed. However, the fog has not entirely cleared yet. Chemists succeeded to mimic part of the dinitrogen reduction with purely chemical complexes at ambient conditions (Leigh 2003). In these complexes, molybdenum and not the iron was the critical catalyst (Yandulov and Schrock 2003). This mode of action is not yet excluded for the enzyme since an alternative model envisions that nitrogen is bound to molybdenum in the MoFe protein, possibly after dissociation of the carboxyl group of homocitrate (Pickett 1996).

Why do I tell you this story? First, it is one of the few fundamental biochemical reactions that make life possible on Earth. Before you can eat your foodstuff, some humble organisms must take care to funnel the basic atoms of life into the food chain. In fact, these organisms are not so humble at all even though this reaction is limited to bacteria and Archaea. None of the so-called higher organisms is able to perform this task although plant life and modern agriculture critically depend on the reaction. Even today, biological nitrogen fixation still contributes about half of the total nitrogen input to global agriculture.

Evolution of Enzymatic Mechanisms

In addition, there are basic lessons one can retrieve from this story. For example, Fe–S centers play a crucial role in redox reactions not only in the nitrogenase but also in respiratory enzymes. It seems that these catalytic centers are still memories from the early steps of the "bio"chemistry in the prebiotic phase. Apparently, enzymes involved in basic reactions in biochemistry still kept the memory of the postulated prebiotic Fe–S world. Fe–S complexes are found in a number of enzymes and come in different degrees of chemical complexity. In the simplest case, a single iron atom is tetrahedrally bound to the sulfur of four cysteines. In [2Fe–2S] clusters, two irons are each bound to two inorganic sulfides and two cysteines sulfurs, while [4Fe–4S] clusters contain four irons, four inorganic, and four cysteine sulfurs. Each iron atom is coordinated with three inorganic sulfides and one cysteine.

Successful inventions in the quest for food were never lost in the evolution of life. Small islets of the prebiotic world can thus still be found in the enzymes of contemporary organisms. We alluded to similar observations when noting the frequent use of the nucleotide cofactors in modern enzymes, which were inter-preted as the relics of the RNA world that preceded the DNA world. Another lesson is the modular character of the basic building blocks. To stay in the example of the nitrogenase complex, the reductase contains a [4Fe–4S] cluster, while the P cluster of the nitrogenase corresponds to two linked [4Fe–4S] clusters. The series of Fe–S complexes represent in some way duplications of a basic principle. The modular character of the nitrogenase complex also becomes evident at the genetic level. The capacity of nitrogen fixation is widely distributed in eubacteria and Archaea, but no association with phylogenetic lineages can be perceived. Apparently, in the evolution of life the genes for nitrogen fixation have many times been horizontally transferred between organisms. Mother Nature uses and reuses its successful inventions. This is not a conservative preoc-cupation of nature. Perhaps some inventions in early biochemistry were so ingenious that during the few billion years of evolution no superior solutions were or could be found. However, one has also to consider the situation that the early biochemical inventions shaped their chemical environment, which made the replacement of the former inventions by newer, but fundamentally different solutions impossible. Only tinkering with the ancient basic tool kit was allowed from a certain stage of biochemical evolution. Like in the famous Lego toys of your childhood, Nature could build ever-bigger forms by new combinations, but you had to use a fixed set of basic building blocks and chemical principles. Actually, evolution has been compared to the repair of a car that is in full driving speed. This is a good picture indicating that some types of fundamental ameliorations are simply not allowed since you cannot stop the car. This conser-vative nature of evolution comes at a price when the environment on Earth fundamentally changes. Ironically, the invention of oxygenic photosynthesis created also problems for its inventor, the cyanobacteria. One of the following sections will show how cyanobacteria fixed this problem by a costly mending exercise.

Nitrification

In the previous sections, we have explored nitrogen fixation, the transition of atmospheric N_2 into ammonium NH_4^+. This is an important part of the cycling of nitrogen in the biosphere. To close the cycle, you need nitrifying bacteria that oxidize ammonia to nitrite (NO_2^-, nitrosobacteria), followed by nitrobacteria that further oxidize nitrite to nitrate (NO_3^-, nitrobacteria). The loop is closed by denitrification, the reduction of nitrate to N_2, a reaction mediated by over 40 genera of bacteria and a single group of Archaea. This bias toward eubacteria was even more pronounced with respect to nitrification, which was thought to be limited to bacteria. In fact, for years Archaea were considered by microbiologists as extremophiles that thrive only in very special and, as the name indicates, extreme environments. The currently cultivated Crenarchaeota are sulfur-metabolizing thermophiles. However, this observation is certainly an isolation bias. Crenarchaeota are with an estimated 10^{28} cells a dominant constituent of the oceans. They thrive not only in cold oxic ocean waters, but also in terrestrial environments. One should therefore suspect that they play an important role in global biogeochemical cycles. But in what cycle? Oceanographers found recently a lead (Könneke et al. 2005). When they enriched microbes by serial passage from a tank of the Seattle Aquarium, they purified a biochemical activity with enrichment for Crenarchaeota, namely the oxidation of ammonia to nitrate. The medium contained only bicarbonate and ammonia as the sole carbon and energy source, which speaks for an autotrophic organism. Eubacteria could be excluded in this fraction. What they found were small rods that stained as peanut-shaped microbes, a typical finding for marine Crenarchaeota. They showed a near-stoichiometric conversion of ammonia to nitrate and seem to use a pathway known from nitrosobacteria: they oxidize ammonia by the ammonia monooxygenase to hydroxylamine, NH_2OH. In nitrosobacteria, this step uses molecular oxygen as oxygen donor and NADPH or ubiquinone as electron donor to drive the reaction. In a second step, hydroxylamine is oxidized to nitrate by hydroxylamine reductase. Water is now the oxygen donor and four protons are generated in the reaction (the acidity produced in this reaction actually destroys old limestone buildings). Since the first step needs reducing equivalents, nitrosobacteria show large membrane systems as an adaptation. Notably, the ammonia-oxidizing Crenarchaeota lack conspicuous membrane systems speaking for a different biochemistry providing the reducing equivalents. As a further difference, these Crenarchaeota fix carbon via the 3-hydroxypropionate pathway in contrast to nitrosobacteria that all use the Calvin cycle.

Closing of the Nitrogen Cycle by Anammox Bacteria

We have encountered several cycles in our survey of the basic aspects in the quest for food: the carbon cycle from CO_2 to organic carbon by various ways of CO_2 fixation and then back to CO_2 by respiration; or the water cycle with water splitting to O_2 in photosynthesis and back to water via the reduction of O_2 in respiration. Likewise in the N cycle, there is one arm of the cycle where atmospheric N_2 is fixed into NH_4^+, so this process must also be reversed where NH_4^+ is again transformed back into N_2. When microbiologists designed the nitrogen cycle at the end of the nineteenth century, the possibility of anaerobic ammonium oxidation (hence the name "an-amm-ox") was not considered. Since 10 years, we know that about 50% of the removal of the fixed nitrogen from the oceans occurs via anammox bacteria. In addition, one major advance in wastewater engineering was the removal of ammonia by the anammox process. This biogeochemical missing link was subsequently associated with an autotrophic member of the bacterial order *Planctomycetales* (Strous et al. 1999). However, the intermediate metabolism of these bacteria was unknown. To get a hand on this organism, a large consortium of microbiologists fed an anoxic bioreactor with ammonia, nitrate, and bicarbonate as nutrients and wastewater as source of microbes (Strous et al. 2006). They had to wait a year until they had enough material for analysis by community genomics. The culture grew only very slowly with a generation time of two weeks, but the DNA sequences told them that is was finally dominated by a single bacterium *Kuenenia stuttgartiensis*. They could compose a 4.2-Mb genome for it, which allowed them to deduce the biochemical pathway of the anammox reaction. In this scheme, nitrate is reduced to NO. With the addition of three further electrons, NO combines with NH_4^+ to give hydrazine (N_2H_4) using a novel enzyme. With the help from another novel enzyme, high-energy electrons from hydrazine are transferred via ferrodoxin yielding N_2. These electrons are then used by the acetyl-CoA synthase, which fixes CO_2 via the CoA pathway. As these biochemical reactions proceed slowly, special demands are made on the membranes of this organism to keep the proton gradient intact, which is necessary for energy production in a complicated branched respiratory chain. Anammox bacteria found an interesting solution to this biochemical challenge: They evolved ladderane lipids as the major component of their biomembranes. More precisely—of the membrane that surrounds the anammoxosome, a specialized intracytoplasmic compartment that fills much of the volume from this small cell (Sinninghe Damste et al. 2002). Ladderanes are used by engineers working in optoelectronics, but Mother Nature has invented them well before human chemists. The scientists demonstrated that the membrane from the anammoxosome is especially impermeable when compared to the cytoplasmic membrane. There is more than one good reason for this extremely dense membrane structure: not only must the proton gradient be conserved in this slow growing bacteria, the organism must also protect its cytoplasm from hydrazine and hydroxylamine (NH_2OH), very toxic intermediates in its energy pathway. The organisms live—as already evident from its slow growth and small size ($<1\,\mu m$)—at the limit of its bioenergetics possibilities. The researchers calculated that only a 10%-hydrazine loss across the membrane would result in a 50% decrease in biomass yield, not to speak of the toxic effects. However, the take-home message in microbial ecology is always the same. Even if the process is inefficient, if it is only marginally energy-yielding and there are no competitor having better solutions to the problem, the ecological niche is yours for exploitation.

The ladderanes, which make up a major part of the intracellular membrane, show a peculiar structure of five fused cyclobutane rings linked to rather special membrane lipids, e.g., glycerol monoether, which were previously believed to be specific to the domain Archaea. Phylogenetic tree analysis demonstrated that *Planctomycetales* or the ancestor of anammox bacteria represents one of the deepest branches in the bacterial phylum, which even suggested to some scientists a nonhyperthermophilic ancestor for bacteria (Brochier and Philippe 2002). More recent phylogenetic analysis with a much larger protein database supported, however, an evolutionary grouping with Chlamydiae (Strous et al. 2006).

Plant Symbiosis for Nitrogen Fixation

This section deals with the acquisition of the atoms of life and how they get from the world of inorganic chemistry into biomass. Animals lack the capacity to assimilate carbon, nitrogen, and sulfur from inorganic precursors like CO_2, N_2, or sulfate. Since we eat bacteria only in limited amounts, we need primary producers like plants as food for us or as feed for the animals, which we subsequently eat. Plants are autotrophs and thus equipped by nature to fix these compounds. To fix CO_2, plants have acquired an ancestor of cyanobacteria as an endosymbiont, which became the chloroplast. Plants do not have the enzymatic apparatus to fix dinitrogen, and nitrogen nutrition is thus a dilemma for plants if it does not come in the soil as either a reduced (ammonium) or an oxidized (nitrate) compound, which its root cells can absorb. Plants have not acquired dinitrogen-fixing endosymbionts, but some plants have developed intimate relationships with nitrogen-fixing bacteria that come close to this goal. With respect to assimilatory sulfate reduction, plants have maintained or evolved their own enzymatic equipment. In the following two sections, we will explore these two assimilatory capacities of plants.

Biological Fertilization

In nonnodulated plants, the root cells can adsorb nitrate from the soil, which is reduced to nitrite and then to ammonium in plastids from the root cells. The ammonium ion is used for incorporation into amino acids. Ammonium ions and a few other nitrogenous compounds can also directly be extracted from the soil. However, nitrogen deficiency frequently limits plant productivity. The symptom of nitrogen deficiency is chlorosis. The older leaves turn yellowish because the nitrogen deficiency limits the synthesis of chlorophyll. Remember that the central Mg ion in the tetrapyrol ring of chlorophyll is held in place by four nitrogens. To remedy this situation, farmers use manure, rich in nitrogenous organic waste products, on their fields long before the biochemical rationale for this procedure was elucidated. Leguminous plants have solved this problem biologically. They team up with nitrogen-fixing bacteria called *Rhizobium* (Figure 4.13). This symbiosis has attracted the curiosity of biologists for its economical value. Grain legumes (peas, beans, and soybeans) and forage legumes (alfalfa, clover) known for their root nodules fix nitrogen at a rate of 100 kg/ha/year. It has been estimated that the overall biological N_2-fixation in terrestrial systems adds up to 90–140 Tg/year (Tg is teragram or 10^{12} g or 10^6 metric tons). The annual fertilizer synthesis is worldwide, only 80 Tg. Actually, 80% of the nitrogen available to plants comes from biological nitrogen fixation, 80% of it from symbiotic associations.

The *Rhizobium* Megaplasmid

The molecular basis for the symbiosis between *Rhizobium* and legumes is provided by a bacterial megaplasmid of 536 kb (Freiberg et al. 1997). Note that this is nearly the genome size of intracellular bacteria like *Mycoplasma*. This large plasmid does not contain genes essential for transcription, translation,

FIGURE 4.13. Nitrogen-fixing bacteria of the genus *Rhizobium* associate with the roots of *Lupinus* forming nodules.

and primary metabolism. In fact, the bacterium can be cured from the plasmid without compromising the survival of the organism. What genes are found on this plasmid? You find neatly ordered the *nifH, D, K* genes encoding the γ-subunit of the Fe protein and the α- and β-subunits of the MoFe proteins, respectively. These genes are followed by the *nifE* and *nifN* genes; they have to cooperate in MoFe cofactor synthesis with the *nifB* gene, which was transferred into the nearby *fix* gene cluster. Some 20 kb apart, you find a second copy of the *nifH* to *K* genes. The *fix* gene cluster encodes proteins involved in the electron transport to the nitrogenase complex (including the ferrodoxin genes). However, nitrogen fixation genes make only a minor but clustered part of the gene content from this plasmid. You find numerous ABC transporter genes, genes involved in protein secretion and export, various enzymatic functions, genes involved in polysaccharide and oligosaccharide synthesis, and last but not least nodulation genes, whose function I will discuss below. The scientists who sequenced the plasmid noted that the megaplasmid is a mosaic not only with respect to gene function, but also with respect to GC content. According to this argument, the nitrogen fixation and nodulation genes have been recruited from different genetic sources. They further noted that it seems to act as a transposon trap and contains an *Agrobacterium*-like conjugal transfer cluster. In their view, the megaplasmid was gathered through transposition with other soil bacteria, and the symbiotic genes were acquired by lateral gene transfer.

Controlled Relationship

This relationship is of mutual interest, but for both sides it comes with costs. The plant has to offer about 15 g carbohydrate to the symbiont to receive 1 g of nitrogen. Not surprisingly, plants have measures to end the symbiotic relationship when nitrate or ammonia is available in the soil at sufficient quantities. Interestingly, most rhizobia bacteria do not fix N_2 outside of their plant host. Apparently, without the carbohydrate offer from the plant this activity becomes too costly. However, in association, the nutritional relationship becomes so useful for both partners that each side has even developed means to locate the other in the soil. The root cells hereto exude flavonoids. These flavonoids are then sensed by free rhizobia in the soil. In fact, different legumes secrete different flavonoids to assure a specificity of the interaction. However, an individual plant can secrete different flavonoids at different developmental stages and can associate with different rhizobia. Systematically, the major symbiotic bacteria of legumes belong to the α-proteogroup of eubacteria: *Rhizobium* and *Sinorhizobium*. They are closely related to *Agrobacterium tumefaciens*. This is an interesting relative since this bacterium forms crown-gall disease in many dicotyledonous plants by transforming plant cells with the T-DNA from its Ti (tumor-inducing) plasmid. This means that the association between rhizobia and plants might even be considered as a mild, controlled "infection." If they allow a too close relationship to a nonnitrogen-fixing bacterium, the plant runs the risk of being exploited. If the purported partner is a plant pathogen, it spells disease. The flavonoids are powerful messengers. Some induce directly a positive chemotaxis toward the source of the flavonoid signal and the rhizobia associate with the root hairs. The flavonoid signal is probably directly sensed by the NodD protein, which is also a potent bacterial transcription factor activating many nodulation genes (*nol* and *nod* genes). The core *nod* genes encode the Nod factor, which is sensed by the plant hair roots. These bacterial Nod factors chemically represent lipooligosaccharides. These consist of three to five *N*-acetylglucosamine residues. The chemical specificity is achieved by chemical modifications with SO_3^- or acetate groups and fatty acid side chains.

The Nodule

The root hair growth is perturbed by the Nod factors, the hair deforms and curls as it grows. Only minutes after contact with the Nod factors, the membrane of the root hairs depolarizes and ion fluxes are observed. Conspicuous are periodic spikes in cytoplasmic calcium. The bacteria are trapped in this curl, penetrate the root hair by forming an infection thread. In this thread, the bacteria proliferate as they invade the plant tissue. The plant–bacterium interaction creates a new morphological structure, the root nodule. There are two forms of it: cyclindrical and spherical. The cyclindrical shows a characteristic succession of tissue. At the apex is a meristem (zone of active cell division), followed by an infection zone containing the infection threads, and next comes an early symbiontic zone, where bacteria develop into bacteroids. The bacteroids are surrounded by a plant-derived membrane, which controls the metabolite fluxes into and out of the bacteroid. What follows then is the nitrogen-fixing zone, which is nearly entirely occupied by bacteroids and finally a senescence zone toward the basis of the nodule. Plants actually assist the nitrogen-fixation process by keeping the oxygen tension low. They achieve that by synthesizing leghemoglobin, an oxygen-binding protein. Leghemoglobin is another fascinating illustration of the conservative attitude of Nature. It shares less than 20% amino acid sequence identity with hemoglobins, but the 3-D structure identifies it clearly as a member of the hemoglobin family. Bacteria assist in the process of reducing the intracellular oxygen concentration by respiration. Actually, bacterial respiration represents another important oxygen sink. At the same time, this provides the energy needed for the nitrogen fixation. To do that at the lowered oxygen tension, rhizobia change to a cytochrome oxidase with a very low K_m for O_2 of 8 nM (compared to 50 nM for the enzyme in free-living rhizobia). As in cyanobacteria, a complicated network of two-component response regulators—that sense the supply of dicarboxylic acids (the wedding present of the plant root) and a low O_2 tension—cooperates with transcriptional regulators. The membrane bound FixL protein senses O_2. In the absence of oxygen, it becomes phosphorylated and in turn phosphorylates the soluble FixJ response regulator. The latter binds to the promoter (appropriately dubbed "anaeroboxes") of the *nifA* and *fixK* genes. The latter two are the transcriptional activators of the *nif* and *fix* gene clusters. The *nifK*, *D*, and *H* genes, for example, encode the protein subunits of the nitrogenase complex. The photosynthate from the plant enters the nodule as sucrose. As the bacteroid membrane cannot transport sugars, the sucrose is first converted into organic acids. These organic acids fulfill two functions. Malate enters the bacteroid where it is oxidized to provide ATP for nitrogen fixation. The nitrogen leaves the bacteroid as ammonium. Another part of the organic acids provides the carbon backbone (glutamate) for the synthesis of nitrogen-containing transport compounds from ammonium (glutamine).

Sulfur Uptake by Plants

Role of Sulfur

Sulfur is an essential macronutrient required for plant growth. Sulfur's primary use in the plant biochemistry is to synthesize the two sulfur-containing amino acids, cysteine and methionine. The activities of several chloroplast enzymes involved in carbon metabolism and the photosynthetic process are regulated by reversible disulfide bond formation. Glutathione, a tripeptide synthesized by enzymes and not the ribosome, is involved in growth and development of the plant by its use as a storage and transport form of physiological usable sulfur. Even a casual look into a biochemistry book shows that sulfur chemistry is used in many coenzymes and vitamins (e.g., Coenzyme A, thiamine, biotin) and at active sites of many enzymes (e.g., Fe–S centers). Sulfur-containing lipids are found in chloroplast membranes, in lipooligosaccharides that function as Nod factors or in phytoalexins (plant compounds produced in response to pathogen attack).

Distribution in the Plant

Plants acquire sulfur as sulfate via the roots by an active transport process, and in polluted areas as gaseous sulfur dioxide via the leaves. Plants have a reductive sulfate assimilation pathway. The root cell membrane contains a high-affinity sulfate permease that cotransports the sulfate together with three protons. When incubated over a range of sulfate concentrations, a multiphase sulfate uptake is measured in roots suggesting the existence of multiple transporters with distinct affinities for sulfate. The transport of sulfate is driven by a proton-pumping ATPase. Within the root cell, sulfate is also stored in the vacuole and is transported across the tonoplast (the membrane surrounding the vacuole) by a uniporter. This transport is powered by the electrochemical gradient created by the highly acidic vacuolar sap. Root plastids also import sulfate, which is then fixed into cysteine. The transport mechanism is probably via an antiport mechanism, which exchanges sulfate against phosphate. From the root, sulfate gets via the xylem transport system to the leaf cells, which extract sulfate from the xylem by a low-affinity sulfate permease. The chloroplast is a major place of sulfate reduction to cysteine and glutathione. Animals cannot reduce sulfur and therefore depend on sulfur-containing amino acids in their diet to satisfy their sulfur needs. Increasing the organic sulfur content of food and feed plants is thus a major biotechnological challenge. Sulfate is generally relatively abundant in the environment and thus not a growth-limiting nutrient for plants. However, oilseed *Brassica* varieties respond to sulfur fertilization with increased harvests. The high-affinity sulfate permease in the root cells is regulated by the sulfate concentration in the soil water. Under sulfate starvation conditions, the level of mRNA for the high-affinity permease increases rapidly. This permease is a single polypeptide with 12 membrane-spanning regions.

Assimilatory Sulfate Reduction

Chloroplasts contain the entire pathway of cysteine synthesis and are probably the primary site for cysteine synthesis in plants. The reason is clear: The reduction of sulfate to sulfide requires a lot of energy—nearly twice as much as nitrate and carbon assimilation. ATP and reductants are at hand in chloroplast since they are produced abundantly by photosynthesis. The first biochemical step is catalyzed by ATP sulfurylase, which replaces the two terminal phosphate groups of ATP by a sulfate group creating 5'-adenylylsulfate (APS). This compound contains a high-energy phosphoric acid–sulfuric acid anhydride bond that prepares sulfate for the following metabolic steps. This reaction is thermodynamically not favored and must therefore be driven by the consumption of APS in the next biochemical steps and the splitting of the leaving pyrophosphate group. The next step in sulfate reduction in plants is not clear. One hypothesis postulates the sequential action of an APS kinase adding a phosphate group to the 3' hydroxyl of the ribose moiety of APS creating phosphoadenylyl sulfate (PAPS). This reaction is followed by a hypothetical PAPS reductase step which reduces the PAPS sulfate group to a free sulfite using thioredoxin as reductant and then finally a sulfite reductase transferring six electrons from ferredoxin on sulfite, reducing it to the sulfide oxidation level. This pathway is the mechanism of sulfate assimilation in cyanobacteria, the cousins of the postulated ancestors of chloroplasts. In an alternative hypothesis, sulfate is transferred from APS to a reduced thiol group (glutathione is a candidate) resulting in a thiosulfonate, which is then reduced to a thiosulfide and then released from the carrier binding by reduction to hydrogen sulfide.

Recently a glutathione-dependent APS reductase was characterized in the plant *Arabidopsis* that could replace APS kinase and PAPS reductase in *E. coli* double mutants. Apparently, this enzyme transforms APS into the free sulfite using glutathione as reductant. The domain structure of this enzyme suggests an N-terminal transit peptide that directs the protein to chloroplasts followed by a large reductase domain and a C-terminal domain which functions as a glutaredoxin. This latter domain contains two nearby cysteine residues that shuttle electrons via dithiol–disulfide interchanges.

The final step in chloroplast cysteine synthesis is the condensation of serine and acetyl-CoA to *O*-acetyl-serine (OAS), mediated by serine acetyltransferase. This enzyme exists in a complex with an OAS lyase that splits OAS with hydrogen sulfide into cysteine and acetate. In plants, sulfate reduction is regulated at several levels. OAS lyase dimers exist in excess over serine acetyltransferase tetramers. If OAS accumulates because enough sulfide is not present to transform it into cysteine, it dissociates the lyase/acetyltransfer complex, thus reducing

OAS synthesis. Another strong regulator of sulfate reduction in plants is the developmental phase. Reduction is high in young tissues and declines markedly in old plant tissues. Plants also maintain a relatively fixed ratio of reduced nitrogen to reduced sulfur of 1:20, the appropriate ratio to maintain protein synthesis at the ribosomes. A key to this regulation might be the fact that APS reductase activity declines in response to nitrogen starvation.

Nutritional Interactions in the Ocean: The Microbial Perspective

Stromatolites and Biomats

Fossil Stromatolites

The earliest appearance of purported photosynthetic bacteria in the fossil record was in strange biological structures of the shallow water called stromatolites. Stromatolites are laminated domes, which are commonly regarded to have formed by the activity of ancient microbiological mats composed mainly of cyanobacteria. Since the record of stromatolites goes back to 3.5 Ga ago, they are considered as a proxy of early life on Earth. However, their association with cyanobacteria becomes thus problematic since—as we have seen in a previous section—unequivocal evidence for cyanobacteria goes back for "only" 2.5 Ga with an estimated age for oxygenic photosynthesis of 2.8 Ga. The case for a biological origin of stromatolites was strengthened by the famous and undisputed finding of microfossils in the Proterozoic Gunflint Formation of Canada by the pioneers of this technique, namely Tyler and Barghoorn. Later on, this view was reinforced by large lithified modern stromatolites of Shark Bay in Australia. However, stromatolites frequently do not contain fossils, and some geologists prefer an abiotic origin of these structures based on a mathematical fractal analysis (Grotzinger and Rothman 1996).

Modern Stromatolites

Stromatolites persisted up to our days, although at much lower frequency and in very specialized habitats not claimed by other life-forms. What can these structures tell us about stromatolite biology? When looking at contemporary stromatolites from the Bahamas, which grow in seawater of normal salinity, a carefully orchestrated succession of microbiological communities participate in the accretion, lamination, and lithification of these structures (Reid et al. 2000). There were three types of structures. First, there is a pioneer community of the gliding cyanobacterium *Schizothrix*, entwined around carbonate sand grains. The second type represents a calcified biofilm. *Schizothrix* excreted surface films of exopolymers that become the food to heterotrophic bacteria. The third and last community includes in addition the photosynthetic coccoid cyanobacterium *Solentia*, which actually bores into the sand grains. Eukaryotic algae can associate with this population. The lithification process occurs by the bacterial decomposition of the exopolymers in the photic zone of the stromatolites.

Accelerated Cycling of Organic Carbon

The advent of oxygenic photosynthesis on Earth may have increased global biological productivity by a factor of 100–1,000. Hydrothermal sources, which might have been the cradle of life, deliver 0.1×10^{12} to 1×10^{12} mol/year of reduced S, Fe^{2+}, Mn^{2+}, H_2, and CH_4 and sustain 0.2×10^{12} to 2×10^{12} mol C/year of organic carbon by microbes. The hydrothermal activity of the young Earth was greater than nowadays. However, it could not reach the estimated global photosynthetic productivity estimated at 9×10^{15} mol C/year (De Marais 2000). Much of this new productivity probably occurred in microbial mats where photosynthetic and anaerobic microbes are metabolically closely associated: 99% of the biomass, which is produced in the mat by photosynthetic organisms, is rapidly mineralized by heterotrophic bacteria. This high figure should not surprise us. Terrestrial photosynthesis releases enormous amounts of oxygen but has little net effect on the atmospheric O_2 level because it is balanced by the reverse process of respiration. Marine photosynthesis, in contrast, is a net oxygen producer because a fraction of the newly synthesized organic matter is not respired but buried in sediments. However, this fraction is small and corresponds to about 0.1% of the total (Kasting and Siefert 2002). *New additions of organisms into the great game of evolution will thus not necessarily cause a rapid increase in one chemical component, they will mostly speed up the turns of the geochemical cycles. Biomats, despite their productivity, will not necessarily explode in biomass because this process is cancelled by organisms that live from biomats.* Modern mats are, for example, excellent food sources for snails, crustaceans, and small invertebrates (Jorgensen 2001). They are so attractive to animal life that the absence of feeding traces on ancient stromatolites is taken as evidence for the absence of animal grazers at the indicated geological period.

Stromatolites in Hypersaline Water

Centimeter-sized mats built by cyanobacteria occur in the hypersaline lagoons in Baja California where they find ideal breeding grounds for their growth. These microbial mats produce CO, H_2, and unexpectedly CH_4 (Hoehler et al. 2001). CO and H_2 are the products of photosynthetic cyanobacteria in the topmost layer of these communities. CO (probably a by-product of cyanobacterial photosynthesis) and H_2 (probably a by-product of cyanobacterial nitrogen fixation) show an alternate appearance: H_2 is high during nighttime and low during the day. The flux of H_2 is substantial and allows the synthesis of methane by methanogens in the topmost layer of the mats. The mat-driven flux of H_2 exceeds the geothermal flux of H_2 by 2–4 orders of magnitudes. This created new metabolic possibilities for biomass production by prokaryotes associated with cyanobacteria. Actually the exchange of reducing power between different metabolic groups of prokaryotes was not the only consequence. Since the microbial mats were a dominant form of microbial life for such long geological periods, they also profoundly influenced the atmosphere. The large fluxes of H_2 from the mats, followed by the escape of H_2 into space (molecular hydrogen is so light that it cannot be held by the Earth's gravity field), paradoxically contributed to oxidation of the primitive ocean and atmosphere. We clearly see here a biogeochemical impact of cyanobacteria at a global level.

The Sulfuretum in Salt Marshes

Cyanobacteria were not only cooperating with other bacteria; they probably also used the invention of oxygenic photosynthesis as a chemical club leading to the observed stratification in ancient stromatolits. Similar stratification is also seen in modern salt marshes, flat coastal areas that are flooded and that fall dry with the tides. Over the first centimeter, they show an intensive color change. This very productive ecosystem contains primary producers like the nitrogen-fixing and oxygenic photosynthetic cyanobacteria. In the upper 1 mm, the cyanobacteria are associated with diatoms (gold-brown algae of the *Chrysophyta* genus), which produce the yellow-brown color of the sand. Over the next 2 mm cyanobacteria solely dominate giving the sand a blue-green hue due to the maximum of chlorophyll a absorption at 680 nm. The next 3 mm is dominated by purple sulfur bacteria using first bacteriochlorophyll a and, in deeper layers, bacteriochlorophyll b with absorption maxima at 850 and 1,020 nm, respectively, which gives the mat a pink and then a peach color. As in stratified lakes, green sulfur bacteria live underneath the purple sulfur bacteria, changing the color to olive green. At 7 mm depth, one can find black layers dominated by the sulfate-reducers like *Desulfovibrio*. Due to the intensive turnover of sulfur compounds in this ecosystem, it is sometimes called a "sulfuretum."

Read my Lips: Cyanobacteria at the Ocean Surface

Two Small Bacteria

Current estimates are that about half of the global photosynthesis and oxygen production is achieved by the phytoplankton, single-celled organisms that live in the top layer of the ocean where sunlight can drive the light reactions of photosynthesis. Cyanobacteria take the greatest share with essentially two basic types: 0.9-μm-large *Synechococcus* strains (which is not large for bacterial standards) and the even smaller and also more abundant 0.6-μm-large *Prochlorococcus* strains. Several of these cyanobacteria had their genome sequenced, which allows now an *in silico* insight into their nutritional lifestyle (Fuhrman 2003). Before getting to this topic, we need to spend some words on living in the sea.

The Ocean: Nutrients

"*Omne vivum ex mare*"—All life comes out of the sea. Life has a peculiar relationship with water as a solvent such that the search of water on other planets becomes a proxy measure for searching life. One should therefore suspect that it must be easy to make a living in the oceans. And this is indeed true: the ocean offers an enormous stability against desiccation, protects against large temperature fluctuations and against UV light as a harmful radiation to the genetic material. However, in analogy to the American slogan "There is no free meal," oceans are also difficult environments for life. In fact the lack of food is the major problem in the marine environment: Oceans have been compared to huge stable deserts. Offshore oceans correspond nutritionally to oligotrophic lakes and depend strongly on nutrient supply through water currents. Areas that experience upwelling of nutrient-rich deep-sea water can reach, therefore, productivity levels of eutrophic lakes. Some of those represent the richest fishing grounds on Earth, e.g., the Humboldt current at the Peruvian coast of South America.

The Ocean: Light

The productive zone of the oceans just extends to 50 or 100 m depth. In this upper layer, one can find the primary producers. The reason is simple: If nutrients are scarce, organisms are favored that can use light as energy source to fix CO_2. Because of its high alkalinity, seawater can maintain CO_2 at 2 mM at its surface compared with pure water that maintains only 15 μM CO_2 in equilibrium with the atmosphere. CO_2 is thus not a limiting factor. Light can penetrate a water body in a wavelength-dependent way. Infrared is immediately adsorbed by water molecules and light with wavelength shorter than 400 nm and longer than 700 nm does not penetrate into water to great depth, e.g., the intensity of red light falls to 1% of its incident value within less than 3-m water depth. These physical and chemical constraints determined the possibilities of life in the ocean. Not surprisingly not only photosynthetic bacteria but also small photosynthetic eukaryotes comprise a prominent part of this environment. Indeed the upper 100 m of the oceans account for nearly 50% of the net primary productivity of the biosphere. Due to their tiny size, cyanobacteria comprise only 1% of the total photosynthetic biomass, yet account for the majority of the marine nitrogen fixation (Bryant 2003). In addition these cells represent an important biological pump, which traps CO_2 from the atmosphere and potentially stores it in the deep-sea sediment. One primary problem for the photosynthetic CO_2 fixation is how to deal with the differing light intensities they meet. How to react toward a damaging too much of light in the topmost layer and how to capture the decreasing light intensity with increasing water depth? The microbial solution is straightforward and demonstrates the beautiful rationality of biological systems. For example cyanobacteria in the upper layer have a photolyase gene that helps them in repairing ultraviolet damage to their DNA. This gene was lost in the low-light ecotype because it became useless in this environment. In contrast, the low-light-adapted form showed more genes encoding chlorophyll-binding antenna proteins than the high-light-adapted forms. Eighteen monomers of this antenna protein form a ring around the trimers of the PSI reaction center. This is a clever adaptation that maximizes the capture of more photons in the dim light conditions at the lower photic zone (Bibby et al. 2001a,b). In the case of low light adaptation, the antenna is not expressed from *isiA*, which is used in low-iron adapted strains, but from another gene: *pcb*, for chlorophyll-*binding* (Bibby et al. 2003). Very low light adapted strains have up to seven *pcb* genes, which allow growth under the very low light intensities found at the bottom of the euphotic zone.

Furthermore, *Synechococcus* uses a different light-harvesting system than *Prochlorococcus*, namely phycobilisomes. They are composed of a system of core and radially arranged rods. The core and rod cylinders consist of discs containing phycocyanin and phycoerythrin as pigments. These pigments have the remarkable property to adsorb light in those frequency bands not covered by chlorophyll a and β-carotene. *Prochlorococcus* uses divinyl derivatives of chlorophyll a and b, which are unique to this genus thus allowing complementary light use by different cyanobacteria and the optimal use of those frequency bands still reaching deeper water layers.

The solution to the light problem is thus differentiation of marine cyanobacteria into two major genera, *Synechococcus* and *Prochlorococcus*, each of which come in many ecotypes. Sequencing of their RNA polymerase genes revealed a division of the former in more than eight clades, while the latter showed a splitting in two clades that separated according to the depth of the clones' collection (Ferris and Palenik 1998). This observation suggests indeed physiological specialization of the dominant marine cyanobacteria according to nutritional needs.

Biology has a Few Basic Principles

Before we investigate these physiological adaptations in cyanobacteria, let me lean back for a short moment. You will have certainly remarked recurring subjects in this survey of the natural history of eating. In the past, biology was set apart from chemistry and even more from physics because it lacked unifying principles and general laws. Biological research over the last decades has demonstrated a lot of recurring principles in many biological phenomena ranging from pattern determination in embryogenesis to the interaction of organisms in complex ecosystems. Biology is thus not an endless enumeration of special phenomena and hard to understand peculiarities. Biology has its clear-cut theoretical foundations and principles. For example one of the unifying principles in biology is diversity. With much more foundation one could claim a "horror vacui" (fear of the void) in biology than in physics: Biological systems want to fill each and every place on this planet, as long as even the most basic constraints of biochemistry are satisfied. But in order to do so, organisms need to be diverse since not even a seemingly uniform environment like the ocean is spatially homogenous. In addition as Greek natural philosophers expressed with the statement "panta rhei" (everything flows), the only constant aspect in Nature is change. In fact ecologists state that there is no such thing as an organism being adapted to its present environment. Any living organism was adapted to an environment that existed in the past. Any individual must again find its place in the currently existing environment in this eternal fight for survival of the fittest. Species that were too much adapted to the conditions that prevailed yesterday might lack the flexibility to adapt to the environment of today and the changes that will occur tomorrow. Species must therefore be able to adapt—those that have lost their genetic diversity and thus the adaptation potential are condemned to die out.

Prochlorococcus: Small is Beautiful

Prochlorococcus is the smallest known oxygen-evolving autotroph. It dominates the phytoplankton communities in most tropical and temperate open ocean ecosystems. It comes in different ecotypes that differ in growth rate in high and low light (Moore et al. 1998), and three have been sequenced allowing metabolic inferences (Rocap et al. 2003; Dufresne et al. 2003). The high-light-adapted ecotype found in the upper parts of the euphotic region of the ocean has, with 1.6 Mb, the smallest genome of any known oxygenic phototroph. It is somewhat astonishing that a mere 1,700 genes are sufficient to build a bacterium, able to create globally abundant biomass from solar energy and inorganic compounds. Genetic complexity is thus not a necessity for evolution. Small bacteria, which made some key inventions early in the evolution of life, could apparently maintain their position on our planet over billions of years. In the script of evolution, there is apparently no inescapable rule toward increasing complexity of the evolving organisms. Cyanobacteria from billion-year-old fossils and from those cyanobacteria living today cannot be distinguished by a nonspecialized microbiologist. In fact being small can be an asset. In the top layer of the ocean where light can drive photosynthesis, nutrients such as nitrogen and phosphorus are frequently extremely diluted. Having a small genome means reducing the need for N and P, both of which are essential chemical ingredients of DNA. Being small means that you need substantially less food to build biomass and to power the smaller body.

However, one should not underestimate these bacteria. The low-light-adapted ecotype of *Prochlorococcus*, which dominates the deeper waters, shows a significantly larger 2.4-Mb genome with nearly 2,300 genes. Only 1,350 genes are shared between both strains demonstrating highly dynamic genomes that have changed in response to myriad selection pressures. Most of these shared genes are also shared with the 2.4-Mb genome of the other dominant, motile cyanobacterium *Synechococcus* (Palenik et al. 2003). Not surprisingly major differences between the three *Prochlorococcus* genomes were found in genes that mediate light acclimation (Bibby et al. 2003).

Other important metabolic adaptations occur in the nitrogen metabolism. Each ecotype of *Prochlorococcus* uses the N species that is the most prevalent at the light, i.e., water depth, levels to which they are best adapted: ammonium in the surface water and nitrite at deeper levels. Both sequenced genomes lack the nitrogenase genes, which would be essential for dinitrogen fixation. In contrast, the *Prochlorococcus* genomes showed various genes for oligopeptide and sugar transporters genes suggesting the potential for partial heterotrophy. Indeed some strains were found in zones so deep that photosynthesis alone could not sustain their survival. Taken together, we see that *Prochlorococcus* diverged into at least two ecotypes, one that inhabits the upper, well-illuminated, but nutrient poor

100-m layer of the water column and another that thrives at the bottom of the light zone (80–200 m) at dimmer light but in a nutrient-rich environment.

Synechococcus

Synechococcus is less abundant in very oligotrophic environments than *Prochlorococcus* but has a broader global distribution. Its genome identifies it as being nutritionally more versatile compared to its *Prochlorococcus* relatives. It can use more organic compounds as nitrogen and phosphorus sources. It is motile, which allows it to search nutrients released by heterotrophic bacteria. However, this mobile lifestyle probably exposes it to some dangers like attack by phages and grazers and toxins released by bacteria that want to defend their niche. Fittingly, *Synechococcus* has also many efflux systems, some of which seem to be responsible for toxin efflux.

Exotic Niches

Cyanobacteria are not only major contributors to the global carbon and nitrogen budgets, they are also a very flexible group, which has footholds in salt and freshwater. They also form symbiotic relationships with animals and plants. In contrast to the oceanic generalist, one also finds many niche specialists in this bacterial group. I will illustrate this with a salt-water specialist and a freshwater specialist. Not all cyanobacteria absorb at 680 nM, there are also forms like *Acaryochloris*, which sport chlorophyll d with an absorption maximum at in the far-red spectrum (700–720 nm; Miyashita et al. 1996). This is apparently an adaptation to their habitat, which is strange enough. They live as a photosynthetic symbiont underneath of an ascidian (sea squirts, a primitive chordate). This ascidian apparently likes photosynthetic bacteria as supplementary power houses since it harbored an additional—this time conventional chlorophyll a-containing—cyanobacterium in a light-exposed body cavity (Kühl et al. 2005). Far-red in contrast to visible light penetrated much deeper into the ascidian tissue, which allows this chlorophyll-d-containing cyanobacterium to sustain photosynthesis and thus to thrive in extreme shade. This interpretation is consistent with the occurrence of other species of this bacterium, like the epiphytic *A. marina*, which is associated with the underside of red algae (Murakami et al. 2004).

Green Manure

Still other cyanobacteria use nitrogen fixation in symbiosis with plants. Most cyanobacteria practice extracellular symbiosis. An especially well investigated case is that of *Anabaena azollae*, which associates with the water fern *Azolla*. The sporophyte of this fern floats on quiet water where it forms a dense surface mat in tropical countries. It is actually an important co-crop in rice paddies in South East Asia, where it controls weed growth, prevents insect proliferation and contributes fixed nitrogen exceeding 50 kg/ha per year and thus allows sustainable rice cultivation. *Azolla* has bilobed leaves that float on the water surface. The dorsal aerial lobe contains chlorophyll and air cavities containing cyanobionts, the symbiotic cyanobacterium. Interestingly, here, the plant has succeeded to take control of the symbiont: Every second cell instead of every 10th cell of the filament is differentiated into a heterocyst and the heterocyst induction is no longer controlled by the nitrogen status in the bacterial cell. Notably a third partner still belongs to this community: *Arthrobacter*. This bacterium apparently creates a microoxic niche necessary for the nitrogen fixation by increased respiratory activity. This observation leads us to the next section.

Problems with Nitrogen Fixation for Cyanobacteria

The Missing Bacterium for Oceanic Nitrogen Fixation

Nitrogen is an essential atom of life—all living beings need it for building biomass. Fixed nitrogen, in contrast to the N_2 gas, is a limiting nutrient in much of the sunlit layers of the ocean. For an ocean organism, it would be highly advantageous to have the biochemistry to fix nitrogen. Surprisingly this is not the case for abundant cyanobacteria like *Prochlorococcus*. They reach concentrations of 10^8 per liter in warm ocean water and account for much of the primary production in tropical and subtropical oceans. They are great in CO_2 fixation, but N_2 fixation is the job of other cells. It was clear to marine biologists that the availability of nitrogen is important in regulating biological productivity in the ocean. Therefore deepwater nitrate has long been considered as the major source of new nitrogen, which supports the primary biomass production in oligotrophic regions of the world's oceans. Today we know that nitrogen fixation by cyanobacteria provides an important input into the N budget of the surface layers of the ocean. Before discussing the case of the missing nitrogen fixers, I will present some data on an extensively investigated nitrogen-fixing freshwater cyanobacterium.

Heterocysts in *Anabaena*

In filamentous cyanobacteria like *Anabaena*, many freshwater cyanobacteria, or cyanobacteria of northern seas, the adaptation of the nitrogenase to oxygen stress has a morphological correlate, the heterocyst. When exogenous sources of fixed nitrogen (NH_4^+ or NO_3^-) are depleted in their environment, about every 10th cell in the filament develops into a heterocyst. The differentiation of a normal vegetative photosynthetic cell into a nitrogen-fixing heterocyst takes about 20 h. The heterocyst is surrounded by three layers, while the vegetative cell shows only one layer. The extra layers consist of a laminated glycolipid and a fibrous polysaccharide layer, and form a barrier against the diffusion of O_2 into the cell. In this way, the nitrogenase is protected against intrusion of external oxygen. However, there is still a problem: Cyanobacteria acquire energy by photosynthesis that develops oxygen inside the cell by PSII. There is thus an internal oxygen source that needs to be neutralized. *The cells deal with this problem by a strategy that is a blend of cellular differentiation and cellular cooperation, which is not a very common event in most bacteria and points already toward a path, which multicellular organisms will take.* The differentiation process can be followed microscopically. In the proheterocyst, the thylakoids are rearranged and become contorted. Biochemists observed that at the same time the oxygen-developing center in PSII is lost; this also applies to the electron transport from PSII to PSI. PSI, however, cannot be thrown away: The cell is still in high need of energy for nitrogen fixation. However, PSI alone, separated from PSII, cannot deliver reducing power. The electrons are provided to the nitrogenase complex by a heterocyst-specific ferredoxin, FdxH, which is reduced by a pyruvate: ferredoxin reductase. However, PSI still produces ATP by a process that has aptly been called cyclic photophosphorylation. During cyclic electron transport, the conventional ferredoxin is produced as usual by PSI, but instead of transferring the electron to $NADP^+$ it interacts with a membrane-bound oxidoreductase that allows the transfer of the electron back into the quinone pool and from there into the cytochrome b_6f complex back to PSI via plastocyanin. As no new electrons are fed into the chain via PSII and as electrons cycle around this loop, powered by the light-driven reaction in PSI, the name cyclic photophosphorylation is explained. The important feature of this electron-turning wheel is that protons are still transported vectorially across the thylakoid membrane by the plastoquinone oxidation–reduction cycles. The energy for this uphill process is provided by the light captured by PSI. Thus this proton gradient still allows the production of ATP by the ATP synthase complex. Of course due to the absence of PSII, less protons are transported by cyclic electron transport than by the complete PSII–PSI system. The less-efficient ATP production is compensated by oxidative phosphorylation. However, this should lead to problems. We said above that the three-layered cell wall of the heterocysts was directed against intrusion of exogenous oxygen. The intracellular oxygen partial pressure will thus be much lower than in an adjacent vegetative cell. To continue with respiration under reduced oxygen concentration, the heterocyst expresses a cytochrome oxidase with much higher affinity for oxygen. In this way, heterocyst-specific respiration reduces still further the intracellular oxygen concentration. The thick wall of the heterocyst is also a barrier against the import of oxidizable food molecules. Therefore neighboring cells have to provide the substrates for cellular respiration. Heterocysts receive sucrose from the adjacent vegetative cells through small cellular connections (microdesmota).

Heterocysts: Regulation and Differentiation

Bacterial cells are always very economical with their resources, and so are cyanobacteria. Since PSII in heterocysts is rendered nonfunctional for electron transport to PSI, it does not need its light-harvesting apparatus. Early in heterocyst development, a protease is expressed that digests the proteins composing the phycobilisome, the antenna complex of PSII. As the reducing equivalents are now needed for nitrogen fixation, competing consumers like CO_2 fixation need to be eliminated. Consequently another heterocyst-specific protease degrades Rubisco.

Nitrogenase is expressed only very late in this differentiation process. The reason is clear: The dinitrogenase reductase subunit of the nitrogenase complex is highly oxygen sensitive and irreversibly inactivated within a minute. Overall, a lot of genetic regulation occurs during this differentiation: A cascade of two-component sensor-regulator systems, "alarmones" like polyphosphorylated nucleotides (ppGpp), new sigma factors (sigB and sigC) directing the RNA polymerase to the promoters of heterocyst-specific genes are involved. Overall more than 1,000 genes are differently expressed during heterocyst formation. We deal here with a very large stimulon. A stimulon comprises a group of genes that respond to the same stimulus—in this case, the decrease of fixed nitrogen in the environment. As if to demonstrate its versatility, cyanobacteria use even regulation systems that operate at the level of the DNA information. The nifD gene encoding one of the MoFe protein subunits of the nitrogenase complex is actually interrupted by a longer DNA element flanked by an 11-bp DNA repeat. This element is excised by a XisA enzyme, which is induced during the heterocyst development.

The nitrogenase delivers NH_4^+ as end product. As glutamine synthetase and transport systems for glutamine are also induced in the heterocyst, this specialized cell exchanges glutamine against sucrose with the vegetative cell. We have here an early stage of cellular differentiation in a versatile prokaryote. This process announces the path to ever increasing differentiation of cell functions in multicellular organisms, which is also frequently accompanied by nutritional differentiation where different cells capitalize on the execution of distinct metabolic pathways. Another aspect demonstrates the flexibility of cyanobacteria: Instead of using spatial cell differentiation, nonfilamentous cyanobacteria use temporal differentiation, performing oxygenic photosynthesis in the light and nitrogen fixation in the dark.

The heterocysts story fits to the Rubisco story. Basically dinitrogen reduction requires a strongly reducing active enzyme center and is inhibited by oxygen. Oxygen was not a problem when the nitrogenase was invented—the atmosphere was still strictly anoxic. When this changed, Nature found several solutions to the problem. As in other comparable situations instead of redesigning a new enzyme, Nature preferred to develop new layers of control or complexity overlaying the existing system. Either there were no other suitable solutions for the N_2-fixing enzyme at hand or the task had no more than one chemical solution (there is pretty much bioinorganic complexity around the active site of the enzyme) or the organisms could simply not allow playing around with a vital function to their survival.

Trichodesmium

Now back to the missing nitrogen-fixing bacteria. The large colonial cyanobacterium Trichodesmium has traditionally been considered as the dominant nitrogen fixer of the ocean (Sanudo-Wilhelmy et al. 2001; Staal et al. 2003). It occurs as filaments of cells that can reach a length of up to 0.5 mm. Yet this bacterium cannot thrive in colder seas, and other bacteria must here fill the gap in the N budget (Fuhrman and Capone 2001). Even in the tropics and subtropics where *Trichodesmium* is thought to be the dominant N_2-fixing organism, *Trichodesmium* cannot account for the entire nitrogen fixation. It provides about half of the new nitrogen. In the meanwhile, small unicellular cyanobacteria have been identified that fill the gap in the subtropical sea (Zehr et al. 2001; Montoya et al. 2004). With probes designed on the nitrogenase gene *nifH*, the oceans were searched for transcripts of this gene. The marine microbiologists had not to search long for it, they found many expressed nitrogenase genes. The gene sequence attributed them to two clusters of cyanobacteria related to *Anabaena* (group A) or *Synechococcus* (group B) (Zehr et al. 2001). Small unicellular N_2-fixing bacteria were quickly cultivated.

There are other turns to this recent discovery. The unicellular N_2 fixer are probably more important for the nitrogen budget of oceans because they are more uniformly distributed in contrast to *Trichodesmium*, which occurs mainly in a shallow portion of the upper euphotic zone. Furthermore *Trichodesmium* is rather toxic and thus avoided by grazers. The nitrogen fixed by them therefore does not easily enter the food chain.

Notably *Trichodesmium* does not form heterocysts, while in temperate and polar regions, heterocyst-forming cyanobacteria dominate the nitrogen input into lakes and oceans. Apparently there is a complicated trade-off between two processes. The decisive factor, which explains this geographical difference, is temperature. O_2 flux, respiration, and N_2 fixation all depend on temperature. Beyond a certain temperature, the heavy metabolic investment in synthesizing the components needed for the heterocyst does not any longer pay off. In fact the cell will thereby only limit its N_2 influx (Staal et al. 2003).

A World of Iron

Iron Age in Mythology

Mythology attributed an important role to iron. The ancient Greeks kept the idea of an evolution of life through time. They imagined a Golden Age not unlike the Judeo-Christian Paradise, followed by a Silver Age and—alas—an Iron Age. The latter is our current period, the world of hard work, constant warfare, disease, and death. The Judeo-Christian theory of evolution described in the book of Genesis is not far from this view when it imagines us as expellees from the Paradise. The chasing from the garden of Eden marks the transition from a life, which does not know food shortage, competition, predation, work, and war to a world full of hardships. Interestingly the pain comes in two forms linked to the two major forces in biology. The quest for food became hard labor for Adam and the quest for sex was also linked to painful labor to Eve in childbearing. Biologists do not hypothesize a Paradise or a Golden Age in a distant past.

Abundance of food was, is, and will remain a dream in the world of biology; we do not expect there will ever be a relief from the quest for food even for the self-declared "crown of the creation." Anyway perhaps the Golden Age is not a dream, but a nightmare since one cannot eat gold as already experienced by king Midas in Greek mythology. Actually chemists would argue that gold being a precious metal is too idle an element for feeding life. For that we need rough elements like iron capable of redox chemistry. As stated by the Greek mythodology, we were and are like all other organisms children of the Iron Age. Actually from the beginning of biological time to our days a group of reactive metals played a crucial role in the evolution of life, and iron played a key role in the quest for food.

This section is a reformulation of a basic discovery going back to Justus Liebig in the early nineteenth-century Germany, who discovered his law of limiting nutrients in plant growth and thus became the father of agricultural fertilization. This observation leads us to one of the boldest ecological concepts, the "fertilization" of the sea with iron. The idea has nothing to do with agriculture but a lot to do with global warming.

Another Problem in Cyanobacteria: Iron Limitation

As we have seen in a previous section, the growth of the cyanobacterium *Trichodesmium* is either phosphorus or iron limited (Sanudo-Wilhelmy et al. 2001). The case is clearer for the cyanobacterium *Synechococcus*—it showed clear evidence for iron limitation. In fact it possesses a physiological adaptation to low iron conditions. In contrast to other bacteria that secrete siderophores, i.e., iron scavenger proteins, *Synechococcus* solves the problem by economizing iron: It replaces many iron-containing enzymes by alternative versions that rely on copper or nickel as metallic cofactors. In addition the sophisticated antenna system ("phycobilisome") is sacrificed because its synthesis is a very costly iron investment. Giving up a light antenna is not a healthy idea for a phototroph. Therefore this cyanobacterium has a functional replacement. It responds to iron deficiency by expressing the *isiA* gene, where *isi* stands for "*i*ron *s*tress *i*nduced." This protein has significant aa sequence identity with CP43, the light antenna protein built into PSII. In fact IsiA forms a beautiful light-harvesting ring consisting of 18 monomers around PSI and not PSII. This is quite surprising since PSI is already very rich in chlorophylls; the new ring adds now about 200 additional chlorophylls to the already 300 chlorophylls, which decorate PSI naturally (Bibby 2001a,b; Boekema et al. 2001). These observations on iron limitation in the ocean led to a very ambitious series of ecological experiments: the seeding of the sea with iron. The stakes are high because the researchers expected that increased photosynthesis in the ocean would lead to a draw down of atmospheric CO_2 levels. As this is one of the major greenhouse gases discussed in current global climate models, the hope was that this could be a mean to counteract the relentless industrial increase of CO_2 emission by ecological interventions. But did it work?

Sowing the Sea with an Iron Plow: Where Feeding Impacts on Global Climate

Carbon Cycles

Before I present you data on these exciting field experiments, I will provide you some background information on the carbon cycle and iron. Carbon comes in six major pools. The largest one with an estimated 7×10^7 Pg ("P" read "peta," 10^{15}) is bound as inorganic carbonate in marine sediments and sedimentary rocks. This is a relatively inert pool, which is augmented by a small influx of 0.2 Pg from the ocean. The carbon in the ocean comes in two pools: (1) carbonate, bicarbonate and dissolved CO_2 in the seawater with an estimated 4×10^4 Pg and (2) a smaller DOC plus biomass pool of 700 Pg. Both pools exchange annually 10 Pg. The ocean CO_2 and the atmospheric CO_2 pool have a greater exchange rate of 90 Pg, but the overall air CO_2 pool is with 700 Pg much smaller than the ocean CO_2 pool. The atmosphere gets about 5.5 Pg CO_2 from the large fossil carbon pool (with a total pool size of 5,000 Pg of reduced energy-rich carbon such as coal, natural gas, mineral oil, which are the result of past biological activity), 15 Pg from the organic carbon deposited in soil (humic compounds amount to 1,500 Pg) and 45 Pg from dead biomass (120 Pg total pool). The atmospheric CO_2 pool loses annually 60 Pg CO_2 to the living biomass, which represents globally a pool of 600 Pg. The same amount of 60 Pg leaves the living biomass as dead biomass. The living biomass is nearly exclusively made up of plant material: consumers such as animals and humans and mineralizers like prokaryotes add up to only 2% of the overall biomass ("standing crop").

The Missing Carbon

In the iron seeding experiments, oceanographers tried to pump atmospheric CO_2 into oceanic biomass in the hope to increase the fraction of the marine carbon forced into oceanic sediment, which would become—at least for climate considerations—immobilized. If all processes in the carbon cycle are in equilibrium, the CO_2 levels would not change. This is, however, not the case for the current terrestrial/atmospheric system. We see a relentless increase of atmospheric CO_2 concentration since the onset of the Industrial Revolution. The additional influx of CO_2 comes from human activities. A chief cause is the burning of fossil fuel, which contributes photosynthesis-derived reduced carbon from past geological periods as CO_2 into the atmosphere. In a way, this reestablishes another equilibrium, photosynthesis has in the past produced more oxygen from water splitting than respiration could consume. Heterotrophs took a while until they learned to digest cellulose associated with lignin, which were invented by the early land plants. In addition substantial parts of the reduced carbon created by photosynthesis was secluded from respiratory degradation by geological processes that buried the biomass leading to coal and oil deposits.

Returning Fossil Carbon

Actually by burning fossil fuel, we reestablish old balances in decreasing an oxygen surplus and remineralizing the reduced carbon to its fully oxidized form: $CHO + O_2 \rightarrow CO_2 + H_2O$. With this process, we replace an O_2 molecule by a CO_2 molecule in the atmosphere. If this CO_2 increase is not consumed by other biological or chemical processes, the atmospheric CO_2 concentration will increase. Since CO_2 is a greenhouse gas, the global temperature of our planet will also increase (Meehl and Tebaldi 2004; Kerr 2004). We change the CO_2 budget with other activities as well, e.g., deforestation, which increases CO_2 by burning the wood and by the lesser sequestration of CO_2 due to diminished photosynthesis by the decreased plant cover. Agricultural activities also add to the positive atmospheric CO_2 budget (Janssens et al. 2003).

Curbing the apparently relentless increase in atmospheric CO_2 concentrations would thus be a highly welcome event. Political agreements like the Kyoto treaty are intended to achieve this. *It is doubtful whether the current international political situation allows an efficient control of this process, however important it might be for the future of the human civilization. Therefore scientists were thinking on alternative solutions as to how the CO_2 concentration could be maintained or even decreased despite an unabated fossil fuel burning.* This led to an ambitious project, which one could call global geoengineering. Oceanographers are currently realizing how frustratingly small our knowledge basis is if we deal with global ecological problems. *However, scientists are the daughters and sons of Prometheus and are thus not easily deterred by obstacles.*

The Martin Iron Hypothesis

The basic observation of the oceanographers is a paradox. About 20% of the world's oceans are replete with major bacterial and plant nutrients (nitrate, phosphate, and silicate); they receive enough light for photosynthesis and CO_2 for carbon fixation, but they still show only low biomass production. These enigmatic zones are known as HNLC (*h*igh-*n*itrate-*l*ow-*c*hlorophyll areas) to oceanographers. Apparently there is a limiting factor that prevents photosynthetic bacteria and phytoplankton from exploiting these resources. The iron hypothesis was proposed by J. H. Martin in 1990 and led to numerous large-scale field trials. He pointed to iron as the likely limiting factor for plankton growth. If this hypothesis of iron limitation is correct, one could even envision sowing the sea with iron to draw CO_2 from the atmosphere and thus reverse the secular trend of increasing CO_2 concentrations. Iron availability is a problem in many aqueous environments. In anoxic water, Fe^{2+} is soluble up to $0.1–1\,\mu M$, depending on the carbonate concentration, and can thus satisfy the needs of living cells. However, in the presence of oxygen and at neutral pH, Fe^{2+} is oxidized to Fe^{3+}, and the solubility of iron drops dramatically to a concentration as low as $10^{-18}\,M$.

Iron Limitation

Several observations pointed to iron as a limiting factor. For example flavodoxin was identified as a biochemical marker of iron limitation in the sea. Flavodoxin is a redox protein that contains riboflavin $5'$-phosphate as cofactor and replaces the iron–sulfur protein ferredoxin in many microorganisms, including diatoms, a major photosynthetic protist of the ocean. The flavodoxin level in the chloroplast of diatoms was inversely proportional to the ambient iron concentration and its synthesis could be decreased by iron addition (La Roche et al. 1996). Since addition of iron to seawater samples in bottles also stimulated the growth of phytoplankton, iron limitation in the ocean was not an unreasonable working hypothesis. Observational studies confirmed the link between natural iron concentrations, plankton blooms, and CO_2 draw down (de Baar et al. 1995).

The Iron Fertilization Experiments IronExI and IronExII

Thus encouraged, oceanographers made in the 1990s bold steps in a series of iron fertilization experiments. In the IronExI mission, a 64-km^2 patch of the equatorial Pacific Ocean was literally seeded with 450 kg of iron in an acidic solution. This intervention resulted in a doubling of the biomass both in cyanobacteria and algae, and chlorophyll concentrations tripled, but the effect was short lived. Four days after the addition, the seeded patch was subducted into a deeper zone and no draw down of CO_2 or nitrate could be measured (Martin et al. 1994). In IronExII, the same amount of iron fertilizer but subdivided into three doses was given over a week. When the iron concentration increased from the preintervention levels of <0.2 nM to 2 nM, the chlorophyll concentrations increased in parallel. Now a decrease in nitrate and a decrease in CO_2 fugacity were measured. Actually the equatorial Pacific was still a source of CO_2 to the atmosphere, but its contribution decreased under the iron fertilization experiment. Cyanobacteria, under normal conditions one of the major contributors to CO_2 fixation in the ocean, increased in biomass by a factor of two, but the great winner was diatoms showing an 85-fold increase. The reason was simple: The microzooplankton predators of the smaller cyanobacteria, namely ciliates, increased in parallel with their photosynthetic prey and could keep the cyanobacteria increase in check. In contrast, the larger diatoms escaped this grazing pressure since they are eaten only by larger mesozooplankton, calanoids, and copepods (both small crustacea), which have longer generation times than diatoms (Figure 4.14). In fact copepods are already growing at the maximal achievable rate limited by mortality due to their own predation (Coale et al. 1996). Under the enrichment experiment, a changed photochemical quantum efficiency was measured. Since the species composition of the photosynthetic organisms was not detectably changed, physiological changes in the existing species must have occurred (Behrenfeld et al. 1996). The map of the transiently iron enriched zone matched also the map of a 60% decrease in the CO_2 flux from the ocean to the atmosphere (Cooper et al. 1996). On the other hand, the iron effects were too transitory to affect a draw down of CO_2 to the sea bottom. Three days after the last iron addition, the

seeded zone had returned to preintervention iron levels. The greatest effect of CO_2 uptake was calculated for the polar southern ocean, the site for the third large-scale iron fertilization experiment (Boyd et al. 2000). The seeding of the sea with 1.7 tons of iron resulted in an extra 800 tons carbon fixation mainly by a diatom-dominated algal bloom. Levels of chlorophyll increased sixfold and flavodoxin levels, the abovementioned marker of algal iron stress, decreased. Increased chlorophyll levels were still measured 40 days after the intervention, but despite that sustained effect the second tenet of Martin's iron hypothesis, which is about the link between iron supply and subsequent downward particulate carbon export, could not be confirmed. This field experiment demonstrated that a modest sequestration of atmospheric CO_2 by artificial addition of iron to the Southern Ocean is in principle possible. However, it falls short of the hopes of those who applied for patents on ocean fertilization as a mean of carbon emission trading after the Kyoto treaty.

The Southern Ocean Iron Fertilization Experiment

However, oceanographers were not deterred and continued with SOFeX, the southern ocean iron enrichment experiment (Coale et al. 2004). This experiment was conducted in two parts: a northern and a southern intervention. Both zones were high in nitrate but differed in silic acid concentration being low in the northern intervention zone. Silicon is a major nutrient for diatoms because it is required for the biosynthesis of their exoskeleton, which is called a frustule. The results were very revealing: As expected, the iron concentration increased in the intervention zone and were paralleled by an increase in chlorophyll and particulate organic carbon concentration, and a decrease in nitrate concentration and partial pressure of CO_2. The enhanced growth in the northern silicon-limited region was mainly attributed to nonsilicious flagellated phytoplankton groups (e.g., dinoflagellates), whereas the silicon-sufficient southern region became dominated by a 20-fold increase in diatoms. Surprisingly diatoms in both regions remained thin-walled.

FIGURE 4.14. Pelagic life. The pelagic zone consists of the entire ocean water column. It contains three forms of life: the phytoplankton, which provides via its photosynthetic capacity the food basis for all marine animals; the zooplankton, which comprises marine animals that rely on water motion for transport. They feed on phytoplankton and smaller zooplankton. Numerically, the zooplankton is dominated by crustacean copepods and euphasiids. The third form of life, the nekton, constitutes the free swimmers and is numerically dominated by fishes, mollusks, and decapods. The *top quarter* shows a decapod (*Panulirus*) larva, three fish eggs, a bony fish (*Coryphena*) and the larva of another decapod (the lobster *Homarus americanus*). Below them, the figure is dominated by the spindle-shaped bodies of five different copepods (*Setella, Calocalanus, Copilia, Oithonia*), all ending in fancifully shaped appendices. The *bottom* row shows at the *left* two further copepods from the Gulf of Naples. The largest animal in the lower part is a crustacean.

This observation might be explained by observations with iron-limited diatoms. When iron was added to bottled diatoms taken from the ocean, the diatoms answered with an increased nitrate and phosphate consumption compared to iron-limited controls, while iron addition had no effect on the silicon uptake (Takeda 1998).

Problems with the Downward Particle Flux

Highest primary production was seen in the surface layer. As the bloom developed, subsurface production decreased below preintervention levels. This process was attributed to self-shading by the phytoplankton since the 1% residual light level was attained at lower depth. But how much of the decreased CO_2 pressure is translated in actual draw down of carbon? The carbon sequestration into deeper layers was measured by observing the depth profile of particulate organic carbon with radioactive tracers adsorbing to the sinking material (Buesseler et al. 2004). As the experiment progressed, the particle flux extended into deeper waters, the most marked effect was a sixfold increase in particle flux at 100-m depth. This is a relatively modest effect since it means that the ratio of carbon sequestered to iron added was 1:1,000. Another group used floats profiling with robotic observation, and estimated molar ratios of iron added to carbon exported to deeper ocean layers to be between 10^4 and 10^5 (Bishop et al. 2004). The scientists concluded that it is difficult to see how ocean fertilization with such low export efficiency could be scaled up to solve the global carbon imbalance problems.

Only the most recent iron fertilization experiment conducted offshore of Alaska documented the rise and fall of a diatom bloom (Boyd et al. 2004), while the previous experiments were mostly terminated by subduction events of the enriched areas. Diatoms increased in mass in parallel with the particulate organic carbon fraction. Nutrient limitation by the dilution of the added iron and the exhaustion of silicic acid terminated the bloom, top-down control by predators played no role. Sediment traps were installed at various depths and intercepted at 50 m, an increased diatom and aggregate count compared to mesozooplankton fecal pellets, but this increase was less evident at lower depths. Below 50-m depth only a transient and relatively small increased particle rain was observed. The authors concluded that secondary silicic acid limitation and the inefficiency of the vertical carbon transfer compromise the iron fertilization strategy. What is the problem with the vertical carbon flow? The answer was provided by recent analysis of the cycling of organic carbon in the ocean, the subject of our next section.

Photosynthesis Versus Respiration in the Ocean: The Closing of the Carbon Cycle

Digestion of "Marine Snow"

In principle, large, rapidly sinking organic aggregates are an important component of the carbon flux from the ocean's surface to its depth. However, in the early 1990s, it was discovered that large (>0.5 mm) aggregates ("marine snow") are heavily colonized by bacteria already in the surface layer of the ocean. Initially it was thought that the carbon demand of these bacteria is so small that it would take months to consume the aggregates' carbon. Video cameras documented a sinking rate of 50 m/day or more (Smith et al. 1992), which should assure an efficient transfer of particulate carbon to the deeper ocean layers. However, the aggregates are associated with high hydrolytic enzyme activities that solubilize particulate amino acids with a turnover time of sometimes less than a day. As the associated heterotrophic bacteria are unable to take up these nutrients, this "floppy feeding" or "uncoupled hydrolysis" by extracellular or cell wall–bound bacterial enzymes transfers carbon from the particulate to the dissolved organic carbon (DOC) pool. This observation is in agreement with the iron enrichment experiments that showed a selective loss of carbon over silicic acid from the sinking diatoms. In some way, the idea of sequestration of photo-sythetic organic carbon with sinking photosynthetic organism is naïve since it does not account for nutritional equilibria in nature. Where primary producers thrive, there will be organisms that make a living from the newly synthesized biomass. At first, predators may come that eat the photoautotrophs. However, as these photoautotrophs decay, they also become food to decomposer bacteria as those colonizing marine snow. Since these bacteria apparently release substantial amounts of undigested organic carbon and nitrogen, they provide a food basis for numerous other heterotrophic bacteria that remineralize the organic carbon to CO_2. The extent of bacterial respiration in the ocean was until quite recently underestimated (del Giorgio and Duarte 2002). In fact the released DOC is such a rich nutrient source in some oceanic regions that even phototrophs exploit this food. For example anoxygenic phototrophic bacteria were initially supposed to be competitive only in anoxic illuminated regions of the sea. Today they are known to be abundant in the upper open ocean, where they represent up to 10% of the total microbial community (Kolber et al. 2000, 2001). These α-Proteobacteria of the *Erythrobacter* cluster contain 10-fold less bacteriochlorophyll than purple bacteria and satisfy only about 20% of their cellular energy requirements by photosynthetic electron transport. They are facultative phototrophs, which switch to a mostly heterotrophic respiratory metabolism in organic-rich environments where they rely on exudants produced by oxygenic photoautotrophs. These bacteria fix only relatively small amounts of CO_2, but since they support their heterotrophic metabolism by photosynthetic ATP synthesis, they release less CO_2 per unit biomass synthesized from the DOC than pure heterotrophs.

Siderophores

As could be suspected from the low draw down of photosynthetically fixed carbon, heterotrophic bacteria play an important role in the carbon cycle already in the upper ocean layer. They constitute not only about 50% of the total partic-ulate organic carbon in the ocean but represent the largest fraction of biogenic iron in this system. In iron-depleted water the Fe:C ratios of the heterotrophic bacteria were twofold higher than those of phytoplankton (Tortell et al. 1996).

In iron fertilization experiments, these heterotrophs will thus directly compete with phytoplankton for iron by the release of high-affinity iron uptake ligands ("siderophores"). In fact dissolved Fe(III) in the upper oceans occurs almost entirely in the form of complexes. To cope with the extremely low marine iron concentrations, *Alteromonas*, an open ocean bacterium, has developed two siderophores that coordinate iron via catecholate and β-hydroxy-aspartate moieties with the exceptionally high association constant K_a of 10^{-53}, which exceeds the most tightly binding siderophores of terrestrial bacteria (enterobactin; Reid et al. 1993). The heterotrophic marine bacterium *Halomonas* elaborates a siderophore called aquachelin. In contrast to alterobactin 1, which is a complex heterocyclic ring system, aquachelin is a linear molecule consisting of a long fatty acid tail and a longer peptidyl chain. The side chains contain one β-hydroxy-aspartate and two hydroxamate groups, which each bind one Fe(III) ion. This siderophore undergoes a photolysis reaction in sunlight that results in an oxidative cleavage of the siderophore and a concomitant ligand-to-metal charge transfer reaction resulting in the reduction of Fe(III) to Fe(II) (Barbeau et al. 2001). This reduction increases the bioavailability of iron for the bacterium, which binds this siderophore and explains also the diel (i.e., in daily rhythm) Fe(II) cycling in oceanic surface water.

Are Plankton Respiration and Photosynthesis Balanced?

The ecological importance of the bacterial heterotrophs became clear when the German research ship *Polarstern* cut a north–south transect through the Atlantic by taking 170 water samples from 11-m depths. The oceanographers determined chlorophyll a concentrations as a measure of primary production. They observed high chlorophyll concentrations in the northern and southern cold seas and low chlorophyll values in the oligotrophic tropical regions (Hoppe et al. 2002). Then they measured the bacterial growth by radioactive leucine incorporation, which showed high values of bacterial growth in some regions of low primary productivity. The comparison of both data sets showed an alternating pattern of prominence of autotrophic–heterotrophic–autotrophic regimes, which correlated with northern cold-tropical warm temperatures and southern cold surface temperatures. High temperature is correlated with increased respiration rate and also frequently with oligotrophic ocean conditions. However, the plot of the ratio of heterotrophic bacterial carbon demand versus primary production led to a paradox: Over a broad belt around the equator, more carbon was consumed as biomass and in respiration than was locally produced by phototrophs. The system is clearly not in equilibrium: Not only are these oligotrophic ("unproductive") regions of the ocean net CO_2 producers (del Giorgio et al. 1997), the reduced carbon demand must be covered by other processes than marine photosynthesis. What sources fill this reduced carbon gap? At one side, there are equatorial upwellings, which provide aged DOC. When looking at the map, there was a conspicuous high bacterial growth opposite to the inflow of the Amazonas into the tropical Atlantic. Big rivers import large quantities of terrestrial organic matter into the ocean derived from the photosynthetic activity of land plants.

A priori, one would expect recent terrestrial dissolved organic material in this input—this is paradoxically not the case. The older dissolved organic material is enriched in the riverine carbon influx into the ocean. The dating of the DOC fraction is facilitated by the atomic bomb tests over the last decades, which result in ^{14}C-enrichment in recent material. The paradox is solved by the observation that this recent ^{14}C-enriched material is biologically labile and thus prone to bacterial degradation. The smaller pool of older ^{14}C-depleted biologically refractory material is thus preferentially available for export into the ocean (Raymond and Bauer 2001). However, rivers cannot fill the organic carbon deficit of the equatorial Pacific Ocean. On the basis of quantitative comparisons of gross primary production and net community production, some oceanographers claim that the open oceans as a whole are not substantially out of organic carbon balance with respect to plankton respiration and photosynthesis (Williams 1998).

Food Pulses to the Depth

Comparison of the DOC concentrations and the apparent oxygen utilization at different depths of the ocean demonstrated that the DOC flux supports only about 10% of the respiration in the dark ocean (Aristegui et al. 2002). The researchers concluded that the particulate organic carbon flux was severely underestimated as a source of carbon for respiration in the dark ocean since it must provide 90% of the energy supply. In fact more than 50% of the Earth's surface is sea floor below 3,000 m of water. This region represents a major reservoir of the global carbon cycle and the final repository for anthropogenic CO_2 as targeted by iron fertilization experiments. Marine biologists knew that these areas are characterized by severe food shortage; phytodetritus is the major food source for the abyssal benthic community. However, until quite recently not much was known as to how this community deals with the food pulses that arrived at this depth. Bacteria usually dominate benthic biomass in deep-sea sediments, but when marine microbiologists offered experimental food pulses, bacterial respiration and growth were not quickly induced. The radiolabel from phytodetritus entered first the numerically and biomass-wise much sparser metazoa: The macrofauna in the top 5–10-cm floor layer became labeled within days, the smaller nematoda only within weeks. The food material passed first through the gut of larger animals, which rapidly subducted the labile food down to 5–15-cm depth where it was then quickly degraded by bacteria (Witte et al. 2003). Estimates of the contribution of metazooplankton to respiration in the ocean are highly variable and range from 1 to 50% depending on the ocean region. Vertebrates, which occur three trophic levels above the primary producers, account for less than 1% of the respiration in the oceans.

Marine $CO_2/HCO_3^-/CO_3^{2-}$ Equilibria

There are still other marine processes that impact on the CO_2 release by oceans, which have nothing to do with respiration and which make balancing calculations a complex business. Take the example of coccolithophorids, photosynthetic protists that build calcareous skeletons. After their death they sink to the depths generating a continuous rain of calcium carbonate to the deep ocean. This would be a direct way to bury surplus atmospheric CO_2 in the ocean sediment, but things are generally not that simple as geoengineers would wish in their fight against the increasing atmospheric CO_2 levels. CO_2 is involved in notable equilibria (Gattuso and Buddemeier 2000). For example in terrestrial systems, extra atmospheric CO_2 may have a direct fertilizing effect resulting in increased photosynthesis. Not so in marine environments, where most algae use the bicarbonate ion (HCO_3^-) rather than CO_2 as photosynthetic substrate. Yet CO_2 is in equilibrium with bicarbonate, which could then still push photosynthesis. However, the dissolution of CO_2 acidifies the water according to the equation $H_2O + CO_2 \rightarrow HCO_3^- + H^+$. It was calculated that the increased atmospheric CO_2 concentrations will have caused, by the end of this century, a drop in the ocean pH by 0.35 units. However, the ocean surface is not a pure water solution, it contains many ions, including carbonate ions. CO_2 combines also with carbonate according to the equation $CO_2 + CO_3^{2-} + H_2O \leftrightarrow 2HCO_3^-$. Increased CO_2 concentration thus leads to decreased carbonate concentration, which should result in decreased calcification. This is already observed in the field with reef-building corals and coralline algae and in experiments with coccolithophorids exposed to increased atmospheric CO_2 (Riebesell et al. 2000). Not only is the calcite production decreased but also the protists showed an increased proportion of malformed coccoliths and incomplete coccospheres, which will affect their sinking rate. Owing to the second equation read in reverse, calcification is a source of CO_2 to the surrounding seawater. Decreased calcification would thus diminish the release of CO_2 by the ocean and provide a negative feedback. While the prediction of rising CO_2 levels on the ocean chemistry is reasonably straightforward, things become rather complex due to the interaction of several, altogether not that well understood biological systems.

CO$_2$ Levels and Climate Record

In the last two sections, we have discussed the possibility to restrain the future rise of atmospheric CO_2 concentration and thus a further warming of the globe by iron fertilization of the ocean. Excessive iron fertilization by strong winds is also discussed for an alternative scenario: ice ages. The experimental evidence came from ancient air samples encaged in air bubbles trapped in the ice core. Ambitious drilling projects were done both in the northern (Greenland GRIP project) and the southern hemisphere (Antarctic Vostoc project). In 1998 the latter reached the record depth of 3,600 m and provides us with a climate and atmospheric history for the past 420,000 years. This period covers four transitions from glacial periods to interglacial warm periods. We live currently in a more than 11,000 years long stable warm period, the Holocene, which is by far longer than the previous, about 4,000 years long warm periods. Some researchers even suspect that the winter sport pleasures documented in Dutch paintings and reported in English literature commonly called the Little Ice Ages were a hint to a cooling trend, which was only prevented by the CO_2 warming effect. Whatever hypothesis is sported, the ice cores document a major, about 100,000 years, periodicity and a minor, 41,000 years, interval. This regularity was interpreted as a pointer to changes in the orbital parameters of the Earth (eccentricity, obliquity, and precession of the Earth's axis; Stauffer 1999). The ice core analysis allowed establishing a temperature record describing amplitudes of temperature change of about 8 °C. The temperature record is mirrored by the changes of two greenhouse gases, CO_2 and CH_4. The correlation coefficient for both gas concentrations and the average temperature is remarkable. Detailed analysis of the Vostoc ice core revealed that the CO_2 decreases lagged the temperature decreases by several thousand years (Petit et al. 1999), while the Greenland ice core showed that the CH_4 changes were in phase (± 200 years) with the Greenland climate (Chappellaz et al. 1993). However, the radiative effect of methane is too small to account for the observed temperature changes and also the combined CO_2 and CH_4 changes explain perhaps 2–3 °C temperature differences and thus only half of the overall effect.

Intervention in an Unknown System?

Some oceanographers expressed concerns (Chisholm 2000) that we should not intervene, e.g., with iron fertilization at a massive scale, in a system that we understand so little, namely the population structure of the microbial cells in the ocean surface layer.

A vivid illustration of our ignorance is provided by a shotgun sequencing project in the Sargasso Sea (Venter et al. 2004; Figure 4.15). Two-hundred-liter surface-water samples were filtered, and the genomic DNA from an intermediate size fraction (greater than bacteriophages, smaller than algae) was sequenced. It yielded about 1 Gbp of DNA sequences, the equivalent of about all sequenced bacterial genomes in the public database. Not surprisingly the abundant cyanobacterium *Prochlorococcus* dominated the sequences. *Shewanella* and *Burkholderia* DNA represented also a major part in the sequenced DNA. However, with low prevalence, DNA from an estimated 1,000 further bacterial species were detected in this sequencing tour de force pointing to an enormous diversity of microbial genomes, which we had completely ignored before. The stage is thus clearly not set for a targeted climate engineering of the oceans— much remains to be understood before such approaches become feasible.

Should we stop with this form of targeted geoengineering? We already do interventions of this type on a massive scale. Examples are the drainage of fertilizers used in agriculture, containing N and P via rivers into the oceans or the transformation of organic C sources into CO_2 via burning of fossil fuels. Nevertheless some scientists are skeptical since iron fertilization might also influence other biological systems in an unpredictable way. Our knowledge of marine microbiology is still dramatically fragmentary as revealed by the recent discoveries about the most abundant cells in the ocean, presented in the next section.

FIGURE 4.15. The Sargasso Sea is a free-floating mass of seaweed, the most frequent being *Sargassum natans*. The brown alga merits its swimming name: air-filled floaters as seen in the picture keep the algae near the surface to allow photosynthesis.

The Most Abundant Cells on Earth are on a Small Diet

The SAR11 Clade

I will illustrate this point with bacteria that were until quite recently only known by their code name SAR11 (Rappé et al. 2002). In some way, these bacteria are not exotic: They form a clade in the α-Proteobacteria, the nearest cousins are *Rickettsia*, they belonged until recently to the large group of uncultivatable bacteria. This property is the rule in bacteriology rather than the exception. The fraction of prokaryotic species known to exist in the environment, which have not yet been cultured, is estimated to be as high as 95–99.9%. Some of the limitations to cultivation are certainly physical in nature. Obviously, bacteria living at 3,000-m depth in the ocean are adapted to the high-pressure conditions (obligatory barophilics) and they will not survive decompression. Other limitations are certainly nutritional as can be seen with the SAR11 clade. These bacteria could be cultivated by dilution into sterilized natural waters or diluted media. Growth was only observed at nutrient concentrations that were three orders of magnitude less than in common laboratory media. Addition of even small amounts of dilute proteose peptone (0.001%) inhibited growth (Rappé et al. 2002). What makes this clade special is its abundance? These 1-μm small curved rods account for a quarter of all ribosomal RNA genes that have been identified in seawater. The results were confirmed by in situ hybridization techniques using fluorescence-labeled probes to its 16S rRNA in the ribosomes. SAR11 clade accounted for 30–40% of cell counts within the euphotic zone (a maximum was reached at 40-m depth) and 16–19% between 250- and 3,000-m depth. By extrapolation it was estimated that globally there are about 10^{28} SAR11 cells in the oceans (Morris et al. 2002). If this figure is correct, SAR11 is among the most successful organisms on Earth. However, we have no data on their physiology except that they are heterotrophs, and genome sequencing efforts are currently our only guides to a deduced metabolism. It is of course risky to intervene in a system with iron fertilization when we know its microbial constituents so poorly even if they represent major players in the biosphere.

Reading the *Pelagibacter ubique* Blueprint

If you lack physiological information on a bacterium, one can nowadays obtain a first insight by looking into its genome sequence. This was done with the cultivated strain from the SAR11 clan, which, in the meanwhile, got the fitting name *Pelagibacter ubique* (Giovannoni, Tripp et al. 2005). The first part refers to its occurrence in the pelagic zone, i.e., the ecological realm that includes the entire ocean water column; the second part alludes to its ubiquitous distribution. It has an astonishingly small genome: With its 1.3 Mb, it represents the smallest genome for a free-living microorganism. In contrast to parasitic bacteria, *P. ubique* has all it needs for independent life: It encodes genes for the biosynthesis of all 20 amino acids; catabolism and energy supply is assured by the Entner–Doudoroff, TCA cycle (plus glyoxylate shunt), and a respiratory chain. Key glycolytic enzymes were, however, lacking. According to the transporter genes detected, it makes its living by assimilating organic compounds from the oceans's DOC reservoir. The genomes show that this extremely successful bacterium has only few extra genes like a light-driven proteorhodopsin proton pump. Otherwise the organism is putting its strength in economy. The GC content is low, possibly to spare N (C, cytosine, contains one N atom more than T, thymine); the genome size is small, which spares N and P; the cell is small (the small genome occupies already 30% of the cell volume), which assures a large surface to volume ratio. This is a common strategy in starving bacterial cells and in cells from oligotrophic environments. Regulation is minimal, only four two-component regulatory systems deal with responses to P and N limitation and osmotic stress. The bacterium maintains its basic rhythm with a low growth rate (one division every 2 days) and does not respond to nutrient addition. *Pelagibacter ubique* has an extremely streamlined genome, with practically no noncoding DNA; on average a mere 3 bp separate genes.

Roseobacter

Recently further phylotypes of α-Proteobacteria were detected as abundant members of the bacterioplankton. Since they are phylogenetically related to *Roseobacter*, they were dubbed the *Roseobacter* cluster. Interestingly the SAR11 clade and the *Roseobacter* cluster showed distinct geographical distribution pattern. The former was found as a dominant species in tropical and subtropical oceans, while the latter was only detected in temperate and polar oceans. As in the SAR11 clade, different phylotypes were also detected in the *Roseobacter* cluster that followed a north–south gradient (Selje et al. 2004).

Silicibacter Genome

Silicibacter follows a different strategy. The sequencing of a member of the *Roseobacter* clade now allows a comparison of the feeding strategies of abundant oceanic cells as revealed by genome analysis (Moran et al. 2004). In contrast to *P. ubique*, *S. pomeroyi* is not a minimalist. It has a 4.1-Mb genome and a 0.5-Mb-large megaplasmid. It combines two nutritional strategies. One is lithoheterotrophy: It gains energy, but not carbon by oxidizing CO to CO_2; likewise it oxidizes reduced inorganic sulfur compounds. The key genes for both processes, *coxL* and *soxB*, were represented in the Sargasso gene library from the Venter lab at an abundance of one per ten bacterial cells. CO is ubiquitous in marine surface waters since it is a photooxidation product of dissolved organic material (DOM) . In addition *Silicibacter* has numerous transporters for organic compounds including peptides, amino acids and algal osmolytes like glycine betaine and DMSP (more on this fascinating compound, which can represent 20% of the cytoplasm of algal cells in the marine food chain section). *Silicibacter* is motile and shows quorum-sensing, i.e., the capacity to change its metabolism according to cell numbers in the environment. This allows this bacterium to switch metabolically from a particle-associated state, when sticking to marine snow or algal debris (characterized by high population density and high substrate availability), to a free-living state, defined by low cell numbers and low substrate concentrations. The authors dubbed it an "opportunitroph" able to switch from lithoheterotrophy to the rapid exploitation of pulses of nutrients. *Plegibactor ubique* is in comparison a dull, but efficient nutritional "long-distance runner."

Archaea

Another gradient was detected along the depth profile. While bacteria clearly dominated the upper 500 m of the ocean water column, a subgroup of the Archaea, the Crenarchaeota, became as common as bacteria in the subtropical North Pacific at a depth of 2,000 m and remained prominent down to 5,000 m and the sea floor (Karner et al. 2001). These cells were metabolically active, and they were also estimated to 10^{28} cells globally. The data demonstrated that archaea are far from confined to very specialized niche habitats occupied by extreme thermophiles, halophiles, and methanogens. We still need to learn much about the metabolism of these abundant prokaryotic cells and their interaction before we can realistically design ecological engineering at a global level.

Depth Profile

Diving Deep in Hawaii

As already demonstrated by the Sargasso Sea sequencing, a new trend in microbiology is to sequence an entire ecosystem. In an especially informative report, genomics specialists teamed up with oceanographers and investigated prokaryotic genome sequences along a vertical transect at the ALOHA ocean station near Hawaii (DeLong et al. 2006). Samples were retrieved in a depth profile covering the upper euphotic zone between 10 and 70 m, the bottom of the chlorophyll maximum (130 m), below the euphotic zone (200 m), the mesopelagic (500 m), the oxygen minimum at 770 m, and a layer 700 m above the seafloor, which corresponded approximately to the average depth of the oceans on Earth of 3,800 m. About 10,000 sequences were obtained for each zone. Sequencing of the small subunit RNA allowed an inventory of the bacterial taxa at the different levels. As expected the surface layers were dominated by organisms like *Prochlorococcus*. The lower layers were dominated by Delta-Proteobacteria (*Desulfo…* species) and the SAR11 bacteria. A surprise was the heavy contribution of viral sequences in the photic zone, which represented with 21% the largest fraction of the sequenced DNA. The result was not expected because filters were used that excluded viruses from the sampling. The fraction was dominated by T7- and T4-like Podo- and Myoviridae, respectively, which were apparently replicating inside of cyanobacteria. The authors estimated an infection rate of cyanobacteria of 10%, demonstrating that the photic zone is a battlefield between phototrophic bacteria and their phages. Even more interesting, this community genomics approach allowed to identify specific genes and thus to associate specific metabolic traits with different depths. In the photic zone, sequences associated with photosynthesis, porphyrin, chloro-phyll, and carotenoid synthesis, maltose transport, lactose degradation, type III secretion, vitamin B6 metabolism, and heavy metal export were prominent. In the deeper layers, sequences were more associated with protein folding, methionine, glyoxylate, dicarboxylate, thiamine, and methane metabolism than in the surface layer. The data now allow painting a portrait of the different depth layers. The photic zone shows the importance of cyanophage predation and the need for restriction–modification systems. In the light, energy is gained by photosynthesis and by proteorhodopsins. Motility and chemotaxis are important here for heterotrophic bacteria to swim toward nutrients associated with particles and algae. Photolyases and carotenoids protect against photodamage. The sequences paint a different picture for the deeper layers. Here a surface-attached lifestyle dominates with pilus synthesis, protein export, polysaccharide, and antibiotics synthesis as important traits. Metabolically the glyoxylate cycle and urea metabolism play a great role. HGT is important in both layers. Photosynthesis genes like *psbA* and *psbD* and transaldolases are horizontally transmitted by cyanophages. Transposases and phage integrases become prominent in the lower layer.

Pressure-Adapted Bacteria

This exciting exercise in community genomics is complemented by the sequencing of bacteria adapted to high pressure, characteristic for life at depth. If these bacteria have sequenced relatives living near the surface of the biosphere, comparative genomics can become especially revealing. This is the case for *Photobacterium profundum*, a piezophilic ("pressure-loving") Vibrionaceae. As its relatives in the *Vibrio* genus, it shows a bipartite genome. One part is a 4-Mb stable genome with established genes, which is active in transcription. The other part is smaller (2 Mb) and looks like a genetic melting pot (Vezzi et al. 2005). The genome sequence revealed that at high pressure energy conservation becomes a problem: The organism needs two F_1F_0 ATP synthases and three cytochrome oxidase genes. Microarray expression analysis showed an interesting pressure-regulated pathway: the Stickland reaction. In this pathway amino acids (which have the redox state of sugars) are used for fermentation. Characteristically one amino acid is oxidized, which looses thereby the amino and carboxy group, and two hydrogens are abstracted that are used to reduce the second amino acid. The oxidative branch of the Stickland reaction delivers energy via substrate-level phosphorylation. Pathways that degrade complex carbohydrates like chitin, pullulan, and cellulose are activated under high pressure, which fits with the observation that they represent an important source of carbon, sinking into the abyssal environment from the photic zone. Decompression is survived by the organism, but leads to an important activation of stress genes, demonstrating its adaptation to high depth. Another important nutrient input to the deep-sea ecosystem—until recently overlooked—is DNA. DNA concentration is with $0.3 \, g/m^2$ seafloor extremely high and 60% of the DNA pool is extracellular and thus enzymatically digestible. It provides 47% of the daily P demand in this ecosystem. DNA represents thus a key trophic resource and contributes to the biogeochemical P cycle (Dell'Anno and Danovaro 2005).

Sediments

Sediment Formation

At any one moment, some 10 billion tons of particulate organic matter is sinking down into the world's ocean (Parkes et al. 1994). These particles settle finally at the bottom of the sea and build huge sediments. We should therefore not be surprised that we find in the sediments a numerous bacterial community. However, when calculating the annual deposition of material in the deep sea, one gets a meager upper limit of 0.02 mm. Even ocean areas of high productivity like marine upwellings produce annually not more than 0.3 mm, which is small even in comparison with oligotrophic lakes that settle up to 2-mm sediment per year. One of the reasons for this low sedimentation rate in the deep sea is that the sinking material has to feed many mouths during its crossing of several thousand meters of water column. In fact all easily digested material (carbohydrate polymers, proteins, lipids, nucleic acids) is eaten up during this voyage such that the organic carbon content is only 0.5% in the deep-sea sediment as compared with 6% in sediment from an oligotrophic lake. What arrives at the deep-sea floor is mainly decay-resistant material, some of which is so resistant that it can serve as paleobiological tracers (we already discussed the membrane lipids hopanoids as geological marker for cyanobacteria). Other resistant material includes hydrocarbons, branched fatty acids, and carotenoids. The organic material in sediments is transformed over long periods of time and leads in a process called diagenesis to organic derivatives that become more and more recalcitrant to digestion. Organically rich deep sediment then develops into a paste-like material of oily character called kerogen, which resembles components of mineral oil.

Terminal Electron Acceptors

Beyond the problem of the food source material, microbes meet another problem in sediments, namely that of the terminal electron acceptor. Redox reactions that energetize microbes show a clear hierarchy with respect to the acceptor redox pair, regardless of what ecosystem is investigated and what organic food material is considered. The preference list of microbes is easily understood since it follows mainly the decrease of the redox potential. The highest biological redox potential is that of the O_2/H_2O redox pair with a standard potential $E0'$ of $+810$ mV. However, oxygen penetrates into the sediment of a eutrophic lake less than 1-mm deep. Oxygen penetration is several centimeters in the sediment of an oligotrophic lake or a productive marine sediment and up to 1 m into an offshore deep-sea basin sediment. If oxygen is consumed, microbes use the next efficient electron acceptor, nitrate. Its redox potential varies from $+751$ to $+363$ V depending whether it is reduced to N_2 or NH_4^+. The separation between successive biological redox partners is not always that sharp because the redox potential is a thermodynamic value and describes the maximal energy gain which can be obtained by using the indicated electron acceptor. Since biological redox reactions are catalyzed within organisms, the efficiency of the process also depends on kinetic factors determined by the substrate affinity and the specificity of the substrate uptake systems. Thus some bacterial strains belonging to conventional genera (e.g., *Pseudomonas*) show aerobic nitrate reduction, i.e., they use nitrate as electron acceptor even if oxygen is present. Reduction of oxygen and nitrate can be coupled to the complete oxidation of organic substrates to CO_2 and a single bacterium can complete the mineralization of organic carbon. After exhaustion of these electron acceptors, i.e, at a given depth of the sediment, the mineralization of organic carbon becomes the cooperative effort of different microbial groups. Primary fermenting bacteria convert polymeric or monomeric substrates into the classical fermentation products, which are then oxidized to CO_2 using Fe(III) or Mn(IV) as electron acceptors. *Shewanella* is a well-investigated organism that can oxidize a broad spectrum of organic compounds using iron as electron acceptor. Mn-reducers are geographically more limited (e.g., Baltic Sea). The next preferred electron acceptor is sulfate, but the redox pair sulfate/hydrogen sulfide has already a pretty negative redox potential of $-218\,mV$. Organic carbon mineralization is thus the job of two communities: primary fermenting bacteria and sulfate reducers. However, sulfate is dissolved in high amounts in the seawater and reaches thus rather deep into the sediment. The reduced hydrogen sulfide eliminates the last traces of oxygen in the sediment and lowers the redox potential drastically. The sediment is now ready for a carbon mineralization by methanogens, which use CO_2 as electron acceptor to produce methane relying on a reaction with a marked negative redox potential $(-244\,mV)$.

Viable Cells in the Sediment

Recent profiling of subsurface metabolic life in ocean drilling cores has largely confirmed these general concepts (D'Hondt et al. 2004), but demonstrated major differences between a sulfate-rich open ocean province and a sulfate-depleted ocean margin province found along both coasts of the American continent north of the equator (D'Hondt et al. 2002). The low sulfate regions were also regions of high subsurface methane concentrations and an earlier suspicion that anaerobic methane oxidation may be the dominant sulfate sink in the sediment were in the meanwhile confirmed (see below).

Remarkably, neither the difficult carbon nature of the food nor the problems with the electron acceptors deterred prokaryotes to conquer the sediment for the biosphere. This realm of the life was discovered relatively recently (Parkes et al. 1994). Total bacterial cell counts were as high as $10^9/cm^3$ at the sediment surface, it dropped to $10^7/gram$ several meters below sea floor and maintained this level to the highest depths measured at 500 m below the floor. British marine geologists asked whether they dealt with viable bacteria. Two lines of argument gave them confidence: They found a constant level of dividing cells (about 5% of the total) across the depth profile, and they could cultivate between 10^5 and 10^2 fermentative bacteria per cubic centimeter for the upper and lower sediment layers, respectively. The cultivated bacteria were not exotic. Sequencing of 16S rDNA and metabolic tests identified many sulfate-reducing bacteria related to *Desulfovibrio*, but the sediment isolates showed as expected metabolic activity at much higher pressure and temperature than comparable terrestrial isolates. Due to these characteristics the authors suspected that these bacteria will colonize the sediments, which can be more than 1-km thick, to even greater depth. Since the oceans cover 70% of the Earth's surface, the sediment biota is likely to represent a substantial part of life on Earth. These early data were confirmed in a number of subsequent studies. For example a recent study used an enumeration technique that counted only cells with active metabolism (i.e., cells containing high numbers of ribosomes). The counts from the German microbiologists were only one log lower than those of the British scientists (Schippers et al. 2005). Interestingly the number of bacteria was almost identical to the number of total prokaryotes, Archaea contributed at least one log less viable cells. This is a remarkable result since methanogenesis is the exclusive domain of Archaea. Drilling cores from the Peruvian ocean margin and open ocean sites were investigated for the population structure with extensive ribosomal RNA sequencing. The most frequent isolates were related to Firmicutes (*Bacillus*), α-Proteobacteria (*Rhizobium*) and γ-Proteobacteria (*Vibrio*; D'Hondt et al. 2004). The estimated mass of the subsurface microorganisms was calculated to a global 10^{17} g. With that addition to the prokaryotic world, the prokaryotes make up to 40% of the total amount of carbon bound in the biosphere. The vast majority of terrestrial carbon is bound in plants where the extracellular carbon exceeds by far the protoplasmic carbon. Previous approximation that half of the protoplasm carbon in the biosphere is found in prokaryotes is therefore likely to be too conservative.

Ocean Drilling

Bacteria living in the dark of deep sediments with hardly any energy supply, where they survived for millions of years (sediments located at 90 and 190 m below the sea floor were dated to 0.8 and 2 My ago), were understandably met with disbelief. How can prokaryotic processes operate on geological timescales under such conditions? Data on their metabolism will influence our ideas on fossil formation, subsurface life on other planets, and theories on the origin of life. The Ocean Drilling Program off the Peru ocean margin and at a Pacific open ocean site provided important data (Parkes et al. 2005). Deep brine incursions into the sediment provided sulfate for anaerobic sulfate reducers. The high content of organic matter in the sediment of up to 8% allowed—against previous hypotheses—active methanogenesis to coexist with sulfate reduction. At 30 and 90-m depths, you find the upper and lower sulfate/methane interfaces where bacterial populations showed marked rise. Near the surface H_2 and CO_2 were the substrates for methanogenesis, while in the depth acetate became the substrate. That the cells are not only numerous (up to 6×10^8 cm^3) but also metabolically active was demonstrated by thymidine incorporation into dividing cells.

Prokaryote Numbers in the Biosphere

Whitman and colleagues (1998) dared to estimate the relative number of prokaryotic cells in the biosphere. The big numbers were contributed by four ecosystems: The front runner was ocean subsurface with 3×10^{30} total cells, followed by terrestrial subsurface with less well constrained but substantial cell numbers (0.2×10^{30} to 2×10^{30}), then came the global soil with 2×10^{29}, and then the ocean with 1×10^{29} cells. All other sources of prokaryotes mustered substantially lower cell number, although many of them remained substantial (prokaryotes in animal guts, e.g., mammals and termites; on leaves or associated with roots), while still other were negligible in total numbers despite the large size of the system like the atmosphere. The error range of these estimates is still substantial, Nature does not cease to surprise us with discoveries. The high cell numbers in the oceanic subsurface should be seen in perspective to their metabolic activity. Marine microbiologists have calculated the turnover time in this community by dividing the carbon flux available for these bacteria by the total number of living bacteria. This gave values in the range from 0.2 to 2 years (Schippers et al. 2005): This means that the average bacterium in this environment divides only once a year. The metabolic activity in the subsurface is thus very low in comparison with that in the overlying ocean. However, quite comparable values were determined for bacteria in soils.

Sediment–Soil Comparison

In the planning of this book, I wanted to illustrate the path of knowledge acquisition with recent research results. You might therefore ask why I have not chosen the soil instead of the sediment as an illustration for a microbial community. Due to its primordial importance for agriculture, the soil determines our own food basis directly while the microbiology of the ocean sediment is more of geological interest. There are two main reasons for this choice, the presented ecospheres were selected to demonstrate a progressive time line through evolution, and I have in the previous sections concentrated the discussion on the prokaryotes and their feeding habit. Prokaryotes have a much bigger impact on the food chain in sediments than in soil. In the soil, bacteria make only 7% of the biomass comparable to the share of various soil animals. The big share in the soil biology is taken by fungi, which constitute the remaining 86% of the soil living biomass. However, there is also a scientific difference between sediment and soil. While only 80 different microbial genomes were detected in a water sample from a fishpond, you find already 1,000 different prokaryotic genomes in a coastal marine sediment. This figure increases to at least 10,000 different genomes in a soil sample. Furthermore ocean sediments change perhaps at the centimeter to meter scale, while it can be argued that soil samples change at the millimeter level or smaller. Soil is also much more variable geographically than ocean sediment. When we leave laboratory biology, we quickly realize how limited is our current knowledge basis. It is thus understandable when scientists, who are trained by a reductionist approach, prefer sediment over soil research. This is also a question of practicability: You choose your object of research where your current technology is the most likely to yield insights into the functioning of the system. As soil research is confronted with systems that are somewhere between one to three orders of magnitude more complex, it is understandable that microbial ecologists concentrate on systems where they are more likely to succeed with their analysis. There are thus much more high profile reports on ocean sediments than on soil ecology in the nonspecialized scientific literature. One prominent microbiology textbook referred to this situation with the judgment "that microbial ecology in the future will have to deal with soil as its most important subject, for general scientific reasons as well as for economical and agricultural needs" (Lengeler, Biology of the Prokaryotes, Thieme, *1999).*

Early Steps in Predation

Paradise Lost?

Arms Race

Until now we have seen a plethora of honest strategies in the quest for food. Microorganisms developed innovative enzymatic systems to make a living with the offered nutritional sources in a given environment. However, there are alternatives: Instead of investing into costly metabolic pathways, a cell might choose an easy way—you steal the nutritional resources from another cell; in the most drastic case you eat the other cell. When this invention was made, life became very dangerous on Earth. As part of the honest guild of the prey, you had to run or protect yourself with physical or chemical armors to escape predation. This trend obliged the predator to run even quicker or to develop tools to deal with the different types of defense strategies mounted by the prey. In sum, the arms race was on and once started, evolution did not offer the option to return to an earlier level of competition.

A Horticultural Eschatology

Humans occasionally managed to deal with this dilemma, they invented the truce in their wars or in exceptional cases bilateral arms reduction. A lot of philosophical and religious thought went into utopias that existed before the time or after the time. Man imagined the Paradise as a safe place from all the labor and danger implicated in the quest for food. Costless food is offered in a horticultural eschatology, later Jewish prophecy painted a vision where the sheep lays next to the lion—a powerful picture that the quest for food is suspended. Interestingly our quest for sex was felt and still satisfied in the Paradise by the Judeo-Christian creator, while our quest for knowledge led to our expulsion from the Paradise before we could taste the fruit of the tree of the everlasting life, which thus became a vain desire.

Food and Temptation

The snake-tempter came with a fruit and the promise "Eritis sicut deus scientes bonum et malum" *(You will be like God, knowing good and evil). Actually it is not so clear why God should deny or envy us this knowledge, but beware the Hebrew God is described as a jealous god in the Old Testament. The pursuit of knowledge (the more if it has ethical undertones) is one of the nobler instincts of man (at least in the view of scientists). This view found material expression in a thought-provoking sculpture on the campus of the University of Mexico City where a small symbolic tree carries a single fruit, which looks like a hybrid between an apple and crystal structure of a virus, highlighting both "poison" (the Latin meaning of virus) and knowledge. The tree is small as if the artists wanted each student to grasp for the scientific "fruit". Perhaps the text wants to express the idea that as early as we understand our living condition, we recognize that the Paradise is nothing but a dream. Psychologically it is probably significant that the temptation and interestingly the redemption came with food, the fruit in the paradise and the bread in the Host (Christians are eating their salvation in the Holy Supper, called theophagy by historians of religions). In the Latin Vulgata Bible the promise of the snake must have sounded ironical to the reader. In Latin "malum" means "apple" and at the same time "sin, evil," eating "malum" created "malum." This is perhaps the reason why Eve offers in many classical paintings an apple, despite the fact that the fruit is not specified in the Bible. These homonyms derive from distinct etymologically sources:* Malum-*apple derives from the Dorian Greek* μαλον, *while* malum-*evil has the same root as the English word "small." The identical sound pointed the Latin reader perhaps to a mysterious union between a property, here a sin (or in other cases a virtue or a strength) with a food item and was perhaps extended to the bystander-tempter offering this food item, creating the link between sin and sex/woman. The ingestion of the food endows the eater with a property associated with the food source (apple with the knowledge because it comes from the tree of knowledge, the consecrated wafer with salvation because it is mystically derived from the body of the Savior). I am not sure whether this is just a fancy idea or has a meaning in the collective subconsciousness of humans. The latter interpretation is suggested by many fairy tales (the evil comes to Snow-white when eating an apple; eating a snake endows in other fairy tales with the knowledge to understand the language of animals).*

Paradise as a Universal Dream

The dream of a Paradise belongs to many people and many times. The Greeks kept the memory of a Golden Age in a distant past, Muslims adopted the Judeo-Christian paradise where the horticultural dream of an oasis was supplemented by female attractions. As late as at the eve of the French Revolution artists painted the dreamland of Arcadia, interestingly not any longer as a horticultural but a pastoral idyll as if they went even further back in time (pastoralism preceding agriculture). Also projections into the future like the classless society of the communists alluded to the ideas of the paradise with the quest for food suspended (everybody to its needs).

Only a Dream?

When reading S. Mithen's fascinating book "After the Ice" (Harvard, 2003) I got the impression that the Biblical record of the Paradise simply kept a memory of the Fertile Crescent during the warm and wet period of the Late Glacial Interstadial (10'800 to 12'700 BC), when humans lived a comfortable life from hunting the large gazelle herds roaming the grasslands. This period was followed by the cold and dry period of the Younger Dryas (9'600 to 10'800 BC), where this food resource suddenly collapsed. Humans lived a period of food shortage leading finally to the "invention" of agriculture by the Neolithical Revolution. Climate change was then interpreted by the religiously inspired authors of the Old Testament as banishment from a food paradise into the strenuous life of early farmers and pastoralists. How should prehistoric humans interpret the disappearance of their gazelle food source if not by an insult to their god? The authors of the book of Genesis claimed that humans ate a forbidden food – this interpretation does not lack logic for humans governed by the quest for food and sex. In fact, the authors of later books of the Old Testament came back to this interpretation by structuring human life by a wealth of food regulations as religious obligations.

Bacterial Predators

Let's come back to the realm of biology and ask whether there is any biological underpinning to this dream of a predator-free world? Over a long time period, prokaryotes were alone. What limited the life of bacteria at a time when ciliates, flagellates, and copepods were not yet present to sieve bacteria out of suspension and amoeba, rotatoria, and nematodes did not exist to suck bacteria from surfaces? Was there a type of prokaryotic Arcadia? In the framework of biological thinking, a predator-free world is but a dream. The idea of stealing the resources from other cells was too obvious to be overlooked. This stealing probably took different forms. One form is a predator bacterium that relies on other bacteria as food.

Bdellovibrio

Different forms of bacterial predation were described in morphological terms. The first was the feeding behavior of a Gram-negative bacterium belonging to the δ-Proteobacteria group carrying the telling name *Bdellovibrio bacteriovorus* ("a bacterium-devouring leech"). It is a curved rod with a flagellum at one cell pole. The flagellum is unusually thick since a sheath covers it. Its life cycle is complex even though it takes only 3 h for completion. The predator bacterium swims around vary rapidly (100 cell lengths per second) until it bangs on its prey, which are other Gram-negative bacteria. This is attack phase, stage I. Stage II is a recognition period of reversible binding followed by irreversible anchoring. In stage III the bacterium begins to rotate at the surface of the prey at a rate as high as 100 revolutions per second and bores a small hole through the host outer membrane and peptidoglycan cell wall in about 10 min. A mixture of hydrolytic enzymes is applied locally to keep the hole small to prevent excessive damage of the prey cell and leakage of its content. In stage IV, the predator sneaks into the hole and takes residence in the periplasmic space loosing its flagellum. In stage V the preyed cell changes form, from rod-shaped to a rounded form, and the predator begins to extract solutes from the prey's cytoplasm.

Bdellovibrio's Genome and Life Stages

Despite its small size (some forms are a mere 0.2-μm wide and 0.5-μm long), the predator has a genome of 3.7 Mb comparable to that of such metabolically versatile bacteria like *E. coli*. The recent sequencing of its genome revealed some of its secrets (Rendulic et al. 2004). According to that sequence analysis, it can only synthesize 11 of the amino acids needed for protein synthesis. Apparently it imports the lacking amino acids directly from the prey's cytoplasm into its own, despite the double barriers of the prey's plasma membrane and its own surface. Not surprisingly its genome encodes many ABC transporters that outrival the numbers found in most of the other sequenced bacterial genomes. Also ATP is directly obtained from the prey's cytosol even though *Bdellovibrio* can generate its own ATP by gylcolysis, via the tricarboxylic acid cycle and through fatty acid degradation. The large number of hydrolytic enzymes encoded in its genome suggests that it degrades the prey molecules to its constituent bases, sugars, and acids before their reuse. In stage VI *Bdellovibrio* transforms its shape: it gets elongated and convoluted, and synthesizes DNA. The genome tells us that it synthesizes all bases from scratch instead of relying on the host's bases, probably because the need exceeds the supply. When grown to multiples of its normal length it starts to septate. Such multiple fissions are rare in prokaryotes. Stages VII and VIII see the development of flagellated cells and the lysis of the outer cell membrane and cell wall of the prey, respectively. The predator has multiplied and it sets out to find a new prey. *Bdellovibrio* is now a pet bug in a number of laboratories since there is some hope to develop it into a living antibiotic for future pharmaceutical use. Some microbiologists argue that the prey invented physical armor against *Bdellovibrio* in the form of the S-layer. This sheet consists of protein or glycoproteins showing a *para*-crystalline appearance that adheres in Gram-negative bacteria directly to the outer membrane. It protects not only against pH fluctuations and osmotic stress but apparently also against enzymes and *Bdellovibrio* attack. Here is a recurring theme in evolution of increasingly better armors and better drills.

Other Predator Bacteria

These bacteria are mainly known from morphological studies conducted in fresh-water sulfurous lakes. One bacterial predator was called *Vampirococcus* since it adsorbs to the surface of the phototrophic purple sulfur bacterium *Chromatium*. As its name proposes, it develops a peculiar attachment structure and sucks out the cytoplasm of its prey. All what remains from the prey is the cell wall, the cytoplasmic membrane, and inclusion bodies. *Vampirococcus* persists freely suspended in these lakes, but was seen dividing only when attached to its prey defining it as an epibiont. In the same lakes, there is another bacterial predator called *Daptobacter*; it penetrates the prey cell and multiplies within the cytoplasm of the prey cell yielding at the end of the cycle again an empty cell consisting only of cell wall and membrane plus storage particles (Guerrero et al. 1986).

Myxococcus

There also exist bacterial predators that attack bacteria growing on a surface, the most thoroughly investigated are myxobacteria. These are Gram-negative aerobic soil bacteria that secrete an array of digestive enzymes that lyse insoluble organic material. They also digest bacteria and yeast that they find on their way using an abundance of hydrolytic enzymes. Many myxobacteria also secrete antibiotics that kill their prey. They digest the food cell and adsorb primarily small peptides from them. Amino acids are the major source of carbon, nitrogen, and energy for myxobacteria. The feeding occurs in dense populations containing thousands of cells that exhibit gliding motility. The advantage for this collective feeding is twofold: Secretion by so many myxobacteria assures a high concentration of extracellular enzymes. In fact the growth rate of myxobacteria increases with cell density demonstrating clearly the selective advantage of collective feeding behavior. Secretion of hydrolytic enzymes is a risky exercise and many other bacteria have opted for cell wall-bound hydrolytic enzymes. The digestion by cell-bound enzymes assures that the cleaved substrate remains in the vicinity of the cell and does not get stolen from bystander cells or gets passively drifted away from the feeding cell. When a large number of cells belonging to the same bacterial clone glide along a surface, no risk is associated with enzyme secretion. Cooperative behavior becomes thus a strong selective advantage and drives these prokaryotic cells to a multicellular structure. In fact *Myxococcus* also capitalizes on this collective behavior in the inverse situation when the food is exhausted and starvation sets in. Amino acid depletion coupled with high *Myxococcus* cell density on a solid surface induces one of the most complex development processes in prokaryotes, the building of a fruiting body. These are nicely colored, and in the case of the myxobacterium *Stigmatella* these are beautifully structured bouquet-like bodies measuring nearly 1 mm in length containing more than 10^5 cells. One day later, some cells in the top sporangiole of the fruiting body differentiate into myxospores, while others lyse. The elevation of the spores above the ground assists their dispersal. "Sticking together" provides a distinct selective advantage and this primitive community aspect of bacteria point the way to multicellular life cycles. Motility is part of the success of *Myxococcus*. Two systems underlie this gliding movement: S-motility depends on cell–cell proximity, pili, and an extracellular matrix of cohesive fibrils. In contrast, the A-motility allows individualistic movement driven by slime extrusion through nozzle-like pores. It is notable that A+S– "asocial" mutants quickly evolved new swarming capacities by the enhanced production of an extracellular fibril matrix that binds cells together. Apparently the advantages of social life are so great for this organism that transitions to primitive cooperation occurs spontaneously, even if it is costly to the individual cell (Velicer and Yu 2003).

It is not yet clear whether these forms of bacterial predation really play a role in controlling natural bacterial populations in the environment or are a microbiological curiosity. However, there are professional predators of bacteria in nature. They belong to two classes. One class is bigger than bacteria and belongs to the protists. These predators are known for a long time and play an ecologically important role in transferring bacterial biomass up into the food chain. The other class of bacterial predators is bacteriophages, they are smaller than bacteria and are also known for a long time, but their ecological role became clear only relatively recently. Bacteriophage literally means bacterium eaters. We will first explore their lifestyle, before reviewing their role in the ecosphere.

The Phage Way of Life: Bacterium Eaters

Bacteria can be quite abundant if you happen to be in the right environment. The right environment is not a major problem for the phage. As each phage is produced by the previous host cell, its next target cell is in general not far away. Therefore phages have not developed motors to propel them to the next target. This sounds somewhat ridiculous when speaking of viruses, but we have seen that substantial genomes can be associated with viruses—so developing a propeller is not necessarily excluded. Phages rely on diffusion and convection currents and meet their target cells by random encounters. If the number of target cells becomes low, diffusion becomes limiting. In the laboratory, phage infection does not occur below 10^4 cells/ml (that number depends somewhat on how many phages are added). Ecological observations in the ocean generally confirmed this value, although the limiting bacterial concentration was estimated as high as 10^5 and as low as 10^3 cells/ml. This refers of course to the number of susceptible target bacteria, not the absolute number of cells. Phages are actually quite choosy with their prey. Since they depend for their metabolism entirely on the host cell, they must be careful to match their own metabolic and nucleic acid handling machinery with the systems of the cell. It is generally said that viruses respect the species barrier. Cross-species infections occur relatively frequently, but they do not lead to a new infection cycle—the heterologous species is mostly a dead-end host. The above-mentioned concentrations of specific host bacteria required to sustain viral infections are actually quite high when considering that the average concentration of bacterial cells is about 10^6 per ml in the ocean. In the natural environment like a lake, you generally count with 10–50 different bacterial species per sample taken. This means that the average concentration of specific bacteria is quite close to this limiting minimal concentration of bacteria necessary for efficient phage infection. This has two important ecological consequences. Phages will mainly propagate on the dominant bacterial population—a concept described in the literature as "killing the winning population" phenomenon. In fact careful ecological time series of phages and their host bacteria in a pre-alpine lake demonstrated that seasonal bacterial blooms were terminated by a collapse due to phage lysis. The observed kinetics of the appearance of bacteria, infected bacteria, and free phage fit the theoretical model (Hennes and Simon 1995). At lower cell densities, the ecological consequence is equally important. It explains why infectious phage coexists with a susceptible bacterial cell. *The diffusion–limitation of the phage infection process is a type of ecological safety valve for diversity in the environment: It trims back overshooting bacterial clones that would consume the resources of the entire community and spares low-abundance bacteria. Thus phages cannot even drive fully susceptible bacteria into extinction. In the laboratory, where the interaction of a phage with its host bacterium was studied in isolation, a different picture was painted for phage–host interaction. It was seen as a type of arms race with bacteria developing resistance to the phage (frequently by changing the receptor molecule on the cell used by the infecting phage) and the phage taking countermeasures such as developing new receptor affinities. Phages and bacteria are thus locked in a war of attrition over long time periods.*

Numbers

In oceans and lakes, the total number of phages exceeded bacteria generally by a factor of 10. Ecologists conclude from that observation that phages are one of the major factors controlling the bacterial population size in the wild. When these data were published for the Northern Atlantic (Bergh et al. 1989) and quickly confirmed for other environments (Suttle 1990), it suddenly changed our ideas on the abundance of life-forms on our planet. Phages were in fact not limited to surface water, but also found in deeper water layers as well as in many soils, albeit at lower titers. As we inhabit a blue planet, where most of the surface is water, a conservative estimate calculated 10^{30} tailed phages in the biosphere. In sheer numbers tailed phages represent probably the most abundant "organism" on Earth. Of course this is genetically not a monolithic population, ecologists estimated in excess of 100 different phage strains in any environmental sample, genomics approaches working with 100-l water samples approximated up to 7,000 different viral sequences in such samples (Breitbart et al. 2002). There are data, which suggest that the DNA sequence space for viruses is as large as for cellular life-forms on Earth (Pedulla et al. 2003). This figure is counterintuitive when considering the small size of viral genomes, but might be compensated by the fact that practically all investigated species are infected by viruses and most are actually infected by many different viruses—take the table of contents from a textbook of medical virology as a witness. There are only few exceptions like yeasts, which sport only intracellular viruses.

Choosing the Target

If bacteria are numerous, the problem of the phage is not the diffusion-controlled encounter, but the specific recognition of the target cell. How can the *E. coli* phage T4 attack the right cell if in the human gut statistically only one in a million is a suitable cell? If the phage injects its DNA into the wrong place—either a nonpermissive cell or a cellular debris—it is gone. The phage has only a single shot and therefore it uses multiple sensors that need positive feedback before the attack is launched. The T4 phage initially recognizes the host cell via reversible interactions of the tips of its long tail fibers with surface receptors like the lipopolysaccharides (LPS) on the cell surface of *E. coli*. T4 has six long tail fibers fixed to the baseplate, a type of landing module or platform near the most distal part of the tailed phage when seen from the phage head. One can imagine how T4 is sweeping with its tail fibers across the cells in its vicinity to get a feedback for the location of a prey. Binding of a single fiber to LPS is not sufficient. Multiple fiber binding, probably by at least three of the six fibers, is a necessary next step, but this is not a problem since LPS covers the entire bacterial cell surface. There is a further trick with the tail fibers: They consist of a proximal and a distal part separated by an elbow structure. The arms of the tail fibers are so long that their elbow can bind to the whisker fibers located at the neck of the phages—between the head and the tail. Apparently under some adverse conditions it might be advantageous not to get engaged into an adsorption process. If everything is fine and the fibers got bound, the baseplate, which is initially 100 nm away from the cell surface, moves closer. Here again the elbow of the tail fiber allows some flexibility since the tail fibers remain bound to the LPS during this approach. The flexible tail fibers are used as levers to move the baseplate toward the cell surface (Leiman et al. 2004). The baseplate contains short tail fibers, which then bind to the cell irreversibly.

Injecting the DNA

The phage particle then behaves like a spring under tension. The baseplate, which has a hexagonal shape in mature virus, changes to a star shape after adsorption to a host cell. This initiates sheath contraction, which propagates in a manner analogous to falling dominos. The sheath contracts to less than half of its original length and drags the baseplate along the tail tube. The tail tube protrudes and penetrates through the outer cell membrane. This nanosized DNA injection machine has been investigated at 17-Å resolution by cryoelectron microscopy and partly also by X-ray crystallographic analysis. The central hub of the baseplate was resolved at a 3-Å resolution (Kanamaru et al. 2002). It appears like a torch. The bigger head is formed by gp27 of T4, a trimer arranged in a torus large enough to assure the passage of the dsDNA genome. The handle of the torch is built by a trimer of gp5. Gp5 experienced a maturation cleavage into an N-terminal gp5* with a lysozyme domain and a C-terminal gp5C, which forms a hollow needle pointing to the host surface. The tail sheath contraction forces the tail tube toward the cell membrane. The gp5C needle punctures the outer cell membrane, the externally located lysozyme domain then digests the peptidoglycan layer between outer and inner membrane, allowing the penetration of the tail tube into the inner membrane for injection of the phage DNA into the host. The phage genome, and only the phage genome, enters the cell; the protein shell is left on the cell envelope.

Wait and See: Lysogeny

Not all problems are solved when the T4 DNA is inside the right cell. Within an hour, a productively infected cell produces up to 300 progeny T4 phages, each with a genome of 170 kb. Without considering all other biosynthetic activities redirected by the phage, DNA synthesis alone amounts to 50 Mb—a substantial synthetic activity for a cell that normally replicates just a 4-Mb genome during this time period. It is apparent that many cells are not in a nutritional state to allow such a spree of metabolic activity. How does the phage react? T4 can follow a policy of wait and see: The genome remains relatively dormant, and the phage process gears up when the cell resumes growth after encountering a new food-rich environment. However, a phage cannot wait too long and many phages infecting bacteria from nutrient-poor environments can become a highly appreciated source of phosphorus and nitrogen to the starving cell. An interesting class of phages called temperate phages evolved a sophisticated strategy to cope with the problem of how to infect a starving cell. These phages evolved a genetic switch that allows a change in lifestyle. They choose between lytic infection of a well-fed cell or lysogeny in a nutritionally deprived cell. Phage lambda is the prototype of this viral group. Lysogeny is a type of dormancy where the phage incorporates its genome into that of the host bacterium. If your host cell is starving, the cell in the neighborhood is most likely also starving. Amplifying new phages makes little sense, and they will not find well-fed cells to maintain a replication cycle. Released phage would only succumb to the process of phage decay. Phages have been likened to crystals, but in the ecological context, they are not spared from decay processes to which all biological material is subjected. The sunlight at the ocean surface is one of the killers of phages, marine nanoflagellates (a protist) were also reported to graze viruses. It is thus a safer strategy to hideout in the genome of the host as a prophage and to leave the cell the task to replicate the prophage DNA. The prophage can occasionally be induced and when in a well-fed cell, it leads to a full multiplication cycle of the phage in an environment that will sustain further infection cycles. However, prophage induction will also occur in a cell that goes from bad to worse—here the prophage has a sensor (actually the same protein which silences the prophage gene expression) that captures the message, when the cell is likely to die. Instead of dying with the cell, the prophage gets induced and tries a desperate last chance operation.

Conflict and Cooperation

The close association of prophage DNA with the bacterial genome creates a fascinating mixture of conflict and cooperation with the cell. Conflict because, as all selfish DNA, the cell has to watch that it replicates only its own DNA. Prokaryotes have a very efficient DNA detection and destruction system that distinguished self from nonself DNA (restriction–modification systems). As temperate phages literally bombard all bacterial cells, the bacterial genome would be diluted out by integrating phage sequences. About half of the bacterial genomes carry prophage sequences, many even carry multiple sequences. The most extreme case is currently a pathogenic *E. coli* strain that carries 18 prophages, which cumulatively adds up to 0.5 Mb—a lot for a 4-Mb genome. *Bacteria keep small genome sizes, and they apparently mount genetic mechanisms that purge the genome from foreign DNA elements. At the same time, bacterial adaptation relies on external DNA capture—the acquisition of plasmids or transposons carrying antibiotic resistance genes is a lively reminder of that need. Phages apparently take account of this dilemma of the bacterial cell and frequently come with genetic gifts to the cell. This cooperation pays out for both parts. Bacteria, which acquire prophages containing genes that confer a selective advantage, prosper in their environment. The prophage also profits by passive propagation in a successful bacterial clone.* In fact bacterial pathogens are known to human medicine, where the changing pathology can be correlated with the sequential acquisition of prophages offering specific combinations of virulence factors, e.g., *Streptococcus pyogenes* (Beres et al. 2002). Prophages are thus exploited by the bacterial cell, which then becomes a successful human pathogen such as *Vibrio cholerae* (Waldor and Mekalanos 1996). In fact possession of prophages is more the rule than the exception in bacteria as revealed by bacterial genomics (Canchaya et al. 2003b; Brüssow et al. 2004). Apparently prophages do not only influence short-term bacterial adaptation but they also affect long-term bacterial genome evolution (Canchaya et al. 2004).

Phages in the Microbial Loop of the Food Chain

Lysis by Phages

Even if phages can be deadly devices, their effect on bacteria is not exclusively detrimental. This is one of the remarkable lessons from broader ecological considerations: What appears at the individual level a nuisance or a nemesis can at a higher ecological level even represent a motor of evolution or a stabilizer of biodiversity. I will illustrate this for the lytic infection in which the bacterial cell dies (Figures 4.16 and 4.17). The characteristic effect in the laboratory is the sudden clearing of a test tube. The transition from a turbid to a clear tube is achieved by a combination of two proteins (phage holin/lysin), activated at the end of the infection cycle. If the phage does not find a way to destroy the cell's structure, its progeny phages remain encased in the bacterial cell. A holin inserts into the bacterial cell membrane and forms pores. The phage lysin gets exported via these pores and nibbles away the bacterial peptidoglycan layer. As bacteria are under osmotic pressure, they literally explode when the cell wall gets digested. If you look at the cleared bacterial suspension with the microscope, you hardly see any bigger cellular debris. This bacterial death has an important ecological consequence.

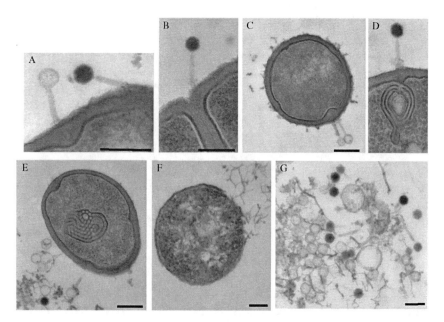

FIGURE 4.16. The attack of the *Lactobacillus* phage LP65 on its host cell illustrates the lytic phage way of life, probably one of the earliest forms of predation in the biosphere. **A** and **B** show the attachment of the phage with its base plate–tail structure to the cell. Darkly stained capsids still contain DNA; the empty capsids have their DNA already injected into the bacterial prey. Details of the phage landing module on the cell wall and the syringe-like DNA injection apparatus are shown in **C** and **D**. Early in infection the cell shows stress symptoms (**E**), later the cell undergoes phage-induced lysis of the cell wall (**F**). Finally the lysed cell releases its content into the medium, including the progeny phage (**G**). The images are produced by ultrathin section electron microscopy.

Microbial Successions in a Lake

To understand this connection, we need a short excursion into a lake in early spring. It is getting warmer, light is getting more intensive, essential nutrients such as silicon, nitrogen, and phosphorus are available. Silicon? You might not be aware of silicon as a nutrient, but the fast growing diatoms and *Chrysophycea* dominate the water initially—and both depend on silicon as an essential nutrient. Once the silicon resources have been used up, the next generation of primary producers takes over. The population then shifts to green algae as the dominant organism. The algal biomass contains the elements carbon, nitrogen, and phosphorus in a characteristic ratio, named after the aquatic biologist A. C. Redfield. The Redfield ratio sets C:N:P as 106:16:1. Many lakes are now eutrophied since they are supplied with ample amounts of phosphorus. This means that the next element, which becomes limiting is organic nitrogen. When this happens, the next generation of primary producers follows; these are cyanobacteria like *Anabena* or *Microcystis*. They have learned to harness atmospheric N_2 into organic nitrogen by

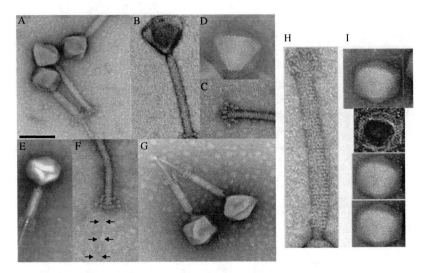

FIGURE 4.17. The tailed *Lactobacillus plantarum* bacteriophage LP65 seen with negative staining electron microscopy. Phages are depicted with extended (**A**) and contracted (**E**, **G**) tail, **B** and **H** visualize the tail, **D** and **I** the capsid, **C** and **G** the base plate, **F** the tail fiber in greater detail. The bar marks 100 nm.

their nitrogen-fixation apparatus. The photosynthetic phytoplankton exudates organic molecules (glycolate or glycerol). This material can make up an astonishing 5–35% of the total photosynthetic production. This material enters the dissolved organic matter (DOM) pool and feeds heterotrophic bacteria. They decompose this DOM into CO_2 and minerals, which can again feed the photosynthetic organisms. This literal feedback is called a microbial loop.

Food Chain

Another part of the organic material released by the primary producers is accumulated as biomass in the heterotrophic bacteria. The biomass of the photosynthetic and heterotrophic microorganisms makes up the particulate organic matter pool (POM). This pool feeds now a series of increasingly larger predators. The first line of grazers is protozoa, which are eaten in turn by metazoan zooplankton. In a lake fish is frequently the top predator (if no human beings are around to eat the fish). The grazers are heterotrophs and will recycle part of the consumed biomass as CO_2 through respiration. Another part enters the pool even before digestion via sloppy feeding. Bacteria and protozoa are the most important actors in the classical food chain of a lake when considering their metabolic activity and their total biomass. Fluorescence staining estimates prokaryotic cell numbers for lakes in the range of 10^4–10^6 cells/ml. These numbers vary by one log through the annual seasons, but much less with the nutrient content of the lake. Ecologists deduced from this observation that the bacterial cell number is more determined by the grazing efficiency of the zooplankton and protozoa ("top-down control") than by available food resources ("bottom-up control").

The fact that the eaters increase in size has an important consequence in aquatic environments, being lakes or oceans. Particulate detritus is formed either as fecal pellets or as dying animals. Due to their density and weight, they sink relatively quickly to the bottom of the water column and are therefore lost to the nutrient cycle in the sunlit zone of the water. Some digestion products are smaller but still large enough to sediment slowly to the bottom. Particularly interesting are the fragile aggregates of several centimeter size known as "marine snow." This snow shows enhanced microbial activity when compared with the surrounding medium. Interestingly this applies both to the bacterial and the phage part. While in the surrounding water, the phage to bacterial counts behaves like 6:1, this ratio was up to 40:1 on the marine snow flakes. The increase in ratio probably directly reflects the higher growth rate of the bacteria on the "snow."

Phages in the Microbial Loop

Where are the phages or, more general, the viruses in this scenario? Actually, viruses were relatively long overlooked even by microbial ecologists and came into focus only relatively recently (Fuhrman 1999; Wommack and Colwell 2000). As ecologists like model building, they first incorporated viruses into their models and then checked by surveys and mesocosmos experiments whether the prediction fit with the reality. Overall, these investigations added a new loop into the food web. Viruses infect the phytoplankton as well as the heterotrophic bacteria. This leads to the lysis of the cells and two components are released, DOM from the cellular remnants and the progeny phages/viruses called virio-plankton. Both fractions are too small in size to sink—consequently they remain in zone of the bacterial activity. The viral lysis of the bacteria then again feeds the bacteria. This would not be the case when the bacteria were not lysed by phages but eaten by protists in the food chain. This organic matter would be sequestered into another compartment of the ecosystem. The net result of viral infection was always the same, irrespective of whether bacterial mortality was set as 50 or 10% as virus-induced, viral lysis resulted in an about 30% increase in bacterial carbon mineralization and bacterial production rates. Phages are thus not rogue killers. They play an important role for bacteria: They maintain biodiversity, enhance the food resources, mobilize bacterial genomes, drive thus bacterial evolution, and direct microbial succession patterns.

On Starvation, Sporulation, Cannibalism, and Antibiotics: Near Death Experiences

Copiotrophic Versus Oligotrophic Bacteria

It might sound like a contradiction to include starvation strategies into a history of eating. However, to be eaten is just the other side of the coin of eating, so how to deal with starvation is perhaps one of the most important strategies of eating. In fact exponential growth of bacteria with short doubling times is even in the laboratory broth only a short period in the life history of a bacterium. As early as the first nutrient becomes limiting or waste products exceed a certain level, growth of the bacterium ceases. However, if the bacteria are transferred into a fresh medium, growth resumes after a certain lag phase. If no fresh medium arrives, the bacterial culture goes into the death phase and the viable bacterial cell counts start to drop. Bacteria have designed a number of solutions as to how to cope with the problem of starvation and dying. Indeed bacteria showed all their creativity in facing this problem. It was frequently stated that bacteria are confronted with a life between feast and famine, ecological situations that oscillate between copious supply of nutrients (e.g., life in the gut of an animal, in milk, or on decaying plant material), alternating with a more or less long period to reach the next rich food source. These bacteria have been called copiotrophic, and *E. coli* is one of them. These organisms are fast growing casual workers during the feast periods, but they show little ability to endure starvation over prolonged periods of time, and they die fast. However, not for all bacteria applies the feast/famine comparison. For example oligotrophic bacteria have opted for another strategy, they are specialized to live under low substrate supply conditions. Bacteria living in the deeper ocean layers or the water in the rock beds are examples. They are typically slow-growing, they do not reach high population sizes, but their environment has nevertheless a decisive advantage: It is extremely stable. One would therefore predict that these bacteria can sacrifice complex regulation systems to cope with rapidly changing environmental conditions.

Strategies of Starvation

Still other bacteria have tried to combine the capacity to achieve rapid growth and large population size with an extreme starvation resistance by differentiation processes (spore development). Even if this trait is not widely distributed in bacteria, its theoretical implications for the dissemination of bacterial life are enormous. Other bacteria, *Caulobacter* is a representative, increase their cell surface when starving, and they build stalks that apparently allow the efficient extraction of decreasing nutrient concentrations. Many Gram-negative bacteria follow the opposite strategy: They decrease their energy needs by decreasing their cell size. Dwarfism is a widely distributed observation in marine organisms that are frequently much smaller in their natural environment than when grown under laboratory conditions. This process has been studied in some detail in a *Vibrio* strain. It undergoes an ordered sequence of events: In the first phase, intracellular proteins are degraded, while new starvation-induced proteins are synthesized and respiration decreases. In the next phase, intracellular storage material like polyhydroxybutyric acid is mobilized and high affinity transporters try to catch the remaining substrates in the environment. If that does not help, further reduction is started in the third phase: The cell looses its flagellum, decreases in size, and reduces its metabolic activity still further except for an important production of exoproteases. At the end of this series stands a dwarf cell.

Spores

The bacterial spore is, in some bacterial groups, the last response to a harsh environment including prolonged starvation. Spores have not only a remarkable resistance against physical and chemical stress but also a remarkable capacity of dispersal because they are easily carried by air currents. In fact spores have stimulated the fantasy of so prominent astrophysicists as Fred Hoyle, who revived the ideas of natural philosophers from the nineteenth century speaking of panspermia. According to this school of thought, life did not originate on Earth, but was seeded by spores coming from space. Hoyle pretended that his studies on the absorption properties of interstellar dust allows the interpretation that part of the dust grains were derived initially from viable bacteria. Indeed, laboratory experiments showed that these spores can survive conditions of very high vacuum, low temperature, and intensive UV irradiation prevailing in space. It was calculated that in interstellar clouds spores might survive up to 45 My.

Bacillus Sporulation

Spore formation in *Streptomyces* and *Myxococcus* leads to complex multicellular mycelia and fruiting bodies, respectively. The sporulation process in *B. subtilis* implicates less multicellularity. Only two cell types are built, the mother cell and the spore. However, a rather complex interplay of genes and morphogenetic process is observed in *B. subtilis* sporulation. Not less than four different sigma factors are induced in a distinct spatiotemporal way during *B. subtilis* sporulation that redirect the promoter recognition capacity of the RNA polymerase in the two cellular compartments and thus the local transcription program. The last sigma factor expressed in the mother cell, σK, has an especially striking origin. Its gene is the result of a genetic recombination between two different genes, *spoIVCB'* and *spoIII'C* physically separated on the *B. subtilis* chromosome. The process is mediated by a specific recombinase, SpoIVCA. This uses a repeat sequence at the end and the beginning of the two gene halves for recombination, resulting in the discarding of the intervening DNA segment. The process is thus linked to an irreversible genome change, which is, however, not so dramatic since it occurs in the mother cell and not in the spore. After further proteolytic processing of the pre-σK protein by another sporulation protein, σK directs the transcription toward genes involved into the synthesis of cortex and coat structures of the prespore. This excursion must suffice us as a glimpse into bacterial cell differentiation. The very fact that two different cell types developed in this process gave rise to a phenomenon, which is widely distributed in multicellular life, namely programmed cell death. The mother cell commits suicide when its task has been achieved, and the spore is formed. This job is morphologically rather complex and goes through different stages regulated by numerous genes classified into different groups: *spo0* to *spoVII*, according to the morphological stages 0 to VII affected by the mutant phenotypes. Let us concentrate just on the nutritional aspects of this differentiation process, which is most evident in the early phase.

Bacillus subtilis cells undergo sporulation during the stationary growth phase when the cell density is high and the nutrients are low. Experimental evidence links both parameters to the induction of sporulation. Depletion of either a carbon or a nitrogen source is necessary for the activation of sporulation. One identified nutritional signal is the reduction of the GTP level. Indeed addition of purine analogues inhibited GTP synthesis and induced sporulation even when the best carbon (glucose) and nitrogen (glutamate) sources were abundantly available. Interestingly several proteins are guanylylated when the cell enters the stationary growth phase. Intriguingly this includes EI and HPr known from the carbohydrate: phosphotransferase system, although the causal relationship, has not yet been deciphered. Here we see a clear advantage of research with prokaryotic organisms. Even if each process is very complex in its regulation, we have only a rather limited number of partners around, which can enter the game. Mother Nature is obliged to reuse the same elements to the delight of the student of biochemistry, who after a while perceives structures in the chemical fog of the bacterial cell. The cell density is apparently sensed by a proteinaceous factor excreted into the extracellular space. Indeed a "preconditioned" medium from late exponential phase cultures can induce sporulation.

Regulation of Sporulation

We touch here the area of "quorum sensing" and how bacteria monitor their population density and adjust their behavior accordingly. Bacteria use, of course, chemicals elaborated by them to assess the population size. *Bacillus subtilis* recently revealed a fascinating story of cannibalism. To understand these new observations, we need some background information. First, the different signals that initiate sporulation must be integrated into a pathway. This signal transduction pathway is known as Spo0-phospho-relay. We have here the familiar theme of a somewhat reticulate two-component regulator protein system consisting of two sensor systems, KinB and KinA, reporting extra-cellular and intracellular stimulator molecules, respectively, to the response regulator Spo0F. The Spo0B protein phosphotransferase passes the phosphate group then to the central response regulator Spo0A. This transcription regulator affects the expression of genes in various operons negatively and positively. The multistep nature of this phosphorelay with different kinases and different phosphatases is thought to facilitate the integration of various physiological signals. This integration is apparently important since sporulation is an energy demanding process, which once initiated can no longer be reversed. The cells are confronted with a dilemma: if they wait too long on the arrival of new nutrients, they might starve to death. If, however, they start too early to commit to sporulation, cells that waited are favored since they can profit from the newly emerging food source, while the committed cells have to complete their sporulation cycle first. This dilemma created a curious wait-and-see strategy.

Cannibalism in *Bacillus*

Recently, two new operons were added to the list of SpoOA regulated genes: *sdp* (stands for *s*porulating *d*elay *p*rotein) and *skf* (*s*porulating *k*illing *f*actor). The *spdC* gene encodes a small extracellular protein signaling factor that stimulates the expression of *yvbA* gene. The *YvbA* protein stimulates the transcription of two further operons involved in lipid catabolism and ATP production. This internal supply of energy delays the initiation of sporulation, allowing the cell to observe the scene. At the same time, what does the *skf* operon do? As the name indicates, this operon elaborates a killing factor; this factor is released into the medium and readsorbed by the cells. *Bacillus subtilis* cells that have an active *skf* operon have also produced two other proteins, *SkfE/F*, which apparently function as export pumps that escort the killing factor again out of the cell. Cells which do not have an activated SpoOA have not taken the precaution to transcribe the *skf* operon. Consequently the killing factor does its job, it kills the cell, and this killing releases the cell nutrients into the medium, which is taken up by the active SpoOA cell, which thus gets an extra nutrient input to continue with its wait-and-see position. As this process occurs between *B. subtilis* cells, the group of biologists describing it called it cannibalism (Gonzalez-Pastor et al. 2003). Other biologists preferred a different terminology and speak of self-digestion (Engelberg-Kulka and Hazan 2003) and see this as an act of a multicellular organism, like the self-digestion of the mother cell during sporulation. The critical question now is how related are the bacteria which are involved in this interaction. Is the killing factor only directed against cells of the same lineage or also against other bacterial cells.

Antibiotics

Chemical warfare is not rare in bacteria fighting for limited food resources. Bacteriocins are proteins that are produced by plasmid-carrying bacteria; these proteins kill the same or related bacteria lacking the plasmid and thus the immunity function against the bacteriocin. The target of these bacteriocins can be variable: Colicin E1 from *E. coli* permeabilizes the cell membrane, E2 degrades DNA, and E3 leads to rRNA degradation. Many bacteria produce antibiotics. Nutritional limitation seems to play a key role in their production. Antibiotics belong to the secondary metabolism of bacteria. They are synthesized after the growth phase (trophophase) in the so-called idiophase and need to be triggered by nutritional stress. Antibiotic production is seen as a chemical differentiation where the chemical individuality of the specific cell clone is expressed. A large part of natural isolates of actinomycetes produce antibiotics when freshly isolated, and they use them as chemical clubs to defend their position especially in oligotrophic environments such as soil and water. However, microcins are produced by enterobacteria and are thought to be important in the colonization of the human gut early in life. Yet, antibiotics are not only weapons in the bacterium–bacterium competition for limited food resources. Bacteria like *Serratia marcescens* use antibiotically active pigments in their fight with predator amoeba.

Increasing Complexity

The Birth of the Eukaryotic Cell

Giardia at the Root?

The origin of the eukaryotic cell is as unclear as the origin of the prokaryotic cell and numerous competing hypotheses have been formulated to explain its creation. In contrast, only few experimental facts have been accumulated. The rRNA evolutionary tree pegs parasitic diplomonads and microsporidia as the most primitive eukaryotes living today. Diplomonads got their name from two nuclei they contain in a kite-like body displaying a characteristic face-like appearance. Actually, the diplomonad was one of the first microbes that van Leeuwenhoek observed in 1681 in his own stool by using his microscope. He marveled about the Giardia's eyes (nuclei) gazing at him. Until quite recently, Giardia was believed to be one of the most primitive eukaryotes—and much hope was fixed on it for unraveling the origin of the eukaryotic cell. Giardia lacks mitochondria, smooth endoplasmic reticulum, Golgi bodies, and lysosomes, in fact much of the defining membrane-enveloped organelles characterizing a eukaryotic cell. At the biochemical level, *Giardia* lacks the tricarboxylic acid cycle and the cytochrome system. It uses glucose and stores glycogen but can live without glucose. Unlike other diplomonads that possess a cytostome (cell mouth) through which they endocytose bacteria as food, *Giardia* feeds by pinocytosis on mucous secretions of the host's intestinal tissue. *If we take the cell biology and biochemical data on Giardia at face value, we might be tempted to speculate that the earliest eukaryotes were protists lacking organelles such as chloroplasts and mitochondria, they were heterotrophs and not autotrophs, anaerobes and not aerobes.*

Fossils and the Size Argument

The fossil record of early eukaryotes is relatively unrevealing. As the nucleus left no fossil traces, arguments for the eukaryotic nature of a fossil reside mainly on a size criterion with a 60-μm boundary once accepted as a safe limit for excluding a prokaryotic cell. The earliest hints of large-celled eukaryotes were found in 2.1 billion-year-old rocks and demonstrate spirally coiled millimeter-wide ribbons called *Grypania*. However, safer eukaryotic fossils are balloon-like cells dated to 1.8 billion years found in the Jixian valley in China. From about 1 billion years ago, acritarchs are reported: Their morphology suggests something like algal phytoplankton. One might therefore suspect that these cells already contained mitochondria and chloroplasts, the two powerhouses of the eukaryotic cell. However, the impressive scientific name of these creatures should not instill too much confidence: acritarchs derive their name from the Greek word *akritos*, meaning confusing, uncritical. The size criterion is somewhat tenuous as giant prokaryotes have been reported recently. The first was a 500-μm-long relative of *Clostridium* called *Epulopiscium* that dwarfs many protists. The Latin name means "feast in fish," and this describes the lifestyle of this fish gut commensal. It is motile, swims with bacteria-type flagella at the decent speed of 2.4 cm/min, and shows a highly convoluted plasma membrane, which helps to overcome the size limits set by diffusion processes and for nutrient transport. Previously, such large prokaryotes—in fact a million times larger than *E. coli* on a volume basis—were not believed to be possible. Now even greater prokaryotes have been described. *Thiomargarita*, which we discussed in the section Sulfur Worlds, is 100-fold greater than *Epulopiscium*. Prokaryotes now reach far into the sizes of eukaryotes and disqualify the size criterion (Sogin 1993). In addition, tiny, but truly eukaryotic microorganisms have been described. One example is *Nanochlorum*. While only measuring 1–2 μm in diameter, it still has the place for a nucleus and a single chloroplast and mitochondrion.

Endosymbiont Hypothesis

Many of the hypotheses on eukaryotic origins postulate a critical fusion event between two or more prokaryotes, endowing the new cell with new metabolic capacities that propelled the rapid evolution of cell systems of increasing complexity. Perhaps the most stimulating hypothesis on the origin of the eukaryotes is Lynn Margulis' version of the "endosymbiont hypothesis" for the origin of mitochondria and chloroplasts. This theory goes as follows. There was first an anaerobic prokaryote that invented typical eukaryotic endomembranes containing cholesterol and leaving steranes as a fossil trace. These cells made a living by engulfing and digesting bacteria. Then came a critical event. Instead of being digested, a bacterium survived in the predator cell. And the prey turned out to be of a major selective advantage to the predator cell as it showed the capacity to cooperate with the metabolism of the cell. On the basis of a wealth of biochemical data, for example, the striking similarity in the respiration chain organization of the bacterium *Paracoccus* and the corresponding mitochondrial enzymes, it was postulated that an α-Proteobacterium became the first endosymbiont of this early eukaryote. Over time the endosymbiont lost most of its genes that were transferred to the nucleus of its host cell. The relationship thus became genetically fixed and was evolutionarily stabilized through mutual benefit (others would speak of slavery): Respiration-derived ATP was traded against organic substrates for oxidation. Oxygen was not an essential partner in this deal as alternative electron acceptors like environment-derived nitrite or metabolism-derived succinate could take the role of oxygen. In fact, a long list of eukaryotes ranging from protists and algae to flatworms and nematodes, mussels and snails possess anaerobic mitochondria. This process was repeated a second time when bacteria closely related to contemporary cyanobacteria became endosymbionts in an already mitochondria-containing cell. Cyanobacteria endowed the cell with photosynthetic capacity, neatly explaining the striking similarity of the PSI and PSII in cyanobacteria and chloroplasts. As in the case of mitochondria, cyanobacterium genes were transferred to the host, a process that can still be studied as the transfer of chloroplast genes to the cell nucleus (see the next section). *This theory satisfied many scientists as it could explain a wealth of genetic and biochemical data, it concurs with the basic principle of the conservation of basic and successful biochemical mechanisms that become integrated into complex systems. Evolution in this scenario is mainly the innovative combination of existing parts and only occasionally the invention of a new biochemical reaction.*

From Mitochondria to Mitosomes

Phylogenetic evidence derived from both ribosomal RNA and protein sequence data tells us that all the mitochondrial genomes we know today are derived from a common protomitochondrial ancestor. This interpretation implies that mitochondria are monophyletic and originated only once in the evolution of the eukaryotic cell (Gray et al. 1999). The mitochondrial genomes in different eukaryotic lineages are the end result of independent reductions (<6kb in *Plasmodium*, 360 kb in the land plant *Arabidopsis*) of a much larger eubacterial genome. While all members of the so-called crown group of eukaroytes contain mitochondria, this is not the case in the stem group. At this position some amitochondrial organisms are found: these group include entamoebae, microsporidia, and diplomonads. As the latter two amitochondrial organisms represent one of the longest (deepest?) branches in the eukaryotic phylogenetic tree, *Giardia* was a hot candidate for being closest to the origin of the eukaryotic cell. But there are cracks in this appealing scenario and the *Giardia* link for the earliest eukaryote recently lost a lot of supporters (Henze and Martin 2003; Knight 2004). Mitochondria comprise a diverse family of organelles and *Giardia* possesses one highly reduced form of it called a mitosome. These small organelles are surrounded by a double membrane, but do not produce ATP. They are instead factories for the assembly of iron–sulfur clusters as suggested by the presence of several enzymes involved in this pathway (Tovar et al. 2003). As we have alluded several times in our survey, these bioinorganic clusters are relics of the earliest times of the biochemical evolution and provide the critical prosthetic groups to a number of enzymes. As these reaction centers were designed before the advent of oxygen in the atmosphere, they had to be protected from oxidation during their synthesis. A paradoxical, but finally logical location for these Fe–S cluster assembly proteins is the mitochondrion. As these organelles avidly reduce oxygen to water, a low oxygen partial pressure is found in this organelle. Genetic data from yeast suggest that Fe–S cluster biosynthesis is a more essential task of mitochondria than is respiration. As *Giardia* lives as a parasite in the intestine of vertebrates (fish, birds, mammals including man, where it causes either important economic losses or health problems), its environment contains only low or no oxygen, making oxygenic respiration obsolete. To be protected during transmission from one host to the next *Giardia* gets a protective shielding (encysted) before leaving the intestine and the protective shield is lost (decysted) only in the intestine of the next host. Thus there is no need to maintain the respiration part of the mitochondrion, and these observations are strong arguments for a reductive evolution from mitochondria to mitosomes in *Giardia*.

Microsporidia

A very similar story can be told for microsporidia, intracellular parasites also found in humans. Ribosomal RNA placed them among the most deeply branching eukaryotes. As they lacked mitochondria, they were thought to be relics from a eukaryotic past before the endosymbiosis event led to mitochondria. Later on microsporidia were identified as specific relatives of fungi (Roger and Silberman 2002). Apparently, parasitism led to economies in every aspect of life, and the genetic information was compacted into a minute eukaryotic genome (Katinka et al. 2001). This genome still contained genes for five proteins from mitochondrial descent, including one involved in generating iron–sulfur clusters. Antibody directed against this protein detected in microsporidia a membrane-bound organelle sometimes called a crypton to denote its still unknown functions (Williams et al. 2002). Many microbial eukaryotes that were previously thought to lack mitochondria have apparently retained a reduced organelle.

The Hydrogenosome

Another organelle, the hydrogenosome, was discovered 30 years ago by Miklos Müller and plays a prominent role in a fascinating hypothesis on the origin of the eukaryotic cell. These are hydrogen-producing organelles in protists (specifically ciliates and flagellates) living in anoxic environments. These ciliates ingest bacteria as foodstuff, and they degrade their macromolecules into pyruvate, which is transferred to the hydrogenosome. Pyruvate is decarboxylated and oxidized to acetyl-CoA, leading to the release of H_2 (hence the name of the organelle) and CO_2. ATP is formed in the hydrogenosome when acetyl-CoA is transformed into acetate, which is exported out of the organelle. CO_2 and H_2 are not waste products when the producing cell can cooperate with methanogens that transform it into methane in an exergonic reaction according to the equation $CO_2 + 4H_2 \rightarrow CH_4 + 2H_2O$. This is exactly what the ciliates are doing. Ciliates from the rumen of farm animals are tightly associated with methanogens at the cell surface. In fact, ruminants are an environmentally important source of methane, which is released during the belching of cows. This recently led to a project to vaccinate cows against methanogens to reduce this important source of greenhouse gas production. Ciliates living in strictly anoxic, eutrophic sediments go one step further. They carry the methanogenic partner even inside the cell. Feeding experiments have shown that methanogen-containing ciliates grow faster and give up to 35% higher yields than methanogen-free controls. The hydrogenosome and the methanogen form a functional unit, sometimes even a morphological correlate in the form of hydrogenosomes and methanogens organized into alternate sandwich arrangements. As hydrogenosomes were linked to mitochondria we see here another important function of primitive mitochondria that is not linked to aerobic respiration.

But how good is the evidence for the mitochondrial link? This point was until recently undecided. Hydrogenosomes from the human parasite *Trichomonas vaginalis* do not contain a DNA genome. Its ancestry had to be deduced from protein sequences. However, previous investigations on a hydrogenosome-specific protein led to divergent conclusions. One group placed it next to the mitochondrial lineage (Hrdy et al. 2004), another declared hydrogenomes as more distant relatives of α-Proteobacteria (Dyall et al. 2004b). This unpleasant visitor in the vagina, which causes vaginal discharge, malodoration, and irritation, was apparently not willing to reveal its murky evolutionary past. Clarification came finally from the anaerobic *Nyctotherus ovalis*, which inhabits another very special niche—it lives in the hindgut of cockroaches. In contrast to hydrogenosomes from flagellates, some fungi and ciliates, the organelle from *Nyctotherus* contains a small genome. This rudimentary genome encodes components of a mitochondrial electron transport chain. If complemented by some nuclear genes, functional mitochondrial complexes I and II could be reconstituted, while complex III and IV lacked entirely (Boxma et al. 2005). This hydrogenosome is closely associated with an intracellular methanogen archaeon. With its cristae, this organelle looks definitively mitochondrion-like. Phylogenetic tree analyses of the DNA sequence associated the genes of this organelle with the mitochondrial genes from aerobic ciliates. The *Nyctotherus* organelle is thus a missing link between mitochondria and hydrogenosomes. Apparently, hydrogenosomes are derived secondarily from mitochondria. Tracer experiments demonstrated that decarboxylation of pyruvate occurs via pyruvate dehydrogenase, but the enzymes of the citric acid cycle are lacking. Fumarate is used as a terminal electron acceptor like in some anaerobic mitochondria.

Syntrophy

Only acetate, CO_2, and other C1 compound plus H_2 serve as substrates for methanogens. Fatty acids larger than acetate like propionate and butyrate would persist in nature under anoxic conditions if there were not bacteria like *Syntrophobacter*, *Syntrophomonas*, and *Syntrophospora* that transform these substrates into products suitable for methanogens according to the equation: butyrate + $2H_2O \rightarrow$ 2acetate + H^+ + $2H_2$. However, there is a problem: Under standard conditions, free energy input is needed to drive this reaction. How can these bacteria derive energy from this reaction? The answer is straightforward: These bacteria associate with methanogens that capture acetate and H_2. The removal of the reactants pulls the reaction and ATP synthesis becomes possible.

The Martin–Müller Hypothesis

Syntrophic relationships between a Eubacterium and an Archaeon play an important role in a recent metabolic interaction model for the origin of the eukaryotic cell. According to this new hypothesis, it is suggested that eukaryotes have arisen through a symbiotic association of an anaerobic, strictly hydrogen-dependent autotrophic Archaeon (the host) with a Eubacterium (the symbiont). The latter was able to respire but generated molecular hydrogen as a waste product of anaerobic heterotrophic metabolism. Explicitly, the model postulates a symbiosis between a free-living H_2 and CO_2-producing Eubacterium and a methanogenic Archaeon (Martin and Muller 1998). If the symbiont escapes, the host starves immediately. An efficient hydrogen transfer calls for a large surface area that surrounds the symbiont; engulfment and maintenance were made possible by the build-up of intracellular membrane systems.

The Archezoa Alternative

This model has several definitive advantages. *One is its purely logical appeal. This hypothesis does not postulate the creation of a fundamentally new eukaryotic creature de novo. The eukaryotic organization emerged from the fusion of two known prokaryotic elements according to the conservative nature of evolution that prefers to reuse all its previous successful inventions before it goes for new inventions. Novelty is frequently the result of new imaginative combinations of otherwise well-known elements. In fact, the hypothesis does not request the existence of a primitive eukaryote lacking endosymbionts as formulated in the "Archezoa hypothesis." The elusive character of this endosymbiont-free eukaryote has been demonstrated by the studies with Giardia: It contains mitochondria-like organelles.* Comparative genomics also questions the root position of "primitive" protists: Microsporidia are now considered a likely sister group to fungi, and the deep divergence of diplomonads and parabasalids (their nearest phylogenetic neighbors) obtained by several molecular data sets is now seen as a "long branching artifact" (Dacks and Doolittle 2001). The parsimony argument for all these molecular data is that the last common ancestor of all currently known eukaryotes had mitochondria. What is then closer than to reject the archezoa concept and replace it by a hypothesis where the origin of the eukaryote is the fusion event of two distinct types of prokaryotes?

The Dual Ancestry of Eukaryotic Genes

Another element is plausibility: The model is so close to the known syntrophic relationships described between extant prokaryotes and protists that it does not request for improbable events. Finally, the eubacterium–archaea fusion can explain a curious observation with prokaryotes: The information-processing systems of Archaea (e.g., DNA transcription) resemble those of eukaryotes more than those of eubacteria. In contrast, eukaryotes have a more eubacterial than an archaea-type intermediary metabolism. The hypothesis that the eukaryotic cell is the consequence of an archaea host with a eubacterial symbiont explains this observation easily.

Further Partners?

The prokaryotic fusion hypothesis dominates now the eukaryotic origin discussion but it still comes in different forms. Critical differences between the hypotheses are the nature of the eubacterial partner(s). Lopez-Garcia and Moreira (1999), for example, postulate two successive events: first, a sulfate-reducing δ-Proteobacterium, which also produce hydrogen from fermentation and form syntrophic consortia with methanogens. Second, either at the same time or shortly later came an α-Proteobacterium into the consortium. It fed on methane produced by the archaea partner (i.e., they were methanotrophs) and produced CO_2. The hydrogen produced by the δ-Proteobacterium and the CO_2 produced by the α-Proteobacterium together stimulated the methanogen, further increasing the metabolic link between the different elements of this initially prokaryotic consortium, which quickly developed new characteristics transforming the ancestor of all eukaryotes.

Metabolic Interaction

Whether two or three partners, both schools of thought stress metabolic interaction as the key to the evolution of complexity. The reason seems clear: The cycling of matter is the key to sustainability in biochemistry. Organisms that create methane need to be associated with organisms that feed on methane to close the loop. Provided that you dispose of an external source of energy, the cycle can go for an undefined time. In prokaryotes that are only minimally compartmentalized, these competing processes have to occur in different organisms. If they get separated, the cycle gets interrupted and both partners start to starve. In the primitive eukaryote the cycling got permanent because the partners conducting opposing biochemical processes were closed into a loop and could thus not be lost easily. This does not mean that eukaryotes developed a more complex biochemistry than was ever achieved in prokaryotes. The fusion event sampled only two or three metabolic types of prokaryotes out of the myriad of possibilities. This intertwined process perhaps got more efficient and later experienced many variations, but except for those eukaryotes that later still acquired photosynthetic symbionts, the eukaryotes had only a limited possibility of metabolically exploiting their environment and had to become efficient predators. They ate the biomass and fuel created by the multitude of prokaryotes in a diverse set of environments with an incredible variety of biochemical reactions.

Alternative Models

Not all eukaryotic origin hypotheses are based on metabolic advantages. Margulis et al. (2000) postulate a fusion between an Archaeon resembling *Thermoplasma* (living in warm, acidic, sporadically sulfurous waters using elemental sulfur as an electron acceptor creating H_2S) and a spirochaete-like eubacterium. The spirochaetes associate with the surface of the archaeon creating a motility symbiosis. The spirochaetes were the swimmer structure that became the mastigon (a cell whip, an eukaryotic flagellum) in this postulated system. When attached to a nuclear structure, which developed in the *Thermoplasma* partner, a structure developed, which is called a karyomastigont—well known from extant protists. Yet a syntrophy idea centered on a sulfur cycle also forms a part of this hypothesis.

The Ring of Life

That the origin of the eukaryotic cell is enigmatic should not surprise us as it goes back to an event that occurred at least 1.4 billion years ago (Javaux et al. 2001). Fossil finds are some help for dating the event, but they cannot clarify the underlying genetic processes leading to the eukaryotic cell. In contrast, genome comparisons should shed some light on this process. Depending on the methodology, different results were obtained. When the ribosomal RNA sequence was used as a yardstick, Eukarya showed a closer relationship to Archaea than to Bacteria. Even if Archaea appeared there as sister groups of Eukarya, the deep connections between eukaroytes and prokaryotes remained unresolved. The use of a single gene for evolutionary inferences of this importance was, however, criticized. When many genes were taken for comparisons, the picture did not become much clearer. Informational genes (i.e., genes involved in transcription, translation) were mostly related to archaeal genes, whereas operational genes (i.e., genes involved in cellular metabolic processes like amino acid, lipid, and cell wall synthesis) are mostly closely related to eubacterial genes (Rivera et al. 1998). Furthermore, pervasive lateral gene transfer was discovered between organisms, which discouraged the initial hopes that the origin of the eukaroytic cell could be deciphered from the genomes of extant organisms (Doolittle 1999). In fact, when 10 complete genomes from organisms of the three domains of life were compared by using a new statistical method ("conditioned reconstruction," which uses shared genes as a measure of genome similarity irrespective of whether these are horizontally, i.e., brother–sister, or vertically, i.e., mother–daughter, inherited genes), neither the root of the tree nor the origin of eukaryotes could be determined (Rivera and Lake 2004). However, the authors observed a repeating pattern in the five trees they obtained, indicating that the trees are simply permutations of an underlying cyclic pattern. This observation has an important implication: The eukaryotes are the result of the fusion of a bacterial and an archaeal genome. The tree of life became a ring of life. This genome analysis fits remarkably well with metabolic hypotheses on the origin of the eukaryotic cell (Martin and Muller 1998), but it does not reveal whether this event preceded the symbiotic acquisition of mitochondria. As evidence for an earlier endosymbiosis event is entirely lacking, one might suppose that the eubacterial genes in the eukaryotic genome stems from the mitochondrial endosymbiont, which transferred most of its genes into the genome of its host (Martin and Embley 2004).

Molecular Fossils

What is the time horizon for the emergence of eukaryotes? Molecular phylogeny and biogeochemistry indicate that eukaryotes differentiated early in the history of life on Earth. C27 to C29 sterans derived from sterols synthesized only by eukaryotes push the origin of eukaryotes up to 2 Ga or more ago (Brocks et al. 1999). This early date might cause raised eyebrow, but fossil data from 1.5 Ga-old shale from Australia demonstrate an already well-diversified protist life

(Javaux et al. 2001). One cell type is called *Tappania* and shows large vesicles with hollow processes that branch dichotomously. They were interpreted as germinating algal cysts. Another form is called *Satka*, characterized by many rectangular scales on the cell wall. The geologists could reconstruct the paleoenvironment and found that these early eukaryotes were not only morphologically diversified into a handful of organisms but they had already formed an ecologically differentiated community. Species richness and abundance were greater at the ocean margin than in the storm-dominated shelf.

The Story of O and the Malnourished Ocean

Stasis

With the evolution of the eukaryotic cell, the way was opened to organisms that are more familiar to us than prokaryotes. Why I spent so much space on basic biochemistry and prokaryotes is easily explained: More than half of the total time of life on earth is dominated by prokaryotes. There is another reason for the focus on prokaryotic life: The new eukaryotic life explored the planet relatively slowly and for a long time was restricted to unicellular life-forms. In fact, what happened between the purported origin of eukaryotes 2 Ga ago and the appearance of the first animals? Fossil evidence dates the emergence of algae to about 1.2 Ga (Dyall et al. 2004a), in other estimates 1.5 Ga (Javaux et al. 2001), in more daring estimates even 1.9 Ga before the present. No clear evidence for tissue-organized life exists before the appearance of the first animals at about 0.6 Ga. *In previous sections we saw that time periods much smaller than 1 Ga were sufficient to lead from prebiotic levels of organization to the prokaryotic cell. What explains this stasis in the evolution of the early eukaryotic cells? What was the likely brake for the evolution of higher life-forms?*

The Story of O

Geologists gave firm statements with respect to the early atmosphere: Earth started out with no free oxygen. All oxygen was bound in rocks and water. You can then follow up minerals until about 2.4 or 2.2 Ga ago, none shows evidence that it was ever exposed to oxygen. However, at 2.4 Ga ago something changed, and this something became known as the great oxidation event. The most likely source of this event was cyanobacteria. We already discussed the controversy of their fossil appearance, but 2.7 Ga ago might not be too daring a hypothesis. This leaves, however, a gap of perhaps 300 million years from the appearance of oxygenic photosynthesis to measurable increases in atmospheric oxygen concentrations. What was slowing the process of oxygen accumulation in the atmosphere? Some geologists argued that cyanobacterial photosynthesis was in the beginning not oxygenic. Other geologists suspected that reducing volcanic gases reacted with oxygen such that it was constantly produced, but as quickly consumed. These processes ended sometime before 2.4 Ga ago, leading to an increase in atmospheric oxygen. Fittingly, with this change came biological reactions. At an estimated 1.9 Ga ago conspicuous spiral structures were interpreted as first eukaryotes, and they even got a species name: *Grypania spiralis*. You would now expect that the Proterozoic ocean would soon teem with eukaryotic life-forms. However, this was not the case. Nucleated algal cells were documented at about 1.4 Ga ago, but algal evolution made no leaps, it crept over nearly a billion years. Evolution got on the fast lane only with the second oxidation event, which occurred 0.7 or 0.6 Ga ago. In a later section we will encounter the Ediacaran fauna of this period and from that time point on, animal evolution was getting on gear, apparently fuelled by the new supply of oxygen. *Scientists have not yet found a good explanation for the second oxygen rise: Was it induced by lichens creeping on land, starting to weather the rocks and thus providing nutrients to the ocean? Or was the gut invented in animal evolution, perhaps still at the zooplankton level, which produced fecal pellets that could sink into the deep sea before it was decomposed by oxygen-consuming heterotrophs.* The essence is that primary productivity, which releases oxygen in photosynthesis, was buried in the ocean sediment before it could be annihilated by respiration, which consumes oxygen (Kerr 2005).

The Canfield Ocean

Something was odd with the ocean, but what? Geologists state that the preceding Archean ocean was anoxic and iron-rich. Our current ocean is oxygen-rich to its deepest bottom, but as we have seen, it is iron-poor. Where is the Proterozoic ocean with respect to these conditions? The classical argument for the oxygen enrichment of the deep oceans was the disappearance of the banded iron formation at about 1.8 Ga ago. Reduced iron, Fe^{2+}, arrives from hydrothermal sources, gets oxidized by oxygen in deep ocean water to Fe^{3+} and forms then in the presence of oxygen insoluble iron oxyhydroxide, which precipitates in global layers as BIF. This scenario fits with the oxygen increase in the atmosphere, known as the great oxidation event. According to this model, the end of the BIF period signals an oxic ocean. This should have spurred the evolution of the algae, but nothing like this was observed in the geological record. D. Canfield searched for an alternative scenario, which could explain this biological stasis (Canfield 1998). He postulated that atmospheric oxygen increases at 2.3–2.0 Ga were sufficient to lead to some weathering of surface rocks, oxygen would react with sulfide minerals and export them as oxidized sulfate into the ocean. However, in his scenario the produced oxygen quantities were not high enough to create an aerobic deep ocean. The increased inflow of land-derived sulfate into the ocean would feed sulfate-reducing bacteria in the ocean. As an end product of their metabolism these bacteria will produce sulfide. As the solubility of iron sulfides is also low, it would also precipitate in the ocean. The result would be an anoxic, sulfidic, and iron-poor ocean over most of the Proterozoic era. Only in the Neoproterozoic era (starting at 1 Ga ago), the oxygen concentrations became so high that they could sweep the deep ocean of sulfide.

The Malnourished Ocean

The consequence for the chemical composition of the Proterozoic ocean would be dire: Fe and Mo (molybdenum) were removed from the ocean. In the Archean ocean, the Fe concentration might have been as high as $50\,\mu M$; this concentration probably decreased 1,000-fold in the Proterozoic ocean. In the Black Sea, currently our closest analog of a sulfidic ocean, the iron concentration in the oxygenated surface water is $40\,nM$ and this falls to $3\,nM$ below the chemocline, when reaching the sulfidic layers. In the malnourished ocean hypothesis (Anbar and Knoll 2002), life was not well prepared for this change. Fe and Mo are critical elements in the enzyme nitrogenase of nitrogen-fixing prokaryotes like cyanobacteria. One can therefore suppose a decline in the biological productivity of the ocean. Algal cells were hit especially hard as they were not able to fix nitrogen from N_2 and thus depended on cyanobacteria. Under nitrogen-limiting conditions, algae are easily outcompeted by cyanobacteria. Even non-nitrogen-fixing cyanobacteria like *Synechococcus* are more flexible than algae, thanks to a sophisticated acclimation to macronutrient deficiency (for a recent review: Schwarz and Forchhammer 2005). *The flexibility is still documented in biochemistry textbooks: We know of bacterial nitrogenases that use pure Fe cluster or V (vanadium) and Fe cluster instead of the Fe_7MoS_9 catalysis cluster, probably to economize Mo. Algae responded as well: In the next section you will hear of the "red" and "green" lines of algal plastids. The green lineage prefers Fe, Zn, and Cu as cofactors, whereas the red lineage prefers Mn, Co, and Cd. One might argue that algae desperately tried different combinations of redox-active metals as cofactors for photosynthesis, for nitrate transport during N assimilation, and for the electron transport chain.* This period of Proterozoic marasm probably ended with the Grenvill orogeny at 1.2 Ga ago when mountain-building processes led to an increased weathering and thus influx of the missed metals into the ocean. This led to an increased productivity of the oceans and a diversification of algae in the Neoproterozoic Ocean.

Before resuming the thread of action with algae, we must explore the feeding modes of nonphotosynthetic organisms at the root of the Eukarya tree. Many of them lack mitochondria, and biologists expected from them insights into eukaryotic life before the acquisition of mitochondria.

Vita Minima: The Reductionist Lifestyle of Protist Parasites

Protista

The eukaryotic organisms at the root of the Eukarya tree are collectively called Protista. The definition is essentially negative, covering unicellular eukaryotes. If they occur as colonies of cells, they lack true tissues as seen in plants, animals, and fungi and are smaller in size than $5\,\mu m$. Even a casual look at a eukaryotic 18S rRNA tree will reveal that the long branches of the protista cover a diversity of life-forms that far exceed what animals and plants will develop later in evolution. These protista are also much further down to the Eukarya tree, suggesting their greater antiquity over animals and plants. In this group, you find so many diverse organisms that they defy a common denominator. Many parasites belong to these clades, which makes it difficult to differentiate genuinely primitive characteristics from secondary reductions due to the parasitic lifestyle. The genomes from a few protists of medical importance have been sequenced. I will discuss two of these to explore what can be gleaned from the sequences with respect to nutrition.

Entamoeba: Life Cycle

The first is *Entamoeba histolytica*, a pseudopod-forming rhizopod (Figure 4.18) and the causative agent of amoebiasis. Its life cycle is simple. Ingestion of a cyst contained in drinking water or vegetables assures a safe passage of the parasite through the stomach. In the small bowel, the parasite leaves the cysts and the amoebal trophozoit feeds on bacteria in the colon. With excreted glycosidases and a membrane-bound neuraminidase, it can also use the intestinal mucins as a food source. However, *E. histolytica* can likewise attack the epithelium of the colon. It kills the gut cells only upon direct contact by inducing an increase in the intracellular Ca^{2+} level in the target cell. It can also invade the colonic epithelium directly, where they form ulcers. The invading parasite frequently contains ingested red blood cells. Following the portal vein, the parasite reaches

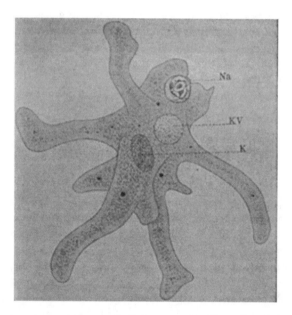

FIGURE 4.18. *Amoeba proteus*, a unicellular protozoan from the phylum Rhizopoda. One can differentiate the clear ectoplasm from the granular endoplasm, the pseudopodes, and inside the cell, the nucleus (K), the food vacuole (Na), and the water expulsion vesicle (KV).

the liver, where it causes an abscess. When the intestinal trophozoit reaches the terminal part of the colon, it gets encysted and once defecated, the cyst remains infectious for weeks or even months in an appropriate moist environment. This simple but efficient parasitic lifestyle allows the parasite to infect an estimated 10% of the world population, causing 50 million cases of invasive diseases per year.

Metabolism

The genome sequence was anticipated with great interest as it allowed insights into metabolic adaptations shared with other amitochondrial protist pathogens, namely *Giardia lamblia* and *T. vaginalis*. It showed a 23.7-Mb-large genome encoding nearly 10,000 genes (Loftus et al. 2005). This is only slightly larger than the largest currently sequenced prokaryotic genome, namely *Streptomyces coelicor*, which encodes somewhat more than 7,000 genes in a 9-Mb genome. The metabolism of *E. histolytica* is shaped by secondary gene loss and lateral gene transfer: It is an obligate fermenting organism using bacteria-like metabolic enzymes and bacteria-like glucose transporters. Glucose is the main energy source, which is degraded by glycolysis. Also the pentose phosphate pathway contains many bacteria-like enzymes. The TCA cycle and the respiratory chain are lacking. However, the genome data support the presence of an atrophied mitochondrion. The biosynthetic capacities are very restricted and from the 20 amino acids only those of serine and cysteine are maintained. The high cysteine levels functionally replace the lack of glutathione as buffer against oxidative stress. The genome contains about 100 prokaryote-like genes, half of them are involved in metabolism to increase the range of substrates available for energy generation. Notably, the major donor for these genes seems to come from the Bacteroidetes group of bacteria, which includes *Bacteroides*, the predominant commensal of the human gut microbiota. As a phagocytic and pathogenic resident of the human gut, *E. hystolytica* has a number of virulence genes that range from numerous lectins for adhesion to the host cells to pore-forming peptides to lyse the cell and many secreted hydrolytic enzymes.

Encephalitozoon: Life Cycle

The second parasitic protist, which I will discuss, has the not very inspiring name *Encephalitozoon cunuculi*. It was the first of the pathogenic parasites that was sequenced—the main reason was the small size of its genome. It belongs to the Microsporidia clade, a unique group of obligatory intracellular, spore-forming protozoa. Historically, it was known to cause serious economical problems: it infected silkworms, honeybee, and commercial fish. Medically, it rose to prominence as a common secondary infection of HIV patients. In HIV patients, microsporidia infections manifest as ocular disease. Also extensive tracheal, renal, and intestinal infestation were reported. The transmission is via spores of 2-μm diameter that resemble superficially a nematocyst of jellyfish. Commonly, the spore is inhaled, the change of environment leads to an eversion of the coiled tube, which fills the spore. This combination of a harpoon and hypodermic needle punctures the host cell, and the sporoplasm of the parasite is forced through the tube into the host cell. In the cell, the sporoplasm develops in proliferative stages, leading to fission processes called merogony. In the third phase of its life cycle, the meront is thickening its membrane and the cell transforms into a mature spore.

Genomics and Metabolism

Microsporidia are also of biological interest (Keeling 2001). Their simplicity and lack of mitochondria gave rise to the idea that microsporidia might be a primitive eukaryotic lineage that evolved before the acquisition of mitochondria (Cavalier-Smith 1987). Indeed, early tree building with the ribosomal RNA sequences suggested that microsporidia are extremely ancient eukaryotes (Vossbrinck et al. 1987). The genome sequence was therefore of great interest for biologists, and the results were really astonishing (Katinka et al. 2001). It presented a eukaryotic genome that was so small that its 11 linear chromosomes ranged in size only from 200 to 300 kb, the size of greater bacterial plasmids. Cumulatively, this remarkably reduced genome tallies only 2.9 Mb—well near the median size of bacterial genomes. The genome was compacted: 90% of the DNA is used for coding the 2,000 genes, again resembling the gene density of a prokaryote. When compared to the yeast genome, even the average protein size was significantly reduced in *Encephalitozoon*. Many genes deal with the transmission of genetic information and little coding capacity is left for metabolism. A complete glycolytic pathway was detected, which was fed by the disaccharide trehalose, the major sugar reserve in microsporidia. ATP synthesis is only by substrate-level phosphorylation. TCA cycle, respiratory chain, and ATP synthase are all lacking. However, when actively growing inside the host cell, it actively recruits host mitochondria near their plasma membrane (the parasite is, like in plasmodia, still surrounded by a membrane of the parasitophorous vacuole of the host). Four carriers then import mitochondrial ATP into the intracellular parasite. Phylogenetic analysis of the sequences supported the hypothesis that microsporidia are in fact atypical fungi that lost mitochondria during evolution. The genome sequence revealed still 22 genes of putative mitochondrial origin, including a Fe–S cluster assembly machinery, a hallmark of mitosome function. From the genome data, the authors reconstructed a metabolic function for this mitosome. This scheme would allow the use of pyruvate and NADH produced in glycolysis, which would otherwise accumulate in the parasite.

Primary Endosymbiosis: The Origin of Chloroplasts

The Second Takeover

Some of the new eukaryotic cells are interested in cyanobacteria. A newly created eukaryotic cell would find an ideal partner in a cyanobacterium. A captured bacterial phototroph would deliver homemade oxygen as an electron acceptor for the eukaryotic mitochondrion. In addition, it would deliver ample reducing equivalents and ATP for any biosynthetic activity. CO_2 set free during food oxidation could be refixed by the Calvin cycle in organic carbon. This looks like an ideal *partnership where the waste of one bacterial cell becomes the food for the other bacterial cell and the bystander eukaryotic host would reap the metabolic benefits of both systems. This is too good to not be used. And it was used, not by all of the new eukaryotic cells, probably because this new endosymbiosis came with one major constraint—to power this self-sufficient cellular economy, you need light as a primary energy source. Photosynthetic symbiosis restricts life to the sunlit zones of the biosphere. From the distribution of mitochondria and chloroplasts in eukaryotes, it is deduced that the arrival of mitochondria predates the advent of plastids.* To provide a minimal time horizon for these events, fossils of red algae-like organisms were dated to 1.2 Ga, some even to 1.5 Ga ago.

Distinct Steps in the Takeover

Chloroplasts are found in algae-like eukaryotes and land plants representing five phylogenetic distinct groups: Euglenozoa, Alveolates (dinoflagellates), Stramenopiles (brown algae, diatoms), red algae and green algae, plus the land plants derived from them (Figures 4.19 and 4.20). From the scattered phylogenetic distribution of chloroplasts, one might suspect independent takeover events of cyanobacteria in these lineages. As we will see, this second major endosymbiosis does not go back to a single event in the history of life but can be differentiated into primary, secondary, and even tertiary acquisition steps.

Even early phases of the takeover process were documented as the startling observation of an endosymbiotic cyanobacterium that inhabits the biflagellate protist *Cyanophora paradoxa*. Here it apparently acts as a chloroplast, but this cyanelle, as it is called, is still surrounded by a peptidoglycan layer and thus has a striking resemblance to a bacterium. However, it has lost the LPS and the outer membrane, indicating a bacterial endosymbiont on its way to a chloroplast.

Red and Green Plastids

There is growing evidence that all current chloroplasts are derived from a single cyanobacterial ancestor. This is not obvious at first glance. If you look at photosynthetic protists, you have a wide and scattered distribution of chloroplasts between groups of organisms that are phylogenetically not closely related. The classification of photosynthetic organisms was based on chemical details of their accessory pigments and the structure of the light harvesting system. Even if you limit the comparison to chloroplasts enveloped by two membranes, the result of the primary endosymbiosis event, you already have two major lines: On one side the "green" plastids in green algae, which later gave rise to land plants. These Chlorophyta contain chlorophyll a and b, the core antenna complex and the light harvesting system I. On the other side are the red algae (Rhodophyta) containing "red" plastids, characterized by the possession of chlorophyll a-containing antenna complex and phycobilisomes containing phycobilins as pigment associated with the core complex. These green and red lineages are also differentiated by other nutritional specializations. Not enough with this diversity, there are photosynthetic protists sporting chloroplasts that

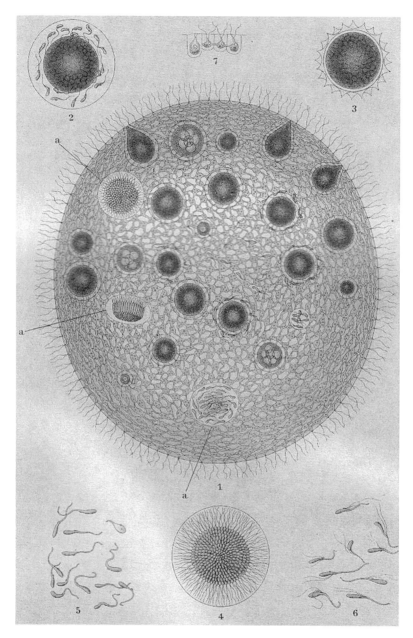

FIGURE 4.19. The green algae *Volvox* marks the transition from a colony of identical cells to a multicellular individual. We will also see that *Vibrio cholerae* feeds on the mucin layer of *Volvox* when persisting in environmental water.

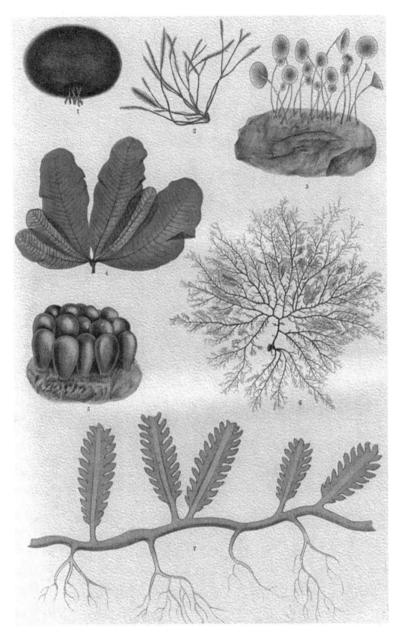

FIGURE 4.20. Green and red marine algae. Green algae/Chlorophyceae, order Siphonales:
(*1*) *Codium*, (*2*) *Bryopsis*, (*3*) *Acetabularia*, (*5*) *Valonia*, (*7*) *Caulerpa*; Rhodophyceae:
(*6*) *Plocamium*; (*4*) *Hydrolapathum*.

are surrounded by three and even four membranes. In the latter class come, for example, diatoms containing fucoxanthin–chlorophyll a and c as pigments (*Chromophyta*; Falkowski et al. 2004). At first glance, one might anticipate a polyphyletic origin of chloroplasts.

A Monophyletic Origin?

However, a number of data speak in favor of a monophyletic origin of modern plastids. One piece of evidence was immunological cross-reactivity between the proteins carrying the different pigment systems. Antibodies raised against a barley PSI complex cross-reacted not only with all chlorophyll a/b-binding proteins from the green lineage, but also with LHC I proteins from several red algae and even a diatom. In contrast, this antiserum did not react with pigment-associated proteins from prokaryotes, namely cyanobacteria (which have phycobilin-containing phycobilisomes associated with the core complex), or with prochlorophyta (with chlorophyll b-binding proteins associated with a core lacking phycobilins; Wolfe 1994).

Subsequently, these immunological data were corroborated and extended by phylogenetic tree analysis performed with a concatenated set of more than 40 proteins shared between a wide range of photosynthetic organisms. At the basis of the tree the researchers placed the above-mentioned *C. paradoxa*; this "cyanobacteria carrying paradox" is a eukaryotic alga (Glaucocystophyta). The next branch is occupied by rhodophytes and diatoms. This relationship was later confirmed when the researchers learned more about the process of secondary endosymbiosis. Then came the branching of the green lineage with *Euglena* as an early branch and then green algae plus all land plants.

A Unifying Model

The chemical taxonomy based on chlorophyll relies on minor changes in the pigments. For example, chlorophyll a and b differ by a methyl or a formyl group as a substituent at pyrrole ring II. A single enzyme is needed to transform chlorophyll a into b, namely chlorophyll a oxidase. The phylogenetic tree analysis of this enzyme showed that the ability to synthesize chlorophyll b did not arise independently several times. The extant prokaryotic prochlorophyta are thus directly linked to chlorophyta (Tomitani et al. 1999). The Japanese researchers who elaborated these data also offered a model to reconcile the phycobilin–chlorophyll b dichotomy that separates what should belong together both in the prokaryotic and in the eukaryotic realm. The solution is the hypothesis of an ancestor that contained both phycobilin-associated antenna and chlorophyll b-associated antenna. The extant bacterium *Prochlorococcus marinus* is currently the only known organism combining chlorophyll b with a type of phyco-erythrin (Hess et al. 1996) and could thus serve as the closest proxy of this ancestor. The Japanese scientists further proposed that this ancestor was swallowed by a nonphotosynthetic eukaryote. In this ancestral chloroplast, the prokaryotic chlorophyll carrier protein, belonging to the IsiA family, was exchanged by an antenna protein, belonging to the LHC superfamily (La Roche et al. 1996). Now the stage was set, and the somewhat bewildering diversity observed in current organisms is a story of selective losses creating the sister groups Chlorophyta/Rhodophyta and more distantly Glaucophyta (Moreira et al. 2000) in primary endosymbionts and subsequent forms of chloroplasts by secondary endosymbiosis. A similar story of selective loss separated the cyanobacteria from the prochlorophyta lineages in the prokaryotes (Tomitani et al. 1999).

The Dynamic Plastid Genome

Gene loss is actually a common denominator of plastid genome evolution. Plastids of higher plants contain only about 120 genes, whereas their prokaryotic ancestors might have contained more than 3,000 genes. Chloroplast genomes in basal groups like the flagellate *Mesostigma* (at the base of the green algae lineage) contain a marginal higher number of 135 genes (Lemieux et al. 2000). The chloroplast from the red alga *Porphyra* figures with a genome size of 183 kb, and 210 genes at the upper end. Most chloroplast genomes encode part of the prokaryotic-like transcription and translation apparatus and part of the photosynthesis proteins. *Porphyra* chloroplasts still contain about 130 genes with metabolic function, more than twice the number of such genes in chloroplasts of land plants. These genes comprise a number of enzymes of intermediary metabolism (nitrogen assimilation, biosynthesis of amino acids, fatty acids, and pigments; Palmer 1993). Proteome analysis led to the approximation that plastid function depends on some 3,000 genes. Where are they? They are now residing in the nuclear chromosome. And there is some good evidence for it. Recall Rubisco, its L-subunit is encoded on the chloroplast genome whereas its S-subunit is now found in the nuclear genome. However, if you screen the tobacco nuclear genome, you come across large tracts of plastid DNA. If you screen for a given gene, for example, that encoding the large subunit from Rubisco, you identify 15 fragments of different sizes, the largest being 15 kb long. Notably, the gene transfer is a one-way process: No nuclear gene fragments were detected in the chloroplast genome. However, this gene process is not limited to plastids—the other organelle, the mitochondrion also experienced a massive transfer of mitochondrial DNA to the nuclear genome. The model plant *Arabidopsis* contains, for example, 13 small inserts and 1 large insert of more than 600 kb of putative exmitochondrial DNA in its nuclear genome (Maliga 2003). This organelle-to-nucleus gene transfer was not a distant event at the origin of the endosymbiotic relationship but is a still ongoing process. Australian scientists introduced an antibiotic resistance gene into the tobacco chloroplast and screened 250,000 progeny seedlings for the expression of the resistance gene and, indeed, they identified 16 plants with the selected phenotype (Huang et al. 2003). Actually they used a few genetic tricks to obtain the result: They introduced a nuclear-type intron into the gene to prevent its expression in the chloroplast, and they placed the selection marker downstream of a strong, constitutive plant viral, i.e., the eukaryotic-type promoter to allow its expression in the nucleus. A control resistance gene without a eukaryotic promoter was not expressed. These constructions were necessary to visualize the transfer, but the transfer efficiency of 1:16,000 pollen (which do not contain chloroplasts) indicates that the transfer efficiency is in reality much higher, but not visualized as the transferred gene fragment remains silent. This observation highlights the difficulty in a functional transfer as the transferred gene changes from a prokaryotic to a eukaryotic genetic apparatus with respect to transcription and translation. Furthermore, the nuclear-encoded gene product must then still travel back from the cytoplasmic site of translation to the chloroplast. The transferred genes thus must also acquire a tag sequence that redirects the protein back to the site of its need—the chloroplast. Functional transfers will thus be very rare events, transfers of organelle DNA to the nucleus in contrast seem to be relatively frequent. This process was investigated in yeast with mitochondria DNA. Interestingly, all transferred DNA turned up in nontranscribed regions, half inserted in the vicinity of retrotransposon long terminal repeats or tRNA. The transferred DNA originated from all regions of the mitochondrial genome (Ricchetti et al. 1999).

Reasons for Gene Transfers?

Why should plastids give their genes to the nucleus? The loss of critical genes means loosing the genetic independence because escape from the symbiotic relationship becomes impossible. It might be too anthropomorphic to see the nucleus as a slave master or as a genetic black hole sucking in all genes from smaller satellite systems that come into its cytoplasm. Also, intracellular bacterial pathogens like Chlamydia and Mycoplasma show spectacular gene losses, which leads to genome reductions approaching the 500-kb genome size limit. As the intracellular bacterium gets food from the surrounding cytoplasm of its host, it might be evolutionarily advantageous to reduce the genome size because many metabolic functions are now provided by the host. In some bacteria you can actually see the ongoing gene inactivation process (e.g., Mycobacterium leprae). Perhaps the nucleus is not to be blamed for this process, and the organelles might only "park" their genes in the nuclear chromosome. There is a powerful population genetics argument for this scenario known as Muller's ratchet—the rapid accumulation of deleterious mutations in asexual populations. Transfer of the chloroplast genes to the nucleus submits them to frequent recombination, which can purge the genetic load of deleterious mutations. Yet, then one might ask why the process of gene transfer did not go to its logical end—the complete transfer of the entire plastid genome to the nucleus. In the end a large proportion of the extant chloroplast genomes are now dedicated only to maintain the prokaryotic translation apparatus, part of the photosynthetic apparatus (Rubisco and complexes of the thylakoid membrane system) and NADH dehydrogenases. One might argue that these functions cannot be transferred because they would disturb the differently organized nuclear processes in translation and division and must remain with the organelle.

Predator Protozoa...

Ciliates

Protists are not only characterized by a wide genetic diversity, they also show a substantial morphological variability. Even within a single group of protists like the Ciliates you find many forms and feeding habits. *Paramecium* is a typical ciliate (the name derives from the many cilia organized in rows on the surface of this unicellular organism, which are the locomotory organelles). As a particularity, it shows a strange division of labor: A polyploid macronucleus containing many genome complements and fulfilling metabolic functions and the general operation of the cell and then a diploid micronucleus, which has exclusively a reproductive function. Paramecia, even when they consist only of a single cell, have a complicated inner structure. When one stains their food, e.g. yeast cells, with a pH indicator dye, one can follow the fate of the prey during digestion. Many ciliates have a specialized cell surface area that functions as cell mouth (cytostome). This organelle leads into a gullet, which transfers the prey into the food vacuole. During the digestion process, the vacuole cycles through the cytoplasm. The food vacuole first becomes acidic and then neutral and then alkaline. The digested food material is absorbed, the vacuole shrinks, and the residual nondigestible material is then extruded. This sophisticated and stereotypical organization of cellular digestion ("cyclosis") underlines what care even single cells attribute to the capture and digestion of prey.

Different Feeding Modes

The unicellular ciliates have already developed many ingenious eating forms. Take *Didinium*, which attacks fellow ciliates like the above-mentioned *Paramecium*. The prey is twice as large as the predator, therefore the cytostome had to adapt for taking greater bites. It is actually everted as a type of projection that captures the prey and then it is inverted back into the cell, slowly sucking in the paramecium. In *Nassulopsis* the cytostome is reinforced with a complex meshwork of microtubules that allows this ciliate to suck up filamentous cyanobacteria like spaghettis. Then you have ciliates like *Stentor*, which resembles a funnel where the upper rim is decorated with a circular arrangement of compound cilia (cirri, membranelles) that create water currents and the organisms filter small food particles from the suspension into the funnel orifice (peristomium), leading into a buccal cavity and then into the cytostome. Still another feeding type of ciliates is seen in the suctorians. *Acineta* is such a "sucker," the body of this unicellular organism is calyx-like and decorated at the top end with bundles of feeding tentacles ending in a terminal swelling. When they make contact with a potential prey, the tips of these tentacles discharge haptocysts into the body of the food organisms. Enzymes released from the haptocysts paralyze the prey and start its digestion. Then a ring of microtubules forms a tube within the tentacle and the cellular content of the captured prey is sucked into a food vacuole of the suctorian. Other ciliates developed toxicysts, mucocysts, and trichocysts that discharge aggressive chemicals or protective coatings. Some show nail-like structure, used likewise for attack and defense. Not just that, ciliates also comprise in their feeding range parasites that live in the cytoplasm of other ciliates, on the gills of freshwater fishes, in the intestine of pigs (*Balantidium coli*). Ciliates adhere as ectoparasites on crustaceans and belong as symbionts to the rumen population of cattle. Here I again stress that all these different eating forms are realized by organisms that consist of a single, admittedly relatively large cell. We do not know whether protists were so diversified from the beginning of their evolution, but some ciliates that secrete an external skeleton ("loricae") can be traced in the fossil record until the Ordovician (500 My ago).

... And How Bacteria Get off the Hook

Arms Race

Here we must look back. In one of the previous sections, I told you that bacterial mortality has two major causes: phage lysis and protozoa grazing. Bacteria are thus an important food source for heterotrophic protozoa. In view of what I told you about the rules of the quest for food game, it would be surprising if bacteria did not respond to this grazing pressure. On a theoretical basis, predators are potent agents of natural selection in biological communities. The effect can be seen when putting food bacteria and grazing protozoa in a vessel. Bacteria will grow to a maximum, followed by a collapse of the bacterial population. This collapse is caused by the outgrowth of the grazers, which feed on the bacteria thereby decreasing their population size. This will likewise be followed by a collapse of the grazer population due to bacterial food getting exhausted. This process is accelerated by the appearance of grazing-resistant bacteria, which have developed some antipredator traits and resume growth by replacing the initial bacterial population (Matz and Kjelleberg 2005). This is not a particularity of this system. Very similar data were already reported by the founding fathers of molecular biology, when Salvatore Luria and Max Delbrück (Luria and Delbrück 1943) "played" with *E. coli* and its phages in a vessel. Repeated cycling of both partners was observed, suggesting an arms race between predator and prey. In the bacteria–protozoa war, bacteria developed a number of tricks, which can be divided into predigestion and postdigestion adaptations.

Oversize

One possible strategy is to develop long filaments: Small protozoa prefer prey in the size range of 1–3 μm. Longer prey is avoided when smaller prey is present because its digestion takes longer time. This selection pressure can explain the appearance of blooms of filamentous bacteria, which represent half of all bacterioplankton in freshwater environments under grazing pressure. The bloom is only short-lived as metazoan predators like *Daphnia* take over because this filter feeder has a larger size limitation (Pernthaler et al. 2004).

Masking

Another strategy is surface masking. It was long known that amoeba need to recognize surface structures on the bacterial prey to start the phagocytosis process. J. Lawrence and colleagues recently came up with a thoughtful hypothesis backed by experiments. *Salmonella enterica* covers about 70 O types, chemically distinct forms of LPS, which decorate the surface of the bacterium. The classical interpretation is that this is a virulence factor important for escape from immune surveillance by the vertebrate host. However, *Salmonella* is not much exposed to the immune system. It is either in the gut or inside a cell. Their experiments showed selective feeding of gut amoeba with respect to *Salmonella* belonging to different O serotypes. The authors argue that the "ecological reason" for variability of the *Salmonella* O serotypes is likely surface masking to avoid protozoa feeding (Wildschutte et al. 2004).

Motility

Still another avoidance strategy is speed development in motile bacteria. A common bacteriovorous protozoon is the citiate *Paramecium* (Figure 4.21), which itself frequently becomes the prey of other ciliates (Figure 4.22) as a lively illustration of the eat and be eaten. High-speed versions of bacteria bang more frequently on bacteriovorous protozoa, but these bacteria are not caught in the feeding current created by the flagellate protozoon and thus escape the drag into the food vacuole. Speed is thus not only a strategy to reach new food patches more quickly, but also an antipredator response. In fact, in the presence of the predator, the bacterial population experienced a selection for higher speed and smaller bacteria size (Matz and Jurgens 2005).

Fight from Within

Once inside the food vacuole, bacteria keen to escape from predation have to opt for other strategies. One option is resistance to digestion. For example, a few minutes after ingestion of the cyanobacterium *Synechococcus*, flagellates reject the prey. They are apparently unpalatable probably because of the proteinaceous S-layer, with which the cyanobacterium surrounds itself. Other bacteria place more trust on active measures against the protozoan predator. Some bacteria elaborate toxins that have potent antiprotozoal activity. One of these compounds is the pigment violacein produced by *Chromobacterium*. In feeding experiments, the predator flagellate *Ochromonas* does not distinguish between pigmented and nonpigmented bacteria. However, after digestion of as few as two pigmented bacteria, the predator's fate is sealed. After 30 min, the predator lyses and releases its cell content. The ingested bacterium is dead (the toxin is not released from the intact cell), but the cytoplasm of the lysed protists now feeds the bacterial population (Matz et al. 2004).

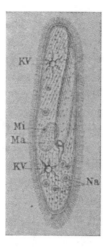

FIGURE 4.21. *Paramaecium caudatum* is a ciliate protist; the cell is covered by numerous cilia, the locomotory organelles. Food is taken up by a region of the cell differentiated into a cell mouth, the sac-like structure, from which food vacuoles (Na) enter the cell body. Inside the cell you can see water expulsion vesicle (KV), the mironucleus (Mi), and macronucleus (Ma).

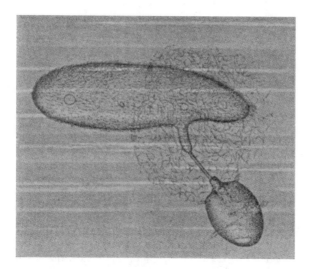

FIGURE 4.22. Fight between ciliates, phylum Ciliophora of the protozoa. *Didinium nasutum* (*bottom*) attacks a paramecium with its feeding tentacle. The latter defends by retaliating with trichocysts.

Origin of Bacterial Pathogenicity?

Actually, the number of bacterial toxins that act on eukaryotic competitors and hosts speaks in favor of an early coevolution. Actually, one might even suspect that measures deployed against protozoan predators were later used by bacterial pathogens that target animal and human hosts. The common denominator would be the formal similarity between phagocytes and amoeba. In fact, in the context of human bacterial pathogens, we should mention that many pathogenic bacteria survive within the vacuole of phagocytes. These vacuoles are no longer food vacuoles, they have become means of pathogen destruction. Take the example of Listeria monocytogenes, which survives in phagocytes. L. monocytogenes is probably erring only in humans; it is so closely related to L. innocua, which lives in soil, that the conclusion might be permitted that L. monocytogenes is a soil bacterium that had somehow acquired a few virulence islands important for its survival in food and the human body. In the soil, survival within the food vacuole of amoeba might be a useful survival strategy. This is not a rare strategy: *E. coli* survives in food vacuoles from amoeba in drinking water. Ingested bacteria were resistant to chlorination, whereas free-living bacteria were killed. Protozoan-digested pathogenic bacteria are thus occasionally associated with residual infectivity in chlorinated drinking water (King et al. 1988). *Taken together the data suggest that a good deal of bacterial pathogenicity in humans might have originated as antipredator measures of bacteria against protozoa.*

Algal Slaves

Mesodinium

You might suspect that I have searched entertaining and refreshing titles to compensate for all the complicated biochemistry of the first half of the book, but the link to the section heading comes from a scientific paper, which I want to present next. Unusually, the title of this *Nature* paper is "*Algae are Robbed of their Organelles by Mesodinium*" (Gustafson et al. 2000). *Mesodinium* is a common photosynthetic marine ciliate. This protist has lost its cytostome and lives with an algal endosymbiont. This is not an exotic ciliate; many harbor endosymbionts and are sometimes quite green. During the summer months, ciliates with plastids can become functional phytoplankters and dominate the ciliate fauna. *Mesodinium* is a vigorous ciliate showing extremely high rates of primary production in fjords. Its cellular metabolism is apparently heavily supported by high photosynthetic rates. Pale, plastid-lacking *Mesodinium* do not show sustained growth. However, happiness comes back when you offer algal prey. The ciliate ingests these free-living algae, but then a peculiar effect is observed. Algal nuclei first increase in number, but are not retained. In contrast, chloroplasts are maintained and not digested. Apparently, *Mesodinium* steals the chloroplasts, which become temporary working slaves in the ciliate, whereas the algal cytoplasm, nucleus, and mitochondria are treated as fast food.

The chloroplasts synthesize sugars for the ciliate. This happiness is, however, not enduring: The algal chloroplasts can maintain chlorophyll synthesis and photosynthesis for several weeks, but then the chloroplast fails because its algal nuclear command center was lost to digestion. The plastids start to age and the ciliate must replace them by new algal preys. Yet, despite this complication, the photosynthetic support to *Mesodinium* is so important that this ciliate can form marine blooms known as nontoxic red tides.

Secondary Symbiosis

Stealing of plastids is a widely used practice in the quest for food. Actually, a number of protists found it cumbersome to renew the aging chloroplasts on a regular basis. They did something, which is called secondary symbiosis. A heterotrophic protist engulfs not a cyanobacterium, as in primary endosymbiosis, but a eukaryotic alga. You play here the game of a simple Russian doll. If you destroy the nucleus of the alga as does *Mesodinium*, you cannot maintain the chloroplast as long-term companion and synthetic powerhouse to the cell. Therefore you leave the entire algal cell in your food vacuole, but with the directive not to digest your slave. This leads now to a peculiar cytological structure. The chloroplast from the secondary symbiosis event is now enveloped by four membranes: Two are the heritage of the cyanobacterium's inner and outer membranes, the third is the plasma membrane of the swallowed algal slave, and the fourth is the membrane of the food vacuole of the secondary host. Many prominent organisms of the phytoplankton are the result of this secondary symbiosis: euglenoids, cryptophytes, coccolithophores, diatoms, and dinoflagellates. The latter are an especially intriguing group, some of them derive their photosynthetic slave from the green algal lineage, others from the red algal lineage, and still other dinoflagellates are even the result of a tertiary endosymbiosis. In the latter case the cryptophytes, a secondary symbiosis product, now live within another dinoflagellate host cell.

Guillardia

Despite the wide distribution and the ecological success of these secondary symbionts, the genetic organization of these cells is strange. Take the cryptomonad *Guillardia theta*, which was investigated by sequencing approaches (Douglas et al. 2001). It contains 121- and 48-kb prokaryotic-type genomes in its chloroplast and mitochondrion, respectively. Then it sports a 551-kb genome in the so-called nucleomorph, which is a strange cytological structure apposed to the chloroplast in the food vacuole that wraps the chloroplast. The fourth and outermost membrane of the chloroplast is actually continuous with the membrane of the cryptomonad nucleus, which musters a not yet sequenced sizable genome of 350 Mb. But what is the nucleomorph? *In fact, this structure is nothing other than the rest of the nucleus of the algal slave. We are sometimes told that engineers look at solutions, which Nature found for technical challenges, to get new ideas. Indeed, the tensile strength of plant fibers or spider webs might be impressive to such an engineer. However, Nature's solutions are not designer made. You see this clearly with cryptomonads. As they are a mixture of four different genetic systems—two prokaryotes and two eukaryotes, each with its own cytoplasm—cryptomonads need to maintain four different protein translation systems. The nucleomorph has undergone spectacular genome shrinkage as it shows only the genome size of an intracellular prokaryotic organism and is now a genetic shadow of its algal past.* The shrinkage was accompanied by a compaction of the genome, the gene density is extremely high for an ex-eukaryote: 1 gene per 1 kb, like in prokaryotes. The function of half of the 460 genes is totally unknown. Nearly all genes for metabolic functions were lost. However, when studying *Mesodinium* we realized that chloroplasts could not be maintained because the chloroplast genes transferred to the algal nucleus were lacking. Consequently, you find that the nucleomorph of cryptomonads still maintained 30 genes for chloroplast-located proteins. To allow expression of these proteins, the nucleomorph retains 100 genetic-housekeeping genes. Less than10% of the genes encode functions that are useful to the rest of the cell, like those linked to starch synthesis in the periplastid space. However, most of the identified genes are simply needed for the self-perpetuation of the nucleomorph and its periplastid ribosomes.

Cyanidioschyzon

Yet, there is no absolute necessity for maintaining a nucleomorph. Apparently, you can transfer the genes needed for chloroplast function to the nucleus of the secondary host. Actually, this solution was followed by practically all other secondary symbionts except in cryptomonads. The recent sequencing of the 16-Mb genome of the red alga *Cyanidioschyzon* (Matsuzaki et al. 2004), a primary symbiont, and of the 34-Mb genome from the diatom *Thalassiosira* (Armbrust et al. 2004), a secondary symbiont, was quite revealing for the symbiont hypotheses in photosynthetic phytoplankton. *Cyanidioschyzon* is a eukaryotic alga with the size of a bacterium ($2\,\mu m$). It inhabits acidic, sulfate-rich, moderately hot springs. It is actually not only ultra small, but—necessarily—also the ultimate minimalist. It shows a nucleus, a single plastid (150 kb), mitochondrion (32 kb genome), microbody, Golgi apparatus, and endoplasmic reticulum. It lacks a cell wall and contains only a difficult-to-detect cytoplasm. Most of the cell volume is taken by the plastid and when the plastid divides by the fission way of its prokaryotic ancestors, the alga strangely resembles a molecular model of the water atom. The chimeric nature of the genetic heritage of this cell is nicely revealed during plastid division: a bacterial FtsZ ring, a plastid PD ring, and a eukaryotic mechanochemical dynamin cooperate in this process. Cooperation is also the motto for the phycobilisome components of this red alga, its components are mostly encoded on the plastid genome, but many were actually also transferred to the nuclear genome. Also the enzymes of the Calvin cycle reveal a mosaic origin, Rubisco is the product of HGT, whereas the other enzymes of this cycle are essentially identical in *Cyanidioschyzon* and *Arabidospsis*, the model higher plant. This is strong genomic evidence for a single event of primary plastid endosymbiosis. Green plants and red algae derive their plastids from the same fusion event of a cyanobacterial-like ancestor and a eukaryotic host.

Thalassiosira

The genome of the diatom also provides evidence for its chimeric origin marked by the fusion of different phylogenetic lineages. A total of 11,000 genes are predicted for *Thalassiosira*. Almost half of the diatom proteins have alignment scores equidistant to the green plant *Arabidospsis*, the red alga *Cyanidioschyzon*, and the mouse, underscoring the evolutionarily ancient divergence of Plantae (red algae, green algae, and plants), Opistokonta (animals and fungi) and the unknown secondary host that gave rise to the diatom lineage. This diatom showed about 800 exclusively animal-like genes, as much as plant-like genes, but much less genes resembling exclusively red algae genes (200). Interestingly, a number of genes from the cryptophyte *Guillardia* nucleomorph were detected in the diatom nucleus, thus demonstrating the successive stages in gene transfer during the multiple endosymbiosis events. In the next section, we will investigate diatoms in greater detail. *This will not only illustrate the nutritional way of life of the major carbon-fixing organisms of the ocean, but it will also introduce the measures the phytoplankton has taken against unfriendly takeovers by grazing animals. It will show how poisons were developed to discourage predators. In a later section, I will tell you how even more poisonous protists, namely dinoflagellates, encounter predators that evolved countermeasures against these dinoflagellate poisons, showing that the quest for food is also an eternal biological arms race.*

Diatoms and the Marine Food Chain, on Toxins and Armors, Art, and Purpose

The next organism I have chosen for discussion needs no justification. Its ecological importance is so overwhelming that omitting it from our survey cannot be excused. The organism in question is a unicellular, photosynthetic eukaryotic alga called diatom.

Form

These are gracious microorganisms (Figures 4.23) that have fascinated light microscopists since the late nineteenth century. Their name is very revealing, it means "cut in two" in Greek. Semantically, this is not a very intelligent word coinage because it contains a contradiction. "Atomos" means something that cannot be cut in Greek, which, nevertheless, gets separated in two subunits. However, morphologically the word diatom is very fitting. One major class of diatoms, the centrics, resembles the Petri dish of the microbiologist. The larger half is called an epitheca, the smaller half a hypotheca. Like in a Petri dish, during fission of the cell, you can literally lift the top lid and separate both parts of the dish and each daughter cell will synthesize a new theca within the

FIGURE 4.23. Diatoms: Green algae from the order Conjugales. *Micrasterias* (*1, 2*), *Cosmarium* (*3, 11*), *Xanthidium* (*4, 9*), *Staurastrum* (*5, 10*), *Euastrum* (*6*), *Penium* (*7*), *Closterium* (*8*), *Aptogonum* (*12*). Diatoms are discussed in this book as part of the marine food chain, in salt marshes, in microbial succession in lakes, during chloroplast take-over, in blooms, as producers of toxin and armor, in their fight with predator copepods, in discussions on their peculiarities of C and N metabolism, in giant mats and during iron fertilization experiments and with the genome sequence of *Thalassiosira*.

old one. Necessarily, there is always a loser, which gets the smaller straw and decreases in size. Actually, if I promised you sex with eukaryotes, diatoms use it cautiously. Only when the losers get down to about 30% of the original size do the diploid cells go into meiosis, form gametes, which fuse into a zygote. The zygote develops into an auxospore, which grows and builds a new organism, which resumes mitotic divisions until the frustules (its silicon shell) go down again in size. However, these pretty algae with such a variety of artistic shells are not just a fancy of Nature that plays with forms in endless variations; diatoms are major workers in the global carbon cycle. They generate about 40% of the organic carbon of the sea. This is a sizable quantity of something like 20 billion metric tons of organic carbon—a single algal genus produces as much biomass as all terrestrial rainforests combined (Field et al. 1998). Diatom's productivity is the basis of a short and energy-efficient food chain in the ocean, which leads quickly to biomass of direct interest to mankind.

Evolution

Actually, diatoms are quite young algae and we allow here a premature appearance in our evolutionarily guided account. The fossil record of marine diatoms documents that they only evolved in the Jurassic and became more common in the Late Cretaceous (Falkowski et al. 2004). This approximately "only" 100 Ma time horizon of diatoms was also confirmed by molecular fossils in the form of highly branched isoprenoid alkenes, which come in a characteristic chemical T-form (Damste et al. 2004). These compounds are so stable toward digestion that they survived the last 100 Ma in sediments and in petroleum. These tracer molecules apparently evolved twice in both major groups of diatoms, which can be distinguished by rRNA sequence and morphology, namely the centrics and the pennates. The latter are, as the name indicates, feather-like elongated frustule structures with long grooves and side ribs. The ecological dominance of diatoms thus occurred relatively recently and marks a major event in the world's ocean that caused a shift in the relative importance of the different phytoplanktons, namely from calcareous nannoplankton and dinoflagellates to diatoms. A major anoxic event and the following mass extinction might have created possibilities for newcomers. Furthermore, the evolution of grasses on the land forced by the coevolution of hypsodont (high crown) dentition of grazing ungulates might have provided the necessary silicon to the ocean. Grasses are major agents of rock weathering and up to 15% of their dry weight consists of opal phytoliths (silica incorporated into their cell walls). Only this mobilization of terrestrial silicon for the ocean probably allowed the dominance of diatoms in the phytoplankton (Falkowski et al. 2004).

Blooms

The productive regions of the ocean are characterized by seasonal blooms of phytoplankton, which are commonly dominated by diatoms. Surprisingly, diatom blooms did not end with an explosion of its major predator, the copepods (Figure 4.24). Instead, the bloom ends in mass sinking of cells and phytodetritus. The bloom is limited in time to two weeks and is frequently followed by another outgrowth of a different diatom species (Scholin et al. 2000). It is not clear what limits the bloom. Is nutritional exhaustion of the resources the cause? But do different species of diatoms have such different and complementary nutritional requirements? Do algal viruses always kill the winning population and the given strain or species is only relieved from viral pressure when the cell number passes a lower threshold under which viral predation is diffusion-limited? The genome of diatoms contains more than 20 different chitinases (Armbrust et al. 2004). Are they required to regulate the length of the chitin fibers that diatoms extrude through the pores in their frustules to influence their sinking rate or are these chitinases defense systems against fungal attack? The reason for the ending of the bloom is unknown, but copepod grazing is not the cause. Apparently, copepods are unable to track the diatom blooms, which tend to occur too early in the season when copepod populations still recover from overwintering. Italian scientists exploring the Adriatic Sea came up with a fascinating observation (Miralto et al. 1999). They observed that diatom blooms caused, as expected, a higher fecundity of copepods as measured by egg production rates, but the hatching rate fell dramatically.

Copepods

What did happen? The researchers succeeded in isolating simple aldehydes with a polyunsaturated C10 chain length and various *cis–trans* conformations that reduced the hatching success of copepods by inducing apoptosis (programmed

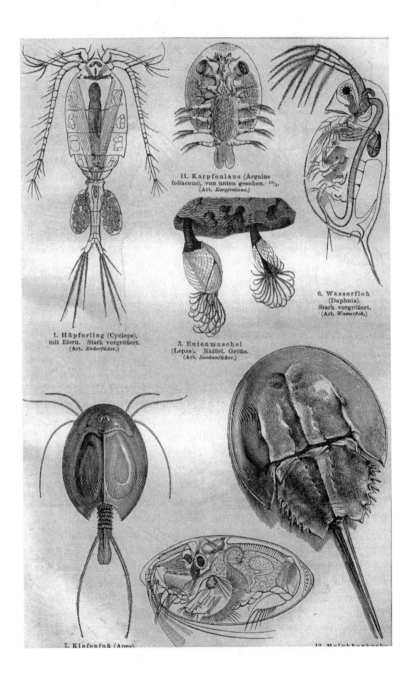

11. Karpfenlaus (Argulus foliaceus), von unten gesehen. ¹⁰/₁. (Art. *Karpfenlaus.*)

6. Wasserfloh (Daphnia). Stark vergrößert. (Art. *Wasserfloh.*)

1. Hüpferling (Cyclops), mit Eiern. Stark vergrößert. (Art. *Ruderfüßer.*)

3. Entenmuschel (Lepas). Natürl. Größe. (Art. *Rankenfüßer.*)

7. Kiefenfuß (Apus)

cell death). Apparently, a primary producer mounts a chemical fence against a herbivore. This strategy is not unusual and is known from land plants that produce phytoecdysone (Ribbins et al. 1967) or phytoestrogens (Leopold et al. 1976) that negatively affect the reproduction success of grazing insects and vertebrates.

When the diatom is eaten by the copepod, its shell is crushed in the foregut. Within seconds enzymes are activated, which transform fatty acids into the toxic aldehydes. Animals fed exclusively with diatoms experienced in their progenies a stop in the development of the nauplius larvae. Spectacular malformations of the nauplius' appendages were observed (Ianora et al. 2004). The diatoms and the purified aldehydes elicit the same teratogenic effect. The bottom line of these laboratory observations seemed clear: a transgeneration marine plant–herbivore interaction prevents an efficient exploitation of the primary producer by its chief predator, allowing the development of a bloom followed by a sinking of the produced biomass, which prevented the produced organic carbon to climb to higher trophic levels.

Food Choices

These observations challenge the classical view of marine food web energy flow from diatoms to fish by means of copepods. Calculations on a sustainable fishery are based on 10% energy transfer efficiency from the first to the second trophic level and would thus be critically flawed. However, a multinational group of marine biologists argued that the laboratory observation of working with a diet consisting of only two protists is not representative of the situation in the field where copepods have the choice between many diets. Actually, there are data that show that copepods prefer motile protists (ciliates and dinoflagellates) on the immobile diatoms. These scientists argued that diatoms might only be

---

FIGURE 4.24. Crustacean diversity. *Top left*: Copepods are probably the most abundant metazoan on earth. They belong to the crustacean subclass Copepoda. The picture shows a female *Cyclops* with many eggs (*1*). *Top right*: *Daphnia*, the water flea from the order Phyllopodia (*2*). In between is below *Lepas*, a true barnacle, a Cirripedia crustacean (*3*). This sessile animal is attached to floating objects. Its feathery thoracopods ("breastfeet") are used for suspension feeding. Above is *Argulus*, a parasitic crustacean seen from below, belonging to the Brachiura group of crustaceans. With its suckers and hooks it attaches to host fish and feeds by piercing skin and sucking blood from its victims. In the *bottom row* from *left*: the tadpole shrimp *Apus*, a Triopsida belonging to the Branchiopoda group of crustacean (*7*). It feeds on organic material, which it stirs up from the ground when moving. However, this quickly growing animal is also a cannibal. It is a living fossil, which has not changed much over evolutionary periods. On the *right* is another living fossil, the horseshoe crab *Limulus*, a Xiphosura. It is a bottom crawler, preying and scavenging on other animals. The late-nineteenth-century authors providing this picture got it wrong: *Limulus* is not a crustacean, it is a Chelicerata and thus closer to true spiders and the dreadful extinct water scorpion *Eurypterus* than to crustaceans.

nutritionally deficient and not necessarily teratogenic. In fact, field observations from 17 sites covering many areas of the oceans did not demonstrate a
negative relationship between copepod egg hatching success and diatom biomass
or dominance (Irigoien et al. 2002). If the inadequacy of a diatom monodiet
for copepods results from the absence of a dietary factor (e.g., a fatty acid),
the ecological consequence would be limited as even in a dense diatom bloom,
alternative preys are available. To test the nutritional hypothesis, UK scientists
offered diets consisting of only diatom or dinoflagellates or a graded mixture
of both. Even a small proportion of dinoflagellate food had a beneficial effect
on copepod growth efficiency. Copepods apparently "knew" about the better
nutritional value of dinoflagellates. When they constituted only 25% of the total
population, they represented >70% of the biomass ingested by copepods (Jones
and Flynn 2005). *This is the last word in this debate because it was published
just before the writing of this book, but it will certainly not close the discussion.
Science is an endless story and a major fun of scientific argument is the exchange
of arguments based on experiments or field observations.*

Toxins

Other toxins of diatoms go through the food chain and come to the attention of
marine biologists only when top predators get killed. This occurred when there
was a mass mortality among sea lions (Figures 4.25) stranded on the Californian
coast over a 2-week period (Scholin et al. 2000). The cause of death was the
neurotoxin domoic acid that was produced by diatoms. It got along the food
chain accumulating in anchovis, a planktivorous fish, and ending with the death
of marine birds and mammals. The blue mussel *Mytilus edulis* (as the name
indicates an edible and widely distributed human food item) did not accumulate
this toxin despite its suspension filtering way of food acquisition. Humans were
thus spared this toxic wave, which caused neuronal necrosis and vacuolation in
the brain of the autopsied sea lions.

Toxic algal blooms are not rare—thus chemical warfare seems to be common
in algae. However, the response is different: the diatom aldehydes kill copepods
indirectly and insidiously, whereas copepods are not affected by a potent
neurotoxin produced by dinoflagellates (saxitoxin). Copedods accumulate these
neurotoxins in their tissue to levels that kill their predator fish and are thus
directly beneficial to the copepods. Diatoms are interesting not only for their
chemical defenses but also their armors. As these armors of diatoms are of
esthetical appeal, I will tell this story from a somewhat philosophical viewpoint.

Art Forms in Nature

*Hypothesis formulation has an important role in science but is sometimes under-
rated in biology. It is frequently argued that biological phenomena are so
complex that concept building is a rather futile exercise. To a certain extent
this is true but without a drive to theory, it might be difficult to design experiments in biology that go beyond phenomenological description. I will illustrate*

FIGURES 4.25. Pinnipedia are aquatic fin-footed mammals in the Carnivore order. Pinnipedia come in three families: Odobenidae (e.g., the walrus *Trichechus rosmarus,* (bottom left) Otaridae (the eared seals, e.g., the sea bear *Otaria ursina* (top left) or the sea lion *Otaria stelleri* (center left)), and Phocidae (the ear-less seals, e.g., the seals *Phoca vitulina* (top right) and *P. groenlandia* (also top right). The hooded seals *Cystophora cristata* (bottom right), and the elephant seal *Mirounga angustirostris* (center right), also belong to the Phocidae family. Pinnipeds are aquatic with respect to food search and terrestrial for mating. Their diet consists of fishes, cuttlefishes, octopuses, and crustaceans. The walrus is a bottom feeder of sessile mollusks. Pinnipedia occur in this book both as predators and as prey.

this conflict with diatoms. E. Haeckel, zoologist, philosopher of the evolution theory in the late nineteenth century in Germany, frequently marveled about the various and artful forms of the shells from protists (Figure 4.26). He wondered about the art forms nature has developed. His feeling is totally justified from a psychological aspect—even modern general microbiology textbooks frequently sport tables where many diatoms are exposed and inspire admiration in the reader. Of course, if you look at animal and plant life you cannot easily escape the impression of a highly esthetical show. We enjoy the colors of flowers and admire the plumage of birds. However, the contemporary biologist comes here with the question: What is its biological function? What is the evidence for the usefulness versus artfulness of diatom forms in nature?

FIGURE 4.26. Radiolaria, one of the four major groups of the Protist phylum Actinopoda. The siliceous skeletons, which preserve very well in the fossil record, were painted by nineteenth-century zoologist and natural philosopher Ernst Haeckel. These organisms led him to a discussion on art forms in nature. The central star is *Hexacontium drymodes*.

Smetacek's Ocean

In an article entitled *A Watery Arms Race* Viktor Smetacek (2001) compares the phytoplankton with the terrestrial plants and notes the fleeting appearances of photosynthetic life in the ocean. Species that possess chloroplasts do not look different from those that don't. Even marked algal blooms last at most for only a few weeks. Space-holding plankton, apparently, has never developed in the sea. Space does not seem to be a limiting factor and, according to Smetacek, competition between organisms that target the same niche is not the factor that rules the evolution of plankton. *However, this does not mean that life is cosy in the sea. We have already heard about the comparison of the open ocean with a huge stable desert. And this desert is full of snakes. Predation is omnipresent, predation comes from the top where organisms want to eat you and predation comes from the bottom, where others want to infect you. You are constantly hunted.* Smetacek developed a theoretical concept on the microbial way of life in the ocean (Smetacek 2002). He explains the remarkably stable bacterial number of around a million cells per milliliter of ocean water by the heavy predation pressure of a few genera of flagellated protists that hunt bacteria individually. Bacterial populations that succeed in escaping from this pressure are decimated by viruses. Marine bacteria are thus permanently locked within close boundaries by the double threat of predators and pathogens. Bacteria are thus better off when roaming the ocean individually for food as diffusion will immediately obliterate their chemical tracking. If they assemble with nutritional hot spots like larger sinking particles, predators and, apparently, phages can track bacteria much more easily. Many predators and pathogens feed and infect selectively. Small predators hunt individual cells, like copepods devouring diatoms. Captured cells are pierced, ingested, engulfed, or crushed.

On Teeth and Armors

Here we need a short stop in basic zoology. Copepods belong to the subphylum Crustacea of the phylum Arthropoda. Even nonzoologists know Crustacea as the world's most appreciated gourmet food is made with them (lobsters, crabs, and shrimps; Figure 4.27). They show six thorax segments, where the first is always fused to the head segment. The head segment carries two whips called an antenna and an antennule. The thorax has a few legs, which are kept locked together for swimming. At the rear you find an abdomen without appendages but two large egg sacs. The end segment (telson) is frequently luxuriously decorated with featherlike structures. The basic body structure is still that of a miniature lobster. The copepods comprise 12,000 species that are quite diverse. The calanoid subgroup is planktonic and as a major primary consumer in marine food webs, it is of extreme importance. The calanoids have a short foregut, which leads into the midgut accompanied by an anterior and a posterior cecum. The hindgut is a thin tube that ends in an anus near the telson. After ingestion, the food is handled mechanically in the foregut. Sometimes you find even a type of gastric

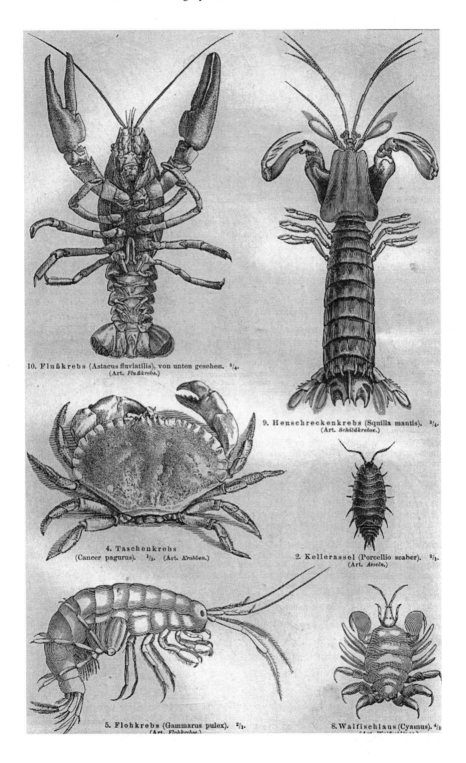

10. Flußkrebs (Astacus fluviatilis), von unten gesehen. ³/₄.
(Art. *Flußkrebs.*)

9. Henschreckenkrebs (Squilla mantis). ³/₄.
(Art. *Schildkrebse.*)

4. Taschenkrebs
(Cancer pagurus). ¹/₃. (Art. *Krabben.*)

2. Kellerassel (Porcellio scaber). ²/₁.
(Art. *Asseln.*)

5. Flohkrebs (Gammarus pulex). ²/₁.
(Art. *Flohkrebse.*)

8. Walfischlaus (Cyamus). ⁴/₁

mill, which carries sclerotized, sometimes even siliconized, teeth, grinding the food into smaller pieces.

Smetacek's thesis is that planktonic evolution is ruled by protection and that the many shapes of plankton reflect defense responses to specific attack systems. This concept can be tested and to do that he teamed up with physicists. They tested the physical strength of diatom's shells with calibrated glass microneedles to load and to break the shells with defined forces (Hamm et al. 2003). The diatoms turned out to be stable and lightweight like the best man-made constructions. They resisted forces of up to $700 \, tons/m^2$ and demonstrated elastic deformation under stress. They also showed favorable stress distribution when challenged on several small patches simulating a copepod's mandible bite. We see here an obvious arms race as copepods also have developed silica-edged mandibles and gizzards lined with formidable arrays of sharp "teeth." There are several thousand species of copepods. Do the many forms of frustules from diatoms represent the optimization of resistance against the mechanical stress exerted by their specific mandible morphology? Is artfulness a pure illusion in biology and is everything the product of selection?

Diatom Nutrition

Gas Solubility

It is generally considered that inorganic carbon supply does not limit the growth of phytoplankton. However, there might be problems: The molecules of the biologically important gases CO_2, O_2, and N_2 are nonpolar. In CO_2 each C=O bond is polar, but the two dipoles are oppositely directed and cancel each other. In the gas phase, the movement of these molecules is unlimited, whereas in the aqueous phase, their motion and that of the water molecules are constrained: Solution of gases in the water phase decreases the entropy. Lack of polarity and decreased entropy together make that these gases have a low solubility in water: for O_2 this is 0.035 g/l, while for CO_2 it is 0.97 g/l. Since the last glaciation the atmospheric CO_2 concentration has increased from 180 to 355 ppm in the

---

FIGURE 4.27. Crustacean diversity II. *Squilla mantis* is a typical Stomatopoda crustacean (9). The animals are found in shallow marine environments. They live in burrows and are raptorial carnivores, preying on fish, mollusks, cnidarians, and other crustaceans. Better known from this class of animals is the crayfish, here represented with a freshwater specimen *Astacus fluviatilis* (10), the clawed lobster. Below is *Cancer pagurus*, a true crab (4). Both animals belong to the Decapoda order, the scientific name refers to its 10 feet. *Bottom*: *Gammarus pulex*, the sand flea (5) and the whale lice *Cyamus* (8), they belong both to the Amphipoda order of crustaceans. The latter is a parasite of dolphins and whales. Also the pillbug *Porcellio scaber* (2) is a crustacean, it belongs to the order Isopoda (referring to the fact that all feet are similar in form). Pillbugs are the most successful terrestrial crustaceans. They are herbivorous or omnivorous scavengers or detrivores.

atmosphere, but in absolute terms the CO_2 concentration is still small when compared with what prevailed when oxygenic photosynthesis evolved.

Is CO_2 a Limiting Nutrient?

Dissolved inorganic carbon exists in the seawater in three interchangeable forms: CO_2, HCO_3^-, and CO_3^{2-}. In the slightly alkaline seawater (pH 8.2), less than 1% of the inorganic carbon exists in the CO_2 form, which is the only inorganic carbon form accepted by Rubisco for carbon fixation. This value corresponds to a CO_2 concentration of $15 \mu M$. In upwelling regions of the ocean, where high-latitude phytoplankton spring blooms are observed, the nitrate and phosphate concentrations amount to 35 and $2 \mu M$. This gives a $CO_2/NO_3^-/PO_4^{3-}$ ratio of 28/38/1. However, the Redfield ratio defines the C/N/P nutrient requirements of phytoplankton as 106/16/1, which resembles the ratio of the major dissolved nutrients in the deep ocean. Before we use this anomaly as an argument, I should mention that not all odd Redfield ratios signal a problem. Non-Redfield ratios are now increasingly being reported (Arrigo 2005). At the most basic level, this stoichiometry reflects the elemental composition imposed by evolution on phytoplankton. The green and red sublineages of phytoplankton show with 200:27:1 and 70:10:1, respectively, substantial deviations from the standard ratio. Since proteins and chlorophylls as resource machinery are high in N with respect to P, whereas ribosomes as growth machinery are high in both N and P, organisms adapted to different physiologies also demonstrate distinct Redfield ratios (Klausmeier et al. 2004). The "bloomer" adapted for exponential growth with a lot of ribosomes will show N:P ratios <10. The "survivalist," which bets on a copious resource-acquisition machinery in an oligotrophic environment, will have an N:P ratio >30. Only the generalist has the standard ratio of 16. Anthropogenic inputs (e.g., nitrogenous fertilizers from agriculture) in some regions will sensibly change this ratio.

With this critical remark on the Redfield ratio, we will go back to the above-mentioned bloom figures. CO_2 could thus, under upwelling conditions, become the diatom nutrient in shortest supply. To test this prediction, marine microbiologists from the Alfred Wegener Institute investigated the growth rate of temperate (*Thalassiosira*) and polar (*Rhizosolenia*) phytoplankton under different CO_2 concentrations. In fact, below $15 \mu M$ CO_2, the cell division rate decreased with decreasing CO_2 levels (Owen et al. 1974). Because CO_2 is in equilibrium with much larger concentrations of bicarbonate in the seawater, the CO_2 supply limits the growth rate but not the final biomass of the diatom bloom, which is defined by a complex interaction of three factors: nitrate/phosphate nutrient exhaustion, sinking rate of dead diatoms, and grazing by zooplankton. However, US oceanographers showed that in the natural assemblage the growth of temperate diatoms was not limited by CO_2 concentrations. In fact, the observed growth rate could not be sustained by pure CO_2 diffusion and they explained the rapid uptake of radiolabeled CO_2 by an active transport of bicarbonate or CO_2 across the plasma membrane (Boklage 1997). Aquaporins, transmembrane proteins that form water-permeable channels across the membrane, can also actively transport CO_2 gas. Plants where the aquaporin expression level was decreased by an antisense RNA or increased via a tetracycline-inducible promoter showed decreased and increased CO_2 transport and leaf growth (Uehlein et al. 2003). The *Thalassiosira* genome showed a bicarbonate transporter (Armbrust et al. 2004).

Carbonic Anhydrase and Zinc

However, this creates a problem for the cell as bicarbonate is not accepted as a substrate by Rubisco. Here another observation was helpful: An inhibitor of carbonic anhydrase had no effect on the intracellular accumulation of bicarbonate, but decreased the carbon fixation by diatoms (Boklage 1997). Carbonic anhydrase catalyzes the reaction $CO_2 + H_2O \leftrightarrow HCO_3^- + H^+$ a million-fold over the chemical equilibrium. This enzyme could thus liberate the necessary CO_2 for carbon fixation when bicarbonate is the actively transported species. A look at its active site is revealing: it contains a zinc ion held in place by imidazole groups of three histidine residues. The zinc ion complexes, in addition, one CO_2 molecule and one water molecule. The bound water can rapidly be converted into a hydroxide ion, which attacks the C atom of CO_2 leading to the formation of HCO_3^-. This structure explains other observations of diatom nutrition. Most of the zinc in diatoms is bound to carbonic anhydrase. At low zinc concentrations, the activity of the enzyme fell and under these conditions the CO_2 concentrations became a limiting factor for the growth of diatoms (Morel et al. 1994). Interestingly, the enzyme expression was only induced under low CO_2 levels. This led the researchers to the formulation of a zinc–carbon colimitation hypothesis of marine phytoplankton growth. Colimitation is nowadays a very popular concept in oceanography and is about to replace Liebig's law of the minimum, which states that only a single resource limits plant growth at any time (Arrigo 2005). With zinc we have a competitor to the iron fertilization hypothesis (Stanojevic et al. 1989) for the draw-down of CO_2 into organic biomass. Another aspect of diatom nutrition is of interest in this context. Inorganic zinc levels are with 2 pM quite low in the sea and zinc will thus frequently limit the growth of diatoms. Their carbonic anhydrase shows here an interesting adaptation: At low Zn levels, its activity can be restored by the addition of cadmium or cobalt, which replace the zinc ion at the active center (Morel et al. 1994). *In terrestrial systems, cadmium is a serious poison—in the sea cadmium is actively sought by diatoms and many red algae. In this context, it becomes an important metallic cofactor. This observation is of some importance in ecological discussions, for example whether it is better to pull defunct oil platforms to special waste deposits on land or to dump them into the sea.* Other algae, for example, coccolithophores, do not possess carbonic anhydrase activity. The absence of this activity can be understood directly from the physiology of this cell. As it builds its shell from carbonate instead of silicate, it can derive CO_2 for carbon fixation from the calcification reaction: $2HCO_3^- + Ca^{2+} \rightarrow CaCO_3(\text{solid}) + CO_2 + H_2O$.

Trading Iron for Copper

Diatoms are quite flexible organisms when it comes to metal use. In higher plants, green algae, and cyanobacteria, the Cu-containing redox protein plastocyanin transfers electrons from the cytochrome b_6f complex to PSI. Algae containing chlorophyll c use for this task a functionally equivalent Fe-containing protein, cytochrome c_6. Diatoms were thought to be members of the same club. This is also definitively the case for coastal species of diatoms like *Thalassiosira weissflogii* that lacks the gene coding for plastocyanin. However, Canadian oceanographers have now shown that the oceanic diatom *T. oceanica* uses plastocyanin for this critical step in electron transport (Peers and Price 2006). Consistent with this finding, the oceanic form had a 10-fold higher requirement for copper, and copper deficiency limited photosynthetic electron transport regardless of iron status. What could be the reason for this switch of electron carriers? Offshore oceanic water contains much lower concentrations of many dissolved metals than does coastal water, but this includes copper with 0.4 and 50 nM concentrations in both waters, respectively. However, the gradient for iron supply is even steeper and therefore it makes sense to trade iron for copper. The flexibility in electron carrier use within two ecotypes from the same diatom genus underlines again the intrinsic flexibility and apparent modular structure of the electron transport chain, which we have already seen with other examples.

C4 Metabolism

Carbon fixation by diatoms offers another fascinating twist. Rubisco from diatoms has half saturation constants of 30–60 μM for CO_2, much higher than the 15 μM concentration in the seawater. As diatoms contribute so much to the marine carbon fixation, it is unlikely that diatom's Rubisco works at these low concentrations. Bicarbonate transport is only part of the explanation for the ecological success of diatoms. Some data indicate that diatoms show C4 and not the usual C3 photosynthesis. What does this mean? When Melvin Calvin and coworkers labeled algae with a short radioactive CO_2 pulse in the <1 min time range, they isolated the C3 compound phosphoglycerate as the first labeled compound. When John Reinfelder and colleagues repeated this experiment with the diatom *Thalassiosira*, 70% of the early label was found in the C4 compound malate. After a chase period of 1 min, the label disappeared from malate and appeared in phosphoglycerate and sugars (Reinfelder et al. 2000). What had happened? The authors argued for a model like in classic multicellular C4 plants. In the cytoplasm, the enzyme phosphoenolpyruvate carboxylase uses bicarbonate as C1 donor to transform the C3 substrate to the C4 product malate via the intermediate oxaloacetate. Malate is then transported into the chloroplast of the diatom where the enzyme phosphoenolpyruvate carboxykinase generates CO_2 for Rubisco and PEP for recycling back into the cytoplasm. They argued that this C4 metabolism was previously overlooked as it is only observed under moderate zinc deficiency when carbonic anhydrase activity is low. In accordance with this model, the researchers found decreasing levels of PEP carboxylase activity with increasing CO_2 concentration, and PEP carboxykinase activity was mainly associated with the chloroplast. I have already stressed that diatoms are not particularly old algae; they radiated during the Mesozoic and came to marine dominance just over the last 40 Ma. However, even then the C4 metabolism in diatoms would precede that in the terrestrial system where the C4 and CAM plants evolved only about 7 Ma ago. Yet, the last word was not yet spoken on this issue as other biologists doubted whether the presented evidence suffices for diagnosing a C4 metabolism (Johnston et al. 2001), and genomics scientists could not reconstruct the C4 pathway from the predicted genes of *Thallassiosira* (Armbrust et al. 2004).

Nitrogen Limitation

There is no doubt that the nutritional way of life of diatoms keeps the secret of the ecological success of this algal group. The versatility of diatoms becomes also evident when looking at their nitrogen metabolism. Like all eukaryotic cells, diatoms are unable to fix molecular nitrogen, but diatoms have developed a very efficient system of nitrate acquisition. The basic problem for photosynthetic organisms is a dilemma; they must stay in the sunlit zone of the ocean to achieve photosynthesis, and this zone extends, at maximum, to 100-m depth. However, this zone is at the same time an extremely nutrient-poor area; nitrate concentrations are very low there, but rise steeply below the nutricline at 100-m depth (Hayward 1993). There is a 1,000-fold concentration gradient between both layers, but diffusion is considered to be inefficient to provide the necessary influx of nitrate. Of course, upwelling of nutrient-rich colder and denser deeper waters is a potential solution to the problem, but upwelling is not a regular feature in many open ocean areas. A rational solution to this problem would be migration of an organism in a day-and-night rhythm between both layers. This is actually done by some photosynthetic protists: In coastal waters, dinoflagellates sink at night to feed on nutrient-rich water. But the distances traveled are only a few meters and dinoflagellates are—as the name already suggests—motile protists. Diatoms are nonmotile and they would have to bridge 100-m-deep distances.

Rhizosolenia

Nevertheless, some diatoms do just this as demonstrated by scuba diving oceanographers (Villareal et al. 1993). They detected macroscopic, buoyant diatom mats composed of *Rhizosolenia*, which occur over broad expanses of the oceans. These up to 30-cm-large mats showed an impressive ascent rate of up to 6 m/h and a healthy physiology in a low nutrient environment with a cell division rate of 0.6 per day. A key observation was that floating mats showed a 10 mM intracellular nitrate concentration, whereas sinking mats sported only 2 mM concentrations. Isotope investigation provided a further hint to the origin of the intracellular nitrogen: The mats have a higher ratio of the ^{15}N to ^{14}N

isotope than the nitrogen that is recycled in the upper sunlit waters. Nitrate from below the nutricline showed the same signature. As scuba diving was limited to a 20-m depth, the US oceanographers came back to the floating *Rhizosolenia* with remote video recorders that could detect the mats sinking just below the nutricline at 150 m (Pergams et al. 2003). The video images documented mats at much higher density, the majority was smaller than 1.5 cm in diameter. The researchers deduced a vertical migration as a vegetative growth strategy. *Rhizosolenia* mats mediate a substantial vertical nitrate flux to the sunlit zone, and they were compared to a food conveyer belt in the ocean. Another observation is notable: The diatom genera *Rhizosolenia* and *Hemiaulus* contain the endosymbiontic N_2-fixing cyanobacterium *Richelia intracellularis*, which is capable of extremely high N_2-fixation rates. In blooms these diatoms contribute up to 70% of the nitrogen demand in surface waters (Arrigo 2005). The dominance of some diatoms has thus an understandable nutritional basis. The productivity of these diatoms can amount to breathtaking quantities. Algal mats were probably seen on several occasions in the late 1980s by NASA astronauts at the Space Shuttle Atlantis. A particularly spectacular case was published under the heading *A Line in the Sea* (Yoder et al. 1994). Meteorologists and oceanographers were on the spot with aircrafts and a research ship to investigate a several hundred kilometers long line that looked like a fracture zone in the water. It turned out to be an open-ocean front between a down-gliding cold water front opposing a warm water front. At the front line a very dense culture of *Rhizosolenia* mats discolored the sea.

Diatom Mats as Large as Australia

This current exuberance regarding diatoms is only a weak mirror of its activity 10 million years ago. The Pacific Ocean was apparently covered by floating mats of diatoms, each tens of square centimeters in size. These mats were probably grouped into massive patches the size of Australia by wind action (Sancetta 1993). When the cells died, they lost buoyancy and the mats settled rapidly to the sea floor, where they deposited large quantities of organic matter. Geologists found decimeter-scale sediment beds, which were deposited in a few thousand years. The sites extended over 2,000 miles (Kemp and Baldauf 1993). The surprising observation was that the laminated sediments still contained microscopically detectable diatoms that could be classified as diatoms up to their species level, *Thallassiothrix longissima*. In the modern sea they occur with cell densities of up to 10^6 cells/l, a remarkable concentration in view of their length of up to 3 mm. In the sediment, the individual frustules formed an impenetrable meshwork of great tensile strength that physically suppressed all attacks of burrowing animals in the deep sea. The rapid sedimentation of diatoms caused a major draw-down of carbon, silica, and other nutrients. The most surprising feature is that this organic material remained in situ and geologists concluded that anoxic conditions prevented their reminer-alization.

Sapropels

A similar phenomenon of mass deposition of diatom mats was observed in the Mediterranean Sea, where the deposits became known as sapropels (Kemp et al. 1999). Sapropels are sediments rich in organic carbon formed in low-oxygen conditions. The Mediterranean sediments consist of alternating layers of two diatom assemblages. One layer consists mainly of rhizosolenids accumulated during a summer bloom, when the sea is well stratified; one possible reason was massive freshwater input rich in nutrients from the Nile. Owing to their lifestyle, rhizosolenids are very sensitive to water-column agitation. When the autumn storms mix up the seas, a massive sinking occurs. In winter/early spring, another, but this time rather mixed, diatom population takes over and becomes the dominant phytoplankton contributing the next layer to the sapropel. The diatoms can still be microscopically determined to the genus level in the Quaternary sediments. The maintenance of organic-rich sediments is crucial for the genesis of petroleum source rocks.

Dinoflagellates

Evolution History

Dinoflagellates represent an important part of modern phytoplanktons in the ocean. They are second only to diatoms, but this has not always been so. The rise of the modern eukaryotic phytoplankton community began in the Triassic and it was an important response of life to the massive extinction at the end of the Permian. Dinoflagellates led this repopulation process, followed by coccolithophores and only from the Cretaceous diatoms grew in importance to outpace dinoflagellates in the Tertiary, when dinoflagellates experienced a decline. The Mesozoic radiation of the eukaryotic phytoplankton was the response to sea level rises due to substantial warming, which created expanded flooded shelf areas and thus more niches for these photosynthetic algae (Falkowski et al. 2004). According to molecular phylogeny data, dinoflagellates must be older than Foraminifera and Radiolaria. These protists had durable skeletons made of carbonate and silicate, respectively, which allowed tracing their fossils into the Precambrian.

Morphology

Also dinoflagellates have an exoskeleton that gives them characteristic forms. *Gonyaulax* somewhat resembles a walnut, *Ceratium* has an elongated structure due to the possession of characteristic spikes. These armored forms are called thecate in contrast to naked forms like *Noctiluca* (the nice name refers to the night glow of the ocean caused by them). The theca is divided by two perpendicular girdles, each containing a flagellum. Two flagella positioned like this gives them a characteristic whirling spin, which is the meaning of the Greek word "dinos." However, in contrast to the protists having inorganic skeletons, dinoflagellates use cellulose as construction material. Organic walled protists have been described as acritarchs, but their identity with dinoflagellates is contested. However, molecular fossils from dinoflagellates like the characteristic membrane lipid dinosteran (a cholesterol derivative) became abundant in the Early Cambrian and can still be traced to the Riphean (800 Ma ago) (Moldowan and Talyzina 1998).

Most dinoflagellates belong to the red photosynthetic lineage that acquired their chloroplasts from red algae.

DNA

With respect to genomics, dinoflagellates are special: Their nuclear DNA contains unusual bases, lacks typical histones, and the DNA is permanently condensed, which occurs only during cell divisions in other eukaryotes (McFaden 1999). Also their chloroplast genome is special: Not only is it the smallest known chloroplast genome only coding for two and three proteins from PSI and PSII, respectively, cytochrome b_6f, a subunit of the ATPase, and a 16S and 23S rRNA (and no Rubisco gene); to mark its exotic genomic nature, each of the few remaining chloroplast genes comes on an own DNA minicircle each sporting conspicuous noncoding DNA repeats and thus own replicon functions (Zhang et al. 1999). The rest of the chloroplast genes have been transferred into the dinoflagellate nucleus.

Symbiont

Dinoflagellates make major contributions to the CO_2-fixing budget of the ocean and in a change of role this secondary symbiont itself became the most important endosymbiont in corals and thus a unicellular protist contributes to the constructions of organisms that can be seen from space. I am referring to the Great Barrier Reef in Australia. *In the corals, dinoflagellates apparently have found a rather liberal employer. Corals did not enslave their photosynthetic helpers. In fact, as we will see in a later section on coral bleaching, reef corals have a dynamic symbiotic relationship with dinoflagellates.*

Gonyaulax: Toxins

Like diatoms, dinoflagellates are not the benign organisms one would expect from a cell endowed with the gift of photosynthesis. However, we should not forget that we have here a heterotrophic wolf in photosynthetic sheep clothes. The predator character of the heterotroph, which tamed a red alga for its purpose, is more than evident when looking at some dinoflagellates. Take *Gonyaulax tamarensis*: on the bright side, it is known for its bioluminescence in the sea using a luciferin–luciferase system; on the dark side, it is dreaded as the cause of a major outbreak of paralytic shellfish poisoning, where it produces the potent neurotoxin saxitoxin. The sequence of events is straightforward as it relies on toxin transmission via the food chain as a conveyer belt. Marine suspension feeders such as mussels and clams filter dinoflagellates from the water and store the toxin in their tissue. Fish eating shellfish get intoxicated. Indeed, massive fish kills in the mid-Atlantic involving several million menhaden (*Brevoortia*) have been attributed to dinoflagellate's toxin. This toxin even reaches the top consumer: more than 300 fatalities have been reported in human beings where the major symptoms are muscular paralysis and respiratory failure. This trend for shellfish poisoning is increasing, and we will return to this neurotoxin in a later section, not for its medical and economical importance, but because it represents a fascinating case for predator–prey coevolution.

Gonyaulax: Red Tides

Algal blooms with dinoflagellates ("red tides") show a characteristic seasonality. Part of this rhythm is explained by the life cycle of *Gonyaulax*. It oscillates between motile, vegetative cells and resting cysts, which overwinter in bottom sediments. Newly formed cysts have a mandatory months-long dormancy period during which germination is not possible. Plants use mostly environmental cues to retrigger growth after over wintering. This is, however, difficult for *Gonyaulax*. Its resting cysts at the sediment bottom experience constant temperature and do not see light. Therefore, this species has evolved an endogenous annual clock, which triggers the release of motile cells (Anderson and Keafer 1987). Dinoflagellates are not just motile, many have eye spots (stigmata), some even with lens-like structures that can focus light, which could help them searching the sunlit part of the ocean to exploit their photosynthetic capacities.

Fish-Eating Dinoflagellates

Toxin production in dinoflagellates was seen as a defense measure to discourage herbivores as done by diatoms. Therefore, it was as a surprise when it became clear that dinoflagellates use the toxin for their own nutritional means. Fish pathologists described a *Dinamoeba* that excysts (germinates) after live fish or its excreta are added to the aquarium. It completes its sexual cycle (fusion of a female and male gamete looking like small dinoflagellates, the latter with a specially long flagellum) while killing the fish. Additional vegetative cells are formed when live fish is present. The cells search dying fish or flecks of sloughed tissue and digest the cell debris by forming a peduncle, a type of sucking device. If no new fish is left, they build new cysts. Notably, cell-free filtered aquarium water still induced neurotoxic signs in fish pointing to a neurotoxin (Burkholder et al. 1992). I mention this point specifically because recent aquarium experiments with the dinoflagellate *Pfiesteria* have challenged the toxin hypothesis. Fish mortality was only observed when the dinoflagellate was in direct physical contact with the fish, but not when separated by a membrane that would allow the passage of toxins. In direct mixing experiments the observations were clear: *Pfiesteria* exhibits rapid chemotaxis, attaches to the epidermis of the fish larvae by the peduncle, and then literally sucks the life out of the fish. *Pfiesteria* fed for 1 min showed substantial swelling due to the ingestion of the fish epidermal cytoplasm via the peduncle (Vogelbein et al. 2002). Apparently, not all dinoflagellate fish pathogens are toxigenic, some are a type of ambush predator.

The First Animals

The Origins and the Sponges

You might expect that the subject of eating will get more fascinating with the advent of animals on earth. This great expectation partly reflects our anthro-pocentric view: We are finally animals and we feel with them more than with plants or bacteria. Their cycles of eat and be eaten are much more familiar to our view of the world than the remarkable biochemical deeds of micro-scopic organisms or the brave world of photosynthesis. Actually, I stretched your patience before reaching animals. This delay is justified because animals arrived relatively late on earth. When did metazoans (multicellular animals) actually appear in the fossil record?

Metazoan Fossils

Currently the oldest finds of body fossils in contrast to trace fossils are from the Early Vendian period and have an age of about 580 million years. This is the last period of the Proterozoic (literally: before animal period). The phospho-rites of the Doushantou Formation near Weng'an in southern China opens an amazing view into multicellular life just before the Ediacaran radiation of macroscopic animal life. The marvelously conserved fossils show soft-bodied microscopic remains from animal embryos, a material paleontologists would not have dared to dream until recently (Xiao et al. 1998). The interpretation of the cellular balls as *Volvox*-like colonies of protists was quickly rejected and the spheroid material is currently interpreted as successive binary divisions of animal eggs.

Blastea?

One, two, and four blastomeres arranged in a modified tetrahedron were identified as well as later cleavage stages of embryonal development. Is this the blastea predicted more than hundred years ago by E. Haeckel? Haeckel was explicit and as we will see not far off from current thinking, when he derived the metazoan ancestor from colonial flagellated protists. Actually, these cell balls are only the embryos and not the microscopic animals themselves. In fact, embryos like those found are rare observations in living animals, but they resemble those from some crustaceans. If this interpretation is correct, Bilaterians (bilateral symmetric animals with a major head–tail body axis) were already developed well before the Cambrian Revolution, which opens the Paleozoic era (literally: the period of the old animals). However, fossil evidence from larger animals was not yet unearthed from a period preceding the Vendian. The evidence for this is only indirect and has thus been disputed. In fact, these startling structures were reinterpreted as giant sulfur bacteria (*Thiomargerita*) under nutritional stress (Bailey et al. 2007).

Trace Fossils

Two independent lines of research suggest that animals substantially predate their first fossil finds. One part of the evidence is trace fossils. The paleontologist Adolf Seilacher suggests that the conspicuous traces in the Meso-Proterozoic Chorhat sandstone from India are the burrows of worms that wedged themselves through the sediment by peristalsis. According to this interpretation, these animals used microbial mats both as a food source and as an oxygen mask to power their locomotion (Seilacher et al. 1998). The traces suggest 5-mm-long worms, which he interprets as triploblast metazoan that already lived 1.1 billion years ago. Triploblasts are animals with three germ layers, in contrast to Diploblasts, i.e., animals starting from two germ layers like modern Porifera (sponges) or Cnidaria (jellyfish and the like). The interpretation of trace fossils is, however, a risky business and would be met with more disbelief if it were not corroborated by data from molecular biologists.

The Molecular Clock

Their argument is that the sequence differences that separate the extant organisms should allow a back calculation to the time period when these lines actually diverged in the past. The method is of course not without risk either as it must make some a priori anticipations on constant mutation rates over geological timescales and in different lineages of animals, which cannot be tested directly. However, careful sequence data analysis from 18S ribosomal RNA genes and a panel of seven protein coding genes yielded a divergence time of about 1–1.2 billion years ago for vertebrates, echinoderms, arthropods, annelids, and mollusks. The divergence time determined for the separation of Agnatha and Gnathostomes (vertebrates without and with jaws) was calculated by this method to 600 million years ago. The latter time estimation predates the fossil record, but not dramatically so (Wray et al. 1996).

Slow Burn?

These different data were summarized in two hypotheses: The slow burn interpretation that suggests that animals developed very slowly and for long periods remained morphologically restricted to minute larviform planktics, animals floating in the water that did not exceed 1 mm in size. Opposed to this view is the belief of paleontologists who rely on the body fossils of animals and who advocate the theory of a Cambrian explosion. This revolutionary period produced in this manner a multitude of body plans within a short period of time that covered essentially all extant animal phyla and even more phyla that disappeared in the depth of time. The Cambridge paleontologist Simon Conway

Morris (2000) brought these divergent views on the point when asking, what was the "Cambrian explosion: a slow-fuse or a mega-tonnage"?

The Speed of Major Inventions: The Animal "Big Bang"

The perceived flaws of the evolution theory are in fact not an argument against the theory, but only testimony of our ignorance. I will illustrate this with two points and articles that I read at the time of the writing of the book. In the typical semantics of research journal titles, one recent heading reads *Animal Evolution and the Molecular Signature of Radiations Compressed in Time* (Rokas et al. 2005). The problem is well known to biologists and we spoke about this problem in a preceding section. All major animal body plans were present in Cambrian fossils but mostly lacking in the Precambrian record. Molecular biologists put much hope in phylogenetic tree analysis by using protein sequences of extant animals to reconstruct their ancestral relationships. The authors of the quoted article used the sequences from 50 genes obtained from 17 different animal taxa covering choanoflagellates to humans. This ambitious 50-gene data matrix did not resolve relationships among most metazoan phyla. Calcareous and hexactinellid sponges, desmosponges, and cnidarian all projected with long branches to the basis of the tree. Likewise, the protostomes were poorly separated. Many trees were derived and none was backed with much statistical support. The authors then applied the same type of analysis to a sister kingdom of Metazoa, the Fungi. Both kingdoms originated within approximately the same geological time frame as suggested by their fossil records (Yuan et al. 2005). In contrast to Metazoa, the Fungi were well resolved in a phylogenetic tree analysis. The authors concluded from this difference that this molecular method works in principle even with this distant time horizon, but that there is something special with the tempo and mode of early metazoan evolution. The major metazoan lineages were according to them characterized by closely spaced series of cladogenic events. They placed the origin of Metazoan to approximately 600 Ma ago with poriferans, cnidarians, and the earliest bilaterians appearing in a short 50 Ma time span. They argued that even fossil evidence sometimes supports very short time spans for the appearance of major lineages like 20 Ma for the lobe-limbed vertebrates (lungfish, coelacanths, and tetrapods), which occurred 390 Ma ago. Actually, these molecular data are consistent with the interpretation of the Cambrian explosion hypothesis of animal evolution. If you would like to tease biologists you could now argue that the data are also compatible with the creation of animals perhaps not in a single event of divine decision to animate creation with livelier organisms, but over a surprising short period of time. This is not a vain point for biologists as the invention (to use a neutral term also acceptable to biologists) of new structures leading to new taxa and at a lower level, the speciation, is and remains a fundamental problem in biology. How is this discussion settled in biology? First, the topic is not settled with a single paper, especially as previous publications (e.g., Peterson and Butterfield 2005) concluded that the divergence of animal phyla occurred gradually over a period stretching hundreds of millions of years into the Precambrian. Furthermore, there is nothing like direct evidence in science, all conclusions are based on other assumptions or facts taken as granted, which, however, also rely on indirect conclusions. This scepticism applies especially to branches of biology that occurred only once in the history of life on Earth and cannot be repeated in controlled laboratory experiments. The paper of Rokas et al. (2005) was therefore heavily criticized in a comment appearing in the same journal on the basis of perceived flaws in the statistical analysis of the sequence data (Jermiin et al. 2005).

Sponge Fossils

Now back to the oldest real animal fossils and the implications of these findings for the "quest for food" subject. The Doushantou phosphate deposits contain 580 million old sponges (Porifera), which display cellular structures. The fossils show typical sponge cells like porocytes, sclerocytes, and amoebocytes. Even cell nuclei can be discerned. Highly diagnostic spicules classify these sponges as close relatives of modern monoaxonal *Demospongiae* sporting silicon-based needles. In addition, developmental stages of morulas, blastulas, and larvae with peripheral flagellae were identified (Li et al. 1998).

Sponge Anatomy

To appreciate these animals and to understand their mode of feeding, we need a short excursion into invertebrate zoology. Sponges show a very simple body plan. In its simplest form sponges resemble a flower vase (Figures 4.28 and 4.29). The outside layer is the pinacoderm, a one-layer epithelium. The inner layer is the choanocyte layer, containing the actual feeding cells. Between both layers you find a mesohyl, a type of mesenchyme, which contains a number of crucial cells. These are sclerocytes, cells that secrete the crystalline spines of the sponges. These spikes give sponges a fixed form and allow them to withstand tidal waves. However, the spicules also have a discouraging function toward predators. They make sponges rather unpalatable (the familiar bath sponges have lost the spicules and its leathery structure is provided by collagen fibrils called spongin, the matrix of the mesohyl intermediate layer). Another important cell is the porocyte, a cell, which rolls up into a tube-like structure around a central opening. They build the water inlets (ostium). A water stream is pulled into the inner part (atrium, the inner part of the vase) of the sponge by the coordinated action of the flagella beats from the choanocytes. They maintain a flow of water across the chambers filled with a choanocyte cell layer and push the water out of the sponge through an outlet channel called osculum.

Feeding

Sponges are suspension filtering animals, they exclude broader particles from their inlets and capture particles in the 2- to 5-μm size range (bacteria, small

FIGURE 4.28. Different sponges, *top right: Euspongia officinalis*, the commercial bath sponge, *bottom right: Sycon raphanus*, a sponge at the syconoid condition (folded choanoderm); *top left: Aplysina aerophoba*, nicely colored sponges that change their color on contact, *bottom left: Suberites domuncula*, associated with a hermit crab.

protists, unicellular algae, organic detritus). When the suspended material in the water flow becomes too concentrated and would thus clog the water channels, the sponge can switch from full pumping activity to closure of its ostia within minutes. True muscle and nerve cells are searched in vain in sponges. The feeding cell is actually the choanocyte: it consists of a highly vacuolated cell body, which ends in a collar consisting of up to 50 cytoplasmic extensions (microvilli). In the center of the collar swings the flagellum and it pushes suspended food particles against a mucous layer between the collar villi. They become trapped in this sticky material and are transported by undulations of the collar toward the cell body. Here the food particles are ingested by phagocytosis and liquid droplets are taken up by pinocytosis. The first partial intracellular digestion takes place in the choanocytes, but then the food particle is passed to the archaeocytes located in the mesohyl. Archaeocytes are amoeboid cells and play a major role in digestion, thanks to enzymes like protases, amylases, lipases, and phosphatases. Archaeocytes are also highly mobile cells, which are committed to distributing the nutrients through the body of the sponge. Fluorescent-labeled bacteria are digested within a day, and the waste is expelled with the outflow from the sponge. The water flow in the sponge uses simple but efficient principles of streaming dynamics (diameter increases of the conducts slow the flow over the feeding chambers containing the chaonocytes, while diameter decreases in the outlet accelerate the outflow from the sponge and prevent that waste is recycled back into the sponge). Excretion (primarily of ammonia and CO_2) is mainly by

FIGURE 4.29. Glass sponges (class Hexactinellida of the phylum Porifera) from the deep sea illustrate the extraordinary beauty of these most primitive animals. They show only the parazoan grade of body construction, that is, they lack embryological germ layers. *Euplectella* (*1*), *Lophocalyx* (*2*), *Sclerothamnus* (*3*), *Perifragella* (*4*), *Parrea* (*5*), *Semperella* (*6*).

simple diffusion and water flow, but some sponges have even developed fecal pellets where undigested material is packaged in mucous capsules.

Evolutionary Success

There is a widely distributed belief that Nature pushes for increasing complexity. While it is undeniable that with ongoing evolution organisms with more sophisticated structures and functions appeared on earth, there is no inherent drive for complicated structures. Bacteria are not primitive bystanders of evolution that were sitting on a dead end that did not allow them to develop into complex multicellular organisms. In fact, they might still be the masters of the earth not despite their small structure but perhaps because of their small structure. The same reflection applies to sponges. The fossils of the above-mentioned demospongiae demonstrate that their architecture essentially remained unchanged over half a billion years. This indicates that the basic body plan of sponges was an enormous success story. The success of sponges can also be seen from other figures: Sponges are the dominant animals in many benthic marine environments, sponges are found in many geographical regions ranging from tropical reefs to Antarctica, and finally sponges are found as well in shallow waters as at depths of 5,000 m (Vacelet and Boury-Esnault 1995). The secrets of their wide distribution are the successful solutions, which sponges found for the eternal challenges of the eat and be eaten.

Photosynthetic Sponges

I will illustrate the inventions of sponges with a few examples. First let's have a look at coral reefs where sponges are frequently the second largest biomass component after the corals. Many tropical waters are nutrient deficient— evidently a problem for a suspension feeder. However, sponges found a solution: About 80% of the reef-associated sponges are associated with cyanobacteria or eukaryotic algae (zooxanthellae) sitting in a thin layer of sponge "tissue" exposed to light. In this symbiosis, sponges receive photosynthetically fixed organic carbon from their associated symbionts; the sponge *Verongia gigantea* saturates only 17% of its energy needs over filter feeding. Its species denominator "gigantea" indicates that you can grow to a comfortable size with photosynthetic helpers.

Defense

Now to the defensive side: If a sponge is associated with corals, it must look that its osculum, the water output channel, is not overgrown by corals. This is achieved by the secretion of mucus from the outlet chimneys of the sponge, which contains a toxic compound that strongly inhibits the growth of corals around the oscular chimney. Two closely related coral-burrowing sponges produce two chemically very different toxins suggesting a substantial synthetic activity of sponges. In fact, sponges are today a highly searched source of antiinflammatory, antitumor, and antibacterial compounds. The latter compounds explain why sponges rarely suffer bacterial infections and putrefaction. In principle, sponges as sessile animals are an easy prey to other animals. However, sponges are generally long lived and show only slight eating damage. Two factors explain these observations: First, the calcerous and siliceous needles mechanically discourage many attacks. Second, sponges concentrate toxic chemicals that apparently do not harm an animal lacking a nervous system, but could poison most predators.

The Lost Battle for Silicon

Successful food adaptations are also found in sponges in the deep sea. Before telling these stories, I want to recall a failure where sponges lost an important food battle with diatoms. The scene was reef-building sponge that had developed large hypertrophied spicules (desma) during the Mesozoic. Desma-bearing sponges virtually disappeared from the reefs during the Cretaceous. The reason was silicon limitation in the upper layers of the ocean. The reason was probably twofold: a lesser offer of biological silicic acid, $Si(OH)_4$, and the radiation of diatoms that competed more efficiently for the limited silicon sources. Desma-bearing sponges did not have enough of this essential nutrient to synthesize the hypertrophied needles and had to descend into the deep sea where higher silicon concentrations are found (Maldonado et al. 1999).

Methanotrophic Sponges

In the deep sea, sponges developed new lives. The French submersile Nautile discovered large bushes of *Cladorhiza* sponges in 5,000-m depth. Filter feeding is problematic at this depth because usually there is not much to filter in these low nutrient environments. Notably, this sponge grows on the flanks of an undersea volcano, which supplies the food substrate for a methanotrophic bacterium that grows within the body of the sponge. Two types of bacteria were found: Intercellular bacteria, which were apparently healthy and enjoyed the growth medium "sponge mesohyl" enriched for volcano methane. On the other hand, there are bacteria inside of sponge phagocytes, which were just eaten by the host. In this sponge, the bacterial symbiont was already incorporated into the sponge embryo to assure its transmission (Vacelet et al. 1995).

Ambush Hunter in the Deep Sea

Even more spectacular are real carnivorous sponges discovered in the deep-sea. Its cousins happened to live in a cave in the Mediterranean Sea, which facilitated observation because it only required diving to 20-m depth. In the extremely food-poor environment of the deep sea, macrophagy becomes a better strategy than filter feeding. The record holder of the deep-sea sponges is *Asbestopluma*, reaching down in the abyssal zone of 8,800-m depth. This sponge gave up its entire aquiferous system and choanocytes. Instead it developed tentacles fixed to a long stalk. This weapon with a "Velcro"-like adhesiveness is held in ambush position until touched by quite sizable prey. The prey struggles for hours, but becomes increasingly engulfed by thin filaments that overgrow the prey (Vacelet and Boury-Esnault 1995). As large prey is rare in this environment, the sponge has a back-up food system of symbiontic bacteria that are occasionally digested in bacteriocytes of the sponge. The quest for food in this difficult environment has led to a body plan that would be classified as a new animal phylum if the siliceous spicules do not place them firmly in *Demospongiae*. These sponges have developed a food-induced lifestyle that already points the way taken by the next group of animals, which we will meet on the ladder of complexity, Cnidarians.

Choanoflagellates

Before we get to them, we should, however, take a look back on the possible origin of sponges. Sponges are multicellular animals, but they still lack true tissues, reproductive organs, nervous system, and a body polarity. They clearly contribute to our understanding of the transition of unicellular to multicellular life. Current opinion suggests a direct protist ancestry to sponges. Morphologically, choanoflagellates (they look like a colony of choanocytes inserted into a gelatinous mass that floats in the water) are so similar to the choanocytes of sponges that more than 100 years ago E. Haeckel suggested that these protists are either highly reduced sponges or the direct ancestors of sponges. Tree building with the rRNA sequence from choanoflagellates yielded only ambiguous results. In contrast, four protein sequences including the highly conserved elongation factor 2 and tubulins firmly placed the choanoflagellate sequences with the animal protein sequences at the exclusion of yeasts and fungi. In addition, a gene library from the choanoflagellate *Monosiga* yielded an animal-like receptor tyrosine kinase with an extracellular ligand binding domain. In animals this protein is used for cell signaling and cell adhesion (King and Carroll 2001). *Taken together with the suggestive morphological similarities, the choanoflagellates should be regarded as a candidate organism at the root of the metazoan tree.*

The Ediacaran Fauna

Where are the Animals?

Evidence from molecular biology tells us that major diversification of eukaryotes occurred perhaps 1 Ga (giga anni, i.e. 1 billion years) ago and the major lineages of metazoa were distinct for 700 Ma (mega anni, i.e. 1 million years) . However, the fossil evidence is tenuous. The decline of stromatolites at 800 Ma has sometimes been attributed to grazing. Fossil evidence for this hypothesis is, however, lacking and there are alternative explanations for the decline related to changes in the ocean chemistry. Trace fossils like putative metazoan fecal pellets are quoted as evidence for a metazoan gut, but they might also result from protists. Early metazoans were probably soft bodied and thus unlikely to leave fossils. However, one would then expect trace fossils from burrowing (bioturbation). Nothing like this has been reported in the pre-Ediacaran period. It is thus likely that the early metazoan did not grow beyond the millimeter size.

The low concentration of atmospheric oxygen and the difficulty in oxygen supply to deeper cell layers and the late invention of collagen as a tissue support were proposed as possible explanations.

Why did animals take so long to appear on earth? Eukaryotes developed perhaps 1.5 Ga ago, algae were clearly present 1.2 Ga ago. What prevented the evolution of metazoans (multicellular animals)? The availability of food was not a limiting factor. Stromatolites were easy and nutritious prey. Was the climate too rough?

The Earliest Animal Fossils

The first clear-cut fossil evidence for animals dates back to the Precambrian, to a period called Ediacaran, named after hills in South Australia explored by the geologist Sprigg. The onset of the Ediacaran coincides with the last global glaciation 580 Ma ago. Global means here "Snowball Earth," where nearly the entire planet was covered by an ice shield. The animals were already quite sizable reaching from 3- to 50-cm sizes. Their appearance is rather strange: *Spriggina* has been interpreted as an early annelid worm. But if you put this animal upright, it looks like sea pens, a cnidarian. Others like *Charniodiscus* have a clearer morphology. The body ends on one side with a bulbous holdfast, which was probably tethered to the ground of the sea. From this structure starts a stalk continued into a central axis. From this axis emerges a feather or fern-like frond. Each of these subunits of the fronds shows a quilted mattress appearance typical for these animals. G. Narbonne recently found marvelously conserved specimens at the Mistaken Point (*nomen est omen?*) assembly in the Canadian Newfoundland. Mistaken Point merits its name, it belonged at the time of the fossil to the continent Laurentia, which straddled the equator, one of the warmest places of the Ediacaran world. These soft-bodied fossils gave a nearly 3-D vision of these animals, the fronds exhibited three orders of fractionality in their branching: major branches in the centimeter size range and tertiary branches in the submillimeter range. Some organisms showed a zigzag central axis, others were bush or spindle shaped (Narbonne 2004; Brasier and Antcliffe 2004). The scientists interpreted the spindle forms as recliners that lay on the sea floor, while the bush and frond-like organisms were elevated above the sea floor. Both forms were probably suspension feeders. The early Ediacaran ecosystem shows no sign of burrowing, diagnostic for mobile animals. Tracks of such activities are only known from the later Ediacaran period, while the probably soft-bodied animals themselves left no trace. The furrows suggest animals that grazed microbes on the surface and the burrows indicate animals that ate bacteria and organic material contained in mud and silt.

Vendobionta?

One school of paleontologists sees the story like this: The Ediacaran animals disappeared as suddenly as they appeared in the fossil records. The Ediacaran fauna were not the ancestors of the Cambrian fauna. In fact, they were ancestors to nothing that lived later on the planet. Apparently, they were an experiment of evolution that led to nowhere and vanished from the scene. This is the Vendobionta hypothesis (after the Vendian an alternative name for the Ediacaran period): Animals lacking an alimentary tract, muscles, and nervous system consisting exclusively of a tough cuticle and a quilted mattress-like body. Other prominent paleontologists see, however, the Ediacaran fauna as diploblasts (animals with two germ layers like the modern coelenterates; Conway Morris 1993). *Charniodiscus*, for example, moves thus into the vicinity of present-day anthozoa. The Ediacaran fossil *Tribrachidium*, a disk-like fossil with a central three arm structure, is seen as a protoechinoderm before the stabilization of a pentaradial arrangement found in all modern echinoderms. The same discussion is entertained with the Burgess fauna of the following Cambrian period. On one side is the position of Stephen Gould, formulated in his book *Wonderful Life*, who advocates the position of a multiplicity of body plans in the Cambrian far exceeding that of present-day life. On the other side is the position of S. Conway Morris (1993), who perceives these forms of life as precursors of present-day triploblasts (animals with three germ layers like the modern protostomes and deuterostomes).

Mollusk-like Kimberella?

A well-preserved late Precambrian fossil from the White Sea in northern Russia, called *Kimberella quadrata*, was interpreted as a bilaterally symmetric, benthic animal with a nonmineralized firm univalved shell and soft undulated parts that extended beyond the shell (Fedonkin and Waggoner 1997). These soft parts were interpreted as a type of ventilatory flaps. The largest animals were 14 cm long and the researchers found hints for a creeping foot and an anterior bulge was interpreted as a mouth or stomach. This report contradicts the interpretation that the Ediacaran fauna represents an extinct grade of nonmetazoan life like vendobionts sui generis, protists, or lichens. The animal is, according to this article, clearly a triploblastic animal, which traces its origin and diversification before the beginning of the Cambrian. For the authors the fossil resembles monoplacophoran mollusks. When looking at these extant animals from the underside, a certain resemblance cannot be denied. The finding of a living *Neopilinia galathea* by a Danish marine expedition in 1952 was a zoological event in itself as these animals were only known from fossil shells dated between the Cambrian and the Devonian. A mouthpart is covered with cuticular plates and leads into a pharynx. Two pairs of digestive glands secrete into the anterior gut, followed by a long, rolled middle gut ending with a rectum. *Neopilinia* lives on a muddy seafloor at 3,000-m depth and its gut content revealed detritus as well as remains from radiolarians and diatoms. Herbivore and predator mollusks contain a characteristic and uniquely molluskan structure, the radula (Figures 4.30 and 4.31). Commonly, this is a ribbon of chitinous teeth that project from the buccal cavity. The radular membrane moves back and forth by protractor and retractor muscles. The teeth are continually replaced from odontoblasts and are further hardened by the incorporation of iron compounds. However, the basic plan

FIGURE 4.30. Chitons are marine mollusks from the class Amphineura (Polyplacophora). The picture shows *Chiton elegans*; on the back you can see eight plates held together by a girdle. The animal creeps on a large muscular foot and scrapes algae from the rock with a mollusk-specific file, the radula.

experiences many variations to adapt to different feeding modes. In some the teeth rows are used as scrapers to remove food particles from solid surfaces, in others they function more like brushes. Combined with powerful jaws, the radula allows strong rasping, tearing, or pulling action. In still other mollusks, the radula is used in combination with an acid-secreting gland as a drill to penetrate calcerous protective coats from prey animals or the teeth are transformed in stylets specialized for blood sucking. In cone snails (*Conus*) the radula became transformed into a poison-injecting harpoon fixed at the end of a proboscis, which can be ejected to capture fish. However, *Kimberella* does not show a radula, the molluskan affinity of this fossil thus remains in a limbo.

Stromatoveris and Comb Jellies?

Another Ediacaran-like fossil was recently reported from the lower Cambrian (Shu et al. 2006). The body is leaf-like with 15 branches, which probably carried fluid-filled canals in the living organism. The animal was benthic and embedded in the seafloor by a stalk. The authors deduced that the branches were ciliated and served to transport food particles via narrow grooves between the branches. Overall, the fossil is frondlike as many Ediacaran creatures, but the Chinese discoverers did not favor a cnidarian affinity because no zooids could be identified, excluding in their opinion a pennatulacean (sea feather) interpretation. Instead, they attributed this creature to Cambrian ctenophores. If this interpretation is correct, the Ediacaran fauna is not a separate and abandoned experiment of evolution, but firmly embedded into the early phylogeny of metazoa as we know them today. Ctenophora, which carry the trivial but very descriptive names comb jellies (Figure 4.32) or sea walnut, resemble cnidarians in many respects,

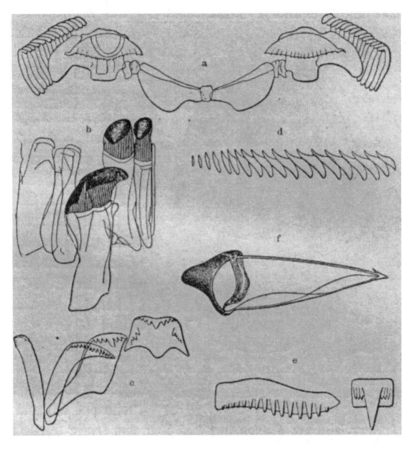

FIGURE 4.31. The buccal region of the molluscan foregut typically bears a unique structure, the radula. It is a toothed, rasping, tongue-like strap used in feeding. The radula is usually a ribbon of recurved chitinous teeth, which varies extensively according to the feeding needs of the specific mollusk. Six different radula forms are shown in the figure for the following mollusks: *Neritina* **a**, *Patella* **b**, *Bythinia* **c**, *Scalaria* **d**, *Mitra* **e** and *Conus* **f**. The latter is a single poison-injecting tooth ("toxoglossate radula") from a predatory cone snail that even attacks fish. The Prosobranchia order of the Gastropoda are actually systematically classified according to their radula type into Docoglossa (**b**), Rhipidoglossa (**a**), Taenioglossa (**c**), Rachiglossa (**e**), Toxoglossa (**e**, **f**), Ptenoglossa (**d**) and finally those lacking this structure, Aglossa, where the scientific name describes the form of the radula.

but most zoologists see these similarities as convergences, whereas other zoologists see ctenophora—with no more evidence—as distant relatives of flatworms. Currently, ctenophora are interpreted as a separate group that arose early in the evolution of the metazoa (Shu et al. 2006). These animals are ovoid in shape, where one pole of the body carries a mouth leading into an elongated pharynx. The ingested food then passes via a small stomach into a complex system of radiating gastrovascular canals. This canal system finishes the digestion process

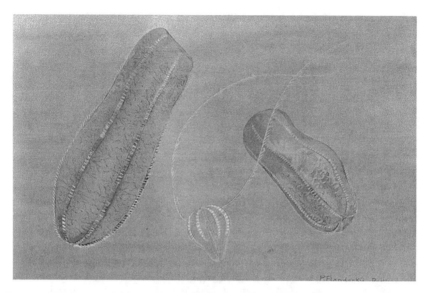

FIGURE 4.32. Two different comb jellies, phylum Ctenophora (*Hormiphora plumosa* and *Beroe ovata*). Two long tentacles, four radial comb plates (ctenes) and at the bottom a mouth extending into a paragastric canal, a pharynx and an aboral canal can be distinguished in the central animal. The outside animals show fluorescent comb plates.

and distributes the nutrients through the body. Gastrodermal cell rosettes regulate the flow of the digestive soup. No circulatory system has evolved in these animals and the indigestible food parts and metabolic waste is expulsed from the anal pole opposite from the mouth. Despite their translucent graceful marine appearance, ctenophora have a predatory lifestyle. The cydippid ctenophora possess long tentacles, which emerge from tentacle sheath. The tentacles contain the interior muscle fibers in a mesenchymal layer, whereas the epidermis is covered by so-called colloblasts. This intricate cell develops into a straight central filament surrounded by a spiral filament and ends in a cap of secretory granules. The tentacles trail passively in the sea. However, upon contact with the prey, the colloblasts explode and release an adhering glue, which fixes prey-like copepods. The muscles in the tentacles are then used to wipe the food-laden arms across the mouth. The animals derive their scientific name from comb rows composed of comb plates or ctenes. Each ctene consists of traverse bands of fused cilia. Their beating is used for vertical movements of the animal in the water column (Figure 4.33).

You have seen that I used weak resemblance of early fossils with extant organisms to provide at least some ideas about their feeding modes. The links between *Stromatoveris* and jelly combs and *Kimberella* with mollusks are not too apparent for a casual observer, whereas the resemblance between *Charniodiscus* and sea feathers is clearer. I therefore use the opportunity to explore with you the cnidarian feeding mode before we move on with the fossil record of the Cambrian Revolution.

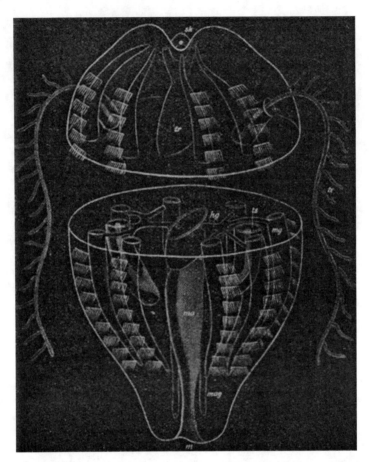

FIGURE 4.33. Scheme of the comb jelly (Ctenophora): the mouth (m) is served with food by the tentacles (te, containing smaller tentilla) and the food is then passed to the pharynx (ma), which is accompanied by parallel pharyngeal canals (mg). Food is then distributed via an infundibulum or stomach (tr) into a complex system of radiating gastrovascular canals; the figure shows transverse canals (hg) and meridional canals (mg). Further annotated structures are apical sensory organ (sk) around the anal pore and tentacle sheaths (ts). The brushes are the ctenes of the comb rows.

Cnidarians: From Sea Pens and Different Feeding Habits...

It is of course relatively difficult to write a history of eating when fossils provide only few hints on the feeding behavior of the investigated organism. I will therefore try a description of the feeding modes from extant cnidaria to provide an idea as to how the Ediacaran fauna might have subsisted. Of course, several hundred million years separate these organisms and we have to adopt the conservative hypothesis that Charniodiscus is the ancestor of current sea pens.

Systematics

Sea pens, corals, and anemones form the class Anthozoa in the phylum Cnidaria (Figure 4.34). The other classes in this phylum are Hydrozoa (an example is *Hydra* (Figure 4.35), where small asexual benthic polyps alternate with sexual planktonic medusae), Cubozoa (the jellyfish of the tropical sea called sea wasps because of their toxic stings), and Scyphozoa (the harmless jellyfish like the medusa *Aurelia*). Cnidaria, despite their morphological and dietary diversity, are still a primitive form of animals. They are diploblastic metazoans, i.e., they derive their tissues from two true embryonic germ layers in contrast to sponges, which are still at a more primitive parazoan grade of body construction. Parazoa are metazoa, which lack a true embryological germ layering. Cnidaria possess muscle and nerve cells—where they make their first appearance in animals, which gives them a higher degree of mobility than in sponges. We see here for the first time real animal locomotion, attacking and fleeing now become a standard part of the daily activities in the quest for food. Many cnidarians show a dimorphic life cycle, which includes two totally different adult forms: the sessile polyp and the planktonic medusae. Most cnidarians are carnivorous, but the 11,000 extant species have developed many feeding habits. If *Charniodiscus* is really the ancestor of the current sea pens, cnidarians would have already achieved in the Ediacaran the most complex body organization of current Anthozoa.

Sea Pen Anatomy

Sea pens are actually a colony built around a primary polyp, which differentiates into a main supportive stem. The base of the primary polyp is anchored in the sediment, and the upper portion (the rachis) produces secondary polyps in rows sometimes united in what appears as "leaves." These secondary polyps come in two forms: autozooids bear tentacles and function in feeding; siphonozoids are smaller and create water currents through the colony. Thus sea pens swell by water intake in the night, get erect, and go for food catching. With the beginning of day, the siphonozoids expel water, the animal shrinks to a third of its night size and lies nonconspicuously on the ground. In this state, they can creep slowly on the ground and when they have found a suitable place they can drill their stem into the sediment. Intake of water fixes the stem in the ground and the animal gets erect again, sea pens thus have an efficient hydrostatic skeleton. The tissue organization of the polyp is still primitive: It is essentially a central empty sac ("coelenteron") limited by an epithelial lining. At the top is the mouth opening, surrounded by tentacles. At the bottom is an opening connecting the individual polyp to the colony. The cnidarian epithelia come as an outer epidermis, separated by an acellular mesogloea from an inner gastrodermis. Near the mouth ("hypostomium"), the epithelium shows the most differentiated structure. On the gastrodermal side, you find nutritive-muscular cells overlaid by mucus gland cells. On the epidermal side, you detect in the epidermis sensory cells, nerve cells, and cnidae (nematocysts in the older literature), which are

FIGURE 4.34. The sea pen *Pennatula phosphorea* is a colonial animal of the Anthozoa class (phylum Cnidaria). At the bottom is the peduncle, which fixes the animal to the ground; it continues at the top into a rachis carrying branches, which in turn carry polyps.

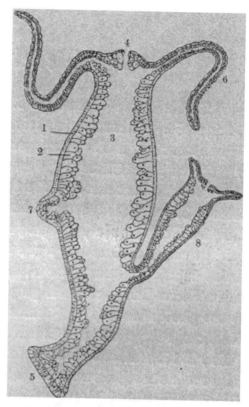

FIGURE 4.35. Cut through *Hydra*, with ectoderm (*1*), entoderm (*2*), coelenteron (*3*, gastrovascular cavity), mouth (*4*), basal disc (*5*), tentacle with nematocyst batteries (*6*), and buds in different development stages (*7, 8*).

so characteristic for these animals that their phylum name is derived from this formidable weapon, probably the most complex secretory product that the Golgi apparatus has ever developed in nature.

Feeding

In the gastric region of the coelenteron, the gastrodermis is enriched by enzymatic gland cells and cells with food vacuoles, which mix with nutritive-muscular cells. The task of the nematocyst-laden feeding tentacles is the capture of prey (Figures 4.36–4.38). Many nematocysts discharge a toxic mixture of phenols and proteins into the flesh of the prey to paralyze it. The tentacles then carry the prey to the mouth and the prey is ingested as a whole. The first phase of digestion in the coelenteron is extracellular. Mucus and digestive enzymes are secreted from the gastrodermis, transforming the food into a soupy broth. The digested nutrients are taken up by the nutritive-muscular cells via phagocytosis and pinocytosis. Their digestion is finished intracellularly in their food vacuoles.

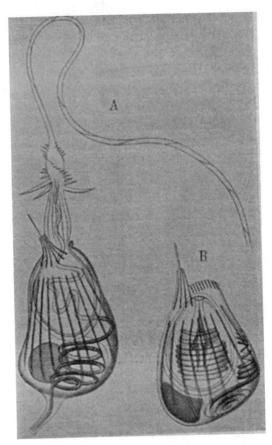

FIGURE 4.36. A nematocyst from the polyp *Hydra*, the feared weapons of cnidaria used for foraging and defense. *Right*: a resting nematocyst, *left*: an exploded nematocyst.

Nondigestible material is expelled via the mouth. The coelenteron serves multiple functions: It serves as a circulation system as well as a digestion and food distribution system. Basically, cnidarians are characterized by radial symmetry. As the gastrovascular cavity is frequently subdivided by longitudinal, ridge-like mesenteries, more complicated and even bilateral symmetry can result. In Anthozoa the mesenteries prolong into long filaments ("aconita") that hang free into the coelenteron. They serve both feeding and defense functions. They are covered by cnida, which are used to subdue live prey in the body cavity or they are expelled out of the mouth as defense organs when the Anthozoa contract violently.

Nematocysts

Nematocysts sit in capsules from cnidocytes. The upper end of the capsule is turned inside and forms a coiled tube. When triggered by a prey, this weapon

FIGURE 4.37. The green hydra (*Chlorohydra viridissima*) and the brown hydra (*Pelmatohydra oligactis*, *top*) are cosmopolitans, they belong to the order Hydroida, class Hydrozoa, phylum Cnidaria. Fragile as it appears, it can subdue with its tentacles even greater prey than the *Daphnia* depicted. *Daphnia* or the water flea is a crustacean from the order Cladocera.

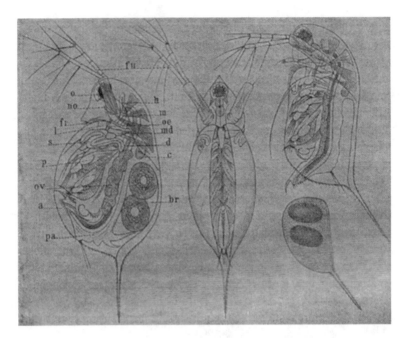

FIGURE 4.38. *Daphnia longispina*, a water flea (Cladocera), from the order Phyllopoda of the Crustacea. *Left*: a female with summer eggs is seen from lateral and in the center in frontal position. At the *right* you see a male animal with an ephippium (*bottom right*). O: compound eye, fII: antenna, br: brood chamber with eggs, ov: ovary. The alimentary tract consists of a rostrum (fI), md: mandible, l: labrum, oe: foregut, h: digestive cecum, d: a midgut with a food string, a: anus.

explodes. The kinetics of this nematocyst explosion is so quick that it requires high-speed cinematography to follow the process (Holstein and Tardent 1984). The first step is a swelling of the cyst, then the lid ("operculum") on the capsule is thrown open and the cyst's content everts and three stylets rush out to form an arrowhead that punches a hole into the prey's integument. This process occurs at the speed of 2 m/s. In the next phase, the long coiled tubule is evaginated at a 10-fold lower speed. The tubule is armed with spines that anchor the organelle in the victim. Toxin is then injected through the tubule into the prey. The overall process is too fast to be mediated by the fastest known striated muscles and it is likely that the discharge is powered by the release of mechanical energy stored in the capsule wall (tension hypothesis in contrast to the osmotic and contractile hypothesis). The nematocyst wall shows indeed a structure that might equal the tensile strength of steel. Atomic force microscopy revealed a fibrous supramolecular organization of the inner wall with periodic cross-striation consistent with bundles of collagen-like fibrils (Holstein et al. 1994). This structure can withstand the high osmotic pressure prevailing in the resting capsule.

Senses

On the sensory side, Cnidaria are still very simple animals; they lack, for example, any visual system to locate their prey. But these predators are not primitive, they can fine-tune their weapons. They possess a single cilium on the cnidocyte, which functions as a mechanoreceptor. The discharge of the nematocysts occurs only at certain frequencies of vibrations in the water. In addition, the cnidocyte is flanked by supporting cells that carry chemoreceptors. They detect mucins and N-acetylated sugars from the food source in a dose-dependent manner. Increasing the sugar concentration increases the discharge rate, but beyond a certain level it causes desensitization. The signal is transmitted to stereocilia on the supporting cell, which then tunes the cilium in a way that shifts the discharge signal to other vibration frequencies. Notably, the power spectrum of their favorite swimming prey, for *Hydra* these are copepods, fits nicely to the vibration spectrum causing maximal nematocyst explosion (Watson and Hessinger 1989). The war of the senses already begins before any cephalization in the body plan of animals becomes morphologically visible. The discharge of these cnidae is not only under sensory control but is also modulated by the nutritional state of the animal: In starved anemones, the threshold for nematocyst firing is lower than in satiated animals.

Hunting

Even more remarkable are forms of aggressive mimicry in these blind and simple animals that lure visual animals into their webs. Siphonophores develop 3-D webs, built by fine tentacles armed with millions of nematocysts that they use like spider webs. Strong swimmers hunt their small prey, while siphonophores that feed on large prey are slow swimmers. The reason for this somewhat paradox behavior is economy because siphonophores would only waste energy when trying to run behind their quicker food. Some even arrange their tentacle web in a way that resembles a copepod, which could fool a zoologist. Actually, the motion of this cheating structure even mimics the darting swim style of copepods. Consequently, their gastrozooids (stomach polyps) contained crab larvae, large copepods, and krill, all of which have copepods on their menu card. Another siphonophore showed in its net elongated fleshy structures that resemble small fish larvae. They use the extra trick to move this fake prey in the swimming mode of larval fish. Not surprisingly, the stomach of this siphonophore contained fish larvae, chaetognaths (arrow worms), and shrimp larvae, all eaters of small fish larvae (Purcell 1980). Thus these most simple metazoans use tricks that we knew before only from developed marine organisms like the anglerfish. *It appears that with the invention of animals, Nature deploys all its creativity in the fight for food. Personally, I suspect, however, an anthropocentric view. In bacteria evolution used much genius to develop novel feeding strategies. As these strategies are not visible to the naked eye and need some biochemistry and molecular biology to understand it, we are emotionally much less impressed by evolutionary achievements in the prokaryotic field.*

Fight

Tentacles with nematocysts are not only used for feeding, but also for defense and even attack against other individuals of the same species, if they do not belong to the same clone. Some sea anemones have developed acrorhagi (tentacles with white tipped vesicles), which are used in chemical warfare between sea anemones fighting for a substrate under conditions of crowding. When two anemones are confronting, the fight is slow, but fierce. The mutual touching with the acrorhagi results in severe localized necrosis in the enemy. If strips of bare rock separates otherwise crowded areas, different anemones have probably erected a chemical curtain.

Myxozoa

The feeding habit is very varied in cnidarians. You find true parasites in the Myxozoa, which were initially classified as protists because of their tiny size, but which are now on the basis of 18S rRNA classified as cnidarians. They even contain *hox* genes, leaving biologists wondering about their function in organisms that lack a definitive anterior–posterior body axis (Anderson et al. 1998). Myxozoans contain nematocysts at both ends of the "spores," which could have suggested the cnidarian diagnosis if analogous organelles were not previously found in proven protists like the dinoflagellate *Nematodinium*. Myxozoa cause the fish whirling disease and use tubifex worms as intermediate hosts.

Cnidarians as Weapons

On the other side, cnidarians are also involved in many mutualistic relationships. A spectacular example is the relationship of some sea anemones with the hermit crab. This crab searches an empty gastropod shell to protect its vulnerable rear body parts. A sea anemone might take a place on this shell, getting a free ride from the crab, but also offering chemical defense for the crab. The cloak anemone *Stylobates* often dissolves the gastropod shell with time, but provides the crab a living protective cnidarian "shell" that grows with the crab. For the crab this means that there is no need to change the house when it is growing. This "carcinooecium" (in nonscientific vernacular this reads as "crab house") is hard to distinguish from a gastropod shell and was in fact first described as one. Various animals (sea slug, a turbellarian flatworm; Figure 4.39) consume cnidarians and recycle the unfired nematocysts for their defense or for attack. Some crabs carry anemones in their claws and swing it as a chemical club.

...To Reef Bleaching as Expression of Their Dynamic Symbiotic Relationships

Acropora: Mucus Feeding in Reefs

Not all cnidaria use nematocyst-laden tentacles for hunting food. For example, the tropical sea anemone *Amplexidiscus* uses its expanded oral disk like a fisherman's

FIGURE 4.39. Tricladida (*Mesostoma* (*top right*), *Dendrocoelum* (*top left*), *Planaria* (*bottom*)) from the order Turbellaria, class Plathelminthes (flatworms), they derive their name from the forked gut, nicely visible in the nearly transparent *Dendrocoelum*. Turbellarians feed on nearly any available animal matter small enough to be captured and ingested. Between the epidermal cells they possess gland cells that secrete copious amounts of mucus and chemical deterrents. In a particular twist they eat cnidarians to recover unexploded nematocysts and they install them into their own skin as "borrowed" defense structures. Plathelminthes are furthermore mentioned for their anaerobic mitochondria and for their possible evolutionary link with comb jellies.

cast net by closing it over crustaceans and small fishes that fatefully explore the space above its mouth. Furthermore, some reef-building corals (Figures 4.40 and 4.41) are suspension feeders using mucous nets. Thin mucus sheets are spread over the colony where they collect fine particulate material that is raining on the coral reef. This food-enriched mucus is then driven by cilia to the mouth of the polyps (Figures 4.42 and 4.43). Notably, the mucus is not only feeding the coral stock but in cases of the prodigiously mucus-producing coral *Acropora* the entire reef ecosystem (Wild et al. 2004). This dominant hard coral of the Great Barrier Reef exudes up to 5 l of mucus per square meter of reef area per day. Mucus secretion has functions beyond feeding the coral, it protects the coral against fouling and desiccation during air exposure at extremely low tide and prevents sedimentation of sand or mud on the colony. Coral mucus contains proteins, triglycerides, and wax esters and represents thus a valuable food source for the entire ecosystem. The gel-like mucus forms filaments and strings on the coral, but half of it is immediately dissolved into the seawater. After detachment from the coral branches, it forms mucus floats on the sea, where it feeds bacteria and algae and fuels thus the planktonic food chain. Enclosed air bubbles provide buoyancy and the cohesive floats form large mats that trap bacteria and algae. With the loss of air bubbles, they quickly sink to the sandy bottom of the reef lagoon. Here they become a welcome food source for bottom dwellers like benthic invertebrates and fish. The settling mucus feeds also a dense population of benthic bacteria present in very high concentrations of 10^9 cells/ml. This community consumes 90% of the organic material by oxidative respiration. In this way, the corals create a reef ecosystem

FIGURE 4.40. The most important reef builder corals belong to the order Madreporaria, here represented by a species from the suborder Perforata, *Dendrophyllia ramea*, which can build even in the Mediterranean Sea "trees" of 3-m height. This animal impresses with yellow polyps emerging from branches, which are several centimeters thick.

FIGURE 4.41. A reef at the coast of Java. At the *right*, from *top to bottom*: *Tubipora* (organ-pipe corals, disturbed with retreated green polyps), *Flabellum*, and *Caryophyllia*. At the *left Madrepora, Maeandrina* (the latter two are true stony corals), *Fungia* (mushroom coral), and *left corner* again *Tubipora*. A few reef fishes, sea urchins, and a horseshoe crab complete the picture to underline that coral reefs feed and provide shelter to many animals.

characterized by high biomass, animal diversity, and productivity despite a low nutrient content in the nearby seawater. Part of this biomass is recycled to the corals.

Recycling of N

One pathway is via fish, which is particularly numerous, colored, and varied in coral reefs. Some like grunts, squirrelfish, and cardinal fish rest in large aggregates over coral heads during the day and migrate at sunset to surrounding sea grass beds to feed on benthic invertebrates. With full guts, they return to the coral heads in the morning, which they leave with empty guts at sunset. The excretory and fecal products from fish fertilize the corals. Indeed, coral heads with grunt resting grounds showed not only higher ammonia concentrations, but also significantly higher growth rates than coral heads without grunts (Meyer et al. 1983).

Symbiosis with N and C Fixers

Organic nitrogen is a limiting factor in coral reefs characterized by high productivity. The productivity problem was solved by scleractinian corals (stony corals) in an especially creative and successful way: They team up with bacteria and algae. And what a success: Structures like the Great Barrier Reef are visible from space, and stony corals have an uninterrupted record that dates back to the

FIGURE 4.42. Anthozoa, an important class of the phylum Cnidaria, here represented by *Corallium rubrum*, an Octocorallia (it carries eight tentacles and eight mesenteries that divide the interior of the polyp), an appreciated finery. The polyps are the white stars on the red coral stock. More prominent in our account are Hexacorallia (they carry multiples of six of mesenteries) and comprise the reef-building corals.

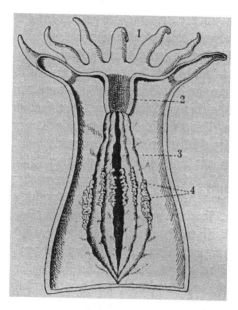

FIGURE 4.43. Cut through an Anthozoa-polyp, with tentacle (*1*), mouth (*2*), radial mesenterial muscle (*3*), and mesenterial filament (*4*).

Triassic. The success is based on mutualistic symbiosis with nitrogen-fixing and carbon-fixing autotrophs. Half of the stony corals are associated with photosynthetic endosymbiontic dinoflagellates (zooxanthellae) located in the gastrodermal cells. These endosymbionts provide via the Calvin cycle more than 100% of the carbon requirement of the animal partner. Actually, zooxanthelles are at the basis of the exuberant mucus production of stony corals. They can actually afford to feed a whole ecosystem because their symbionts are nearly pathological over producers of fixed organic carbon. However, their productivity is limited by the availability of fixed nitrogen. Hence the importance of the fecal fish pellets and the recently described association with nitrogen-fixing cyanobacteria (Lesser et al. 2004). The photosystem of these cyanobacteria is in some disarray, morphologically detectable as random crossing of their thylakoids and biophysically as high fluorescence due to inefficient energy coupling in the primary photochemistry. Actually, the stony corals attracted the interest of marine biologists by their daytime fluorescence. You will recall that the nitrogenase is oxygen sensitive—photosynthesis in the cyanobacterium would only impede the nitrogen fixation. Therefore, the zooxanthellae provide high glycerol concentration not only to the coral but also to its bacterial symbiont. The latter works as heterotroph and uses the glycerol in respiration. This delivers the reductant and ATP needed for nitrogen fixation and at the same time quenches oxygen via respiration. Even more favorable conditions govern at night when coral tissue experiences extreme hypoxia due to the inefficient circulation system and the stop of photosynthetic oxygen production by zooxanthellae.

Coral Bleaching: Possible Causes

The association of zooxanthellae with corals gives the reef their colored appearance. The loss of the symbiontic dinoflagellates turns the stony corals pale. This phenomenon is known even outside of scientific circles as "coral bleaching." The consequences are grave for the stony corals as the zooxanthellae contribute so much to the carbon metabolism of the corals. Alarming reports came from the Caribbean, where the hard coral cover on reefs has been reduced by 80% essentially around the entire basin—an unprecedented case within the past few millenia (Gardner et al. 2003). Their place is now taken by macroalgae, especially fleshy seaweeds, which will choke this formerly productive ecosystem. This is no longer a local problem in the Caribbean, reefs are now threatened globally, an estimated 30% is already severely damaged and close to 60% might be lost by 2030 (Hughes et al. 2003). The reason for the decline of the reefs is not yet clear. Hypotheses range from disease over global warming to direct anthropogenic effects like overfishing or eutrophication by agricultural pollution. Other researchers stressed that the decline of reefs predates the current coral bleaching episodes and began centuries ago. In their comparison of different reef systems, the scientists documented that large animals declined before smaller animals, free-living animals before architectural builders. Human intervention due to fishing was in their eyes the major culprit (Pandolfi et al. 2003). Global warming is for some researchers a likely cause for the decline because stressed, overheated corals are known to expel most of their pigmented microalgal endosymbionts. The bleaching threshold was in one model only 1 °C above the mean summer maximum temperatures.

The Sumatra Incident

If this model and the narrow temperature margin for this symbiosis is correct, the expected levels of global warming could spell havoc for the reefs. However, another scientific detective story from an Indonesian reef told a different story (Abram et al. 2003). A catastrophic death in 1997 was actually linked with an unusually cool water temperature. Winds in the Indian Ocean dipole (a temporary splitting of the ocean into a warm and cool part) drove strong upwelling to the coast of Sumatra. This cool water brought nitrogen- and phosphorus-enriched deeper water to the surface and created the nutritional basis for a phytoplankton bloom. This would, however, not yet explain the catastrophic effect. If N and P are not limiting, iron quickly becomes growth limiting for phytoplankton. The situation slipped out of the usual ecological equilibrium by another factor: massive wildfires in Indonesia following widespread land clearing and forest destruction. Thick smoke covered Southeast Asia. No rain drove down the smoke such that substantial amounts of terrestrial iron were carried with the winds on the ocean. This removed an ecological break on phytoplankton growth. Eyewitnesses reported a massive red tide over hundreds of kilometers in the region. Some tides kill marine organisms by toxin production, whereas others kill corals simply by asphyxia when the abundant dead biomass of the postbloom is oxidized leading to a consumption of oxygen.

Symbiodinium: Symbiosis Trade-offs

The mechanism underlying coral bleaching was under scientific scrutiny in recent years and revealed a complex interplay between stony corals and their dinoflagellate photosynthetic symbiont *Symbiodinium*. The latter inhabit corals, giant clams, and other marine invertebrates as symbiont. Substantial genetic diversity was already demonstrated by its chromosomal analysis 20 years ago and the splitting into different species was proposed (Blank and Trench 1985). Using molecular techniques (restriction fragment length polymorphism), US scientists could demonstrate that the dominant Caribbean coral *Montastrea annularis* is not associated with one, but three ecotypes (species?) of *Symbiodinium*, disproving the widely held belief of one host–one symbiont relationship. Notably, the composition of the symbiont differed and followed gradients of irradiance. This points to a trade-off between the risk of mortality from extreme temperature or light intensity versus the high costs of thermal or photoprotective mechanisms (antioxidants, heat shock and photoprotective proteins, and pigments). Indeed, following natural bleaching events, scientists observed a preferential elimination of symbionts associated with low irradiance tolerance from the brightest parts of the coral stock. Bleaching is frequently patchy and the researchers observed that some corals were protected from bleaching by hosting an additional symbiont that is more tolerant to high irradiance or temperature. Another researcher did the logical next step with transplantation experiments. Upward (deep-to-shallow) transplantation represents an acute stress, which was followed by significant bleaching in half of the transplants, but no long-term death (Baker 2001). In contrast, downward (shallow-to-deep) transplantation represents a chronic stress. This experiment was linked to no bleaching, but significant death. The researcher concluded that without bleaching a suboptimal host–symbiont combination persists, with detrimental long-term consequences. Bleaching under acute stress is an ecological gamble where a short-term risk (throwing out an inefficient symbiont) is traded against a potential long-term advantage (recolonization with a better adapted symbiont). If you loose the bet you die from lack of any symbiont; however, if you do not act, you will probably die anyway. However, this interpretation is not shared by all marine biologists (Hoegh et al. 2002). Andrew Baker proposed that dynamic symbiosis resolves the paradox of reef corals as ecological fragile yet geologically long-lived associations. This process of recolonization was followed after bleaching of the octocoral *Briareum*. The use of genetically marked exogenous *Symbiodinium* proved that bleached corals were repopulated from the environment. However, the new genotype did not always persist over time. New associations could be established and the newly acquired strain was again lost (Lewis and Coffroth 2004). Even without stressing corals, a dynamic developmental host–symbiont relationship could be demonstrated. Juvenile corals acquired a D type *Symbiodinium*, whereas adults were mainly associated with C1 type symbionts. Paradoxically, juveniles grew quicker with the "adult" symbiont form than with the "juvenile" form. Apparently, there was a delicate trade-off between both types: symbiont C1 offered a greater contribution to the host nutrition through faster growth in the host, whereas symbiont D is associated with greater thermotolerance (Little et al. 2004).

Action Outpacing Understanding

What is the take-home lesson from this excursion into the life of a colonial animal? At first glance you might say that corals are a simple model system: Biologists have to deal with one of the simplest animals and its association with a protist. However, if you look deeper into the relationship you realize that we deal here with a Russian doll of increasing complexity at the cell biology level: (cyanobacterium → primary symbiosis) to (red alga → secondary symbiosis) to (dinoflagellate → tertiary symbiosis) to (dinoflagellate in coral). This doll is situated in a highly dynamic ecosystem, on which a complex nexus of physical and chemical environmental factors act. This only seemingly simple system gets very complex and we are with the latest research still in a rather descriptive phase. I have the feeling that our courage to influence the environment with human activity of all sorts is much greater than our understanding to oversee the consequences of our action scientifically. And the reef decline is much better analyzed than, for example, the forest decline. In view of our small understanding about the ecological consequences of our action, we are probably well advised to tiptoe on our planet instead of roaming with bulldozers through the tropical rainforest or to empty the oceans with fishing factories to quote just two sins of our civilization.

The second part of the chapter follows the evolution of eating through time as an organization principle to come to grip with the endless forms of eating encountered in nature. We thus follow the fossil record, but deviate deliberately with excursions into the feeding habits of extant organisms coming close to the fossils. We took the cnidarians as living illustration of the organization level of life and the quest for food in the Ediacaran period. After this excursion, we have to go into the next epoch of the geological record, which has traditionally called the attention of all paleontologists, as the classical fossil history starts with it.

The Cambrian Revolution

The Burgess Shale Fauna

What comes then is the Cambrian, paleontologists hail the new evolutionary developments as the Cambrian Revolution. Many periods have been called revolution whether they refer to events in human history or the history of our planet. The Cambrian Revolution really merits its name. Within a very short period of time between the Ediacaran fauna at 560 Ma ago and the Cambrian fauna at 540 Myr ago the world dramatically changed from an ocean housing a rich but mostly microscopic biota to one full of macroscopic animals showing an astonishing degree of behavioral diversification. Most of the Ediacaran metazoas were either sessile filter feeders or relatively "harmless" predators floating in the water feeding on planktonic organisms. The Burgess fauna shows a wide variety of feeding habits ranging from filter feeders to grazers, scavengers, detrivores to benthic and pelagic predators. These creatures burrowed, walked, floated, and swam. Their common denominator is locomotion. At the Precambrian to Cambrian border all major phyla of animal life made their appearance including probably our own phylum, the Chordata. These animals are wonderfully preserved in the Burgess Shale found in British Columbia, Canada. The extraordinary site owes its existence to a peculiar mode of gentle burying preserving the animals in fine detail. However, it is not only the morphological detail that characterizes this site, but its richness. More than 100 animal species were diagnosed in the fossils of this site. Again, the most striking feature of this fauna is not its diversity but its exotic nature. Many fossils show outlandish creatures that did not resemble anything that came before or will come after this period in the evolution of life. About 60% of the Burgess shale fauna represents problem organisms, "*incertae sedis*" in paleontological jargon, creatures of uncertain affiliation that defy a grouping in extant phyla. Apparently, the Cambrian was a great experimental workshop where Nature invented, tried out, and discarded many forms of animal life. Paleontologists wondered why it was just this period that was so marvelously creative in evolution. No other period in the history of life, whatever smart creature it developed and whatever part of the planet it conquered, developed so many new phyla of animals. Phyla are animals with a fundamentally different body plan. What followed looks—somewhat generalized—like an elaboration of what was invented in this revolutionary period.

Impact of *hox* Genes

Some biologists suspect genetic factors: Segmentation was invented during this time period. The evolution of multicellular organisms means also that the genetic tools that allowed cells to identify their location, both with respect to space (i.e., other cells) and with respect to developmental time, had to be developed. This task is the job of hox genes. They are organized in clusters and their gene order on the chromosome corresponds both to the anterior–posterior order of gene expression as well as their temporal order of expression during the development of the organism. Why this spatiotemporal colinearity exists is a complete mystery. Early in animal evolution these command modules of animal development might have been simpler in organization and less well controlled. This could have allowed extensive playing, yielding many weird and mostly not very competitive organisms, but evolution could play with many different body forms to find the suitable ones to be retained. Later when the better-adapted forms were selected, it perhaps became necessary to freeze this "bauplan" flexibility to avoid spoiling the acquired. Stricter controls might have been imposed at the genetic level that might have allowed their further step-by-step optimization. However, this could also have meant that Nature gave away the playful attitude and a second Cambrian Revolution became impossible because the genetic network of developmental control of the body plan was hardwired and tight. I should stress that these are just speculations aired in scientific articles.

Opabinia

To be on a safer ground, let's return to the fossil evidence and we will now discuss one of the oddballs of evolution in the Cambrian to get a feeling for the experimental mode of evolution in the Cambrian. Whittington recalls a loud laughter in the audience when he first described *Opabinia* in a seminar at Oxford University. What is so funny about this organism? *As a scientist I like, of course, science fiction and I spent some time with my children looking science fiction films. Frequently, I left these films with an unsatisfied feeling—the creatures I saw were so human- or spiderlike that I deplore the lack of fantasy of filmmakers that created these purportedly alien beasts. Many of them can relatively easily be classified by a zoologist.* If you take *Opabinia* you can drive a zoologist into despair. Whatever hypothesis you try, no proposal really stuck. Actually, if higher forms of life had evolved on Mars, a biologist would not be surprised if it resembled *Opabinia*. Very superficially it resembles an arthropod, a distant relative of the extinct trilobites or the living crustaceans. However, *Opabinia* has no good arthropod links, unlike any other arthropod it has body sections with gills on the top and no head appendages. Instead, it has a frontal nozzle with a terminal claw. This organ shares more similarity with the trunk of an elephant than head appendages of arthropods. In fact, *Opabinia* specimens that show the animal from the side reveal a peculiar rearward bending of the front (mouth) end of the gut. As a lucky chance event, another *Opabinia* fossil shows that the extended nozzle could be bent back and was just long enough to reach the bent rostral gut opening. The mere appearance of the terminal claws on the nozzle suggests that *Opabinia* was everything but a peaceful suspension feeder. *Opabinia* has another morphological trait that sets it apart. On its head it shows five eyes. We do not know what picture of its environment they produced to the animal, but two points are remarkable. First, the very fact of eyes: Light was used until now by organisms as an energy source in photosynthesis or as a rough orientation cue for moving up and down in the ocean layer or to trigger cell division. In *Opabinia* light was probably used to locate its prey or predator. Second, the sheer number of eyes suggests that life entered also into a new phase of evolution. In the Cambrian it became important to see because life turned dangerous. The arms race of senses and countersenses entered a novel phase of complexity.

Cloudina's Lesson

Actually, I painted the preceding Ediacaran period as a too peaceful world of suspension feeders. Anyway, for a biologist the term peace has no real meaning, life is from its very beginning a struggle, the survival of the fittest, the eat or be eaten. It philosophically makes no real difference if a lion eats a zebra or a sponge filters bacteria. In both cases, an animal makes its living by eating another organism, great or small. However, the battle got clearly fierce in the Cambrian with the evolution of animals. To be correct, the first trends were already seen in the late Ediacaran. I have not mentioned another Ediacaran creature: *Cloudina*. This organism resembles a stack of ice cream cones that probably harbored a simple tube-dwelling polyp that used its tentacles to fish for food in the surrounding seawater. At first glance there is nothing exciting about this organism, especially as no polyp actually survived in a fossil record. What was preserved was just the calcium carbonate building the cone. However, this calcium carbonate is the new exciting observation. Animals just evolved in the early Ediacaran to a certain size and the late Ediacaran already saw the first skeletal animal fossil. What is so special about having a skeleton? Of course, a skeleton could just be used to better carry a soft body and confer it a hold. However, extant animals use exoskeletons also as armor against an unfriendly takeover. That this was already the case in the late Ediacaran was demonstrated by a survey of *Cloudina* fossils from China. When inspecting 524 specimens, 17 showed conspicuous rounded holes varying in diameter from 40 to 400 μm. The holes were larger when the skeleton was thicker. Bengston and Yue (1992) suggested that the holes were drilled by a predator and as an illustration of optimal foraging theory, greater predators were attacking greater preys. *In fact, the invention of animals characterized by their mobility and sophisticated senses set in motion an arms race between predators and prey. Such a race had already existed before, but the new, more sophisticated tools of the trade accelerated the pace of this race.*

Anomalocaris

I cannot resist describing another strange creature of the Burgess Shale that illustrates this point. It was coined by Walcott *Anomalocaris*, which literally translates to odd shrimp. In fact it was odd—as Walcott had taken just a feeding appendage as the whole animal. Other parts were also described as separate pieces. For example, the mouth consisting of a ring of plates, sporting several rows of sharp teeth in the central opening, were first attributed—without much foundation—to a jellyfish. It was again Whittington in collaboration with Briggs who literally put the pieces together and reconstructed the largest animal of the Cambrian. This giant predator probably swam like a modern manta ray propelled by the beats of its lateral lobes. Its hungry mouth was permanently open and worked like a nutcracker and the two large head appendages could be curled to bring the food to the opening where it could be crushed by constrictions. The teeth probably extended to the walls of the gullet. In the new interpretation, two large eyes flanked the head. This predator dwarfed all other animals of the Cambrian at least by a factor of three. *One can easily appreciate what selection pressure was now on the other animals. Countersenses, speed, and armors had to be developed by animals that wanted to persist. Evolution has once again changed the gear. The Burgess Shale opens the more familiar sections of the Quest for Food and we are touching firmer grounds. For all interested in this world there is a fascinating book written by the paleontologist Stephen Gould. The book is aptly entitled "Wonderful Life" and the title holds what it promises.*

Vertebrates

Toward Vertebrates: An Inconspicuous Beginning as Filter Feeders

Vertebrate Origins

Vertebrates are defined by a number of evolutionary novelties: internal branchia, a dorsal neural tube, and an internal skeleton set them apart from other organisms. But when did they evolve and from which ancestor? For a while vertebrates were put next to worms. Worms is not a defined term in zoology and if used as a morphological description for an elongated organism, it can hardly be wrong. Then specific affinities with annelids were claimed, while other zoologists saw similarities with arthropods. A major argument that led to the rejection of these hypotheses was the ventral location of the neural tube in these animal groups. Saint-Hilaire, an early nineteenth-century zoologist, proposed a fanciful solution to the problem: The worm or arthropod ancestor leading to vertebrates changed the position of top and bottom (dorsoventral inversion). This was not taken very seriously, but ironically support for this hypothesis comes from recent gene expression studies in insects and vertebrates (Shimizu et al. 1996).

Vertebrates evolved a brain case to shield the central nervous tissue and to set it apart from the rest of the body at the front position of the advancing animal. This is also the most logical position, when you want to sense your environment and send the appropriate locomotion signals to the rear. Arthropods capitalized on a different strategy: they—like mollusks—form a ring of nervous tissue organized as a central ganglion around the gut. The paleontologist M. Benton believes that this was a contradiction in the body plan of arthropods that these animals never resolved. If the arthropod's "brain" would evolve, it would logically restrict the gut and thereby the food intake. This problem was elegantly solved in vertebrates by the separation of a neural and a visceral cranium.

Chordata

Vertebrates did not yet exist during the Cambrian period. However, our phylum "Chordata" was already present with a wonderful specimen called *Pikaia*. Another early Cambrian fossil was initially also defined as a different chordate (Chen et al. 1995), but then reinterpreted as a hemichordate (Shu, Zang and Chen 1996), which thus documented also this sister group of chordates deep down in time. Chordata and Hemichordata are now mostly interpreted as sister groups to Echinodermata (starfish, sea urchin and sea-lily belong in this group) and a group of less well-known animals (e.g., Chaetognatha, the arrow worms).

Amphioxus

The nonvertebrate Chordata still have living relatives. For our current discussion, I will describe one group familiar to all biology students: *Branchiostoma lance-olatum*, better known as lancelet or amphioxus (Figure 4.44). It belongs to the group of the Cephalochordata, the "brainy" chordates, although head and brain are only discernible as vague beginnings in this animal. As I said, amphioxus is a living fossil and lives burrowed in clean sand. However, amphioxus is also a good swimmer. With respect to its feeding habit, amphioxus is quite harmless: it is a suspension feeder. It remains covered in the sand of shallow ocean water and only the head with the cirri surrounding the mouth are visible. Water is driven into the mouth; the cirri and the tentacles in the oral cavity prevent larger particles from entering the animal. At the bottom of the pharynx, you find the hypobranchial groove (a homolog of the thyroid gland?) that secretes mucus, which traps food particles. The food particles then pass upward into the epibranchial groove, and from there into the esophagus. At the entrance zone into the esophagus you find a sac called a digestive cecum. The indigestible food elements are expelled via a pore (atriopore) at the ventral side of the animal just under the caudal fin. One can differentiate an extracellular digestion in the gut lumen from an intracellular digestion in the epithelia of the gut wall and even more importantly of the cecum.

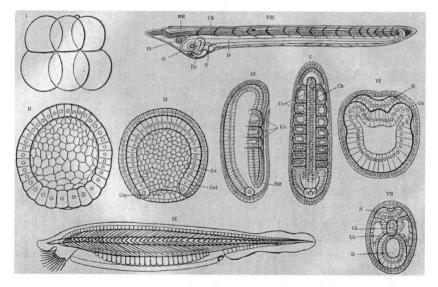

FIGURE 4.44. Amphioxus also known as lancelet (Latin: *Branchiostoma lanceolatum*). The animal is the prototype Cephalochordate. In the book it is linked to early vertebrate fossils, while molecular biologists move it to Echinodermata. Whatever its final position, it is a classic in zoology. The figure shows its embryological stages I–VIII, IX is the adult animal. Ch: chorda, N: neural tube, Ek: ectoderm, Ent: entoderm, UM: Urmund, D: gut, O: mouth, US: coelom.

The dorsal aorta of this small animal sends intestinal arteries to the gut, where they diversify into the first capillary bed around the gut to absorb the nutrients. This blood is then collected into a portal vein that carries this nutrient-rich blood to the cecum. Here the vein splits into a second capillary bed. The nutrient level is regulated in the cecum: This organ functions in lipid and glycogen storage and is very active in protein synthesis. Not surprisingly, it has been likened to the liver of the later vertebrates. The storage of nutrients is done in dorsal and ventral storage chambers. The blood from the cecum flows then into the ventral aorta, through the branchial arteries, and back to the dorsal aorta. The internal branchial gills are important for the attribution of the animal to the chordates although of minor importance for gas exchange, which is actually done in skin folds. The other taxonomically important organs are the notochord, a stiffening rod that gives strength to the animal and that is the anchor point for "V"-shaped muscle bundles (myotomes) and, of course, the nerve cord dorsally of the notochord. Without much exaggeration, here we already see the announcement of the vertebrate design. However, familiar organs like the heart and the brain are still lacking and the environment perception is mainly restricted to tactile senses in the tegument.

Pikaia and Cathaymyrus

In fact, spectacular fossil finds demonstrate that the ancestors of amphioxus can be traced back to the Cambrian period. The Cambrian Revolution thus invented all basic building plans of the animal phyla. Not a single phylum was invented after this explosive period. One of the key fossils is *Pikaia gracilens* as it was called by Charles Walcott, who discovered the Burgess site in 1909. Walcott diagnosed it confidently as a polychaete worm. Fortunately for the history of science, decades later Cambridge paleontologists led by Harry Whittington conducted a careful reappraisal and reinvestigation of the Burgess fauna. Despite their important contributions, Cambrian is not derived from Cambridge, but from Cumbria—the Latin name of Wales. They concluded in the late 1970s that *Pikaia* is not a worm, but probably a chordate. They described a zigzag band of muscles (myotomes) and a notochord, a stiffened dorsal rod that gave the phylum the name. The Burgess creature also shows distinctive features like a bilobed head with narrow tentacles and short appendages. This is why *Pikaia* is now put more in a sideline of the chordate evolution than in the mainstream.

A mainstream Cephalochordate was found in the Chengjiang Lagerstätte in China. In the last two decades China became the new El Dorado of paleontologists. The fossil is therefore fittingly called *Cathaymyrus*. Paleontologists use fancy names, but they are supposed to justify them. Cathay is Latin for China and myrus is Latin for eel. You can really see the morphology of amphioxus in this animal (Shu Conway Morris and Zhang 1996), an amazing feat of evolutionary stasis since more than 550 Ma separate both organisms.

Olfactores?

Before one gets too excited about the deciphering of the vertebrate origins, I should mention that neither cladistic analysis of morphological traits from living and fossil animals nor phylogenetic tree analysis using DNA sequences has yet yielded an accepted tree for the relationships of the great animal groups. This reservation holds also for the nearest relatives of vertebrates. A recent molecular analysis comes up with the heretic view that amphioxus is a closer relative to Echinodermata (sea urchins and the like) than to vertebrates (Delsuc et al. 2006). According to this analysis, which is backed by a genome sequencing project of the appendicularian tunicate *Oikopleura*, tunicates are the closest relatives to vertebrates. Indeed, zoologists had already suspected that the Chordata are not monophyletic and the term Olfactores was proposed to include tunicates and vertebrates. This term refers to an olfactory apparatus in fossils from a precursor of tunicates and vertebrates. In fact, the term is aptly chosen as it underlines the observation that olfactory genes represent the largest group of genes in the human chromosome.

Agnatha

Taxonomy is a creative part of biology and name changes frequently reflect changes in biological perception. Thus the subphylum Vertebrata has in the past also been referred to as Craniota, those possessing a skull. They were in the past separated into two divisions: Cyclostomata (those with a round mouth) and Gnathostomata (those with a jawed mouth). As the possession of the jaw is the dividing line, the former group is now more commonly referred to negatively as Agnatha (those which lack a jaw). Once again, we possess living representatives for the Agnatha, which allow their molecular and behavioral analysis. Hagfish (*Myxine*) is the smaller of the two surviving groups. This 40- to 80-cm-long superficially eel-like animal lives burrowed into soft marine sediment, sometimes living at depth of up to 1,000 m, reaching out only with the tip of the head and locating prey by scent. The prey is soft-bodied invertebrates or dead larger animals. Fishermen frequently complain that they also eat fish immobilized in nets. When locating a prey, it leaves the burrow and coils around the fish and bites into it by protruding and retracting the comb-like horny tooth plate on the floor of the mouth.

Lampreys (*Petromyzon*) are more distinctive and larger (up to 1 m) and lead a more aggressive, parasitic life as adults. They attach to other fish with their mouth and feed on the blood and tissue of the victim to which they remain attached. This allows them to travel large distances despite their own poor swimming capacities, which are explained by the lack of paired fins. A number of other primitive traits, in addition to the lack of a jaw (e.g., single nostril, lack of scales), distinguish agnathans from true fish. Even if the lamprey lacks a jaw, its mouth looks formidable: it is surrounded by a suctorial oral disk bearing several circular rows of horny teeth. To make matters even worse for the victim, the tongue is also equipped with horny teeth. In fact, lampreys became a pest for commercial fish in the Great Lakes of North America, where they had a disastrous killing influence on lake trout.

Ammocoete

Zoologists do not believe that extant agnathans are a good model for the feeding mode of this animal group when it evolved. The argument of such a prominent vertebrate paleontologist like A. S. Romer is that feeding on fish must be a secondary adaptation like the suction mouth. When these animals evolved, fish were simply not around. To settle this point, you have principally two approaches at hand. The first is a principle formulated by E. Haeckel "ontogeny repeats phylogeny." The second is the fossil record.

I realize now that I am frequently quoting this follower of C. Darwin, while most biology textbooks tend to quote his British master. C. Darwin is closer to the contemporary biologists because he was very cautious in his conclusions. E. Haeckel was a more courageous character, who stretched biological imagination into areas C. Darwin would never have touched, like philosophy and politics. This overstretching of biological thinking is blamed by biographers of E. Haeckel. Even if E. Haeckel was a more daring, but also controversial character—he developed his own philosophy ("monism")—some of his biological statements stood the test of time. Take the example of the lamprey. If you follow his principle, you look for the larvae of the lamprey, which is a blind worm-like animal ("ammocoete") that burrows in silt. The larva's mouth is overhung by a hoodlike upper lip that sticks out of the silt. A stream of water passes through this mouth and leaves the animal via the seven pairs of gills. It is a phytoplankton feeder—mucus produced by the endostyle, a pharyngeal gland, traps the food from the respiratory water current. The prediction by the Haeckel principle is that the initial agnathans were suspension feeders, which was in the meanwhile confirmed by fossil finds.

Primitive Fossil Agnathans

The next level of morphological development in Chordata, the true vertebrates, was generally believed to occur not before the Lower Ordovician (475 Ma ago) with the documentation of agnathan (jawless) fish. One find (I spare you the name, they are Chinese–Latin tongue twisters, which will not get popular beyond paleontology afficiados) resembles a modern hagfish, another a lamprey. They are definitively more modern in morphology than *Pikaia* and consorts and sport gill pouches with hemibranches, a heart, dorsal and ventral fins, in one case with a ray structure (Shu 1999a,b). In our context is one specimen, which shows a clear gut structure that is sediment filled as if it was sieving through silt for food particles. Over the last 20 years, a number of new agnathan fossils were described and we possess at least a tentative outline of their family tree (Forey and Janvier 1993). Ironically, according to these data the two living agnathans separated quite early and represent two different and isolated lines in the chordate family tree.

Ostracoderms

The Ordovician (for an illustrated overview of animal life in the Paleozoic era see Figures 4.45I, II, 46–48) also presents vertebrates, which became known as Ostracoderms; they became prominent in the Devonian ocean. They share with the living agnathans a few apparently primitive characters like the single nostril and the absence of a jaw. A characteristic feature is that these animals are covered with a bony armor, either in the form of large plates or as a chain mail-like assembly of small plates or a mixture of both elements. A. S. Romer deduces two facts from the presence of bones in the external armor. First, modern Cyclostomata and Chondrichtyes, which possess cartilage instead of bones in their skeleton, do not represent a primary condition in vertebrate evolution. And, second, life was pretty dangerous for these small fishlike creatures. The Swedish paleontologist Stensiö could literally dissect the head from a creature called *Hemicyclaspis*, he found only a small round mouth orifice without any teeth and large branchial pouches suggestive of a filter feeder. Another about 20-cm-large animal called *Cephalaspis* showed a broad armored head, shaped like a horse's hoof, with a frontal slit of a mouth that could only be used to grub for food trapped in sediments on the bottom (M. Benton). This animal possessed an electric organ probably for locating predators.

Eurypterids

A. S. Romer offers a candidate predator, Eurypterids (Figure 4.49). The laymen are fascinated by sheer size in paleontology and here these predators have something to offer. Eurypterids produced the greatest arthropods that ever populated the world. Specimens like *Pterygotus* reach a length of up to 3 m. These extinct water scorpions belong to the animal class Chelicerata, a group that includes diverse animals like the horseshoe crab, spiders, scorpions, mites, and ticks. These giant scorpions roamed the ancient seas, but also freshwater environments. The larger species were active predators that were probably as good in swimming as in crawling. A. S. Romer suspects that the armor of ostracoderms is the vertebrate's answer to this predation pressure and observes that the disappearance of the ostracoderms is in its timing linked to the demise of the giant water scorpions as if these sea monsters had lost their major food source with the disappearance of the ostracoderms (which they ate up?).

Cephalaspis and Thelodonts

Important innovations were made by these agnathans: *Cephalaspis* (literally who carries a head shield) showed paired pectoral fins and made an important investment in sensory organs to avoid predation. It placed the eyes on the

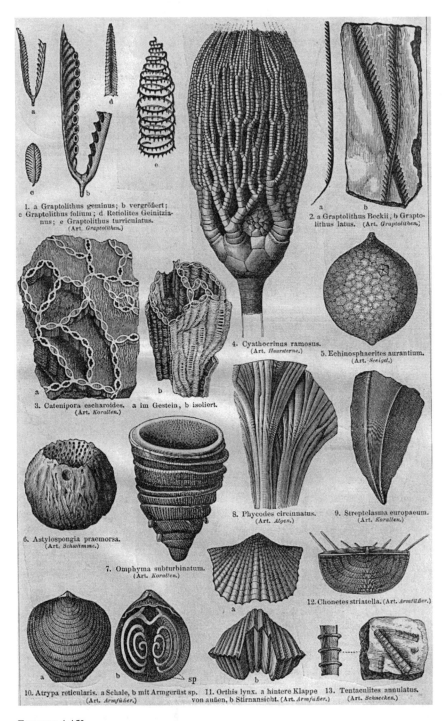

1. a Graptolithus geminus; b vergrößert;
c Graptolithus folium; d Retiolites Geinitzia-
nus; e Graptolithus turriculatus.
(Art. *Graptolithen.*)

2. a Graptolithus Beckii, b Graptolithus latus. (Art. *Graptolithen.*)

3. Catenipora escharoides. a im Gestein, b isoliert.
(Art. *Korallen.*)

4. Cyathocrinus ramosus.
(Art. *Haarsterne.*)

5. Echinosphaerites aurantium.
(Art. *Seeigel.*)

6. Astylospongia praemorsa.
(Art. *Schwämme.*)

7. Omphyma subturbinatum.
(Art. *Korallen.*)

8. Phycodes circinnatus.
(Art. *Algen.*)

9. Streptelasma europaeum.
(Art. *Korallen.*)

10. Atrypa reticularis. a Schale, b mit Armgerüst sp.
(Art. *Armfüßer.*)

11. Orthis lynx. a hintere Klappe
von außen, b Stirnansicht. (Art. *Armfüßer.*)

12. Chonetes striatella. (Art. *Armfüßer.*)

13. Tentaculites annulatus.
(Art. *Schnecken.*)

FIGURES 4.45I.

FIGURES 4.45I, II–4.48. Life's history seen from fossils: a short visual guide through the Paleozoic with illustrations from the end of the 19th century. **Figure 4.45I, II:** In an older definition, the Silurian Period included the Ordovician and the presented fossils date back until about 480 million years ago. This period followed the Cambrian and preceded the Devonian Period. Although fossils from fishes and land plants were found, the fossil record is dominated by marine invertebrates. Graptolites, saw-blade like structures of unknown zoological affinity (I, *1* and *2*) are the most conspicuous and characteristic fossils of this period. Combined with a large morphological diversity of corals (I, *3, 7, 8, 9*), graptolites are used as Silurian time indicators. Simpler multicellular organisms are sponges (I, *6*) and algae (I, *8*). Crinoids (I, *4*) were the most successful Silurian echinoderms followed by sea urchins (I, *5*). Very prevalent in the Silurian are the Brachiopoda (I, *10–12*, II, *3, 5, 7*). They were among the most numerous fossils of the Paleozoic: 12,000 extinct contrast with just 330 extant species. Superficially, they resemble bivalve mollusks, which are also found in the Silurian (II, *6*). This ecological resemblance was probably also the reason for the decline of the Brachiopoda, which lost to the bivalves. In zoological terms, brachiopods belong with echinoderms, chordates, and a few smaller groups to the deuterostomia. Bivalves belong, like the overwhelming part of the animal species, to the protostomes. The difference refers to the fate of the embryological and definitive mouth of the animals. Brachiopoda belong with Ectoprocts (Bryozoa), which were also successful in the Silurian, to the Lophophorates. This name derives from their feeding structure, which filters small organic particles, mainly phyto-plankton across cilia. Mollusks are well represented with bivalves (II, *6*), gastropoda (snails I, *13*, II, *2*), and cephalopods (II, *1, 4*). Within the latter, the progenitors of ammonites appeared and the nautiloids decreased. Within the arthropods, the trilobites dominated the fossils of the Ordovician and diversified in forms, but decreased in abundance during the Silurian (II, *10–13*). Trilobitomorpha are at the base of the Arthropoda group. Most trilobites were benthic crawlers and deposit feeders, some might have been predators, catching passing prey. Some forms developed appendages, which might have kept them afloat for a planktonic life. **Figure 4.46:** The Devonian Period starts at 395 million years ago. The characteristic Old Red Sandstone testifies continent and desert conditions. Land plants formed the first forests. However, marine deposits and their fossils predominate life forms. Two new types of fossils make their appearance. The most spectacular are bizarre, primitive fishes such that the Devonian was dubbed the "Age of Fishes." Many of them were heavily armoured (*5, 8, 9*), suggesting a strong predation pressure. The earliest forms were Agnatha, animals without jaws (*5, 8*), which were mud eaters and scavengers. However, the Devonian also saw the arrival of jawed fishes, the Gnathostomes (*9*). The other significant fossil event in the Devonian was the origin of the ammonoids (*1*, goniatites *2*) from their nautiloid ancestors. Bivalves and gastropoda diversified into many forms (*7, 10*). The trilobites were well developed during this period (*6*, the *left* specimens are enrolled like a pillbug). Arthropoda produced giants like the sea scorpion Eurypterida (not shown). Minute, but geologically more important were Ostracoda, one of the most successful groups of crustaceans. They were so widespread in the Late Devonian that they formed the ostracod-slate or cypridi-nenschiefer (*3*) named after the ostracod *Cypridina* (*4*). With 65,000 species they have the best fossil record of any arthropod group. Surviving ostracods are benthic crawlers or suspension feeders with a planktonic lifestyle. **Figure 4.47:** The Carboniferous Period started 345 million years ago and derives its name from coal-bearing strata. Not surprisingly, it is differentiated from previous geological periods by the abundance and variety of plants. Trees with woody trunks appeared and coal forests built up. No major changes were observed in the marine animal world, the major trends of the Devonian period were maintained. Giant protozoan evolved like the stratigraphically important rhizopod *Fusulina*.

dorsal shield, the shield was fenestrated by enlarged sensory fields connected to the inner ear. In addition, it showed a lateral line system, which was probably used as in modern fish, namely for the detection of water vibrations (Forey and Janvier 1993). Many of these armored animals look rather alien (Figure 4.50). This is not the case for "thelodonts"—a group of animals with laterally compressed bodies ending in an elegant symmetrical forked tail. The body is covered by tiny, hollow, tooth-like scales. They possess paired fins and large eyes—all these aspects appear very modern and suggested to paleontologists an active animal with predatory habit. Some fossil specimens showed a large barrel-shaped organ in the intestine, which they interpreted as a stomach. As these animals do not possess jaws, the possession of a stomach surprised the researcher. Did they swallow the prey in its entity even before the invention of the jaw? However, the muddy gut filling in some fossils suggests suspension feeding in this last step before the invention of the jaw (Wilson and Caldwell 1993).

The Middle Paleozoic Marine Revolution: A Story of Jaws and Teeth

Fossil Record of Jaws in the Devonian: Cladoselache and Acanthodii

Yet what vertebrate group did finally develop jaws? I have to start this section with a deception. No series of fossils documents the development of jaws. The Devonian saw the evolution of jaws in at least three distinct groups

---

FIGURE 4.47. (*Continued*) The coiled mollusks, Goniatites (*1, 4*) and Nautilus (*2*), became useful guide fossils due to their abundance and variability. Trilobites (*3*) and Brachiopods declined. Within fishes, the sharks evolved and left numerous back spines (*5, 6*). Animals followed plants on land. The Carboniferous presents thus fossils from scorpions together with remains of beetles on which they probably preyed (*11*), spiders (*10*), myriapoda and lung snails (*13*). The freshwater contained Phyllopoda crustaceans like *Leaia* (*12*). The Kulm sediment from Germany provided the enigmatic *Bostrichopus* (*8*). **Figure 4.48:**The Permian Period started 280 million years ago and closes the Paleozoic era. With respect to animal life, its upper period does not show dramatic differences from the previous Carboniferous period. However, its end is marked by one of the greatest extinction events in the history of life. The decrease in brachiopods (*10*, see the peculiar *9*), ammonoid cephalopods, and foraminifers was especially marked. Trilobites were not any longer found. Brachiopods were increasingly replaced by bivalves (*2, 3, 5, 14*). Likewise, the *Glossopteris* flora gave way to the Triassic *Dicroidum* flora. Freshwater fishes are now in much greater variety than marine fishes. Typical of the Permian fishes are heterocerc ganoids like *Platysomus* (*1*), *Paleoniscus* (*6*), and *Acanthodes* (*7*). Permian tetrapods are represented on this plate with two Stegocephalia, salamandriform amphibians, namely *Archegosaurus* (*12*) and *Branchiosaurus* (*15*). Notable in this plate is a Bryozoa (moss animal) (*4*); they have a rich fossil record and belong with Brachiopoda to the Lophophorates.

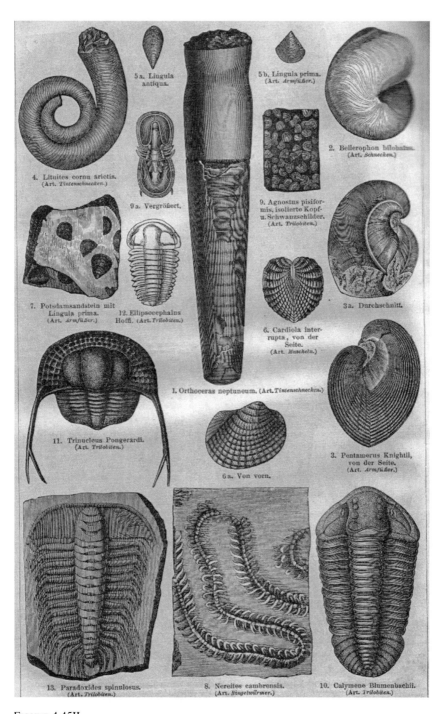

5 a. Lingula antiqua.

5 b. Lingula prima. (Art. Armfüßer.)

2. Bellerophon bilobatus. (Art. Schnecken.)

4. Lituites cornu arietis. (Art. Tintenschnecken.)

9 a. Vergrößert.

9. Agnostus pisiformis, isolierte Kopf- u. Schwanzschilder. (Art. Trilobiten.)

7. Potsdamsandstein mit Lingula prima. (Art. Armfüßer.)

12. Ellipsocephalus Hoffi. (Art. Trilobiten.)

3 a. Durchschnitt.

6. Cardiola interrupta, von der Seite. (Art. Muscheln.)

I. Orthoceras neptuneum. (Art. Tintenschnecken.)

11. Trinucleus Pongerardi. (Art. Trilobiten.)

3. Pentamerus Knightii, von der Seite. (Art. Armfüßer.)

6 a. Von vorn.

13. Paradoxides spinulosus. (Art. Trilobiten.)

8. Nereites cambrensis. (Art. Ringelwürmer.)

10. Calymene Blumenbachii. (Art. Trilobiten.)

FIGURE 4.45II

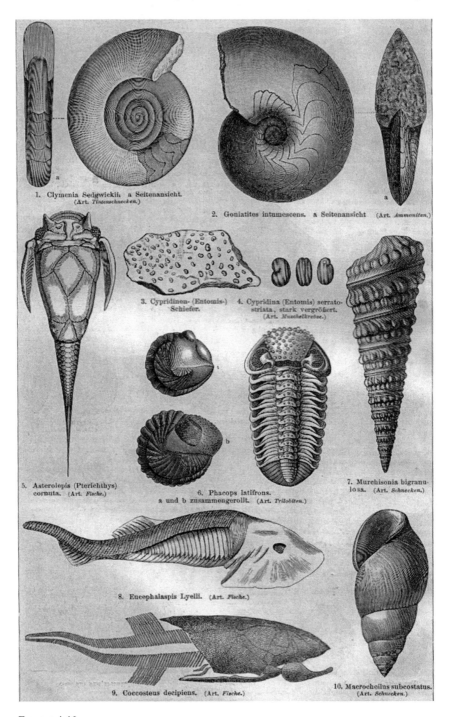

1. Clymenia Sedgwickii, a Seitenansicht.
(Art. *Tintenschnecken.*)

2. Goniatites intumescens. a Seitenansicht   (Art. *Ammoniten.*)

3. Cypridinen- (Entomis-) Schiefer.

4. Cypridina (Entomis) serrato-striata, stark vergrößert.
(Art. *Muschelkrebse.*)

5. Asterolepis (Pterichthys) cornuta.   (Art. *Fische.*)

6. Phacops latifrons. a und b zusammengerollt.   (Art. *Trilobiten.*)

7. Murchisonia bigranulosa.   (Art. *Schnecken.*)

8. Eucephalaspis Lyelli.   (Art. *Fische.*)

9. Coccosteus decipiens.   (Art. *Fische.*)

10. Macrocheilus subcostatus.
(Art. *Schnecken.*)

FIGURE 4.46

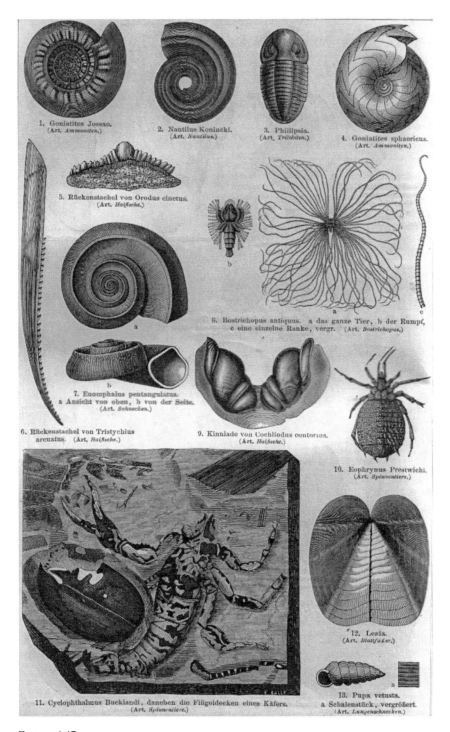

1. Goniatites Jossae.
(Art. *Ammoniten.*)

2. Nautilus Konincki.
(Art. *Nautilus.*)

3. Phillipsia.
(Art. *Trilobiten.*)

4. Goniatites sphaericus.
(Art. *Ammoniten.*)

5. Rückenstachel von Orodus cinctus.
(Art. *Haifische.*)

8. Bostrichopus antiquus. a das ganze Tier, b der Rumpf,
c eine einzelne Ranke, vergr. (Art. *Bostrichopus.*)

7. Euomphalus pentangulatus.
a Ansicht von oben, b von der Seite.
(Art. *Schnecken.*)

6. Rückenstachel von Tristychius
arcuatus. (Art. *Haifische.*)

9. Kinnlade von Cochliodus contortus.
(Art. *Haifische.*)

10. Eophrynus Prestwichi.
(Art. *Spinnentiere.*)

12. Lesia.
(Art. *Blattkäfer.*)

11. Cyclophthalmus Bucklandi, daneben die Flügeldecken eines Käfers.
(Art. *Spinnentiere.*)

13. Pupa vetusta.
a Schalenstück, vergrößert.
(Art. *Lungenschnecken.*)

FIGURE 4.47

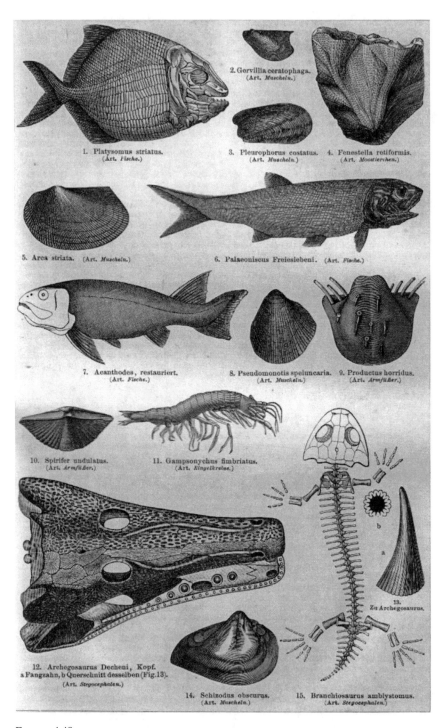

1. Platysomus striatus.
(Art. Fische.)

2. Gervillia ceratophaga.
(Art. Muscheln.)

3. Pleurophorus costatus.
(Art. Muscheln.)

4. Fenestella retiformis.
(Art. Moostierchen.)

5. Arca striata.   (Art. Muscheln.)

6. Palaeoniscus Freieslebeni.   (Art. Fische.)

7. Acanthodes, restauriert.
(Art. Fische.)

8. Pseudomonotis speluncaria.
(Art. Muscheln.)

9. Productus horridus.
(Art. Armfüßer.)

10. Spirifer undulatus.
(Art. Armfüßer.)

11. Gampsonychus fimbriatus.
(Art. Ringelkrebse.)

12. Archegosaurus Decheni, Kopf.
a Fangzahn, b Querschnitt desselben (Fig.13).
(Art. Stegocephalen.)

13.
Zu Archegosaurus.

14. Schizodus obscurus.
(Art. Muscheln.)

15. Branchiosaurus amblystomus.
(Art. Stegocephalen.)

FIGURE 4.48

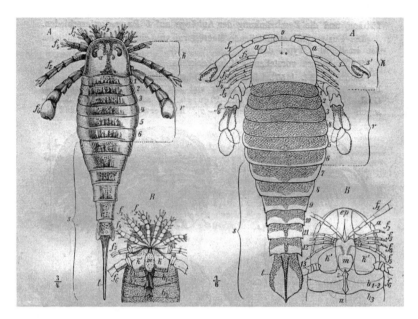

FIGURE 4.49. Water scorpions, an extinct group of Chelicerata. Two extinct forms of Gigantostraca. Left: The 2m-long *Eurypterus* from the Upper Silurian (A seen from above: K: Prosoma; r: Mesosoma; s: metasoma; t: Telson; a: compound eyes; o: median eyes; f2-f5: walking legs; f6: swimming leg; B the head seen from below: m: Metastom; K: Grinding plates). Right: the up to 3m-long *Pterygotus* from the Devonian; s: chelicera

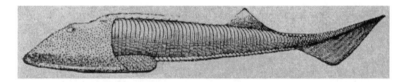

FIGURE 4.50. *Cephalaspis murchisoni* is an extinct early group of vertebrates called ostracoderms, which originated in the Silurian and flourished in the Devonian. They are characterized by an external bony head shield. Eyes are located on the dorsal side of the flat head, the mouth at the ventral side, suggesting a bottom-feeding animal straining organic material from the ground.

of fish: Acanthodii, Placodermi, and Selachii, all three extinct. The fossil record of the first two groups goes back to the late Silurian. The Selachii are the group of cartilaginous fish. Their fossil record starts only in the late Devonian. However, *Cladoselache* already shows all signs of a shark-like animal (Figures 4.51 and 4.52). A streamlined body made for predation: pointed dorsal and tail fins, powerful paired pectoral fins, a long jaw, and the impressive size of 2 m make it a dreaded predator. *Cladoselache* looks like a good swimmer; its predator character is underlined by the lack of any armor. Its prey was probably

FIGURE 4.51. Cow shark (*Hexanchus griseus*). Classical morphologists detected many primitive elements in this several meter-long shark, like 6–7 gills, a persisting chorda dorsalis, a primitive skull architecture, and irregularly shaped teeth. Despite these primitive characteristics, it is a successful predator.

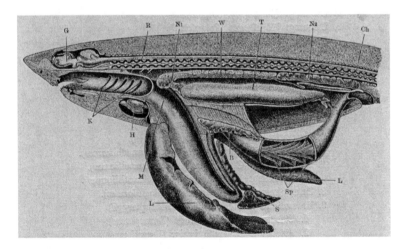

FIGURE 4.52. Cut through the shark *Galeus canis* showing the internal organs. B: pancreas, Ch: chorda and vertebral column, G: brain, H: heart, K. gills, L: liver, M: stomach, N: kidney, R: medulla, S: spleen, Sp: spirals in the gut, T: testis.

acanthodians, the oldest of the jawed fish in the fossil record (current cladograms place them, however, between Chondrichtyes and Osteichtyes, cartilaginous and bony fish, respectively, e.g., Smith and Johanson 2003). The Acanthodii, commonly called spiny sharks, go back to the late Silurian period. They look surprisingly modern, despite the fact that their body combines elements of both sharks and bony fish. They were active midwater swimmers, but their small size of only 30 cm made them an easy prey to *Cladoselache*. They had two protective measures: their fins were reinforced by long spines and they formed large shoals.

Placoderms

In comparison with these two modern-looking forms of fish, the third group of jawed fish, the placoderms, looked rather primitive. They share with the ostracoderms the possession of armor. In placoderms the armor consists of two parts, a head shield and a trunk shield, which are connected by a flexible neck. Most of them were rather small, but a few grew to sizable lengths. Placoderms from the order Arthrodira even showed giant forms like *Dinichtys* and *Dunkleosteus* reaching up to 9 m in length, which made them the most ferocious predators of the Devonian. Even smaller animals of this order, like Brachythoraci, must have been fish predators as proven by the stomach contents of some fossils. Despite the top predator position, these animals maintained heavy armor like the rest of the placoderms. The movable jaws allowed manipulation of the prey; cutting and grinding was made possible by sharp plates in the jaws. The flexibility between the head and the body allowed the Arthrodira to widely open the mouth by lifting—and not by lowering—the head. Not all forms of placoderms were free-swimming predators. The earliest forms were bottom-dwellers and poor swimmers. Sluggish benthic forms had their mouth on the ventral side and probably fed on bottom detritus and invertebrates. The order Rhenanida developed ray-like forms with an anteriorly placed mouth. The order Antiarchi with *Bothriolepsis* was a mud-grubber as deduced by the intestinal content.

Jaws: Comparative Anatomy and Developmental Genetics

If the fossil record cannot clarify the origin of the jaw, biologists must rely on comparative anatomy for insight (Figure 4.53). In a hypothetical scheme the jaw is derived from the skeletal elements reinforcing the gill openings, the pharyngeal arches. The first gill is flanked rostrally by the mandibular arch and caudally by the hyoid arch. In primitive vertebrates the arches resemble each other. Then, however, in higher vertebrates the mandibular arch differentiates into two cartilages and becomes in the gnathostomes the palato-quadrate (upper jaw) and the Meckel's (lower jaw) cartilages. This scheme is widely accepted, but the problem is that this sequence appears to support the homologies between lamprey larval lips and gnathostome jaws (Shigetani et al. 2002). Comparative anatomy does not see a fundamental novelty separating agnathans from gnathostomes. The anatomists sought the help of developmental geneticists. They found that

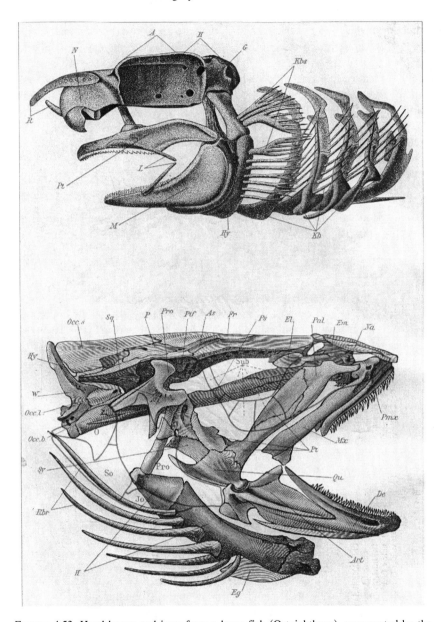

FIGURE 4.53. Head bones and jaws from a bony fish (Osteichthyes), represented by the haddock *Melanogrammus aeglefinus* (*bottom*) and a cartilageous fish (Chondrichthyes), represented by the catfish *Scyllium canicula* (*top*). Note the increasing complexity and sophistication of the head and jaw structure in bony fishes when compared to cartilageous fishes.

in gnathostome embryos, cranial neural crest cells migrate from the mid- and hindbrain into the pharyngeal arches and give rise to the pharyngeal skeleton, including the jaw. Yet, the same immigration of cells is seen in the lamprey. However, instead of forming a dorsal and ventral cartilage, in lamprey embryos a velum, a muscular pump organ, is formed. Geneticists demonstrated the presence of the homeobox genes *hox5-6-7* in lampreys. Jawed vertebrates have three pairs of *Dlx* homeobox genes that are expressed in restricted domains of the branchial arches (Koentges and Matsuoka 2002). Double negative $Dlx5/6^{-/-}$ mutants exhibited in embryonic mice a homeotic transformation of the lower jaws to upper jaws (Depew et al. 2002). *Dlx* genes were also detected in lampreys. Thus, novelty did not come with new genes either. Hence, geneticists suspected that it arrived by distinct regulation of shared genes. This was confirmed in studies where the homeobox gene expression was compared in amphioxus, lamprey, and gnathostomes. The dividing line was a posterior retraction of the *Hox6* gene expression in gnathostomes (Cohn 2002). On the basis of similar expression studies, another group came to the conclusion that the *Hox* gene code and the molecular cascades for interaction in the oral region were already in place in ancestral vertebrates (the *Hox* code even in amphioxus; Shigetani et al. 2002). The acquisition of the jaw is explained by heterotopy, a change in the place of development, here a more caudal restriction of growth factors. *Believe it or not, the Japanese authors from the 2002 Science paper quoted an article from E. Haeckel from 1875, where he formulated this principle.*

The Evolution of Teeth

The Devonian Sea was ruled by placoderms. However, early placoderms still lacked teeth in their jaws. These animals caught their prey with their gums; some had bony cutting blades in the jaws or tusk-like structures, but none of the primitive placoderms sported teeth. Comparative anatomy provided the classical concept for the origin and evolution of teeth. Anatomists deduced a cooption of external skin denticles at the margin of the jaws. Remember the tooth-like scales of the thelodonts with a dentinous crown, a bony base, and a single large pulp cavity (Wilson and Caldwell 1993). Teeth, produced at controlled locations from specific tooth-producing tissue, are a shared derived character (what taxonomists call a synapomorphy) of Chondichtyes, Acanthodii, and Osteichtyes. Therefore it came as a surprise when clear-cut teeth were described in advanced placo- derms belonging to the order Arthrodira (Smith and Johanson 2003). Dentition was seen on the upper and lower dental plates forming an occluding feeding structure. Tooth development was patterned and regulated, starting from tooth primordia resulting in teeth rows. Each tooth was composed of regular dentine formed from cells within a pulp cavity. This observation raises the possibility that teeth developed twice in vertebrate evolution and the similarities are the result of convergent evolution. This would in principle not be too surprising: jaws with high abrasion resistance based on biomineralization were, for example, also developed in totally different animals like the marine polychaete worm

Glycera (Lichtenegger et al. 2002). This jaw exceeds the resistance of verte-brate dentin and approaches that of tooth enamel. However, anatomically it is a totally different structure, which is also chemically based on a totally different biomineral. While the enamel of vertebrate teeth is composed of the calcium-based biomineral hydroxy apatite, that of *Glycera* is composed of the copper-based mineral atacamite. The close similarity between teeth from advanced placoderms and the line of modern fish suggested to some prominent vertebrate paleontologists more a problem of the placoderm/fish tree than evidence for convergent evolution (Stokstad 2003).

Impact on Prey

Whatever the exact position of the placoderms, the development of jaws and teeth in fish contributed to the so-called Middle Paleozoic Marine Revolution, one of the many biological revolutions we will mention in this survey. The impact of jaws and teeth can be directly read from the fossil record and nowhere was the pressure greater as on sessile marine animals like sea lilies (crinoids) (Figure 4.54). In the Ordovician and Silurian about 3% of the crinoid fossils showed arm regeneration as a consequence of predator attack. In the Devonian, this frequency increased to 12% or more (Baumiller and Gahn 2004). The timing coincides well with the marine revolution in the predator guild, but this observation also fits well with the hypothesis of G. J. Vermeji published in the book "Evolution and Escalation," which stated that biological hazards, including that of predation, have increased through geological time. The prey guild had to respond to counter the increased frequency of attacks; as crinoids could not run away, they evolved arm regeneration as an answer to this selection pressure.

Putting Four Feet on the Ground

The First Steps with Four Feet on Land

One of the defining events in the history of life was the emergence of terrestrial vertebrates from early fish. Despite the importance of the event, finally also leading to us humans, the field is still in a flux. Surely, older gaps were closed and new fossil finds were made that qualify as missing links. One of them might become in public perception as famous as the early bird *Archaeopterix*. One major question is why move to the land at first hand? The famous vertebrate paleontologist Alfred Romer gave in the 1950s a perhaps unconventional answer. According to him, they moved on land to stay in water. This hypothesis goes back to an idea from J. Barrell, who linked the periodic aridity exemplified by the Devonian red-bed formations to this crucial event of animal evolution (Ahlberg and Milner 1994). The Devonian was a time of seasonal droughts, and fish had to escape from drying water pools to the next pool. However, there are alternative solutions to the drying pool problem as demonstrated by modern lungfish. This fish buries itself in the mud for a summer sleep at greatly reduced metabolic activity. What are the other possible motivations? Vertebrates

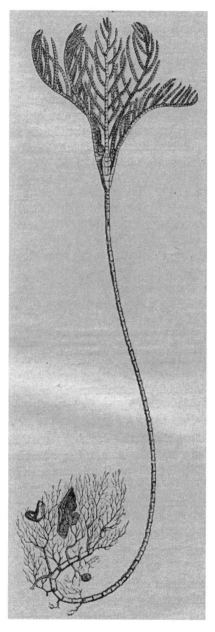

FIGURE 4.54. A sea lily, here *Rhizocrinus lofotensis*, a deep-sea species from the Northern Atlantic, class Crinoidea, phylum Echinodermata. This animal sits on a stalk as its ancestors that go back to the Cambrian. This animal holds its five arms upward and outward and thus captures sinking detritus as if into a funnel. Food grooves line the pinnules on the arms, cilia drive the food from there into the mouth. Bite-injured sea lily arms from the Devonian are the fossil signs of the evolution of animals with jaws.

might have emerged from water to disperse: substantial parts of the globe's surface is terra firma and it would be a waste of opportunities not to use this space if you can do it. For example, oxygen is much more abundant in air than in water–if you are able to breathe it. This expansion follows again the *horror vacui* argument in nature. Finally, the motivation might have been to exploit new food resources offered by the developing terrestrial ecosystem. In the Devonian, plants had already colonized the land and many invertebrates had an even older foothold on land. That the latter motivation is not a futile point is demonstrated by the foraging strategy of the eel catfish *Channallabes apus*, which lives in muddy swamps of tropical Africa (Van Wassenbergh et al. 2006). Aquatic vertebrates commonly suck up their prey by a rapid expansion of the oral cavity. This underpressure creates a vigorous water inflow, which pulls in the prey. Air is nearly 1,000-fold less dense than water—this trick will thus not work. Yet, this catfish has mainly terrestrial insects on its menu card. To get to its prey, the fish propels its head on to the shore, lifts the front part of its body, bends the head down, opens the mouth, pulls down the hyoid bone, and bites with its jaws into the prey. No weight-bearing pectoral fins are necessary for this body-lifting activity. Actually, some paleontologists doubted that the vertebrate limbs developed for locomotion on land. They suspect that limbs were developed as an adaptation for a shallow water predator.

If one looks to the current lineage leading to the modern tetrapods, one gets the following fossil ascension: *Eusthenopteron* → *Panderichthys* → *Tiktaalik* → *Acanthostega* → *Ichthyostega* (Ahlberg and Clack 2006). Of course, this is not a linear sequence as the fossils are side branches from an idealized red thread. The solution of land life obtained with *Ichthyostega* was later abandoned and other vertebrates became the ancestors of tetrapods. In fact, there is still no unanimity about whether Lissamphibia (the living amphibians) and Amniota (reptiles, birds, and mammals) are each monophyletic and whether they share the same common land ancestor. To retrace these lines a bit : as early as 1861 T. H. Huxley associated a group of Paleozoic fish, the crossopterygians, with tetrapods. Thirty years later E. Cope pointed to another group of fish, distant relatives of the crossopterygians, as the source for tetrapods. From the earliest fossil fish in this group, *Osteolepis*, they later got the name osteolepiforms (Janvier 1998). In this paraphyletic group, parallel evolution toward the morphology of large predators was observed. In comparison with their ancestors, the animals showed reduced median fins and an elaborate anterior dentition. The tetrapods thus arose out of one of several similar evolutionary experiments with a large aquatic predator role. *Eusthenopteron* belongs to one of the several clades of osteolepiforms, which became a sister group to the tetrapods. However, this 50-cm-long fusiform predatory fish still shows no terrestrial adaptations and still looks very like a lobe-finned fish (Ahlberg and Milner 1994). *Panderichthys*, the next animal on the ladder, looks like an early tetrapod with paired fins, but in addition it shows a crocodile-like flattened skull with dorsally placed eyes, a straight tail, a flattened body without dorsal fins. The pectoral and pelvic fin bones are even less developed in *Panderichthys* than in

Eusthenopteron, suggesting that the tetrapod-like body morphology developed before the limbs. *Panderichthys'* adaptation is that of a shallow water predator with aerial vision. Its pelvic girdle is 4 cm, small for a 90-cm-long animal (Boisvert 2005). *Panderichthys* is clearly incapable of tetrapod-like hindlimb-propelled locomotion. According to the interpretation of this author the early limb evolution apparently went through a "front wheel drive" stage powered by body undulation, where the pelvic fins were only used as anchors. She concluded that the front part of the animal was lifted from the ground during forward movement.

How right she was became clear 1 year later with a spectacular fossil bridging the gap between the Middle Devonian *Panderichthys* (385 million years ago) and *Acanthostega* and *Ichthyostega* in the Late Devonian (365 million years ago): *Tiktaalik* (Daeschler et al. 2006). It depends a bit on your taste whether or not you prefer a name derived from the first scientific languages Greek and Latin, tongue twisting they are anyway. If you memorize the name of this memorable fossil, you have already learned your first word of the language from Canadian Inuits. Charles Darwin would have loved this fossil, which nearly represents a designer-made animal to fit his evolution theory. This animal follows previous trends (flat body, dorsally located eyes), anticipates future trends (mobile neck, imbricate ribs), and develops a pectoral girdle and forefins, which are nearly forelimbs. They are capable of complex movements and substrate support and can lift the body from the ground (Daeschler et al. 2006). The researchers concluded that the pectoral skeleton of *Tiktaalik* is a fossilized transition from fish fin to tetrapod limb. However, they oppose the view that the tetrapod forelimb (manus) is an evolutionary novelty in tetrapods. It was already assembled in the evolution of lobe-finned fish to meet the diverse challenges of life in the margins of the Devonian aquatic ecosystems (Shubin et al. 2006). These changes included transitions from the pharyngeal and opercular pumps used in fish respiration systems to buccal and costal pumping in the respiratory apparatus of land verte-brates. In *Tiktaalik* the widened skull, expanded buccal cavity, enlarged spiracle, and a more robust hyomandibular bone are apparent adaptations to air breathing. The release of the head from the pectoral girdle allows a greater mobility of the head, which opens new feeding possibilities for a predator of the Late Devonian floodplains.

Evolution is not always logical in its progress, part of this impression might reflect the incompleteness of the fossil record and thus represents an obser-vational bias, which obliges us to compare organisms that do not belong to the exact ancestor series. Another part of this lack of logic is intrinsic to the process of evolution, which does not follow a long-term goal, but seizes short-term opportunities. This notion can be well illustrated with the next animal on the line of the tetrapod's progress: Acanthostega. It retains a true fish tail with fin rays, but has well-developed forelimbs and hindlimbs that bear digits. In fact, *Acanthostega* has eight digits, *Ichthyostega* has seven, *Tulerpeton* has six. It took a while before the pentadactyly (five digits) became fixed in tetrapods and the limb bones cannot be easily equated with the sarcopterygian bone elements in the lobed fins (Ahlberg and Milner 1994). *Acanthostega* has retained fishlike internal gills and an opercular chamber for use in aquatic respiration. It was not an obligate air breather and resembles in that respect extant gill-breathing lungfishes (Coates and Clack 1991). The authors concluded that tetrapod characters such as limbs with digits first evolved for use in water and not for walking on land. The Sargasso frogfish *Histrio* provides a living example of underwater walking (Coates and Clack 1990). In fact, the forelimbs of *Acanthostega* would be too small to be of use for land walking. This trend is extended with a Carboniferous fossil, *Crassigyrinus*, a large predator with a fearful skull and dentition, but tiny forelimbs, which has restricted this tetrapod to a wholly aquatic lifestyle (Ahlberg and Milner 1994). A recent reexamination of the *Ichthyostega* fossil led to a radically new reconstruction of this animal and identifies it as an early but ultimately unsuccessful attempt at adapting the tetrapod body plan for terrestrial locomotion (Ahlberg et al. 2005). This reconstruction was likened to the building principle of a suspension bridge with the trunk during land locomotion lifted by muscles acting like cables running from elevated supports (the pelvic and pectoral girdles) to the neural spines of the vertebrae. Later essays on land locomotion in tetrapods did not build on these neural spines, but reinforced the vertebral centra. As these two vertebral structures are under different genetic controls, competing solutions to land locomotion can be rationalized (Carroll 2005). The elaborate, imbricate rib cage was a design to prevent the collapse of the chest cavity on land. The impressive jaw and dentition of *Ichthyostega* identifies this early, in the literal sense "amphibian", animal as a powerful 1-m-long predator. The long sharp teeth show that it fed on flesh. Land arthropods and worms probably supplemented its fish diet in the water. It did not move quickly on ground, its body construction make frequent stops during its land excursions likely when it probably had to rest its head and belly on the ground. However, this is not an argument against a successful predator. In evolution you need not be a fast runner. To get your meal as a predator you only have to be quicker or cleverer than your prey and probably in the Devonian, *Ichthyostega* exceeded a sufficient number of land invertebrates with respect to speed. If you are not quick enough to run after other animals, you can still specialize on slow prey. In an extreme case you target a prey that cannot run away, namely plants. Later in this book, we will come back to the evolution of herbivory in tetrapods.

Mesozoic Gigantism: The Crown of the Terrestrial Carnivores?

On Truth and Color Pictures

Biologists frequently see animals differently than moviemakers. However, the movies are sometimes so captivating that it is worthwhile asking how realistic they are. Science sometimes rests an abstract game on numbers and facts that cannot compete with the place, which paintings and movies occupy in our mind.

The movie replaces the more abstract reality buried in scientific journals because it is more vivid and more widely distributed in human brains than the message of the scientific reports. Finally, what is reality? Is the mathematical conviction that the Earth revolves around the Sun in the brain of a single person like Copernicus a reality when his fellow citizens believed just the opposite? Of course, there is a deep conviction among scientists that there is only one truth in nature and that this truth is not agreed by democratic elections but by the force of the better arguments. In principle, this structure of nature even exists in the absence of the human mind. However, there is a problem: we cannot recognize the structure of the nature surrounding us without our brain. Evolution has in addition given us a special form of brain, for example, primates are extremely visual animals and human's strong point is color detection. The temptation of colored pictures is so great for us that most scientific journals have now changed to color illustrations. We can immediately grasp extremely complex situations from a color picture, while our brain is very slow with, for example, linear mathematical proofs.

Human Projections

There is a danger with pictures and simplifications—they stick to our mind and cover or sometimes even hide the reality. The paleontologist S. J. Gould has written a thoughtful preface to the picture textbook "The Book of Life" entitled "Reconstructing (and deconstructing) the Past" where he observed that we are "framed on the bias" and where he speaks "on a history of fossil iconography". As an illustration see a picture gallery of Mesozoic life as depicted in late 19th century drawings (Figures 4.55–4.58).

Our brain is very selective when working through observations. The history of science tells us again and again that critical observations need a prepared mind. S. Freud has told us that we have subconscious filters that suppress and magnify observations depending whether or not they fit our interests. I will not work out this idea in the realm of political or religious reasoning, but stick to biology. I would argue that we are preoccupied by gigantism—this phenomenon induces in us a frightful pleasure like children who listen with delight to fairy tales whose content might on a realistic basis abhor an adult. If you look at the numerous legends and myths that tell us about dragons, these fabled animals are so similar to the Jurassic monsters (Figures 4.59, 4.60 and 4.61) that one might be tempted to speculate on genetic imprints of these Jurassic memories in the collective memory of mammals. This idea is, of course, scientific nonsense. A situation where dragons are a mixture of human projections about fear and power, created by our own mind, is more likely. Now paleontologists come with dinosaurs, which fit so neatly with these psychological projections that we transfer all these feelings on dinosaurs. If a moviemaker now takes Tyrannosaurus rex, we do not see just an animal, but our perception is immediately intermingled with collective psychological projections about fear (tyranno-) and greatness (rex)—even its scientific name is psychologically revealing.

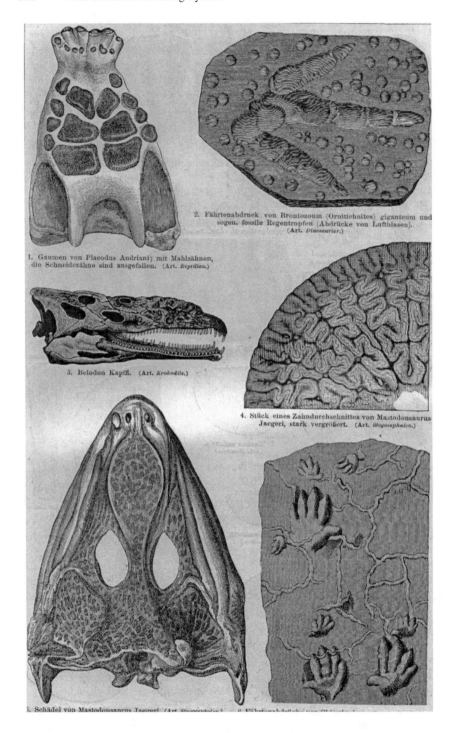

1. Gaumen von Placodus Andriani; mit Mahlzähnen, die Schneidezähne sind ausgefallen. (Art. *Reptilien*.)

2. Fährtenabdruck von Brontozoum (Ornithnites) giganteum und sogen. fossile Regentropfen (Abdrücke von Luftblasen). (Art. *Dinosaurier*.)

3. Belodon Kapffi. (Art. *Krokodile*.)

4. Stück eines Zahndurchschnittes von Mastodonsaurus Jaegeri, stark vergrößert. (Art. *Stegocephalen*.)

5. Schädel von Mastodonsaurus Jaegeri. (Art. *Stegocephalen*.)

The Jaws of *T. rex*

So what do biologists tell us about the jaws of *T. rex*, the most inspiring Creta-ceous dinosaur? In a fruitful interdisciplinary approach paleontologists teamed up with engineers. The biologists provided the fossils showing bite marks of *T. rex* on the pelvis of *Triceratops*. The paleontologists made casts from the tooth marks and tried these tooth replicas on cattle pelvis. The engineers calculated that roughly 6,400 N was required to produce the bite marks observed in the fossil (Erickson et al. 1996). Then they compared the largest maximum bite force that could be developed by *T. rex* (13,000 N) with that of many other animals. We develop as young adults the respectable force of 750 N, 1,400–1,700 N was deter-mined for wolves, sharks, and the orangutan. The lion and the alligator muster 4,000 and 13,000 N, respectively. Slightly disappointingly, *T. rex* produced bite forces comparable to those of *Alligator mississipiensis*, which uses its teeth to procure large prey and to engage conspecifics during confrontation. These activ-ities call for teeth that sustain large compression and bending forces. This could mean that *T. rex* could endure dental stresses associated with prey struggle. It could also mean that *T. rex* could take great bites from scavenged carcasses. Perhaps the king of the carnivores lived from carrion.

T. rex Running After Prey?

A successful carnivore is not just made by a forceful jaw, it also needs legs to run after the prey. Vertebrates had in the meanwhile invested a lot in legs and locomotion. The conquest of land is largely a story of four legs on the ground. In the Jurassic Park movie *T. rex* runs after a jeep. The idea of an alligator's jaw-force-equivalent running behind you with the speed of a car is a nightmare, but was *T. rex* a fast runner? From its stature *T. rex* is certainly the most cursorial of all theropods. However, biologists questioned the speed of this animal. Their argument was indirect, but nevertheless quite convincing. *T. rex* is tall, more precisely the belly and head are 1.5 and 3.5 m above ground, respectively. The arms are small and cannot buffer a fall. *T. rex* running at the speed of a jeep (40 mph) and then falling would not survive this fall when taking data from the car industry simulating car accidents with puppets as reference

◀———————————————————————————————

FIGURE 4.55. The Triassic Period started 225 million years ago and opens the Mesozoic era. Because of the Late Permian mass extinction, the Triassic fossil record contrasts markedly with the preceding world. Marine invertebrates are greatly impoverished. The predominant fossils are from ammonites, followed by bivalves and brachiopods. Other invertebrates are rare. The effect of the Permian extinction was less marked on vertebrates and stegocephale amphibians are still found in the Triassic and are here represented by the skull of *Mastodonsaurus* (*bottom left*). Stegocephales are also called labyrinthodonts; their name derives from the strange internal structure of their teeth that becomes apparent on sections of the fossils (*4*). Reptile fossils from the Triassic are on this plate represented by the palate of the reptile *Placodus* sporting the prominent grinding teeth (*1*) and the skull of the first crocodiles (*Belodon*, *3*). Numerous foot traces of reptiles (*2*) and amphibians (*bottom right*) were also preserved in the Triassic fossil record.

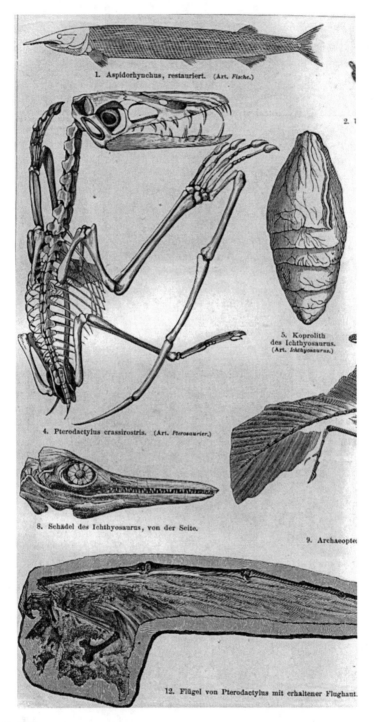

1. Aspidorhynchus, restauriert. (Art. *Fische.*)

2. 1

5. Koprolith
des Ichthyosaurus.
(Art. *Ichthyosaurus.*)

4. Pterodactylus crassirostris. (Art. *Pterosaurier.*)

8. Schädel des Ichthyosaurus, von der Seite.

9. Archaeopter

12. Flügel von Pterodactylus mit erhaltener Flughaut.

FIGURE 4.56

(Alexander 1996). Once again engineers examined the dinosaur and estimated the minimum mass of extensor leg muscle needed for fast running. The chicken, in evolutionary terms a relatively close relative of *T. rex*, runs fast. However, there is a problem: when the body mass increases, the muscle mass must increase more than is proportional to achieve the same running performance. An isometrically scaled-up chicken boosted to the 6,000 kg weight of *T. rex* would have to dedicate 99% of its body weight to the extensor muscle to remain a good runner. Even the most optimistic calculations showed that *T. rex* was limited to walking (Hutchinson and Garcia 2002). This argument does not exclude that *T. rex* was a successful carnivore if it concentrated on large preys like *Triceratops*—according to the same logic this large dinosaur prey must also have been a poor runner. Contemporary crocodiles are sluggish animals—yet you might expect very fast charges on prey animals and vigorous lashings of their tails. However, such bursts of activities are short and are followed by long recovery periods. The reason is that fast running is powered in these large vertebrates by glycolysis. The glycogen stores must be rebuilt in the muscle. A trained athlete needs about 30 min of rest to recover from a 100-m sprint. Giant reptiles would need hours to recover from sprints. In this period, they would become most vulnerable to smaller, more agile carnivores, which are better able to pay back the oxygen debt to the muscle. Thus there is also a good biochemical reason why *T. rex* wasn't a good runner.

Allosaurus

The cranial design of *Allosaurus fragilis*, a smaller and earlier (really Jurassic) cousin of *T. rex* (which came up in the late Cretaceous), was investigated in great biomechanical detail. The name *fragilis* is a misnomer as this 1.5-ton-heavy carnivore occasionally recruited the much larger *Apatosaurus* (*Brontosaurus*) as a meal. Numerous injuries on *Allosaurus* fossils suggest a rough lifestyle for this carnivore (Erickson 2001). Its smaller size allowed it to ambush larger, more dangerous prey. Instead of extracting the fossil from the rock and sectioning the bones, paleontologists have used advanced medical imaging technology like computerized tomography. Not satisfied with that, they converted the data into a

◄——

FIGURES 4.56 and 4.57 The figures show examples of the vertebrate life from the Jurassic Period, which started 190 million years ago. At the *top* are ganoid fishes. The Jurassic presents large marine reptiles like *Plesiosaurus* (*7*) or *Ichthyosaurus* (*8, 13*). The latter is also represented with a tooth (*6*) and a coprolith (*5*), a fossilized fecal dropping, which allows important insights into the feeding of the animals. Other reptiles became airborne like *Pterodactylus* (*4*) for which even soft tissue from the wings were preserved in the fossil record (*12*). Reptiles were not the only animals in the air, insects like dragonflies also belong to the Jurassic life (*11*). The Jurassic Period also yielded *Archaeopteryx* as an early avian and *Amphitherium* as an early mammalian fossil (*10*). The latter belongs to the pantotheres, which are at the basis of all the higher mammals of later times.

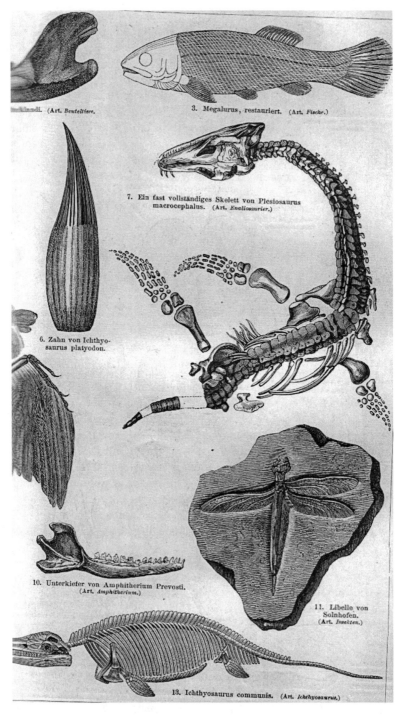

3. Megalurus, restauriert. (Art. *Fische*.)

7. Ein fast vollständiges Skelett von Plesiosaurus macrocephalus. (Art. *Enaliosaurier*.)

6. Zahn von Ichthyosaurus platyodon.

10. Unterkiefer von Amphitherium Prevosti. (Art. *Amphitherium*.)

11. Libelle von Solnhofen. (Art. *Insekten*.)

13. Ichthyosaurus communis. (Art. *Ichthyosaurus*.)

FIGURE 4.57

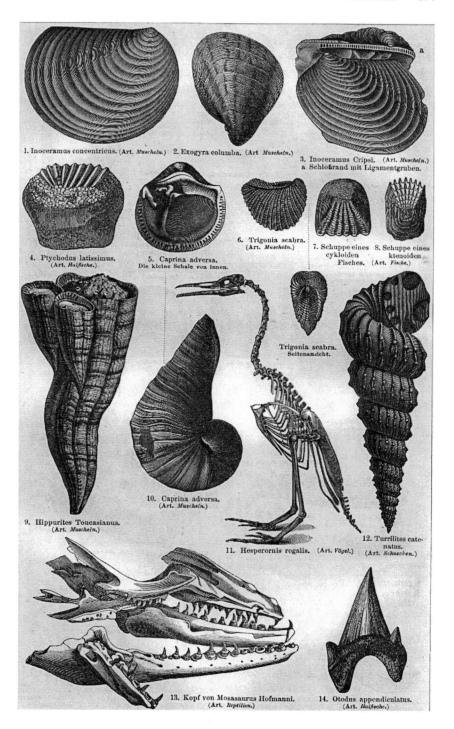

1. Inoceramus concentricus. (Art. *Muscheln.*)   2. Exogyra columba. (Art *Muscheln.*)

3. Inoceramus Cripsi. (Art. *Muscheln.*)
a Schloßrand mit Ligamentgruben.

4. Ptychodus latissimus.
(Art. *Haifische.*)

5. Caprina adversa.
Die kleine Schale von innen.

6. Trigonia scabra.
(Art. *Muscheln.*)

7. Schuppe eines   8. Schuppe eines
cykloiden   ktenoiden
Fisches. (Art. *Fische.*)

Trigonia scabra.
Seitenansicht.

9. Hippurites Toucasianus.
(Art. *Muscheln.*)

10. Caprina adversa.
(Art. *Muscheln.*)

11. Hesperornis regalis.   (Art. *Vögel.*)

12. Turrilites cate-
natus.
(Art. *Schnecken.*)

13. Kopf von Mosasaurus Hofmanni.
(Art. *Reptilien.*)

14. Otodus appendiculatus.
(Art. *Haifische.*)

"finite-element mesh," a mathematic description of the skull usable for computational mechanical analysis like in orthopedic biomechanics (Rayfield et al. 2001). The researchers estimated a maximal bite force of 3,500 N comparable to that of a lion. The skull, however, was clearly "overdesigned" and able to withstand up to 55,000 N before yielding. You can solve this problem more easily: protection from peak stress on the skull can be achieved by the loss of teeth. The team of biologists and engineers suspected that the overdesigned skull was used in a high-velocity impact into the prey somewhat like wielding a large heavy hatchet. Probably, portions of the flesh of the prey were sliced, torn away, and the victim bled to death. *Allosaurus* probably ate only the flesh and viscera. Its dental design was not made for the crushing bite of *T. rex*, which could dismember the carcass and crush the bones.

Coprolite

Further insights into the eating habit of *T. rex* came from what the researchers described as "A king-sized theropod coprolite," in plain English a large piece of fossil stool (Chin et al. 1998). The major fecal content was bone pieces. Histological analysis of the bones allowed the tentative diagnosis of a 200- to 700-kg juvenile ornithischian dinosaur as a prey. Two unexpected observations struck the researchers; they indicated that *T. rex* had a different eating mode than did modern crocodiles, which have a comparable bite force. Living reptiles have a poor dental occlusion and therefore swallow large pieces of the prey. Crocodile have only low enzymatic activity during digestion. In contrast, their gastric juice is practically pure hydrochloric acid, which impacts an aggressive destruction of the inorganic part of the bone and only the organic collagen fibers of the victim's bone remain in the fecal pellet. Just the opposite was seen in the fossil coprolite of *T. rex*: the bones of the prey were literally pulverized to fragments ranging in size from 2 to 34 mm. Birds (in other words avian dinosaurs) use horny gizzards or ingested grit (gastroliths) for food maceration; nothing like this has been observed in theropods. The bone fragmentation therefore must be the result of extensive biting before swallowing. On the other hand, the researchers

◀──

FIGURE 4.58. The Mesozoic era closes with the Cretaceous Period, which started 136 million years ago. Bivalves and gastropods present many forms. The extreme variety demonstrates that many different ecological niches were occupied by these animals. *Trigonia* (6), for example, lived in shallow but open sea. *Exogyra* (2) flourished in the littoral zone. *Inoceramus* (1, 3) is found in offshore sediments. *Hippurites* (9) and *Caprina* (10) are bivalves characteristic of the Cretaceous Period. Gastropods like *Turrilites* (12) show complicated turns of their shells. Cretaceous fishes start to resemble present-day forms. Sharks left characteristic teeth, either pointed (14) or flat (4). Bony fishes are represented with scales of either cycloid (7) or ctenoid (8) form. *Mosasaurus* was a marine lizard that is found abundantly in the Cretaceous fossils. It probably fed on fish and ammonites. Fossils of birds are rare, the figure shows *Hesperornis* (11), a flightless bird characterized by a long jaw carrying many teeth. It was probably diving for fish.

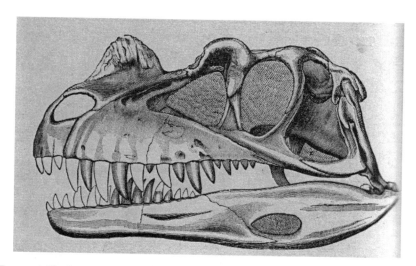

FIGURE 4.59. *Ceratosaurus nasicornis*, a 6-m-sized reptile of the Theropod suborder of the saurischians ("lizard-hips") was one of the heavyweight meat-eaters of the Jurassic. The crest on the top of the snout was probably for sexual display while the teeth leave no doubts about the carnivorous diet of this monster.

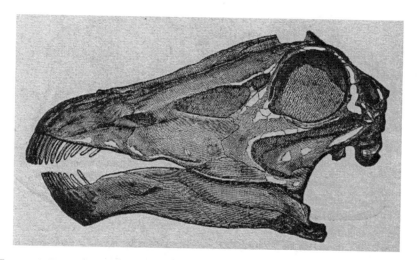

FIGURE 4.60. *Diplodocus* is a giant sauropod (reptile-foot) dinosaur of the late Jurassic. This genus produced, with 26-m size, the longest land animals that ever lived on earth (*Brontosaurus* is a more familiar relative). For its size, it had a small head sitting on a long neck. The eyes were high on the skull and the brain was ridiculously small, making nervous communication with the hindlimbs a slow process. The teeth were pencil-like, dull-edged and located at the anterior margin of the jaws suggesting foraging on soft plant material.

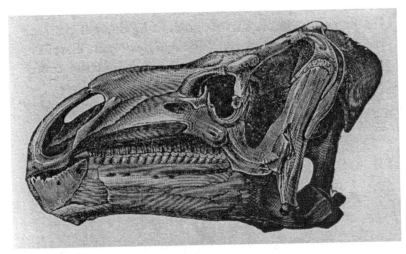

FIGURE 4.61. The *Iguanodon* is a large 10-m long herbivorous ornithopod (bird-foot) dinosaur of the Late Jurassic and Early Cretaceous. The animal has a horse-like skull with long jaws, lined with replacing grinding teeth. As in modern sheep, there are no teeth, but a bony plate at the front of the snout allowing to nip off leaves by a precision grip. The teeth are angled, permitting a sideways tearing action across the teeth. Thus the *Iguanodon* could reduce food plants to a digestible mush and had not to swallow whole plants like *Diplodoccus*, which sported in comparison a primitive dentition. This efficacious feeding capacity allowed the *Iguanodon* to flourish.

were surprised by the high percentage of incompletely digested bone within the coprolite suggesting low stomach acidity or short gut residence time of the food.

Elasmosaurus Screens the Seabed

We spoke about the power of pictures, which can lead to an imaginary existence in the mind of people. Pictures can become part of the cultural heritage, nowhere is this picture-dominated knowledge more apparent than with Cretaceous reptiles. Take the example of *Elasmosaurus*, which was described by the nineteenth-century paleontologist Dean Conybeare as a snake fused to the body of a turtle. On Zallinger's famous paintings *Elasmosaurus* with its head far out of the water looks like an incarnation of the monster of Loch Ness. In other pictures this marine plesiosaurus was shown as an agile hunter chasing cephalopods and teleost fish with its long neck. We will have to modify our ideas on the basis of new findings. The analysis of the fossilized gut content from two 1,000-kg Australian specimens hints at another diet and foraging strategy (McHenry et al. 2005). The gut was filled with the remains of benthic (bottom-dwelling) invertebrates like bivalves and gastropods. These food items made up 92% of its last dinner. Animals from the water column like belemnites, crustacean, and teleost fish contributed only a small percentage to its diet. The paleontologists could even deduce physiological details from the gut content. From the degree of preservation of the mussel shells, gastric acid played no major role in digestion. The researchers found dozens to a hundred of stones in the stomach (gastroliths) that helped to pulverize the food in the stomach and assisted thus the ridiculously undersized jaw of this large animal. The stones were apparently carefully selected for this digestive—and perhaps buoyancy—purpose as they were transported over large distances of several hundred kilometers. So put the audacious air-bound head of Zallinger's Elasmosaurus down and let it screen the sea bottom for bivalve food.

The Invention of the Egg, Brooding and Parental Care

The Triumph of the Egg

As I stressed several times, biologists see the achievements of animals differently than the lay public. I think that many biologists are only moderately impressed by the performance of dinosaurs as the largest terrestrial carnivores. In fact, these animals became extinct proving that their design was not efficient enough to be maintained in the race for the limited food resources on the planet. However, another evolutionary novelty invented by reptiles became an ongoing success story until our days. A. S. Romer called it the triumph of the egg. Some biologists will classify this as an invention in the field of reproduction biology, but the essence of the intervention is nutrition–nutrition of the young to assure the progeny a good start into the race. To fully appreciate the dimension of the invention, we must take a look back at the precursors of the reptiles, the amphibians.

Amphibian Eggs: The Unsolved Problem

Amphibians lay their eggs into water, the egg is fertilized in water, and the embryo is fed by the food reserve laid down in the egg, but when the still small progeny animals hatch from the egg, they have to immediately assure their feeding themselves. This is still the fish way of life. Let's take the 2-mm egg of the newt *Triturus cristatus*. The egg is surrounded by five layers that protect the embryo, the innermost layer liquefies to create a small aquarium for the embryo. The female hides the eggs under submerse plants using a glue-like substance liberated by the outermost layer. Cell division begins after a few hours. After 3 weeks the larvae hatch. Very soon the mouth is opened and the larva must start to feed in order to survive. The animal starts with the metamorphosis into the adult form not before the fifth month of age. During this water-bound stage the egg and the small larva are very vulnerable to predation. Evolution has searched different solutions in amphibians. Some lay very small eggs, the embryo gets very few yolk as reserve, and a tiny larva soon hatches, which has to feed a long time to reach the size appropriate for metamorphosis. Or the animal invests a lot in yolk reserves, which prolongs the embryonic phases. A quite sizable larva hatches, which needs less time for development to metamorphosis. Also this strategy has its price: a large egg is also very nutritious for many aqueous predators and might for them be as tasty as caviar is for us. Some amphibians therefore keep guard next to the water hole, which they used for egg laying. *Ambystoma opacum* lays eggs in holes, which are only filled in the early autumn by rainwater to start development that is out of reach of water-bound predators. Even more extreme is the solution of *Salamandra atra*: the eggs develop in the oviduct, the larva hatches within the oviduct and feeds on undeveloped eggs that liquefy in the oviduct. Thus a live 5-cm-long animal is born for terrestrial life. Another extreme case is represented by *Alytes obstetricans*. As the funny name indicates, the male toad takes care of the egg strings and winds them around the hind legs (Figure 4.62). The father visits the water from time to time to prevent the eggs from desiccating. Evolution has many ideas and the very fact that these animals survived until our days means that they found acceptable solutions. However, all this creativity indicates that there is a basic problem and amphibians did not succeed in finding a generalized solution.

Amniota Brings the Egg Solution

This was actually achieved by reptiles (Figure 4.63). Biologists acknowledged this new invention by classifying reptiles, birds, and mammals as Amniota. A.S. Romer proposed that reptiles did not go on land for living there, they went there for laying eggs. As there were no other larger animals on land yet, there were not many potential predators on land either. Amphibians had to remain in the vicinity of water for reproduction and because they had not found a good solution against

FIGURE 4.62. Midwife toad (*Alytes obstetricans*) is a terrestral amphibian belonging to the Discoglossidae family. The male twists the eggs around its hindlegs.

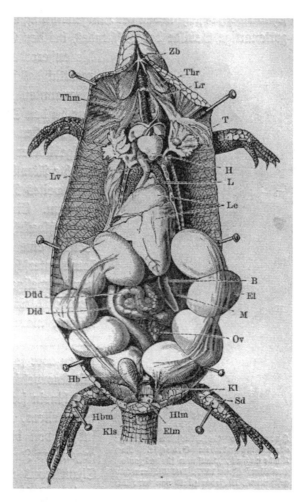

FIGURE 4.63. Anatomy of the reptile *Lacerta agilis*. B: pancreas, Did: large intestine, Düd: small intestine, El: oviduct, Elm: entry of oviduct into cloaca, H: heart, Hb: bladder, Hbm: entry of urinary vesicle into cloaca, Kl: cloaca, L: lung, Le: liver, Lr: trachea, Lv: Vena porta, M: stomach, Ov: ovarium, T: oviduct, Thr: thyreoidea, Thm: thymus.

desiccation of the skin after longer terrestrial excursions. Romer supposed that the earliest reptiles had a lifestyle like sea turtles today, which come to land only for laying eggs. Reptiles invented a small world within an eggshell. The egg is not so uniform as those of birds. Some eggs have a hard calcerous shell, and others have a leathery shell. Under the shell, you find several egg layers and the reptiles differ in detail; not all show what we know from bird's eggs (Figure 4.64), but the general outline is already there. The embryo swims again in a small sea, the amniotic cavity, which is built by layer one—the amnion, which gave the name to the advanced vertebrates. To grow, the embryo needs food, a lot of food in fact. This is provided in the yolk sac that initially fills

FIGURE 4.64. The triumph of the eggs, shown are eggs from various birds (*Struthio* = ostrich (*1*); *Spheniscus* (*2*); *Lophaethyia* (*3, 6*); *Ciconia* = stork (*7*); *Ardea* (*12*); *Anas* (*13*); *Colymbus* (*14*)).

a large part of the egg and gets smaller with the progress of the growth of the embryo. The yolk sac is linked via a duct to the alimentary tract of the embryo. However, the embryo has more needs than just food: it must excrete metabolic waste to avoid intoxication. At the distal end of the embryonic alimentary tract another duct leaves the embryo and opens in still another cavity—that of the allantois. Here the uric acid waste of the nitrogen metabolism is stored. This organ serves still another vital function: the supply with oxygen and the delivery of CO_2. To serve as an efficient gas exchange system, the allantois membrane associates with still another membrane—the chorion, which directly underlies the eggshell. In some reptiles the channel entering into the embryo from the yolk sac joins the channel leaving the embryo for the allantois cavity, nearly forming an umbilical cord. In fact, in the next section we will see that some reptiles evolved a placenta, while some mammals remained at the egg level of embryonic development, so you should not be surprised about the common vocabulary for both animal groups. Indeed, the similarities simply reflect the development of birds and mammals from reptile precursors. I will nevertheless mention a few differences: only crocodiles have an air chamber in the egg, which is a common observation in birds. Birds have two spirally wound strings called chalazae, which always keep the embryo upright. This is important as birds regularly roll the egg during brooding. Reptiles lack these strings and therefore the eggs are fixed in their position in the nest. There are also strong similarities: birds have an egg tooth on the beak, which they use when cutting the egg shell from within. The tooth is part of the horny structure of the beak. Snake embryos also have an egg tooth on the rostral part of the premaxillare bone. It resembles a normal tooth in structure, but is greater and very sharp.

Brooding

What do we know about the fossil record of eggs? The first fossils of reptiles date from the Carboniferous. The first egg fossils stem from Texas and were dated from the Lower Permian. We can thus safely conclude that the egg is an early evolutionary invention in the reptile class. Initially, biologists were quite sceptical whether the fossil record can clarify nonanatomical traits like the behavior of extinct animals. With respect to the eggs, we now discuss exciting fossil findings that demonstrate that behavior, which we commonly associate with birds, was already present in reptiles before birds had evolved. Strikingly, these fossils belonged to dinosaurs and our Jurassic horror vision of the terror reptiles now got maternal undertones. I will illustrate this turn with three research papers. The first is entitled "The nesting dinosaur" (Norell et al. 1995). Ironically, the main actor is called *Oviraptor* (literally egg robber). It was already seen in an early 1923 finding associated with eggs, but as the name indicates, paleontologists believed it died on the eggs when scavenging them. Two further fossils changed the interpretation. The first was the identification of an *Oviraptor* embryo within eggs commonly "preyed" by *Oviraptor* (Norell et al. 1994). The second fossil shows a dinosaur nest filled with 15 eggs arranged in a circular pattern with the broad ends all pointing to the center of the nest.

Above the nest the scientists found the skeleton of an *Oviraptor* in a very bird-like position. The pubis was over the nest, the hindlimbs were tightly folded, the feet lay adjacent to the eggs, both arms were wrapped around the nest as if the animal wanted to protect its eggs against the Mongolian sandstorm that covered the scene with the blanket of death. Probably the animal lay brooding on the eggs. There is a heated debate regarding whether dinosaurs were primitive endothermic animals and one might be tempted to answer this positively by associating brooding behavior with endothermy. This is, however, a precocious conclusion: some extant ectothermic animals like the python are known to brood their nests. Notably, pythons in a zoo were able to achieve a temperature around their eggs that was 5 °C higher than the ground temperature. It was also observed that the muscles of the brooding python twitched and the muscular activity increased when the ground temperature decreased. While this does not indicate endothermy, it nevertheless represents a behavioral adaptation, which heats the eggs and thus accelerates the egg development (Hutchison et al. 1966).

From Brooding to Feeding?

Another well-preserved dinosaur nest allowed a number of inferences. The nest documents primitive and advanced egg laying behavior (Varricchio et al. 1997). From a statistical analysis of the egg position in the nest, it was deduced that two eggs were laid at the same time, indicating two active oviducts. This is a primitive (in the zoological sense, meaning earlier) character as birds have only one active oviduct. Chalazae strings were apparently lacking as all eggs were partially buried and exposed with the large pole at the top. The eggs were numerous and quite sizable: the 50-kg *Troodon* dinosaur produced 24 eggs, each weighing about 0.5 kg. The analysis of the ossification of the embryo suggested that a precocious *Troodon* young left the nest soon after hatching (Horner and Weishampel 1988). Modern birds know both types of young: precocious and altricial behavior, i.e., fleeing from or sitting in the nest (Figure 4.65). The analysis of the embryo from a 5-m-large herbivore *Massospondylus* dinosaur suggests an altricial behavior for the young. The eggs contained only one 8-cm-large embryo that looks unfit for self-feeding (Reisz et al. 2005). It has, for example, only a poorly developed dentition. The head is too great and the pelvis is too small, which probably prevented this young animal from looking feeding-wise for itself. We have indirect evidence here for parental care by a dinosaur after brooding, a remarkable find for a 200-Ma-old animal.

Feeding the Young in Amphibians

As described at the beginning of this chapter, amphibians have evolved many different forms of parental investment, which ranges from hiding, guarding, and transporting the eggs to feeding of their offspring. This fact was already known to zoologists of the nineteenth century. *Ichthyophis glutinosus* (Figures 4.66 and 4.67), which superficially resembles a hybrid between a snake and an earthworm, is a caecilian amphibian (order Apoda, the name recalls

FIGURE 4.65. Naming is a problem in biology. The picture shows a tit (plural: titmice), as known in British English, but "chickadee" in American English, which covers numerous species of small songbirds of the "sparrow" group. In Latin you are on a firmer ground, the depicted animal is *Anthoscopus pendulinus*; it belongs to the family Paridae, order Passeriformes. This bird has quite early attracted human attention for their caring of the young. Eggs are laid into an artfully constructed nest. From Asia to Europe healing properties were attributed to the nest in folk medicine.

FIGURE 4.66. Caecilians are amphibians from the order Gymnophiona (formerly Apoda). They lack legs and eat worms and insects. The picture shows *Ichthyophis glutinosus* from the Indian subcontinent. Females of this amphibian have developed a sophisticated form of skin feeding for the young that resembles in some respects the breastfeeding of mammals.

FIGURE 4.67. The caecilian *Ichthyophis glutinosus* is not a viviparous amphibian, but cares for its eggs, which grow in size on the female, fulfilling one definition of a placenta.

the lack of limbs in this curious tetrapod). Some caecilians are viviparous; this species, however, lays a dozen nearly 1-cm-large eggs. The female curls the body around the eggs and the eggs increase in weight about fourfold during brooding. The eggs are apparently fed by waterous excretions from the skin of the female and the larvae increase in size to 4 cm before hatching. In some viviparous caecilians, the young animal is born with deciduous fetal dentition, which is probably used to scrape lipid-containing secretions and cells from the hypertrophied oviduct of the mother. An even more spectacular case of parental care that approaches mammalian lactation was recently described for a caecilian named after one of the nineteenth-century zoologists, G. Boulenger (Kupfer et al. 2006). The hatchlings of this oviparous caecilian are altricial and remain with the mother until they have substantially increased in size. The adult animal is an active predator of worms and even small snakes. It is equipped for this business with two rows of pointed teeth in the jaws. Even the hatchlings have blade-like teeth, which they use to scrape off the outermost layer of the maternal skin. This tissue is nutritious: The hatchlings grow 1 mm per day. This is, however, a costly exercise for the mother, which looses 14% of its weight. Brooding mothers can easily be distinguished from nonbrooding females by the paler color of the skin, which reflects a hypertrophied epidermis where cells are packed with lipid-containing vesicles. Proteins are also present, but carbohydrates lack in these vesicles. This brooding is also a dangerous exercise for the female as it has—for obvious reasons—to stop the toxin secretion from the skin, which has a defensive function in amphibians.

Mammals: Not so Modest Beginnings?

Cuvier's Early Diagnosis

In 1968 A. S. Romer teased his paleontology colleagues that mammals had a curious evolution "where parent molar teeth gave birth to filial molar teeth and so on down through the age" (Rowe 1999). In fact, the roots of the mammalian family tree were really ill defined. At the bottom was just the famous Middle Jurassic jaw stored at Oxford University. During a stay in England the famous French zoologist and paleontologist Georges Cuvier identified it in 1818 as the bone of a mammal. This brilliant early diagnosis stood the test of time. With an estimated age of 165 Ma it proved that mammals traced their origin deep into the age of the great reptiles.

Hadrocodium

A 195-million-year-old fossil from the Early Jurassic in China represents a new lineage of mammaliaforms that can be clearly differentiated from the nonmammaliaform cynodonts (Luo et al. 2001). This tiny animal fits well with the previous prejudice on modest mammalian beginnings: it weighs a mere 2 g and thus rivals the smallest living insectivores and bats with respect to tiny stature. However, even by Early Jurassic standards of the Chinese Lufeng fauna

Hadrocodium was an outlier. Triconodont insectivores ranged in weight from 27 to 89 g in *Morganucodon* and even more extreme from 13 to 517 g for *Sinoconodon*. Apparently, there was already a substantial trophic differentiation within the insectivorous feeding guild of early mammaliforms. The fossil is of interest to biologists as it shows that the mammalian characteristics were acquired step by step and not by a single episode of rapid evolution. *Hadrocodium* is the earliest taxon that lacks the primitive attachment of the middle ear bones to the mandible. This was a crucial step in the transformation of the mandibular elements from a feeding to a hearing function.

Jeholodens

In the meanwhile a number of spectacular recent findings from China have shed more light on the root of the mammalian tree. A beautifully preserved specimen from the Late Jurassic/Early Cretaceous ($\sim$ 150 Ma ago) documents a triconodont mammal close to morganucodontids (Ji et al. 1999). This tribe can be traced back to the latest Triassic ($\sim$ 210 Ma ago). They are characterized by small size, comparable to an extant shrew (perhaps just 30 g). Their dentition indicates an insectivorous diet. Their postcranial skeleton is a mosaic of ancient characters and a new therian-like pectoral girdle allowing greater excursions of the shoulder. *Jeholodens* was a ground-dwelling animal. The morganucodontid-like animal from China had a comparatively large brain. The small size is a problem because they burn more energy per body weight unit due to the large surface-to-mass ratio. This means that the animal needs proportionally more food to survive. Shrews are known to eat more than their body weight per day to stay alive. They are thus constantly eating. This creates, however, a problem with food processing—in such a small animal you cannot increase the length of the gut. To deal with these large amounts of food, you must increase the efficiency of the food processing by the gut: for example, careful chewing ameliorates the efficiency of the digestion process. An important requirement for good chewing is precisely interlocking cheek teeth. In fact, the fossil shows a nice occlusion of laterally compressed molars that could satisfy this need.

Zhangheotherium

The same formation in Liaoning Province of China that yielded *Jeholodens* provided the next complete mammalian skeleton higher up on the tree—a sister group of the multituberculates (Hu et al. 1997). In the tradition of Romer's joke the name derives from the form of the teeth in these animals—multituberculates refers to the numerous rounded cusps on their teeth. The teeth of the about 10-cm-long animal were more used for crushing and puncturing than for shearing. Despite some similarity with living monotremes (e.g., a pedal spur in the fossil, which resembles the spur associated with a poisonous gland in the modern male platypus), it was diagnosed as an archaic therian. Theria are one of the four subclasses of the class Mammalia to which belong all living mammals (marsupials and placentals) with the exception of the monotremes. The name derives from Greek and means wild beast, not an adequate description for these tiny animals. Interestingly, the animal still showed a straight cochlear, in contrast to the partially and fully coiled cochlea in monotremes and the therian mammals, respectively. Cochlea coiling allows higher sensitivity of the hearing process by increasing the length of the sensory epithelia. Apparently, it became important for the more derived groups of mammals to develop better audition for hunting or avoidance of predation, suggesting a mostly nocturnal activity.

Eomaia and *Sinodelphys*: The Basic Eu- and Meta-Therians

Now it was only a small step to the true therians and this step was passed latest at 125 Ma ago. Two wonderful fossils come from the Lower Cretaceous Yixian formation in China and they were placed by paleontologists at the root of the two living branches of the therians, the marsupials and the placentals. If you look at the reconstructed skeleton you will see two superficially very similar animals. The eutherian (placental) *Eomaia* (Greek for the dawn of a mother) was an agile animal with climbing skeletal adaptations capable of grasping and branch walking (Ji et al. 2002). It was active both on the ground and on the trees or shrubs. It lived in a lakeshore environment. However, this early mother was a tiny animal barely weighing 25 g. The metatherian *Sinodelphys* has, for a naïve fellow biologist, very similar skeletal adaptations as *Eomaia*; I would have difficulty in distinguishing both fossils except for the fact that the root marsupial weighs more than 40 g (Luo et al. 2003). In both animals hairs are preserved as carbonized filaments and impressions around the body. The pelage shows both guard hairs and a denser layer of underhairs. Hairs are thus not only a defining characteristic of living mammals, they appeared early in the evolution of mammals and might be a crucial invention for keeping the body temperature at a constant value. Indirectly, possession of hairs allows a lower food intake for survival. The interchangeable morphology indicates that both groups occupied the same niche, they were small, insectivorous, tree-climbing and—due to the coiled cochlea—probably nocturnal animals. It also indicates that *Eomaia* is a very primitive placental. It has a narrow pelvic outlet indicating a short gestation period and small youngs. Then it shows epipublic extensions from the pelvis—in marsupials these bones fix the pouch in which the young receives milk and grows in size. In short, it reproduced essentially like a modern marsupial (Weil 2002). The nearly identical ecological adaptation in both animals poses a dilemma— how could they coexist when being so similar? The problem might find at least a partial answer from the observation that metatherians dominated during the Cretaceous in North America, while eutherians dominated in Eurasia (Cifelli and Davis 2003).

Monotremes

Where are the monotremes in the fossil record? The monotremes are today represented by only two types, the platypus (*Ornithorhynchus anatinus*, the name refers to the strange duck bill of the animal) and the echidnas or spiny anteaters

(*Tachyglossus* and *Zaglossus*). They are clearly mammals, but show a number of primitive characters like egg laying. Inferences on their ancestry rely on isolated jaws from the Early Cretaceous. One specimen belongs to a platypus-sized animal much larger than the previously reported animals (Flannery et al. 1995). Its teeth suggest that it fed on material that needed crushing, but not shearing. Today, such teeth are found in marine predators like the sea otter, which feeds on hard-shelled animals. The fossil showed a large mandibular canal suggesting that this animal had a bill like the modern platypus. These monotremes dominated during their time the mammalian fauna, at least in the colder, i.e., Australian, portion of the continent Gondwana.

Poisonous Mammals

In fact, all these fossils fit the standard view of the humble start of the mammalian lineage. Yet, the section heading whetted your great expectations. Two recent fossil findings justify this view. One is the jaw of an only 60-Ma-old eutherian mammal from Canada, which still belongs to a small animal. However, its teeth are special: incisor and canine teeth show a venom delivery groove (Fox and Scott 2005). By taking the way of the snakes injecting salivary venoms, the animal can boost its ecological importance. However, like the few extant venom-producing mammals, it probably fed on insects and invertebrates—one should thus not expect sizable prey. It was hypothesized that mammals remained for a long time small, nocturnal insectivores due to the domination of the well-established larger and probably diurnal reptilian carnivores.

Repenomamus

This view must now be corrected with two Chinese fossils from the Yixian formation. One got the telling name *Repenomamus giganticus* (Hu et al. 2005). Well, gigantic here means an estimated body mass of perhaps 14 kg, but this is a lot for a Mesozoic mammal when compared to its tiny cousins. Its dentition characterizes the animal as a carnivore. It is difficult to tell whether it was an active predator or a scavenger, but paleontologists favor the first option. Scavengers are rare in extant mammals—only two species of hyenas are habitual scavengers. *R. giganticus* shows more jaw muscle leverage than these hyenas do, indicating that it was probably capable of catching and handling live prey. Good luck is an essential gift for fossil hunters and the Chinese scientists had it. When they prepared the ribcage of its smaller cousin *R. robustus*, unearthed from the same formation, they found a patch of small bones, which did not belong to *R. robustus*. The placement was suspicious— the supplementary bones were found where one would expect the stomach of the mammal. Actually, the paleontologists could identify the owner of these bones—it was a juvenile individual of *Psittacosaurus*, an herbivorous dinosaur that is common in this formation. The teeth of the dinosaur were used, thus excluding the possibility that *R. robustus* ate an embryo from an egg. The prey was about one third the size of the predator and was dismembered and swallowed as chunks. We now have to conclude that Mesozoic mammals occupied diverse niches and the larger probably competed with dinosaurs for food and territory and did not shy away from attacking small herbivorous dinosaurs.

Early Radiation

Further fossils from the Chinese treasure trove add to the impression of an early diversification of mammals. A 165-million-year-old animal got the nice name *Castorocauda lutrasimilis*, literally an otter-like animal with a beaver tail (Ji et al. 2006). It belongs to the docodonts, Mesozoic mammaliaforms with molars for omnivorous feeding. The recurved cusps of its molars suggest a feeding specialization on fish and aquatic invertebrates like modern pinniped carnivores as seals. This fossil shows well-preserved fur, differentiated into guard hairs and underfurs, suggesting that hairs with their function in thermal insulation and their tactile sensory function were well developed before the evolution of the crown Mammalia. The form of the caudal vertebrae shows a specialization of the tail for paddling and propulsion during swimming. Soft tissue remnants show webbing of the hind feet. Its forelimbs show specialization for digging and for rowing during swimming and diving as in the modern monotreme *Ornithorhynchus*. With its 500-g body weight this swimming mammal is not a giant, but it demonstrates that small insectivores did not represent all of the early mammals. However, except for these aquatic docodonts, most mammals of the Jurassic and early Cretaceous are generalized terrestrial animals. The Late Jurassic *Fruitafossor* shows single and open-rooted molars resembling the continuously growing teeth of armadillos, which feed primarily on insects and small invertebrates (Luo and Wible 2005). The teeth also showed some resemblance to those of aardvark (*Orycteropus*), suggesting specialization for feeding on ants and termites. The forelimbs demonstrated clear-cut specialization for digging activity. The primitive character of this animal is underlined by the observation that the middle ear bones were still connected to the lower jaw. In fact, *Fruitafossor* is not a eutherian, let alone a xenarthran (sloth, anteater, and armadillo (Figure 4.68)), these characters are all the result of convergent evolution.

Mammals: Seamless Nutrition

Monotremes Care for the Young: Platypus

As in the case of the dinosaurs, it was not the eating that pushed the early mammals further in the evolution game, but their feeding of the young. Where did they excel the reptiles? First, it must be stated that mammals are a continuation of the reptile's way of life. This becomes obvious when looking at a basal group of mammals like the monotremes, the only surviving members of the mammalian subclass Prototheria. Their reproductive behavior is a curious blend of primitive and modern elements. The name monotreme refers to the cloaca, the joint exit of

FIGURE 4.68. The giant armadillo (*Priodontes giganteus*) is a mammal, which belongs to the family Dasypodidae, order Edentata. This nocturnal animal is a strong digger and lives in burrows. It feeds on termites and other insects, vegetation, and some carrion. The early mammal *Fruitafossor* shows some analogous features to the armadillo.

the urinary, intestinal, and genital tracts, a typical situation in birds and reptiles. In the breeding season the female platypus digs a burrow and builds a nest from grass and leaves. Into this nest the animal lays two eggs with a leathery shell. After an incubation period of 12 days a 2-cm-long young hatches, which remains in the nest until week 17. During this time the platypus feeds the young with milk. This is a definitively modern character: I am not aware of any of the extant reptiles feeding their young. Some like crocodiles guard the nest and help the young to get out of the eggs, which are buried under a layer of sunbaked earth. Some crocodiles even stay in the vicinity of the freshly hatched crocodiles to deter predators. Feeding the young after hatching is a trait that evolved in the vertebrates only in mammals and birds, an early and a later offspring of the reptile line. Perhaps the living reptiles have lost this trait, which was present in dinosaurs as indirectly suggested by the above-mentioned *Massospondylus* fossils.

The platypus (Figure 4.69) is clearly a mammal—characterized by hairs and the production of milk. However, the mammary gland is still in a primitive stage, it lacks, for example, teats. The young sucks the milk from the skin of the female as the milk emerges from the about 120 openings of the mammary gland. Early investigators like Gegenbaur and Owen identified these ducts in the platypus and their content in the stomach of the young, which convinced them in 1832 to classify the platypus as a mammal.

FIGURE 4.69. Platypus or Duckbill (*Ornithorhynchus anatinus*), sole surviving member of the mammal family Ornithorhynchidae, order Monotremata.

Echidna

Evolution is sometimes strangely modular, and different elements show up in new combinations. Take the anteater, the other living monotreme (Figure 4.70). It was a zoological sensation when Haake reported on an egg in 1884, which he retrieved from the pouch of an echnida female. He was so disturbed and excited by the finding that he crushed the leathery reptile-like egg. Gegenbaur demonstrated in 1886 that the mammary gland was also located in the pouch. However, it differed in the histology from that of typical mammals—the echidna glands looked like derivatives of suderiferous glands. In contrast to the platypus, the echidna mammary glands end in two areolae that open near the midline of the pouch, but the animal still lacks teats. A number of other observations backed the primitive nature of echidna: like reptiles and birds, echnida embryos showed an egg tooth; like birds, echnida females have only one functional oviduct (in both cases it is the left oviduct); like marsupials, echnida have bone extensions from the pelvis, which supports the pouch. For a zoologist this is really a strange mixture of elements.

Parental Care by Marsupials

Paleontologists tell us that by the late Cretaceous, many mammals—except the prototheria—had given up laying eggs and delivered the small young alive. Evolution cannot come with perfect solutions because it has to reorganize the material it found in place and invent a few new tricks. Most new inventions are combinations of already existing elements, which were sometimes used for other purposes. One might ask what was actually won when the transition from the prototheria to the metatheria state was made. The egg was given up,

FIGURE 4.70. Echidna or Spiny Anteater, with only two surviving species in the family Tachyglossidae, order Monotremata, seen from below to visualize the egg pouch, which contains also the mammary gland.

but a placenta was not yet developed as a replacement. Let's take *Didelphis marsupialis*—an opossum about the size of a platypus—as an example for a short description. Not much happens during the first 6 days after the fertilization of the ovum; the embryo grows only slowly, which is astonishing because gestation lasts only 13 days. The longest recorded gestation in marsupials is 40 days in the red-necked wallaby. During that short period there is no development of an intimate mother–fetus connection. In *Didelphis* the about 10 newborn young have just the size of a honeybee. The mother provides some help: it cleans the pouch and licks the way from the birth canal to the pouch, but that is all. The blind newborn has only well-developed clawed forelimbs, and with swimming movement 60% of the newborns reach the pouch within a minute. The rest is lost and the mother doesn't care. Once in the pouch, the young immediately attach themselves to the teats (Figure 4.71). The teat swells in the mouth of the newborn, which has several anatomical adaptations to handle eating and respiration in parallel. In fact, mouth and teat fuse anatomically assuring a fixed position for the young against accidental loss. The *Didelphis* newborn then shows for 6 weeks a milk-driven growth. However, the small young marsupial is not able to suckle—the mother pumps the milk into the young. Then the young starts to leave the pouch, but in case of danger, it returns rapidly to the pouch or later clings to the fur of the mother. Hundred days later, the youngs are weaned. Emotional links between the mother and the young are loose: In case of danger for the mother, adult kangaroos were observed to actively kick out the young from the pouch for an unhindered escape.

Placenta Types in Eutherians

Eutherian mammals are also called placentals because they developed this organ of intimate embryo–maternal relationship in the uterus. The egg is relatively small and not well provisioned. The blastocyst implants into the uterus by an active invasion procedure, which shows some elements of tumor growth. The placenta is the apposition or interdigitation of embryonic (chorion) and maternal tissue (endometrium). In mammals two basic types of placentae are distinguished. The uterine lining may be shed with the fetal membranes in an afterbirth—a condition we find in humans and which is called deciduate. Alternatively, the placenta may be absorbed by the female, which is then called a nondeciduate placenta. Placentae have also been classified on the basis of the relationship between maternal and embryonic tissue. Simple apposition of the two membranes is seen in the so-called epitheliochorial placenta found, for example, in pig, horse, and camel. In ruminants the endometrial epithelium is destroyed leading to a so-called syndesmochorial placenta. In carnivores the trophoblast (the outer cellular layer of the blastocyst) directly contacts the endothel of the maternal vessels in the decidua. In humans, primates, bats, insectivores, and rodents the endothelial lining of the maternal vessel is lost and the maternal blood directly contacts the chorion epithelium of the embryo. One speaks here of a hemochorial placenta. In the advanced stages of pregnancy in rabbits, even the chorionic epithelium is eroded and the embryonic endothelium directly contacts the maternal blood. However, even in this advanced case there is still a membrane separating both blood systems—in no case is fetal and maternal blood mixed in placental mammals.

This histological classification of the mammalian placenta affects physiological properties of the feto-maternal relationship. Just to mention one example: The close contact in the hemochorial placenta allows an efficient transfer of maternal antibodies of the IgG class into the fetal blood circulation. The human baby is thus born under the passive immune protection of the maternal humoral immune system. This has important consequences. For example, you

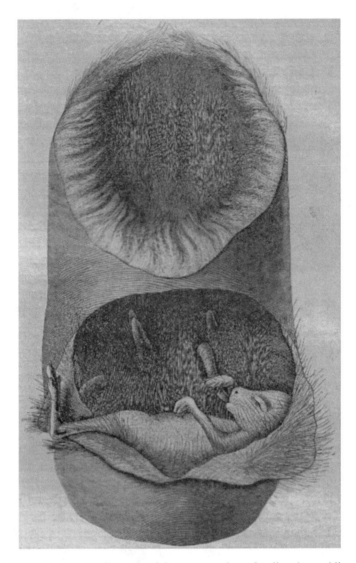

FIGURE 4.71. The pouch of a marsupial was opened to visualize the suckling young kangaroo.

can protect the newborn against tetanus infection (which are in developing countries frequently the consequence of a nonsterile cutting of the umbilical cord) by vaccinating the mother. Immune transfer via breast milk is of lesser importance in humans as demonstrated by the practice of feeding infants with a cow's milk-based infant formula. If you feed a calf with this formula (which due to the industrial processing does not contain active antibodies any longer), it will initially grow well, but will start to show life-threatening, recurrent gastrointestinal and respiratory infections after a few weeks of life. What explains this apparently paradoxical result? The explication is the histological structure of the bovine syndesmochorial placenta. As there are more membranes to cross, evolution decided not to opt for a complicated IgG transport process. The calf is thus born agammaglobulemic, i.e., without maternal IgG antibodies in the blood. The calf therefore lacks passive immune protection at birth. However, evolution is aware of this handicap and it corrects this disadvantage by bovine colostrum (the early milk of the first 2 days), which is extremely rich in IgG. This is quite unusual because in most mammals the predominant antibody in secreted body fluids like milk is IgA. The reason for this difference immediately becomes clear. During the first 2 days of life, IgG from the bovine colostrum is actively transported across the intestinal epithelia of the calf into its blood stream. Bovine milk does not have a feeding function during this time period, but an immune transfer function from the dam to the calf. As IgG is a smaller molecule than IgA, it is easier to transport IgG rather than IgA from the maternal blood stream into the milk and from there again into the blood stream of the calf. As you will realize, the fine structure of the mammalian placenta has important physiological consequences and nature plays a lot with a basic design in a group of phylogenetically closely related animals.

A major, but not exclusive function of the placenta is nutrition of the fetus: the embryo receives oxygen and nutrients from the maternal blood and delivers CO_2 and metabolic waste to the maternal circulation for excretion. The embryo is thus internalized, kept warm, protected from predation, and directly fed by the pregnant mother. It can grow substantially within the uterus. In some mammals precocial young are born that can move with the dam even on the first day after delivery. This is the condition in hares and large grazing mammals. In other mammals the young is born blind, hairless, and essentially helpless (altricial). This situation is found in rabbits, carnivores, most rodents, and humans. However, both types of young animals are fed for an extended time after delivery with milk secreted from the mammary gland of the female.

Pseudoplacentae Occur in Many Nonmammalian Vertebrates

At first glance one would expect that the placenta is *the* evolutionary novelty of eutherians. However, this is not true. The placenta or organs that resemble a placenta have developed independently several times. Within the living marsupials the bandicoots show a chorioallantoic placenta; the difference with the eutherian placenta is that the chorion does not send villi (small projections) into the maternal tissue. Many reptiles are viviparous and show a wide range of nutritive structures for the young within the body of the female animal. At one end is the primitive placenta of the lizard *Lacerta vivipara*. The eggshell is reduced to a thin epithelium and the chorioallantois now exchanges gas and water with the maternal blood system. The embryo is, however, still fed by the large yolk sac in the egg. Early tracer experiments showed only very limited transition of nutrients from the maternal into the embryonal tissue. In some geckos (e.g., *Hoplodactylus*) or some snakes (e.g., *Denisonia*) both chorioallantois and yolk sac seek contact with the maternal blood system resulting in a peculiar placenta consisting of two contact areas. The most developed stage of a placenta in reptiles is, however, seen in the skink *Chalcides chalcides* (Figure 4.72). The egg is reduced to 3 mm and thus contains only low amounts of yolk. To compensate for this lack of food reserve, the skink forms folds in the oviduct into which the egg implants. Over an elliptically demarcated area, embryonal and maternal tissues interdigitate, forming a veritable chorioallantois placenta. Despite the substantial interest that these structures generate with regard to the evolution of central novelties, I did not come across recent molecular investigations of these interesting morphological observations reported in the 1930s and 1940s. However, I found a molecular report on a viviparous fish. The genus of interest is the guppy-like fish *Poeciliopsis*. Mitochondrial DNA analysis separated them into a northern and a southern subclade, which differed in maternal provisioning strategy. One had a "matrophy index" below 1. These were the so-called lecithotrophic species. Stated differently: these animals provided egg yolk to the egg, which loses mass during development because of the inevitable loss of energy in metabolism. In the end, the dry mass of the offspring is lower than the dry mass of the fertilized egg. In contrast, in the so-called matritrophic species the index increased with a progressing stage of development of the embryo and could be as high as 100 (Reznick et al. 2002). In other words, these mothers provided postfertilization provisioning via a structure that was characterized in the 1940s as a "follicular pseudoplacenta." Molecular clock data suggest that the two clades separated about 1 Ma ago. The

FIGURE 4.72. Skinks here represented by *Chalcides tridactylus* belong to the family Scincidae, suborder Sauria of reptiles. They are remarkable for the near absence of limbs and for giving birth to fully developed young.

authors concluded that complex organs like the placenta can develop in less than 1 million years. If this argument is correct, it could explain the multiple appearances of more or less placenta-like structures in fish, amphibians, reptiles, and marsupials reported above. The developmental genetics of the mammalian, say mouse placenta, has been investigated and about 50 genes necessary for its development were characterized (Rossant and Cross 2001). It will be interesting to investigate the developmental program that directs the building of the different pseudoplacentae to see whether they are the expression of a fundamentally similar developmental program in vertebrates like we saw in the evolution of the jaw.

We mentioned the problem Darwin had with the development of a complex structure like the vertebrate eye. Many genes have to cooperate to give a useful structure. If you take out one, you lose the function and no selective advantage results for the carrier of this structure. Perhaps the intellectual difficulty associated with this argument results mainly by looking backward from a given structure. If you take the perspective of the placenta data and look forward, you might see that any of the imperfect placentae fulfill their function, some better, some less so, but all are a solution to the challenges.

Mammalian Versus Avian Feeding of the Young

Lactation was apparently vital to the early mammalian evolution. With a naked altricial young bird in your nest, you cannot look for warmth and food at the same time. The mother bird had to recruit the father for joint parental care. In mammals, females can run the show alone. As an additional advantage, a female mammal can lay down a fat reserve before and during gestation and produce milk and satisfy her own caloric needs by burning this fat depot during lactation. With no need to leave the nest, the young and the mother thus experience a much lower predation pressure. Birds learned to put a fat reserve when preparing for migration and they power long-distance flights from this fat depot. However, with a few doubtful exceptions like the "milk" produced in the gizzard of pigeons, no feeding powered from maternal reserves developed in birds. The bird's way of life and feeding has also maintained a risk, which does not exist for eutherians. There is a vulnerable period when the first eggs are laid, but brooding has not yet started. In this time period brooding parasites can lay their eggs into the nest and exploit the parental investment of other birds. This forced recruitment of a foster mother is not possible in a mammal. Furthermore, when a predator comes, the mammalian embryo has senses and legs borrowed from the dam to run away. No mobility was invented for the avian egg except the distraction maneuver of the parents feigning a broken wing or some female cuckoo carrying the egg in the beak to a foster nest.

The Food Gap between Nutrition via Placenta and Lactation . . .

The heading of this section promised you seamless nutrition of the young. The eggs, pouches, placentae, milk, and food feeding strategies really show that mammals and birds focus heavily on seamless feeding of the young. Here I must add a question, which fascinated me since the birth of my first child. I learned in medical courses that the most delicate moment of human life is at delivery. With the umbilical cord cut, the connection to the maternal circulation is severed. Gas exchange and food provision are no longer assured by the mother; the newborn has to take over these functions. From the food side, the newborn is confronted with a special challenge—the temperature drops from the womb to the ambient atmosphere. The newborn needs more food to maintain the body temperature. However, instead of getting more calories, the newborn gets nothing. Lactation is not yet ready to start. The mother needs the feedback from the vain suckling of the newborn on the nipples of her breast and the vigorous contraction of the uterus expelling the afterbirth to stimulate the breast for lactation. In addition, the newborn is not ready to digest milk because lactase is not yet expressed in its intestine. Obstetricians know that and offer a bottle with a 10% glucose solution to the newborn. Apparently, they judge that the newborn needs calories in addition to rehydration. But why is nature not caring for the newborn during the few days before the onset of lactation? Actually, mothers frequently experience pain during the injection of the milk into the breast, which occurs to a larger extent only 3–4 days after delivery and is frequently accompanied by a short fever pulse. The milk production then steadily increases during the first 10 days after delivery if the newborn suckles. The caloric need of the newborn baby is calculated by the pediatrician according to Finkelstein's rule: number of days minus 1 times 50 = daily needs of milk in grams. If you do this arithmetic exercise, you get 0, 50, 100, 150, 200, 250, 300 for the successive days of the first week. Clearly, the low caloric need of the newborn explains the delayed onset of lactation after delivery. However, this rule does not explain why the newborn needs no food during this most critical period. Actually, after doing the duty by stimulating the breast nipples, the newborn falls asleep and loses any further interest for the breast for at least 1 day. *If the mother has already started milk production, it would be a supply without demand. Intuitively, it is not logical that the newborn goes through this transition period without food support.*

...Is Filled by Autophagy

In fact, it doesn't, and the answer to this conundrum can be found in a recent publication: it is autophagy (Kuma et al. 2004). Autophagy is a well-known process for cell biologists and geneticists. In this process portions of the cell cytoplasm are sequestered into an autophagosome and the bulk of the protein material is degraded to amino acids after fusion of this compartment with the lysosome. Geneticists have identified *16ATG* genes involved in this process, which is the major route for the degradation of long-lived proteins and cytoplasmic organelles. They knew that these genes are widely distributed in eukaryotes. Apparently homologous genes influence the survival of yeast cells during nitrogen starvation, the sporulation in yeast, the fruit body building in *Dictyostelium*, the development of pupal stages in *Drosophila*, and the development of the enduring larval "dauer" stage in *Caenorhabditis*. The importance of these genes in mice development was demonstrated by Japanese scientists. They labeled a protein from the autophagosome with a fluorescence marker. The transgenic mice showed a drastic upregulation of autophagosomes within 30 min after birth; it reached maximal levels at 6 h and returned to the basal prebirth level 2 days after delivery. Particularly marked were the increases in the heart, the diaphragm, alveolar cells, and the skin. Actually, these are all tissues that have dramatically increased energy requests during the transition from aquatic to aerial life. Furthermore, the researchers demonstrated severe hypoglycemia and hypolipidemia in the newborn mice. The amino acids recovered from the degraded proteins had to feed the mammalian organism confronted with the most severe period of starvation during the mammalian life span. The amino acids can be used for the synthesis of new proteins, but the researchers showed that its most important role was as fuel for the energy metabolism. The crucial role of the autophagy process was proven by knock-out mice mutants that lacked the *ATG5* gene. These animals were born normally, but most died within a day even when immediately force-fed with milk. Extrapolating from mice to men, it now becomes clear why we have this curious delayed onset of lactation in humans and the strangely low caloric needs of the newborn infant during the first days. The overall caloric need of the newborn is not greatly different during the first day as at the end of the first week of life. However, the energy demand of important organs involved in the transition to aerial life is not covered by the ingestion of external food, but by digestion of the cytoplasm in physiologically important cells. One might speculate that the seamless feeding of the eutherian young using the placenta, autophagy, lactation, and then solid food feeding in series explains much of the ecological success of this class of vertebrates. It allows, in contrast to the metatherian marsupials, an early development of the brain. In fact zoologists writing 150 years ago were already struck by the systematically lower intelligence when marsupials were compared to their ecological complements in placentals. The very fact that only some birds rival or even outwit "placental intelligence" underlines the importance of this seamless nutrition concept. However, it again demonstrates that the details of the feeding process can be realized with different means; what is important is that it is done, not how it is done. Many ways lead to Rome.

5
The Ecology of Eating Systems

Eat or be Eaten: Anatomy of the Marine Food Chain

Overview

Trophic Levels

There is a phrase from everyday life: "big fish eat small fish." There is much truth in this, but there are also exceptions from the simple idea of big eating small. Take the second biggest fish on earth, the basking shark. It is a plankton feeder and its diet consists mainly of copepods (*Calanus*). Copepods are phytoplankton eaters. The 11-m-long basking shark is thus placed relatively low in the food pyramid. Ecologists classify organisms in eating guilds. Primary producers like the phytoplankton by definition get the attribution to the first trophic level. Copepods, which range in size from 0.5 mm to 1 cm, feed on phytoplankton and are therefore classified in the second trophic level. At first glance, this does not seem to be the food to nourish a giant fish. However, this is not an isolated case. Take euphausians, which like copepods belong to the crustaceans, which reach sizes of 4–15 cm and are also suspension feeders. They are the favorite food choice for the biggest animal on earth, the baleen whales. And there is a good reason for this choice, apart from the fact that it is a protein- and lipid-rich food item. The biomass of one single species of this family, *Euphausia superba*, the Antarctic krill, has been estimated at 500 million tons at any given time. With that biomass it surpasses any other marine animal and even rivals the biomass of the dominant terrestrial animal, namely ants. If you specialize on this food, you have a readily prepared table. Again, despite their size, ecologists place baleen whales only in the trophic level three. This does not apply to other whales. Killer whales prey on mammals like the seal, which is already a predator placed at a higher trophic level. The food chain picture is even more complicated because differences in the attribution to trophic levels not only vary between animals belonging to the same family but also differ for different development stages of the same individual organism. The larvae of large predatory fishes start quite small, and during growth the animal works its way up the food chain until it reaches its final trophic level. As all these complicating factors blur the picture of food chains, ecologists prefer frequently the term food webs to underscore the complex nature of feeding interaction in nature.

In the following, I have selected a number of reports from the recent research literature to illustrate these principles with a few examples from the ocean.

Algae and the Story of DMS

The Chemical Armor of the Primary Producer

It is useful to start the survey with a primary producer. We have met them several times in the book and to avoid repetition, I take now the alga *Emiliania huxleyi* to illustrate a new principle. Like other photosynthetic organisms, it does not like the idea of being eaten. However, it has many enemies that graze on it, one is the protozoan *Oxyrrhis marina*; ciliates and flagellates also belong to its predators. *Emiliania* uses a chemical that deters protozoan herbivores. It contains intracellularly about 100 mM DMSP. In full, DMSP reads dimethyl-sulphonio-proprionate. If you decompose the name, it reveals a relatively simple structure. It is a carboxylic acid consisting of three carbon atoms and has at its other end a positively charged sulfur atom substituted with two methyl groups. One hundred millimolar is quite a substantial concentration stored in algae. The biosynthetic pathway was first deciphered for the green macroalga *Enteromorpha intestinalis* (Gage et al. 1997). It starts with methionine with a transamidation reaction, followed by reduction, methylation, and decarboxylation. Why do I tell you this? It means that the algae invest four enzymes into DMSP synthesis and allot substantial amounts of the sulfur-containing amino acid methionine for its production. Actually, the algae elaborate still another enzyme, DMSP lyase, but they keep this enzyme carefully in a different cellular compartment. Only when the algae are grazed, the enzyme comes in contact with its substrate and does what it was designed for. It cleaves DMSP into DMS (dimethylsulfide), acrylate, and a proton. Acrylate is moderately toxic. Why did algae not design a more toxic compound to deter protozoan grazers? We have seen that diatoms and dinoflagellates synthesize very potent toxins and teratogens. The answer is that the algae were obliged to compromise.

Multiple Purpose Substances

In view of the large amounts of DMSP synthesized, they could not allow to invest in a single purpose compound. DMSP is also a cellular osmolyte, a cryoprotectant, a biochemical methyl donor and, as recently proposed, an antioxidant. However, these arguments would in theory not stop the algae from producing small amounts of a more potent toxin.

Such multiple functions for a single compound are not rare. Marine ecologists know the dual function of diterpene alcohol synthesized by the phaeophyte *Dictyota*, which is a herbivore deterrent and an antifouling substance. Tetrodoxin, the potent toxin from the puffer fish *Fugu*, also serves as a sex pheromone (Matsumura 1995). Despite this multipurpose design, DMSP works sufficiently well as deterrent. Its ciliate and flagellate predators are unable to grow on high DMSP strains, while they grow well on low DMSP strains of *Emiliania*. *Oxyrrhis* grows on both strains, but make a strong feeding discrimination against the high DMSP strain when offered both strains (Wolfe et al. 1997).

Feeding Preferences

Phagotrophic protozoa are highly discerning feeders and use chemical cues, prey morphology, and motility to make their selection. Some ciliates even show an active postcapture prey rejection. The end result is that the grazing pressure shifts to other prey and the high DMSP producer profits from a reduced competition for nutrients. There is more than one ramification to the DMS release argument during zooplankton grazing on algae. In fact, the algae left untouched do not release significant amounts of DMS. However, when they are exposed to copepods, they release increasing amounts of DMS (Dacey and Wakeham 1986). The gut content of copepods and also of penguins contained large amounts of DMS. However, the copepods do not accumulate DMS in their body. When the animals were placed in algal-free water, DMS appeared soon in the water with the elimination of fecal pellets. If one considers further that zooplankton grazing usually matches phytoplankton production, thus keeping phytoplankton biomass at a steady state, one can estimate that 20% of the marine phytoplankton cells are ingested each day.

Meteorology of DMS

In view of the huge amounts of phytoplankton in the oceans, it is then no real surprise that meteorologists became interested in DMS. In fact, they calculated that DMS contributed annually about 10^{13} g of sulfur to the atmosphere. This contribution has enormous atmospheric consequences at the global level. Biogenic DMS is emitted from the ocean and oxidized in the atmosphere to acidic species like H_2SO_4 that form submicrometer aerosol particles. These particles act as cloud condensation nuclei and control then the albedo (reflectance) of clouds. Clouds play a crucial role in meteorology. The surface of the unfrozen ocean, which covers about 70% of our planet, is relatively dark and absorbs thus over 90% of the incident solar energy. The presence of clouds over the ocean decreases the amount of solar radiation reaching the sea surface. Since the transfer of solar radiation energy to the earth is the major driver of both atmospheric and ocean circulation, the role of clouds as radiation screens can thus not be overestimated. Since the solar radiation drives the growth of phytoplankton, it was even hypothesized that marine phytoplankton regulates the production of DMS as part of a global biological system to control the sunlight reaching the Earth. In fact, the observed linearity of the plot of DMS flux against daily solar radiation, underlines that such a feedback mechanism could well be at work (Bates et al. 1987). Variation in the optical properties of maritime clouds could decrease Earth's surface temperature by 1.3 °C

(Charlson et al. 1987) and could thus have an important function to counter-balance global warming by emission of greenhouse gases.

Antioxidant?

Recently, it was demonstrated that DMSP and even more its degradation products DMS and its oxidized product DMSO are substantially more potent antioxidants than ascorbate and glutathione (Sunda et al. 2002). The true function of DMSP could thus be a protection system against hydroxyl radicals formed by solar UV light in photosynthetic organisms living in the upper ocean layer. A number of observations support this conclusion: the seasonal pattern of DMS production (high in the summer, low in the winter), the depth dependence (high at the surface, low in deeper layers), its close association with another antioxidants (carotenoids), its induction by known stressors of photosynthesis like CO_2 and iron limitation, and its increase with high Cu^{2+} and H_2O_2 concentrations. In this view, the deterrent function of DMSP becomes a side show; however, one that still has substantial zoological implications. Actually, these DMS effects go far beyond the primary consumer. DMS became even an important foraging cue for seabirds.

Foraging Cue for Seabirds

Many Procellariiform seabirds make their living flying over vast expanses of seemingly featureless ocean waters in search of food. In this family, you find the albatross, which uses its very developed visual sense to spot and exploit mixed-species feeding aggregates of seals, whales, and conspicuous seabirds. Its smaller relatives, the storm petrels, forage mainly during the night, which excludes visual cues for hunting. The secret of their success was a mystery, but an ability to hunt by smell was suspected from earlier data where petrels were attracted to sponges soaked in cod-liver oil (Grubb 1972). This is not a very likely scent at the ocean surface; therefore, zoologists set out to test other compounds. Petrels forage on Antarctic krill, which as suspension feeders live on phytoplankton. As DMS is released when phytoplankton is grazed, they reasoned that petrels should be attracted to DMS-scented oil slicks applied to the ocean surface. Indeed, a number of petrels showed a significant higher number of sightings over DMS slicks than over control slicks (Nevitt et al. 1995). It was an as strong foraging cue as cod-liver oil, the birds even showed conspicuous turning behavior when flying over DMS-loaded slicks. Petrels show a complex olfactory anatomy, which fits with a night hunting behavior guided by smell. The albatross did not care for the DMS-scented slicks. This observation is surprising since DMS is an ideal foraging cue if your preferred food item is krill. This compound is practically a biogenic aroma identifier for the presence of crustaceans grazing on phytoplankton. However, I was told that the Pacific black-winged albatross has a fantastic sense of smell and can be attracted from very large distances with fish oil (E. Framer, personal communication).

Albatross

The quest for food is not an easy business as can be demonstrated with the albatross (Figure 5.1). The wandering albatross spends 95% of its lifetime in the open ocean, with individual animals showing distinct preferences for specific foraging areas between breeding seasons. With remote sensing methods it could be shown that even a couple separated for over-wintering places as far away as Madagascar and the rim of the Antarctica (Weimerskirch and Wilson 2000). Visual foraging on the sea is not an easy business. Already, nineteenth-century zoologists knew from sailors that the albatross is always hungry demonstrating a certain inefficiency of its hunting technique. The birds regularly dip into water for food or rest. Although their electronically documented path resembles Brownian motion, a mathematical treatment showed that it obeyed a Lévy flight search pattern, a behavior also discussed for foraging ants and flies (Viswanathan et al. 1996). Unfortunately, the albatross spots very efficiently baited hooks from long-line fisheries and tens of thousands are annually killed while scavenging these baits.

Copepods and Krill

Vertical Migration

Interesting marine life histories also occur at the second trophic level. Major insights about the way of life of these organisms came from the documentation of their locomotion. Sophisticated electronics and a lot of mathematics were necessary to track world-traveling seabirds. High-tech equipment is also needed

FIGURE 5.1. The sooty albatross (*Phoebetria fuliginosa*) is a large seabird of the Diomedeidae family, order Procellariiformes, which spends most of its lifetime in the air. Its foraging strategy is explored in the text.

for following the ways of 1-mm-sized zooplankton. Oceanographers placed a multibeam sonar at two coastal sides of the Red Sea (Genin et al. 2005). They could show that zooplankton dominated by copepods swam against the vertical currents at velocities of greater than 10 body lengths per second. In contrast, animals were passively drifting with horizontal currents. If copepods travel vertically, they must sense their depth position. Visual cues could be excluded, but copepods responded behaviorally to small pressure changes. Apparently they possess a "biological barostat." The adaptive benefit of depth retention might be twofold: By swimming against the flow and forming aggregates, the copepods maintain their position in layers of high food concentration where their phytoplankton prey grows. Aggregation also enhances the probability to find a mate in an otherwise sparsely populated ocean. For the predators of copepods, their aggregation behavior is also important. The concentration of copepods is too small outside of these aggregation patches to make feeding on them profitable for such large predators as the basking shark. Invertebrate and vertebrate predators of copepods have thus learned to locate these patches of high prey concentrations.

Predation Risk

The risk of predation is all too evident for copepods. Therefore, it is not surprising that copepods developed countermeasures. One efficient strategy is to exploit the more abundant food resources in the upper strata of the water at night and to descend during the day to a depth where the light intensity is too low for planktivorous fish. A Polish biologist working in the Tatra Mountains found very nice confirmations for this diel vertical migration of copepods (Gliwicz 1986). Since the last glaciation, many mountain lakes lacked fish. In these lakes, no diel vertical migration of the copepod *Cyclops* was found, nor was this behavior apparent in lakes that did not show vertical food gradients. As an adult, *Cyclops* preys on rotifers, while the juvenile nauplia stages of *Cyclops* feed on phytoplankton. In contrast, the noon and midnight position of *Cyclops* differed markedly in lakes, where fish resided for thousands of years (40 vs. 20 m depth, respectively). Fish was introduced over the last decades in many of these lakes and, strikingly, either the copepods disappeared totally or they evolved a diel migration pattern over hundreds of generations. Interestingly, the degree of migration pattern varied according to the length of time since the introduction of the fish. However, this was not the only adaptation that evolved in copepods under predation pressure. In addition, their eggs developed resistance to digestion in the gut of predatory fish.

Diel Migration

For the sake of biological beauty, I have taken here an illustration from a fresh-water environment, but diel migration as antipredation measure was also reported for the marine copepod *Calanus finmarchicus* in the North Sea. This animal is the favorite food for the herring (*Clupea harengus*). *C. finmarchicus* is fat-rich and reaches a size of up to 8 mm. The herring is not a filter feeder, but picks each copepod individually. The herring lives also from other prey like larvae from annelids, small gastropods, and small fish. However, whenever possible the herring prefers *C. finmarchicus* as prey. Since the herring existed in huge amounts in the North Sea, the copepods also felt the predation pressure and developed in the marine environment a diel vertical migration pattern to avoid the risk of the visually hunting herring. In the 1970s, the herring population experienced a population collapse to a historical low before catch quota were introduced that led to a recovery of the herring population in the 1980s. Interestingly, during the 1970s when *C. finmarchicus* was freed from the predation pressure by the herring, they partially lost their diel migration pattern (Hays 1995).

Population Dynamics

Copepods are probably the most numerous metazoans on earth. Their sheer number combined with their importance as a link in food chains makes them crucial elements in all food-web modeling. Small changes in the population structure can have substantial effects on the nutritional interaction in the ocean, ranging from the survival of pelagic fish to the biological pump into the deep sea. An important parameter in marine ecosystem modeling is the mortality rate of zooplankton. Surprisingly few data on zooplankton mortality have been collected. Important data were contributed by copepod counting in the Atlantic near Norway (Ohman and Hirche 2001). Late juvenile (copepodid stage 5) and adult female copepods appeared in late March, they originated from the overwintering population, which survived in deep water. They appear about 40 days before the phytoplankton bloom, the grazing rate is thus low and *C. finmarchicus* lived still largely from stored wax esters put aside for gonad maturation and oogenesis. They also started quickly with egg laying. The initial recruitment rate is high: Up to 60,000 eggs were produced per square meter of sea surface and day. However, eggs produced during the first 20 days of egg laying showed virtually no survival to the first larval stage (the nauplius). A second smaller peak of adult females was observed 10 days before the bloom, it was accompanied by a second peak in egg laying, but now the egg mortality was threefold lower.

Mortality

What is the reason for the high mortality during the first peak of egg production? Loss of eggs by water currents can only account for a quarter of the observed mortality. Diatoms and their toxic metabolites were not around. The male copepods were not short of sperm. The principle prey of copepods, ciliates, and phytoplankton were available. So where is a smoking gun? The most suspicious association was that the egg mortality rate declined in parallel with the decline in abundance of adult females. This observation solved the riddle: The explanation for the density-dependent egg mortality is egg cannibalism by con-specifics. Adult females and C5-stage juveniles ingested eggs with high clearance, particularly in the prebloom period.

Food-web Regulation

This loop introduces a self-limitation device into the population dynamics of copepods that is partially independent from limitation in food resources and independent of predation by animals from the next trophic level. These effects must now be incorporated into food-web models to ameliorate predictions. Predictions are difficult since the food-web interactions are very dynamic, especially in face of external systematic forcing factors like global warming. In the Northern hemisphere, biogeographical northward shifts probably forced by the current global warming trend were already documented in a number of terrestrial animals like butterflies, amphibians, and birds (e.g., Parmesan et al. 1999). Likewise, southward shifts for flies were seen in the Southern hemisphere. Such a trend with a 10° latitude shift for temperate copepods and a decrease in artic copepods was documented in the Atlantic (Beaugrand et al. 2002). This observation has important consequences for predator fish like the cod that relies on an intact food web and which was anyway severely battered by over-fishing. If such an important trophic relay organism as copepods changes its geographical distribution, large-scale ecological changes in the ocean become likely.

Krill or Salp Years

A portrait of life in the second trophic level would be incomplete without some feeding information on krill in the Antarctic food web. Krill despite its small size is a relatively long-lived animal with a life span of more than 5 years (Figure 5.2). Surveys conducted near the Antarctic Peninsula showed large interannual abundance fluctuations between 1976 and 1996, with a significant trend to lower krill abundance in recent years. Marine biologists noted

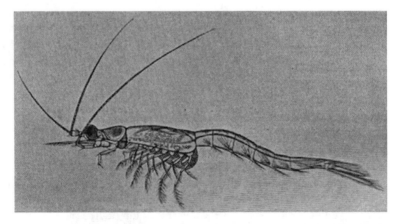

FIGURE 5.2. Krill denotes crustaceans belonging to the suborder Euphausiaceae. *Euphausia superba* is—as the Latin name suggests—a crucial member for the marine food web, providing the food basis for many fishes, birds, and whales.

two interesting correlations. When krill populations were low, salp populations tended to be high. Winters with extensive sea ice cover favored krill and inhibited salp blooms. Salps are pelagic tunicates (Figure 5.3), which form together with the lancelets and the vertebrates the three subphyla of the Chordate phylum.

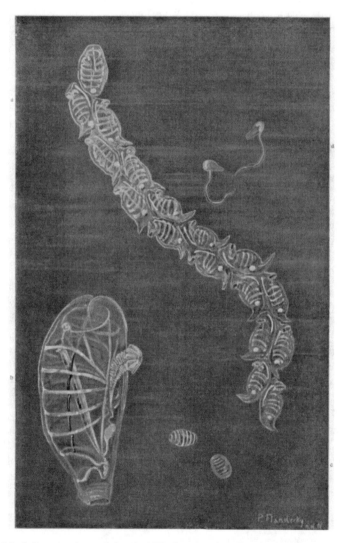

FIGURE 5.3. Salps, members of the class Thaliaceans, belong to the subphylum Tunicata. The two larger salps are *Salpa* and *Cyclosalpa* (*bottom*) and *Doliolum* (*bottom right*). The tailed animal is an appendicularian tunicate (*Oikopleura*). We deal with tunicates in food web regulations (krill vs. salp years) and with respect to tunicate feeding and its close affinity to vertebrates.

Salp Feeding

Salps are rather simple suspension feeders with an oral and atrial siphon at opposite ends of the body and a pharyngeal filtering basket for retaining plankton and organic detritus via various kinds of mucous nets. The linear water flow through the animal provides a means of "jet propulsion" supported by muscular action. Water enters a siphonal chamber, passes into the pharynx, which contains a groove called the endostyle. The bottom contains mucus-producing cells and a row of cilia. Iodine is also enriched from the water and added to the mucus (comparative anatomy links this organ to the later thyroid of vertebrates). The mucus sheets are rolled into cords, which pass into an esophagus, then a stomach equipped with a pyloric gland. Digestive enzymes are secreted into the stomach and an intestine passes the undigested material to the anus, which opens into the atrium near the exit siphon. Most are highly gelatinous and transparent and in contrast to krill not a good food source for higher trophic levels. Salps benefit from open-water conditions, while krill exploit ice algae that develop during the winter months under the ice cover.

Balancing Acts

During blooms, salps can ingest up to 20% of the primary phytoplankton production and can thus deprive krill of sufficient food to support their energy requirements, leading to poor reproductive success of krill (Loeb et al. 1997). Diminished krill production may already have caused a decline in the abundance of one of its vertebrate predators, the penguin *Pygoscelis adeliae*. Krill is one of the most important ecological links between the trophic levels of the primary producers and the vertebrate predators. The carbon from phytoplankton is lost with salp fecal pellets and transported to deeper waters. With autonomous underwater vehicles the relationship between krill and sea ice edges could be investigated. Krill is concentrated fivefold over the open ocean concentration in a narrow band between 1 and 13 km from the ice ridge (Brierley et al. 2002). This is a compromise position: The algal bloom is strongest directly at the ice edge, but receding under the ice offers protection from air-breathing predators. Predation and starvation were traditionally assumed to be the major causes of krill mortality imposing this risk-balancing act. However, data showed that parasitoid ciliates were also a cause of mass mortality of krill (Gomez et al. 2003). As frequently in biology, finding the right strategy is a difficult balancing act. Swarming confers benefits to krill since it improves the capture of motile prey, helps to find a mate and to escape a predator. However, on the dark side, it increases the risk of exposure to parasites.

Planktivorous Fish

Larval Fish

With these animals we do not yet climb up the trophic level, since they are found at the second (phytoplankton eater) or third level (zooplankton eater), but we reach fishes of substantial size or great economical importance. I have chosen three research papers to illustrate this subject. With the first I touch a critical issue of fish ecology with economical relevance for fishery. Because of the prevailing fish reproduction strategy to spawn many eggs, which yield many small larval fish, most fish actually start as plankton feeders before they climb up the trophic level reserved for the adult forms. This is the case for the haddock (*Melanogrammus aeglefinus*, formerly *Gadus aeglefinus*, indicating that it is a cousin of the cod, *Gadus morhua*). The haddock is smaller than the cod, reaches commonly 50 cm length and 6 kg weight, and feeds as an adult mainly on mussels, crustaceans, annelidae, and echinoderms, while the adult cod is a predator of other fishes. The haddock is a commercially valuable fish, but it has suffered from overfishing in recent decades, such that haddock recruitment became a hot topic for fishery researchers.

Year Classes

A long-standing hypothesis in this field contends that the abundance of fish year-classes is determined by the food availability during the critical period of larval development. If food is abundant, one would expect that less larvae die from starvation and the quicker they grow, the earlier they reach larger size classes reducing the predation pressure on them. The relative timing between the spring bloom of phytoplankton and the time of fish spawning should set the pace. To be present before the phytoplankton bloom assures uncompeted access to an abundant food source. However, there is a risk when the spring bloom is delayed a greater proportion of the larvae will starve to death. Actually, the haddock larvae feed mainly on zooplankton, but since the zooplankton feed on phytoplankton, the easier-to-measure phytoplankton can be used as a sentinel. The prediction of the so-called Hjort–Cushing hypothesis is that precocious algal blooms will result in a larger survival index and thus later greater fish stocks. Canadian fishery scientists have surveyed the larvae, juveniles (up to 2-year-old fish), and adults since 1970 off Nova Scotia. The remote-sensing of the ocean color by satellites allowed now a test of the hypothesis. And indeed, 89% of the variance in larval survival could be accounted by the variation of the timing in the spring bloom (Platt et al. 2003). The reproductive success of haddock is thus controlled from bottom-up, it is food controlled. You have more juveniles because less larvae perish from lack of food.

Suspension Feeding in Fish

With the second research report we now turn to a physiological problem of suspension feeding in adult planktivorous fish from the *Clupea* family, comprising such common fish as the herring (Figure 5.4), the sardine, and the anchovy. The reported experiments were actually done with the shad, another member of this fish family. If you look into the mouth of these animals, you see at each side of the lower jaw four gill arches from which many gill rakers branch at a right angle resembling a miniature eel-basket. At the end of the oral cavity, you see the esophagus. The fish takes water up with the mouth

FIGURE 5.4. The herring (*Clupea harengus*), possibly the most abundant of fishes, is shown here in a large school. It eats plankton; and suspension feeding is discussed in the text. Its predator relationship with copepods and its prey relationship with tuna, seal, and sea lion is mentioned, including its echolocation avoidance reaction.

and expels the water again through the gills. However, it was not clear how the gill rakers function during suspension feeding. Two major hypotheses were proposed. In one version, the rakers function like a dead-end filter, they retain particles that are too large to pass the sieve. However, there are two problems with this hypothesis. A lot of food particles would become trapped on the gill rakers and clog the system unless a transport system would collect it from the gill rakers and transport it to the esophagus. The other problem is that these animals consume particles that range in size from $40\,\mu m$ to 1 mm. Many particles would be too small to be retained by the space gaps between the rakers. In the second version some animals solve these problems by a type of hydrosol filtration, they produce large amounts of mucus which traps the food particles (we have seen this above with the salps). Actually neither hypothesis was correct, food particles are caught in the cross-flow (Brainerd 2001). This is a process that is commonly used in the food industry when beverages like wine or beer have to be filtered from particular material to get a clear liquid. In cross-flow filtration, fluid flows parallel to the filter surface and particles become more concentrated as the filtrate (the liquid) leaves through the filter's pores (the gill system). Even if this system is industrially used, we still do not understand the underlying physical mechanisms. The researchers used first some mathematics to investigate the theoretical fluid dynamics by which they excluded the first two hypotheses and then they set out to prove the cross-filter hypothesis by using miniature fiber-optic endoscopes linked to high-tech cameras that they inserted into the oral cavity of the fish (Sanderson et al. 2001). With these medical instruments they measured impressive cross-flows sweeping with a velocity of 55 cm/s across the raker's surface and more than 98% of the particles passed without any contact to the rakers directly into the esophagus.

Basking Shark

With a research report on the third planktivorous fish, I want to introduce a subject of behavioral biology. It deals with the second biggest fish, the basking shark *Cetorhinus maximus*, which reaches a size of up to 11 m. It illustrates a paradox that the biggest animals in the fish and the mammalian category feed mainly on plankton. Apparently, you do not need big prey if you want to grow to large size. What you need is only an efficient suspension filter, but here nature has come up with interesting solutions as we have just heard. However, time is also a problem. The most efficient filter is of no use if the animal forages in an area that does not offer food above a critical threshold. At the end of the day, the basking shark will starve when it has not ingested enough biomass by hunting in a dilute plankton suspension. To succeed, the basking shark must choose the richest, most profitable plankton patches. This is actually a general biological problem with which each mobile animal is confronted. The issue consists of several questions, which can be summarized under the following headings: choosing where to forage; choosing how long to stay; choosing what to eat; maximizing intake rate, while minimizing risk. There is a rich literature on this subject known under the term optimal foraging theory. However, the different animal groups are not homogenously represented in this literature. While birds are favorite targets for animal observation by ethologists, much less literature exists on fish, which is explained by the obvious technical obstacles to observation. Environmental biologists tracked basking sharks visually from a boat, located their exact position, and took plankton samples (Sims and Quayle 1998). The sharks fed in areas where the dominant copepod *Calanus helgolandicus* was 2.5 times as numerous (1,500 organisms per cubic meter) and 50% longer (2 mm) than in areas where they did not feed. When they fed in rich areas, they showed an extensive zigzagging in their traveling, while in poor areas they swam in straight lines. The residence time in a given ocean patch was proportional to the density of the zooplankton. The movement between rich patches took them about 1–2 days where they traveled distances of 10 km. The sharks apparently knew where to feed: They followed fronts where warm water layers met cold layers and aligned their searching with temperature gradients and tidal flows. These are useful rules of thumb, which a marine biologist also would have chosen. The researchers speculated that the sharks sensed the weak electric fields induced from copepod muscle activity or DMS released from grazed phytoplankton. Apparently, the sharks were not alone in sensing these nutrient-rich fronts. The areas, where shark foraged, were also used by shoaling fish, including commercially important fish like the mackerel.

Sardine Versus Anchovy Years

In our survey, we touched several times on the impact of climate on food chains and a spectacular case is presented by two planktivorous fish species, both of substantial economical importance to fishery industries, the sardine and the anchovy, both belong to the herring family (*Clupeidae*). Marine biologists have observed a sharp decline in sardine catches over the last 20 years off Peru, which were one of the richest fish grounds of the oceans. The harvest decreased from 4 million metric tons in the late 1980s to a bleak 40,000 tons in 2001. In view of the economical importance of fishery for Peru this was bad news, especially since the early 1970s a comparable crash in the harvest of anchovies was observed. At first glance this looks like the worldwide decline in fish resources we are currently experiencing as a likely consequence from overfishing. However, two observations did not fit with this diagnosis. One was that the decline of sardines in the 1990s was accompanied by a recovery of the anchovy, although not yet to the catches during the heydays of anchovy fishery in the 1960s. In the 1960s the sardines were not only low, they were practically nonexistent. This raised the hypothesis that we deal here with multidecade changes from a sardine to an anchovy regime (Chavez et al. 2003). The record of fish catches does not extend back beyond 1950 for Peru, so the fishery scientists looked for proxy measures.

On Guano and Haber-Bosch

They looked for the record of guano harvest, which goes back to 1900. There is a good reason for the preservation of this record, which justifies a digression. The German chemist Justus Liebig recognized in the 1840s the importance of fertilizers for agriculture, mainly of a usable nitrogen source for plant growth. Manure was not any longer sufficient as a fertilizer in agriculture to feed the growing world population. South America was during the nineteenth century the source of the most valued animal excretions for readily available nitrogen: guano, the droppings of seabirds, which formed large crusts on the rocks where the seabirds were resting. This business came down since the German physical chemist Fritz Haber convinced an engineer at the BASF chemical factory in the 1910s to reduce dinitrogen from the air with dihydrogen under great pressure and heat. The Haber–Bosch procedure became later actually the detonator of the population explosion during the twentieth century. However, guano remained sufficiently important for Peru that records of guano production were still kept. The Peruvian seabirds are mainly represented by a single species, the cormorant, which feeds nearly exclusively on anchovies. The guano record showed marked fluctuations, too.

El Niño

This observation called the meteorologists on the scene and they defined the anchovy regime by an Eastern Pacific cool phase and the sardine regime by an Eastern Pacific warm phase that varied over periods of about 50 years. The actual connections are more complicated because the El Niño events with shorter periods are disturbing this multidecade pattern. Overfishing and global warming due to anthropogenic CO_2 emission further influenced the analysis. However, without a deeper understanding of these processes, we might be condemned to follow helplessly the coming and going of these important food reserves.

On Cod and Copepods

The collapse of the cod population is commonly attributed to overfishing, in ecological parlance top-down control. However, as is usual in biology, the causal connections are too complicated to be summarized under a single heading. Also bottom-up control contributed to the decline of the cod standing stock after 1980. Warming of the ocean also played here an important role. Survival of larval cod is of critical importance for the later size of the cod population. Survival of the larvae depends on three critical factors: the mean size of the prey, its seasonal timing, and its abundance. The female cod is a prodigious producer of eggs, up to 4 million eggs per animal were counted. Spawning time in the year differs depending on the geographical region. Once the larval fish is in the water, the race is on: The larvae will need appropriately sized prey commensurate with its own size and predator capacities. In the 1960s, the timing was optimal. The larval cod was ready for feeding on copepods between March and July and during this time period its favorite prey, *Calanus finmarchicus*, showed maximal abundance. In the 1970s, the prey showed a shift toward lower and later abundance, but this was not yet critical. Cod had had its hey-day, what fishery biologists speak of as the "gadoid outburst." In the 1980s the situation changed suddenly. *C. finmarchicus* became rare in the North Sea. Of course, copepods did not disappear. There is always this *"horror vacui"* in nature; consequently, its position was taken up by *C. helgolandicus*. The latter grew to even higher concentrations in the sea, but there was a problem. It reached these higher titers only after August (Beaugrand et al. 2003). Ecologists diagnosed a mismatch between prey and predator with lethal consequences on the recruitment of young cod. Since warming caused this regime change in the North Sea, the problem became even more severe for the cod. Rising sea temperature means increase in metabolism and higher energetic costs, all with a thinner diet and unabated top-down mortality by fishing: small wonder that the cod population crashed.

Piscivorous Fish

Tuna

With the next chapter we stay with two other important commercial fish species, the tuna and the cod. I will summarize some recent research on their biology. In a later chapter, I will use the cod to illustrate problems of the sustainability of contemporary fishery. You will also recognize that I use here a journalistic trick. Both fish are herring eaters, we advance therefore in the food chain as the section title promised. You will therefore not be too much surprised to find in the next chapter the cod-eater, then the eater of the cod-eater, then with a slight ironic undertone the eater of the eater of the cod-eater.

The Atlantic bluefin tuna (Thunnus thynnus) has fascinated humans from the beginning of written records. You can quote Aristotele and Strabo from Antiquity literature on the biology of the tuna. The reasons for this fascination are manifold. Humans are generally—a microbiologist would say naively—attracted by the sheer size of animals. Tuna are impressive in this respect: They grow to a length of 3 m and a weight of 680 kg. In many areas of the world, the tuna are prized for their flesh, in Japan a large tuna could command a price of $100,000 per individual fish. One should therefore not be surprised that the stock biomass has dwindled to a mere 20% over the last 30 years. If you want to keep this valued fish on your menu card, a thorough knowledge on its feeding, migration, and spawning is necessary to rebuild the stocks.

Endothermy

However, the tuna has characteristics that make this animal also attractive to biologists (Block et al. 2001, 2005). They are unique among teleosts (bony fish) for their endothermic capacity due to a special subcutaneous bood vessel system. They have thus high rates of heat production. Tuna use this heat to warm their eyes, which facilitates extremely rapid eye movements. The eyes are warmed in preference to other parts of the body and infrared imaging of shoals detects the glowing eyes. In their wide geographical distribution from the Mediterranean Sea over the Atlantic up to the north of Newfoundland to the Caribbean Sea, they experience water temperatures between 3 and 30 °C while maintaining a constant peritoneal temperature of 25 °C. The elevated body temperature of bluefin tuna increases their capacity for rapid migration by enhancing the power output of their muscles. The other side of the coin is that they have a large metabolic demand for food and oxygen. In the Caribbean spawning ground, the warm waters are favorable for the development of the eggs, but the adult animals reach their upper tolerance for heart function.

Migration

The two research papers that I have chosen could be entitled "A year in the life of a western tuna." Actually the fish biologists tagged many tunas electronically and followed them over longer time periods. They documented impressive displacements both vertically and horizontally. In the vertical dimension they saw repetitive diving excursions for food, where the fishes got down to nearly 1000 m, but most of the time they spent in the upper 300 m of the ocean. Most western tuna, which had their spawning grounds in the Caribbean Sea, remained over the coastal shelf of the eastern US and Canada, but some reached out far into the Atlantic coming close to the Spanish coast. The eastern tuna which spawned in the Mediterranean Sea was a more daring character and traveled over the entire northern Atlantic. Apparently, the US/Canadian shelf and south of Iceland were the richest foraging areas, where both populations overlapped substantially. Rapid movements over thousands of kilometers (trans-Atlantic in 1–2 months) are common in tunas. Apparently, the metabolic costs for endothermic fish swimming is low enough in comparison to the foraging and reproduction gain obtained from migratory behavior. However, no mixing of the two populations was observed at the level of the spawning grounds. Tuna smaller than 2 m in length did not enter the spawning area, indicating a late sexual maturation, which is a critical factor for the population biology of fishes. However, tuna grow quickly: fish larvae of few days of age weigh already 50 g, after a month they reach 100 g, while at 3 months they achieve already a weight of 1 kg. Quick growth assures rapid decrease of potential predators. Female tunas invest heavily in egg production, the ovaries increase in weight from 500 g to 6 kg within a month before spawning.

Cod Biology

Cod (*Gadus morhua*) is a smaller fish than bluefin tuna, but still of impressive size. It can reach up to 1.5 m length and 50 kg weight. It reaches 14, 27, 35, and 50 cm length after 1, 2, 3, and 4 years of age, respectively (Figure 5.5). Up to a length of 30 cm crustaceans dominate in stomach contents, then fish starts to appear in its prey and longer cod are nearly purely piscivorous. Spawning starts relatively late when the fish has reached 50 cm length and occurs then in yearly intervals. The cod is a bottom-dwelling predatory fish with populations in the northern Atlantic. Its southern limit was defined by New York and Bordeaux and to the north it reaches into the Artic Sea. As for many other commercially important fish, relatively few data on the foraging behavior of cod is available, which will limit the management efforts of this severely depleted fish stock. Electronic tagging demonstrated striking behavioral differences between North Sea and Irish Sea cod (Righton et al. 2001). The latter was extremely

FIGURE 5.5. The cod (*Gadus morhua, bottom*) and the haddock (*Melanogrammus aeglefinus*), both belong to the family Gadidae.

active at all times, while the former was active during spring time, but stopped physical activity in June. In July they spent most of their time on the seafloor. In August and September they were active only during night and resumed an active life in October. During the summer months the North Sea cod moved less than 1 km in contrast to models that anticipated foraging activity over substantial geographical areas. In accordance with the behavioral differences, microsatellite markers revealed strong genetic differentiation between populations (Nielsen et al. 2001).

Marine Highways

Impressive data obtained with echolocation are available for cod off Newfoundland (Rose 1993). Real cod highways were detected with fish shoals containing more than 10^8 individuals, amounting to 80% of the entire stock estimate in that region. The fish formed a large aggregate of mature animals organized around spawning columns of mature females at the center. Four- and five-year-olds accompanied the spawning core. The migrating cod aggregate was structured by fish size. At the front end were large scouts that led the shoal, the size of the cods decreased toward the rear guard. The aggregates followed a channel of 2–3 °C warm water undercutting a colder shelf water (<0 °C). During the spawning stop, which lasted 10 days, the fish density was 1 animal per cubic meter. When the migration started, a 20 km long column of fish moved forward and upward, keeping an interanimal distance of ten body lengths from the nearest neighbor, apparently maximizing the area swept for prey, while keeping visual contact to the next cods. An impressive echogram showed how cod scouts run into capelin (*Mallotus villosus*), which rouse in apparent panic in front of the hungry steamroller consisting of cod. The capelin searched the same areas for spawning, but one gets the impression that they passed by to serve the cods as fast food.

Piscivorous Mammals

Who eats the cod? Caught cod frequently carry scars from tentative unfriendly takeover, which the fish survived. There must be characters in the sea that do not shy away from this 1.5 m long and strong fish. Such attacks were recently documented for the Weddell seal, animals that weigh as adults 450 kg. If I look with my children at action movies on the TV, I have frequently the impression that you find better scripts in scientific journals than on the TV screen. To prove this point, take the science paper describing the hunting behavior of these marine mammals beneath the Antarctic fast ice (Davis et al. 1999).

The Weddell Seal

The scientists mounted video cameras on the back of the animals such that they could observe what happened just in front of the muzzle of the seal. Recall from a previous section in this chapter how krill explore the underside of the ice for ice algae. This attracts other visitors that are interested in the krill like the small fish *Pagothenia borchgrevinki*. This fish knows that there are others out that have an open eye on them. When hunting for krill, it quickly seeks shelter in ice crevices. Now in the tradition of a good action film, the seal arrives on the scene. The small fish disappears into the crevice, where the seal cannot follow. Now the video documents a surprising behavior—the seals expels a blast of air through its nostrils for 1 s; the fish panics and immediately swims out of the crevice. In the next frame you see via the video camera directly into the eye of the fish, which tries to escape. The seal carried also electronic devices that allowed locating its position during its diving exercises. The scientists followed the hunt in 3D and saw how the seal stalked the large Antarctic cod without startling the fish. Not all attacks were successful, sometimes the cod struck back with a powerful tail thrust and escaped. Sometimes the seal displayed only curiosity, but no aggressive behavior during an encounter with a cod. However, on other days, the researchers observed seals eating cod on the ice. The video documented that the seal hunted under visual control. Some scientists suspected that seals use underwater sonar for hunting, but the electronically surveyed seals were not using calls during the predator–prey interactions. In the wild, blind but well-nourished seals were often described. Apparently, seals can hunt successfully without vision. Zoologists demonstrated that the seals use their whiskers to detect water movements. The spectral sensitivity was well tuned to the frequency range of fish-generated motion. Optical and acoustic cues in bait detection were excluded by placing eye caps and headphones on the animal——in that study handsome harbor seals were used (Dehnhardt et al. 1998, 2001). Seals in a zoo were trained to locate a miniature submarine. From these experiments the biologists calculated that seals detect with their whiskers water movements in the wake of fishes for several minutes. A herring swimming at a sustained speed of about 1 m/s leaves a hydrodynamic trail that remains detectable for the seal when the herring is 200 m away. As is usual in biology, one sense comes seldom alone. There is often a real arms race also at the level of senses between predators and preys. Therefore, one should not be surprised that some teleost fish have lateral line systems, which show hydrodynamic receptor properties that are one or two orders of magnitude more sensitive than the whiskers of seals.

Diving

The Weddell seal dived usually to depths of 100–350 m for 20 min duration, but exceptional dives to 740 m were also measured. Physiologists had a problem with these observations. Swimming is energetically expensive for mammals, resulting in transport costs that frequently exceeded those of fish by a factor of 10. They calculated that the Weddell seal would be unable to complete a 200-m-deep dive using aerobic metabolic pathways. It would have to rely on anaerobic metabolism leading to the accumulation of lactate, and the animal would require prolonged recovery periods. However, no lactate increases were measured even after deeper dives. What had happened? The video film was again revealing (Williams et al. 2000). Diving descents began with up to 200 s of continuous stroking, followed by a prolonged period of gliding to maximal depth. The compression of the air space in the lungs decreased the volume of the animal without a change in mass, thus buoyancy decreases on descent and avoids the energetic cost associated with active stroking. These savings can thus be invested in a prolonged diving time.

Sea Lions

However, when offered a choice between a fish prey closer to the surface and one which necessitates deeper diving, many marine mammals prefer the easy prey even when they are capable of diving. This was the case for sea lions. Engineers used military spying instruments to locate predator and prey off Alaska (Thomas and Thorne 2001). Sea lions feed exclusively on herring schools when they swim at only 10–35 m depth during night. The sea lions lined up side by side with up to 50 individuals swimming along the edges of the school as if they were herding the herring (Thomas and Thorne 2001). The herring stocks are dwindling and food limitation is now the cause for the decline in sea lion populations. Alternative food sources are at hand: four times higher biomass of pollock (*Theragra*) is available in the waters, but the pollock does not figure on the menu card of the sea lion. One reason might be because the pollock swims day and night at 100 m depth. It is not clear why the sea lions do not change the prey since they are well capable of diving up to 250 m.

Killer Whales: Effect of a Top Predator Down the Food Chain

Who eats the seal? The seal is not yet at the top of the food chain, it has still predators. The battle between seals and killer whales (Orcinus orca) is quite interesting since it is also a war of the senses, and I will tell it in one of the next chapters. Otherwise the gradual creeping up of the food chain becomes a bit dull. While it reflects a clear truth and is thus not just a textbook message, it oversimplifies the true ecological situation. Food chains are in reality often food webs. Apex predators often initiate forces that spiral down successive lower trophic levels and have sometimes even a surprising impact on the food base. Such a backfiring story of the killer whale will now be told to highlight the complexity of ecological interactions, this time a top-down control (Figure 5.6).

Killer Whale

Some killer whales prey on harbor seals and sea lions, but these populations have collapsed in the Northern Pacific. The reason for the decline of the pinniped mammals is uncertain, but is probably a response to reduced abundance of their fish prey (Pacific herring) or forage fish stock changes (pollock). The reason for the changes is not very clear; increased fishery, increased ocean temperature, and depletion of baleen whales have all been proposed. If we move from the concept

FIGURE 5.6. Cetacea are a group of aquatic mammals comprising whales, porpoises, and dolphins. The figure represents *Tursiops tursio* (*1*, a porpoise), *Ziphius cavirostris* (*2*), *Grampus griseus* (*3*, a dolphin), *Globicephala melas* (*4*), *Balaenoptera borealis* (*5*, a whale).

of food chains to food webs, the logical connection between the ecological elements will also increase in complexity. However, as in the following story, the food chain can still be followed back as a linear chain of arguments. So with decreasing prey abundance, the mammal-eating killer whales also face population reductions unless they find a replacement for their diet. In fact, killer whales profited from a human intervention, which offered them a new prey, the otter (*Enhydra lutris*).

The Otter

The otter (Figure 5.7) has a colorful history (Estes et al. 1998). Sea otters abounded at the rim of the Pacific until fur trade sent the animal to near extinction. More recently, the sea otter was put under protection and in the 1970s the otter population increased to near maximal densities in many regions, only to decline again in the 1990s. There is a lot of discussion whether infectious diseases have increased in the ocean as a consequence of maritime pollution (Harvell et al. 1999), but no stranded sea otter carcasses were observed. However, marine biologists observed in the 1990s the first attacks of killer whales on sea otters. If the killer whale wants to make a living on sea otters, on a purely energetic basis a single animal would have to consume 1,800 otters per year. Thus a relatively low number of killer whales that discovered this new food source could be at the basis of this tenfold decrease of the Pacific sea otter population. With this story we get into a fascinating scheme, linking an open ocean with a coastal ecosystem.

FIGURE 5.7. Otters are semiaquatic mammals of the weasel family (Mustelidae). The picture shows the common or river otter (*Lutra lutra*), which is a quick runner on land. The text deals with its cousin, the sea otter (*Enhydra lutris*), an exclusive marine otter living in the kelp beds from the North Pacific. It floats on the back with a stone on its chest and opens mollusks by smashing them on the stone. For comparison, the figure shows another Mustelidae: the wolverine (*Gulo gulo*). It is known for its voracity; therefore, it is also called glutton. It eats all kinds of prey and knows, except for man, no predators.

Sea Urchins and Kelp

Sea urchins are the principal prey of sea otters in the Aleutian islands. You have certainly seen these amusing pictures of otters swimming on the back with a flat stone on their breast on which sea urchins are butchered with the help of another stone. If we stay with our linear question "who eats whom?" then we get a simple answer. Sea urchins live on kelp. Kelp is a popular name for large seaweeds found in colder seas (Figure 5.8). Formally they belong to the order *Laminariales* of brown algae. Superficially they resemble higher plants with a root-, stem-, and leaf-like structure, but at a histological level it becomes clear that these are only analogous structures. The kelp *Nerocystis* carries the popular

FIGURE 5.8. Kelp is the popular name for large seaweeds of the order Laminariales, class Phaeophyceae (brown algae). The picture shows several species from the North Sea including *Laminaria saccharina* and *Laminaria hyperborea*, which reach sizes of up to 3 m. Its largest member is Macrocystis with 65-m length. In the text we speak about *Nerocystis*, the "sea otter's cabbage" and top-down control in a marine ecosystem.

name "sea otters cabbage." The name for this up to 40 m long brown alga from the rapid tideways is of course a misnomer. The otter does not eat this cabbage. It searches only its prey there, namely the sea urchins which are herbivores of the kelp (Figure 5.9).

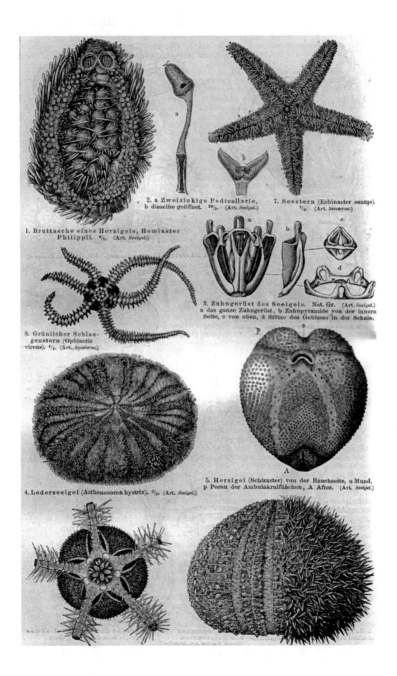

1. Bruttasche eines Herzigels, Hemiaster Philippii. ⁴/₁. (Art. *Seeigel.*)

2. a Zweizinkige Pedicellarie, b dieselbe geöffnet. ²⁰/₁. (Art. *Seeigel.*)

7. Seestern (Echinaster sentus). ¹/₂. (Art. *Seesterne.*)

3. Zahngerüst des Seeigels. Nat. Gr. (Art. *Seeigel.*) a das ganze Zahngerüst, b Zahnpyramide von der innern Seite, c von oben, d Stütze des Gebisses in der Schale.

8. Grünlicher Schlangenstern (Ophiactis virens). ⁴/₁. (Art. *Seesterne.*)

4. Lederseeigel (Asthenosoma hystrix). ²/₃. (Art. *Seeigel.*)

5. Herzigel (Schizaster) von der Bauchseite, o Mund, p Poren der Ambulakralfüßchen, A After. (Art. *Seeigel.*)

Putting all Actors Together

Causal relationships in ecology have a different structure than experiments in laboratory biology. However, the logical links in the killer whale case are very strong: In several islands of the Aleutian archipelago, the 1990s saw a parallel decrease in sea otter populations down to 10% of the 1970s levels. An inverse trend was seen in sea urchins; over the same time period they showed an eightfold increase in biomass. The kelp density, by contrast, decreased by a factor of 12. The kelp tissue loss due to herbivory was 1% in the 1970s and increased to nearly 50% in the 1990s. That the chain of events was actually set in motion by the killer whales was made likely by another observation: Areas that were inaccessible to killer whales for local geographical reasons showed no otter decline, while otters from nearby areas with open access to the ocean were severely affected. In the hindsight, you might argue that one could have predicted this chain reaction as early as the otter appeared on the diet of the killer whale. However, the first event could not be predicted since killer whales and otters coexisted side by side for a while without being involved in predator–prey relationships. Only when the orca's favorite food item disappeared, the otters followed the seals and sea lions on the hunting list of killer whales.

The basic conclusion from this food web was recently confirmed by ecologists who surveyed the Californian kelp forest ecosytem over years to understand the importance of top-down, i.e., consumer-driven, versus the bottom-up, i.e., resource-driven, control of food webs. Coastal kelp forests receive large amounts of fertilizers via anthropogenic runoff. On the other hand, the abundances of large marine top-predators have been substantially reduced over the last decades. A multiyear satellite survey of primary productivity showed that the top-down control explained tenfold more of the variance in the abundance at the bottom trophic level of the kelp than the bottom-up nutrient supply (Halpern et al. 2006).

FIGURE 5.9. Representative extant echinoderms. *Top row*: Sea urchin *Hemiaster* (*1*), which keeps its eggs in a brooding chamber. At the *right* the asteroid *Echinaster* (*7*), with the typical five-rayed symmetry. Between them are pedicellaria (*2*), pincer-like skeleton structures that have a controversial and not yet settled history of interpretation. Some stars use them to capture prey. *Next row*: The ophiuroids *Ophiactis virens* (*8*) and *Ophiothrix* (*bottom row*, *left*); at the *right*, Aristotle's lantern, a complex masticatory apparatus lying just inside the mouth of sea urchins, with five protractible teeth (*3*). *Next row*: Two sea urchins (*Astenosoma* and *Schizaster*) seen from the ventral side showing the mouth as a central opening, A = anus. *Bottom right*: Sea urchin (*Echinus esculentus*, at the left side of this radial arranged animal, the characteristic spines were removed). The sea urchin is the actor in an oscillating marine food chain. They forage on large algae, but are eaten by otters, which fall prey to orcas. You will see how the effect of a top predator can spiral down in a food chain. You will also see Echinodermata as a surprising cousin of *Amphioxus*.

The Fall of the Whales

Humans as Top Predator

The chapter title is on purpose ambiguous. It refers first to the decline in the number of whales due to whaling. Herman Melville's wonderful novel on Moby Dick attracted literature fame to the business of whale hunting, which represented the economical basis for a number of coastal regions in the nineteenth century. Whale hunting is still practiced today for human food consumption. Before human intervention, baleen whales had not much to fear. Although they are situated as phytoplankton and krill eaters at a relatively low trophic level, only few predators have the courage to attack these giants of the sea. Only the killer whales were known to occasionally attack young baleen whale calves. All this changed when baleen whales became the target of the ultimate top predator of our planet. Hunting has thus dramatically reduced all baleen whale populations. Protection measures have led to a recovery of the reduced populations, but to what level? Since we have no reliable statistics of whale numbers before the start of the hunting, it is difficult to estimate the carrying capacity of the oceans for whales. This has practical consequences when, for example, the International Whaling Commission states that catches should not be allowed on stocks below 54% of their estimated carrying capacity. The level of neutral genetic variations increases with population size. This allows suggesting historical population sizes from the observation of genetic diversity in baleen whales. Taken at face value, the genetic data indicate that the current population levels of baleen whales are only a fraction of the past numbers (down perhaps by a factor of between 6 and 20; Roman and Palumbi 2003). This subject has received sufficient media coverage, not to be repeated in this book.

The Death of the Whale

Actually, the story I want to retell starts quite recently with the submarine Alvin in 1987. The researchers crossed by chance the 21-m long skeleton of a blue whale. They found bacteria and worms living on the whale's bones that were previously only known from hydrothermal vents (Haag 2005). All whales die sometime and as they have no natural predators of ecological importance, their carcasses fall to the floor of the sea. Mortality is in fact relatively high in female whales during migration. They are stressed by the birth process, and they accompany the newborns that are sometimes hunted by killer whales. However, the killer whales carry their name with justification since they do eat only few organs of the baleen calves and leave the rest to the sea. A whale carcass is a nutritional feast to the nutritionally desert of the seafloor. Sharks, hagfish, and lobsters prey on the soft tissue for many months. Then the falls feed worms and crustaceans. When anaerobic bacteria start with their work, they create a sulfide-rich environment, which prepares the niche for the invasion of further worms, clams, and mussels. Even when all soft tissue has disappeared, the fat-rich bone marrow of the whale bones will feed a specialized community, which will still live from the whale for up to a century.

Life Histories Between the Land and the Sea

On the Value of Case Reports

Medical research journals publish in addition to controlled clinical trials and large epidemiological surveys also the case reports of interesting medical observations in an individual patient. As an example, I remember a report of a patient that suffered transient paralysis when eating bananas. This fruit is rich in potassium and the patient showed a defect in the absorption of dietary potassium, which led to a short-term benign disturbance of the Na^+/K^+ gradient over the excitable muscle membrane. In this way a single patient might lead to the identification of interesting metabolic mutation and thus contribute new insights into human physiology. In the following, I will report on a series of animal case reports that each touch a nutritional life history of an animal living at the interface between the sea and the land. They are hopefully as entertaining as instructive in our quest for food survey and guide us to the ecology of land animals, which will be the subject of the next section.

The Shrinking Reptile

Some reptiles are special with respect to growth. For example, the Nile crocodile shows an open growth: It grows in length as long as it lives. You can thus actually tell the relative age of these crocodiles by measuring their length. Despite this precedence, it came as a surprise when zoologists learned that reptiles could also reversibly shrink as a response to food shortage (Wikelski and Thom 2000). The animal in question is a marine iguana from Galapagos (Figure 5.10). This long-lived (about 35 cm long) reptile feeds on green and red algae growing in the submerged intertidal zone. During periodic El Niño events the sea surface temperature around Galapagos can increase from 18 to 32 °C due to the disruption of cold nutrient-rich ocean current upwelling. Green and red algae disappear during this regime and are replaced by brown algae that the iguanas find difficult to digest. Until the arrival of the next La Niña conditions, which furthers again the growth of their favorite food, up to 90% of the iguanas can die of starvation. A few animals found an answer: They shrank in size by as much as 20% of the body size between successive years. The degree of shrinkage was positively correlated with the survival of the individual animal. The explanation is easy: As in many grazers, smaller animals are frequently more efficient feeders than larger individuals. The 20% shrinkage cannot be explained by a decrease in cartilage and connective tissue alone, it must also involve bone absorption. Females showed a greater shrinkage than males, probably because they suffer

FIGURE 5.10. Galapagos is a world of its own. The picture shows the reptile *Amblyrhynchus cristatus*, which belongs to family Iguanidae, order Squamata. The book tells the story how this reptile shrinks in body size in times of food shortage.

from the extra energy expenses due to egg-laying. It is striking that this strategy, well known from bacteria under starvation, is so little used in starving animals since it is apparently possible in a large vertebrate.

The Fasting Father Penguin

King penguin partners alternate in staying with their egg in the 54-day incubation period between laying and hatching (Gauthier-Clerc et al. 2000; Figure 5.11). They feed only when at sea and must fast when ashore. The nonincubating partner forages at sea, but has to travel up to 500 km away from the colony to satisfy its food needs. Return to the incubating partner is thus somewhat unpredictable. During the early parts of the incubation period, both parents come back only with a few pebbles in the stomach (as in many birds the stones functionally replace the lack of teeth in the beak). Males try to replace the female near hatching time. The male penguins apparently have an internal clock: They return with about 200 g of fish remains and squid parts in their stomach 20 days before hatching, but increase the stomach content up to 1 kg in a narrow window around the calculated hatching time. During the final male egg incubation period, the males loose 160 g body weight per day. This amounts over a 3-week period to 20% of their initial body weight. If the female does not return at time and the young

FIGURE 5.11. The king penguin (*Aptenodytes patagonicus*) is presented in the book as a caring father to the hungry chick.

hatches, the male starts to feed the chick from its stomach contents. Notably, the regurgitated food shows the same preservation state as the initial stomach filling. The father penguin has not touched the stored food whose energy content covers the energy need of the hatched chick for 10 days, thus providing a buffer against delayed return of the female partner. In contrast, males returning 10 days after hatching come with empty stomach. If the female deserted the egg, the males also returned with empty stomach, they can thus overrule their internal clock.

Growth Limitations for Seabird Colonies

British ornithologists observed that smaller kittiwake colonies tended to grow proportionally faster than larger colonies. A model known as Ashmole's population regulation hypothesis was developed setting an upper limit to the colony growth. According to this model, adult birds forage close to the colony and cause a depletion of local fish prey. Birds from larger colonies therefore have to travel longer to find the same amount of food for their chicks as birds from smaller colonies. The tenets of this hypothesis were tested for the gannet (*Morus bassanus*, Figure 5.12; Lewis et al. 2001). The extensive observation of birds in Britain over nearly a century documented an expansion in size of gannet colonies. Satellite tracking verified that the foraging range is closely related with foraging trip duration. The researchers found a significant

FIGURE 5.12. The gannet (*Morus bassanus*) of the family Sulidae, order Pelecaniformes, is the largest seabird in the northern Atlantic.

correlation between trip duration and colony size. In a densely populated colony with 40,000 occupied sites, the average trip duration was 21 h. This corresponds to an excursion of about 150 km. Birds from smaller colonies had to travel only 50 km to land the same amount of catches. They have to work less hard and achieve thus a better breeding success. The researchers measured that the growth of gannet colonies was density dependent, reaching stagnancy when approaching a limiting upper size. Intraspecific food competition became the limiting factor for further growth. Paradoxically, seabird colonies have only a small influence on local total fish mortality. The seeming paradox was resolved when it became clear that pelagic fish prey of gannets showed an escape response when attacked by gannets. Gannets hunt by plunge diving. In the vicinity of the colonies, fish escaped predation simply by swimming vertically downward.

Diet Shifts in the Great Skua

Fishery produces a large quantity of waste, being offal or the catch of undersized fish without commercial value. This results annually in the discarding of about 30 million tons of fish. A whole population of seabirds feeds extensively on discards and in the past times many of their populations have substantially increased. The current decline of fish stocks, changes in catching techniques, and recovery programs have led in the North Sea to a substantial decline in discarded undersized whitings (*Merlangius merlangus*). The changed discard availability had important effects on the scavenging seabird population consisting of fulmars, gannets, several gulls, and the great skuas (Figure 5.13; Votier et al. 2004). The latter is a top predator in the marine food web of the North Sea. This sizable bird reaches more than half a meter in length and is a restless warrior governed by a nearly insatiable hunger. The bird requires 2,500 kJ/day. The declining fish discards resulted in a shift to alternative prey like the small, shoaling sandeel, a lipid-rich fish. However, the sandeel population also fluctuates in availability and forced the great skua to still other prey and this is notably also other seabirds. In Shetland the skua ate about a third of the adult population of the kittiwakes. It was calculated that with a modest 5% contribution of birds to the diet of the skua, this means that a single colony of 5,000 skuas on the island Foula consumes about 800 kg of birds per breeding season. Prey remains indicate that skuas forage on adult birds between April and mid-July. From mid-July eggs and fledglings from fulmars and puffins are their preferred bird prey. The fish discard decline resulted thus in a substantial pressure on the North Sea bird population and the predation by the skua became a substantial component of the annual mortality of the seabird populations. Only further ecological observation will show whether the skua will drive some seabirds into a population collapse or whether it changes again its diet to avoid its own population collapse due to dwindling food resources.

FIGURE 5.13. Skua (*Catharacta skua*), family Stercorariidae, order Charadriiformes, is an agile predatory bird that forces other birds to disgorge food. When its fish food basis dwindles, it does not shy away to prey on other seabirds as recounted in one of the life stories from the border of the sea.

The Fox and the Gull – Complex Webs of Interaction

The title reads like a fable from Aesop, but this life history was recently told in a scientific journal (Croll et al. 2005). The facts are easily recalled and are seemingly logic. The theoretical implications are, however, important and deal with cascading trophic interactions. Trophic cascades are defined by top-down control of community structure by top predators. These cascading effects can completely restructure an entire food web. The theoretical treatment of food webs is mathematically rather demanding. It is thus gratifying that its basic principles can be illustrated with a simple example. The scene is the Aleutian archipelago. The fur trade in the late nineteenth century led to a collapse of top mammalian predators around the Bering Sea. To revive the fur business, the Artic fox (*Alopex lagopus*) was introduced on some islands where they had an astonishing domino effect. The islands on which foxes were introduced showed a striking transformation from grassland to a maritime tundra ecosystem: The more productive grasses and sedges were replaced by the less productive forbs and shrubs. What caused this change of ecosystem to a nutrient-impoverished regime? Apparently, the introduction of a top predator (Figure 5.14) caused effects that spiraled down to the level of the primary producers. Cascading

1. Fuchs (Canis Vulpes). ¹/₁₂. (Art. *Fuchs.*) — Oben: 2. Schakal (Canis aureus). ¹/₁₂. (Art. *Schakal.*)

trophic interactions were set in motion. The archipelago carried 29 species of seabirds representing a population that exceeded 10 million individuals. On fox-infested islands this bird fauna was severely reduced. Due to their large number the birds were important linkers between the ocean and the terrestrial ecosystem. The birds foraged on the sea and extracted food from the ocean. This food was digested and resulted in the excretion of nitrogen-rich guano on the bird's resting places on the island. The seabirds exported thus nitrogen and phosphorus from the sea onto the land thereby subsidizing the terrestrial ecosystem with ocean-derived nutrients. On fox-infested islands the breeding seabird density decreased by almost 100-fold and the annual guano input decreased in parallel from 360 to $6\,\mathrm{g\,m}^{-2}$, thus reducing the soil fertility. Controlled fertilization experiments on fox-infested islands resulted in the shift back from the tundra to grassland proving the nutrient link for the regime change.

Nutritional Ecology

Trophic Cascades Across Ecosystems

Ecologists have described a number of systems where predation can result in strong effects through a food web. If a member of the food chain shows a change from the aquatic to terrestrial life style, one can get astonishing effects across ecosystems. This was recently shown for dragonflies (Figure 5.15). Dragonflies and damselflies belong to the insect order Odonata. Both larvae and adults are active predators. Larvae consume various invertebrates from their aquatic niche, while adults capture other flying insects. They possess chewing mouthparts with massive mandibles. The labium is modified into a prehensile organ. US zoologists looked into a site in Florida dotted with small ponds (Knight et al. 2005). Four naturally contained fish, four lacked fish, and they observed that fish facilitated plant reproduction near the pond. What happened? Actually, the juvenile forms of dragonflies are submitted to predation by fish. When there are no fishes in the pond, more and larger sizes of dragonfly larvae can develop. This gives rise to more dragonflies around the fish pond. The adults are active flyers, which avoid fish-containing ponds for oviposition since their success in reproduction here will be lower. This increased concentration of dragonflies

---

FIGURE 5.14. The red fox (*Vulpes vulpes*) (*center*) belongs to the dog family (Canidae) like the jackal and the wolf. The fox from the Old World eats mice, rabbits, birds, eggs, and fruits. A lot of folklore surrounds this animal, not the least the fox is the hero in a great verse epos "Reinecke Fuchs" from Goethe. The jackal (*Canis aureus*, at the *top*) is in contrast known to folklore for its cowardice. Its eating habits probably earned it this reputation: the jackal lives from small mammals, carrion, and plant material. It follows lions and finishes the leftovers of their prey. The wolf (*Canis lupus*) (*bottom*) is the largest feral animal of the dog family. This social animal lives and hunts in packs. Its preferred prey is deer, moose, and caribou, but it also attacks mice, rabbits, and birds.

FIGURE 5.15. Dragonflies around a pond. The picture shows various species of Brazilian dragonflies (*top to bottom*: *Chalcopteryx, Hataerina, Euthore, Thore*), family Libellulidae, order Odonata.

around fish-free ponds leads to a decrease in their favorite prey—bees, followed by moths and flies. Bees play an important role as pollinators for plants that grow at the border of these ponds like St John's wart (*Hypericum*). This plant has evolved traits to attract bees as pollinators. However, near fish-free ponds bees are rare and flies have to take up this role, but they are apparently less efficient (Figure 5.16). The consequence is clear: The plants become pollen limited and less seeds are produced. This is not just the effect of lower numbers

FIGURE 5.16. Insects as pollinators. The figure mainly shows flies (Diptera), on the top of flowers *Dioctria, Helophilus*, two species of *Volucella*, at the *bottom left* a Hymenoptera, *Bombus*, a bumblebee. The figure illustrates the prey of dragonfly.

of bees. Pollinators perceive the risk and avoid foraging near dragonflies, they hastily rush from flower to flower and search a quieter place. Animals which change from aquatic to terrestrial ecosystems are not rare; actually an important class of vertebrates carry a name that reflects their split loyalty—Amphibia. Salamanders are similarly predators in both ecosystems.

The War of the Senses: The Example of Echolocation

Killer Whale Versus Harbor Seal

With the killer whales (*Orcinus orca*) we got to a top predator of the ocean. One of its preys is the harbor seal (*Phoca vitulina*) and it needs all its senses to stay alive. In one of the previous chapter, we heard how its bigger cousin was using its whiskers to detect the herring prey. Now the seal needs its ears to detect and interpret the underwater calls of killer whales. Not all killer whales are dangerous for seals. The resident killer whales live in large stable groups and feed exclusively on fish. However, in the waters there are also transient killer whales that live in small groups and feed only on marine mammals. The piscivorous whales emit echolocation clicks in the quest for their prey, and they have in addition a rich repertoire of vocal dialects for communication. Harbor seals have a good underwater hearing and the calls of the fish-eating whales can be heard over great distance. It would be a waste of energy if seals would respond to the calls of all killer whales with an escape response, i.e., leaving the sea surface where the seal is well visible and an easy target to echolocating predators. Therefore, it is not surprising that seals do not care about the calls of the resident whales. To be on the safer side they make the investment to "learn" all local dialects of the resident whales, which is no small memory task (Deecke et al. 2002). The mammal-hunting whales, in contrast, are usually silent, but when they give a call, the seals immediately start with the escape reaction. Probably, this explains why the mammal-hunting whales vocalize less frequently than the fish-eating whales. Then the marine biologists did an interesting experiment. They recorded the calls of fish-eating killer whales living in Alaska, which "spoke" a different dialect, and displayed these sounds to the British Columbia seals. What would they do? Actually, they interpreted the intrinsically harmless Alaska orca calls as danger and showed an escape reaction. Apparently, they did not recognize a specific dangerous killer call, they treated any calls as potentially dangerous except those they learned to associate with harmless neighbors. Only by selective habituation the seals learned what not to fear, a conservative but safe strategy.

Ultrasound Detection and Escape Reaction: From the Herring to the Butterfly

Echolocation is also a hunting strategy used by dolphins. Most fish species do not hear the ultrasonic clicks used by the predator. Cods and clupeid fish like the herring or shad show an escape response. Do they actually hear ultrasound? Zoologists established audiograms for shads by training them to reduce their heartbeat when they detected a sound. They confirmed the sensitivity of shads to ultrasound signals (Mann et al. 1997). In addition, clupeids show a unique ear structure: A pair of thin air-filled tubes project from the swim bladder and terminate in air chambers that are connected with the utricles of the inner ear.

Cetaceans like the dolphin and the killer whale are not the only users of echolocation. The same hunting tool evolved in bats. Also its prey had to evolve counterstrategies to stay alive. Take Lepidopteran as example. One straight-forward strategy is to avoid the bat. As bats are hunting at night, a simple strategy is to be active during daytime. This is the option of butterflies (Figure 5.17), the diurnal Lepidopteran. They rely on a well-developed visual system for predator detection and communication. The first makes them difficult to hunt as you remember from your childhood days and the second makes them beautiful animals. The butterfly was thus "invented" by the bat. Lepidopteran which are active at night are called moths (Figure 5.18). Due to the predation pressure by echolocating bats, moths have evolved five times ultrasound hearing. Escape under the sunlight seems to be a late development of Lepidopteran. Interest-ingly, *Hedyloidea*, the living ancestor of modern-day butterflies, which forages still in the night possess ultrasound hearing. At the basis of their wings, they have ultrasound hearing ears showing a tympanic membrane connected to an ear canal. When they are caught in an ultrasound wave package, they initiate with very short latency of 40 ms—an elaborate flight maneuvre. They increase the flight speed fourfold and make steep dives or climbs mixed with unpredictable loops and spirals (Yack and Fullard 2000).

Parallels and Differences in Echolocation

It is remarkable how a basic problem, namely lack of visibility of the prey to the predator, led to a similar solution. Bats do not see the moth because they hunt at night and dolphins do not see the fish because of the low penetration of light in their marine environment. The very different physical medium, air and water, did not prevent the convergent evolution of very comparable echolocation systems. In detail, there are of course many differences between both echolocation systems. For example, both animals solve differently the problem how to avoid being deafened by their own ultrasound cries when they get very near to the prey. When approaching the prey, fewer echos are lost to the medium, thus the sound becomes louder. In dolphins, the amplitude of the sonar they emitted decreased by 6 dB every time the distance was halved. Thus the echo does not increase in strength. In contrast, bats keep the amplitude of their sonar signals constant, but they can decrease the sensitivity of the hearing process as they get nearer to the prey. The different strategies are dictated by different anatomies in the middle ear. The local muscles are stiffer in the ear of the dolphins than in the bats thereby precluding a down regulation of the hearing sensitivity (Tromans 2003; Au and Benoit-Bird 2003).

Background for the Bat

Other observations, namely those with bats, concur with a food differentiation in coexisting predator species. First, I will provide a short historical note on this fascinating foraging technique. The discovery that bats could "see" with their ears was made by L. Spallanzini more than 200 years ago (Figure 5.19). It was heavily

FIGURE 5.17. Because of their day-flying habits and bright colors, butterflies (superfamily Papilionoidea) are very visible to humans. The figure shows several common examples of European butterflies (from *top*: *Lycaena, Gonepteryx, Vanessa atalanta, Vanessa io, Lycaena arion*). However, the vast majority of the Lepidopteran diversity comes with night-flying and dull-colored moths. Biologists think that the evolution of butterflies is a response to the predation pressure by night-hunting bats.

rejected by G. Cuvier, but after repeating Spallanzini's experiments D. Griffin confirmed the interpretation in 1940. He also coined the term echolocation to describe the phenomenon of how bats use the echoes of the calls they produced to locate objects in their path (Fenton and Ratcliffe 2004). Fossil evidence supported

FIGURE 5.18. The nighttime brethren of butterflies, represented by common European moths, is shown. The three larger animals belong to the order Sphingidae, Heterocera (*Sphinx convolvuli* (*top*), *Sphinx neri* (*center*), *Sphinx ligustri* (*bottom*)). Some moths have developed ultrasound hearing and escape reactions.

the hypothesis that the oldest known bats had already evolved this technique nearly 50 million years ago (Novacek 1985). Echolocating bats can be divided into guilds according to their preferred habitat and foraging behavior. There are "hawks" and "gleaners," "trawlers" and vegetarians within echolocating bats.

Let's start with the trawlers. Five species of the European bat *Myotis* (Figure 5.20) were investigated for their foraging technique. They all screen prey in the aerial edge space near vegetation. This could lead to potential competition

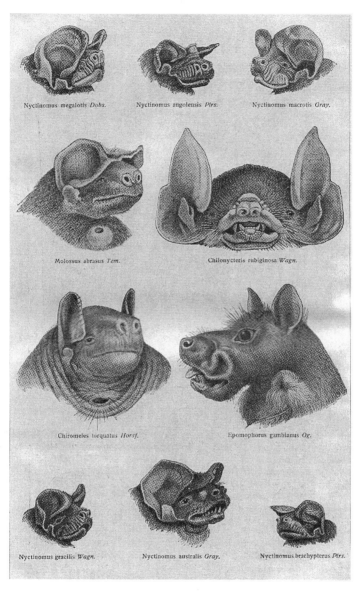

FIGURE 5.19. Portraits of different bats (Chiroptera) showing that in many of them the acoustic orientation dominates over the visual or olfactory senses.

between these animals; however, differences in the echolocation signals contribute to within-guild niche differentiation. These bats catch flying insects and spiders walking close to vegetation. This creates a problem for the bats: How can they differentiate the prey against the background? Experience with mealworms on a clutter screen, which mimicked the vegetation noise, showed that they could not see the prey when sitting on the screen. The prey became

FIGURE 5.20. Brown bat (*Myotis myotis*), a common Chiroptera with worldwide distribution—next to man it could be the most widespread of land mammals. Its echolocation-based foraging covers strategies of hawks, gleaners, or trawlers as discussed in the text.

visible only when removed several centimeters away from the screen. The interesting observation was that the five species differed in their capability to locate the prey according to the distance (Siemers and Schnitzler 2004). Within a 5 cm distance between prey and screen only *Myotis nattereri* showed a 100% capture success rate; the four other species needed a successively greater distance for successful hunting. Visual, olfactory, and passive acoustic cues were excluded for the capture success as well as wing morphology. The difference was in the quality of the echolocating systems. All five bats chose an oblique flight path when approaching the screen. However, the sonar differed substantially: The best performing bat at low distance showed the highest start frequency in the sonar, which then swept also the broadest frequency band from 135 to 16 kHz. Expressed as wavelength it covers 3–22 mm, this is just the size range of arthropods they prey and leave background clutter. In other words, *M. nattereri* illuminates the sonar scene with more "light." It is thus possible that these five bat species are not competing at all for a common food source, but each specialized to an invisible food layer at different distances from the background vegetation.

Bats Structuring the Food Space

Bats are fascinating animals for zoologists. Take the large-eared horseshoe bat (*Rhinolophus philippinensis*), a rare but interesting species from the Wallacea region (*southeast Asia to northwest Australia, named after A. Russel Wallace the co-discoverer of the evolution theory, who did important field work in that region).* This species comes in three morphs: a large, an intermediate, and a small form, where the extremes differ by a factor of two in weight. Each size class was associated with different echolocation call frequencies: the large called at 27 kHz, the small at 53 kHz. The intermediate was also for this property between the extremes (Kingston and Rossiter 2004). The physical consequences are immediately clear: Animals with low frequency calls emit longer wavelengths, which reflect poorly from small prey. Thus for them, prey below 13 mm wing lengths should be difficult to detect. However, as is usual in biology, there is a trade-off: Low frequency calls are less subject to environmental attenuation and allows thus a larger detection range. The large morph samples thus a five times larger volume for large prey compared to the small morph. The latter "sees" nine times as much small insects than the large morph, but its "visual field" range is smaller. The size difference might reflect a secondary adaptation. If you catch bigger prey, you need also a bigger mouth. The impact of this differentiation goes actually further. The bat's sonar is used not only in the quest for food, but also in intraspecific communication, namely mate recognition. The large and the small morphs are functionally deaf to each other's calls. The intermediate form can hear the large and the small morph, but it is unable to establish a reciprocal relationship between either of them. In this way, reproductive isolation can be established in sympatric populations (i.e., those living in the same geographical area without physical barriers).

Bats as Pollinators

You can illustrate many basic biological principles with bats. Blood sucking vampire bats provided fascinating examples for R. Dawkins book The Selfish Gene, where you can deal with the problems of mutual help, cheating, and altruism in mammals. I will use here another aspect of bat's echolocation system. Actually, not all prey dislikes the idea of a visit by a bat. Some like and even need bats and found forms to attract them. This sounds pretty paradox, but is easily understood when realizing that some bats are pollinators. Several hundred species of neotropical plants are pollinated by glossophagine bats. Remember the disrespect for scientific terms, this reads as "tongue-eating" bats. The tit-for-tat is clear; the bat gets nectar as its food and serves as a distributing system of the pollen for the plant. It is one of the great ironies of natural selection that we find the flowers of plants esthetical while they evolved to be attractive for insect pollinators. We and insects see flowers in different wavelengths due to the distinct wavelength of the two eye systems. Bats do not see the flowers when they are foraging during the night, they literally hear them. Plants don't cry, but the bat-pollinated neotropical vine *Mucuna holtonii* evolved a flower structure that functions like a cat's eye on your bike in the optical range. The flower is of the more complicated type of Papilionaceae (e.g., the pea). The inflorescences hang down several meters from the canopy and are located at the edge of the rain forest. When the pollen is mature, the flower lifts its upper petal, which signals a virgin willing to give 100 μl nectar on the first visit and an explosion of pollen on the back of the bat when it presses its snout into the flower. The remarkable observation is that the immature flower is a poor echo reflector, but with the top petal lifted the plant becomes a good reflector and thus very "visible" to the bat (von Helversen and von Helversen 1999). In fact, in the first night bats visited 88% of virgin flowers with the lifted petal. When the researchers cut the top petal, this frequency fell to 21%. To prove that they had only touched the sonar reflector and not other cues (e.g., olfactory signals), they let the flower intact, but put some cotton wool into the top petal reflector. The rate of the bat visits fell to 17% and the reflectance of the flower was only marginally higher than that of the closed flower. *With the latter example we come to a new level of complexity in the sensory structuring of the food space. The senses of the predator are answered by a sensory response of the prey, in this case of a consenting prey. In case of unfriendly takeover, the war of the senses is declared in the "eat and be eaten" scenario.*

Listening Bats

Echolocation is a formidable tool for insectivorous bats, but it cannot serve all purposes. It is well adapted to aerial hawking, i.e., the location and capture of airborne prey. The prey stays clear against a void background. The prey can avoid this dangerous position when hiding in bushes. Bats cannot detect the prey against the reflecting leaves, especially when the prey remains immobile. In such echo-cluttering environments, bats must change their foraging strategy. They change for substrate gleaning. They switch off echolocation and listen instead for prey-generated sounds to locate the insect (Arlettaz et al. 2001). Some prey has to call, for example, frogs that must attract a mating partner. The bat *Trachops cirrhosus* uses these acoustic cues to capture calling frogs. Its performance is poor when the frog is not calling. But if the frog is vocalizing, the bat can distinguish a palatable species from a poisonous one, or a small species from one that is too large to be captured or eaten (Tuttle and Ryan 1981).

Antipredation Strategies

Sleep: The Problem of Putting Senses to Rest

Survival depends heavily on the sharpness of your senses. However, nobody can be on the alert all the time. Higher animals need, like us, a fair dose of sleep. However, the shut down of the senses exposes the animal to substantial danger. Birds have overcome the problem of sleeping in risky situations by developing the ability to sleep with one eye open and one hemisphere of the brain awake. The investigated ducks showed a group-edge effect. Individuals, which are exposed at the edge, kept that eye open that faced the outside of the group. The waking hemisphere was capable of quick predator detection and the ducks initiated escape behavior with a latency of only 0.16 s (Rattenborg et al. 1999).

Alarm Calls

Since predation is the major cause of mortality for most animal species, one should not be surprised that many species have evolved alarm signals that warn conspecifics about imminent danger. Recent studies with *Poecile atricapilla*, a small North American songbird, revealed one of the most subtle alarm systems in animals. Chickadees (Figure 5.21) form groups of eight birds that communicate socially by vocalization. The calls encode information about food, social identity, and also predators. Closer acoustic inspections of the alarm calls demonstrated a sophistication, which ornithologists would not have believed before this study (Templeton et al. 2005). The researchers knew that chickadees distinguished aerial (raptors) from terrestrial (snake, ferret) predators. The bird produce a sound described as a "chick" syllable and one known as "dee" notes. The D-notes differed significantly across the predators and the researchers got a linear relationship when they plotted the number of D-notes against the wingspan of the predator. This is an important information for the potential prey: Birds with large wingspan are less maneuverable, eating few songbirds, while smaller predators are well adapted to hunting songbirds, which thus compose a major part of their diets. The question is, whether the birds were aware of that information in the calls. Playback experiments showed that they modified well their behavior according to the degree of danger. This was not necessarily a flight reaction, mobbing calls recruited other chickadees to harass a predator.

Semantic combinations of two basic calling sounds are now described in *Cercopithecus* monkeys (Arnold and Zuberbuhler 2006). Males call frequently during morning foraging and when seeking sleeping places in the evening. One call warns the group from an approaching terrestrial predator like the leopard (*Panthera pardus*). The group reacts with flight into the canopy. The second call is given when a male had spotted another predator, the crowned eagle (*Stephanoaetus coronatus*). The group does not move into the canopy because this would only increase the predation risk. Psychologists detected now a simple semantic: Combinations of the two calls in series induced a lateral movement in the group. Thus a surprising complex behavior can be induced by using combinations of just two calls.

Flocking

Many vertebrates build large groups of animals, be it shoals of fish, swarms of birds, or herds of grazing mammals. The individual animal experiences a greater protection when hidden in a large group where the predator has difficulties to single out an individual prey than when outside of the group where a predator can easily concentrate on an individual prey. This strategy is widely distributed and also found in invertebrates (e.g., juvenile locusts, Figure 5.22; and migration column of flies, Figure 5.23). The flightless Mormon crickets (*Anabrus simplex*) form enormous migration columns up to 16 km long and several kilometers wide that consist of millions of individuals marching over the countryside, devastating

FIGURE 5.21. Various species of chickadees are shown. From *top to bottom*: *Parus major, Parus caeruleus, Parus palustris* (*left*) and *Parus cristatus* (*right*), *Parus ater*. Their alarm called as antipredation strategy is discussed in the book.

all agricultural areas that they cross. It is commonly anticipated that a number of benefits are linked to this flocking. They range from earlier detection of predators, predator confusion by the sheer mass of prey, and if the predator should nevertheless decide to attack, the individual risk is well diluted and the predator arrives to saturation before the counts of the prey is measurably affected. This phenomenon is known as the selfish-herd effect. Agronomists have measured this protection by marking individuals with lightweight radiotransmitters fixed on the back of the crickets (Sword et al. 2005). Half of the marked crickets were

FIGURE 5.22. Locusts (here the European form *Locusta migratoria*, grasshopper family Acrididae) are known from Biblical times as plagues to humankind. Locusts are also part of the diet from hunter-gatherers.

put back into the marching column, the other half was placed outside of the band. Two days later the radiotransmitters were relocated: All column marchers were alive, while the displaced crickets suffered 50% mortality mainly from rodents and birds.

The desert locust *Schistocera gregaria* is from Biblical time known for its catastrophic effects on human society in economic terms. The animal exists in two forms: a harmless nonband-forming solitary form and an actively aggregating band-forming gregarious form. The solitary phase is the normal state of the species and the gregarious migratory phase is a physiological response to unfavorable environmental conditions or overcrowding. The two stages differ in metabolic state, oxygen consumption, and irritability. An understanding of this transition is crucial to the pest control of mobile swarming insects. Before the adult locusts develop flight, juvenile wingless locusts form like crickets kilometer-long marching bands. Control measures depend on the predictability of the marching direction. Ecological zoologists used a model from physicists and treated the animals as self-propelled particles that adjusted its speed and direction in response to near neighbors (Buhl et al. 2006). They tracked locusts in a video-controlled running arena and varied locust density. The time during which the locusts were aligned increased with increasing density and above a density of more than 70 locusts per square meter, spontaneous changes in direction did not any longer occur. They observed that inactive locusts did not affect the behavior of moving locust and that solitary locusts increased their activity levels within hours when exposed to crowding. The scientists suspected

FIGURE 5.23. *Sciara militaris* from the family Mycetophilidae, order Diptera, is a "midge" that owes its name to a behavior shown by its larvae. The larvae eat fungi and decaying plant materials in forests. Like the crickets described in section "Antipredation Strategies," they form large flocks that march through forests from central Europe, interpreted in the seventeenth century as a harbinger of war (with high predictive value in the times of the Thirty Years' War). In German they are called "Heerwurm," army crowd. The animals probably do not search for food, but they do search for favorable places to transform into pupae.

that marching bands allow a sensitive tracking of weak food gradients in the environment that cannot be sensed by individual locusts.

More on Crickets: A Story of Cannibalism

Joining a large mass of conspecifics is, however, also linked with risks since the migration is apparently forced by starvation. These insects selectively feed on high-protein food sources like seed heads and pods, flowers, carrion, and mammal feces, even soil soaked with cattle urine, suggesting that they suffer from nitrogen and salt deprivation. Protein but not carbohydrate satiation inhibited the tendency to march (Simpson et al. 2006). When crickets could satisfy their protein needs on day 1, they decreased their protein intake and increased their carbohydrate intake on day 2. The starved crickets preferred 0.25 M NaCl strongly over water and they selected food in the path of the marching band according to the protein content. The protein and salt deprivation was so strong that they ingested their own shed exoskeleton after molting. Actually, the richest source of the lacking protein and salts were conspecifics. Some protein-deprived insects consumed another cricket in a single meal. Protein or salt prefeeding reduced that cannibalistic tendency. Actually, the marching is largely imposed by this threat of cannibalism. Consumption of dead animals started within 20 seconds, while experimentally immobilized crickets, which were still able to defend themselves by kicking the attacker with their hindlimbs, largely survived. Cutting one and then two hindlimbs increased the cannibalism rate. Moving in a migratory band is thus a choice of the lesser of two evils, cannibalism over predation threat.

Crickets: A Story of Parasite, Prey, and Predators

Crickets suffer not only from predators, but also from parasites. A gut parasite of crickets is the Gordian worm (*Paragordius*; Ponton et al. 2006). To achieve mating in water, the worm has to convince the cricket to commit suicide in water, which the cricket duly executes. The worm then leaves the gut via the anus of the cricket. In the water, the adult worms are free-living organisms, where they mate as a knotted mass of individuals—hence the name. Few parasites have a predator, but parasites easily fall as victims to predators of their hosts. Some parasites manipulate the behavior of their host such that they avoid zones of predation (or search them actively if their parasites need to reach a different host). During its suicidal action the cricket is active at the water surface and attracts the attention of aquatic predators like fish or frogs. Does it try to get on land? Or does the worm need the next host and the swimming behavior of the cricket is just another trick of the parasite? Such cases are well known in parasitology. The Gordian worm (Figure 5.24) does not need a vertebrate as the next host. Actually, if it would not escape from the attack of the predator, it would be lost together with the cricket. The worm achieves just this escape feat with high efficiency. Minutes after the capture of the cricket prey, the worm escapes from the mouth of the frog or the gill of the trout. Selection has apparently found a solution for the worm's dilemma.

Mimicry

Mimicry comes in many forms and has caused a lot of theoretical discussion since the times of Darwin. Biologists generally distinguish two forms of mimicry. Batesian mimicry is cheating: A prey species uses the look of an inedible species and profits from the protection conferred by the signal evolved by the species warning its predator. In Müllerian mimicry, two species are both unpalatable to a predator and share the cost of teaching a naïve predator their unpleasant nature. However, the teacher pays the lesson with its life and only its conspecifics and the species using the same signals will profit from the deterrence in this educated predator. A recent work with domestic chicks suggested that when the two Müllerian prey species use, for example, two different chemical deterrents, they will heighten the attention level of the predator. When encountering two aversive chemicals, where the predator suspected only one, it will induce a greater caution in the predator (Ruxton and Speed 2005). There are many interesting stories around mimicry—for this book I have chosen two fascinating examples.

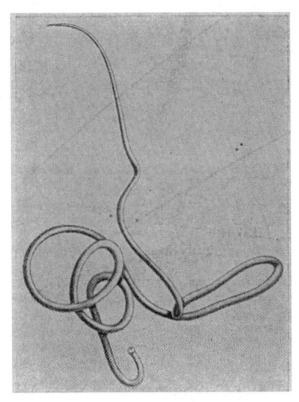

Figure 5.24. Gordian worms belong to the class Nematomorpha of the phylum Aschelminthes. The young animals of this roundworm live as parasites in insects, while the adult is free living in water.

Cleaner Fish

The first example is from the reef. The story has a first chapter, where the model of the mimetic is presented. The model is the cleaner fish *Labroides*, which picks from the client fish *Hemigymnus* parasitic isopods (Grutter 1999). The isopods infest the client fish during both day and night, while the cleaner fish eats its daily meal of 1,200 parasites only during the day. The client fish cycles thus with respect to parasite load between morning highs and sunset lows. Overall, the presence of cleaner fish results after a 12-h period in a more than fourfold lower parasite count for the client fish. The latter benefits from the cleaner and will leave it unscarthed because the cleaning is of mutual benefit for both animals. The story has a second chapter, where the mimetic is presented: the bluestriped fangblenny *Plagiotremus*, which mimics the juvenile cleaner fish. This disguise allows it to approach and ambush other fish tearing away tissue and scales with its large canines (Cote and Cheney 2005). What makes the story special is the third chapter. *Plagiotremus* can change its color at will. Within minutes of a transfer experiment, it changes into a form that easily blends into shoaling fish swimming just above the reef. This does not mean that *Plagiotremus* has changed the character. It conceals only in the fish shoals and will mount attacks from the shoal on nonsuspect passing fish.

Poisonous Frogs

The second story about mimicry is from the Amazonian part of Ecuador. It deals with Batesian mimicry, i.e., the resemblance of an edible mimic with a toxic model. A nontoxic prey animal thus has to deceive its predator by imitating a toxic animal. US zoologists investigated a theoretically interesting scene with four actors (Darst and Cummings 2006). The first two are poisonous frogs: *Epipedobates bilinguis* and *Epipedobates parvulus*, both with a brightly colored back. However, skin extracts injected into mice demonstrated that *E. parvulus* is more toxic than *E. bilinguis*. The third actor is *Allobates zaparo*, the edible mimic. The fourth actors are avian predators, more about them later. The mimic is dimorphic: In Northern Ecuador, *A. zaparo* mimics the there dominant *E. bilinguis*, in Southern Ecuador it resembles the dominant *E. parvulus*. So far, so good—-but what happens in the transition zone where both toxic species overlap? Will the edible frog mimic a polymorphic phenotype or one intermediate between both toxic models or resemble the more prevalent or the more toxic species? In fact, the zoologists were taken by surprise—against their expectation *A. zaparo* mimics the less toxic *E. bilinguis*. To understand this choice we need now the fourth partner in this play, the predator. However, this was a frustrating hunt in the field, even the preparation of plasticine frog models provided no biting marks that could identify the predator. Therefore, the rest of the story unfolded in the Texas garden of one of the zoologists with chicken trained on frog prey. As expected, the lesson with the more poisonous frog was learned quicker and led to generalized avoidance. The lesson with the less poisonous frog, however, was restricted to *E. bilinguis* and its specific mimic. Now it becomes clear why the mimic in the overlap zone mimics counterintuitively the less toxic model. Thereby it reaps the take home messages of the generalized and the specialized lessons from the chick predator and enjoys near complete protection.

Predator–Prey Cycles: From Chaos in the Food Web to Infectious Diseases

It is a major task of ecologists to develop an understanding for the patterns of predator–prey abundance in nature. The first impression when opening an ecology textbook is that of complexity. You find cases where the predator population remains at a rather constant level while its prey shows marked fluctuation in abundance (e.g., tawny owls vs. wood mice and bank voles on the prey side). You find a herbivore population that tracks the abundance of its plant prey, while the plant varies according to other forces than herbivory (e.g., cinnabar moth larvae and ragwort plants). Finally, there are cases where predator and prey populations are linked together by coupled oscillations in abundance (e.g., lynx and snowshoe hare; examples from Begon's Ecology; Blackwell Science 1996). If you read these textbooks, you will see a tendency toward abundance cycles and at the same time you will realize that the mathematical treatment of the predator–prey relationship dominates the discussion. Due to the many factors acting simultaneously on organisms in their environment, the trend for cycles is not so apparent in real populations. Ecologists have therefore increasingly turned to laboratory simulations of predator–prey interactions. In the following, I have chosen a few papers with some recent research results, which document current trends in the predator–prey discussion.

Ciliates and Bacteria in a Chemostat

Even in such simplified systems the outcome of predator–prey interaction can be highly variable (Becks et al. 2005). The ecologists used a simple predator, the ciliate *Tetrahymena*, which had two prey bacteria at its reach, one is the rod-shaped *Pedobacter*, the other the coccus *Brevundimonas*. The ciliate can live on either of the two bacteria as food source, but it prefers the rod on the coccus by a factor of four. In the absence of both bacteria, *Tetrahymena* did not survive the experiment. In the absence of the predator, the rods always outcompeted the cocci documenting their greater fitness in this system. The rules of the game seem pretty clear now. If you are a British, you might find this system worth a bet, but I would recommend that you start with small amounts of money—even this simple system might surprise you. The researchers conducted this three-partner mesocosmos in a chemostate where fresh nutrients for the mesocosm was provided and waste was removed according to a dilution rate determined by the experimentalist. The first run was done with the highest dilution rate $D = 0.9$ per day, i.e., 90% of the medium is exchanged per day. The less competitive coccus is out-diluted within 5 days and the system is reduced to two partners, both the ciliate and the rod bacterium are maintained at high and constant level. The next run was at $D = 0.75$: After 5 days, all three protagonists persisted at constant level with a fixed ratio (rod > coccus). Then the ecologists reduced the dilution to $D = 0.45$ and the system changed its behavior completely. After 10 days you get highly regular abundance oscillations maintained over indefinite time periods. The ciliate and its preferred rod bacterium showed a shift of one-half cycle in their abundance curves, one peaks in the trough of the other. The coccus cycles with lower amplitudes. When you find the system surprising with respect to its variability, you should keep part of your surprise for the final change. The ecologists set D marginally higher, namely to $D = 0.5$. Four replica experiments now showed distinct and aperiodic oscillations, what ecologists call a "deterministic chaos." Theoretical biologists had made many predictions about chaotic behavior in nature, but the surprise was that this chaos can now be studied experimentally in an extremely simple, but real biological system. Microbiologists believe that deterministic chaos is a characteristic of tiny fragmented populations that occur on soil grains or on detritus particles in the pelagic zone of the open ocean.

Rotifer and Algae: Rapid Evolution Affects the Cycles

US researchers studied a predator–prey system with organisms of a higher morphological complexity. Complexity is relative: Rotifers are tiny metazoa— most are less than 1mm long (Figures 5.25 and 5.26). They belong to a group of animals called Blastocoelomata, referring to the retention of an embryonic feature, the blastocoel body cavity, into adulthood. Despite their small size, the body organization is complex and they are divided into head, trunk, and foot. The anterior part bears a ciliary organ, the corona that rotates in the active animal like wheels, hence the English name "wheel animalcules." The pharynx is modified as a mastax, a grinding apparatus with jaws adapted to crushing, grinding, grasping, or sucking. They feed on other small animals and algae. The mastax leads into the esophagus and from there into a thick-walled stomach. Salivary and gastric glands add digestive enzymes to the ingested food. Digestion is extracellular in the stomach where nutrients are also absorbed. The intestine is short and fuses with the "kidneys" to a cloaca. In the experiment I want report, scientists fed the planktonic rotifer *Brachionus* with the unicellular asexual green algae *Chlorella*. As in the preceding experiment, the two organisms were held in a flow-through chemostat. The researchers expected well-behaved cycles as predicted by the conventional predator–prey model. However, what they saw were far longer cycles. Disturbingly, they observed extended periods where algal biomass was high, but rotifer densities remained low. This period was followed by increased rotifer growth, but algal density remained nearly constant (Yoshida et al. 2003). In the classical model, peaks in predator abundance follow peaks of prey by a shift of one-quarter of the cycle. What had happened? The algae evolved under the grazing pressure of the rotifer to a heritably smaller form of lower food value. As the algal population was grazed down by the rotifer, these clones increased to dominance. The rotifer population could not efficiently feed on these clones and consequently crashed. When that happens, the larger algae, which are better in nutrient uptake, outcompeted the smaller clones and rose in prevalence. When they again reached a high density, the rotifers could once again grow out. To test their interpretation the scientists used only a single clone of algae in the next experiment and they obtained the classical cycle with a quarter-cycle delay in predator abundance. This system has important lessons for biologists. One take-home lesson is what was called the "life-dinner" dichotomy. The selection pressure on the prey is greater than on the predator because the former loses in this race its life, the latter only a dinner. The other lesson is that the selection forces are fluctuating: Depending on the presence or absence of the rotifer predator, the algae will experience distinct selective pulls. What might be a selective advantage now, might become a disadvantage a few days later. No optimal organism can develop because selection pulls into one direction only to push back moments later. As an end result, fluctuating selective forces favors genetic diversity in the prey species. However, even if it is only for a dinner, predators can also evolve rapidly, but for fun let's explore the strategies of predators in a different biological system.

FIGURE 5.25. Rotifers, also called Wheel Animalcules, were formerly attributed to the "worm" phylum Aschelminthes, then to an own phylum, which was again contested. The figure shows on top *Melicerta*; bottom left *Noteus*, bottom right *Floscularia*.

Daphnia and Algae: Resource Management via Reproduction

Daphnia are still further up on the complexity scale of the animal reign: They belong to a mainly freshwater order of the Crustacea, the Cladocerans or "water fleas." These 0.5–3-mm-long animals have a carapace that encloses the trunk, which is fused from the thorax and abdomen. Under the carapace is the brood chamber for the eggs. The trunk carries the phyllopodous appendages, which serve in locomotion and in creating currents to get food into the mouth. The

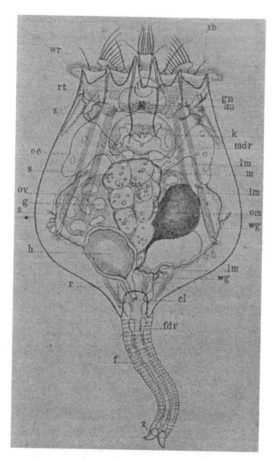

FIGURE 5.26. Section through the rotifer *Brochionus*. Between the ciliary organ (wr: rotating wheels) at the top of the animal you find the mouth, which leads via a buccal tube into a masticating pharynx (k). Between the pharynx (k) and the large stomach (m), salivary and gastric glands (mdr) inject digestive enzymes into the gut. The stomach leads into a short intestine (r) ending with a cloaca (cl) located at the transition to the foot (f).

benthic forms of water fleas scrape organic material from sediment particles; the planktonic species are suspension feeders. The digestive tract is a simple tubing, an esophagus leads into a midgut, that crosses the entire body. This tube is filled with a food string. A digestive cecum is an appendix to the esophagus.

Coming back to mathematical ecology, models predict that in predator–prey cycles the abundance amplitudes will be greatly enhanced by enriching the food source of the prey. However, this enrichment effect was not observed in these *Daphnia*–algae experiments. One explanation for this contradictory behavior could be in analogy to the rotifer–algae situation—the emergence of inedible algae that competed successfully with the edible algae in rich environments and reduced thereby the effective prey-carrying capacity of the system. Indeed, when inedible algae were removed from the system, fluctuations in the daphnia biomass exceeded a factor of five (McCauley et al. 1999). However, some experiments continued to show only small amplitudes. What had happened that created two different situations? There must be alternative attractors that dampened the cycle, other factors than prey characteristics. McCauley and colleagues found the key: *Daphnia* are parthenogenic (they reproduce without sex) and these eggs quickly hatch into juveniles. However, *Daphnia* can have sex and this produces energy-intensive resting eggs (ephippia) that drop to the bottom and do not contribute to population growth. Ephippia are a response to dwindling food resources, their production prevents the over-exploitation of the prey and leads thus to small abundance amplitudes. The researchers replaced in their experiments the ephippia-producing females with asexually reproducing females and—-as a nice confirmation of their predictions—-they obtained especially high amplitude population cycles. We see here a life history feature of the predator that uncouples consumer dynamics from food supply. Many other organisms respond to food scarcity by shutting off reproduction (e.g., secondary amenorrhoea in humans). Natural populations of zooplankton like *Daphnia* are rich in genotypic diversity. Coexisting genotypes frequently show strong fitness differences in the laboratory, yet they are maintained in nature. This phenomenon has been called the "paradox of the plankton." Competition theory offers a solution for this paradox. This theory states that relative fitness is not a constant between genotypes, but a function of the resource abundance, i.e., it is density dependent. To test this hypothesis, Canadian researchers either maintained the algal prey at constant level or induced low or high prey amplitudes. To these systems they added a mixture of *Daphnia* genotypes. Yet, under all externally driven environments, the same genotype became dominant (Nelson et al. 2005). It was a deterministic system. Where was the density dependence? In contrast, when the *Daphnia*–algae dynamics was internally generated by their coupled interaction, selection between the genotypes was reduced. What was now the equalizer of the chances between different *Daphnia* genotypes? Internally generated cycles began with a burst of *Daphnia* fecundity, followed 2 weeks later by a burst of small juveniles, followed later by a burst in large juveniles. However, most juveniles died from starvation during the prey-decline phase before entering the adult stage. The authors concluded that this stage-specific mortality reduced the fitness differences between the genotypes. Survivorship of over-wintering diapause eggs became thus more important for maintaining genotypic diversity in *Daphnia*.

Bacterial Predator and Human Prey: Syphilis

Scientific journals tend to group their research articles according to subjects, but adjacent articles normally do not come in any logical connection. Thus the above *Daphnia* article was followed in the scientific journal *Nature* by an article on syphilis in the USA. However, when I turned the pages I was struck by the similarity of the curves and the scientific approaches used in both articles. When taking a second look on both articles, I realized that the similarities go beyond superficial analogies. Both processes display aspects of predator–prey dynamics. The predator–prey pair is somewhat unusual for ecologists, but the phenomenon can be treated with the same rules of mathematical ecology, even if physicians use the term epidemiology for this form of ecology. Who is the predator? It is a spirochaete bacterium called the "great imposter" by clinicians and *Treponema pallidum* by microbiologists (Figure 5.27). It is diagnosed in darkfield microscopy by its corkscrew appearance and it shares this morphology with cousin spirochaetes like those causing relapsing fever and Lyme disease. *Although not proven by microbiological evidence, historians of medicine believe in their majority that syphilis was imported from the New World by the sailors traveling with Christopher Columbus. When coming back to Spain, they transmitted the new disease, which then caused a pandemic known at the time as the Great Pox to distinguish it from the Small Pox. It quickly moved through Europe in this period of political unrest mainly transmitted by soldiers during the movement of armies through the old continent. Its way can be followed by its names: Spanish, French, and English disease. Its first medical descriptions and its sexual mode of transmission were described in plain English in the Breviary of Helthe in 1547 and in fine Latin hexameters in 1530 by an Italian Renaissance physician. In this poem, Syphilis is a shepherd that angered the god Apollon, who took revenge by giving him this new disease (Grenfell and Bjornstad 2005).* Syphilis remained a major sexually transmitted disease throughout the twentieth century, and the British epidemiologists took advantage

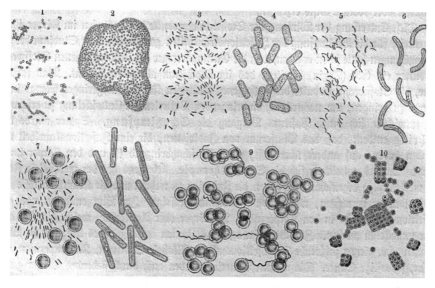

FIGURE 5.27. Bacteria as seen by late nineteenth-century microbiologists: *1, 2 Bacillus* species isolated and in association; *3, 4 Acetobacter*; *5, 6 Vibrio cholerae*; *7, 8 Bacillus anthracis* in low and high magnification, respectively; *9 Treponema pallidum* (with erythrocytes for size comparison); and *10 Sarcina.*

of the US documentation on this notifiable disease to address central questions not only of medical epidemiology but also of population ecology (Grassly et al. 2005). They asked for the role of exogenous environmental factors and density-dependent endogenous biological factors that drove this epidemic. Their numerical basis was a 60-year nationwide documentation of a predator–prey relationship that could make ecologists jealous. In fact, the epidemic pattern of syphilis in the USA has been explained by social and behavioral changes: a high in World War II due to troop movements, another high in the 1970s due to sexual revolution and gay liberation, followed by a 1980s poverty high, and a 1990s decline due to safer sex practices in the wake of the AIDS epidemic. The London epidemiologists did not buy this hypothesis and stated that syphilis is an unusually clear example of unforced, endogenous disease incidence with an 8–11-year periodicity, which is predicted by the infection dynamics and immunity development. They used the simple SIRS model with "susceptible," "infected," "recovered (immune)" states followed by the loss of immunity and return to the susceptible state. By anticipating a reproductive number of $R_0 = 3$ (newly infected cases per index case), the SIRS model predicts very well the period and the amplitude of the epidemic for the USA. The authors provided evidence for increased synchrony of syphilis oscillations across cities over time, pointing to networks between cities, connected by travel and sexual contact. The critical test was made with another sexually transmitted bacterial disease, gonorrhea, caused by *Neisseria gonorrhoea*. This disease is frequently transmitted with syphilis and should therefore show a similar dynamics, but it differs from syphilis by the fact that the *Neisseria* pathogen camouflages itself with different arrays of surface proteins, thereby preventing the development of immunity. In these diseases the simpler SIS model applies because no R state (recovered = immune) exists. Fittingly, no periodicity in the disease behavior was detected in the US data set.

Viral Predator and Human Prey: Measles

If we extend our predator–prey model to include ourself as food prey, many other would-be predators come into focus. I will here mention only one viral predator. In a later section of this book, I will provide some further examples where we become the food of such types of predators. The viral predator that I want to mention here is measles. From a historical viewpoint it is fitting to mention measles directly after syphilis. When the shipmates of Columbus first sailed from Europe to the New World they carried a blind passenger in their ships which might be the key to the startling question of historians as to how so few Spaniards could conquer so belligerent and numerically superior Indian warriors like the azteques. The answer is most likely not the superior European gun power or determination, but the introduction of a new virus into a totally susceptible nonimmune population that had never met this virus. Measles is highly communicable. A single index case can infect up to 15–18 secondary cases and measles is dangerous. It can cause pneumonia, blindness, and can even lead to death. British epidemiologists analyzed measles epidemics in England and they used for their study advanced methods of time-series analysis (Grenfell et al. 2001). Measles epidemics showed seasonal cycles and long-term biennial epidemics, not unlike those we had seen in predator–prey interactions in ecology. The latter cycling became less evident after the introduction of vaccination in 1968. They analyzed the spatio-temporal epidemic pattern with great resolution and got striking results. There was a clear wave of infection moving out of London with a wave speed of 5 km per week. The wave could be followed for up to 30 km out of London. Similar waves spread from large population centers around Manchester–Liverpool to the surrounding hinterland to capture distant small towns. After the epidemic, the recovered individuals become immune to reinfection and are lost from the susceptible pool. In small towns, the infection goes extinct because the concentration of the remaining and replenished (newborns) susceptible subjects became too small. Epidemiologists estimate 300,000 people necessary for measles maintenance. Smaller towns have to wait for a spark from larger population centers to restart an epidemic. In big cities the population concentration is high enough to maintain measles in the population. After a deterministic threshold of susceptible subjects has again been built up, a new epidemic can start.

One could add a number of reflections to infectious diseases as a predator– prey phenomenon. Also microbial predators need a certain prey population size not to go extinct. Some human viral infections might therefore not predate the neolithical revolution. A well-poised predator–prey system will avoid overexploitation (death) of the prey because otherwise the predator might go extinct. Presumably old human infections will thus not do much harm to its prey (e.g., chickenpox, retroviruses). Highly virulent viral infections are likely intruders into the human population coming from other animal species as trans-species infections (AIDS, SARS, Spanish flu, more on that later). Even very virulent infections show a trend for attenuation. Take the Black Death caused by Yersinia pestis. *You can isolate this bacterium from patients in India, but it does not make headlines any longer. Several factors have contributed to this low profile for* Y. pestis. *Certainly, the blood group composition in the European population seems to reflect the selection for more resistant prey genotypes (remember the inedible algae). Antibiotics are also particularly efficient against* Y. pestis. *However, microbiologists would argue that the bacterium has attenuated its virulence and there is some evidence for this hypothesis. In fact, ecology and epidemiology do not only have some formally similar topics, Darwinian thinking might matter in medical research. Actually, the term of Darwinian medicine is already used in the biological research literature.*

Toxic Predator–Prey Arms Races

On Snake's Venom

Many animals have developed venom systems either to their defense against predation or as a tool of predation. We sometimes get into this biological battle line and make painful, occasionally lethal experiences. This experience has shaped our cultural heritage, the most obvious case being the snake which is made responsible for the original sin in the Bible. I regret the shortsighted view of the creationist with respect to the Biblical record, who pulls biologists against the record of the Genesis report. This is a pity because it has—as I alluded at another place in this book—more than one grain of truth. I mentioned the nearly evolutionary report of what Biblical scholars refer to as the P author, who composed in the fifth century BC an attractive cosmogony borrowed from Babylonian mythology. As a scientist, I frequently admire the insight of the authors of the Book of Genesis. Here I will pay my reference to the J author, who wrote in the tenth century BC his more human story of the creation of the paradise from wasteland and the story of Eve and the snake. We see here how animals have taken the role of scapegoats (in fact goats), the very name reminding us how people from the Ancient Near East have used animals to transfer their feeling of sin and guilt to animals. This is certainly an interesting subject to explore for psychologists. However, the Biblical report has also an interesting message for biologists—the snake is punished for its temptation (and what temptation for scientists: knowledge; in this sense, scientists are the most faithful children of Eve. The National Institutes of Health carry the snake in their logo. It might be interesting to look back whether this choice was only taken in the Greek tradition linking the snake to the medical profession by the pharmaceutical activities of snake venom or whether it is also a hidden tribute to the Hebrew snake Najash, the legged Biblical snake). The snake's punishment is the loss of the limbs; the Biblical author thus sees snakes as relatives of four-legged reptiles like lizards. Not enough with that the snake is condemned to eat earth and there is eternal fight between the children of Eve and the snakes because of their poisonous bite (Figure 5.28). I would like to quote two recent research papers which come marvelously close to the insight of the J author of the book of Genesis. Argentine paleontologists found now a mid-Cretaceous snake fossil with a sacrum supporting a nonsutured pelvic girdle consisting of separate pubis, ileum, and ischium bones and robust hindlimbs. Cranial and vertebral bones show an adaptation to subterranean life. The authors speculated on a surface-dwelling terrestrial species that occasionally used tunnels to hunt burrowing prey. The fossil animal got the speaking name *Najash rionegrina* to honor the Hebrew account and the Argentine finding place. Najash is a more basal snake than the other known legged snakes, which carry the funny species names *Pachyrhachis problematicus* (Caldwell and Lee 1997), which possessed hind limb, but no fore limbs, and *Haasiophis terrasanctus*. The latter are marine species from marine limestone at the border of the Tethys sea—intriguingly near Jerusalem—suggesting a marine origin of the snakes, which is now dispelled by the terrestrial Najash. With that new fossil the origin of snakes from Cretaceous marine lizards like mosasaurus is rejected and the pendulum is back to the previous link to fossorial lizard ancestors. *The Bible wins over the Greek goddess Tethys, wife of the god of the oceans Okeanos. Perhaps the NIH should think on a legged snake for its logo?*

FIGURE 5.28. The copperhead snake from North America (*Ancistrodon contortix*) belonging to the viper family is shown in defense position. Its poisonous bite kills rodents within few minutes.

Also the old venomous ancestry of snakes was now confirmed by a large research consortium (Fry et al. 2006). It observed venom not only in advanced snakes, but also in evolutionary relatives like Iguania, Varanidae (e.g., Komodo Dragon), Anguidae, and Helodermatidae lizards (e.g., Gila Monster; Figure 5.29). In fact, the rapid swelling, dizziness, and shooting pain of the bite of the Komodo Dragon is now attributed to bioactive secretions and not bacterial infections as was previously believed. The comparative investigation of the venoms and the anatomical support suggested that venomous functions arose once in the evolution of squamata reptiles at about 200 million years ago. A group of nine toxin types are shared between all these animals, e.g., crotamine is found both in the rattlesnake and the Bearded Dragon. These nine toxins have well-characterized activities that induce hypolocomotion, hypotension, hypothermia, intestinal cramping, paralysis of muscle, bloodclotting disorder, and a marked increased sensitivity to pain. The venomous lizards added later on in evolution only a few new toxins, while the advanced snakes complemented this initial toxin tool set by an impressive array of further toxins. The toxins are stored as liquid venoms in glands associated with the upper jaw in snakes and in the lower jaw in lizards. From the gland a duct leads to a grooved tooth. The evolution of venom is now considered a key innovation driving the ecological diversification in advanced snakes. Beyond this classical snake–venom connection, the venom story has many players and ramifications. I have selected a few recent toxic stories to illustrate the field.

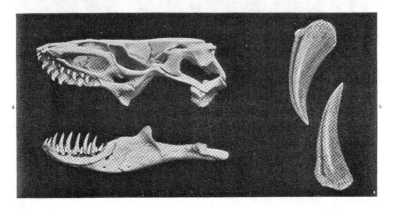

FIGURE 5.29. The Gila monster carrying the ominous scientific names *Heloderma suspectum* and *Heloderma horridum* are venomous lizards from the family Helodermatidae, order Squamata. They feed on small mammals, birds, and eggs. The Gila monster has a strong bite, which is easily understood when looking at the skull of the animal. It shows well-developed teeth especially in the lower jaw. The teeth have groves that conduct the venom from venom glands in the lower jaw into the prey.

Getting Used to Toxins: The Clam Case

Filter feeders like bivalves (Figure 5.30) are especially exposed to toxin-producing phytoplankton like diatoms (domoic acid) and dinoflagellates (saxitoxin). By their nutritional way of life bivalves are also passive carriers of viruses like hepatitis A virus or norovirus. Shellfish surveillance is thus an important public health measure to prevent paralytic shellfish poisoning or the transmission of food-borne viral infections. Bivalves are not targets for the human viruses, which arrive to them via fecal contamination of oyster banks, but how do bivalves escape from the action of saxitoxin? Saxitoxin is an inhibitor of the neuronal Na^+ channel. Marine biologists first thought that bivalves are not susceptible to saxitoxin. This is, however, not the case. They then discovered large differences in the capacity of different bivalves to accumulate saxitoxin or

FIGURE 5.30. Clams are important suspension feeders. The figure shows the mussels *Mytilis edulis* (*1*), *Lima hians* (*2*, *top* swimming), and *Pecten jacobaeus* (*3*, its shell was used for drinking by pilgrims and as decorates in many mediaeval churches) all from the Filibranchia suborder in the bivalve class (phylum Mollusca). The blue mussel (*1*) is raised in Europe as food since the thirteenth century.

tetrodoxin in vivo. Notably, this capacity correlated with in vitro differences in the sensitivity of isolated nerves to these toxins. *Mya arenaria*, a commercially important clam of North America showed a striking geographical distribution with respect to this toxin tolerance. Mollusks from the western and eastern coast of Nova Scotia differed in toxin tolerance; the western population was exposed to regular toxic blooms and was consequently more resistant. This looks like evidence for an evolutionary arms race between predator and prey. Most eastern clams were unable to re-burrow and to retract the siphon after toxin exposure, putting them at higher risk of mortality due to drying and predation. Not so the western races. Explanted cerebrovisceral nerve trunks of the former clams experienced a full block of action potential at 30-μM saxitoxin, while those from the latter were not inhibited by tenfold higher concentrations. The western clams accumulated high toxin levels in the viscera. When the researchers sequenced the voltage-gated Na^+ channels from both populations, they found a single nucleotide change that correlated with the in vivo resistance of clams to the toxin (Bricelj et al. 2005).

The Sodium Channel

Here we need a short backup on the structure of this membrane protein. The Na^+ channel is a multiprotein complex, but the essential part is its about 1,800 aa-long α-subunit. It contains four homologous domains (I–IV), each containing six transmembrane helices (h1–h6). The voltage sensor h4 shifts depending on the polarization state of the membrane toward the outside of the cell membrane and pulls thereby h6, the activation gate opens and allows thus the inflow of sodium ions. To confer selectivity to the ion transport you have as a guardian a "pore region" formed by the protein segment between h5 and h6. The membrane channel rests open only for a moment and the channel is then inactivated via a ball-and-chain mechanism by the inactivation gate, which is formed by a loop connecting domains III and IV.

The resistant clams had an aspartic instead of a glutamic acid in the "pore region" of domain II. The pore regions of the Na^+ channel are conserved over a wide range of animals, from flatworms to mammals. To prove the link, the researchers introduced the mutation into the rat channel protein and they observed a dramatic decrease in saxitoxin and tetrodoxin resistance.

Pharmacologists knew about differences between cardiac and brain Na^+ channels with respect to sensitivity toward tetrodoxin. They localized the difference to a critical amino acid (Tyr 374) in the "pore region," this time in domain I. Electrophysiological work with mutant proteins confirmed that the primary determinant of high tetrodoxin and saxitoxin sensitivity is a critical aromatic residue (Satin et al. 1992).

A Parallel Case: The Garter Snake Versus the Newt

Biologists frequently despair when they are confronted with a sheer endless variation of life strategies and get a gratifying feeling when nature uses and reuses the same basic principles in its manifold emanations. Actually, if you look behind the surface of the organisms and into their inner biochemistry, you don't have to look very far to see the common principle. However, here we have an absolute analogous situation of a toxic predator–prey interaction in two pairs that are evolutionarily widely separated. On one side you have the dinoflagellate–bivalve pair and on the other side you have an amphibian–snake pair. The garter snake (*Thamnophis sirtalis*; Figure 5.31) is among the commonest serpents of North America. It is small, usually not more than 60 cm, and quite harmless. It won't bite you, but may try to deter you by discharging a foul secretion from an anal gland. They live chiefly on insects, earthworms, and amphibians. On this diet they can be quite gregarious especially before breeding and hibernating. Its prey is the Californian newt *Taricha granulosa* (Figure 5.32), this 10–20-cm-long animal belongs to the Salamandridae family. It has a similar diet as the garter snake: earthworms, snails, and slugs. Unfortunately, it is also on the menu plan of bigger predators and newts are an easy prey to snakes. When the newt is attacked, it does not hide, but erects its head and tail to show a warning color. If the predator now bites, the punishment will be severe. The newt secretes the neurotoxin tetrodoxin from skin warts, and virtually all snakes die of muscular paralysis if they mistakenly eat a newt. With one exception: Garter snakes dine readily on newts because they have evolved a remarkable resistance against this toxin. Nevertheless, directly after a newt meal their crawling speed is temporarily reduced (Huey and Moody 2002). Now a fascinating evolutionary playing ground is offered for biologists. If garter snakes developed resistance against tetrodoxin, an untapped food source is opened, unchallenged by competitors. The newt should be doomed. However, if you look on a geographical map of garter snake distribution, the animals vary nearly by a factor of thousand with respect to tetrodoxin resistance (Geffeney et al. 2005). If you look for a molecular correlation of resistance, you again get amino acid replacements in the neuronal

FIGURE 5.31. The garter snake (*Thamnophis sirtalis*) is North America's most widely distributed reptile. This harmless snake for humans (you risk being threatened by a foul discharge from the anus) is involved in a toxic predator—prey cycle with a poisonous Californian newt.

FIGURE 5.32. A Californian newt (*Taricha granulosa*) secretes from its skin a potent neurotoxin, tetrodoxin, to deter its predator the garter snake, resulting in a chemical arms' race in a prey–predator relationship. The newt at the top is *T. granulosa*, the species at the *bottom* are *Molge rusconi* (*left*) and *Molge pyrrhogastra* (*right*).

Na$^+$ channel, this time in the pore region of domain IV. In fact, the molecular data demonstrate that the resistance has evolved at least twice and independently. The investigating biologists were interested in the details of the mechanism to reconstruct a case story for evolutionary interaction. Indeed, the effect of the toxin on the intracellular action potential correlated with the population differences. This was true not only between the populations, but held even for the variation within a given population (Geffeney et al. 2002). Again, the transfer of the mutation into a toxin-sensitive Na^+ channel conferred toxin resistance. Why all this diligence around this observation? In fact, the researchers were interested in tying a point mutation to an ecological relevant physiological phenotype, which can explain much of the evolutionary interplay between the predator and the prey.

Extensions of the Principle

Why do garter snakes vary in tetrodoxin sensitivity. One key to the understanding is that snakes living outside of the geographical distribution range of the poisonous newt do not show this mutation. Then why are not all garter snakes highly tetrodoxin-resistant when they live together with the newt? Here, you apparently have an evolutionary trade-off. The mutation comes with a price. Resistant garter snakes show a slower maximal crawling speed than sensitive snakes. If you introduce a third trophic level, let's say a predator of garter snakes, the reduced speed might result in higher losses due to inefficient flight reaction. Not all details are clear yet in this system, sound ecological data on the third trophic level are, for example, lacking. Nevertheless, we have here a fascinating case of analogous chemical warfare between predator–prey pairs separated by wide evolutionary distances. And this is not the only case where animals use tetrodoxin in self-defense: Fugu pufferfish, Atelopus frogs, blue-ringed octopus, and Phallusia tunicates all use this poison to deter predators. Apparently, also in the ecological context, nature reuses successful solutions to general problems. However, not all animals are likewise successful with this strategy. Fugu became a delicacy in Japanese restaurants without humans developing toxin resistance. The gourmand relies in this case on an anatomical certificate for the chef to remove safely the gall bladder such that his clients escape from deadly food poisoning.

Herbivory

At first glance you might be surprised to see a section on herbivory directly after sections on antipredation strategies and predator-prey interactions. However, if the basic principles of these interactions apply to so unusual pairs as viruses and humans, it would be surprising if the relationships between herbivores and plants were not governed by similar principles. In fact, herbivory is a difficult life style and it needs a lot of education by evolution to make a living from vegetable material. I want to demonstrate this point by three chapters. The first explores the evolutionary origin of land plants and why herbivory took so long to develop in the history of terrestrial life. The second chapter shows what we can do with carbohydrates as omnivores—not too much in fact. The third introduces the surprising observation that even such well-known herbivores like the cow can on its own not do much with plant material. It needs a lot of small helpers to deal with cellulose. Finally you will learn how plants strike back against herbivores.

Terra Firma—Bacteria and Plants Conquer the Land

Problems at Land

Why did life set out to conquer the land? Is the *"horror vacui,"* the fear of the void the guiding principle in biology? Life was born in the ocean because it offered what life needed most, namely water. The land in contrast is often characterized by dryness. Small wonder that life filled first all opportunities offered by the world's oceans, which cover anyway the biggest share of the earth's surface. In the beginning there were other problems, too. Without an ozone shield there were killing conditions at the land's surface—the genetic material of endeavoring organisms would quickly have been destroyed by the intensive ultraviolet radiation. Sufficient amounts of oxygen in the atmosphere might thus also have been a basic requirement for life on land.

Microbial Crusts: Cyanobacteria Again?

The earliest terrestrial communities were probably microbial crusts and mats venturing the land already in the Precambrian and leaving as witness organic traces in paleosols. Good candidates were cyanobacteria living in the intertidal zones. They had already learned to cope with changes from wet to dry conditions and had experienced great fluctuations in salt conditions. Why again cyanobacteria? We encounter them in our survey again and again. Are these cells for all seasons? Probably. I will illustrate their remarkable ecological qualities by one recent publication. Hot and cold deserts support only sparse plant growth because of the surface dessication that they suffer in these environments. In these regions, important primary producers are cyanobacteria. Microbiologists observed a possible key to their success. When they looked at the barren desert soil, nothing betrayed the presence of photosynthetic organisms. Then they wetted the surface with water and observed a greening of the soil, which occurred within half an hour (Garcia-Pichel and Pringault 2001). What had happened? Filamentous cyanobacteria of the *Oscillatoria* genus maintained a population at a depth of about 2 mm below the soil. In the presence of water, the bacteria showed a hydrotactic (water-seeking) reaction. When the soil subsequently dried out again, the cyanobacteria retreated again into the soil. The movements were perfectly reversible. Only high light intensity could prevent the active movement of cyanobacteria to the soil surface. Inhibitors of ATP generation also inhibited the movement. These sturdy cells are thus good candidates for the first pioneers of the continent.

Lichens

After the bacteria, the eukaryotes prepared the assault on the land. Fungi had been on the land at least since the Silurian and some paleobiologists believe that they derive from red seaweeds. Possibly they lost the race with the precursors of the land plants and specialized as saprophytes that lived from sending hyphae into the decaying corpses of the early plants to extract the nutrients. Fungi also discovered the new life style to team up with living photosynthetic cyanobacteria and green algae to form what we call lichens (Figure 5.33). Lichen crusts are still nowadays the outposts of life in the alpine regions. It is well possible that lichen crusts formed the second wave toward the land. Lichen-like symbiosis has a very ancient fossil record. Again, it is the Doushantuo Formation from southern China that provides interesting specimens (Yuan et al. 2005). Filamentous hyphae, which branch dichotomously and carry spore-like terminal structures and loops, surround what looks like cyanobacteria sheathed in a hyaline envelope. The coccoidal cyanobacterial thalli show no evidence of host reaction to the filaments excluding mycoparasitism. The association appears specific since nearby thalli from red algae show no filament association. Since these fossils are dated to 600 million years ago, fungi had in shallow marine environments already evolved symbiotic partnerships before the evolution of vascular plants. Today about one-fifth of all known extant fungal species form obligate symbiotic associations with green algae, cyanobacteria, or with both photosynthetic organisms. The molecular phylogeny analysis of rRNA sequences from 52 species of Ascomycota representing 18 orders revealed that lichens evolved earlier than previously believed and support the interpretation of the fossils (Lutzoni et al. 2001). Only the basal groups comprising organisms like *Candida albicans*, *Saccharomyces cerevisiae*, and *Morchella esculenta* are primary nonlichenized Ascomycota. The acquisition of lichenization followed quickly in one or up to three events. All other nonlichenized Ascomycota are secondary losses. Global weathering of rocks is heavily influenced by lichens as pioneer community since the Devonian.

The first colonists faced a harsh physical environment. Lichens, often associated with bryophytic plants, form still today a biological crust in many

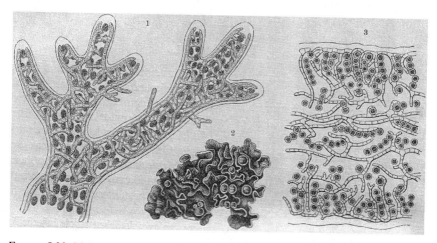

FIGURE 5.33. Lichens are a consortium of algae and fungal hyphae as demonstrated here by microscopical observation of *Ephebe kerneri* (*1*) and *Collema pulposa* (*3*, ×450). *Collema* in original size in (*2*).

harsh terrestrial conditions. Molecular clock estimates support a scenario according to which the major lineages of fungi were present 1 billion years ago and land plants appeared by 700 million years ago, such that a colonization of the land can be anticipated by 600 million years ago despite the lack of a clear fossil evidence (Heckman et al. 2001).

The First Land Animals

These authors also proposed that resistant biological crusts already contained the first animal land colonizers quoting Tardigrades as likely candidates. In zoology textbooks, the Phylum Tardigrada is quoted together with the Phylum Onychophora as distant relatives of Arthropods. In fact, Onychophoran ("velvet worms") resemble caterpillars superficially: The animal shows an inconspicuous head with two antennae and a mouth where circular lips surround a pair of jaws. The body is lined by about 20 pairs of sac-like legs. In fact, many zoologists regard Onychophorans as potential missing link between the Phyla Arthropods and Annelids. Even more exciting from an evolutionary point of view is the interpretation that the Cambrian fossil *Hallucigenia* (*nomen est omen*, a really bizzare spiny creature) is an onychophorian with pairs of elongated legs and pairs of dorsal spines. Likewise, *Aysheaia* from the Cambrian Burgess Shale is a probable onychophoran. These currently exclusively terrestrial animals have thus a prominent family tree. Today they occupy a niche shared with centipedes and are carnivores that prey on small invertebrates. Their feeding strategy relies on slime glands that discharge up to 30 cm of adhesive glue that entangles the prey. The velvet worm then injects the salivary gland contents into the victim, which is then digested extracorporally. The predigested food is then sucked up with the mouth. Fossil evidence suggests that a terrestrial invasion took place in the Ordovician (500 million years ago).

Tardigrades are somewhat more structured and their funny morphology earned them the trivial name "water bears," displaying an eight-legged miniature body in the submillimeter size range with a "smiling" head. Their weapon is an oral stylet, which is pierced into plant or animal cells for sucking the cell sap. Some tardigrades feed on bacteria, algae, and decaying plant material, carnivorous predators of small invertebrates are also known. The mouth opens into a buccal tube, which leads into a muscular pharynx that allows some form of mastication. A small esophagus opens into a large midgut, where digestion and nutrient absorption takes place. Strangely, defecation and molting is synchronized in some species. The association of tardigrades with moss is notorious, a spectacular case being a revival of a tardigrade from a 100-year-old desiccated moss museum specimen. Other characteristics make tardigrades likely candidates for early animal land colonizers. Tardigrades have fossils going back to the Lower Cambrian, are still today associated with early terrestrial colonizers like lichens, mosses, and liverworts, and show metabolic characteristics that allow them to resist physical hardship. During unfavorable environmental conditions tardigrades adopt as cysts a state of dormancy ("anabiosis") with greatly reduced metabolic activity or in the so-called tun stage even a state of cryto-biosis without detectable metabolic activity. This state allows tardigrades to survive extreme temperatures ranging from +140 to −270 °C, extreme dryness, and toxic chemicals like absolute alcohol. Such hardy animals could reasonably accompany photosynthetic organisms in the conquest of the land.

The Next Waves

Lichens as second wave to the land were probably quickly followed by mats of algae and then plants of a moss-like organization (Shear 1991). Notably, nearly all land plants have associations with fungi in their roots, called arbuscular mycorrhizae. In view of their beneficial effect on plant growth and survival via their contribution to nutrient acquisition by the plant, this symbiosis is ecologically important for most vascular plants. The fungal symbionts belong to one order, namely the Glomales in the Zymogomycota division of fungi, one of the four major branches of fungi (Figures 5.34–5.37). Fungi diverged at least 1 Ga ago probably before the divergence of the metaphyta and metazoa lineages in the Eukarya. Molecular clock analysis identified the onset of the diversification in the Glomales at 460 Ma ago in the Ordovician (Simon et al. 1993). This molecular date was later confirmed by a 460 Ma fossil find of hyphae and spores of Glomales fungi from the Ordovician (Redecker et al. 2000). The famous Devonian Rhynie Chert plant fossils like *Aglaophyton*, *Asteroxylon*, and *Rhynia* contain structures resembling vesicles and spores from extant Glomales fungi, demonstrating that latest at 365 Ma ago the arbuscular mycorrhizae was established. Here fungi played again a critical role as helpers for the conquest of the land.

Liverworts, hornworts, and mosses (Figure 5.38) are documented in the fossils from the Ordovician. A number of adaptations are necessary for plants living on land. Conducting strands, cuticles, stomata, and dryness-resistant spores had to develop. Early Devonian plants from the famous Scottish Rhynia Chert like *Cooksonia*, long regarded as the earliest vascular plant, still lacked conducting strands despite its 4 cm height. Phylogenetic studies favor a single origin of land plants from charophycean green algae (Figure 5.39). However, *Cooksonia* demonstrates that not all main adaptations to life on land were made in one phase. In the Devonian, plants with tracheids and cuticles were documented. This line led to quite sizable land plants, which were of 1 m diameter and several meters heigh, the largest land organisms of their time (*Prototaxites*). Tracheids are tubular cells, which are dead when they are functional. To prevent a collapse, their cell walls are strengthened by thick spirals or rings of bands containing lignin. Cuticles appeared first in enigmatic plants called nematophytes. These structures also represent chemical inventions, namely the biosynthesis of lignin and suberin. This is until today a very resistant material with respect to digestion and biodegradation. Since terrestrial plants are not treated in this book with the place they deserve, I offer a figure panorama on the major groups of land plants that shaped the planet Earth to at least visually compensate for this defect (Figures 5.40–5.63).

FIGURE 5.34. Mycophyta I: Phycomycetes. This is a primitive group of fungi that still plays around with different forms of sexuality and cell-wall chemistry (cellulose vs. chitin). The thallus (plant body of lower organisms) consists of siphonal tubes lacking septation. The order Oomycetales is here represented with *Achlya prolifera* (*1*) and *Saprolegnia lactea* (*6*) in the process of release of zoospores (*2–4* and *6, 7*). Chytridiales live as parasites and saprophytes on water plants and water insects, or are as demonstrated by *Polyphagus euglenae* even carnivorous where a single mycelium can contain up to 50 protists in different digestion states. *Rhizophydium* lives on pollen from higher plants. Saprolegiacea live as saprophytes on decaying plants and insects in the water, occasionally also as parasites on fish).

Lignin Synthesis and Degradation

Biosynthesis of Lignin

What is so special about lignin synthesis that makes it so resistant to chemical attack? The start of the pathway to lignin is quite conventional: It starts with the aromatic amino acid phenylalanine. The precious amino group is recovered, remember that photosynthetic organisms are nearly pathological carbon-fixing machines, but they have problems with nitrogen. Therefore, in the final product of the structural tissue of plants you have a lot of carbon, but nearly no nitrogen, which creates problems for herbivorous animals. The phenyl ring undergoes one to three hydroxylation reactions, which determines the chemical identity of the

FIGURE 5.35. Mycophyta II: Peronosporacea are another family of the Oomycetales, order of the Phycomycetes. Most live as parasites on higher land plants and cause substantial economical damage on potato (*Phytophora infestans*), tobacco (*Peronospora tabacina*), or vine (*Plasmopara viticola*). The latter represented in the figure causes a characteristic drying of the grapes (*1*). The fungus lives in the leaves and its sporangia reach the outside via stomata, a gas exchange organ (*2*). It releases spores, which differentiate into zoospores (*3–5*) that infect new leaves by entering again via stomata. If untreated, up to 20% of the wine harvest might be lost to fungal infections, as much as by animal damage.

starting material for lignin and lignan synthesis. According to plants that produce a lot of them, these natural products are called coumaryl, coniferyl, or sinapyl; the carboxyl group of the exphenylalanine is subsequently reduced to an alcohol group. Nothing is special so far. The first complications are introduced when the plant cells synthesize lignans, which are dimers and oligomers formed from primary coniferyl alcohol. It uses for this reaction probably laccase, an enzyme that creates aromatic alcohol radicals with the help of oxygen, but the data are not yet very clear.

A Purposeful Anarchic Synthesis?

Radicals can be created at the eighth position (in the C3 side chain) or at the fourth and fifth positions (the meta and para positions in the phenyl ring). As any of the unpaired radical electrons can combine, a number of chemical bonds can be formed with a single starting compound. In lignans the process is still steered by a

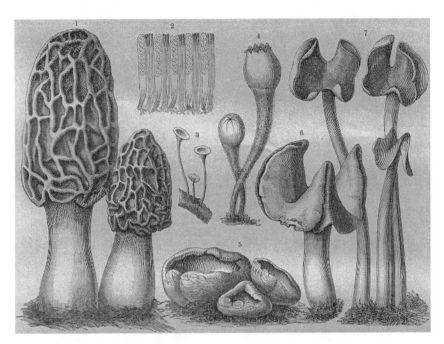

FIGURE 5.36. Mycophyta III: Ascomycetes. Another large group of fungi are the ascomycetes, which get their name from the ascus, a tubular sporangium in which eight spores are created by meiotic division (2). Most of them are terrestrial organisms that live as saprophytes. A few of the ascomycetes represent what the layman associates with mushrooms as the culinary appreciated *Morchella esculenta* (1). The picture represents a collection of specimen from the Discomycetidae: *Helotium* (3), *Anthopeziza* (4), *Peziza* (5), *Helvella infula* (6) and *Helvella fistulosa* (7). These fungi are saprophytes of the forest ground where they live on decaying wood. However, the ascomycetes are a divers group and also include organisms like the yeast *Saccharomyces*, the molds *Aspergillus*, and the plant parasite *Claviceps*, which we will encounter at different places of this book.

"dirigent protein" that yields preferentially 8-8' intermediates. In lignin the matter gets more complicated. The name lignin is derived from the Latin word lignum, which means wood. This is a very appropriate name since wood contains 20 to 30% lignin. After cellulose, lignin is the most abundant organic material of all vascular plant tissues. Lignin is a macromolecular meshwork starting with the three above-mentioned aromatic alcohols. The process is not that anarchic as initially proposed by biochemists. We know now that the tracheid cell still controls this process. Coumaryl alcohols are mainly deposited in the middle lamella between the approximately rectangular cells. Coniferyl alcohols, in contrast, go toward the walls that grow telescope like into the interior of the tracheid cell. Peroxidases working with H_2O_2 as cosubstrate mediate one-electron oxidations leading to radicals that start to polymerize the monomer. The following reactions were initially believed to be nonenzymatically free-radical coupling reactions. *In this chemical view, the second most common compound in the terrestrial ecosystem is not formed under enzymatic*

FIGURE 5.37. Mycophyta IV: Basidiomycetes. The common denominator of this large and divers group of fungi is the basidium, the organ that bears sexually reproduced bodies called basidiospores (7). The picture shows a panorama of Holobasidiomycetidae: *Clavaria* (1), *Daedalea* (2), *Marasmius* (3, 4), *Craterellus* (5), *Amanita phalloides* (6), *Hydnum* (8), and *Polyporus* (9). These fungi have inspired human imagination as an appreciated source of food that could be mixed up with deadly poisonous fungi. The hallucinogenous specimen played a role in many old religious rites. Fungi provide antibiotics and medical and nonmedical drugs and they were used as tinder (*Fomes fomentarius*) for fire making in the neolithic. However, basidiomycetes also contain in the Phragmobasidiomycetidae group important plant pathogens like the rusts (Uredinales) and smuts (Ustilaginales).

control. *This is at the same time astonishing and cute. Astonishing because it leaves such a central cellular process uncontrolled; cute because the anarchic polymerization makes this big macromolecule nearly unassailable to would-be herbivores. As you will see below, this trick worked extremely well during the first*

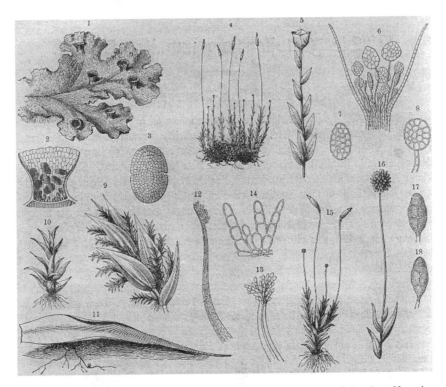

FIGURE 5.38. Bryophyta. Mosses are shown with a representative of the class Hepatica (*1–3, Marchantia polymorpha*) and the class Musci (*4–8 Tetraphis pellucida*) 9, *Leucodon*; 12, *Syrrhopodon*; 15–18, *Aulacomnion*.

100 millions years after the conquest of the land by plants. Herbivory was not an issue. Was it the nonenzymatic growth of lignin that made it so resistant to enzymes of evolving herbivores? Later on, biochemists recognized that the process was not that undirected as initially thought. Some bonds formed with much higher frequency: For example, the 8-O-4' bond represents 50% of all bonds. The synthesis of lignin is still an active research field at the moment as is the degradation of lignin.

Designed Imperfectness in Nature?

In view of its biosynthesis, we should not be surprised that lignin is one of the few natural polymers that can only be degraded with the help of molecular oxygen. Bacteria had their problem with this compound. Small wonder that the earliest land animals were not much more successful in attacking lignin. Under anoxic conditions bacteria could only grab some O-methyl groups from the polymer, but they could not crack the backbone of this formidable macromolecule. The savior was fungi, which were lurking in the zone between the sea and the land and were waiting for their chance. Still the decay of organic plant material in the soil is the job of fungi, which outnumber bacteria in that environment in both biomass and numbers. I said savior with hindsight. Evolution cannot work with perfect

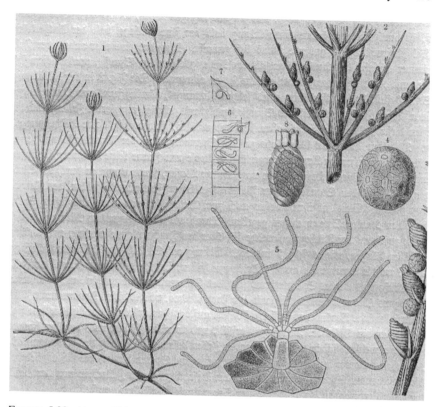

FIGURE 5.39. Algae: Chlorophyceae. *Chara fragilis* (1) belongs to a highly developed group of green algae showing very specialized reproductive organs, oogonia (8) and antheridia (2–4) and spermatozoids (5–7) as depicted in this figure. Some biologists suspect that the ancestor of the extant Characeae gave rise to land plants.

solutions. We heard already one theory which was based on historical arguments and essentially built on an inherent weakness of the evolutionary process that had to use past solutions for the ongoing game. Let's do a Gedanken experiment and imagine that nature, when it worked with lignin synthesis, would not have found the good solution plants possess now, but the perfect one: Lignin cannot be attacked at all, plants have the perfect armor. What would happen? Plants would grow their lifetime and then fall after their death. However, carbon fixed in lignocellulose would not be remineralized. CO_2 would in fact be sequestered in a stable organic form and not recycled into the biosphere. Enormous quantities of carbon would end up on the soil and the carbon cycle would after a while come to a standstill. I am not sufficiently educated in geology to judge to what extent the incapability to cope with fallen plants under anoxic conditions led to the conservation of the Carboniferous forests into coal. If this scenario would be the case, our burning of coal into CO_2 would only pay the debt back accrued by the early problems of bacteria and fungi with decaying plant material. Evolution

FIGURE 5.40. Pteridophyta I: extant Lycopsida. The club moss (*Lycopodium clavatum*) belongs, with the spike mosses (*Selaginella*) and quillworts (*Isoetes*), to the Lycopsida class of the Pteridophyta (fern-like plants). They are spore-bearing vascular plants that range from fossil trees to ground-creeping organisms like the depicted species. Protolepi-dodendraceae like *Drepanophycus*, which closely resembles the depicted living representative of this group, are known from the Lower Devonian and are thus the oldest land plants of Central Europe.

is as usual a complicated trade-off. Elements of recalcitrance are still visible in our world.

Lignin Degradation

Lignocellulose is still today only very slowly and incompletely degraded in soil giving rise to humic acids. *There is also an element of truce: Lignolysis is mainly the job of filamentous white-rot fungi as the basidiomycete* Phanerochaeta chrysosporium. *The curious trivial name refers to the fact that it degrades the brown lignin in wood to gain access to the readily metabolized white cellulose and hemicellulose. Also this fungus cannot support cell growth with an exclusive lignin feed. Where is the truce? White-rot fungi are saprophytes that live from dead plant material; only few are pathogens of living plants (e.g.,* Armillaria mellea). *The solution evolution found is well poised: Lignocellulose is the perfect armor as long as the plant needs protection during life. Once dead, the contract has expired and the fungi get the green light to degrade the plant material to the direct benefit of the fungus, but also to the benefit of the next generation of plants. Ask this from a packaging engineer in the food industry. A bottle that is*

FIGURE 5.41. Pteridophyta II: extinct Lycopsida. An artistic impression of a forest of up to 40-m high Lepidodendrales trees from the Carboniferous; these trees contributed substantially to coal formation. The burial of the photosynthetically produced organic carbon prevented its reoxidation with atmospheric oxygen in respiration, resulting in a net oxygen increase in the atmosphere due to unbalanced water splitting in photosynthesis. Lepidodendron is an extinct relative of the club moss (Lycopsida).

extremely stable when filled with milk or orange juice quickly decomposes into biodegradable compounds when emptied. He would have a hard time to design a solution.

Fungi are well adapted to their degradation job: They possess hyphae that allow fungi to growing into the decaying plant material. This strategy allows release of a true battery of compounds on the spot where it is needed: Numerous enzymes like lignin peroxidase and manganese peroxidase receive H_2O_2 from glyoxal oxidation, metals like manganese and molecular oxygen. The degradation creates, like the synthesis, cation radicals from lignin that then undergo a variety of nonenzymatic degradation reactions. It is curious that a macromolecule that was created via radical reactions is also degraded by radical reactions. However, as the synthesis of lignin has not yet been elucidated in detail, there are still major gaps in our understanding of lignin degradation. For example, lignocellulose is too tight a molecule to allow access to enzymes, at least in sound wood. To what extent are nonenzymatic radical reactions initiated first to make a breach into the substrate, thus allowing access to the enzyme later? *Is the well-known Fenton reaction ($Fe^{2+} + H_2O_2 \rightarrow Fe^{3+} + OH + OH^-$) part of it, creating the highly reactive hydroxyl radical for the pioneer reactions? Our uncertainty about the biochemical details of both the synthesis and the degradation of lignin and its*

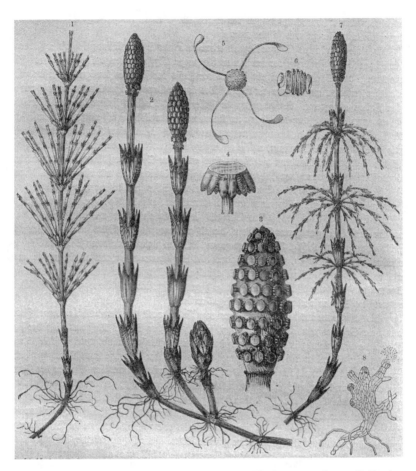

FIGURE 5.42. Pteridophyta III: extant Sphenopsida. The common horsetail *Equisetum arvense* (*1* the summer form; *2* in spring time) grows in meadows and along stream banks where it is fairly abundant, while the wood horsetail (*Equisetum sylvaticum, 7*) grows in cool moist woods. *3, 4* (amplified) sporangium carrier; *5, 6* spores; *8* prothallium. This genus with about 30 species is the only survivor of the class Sphenopsida from the Pteridophyta.

proven resistance in the biosphere is a vivid demonstration of the evolutionary "success" of the lignin design in the plant world.

Taking to the Air: Early Insects

Devonian Insects: Hard Time on Plants

In the Ordovician, there is little evidence for animal life on land. Poorly preserved millipedes and coprolites is all what we have. The Rhynie cherts of the Devonian provide the oldest remains of insects with well-preserved mandibles (Engel and

FIGURE 5.43. Pteridophyta IV: extinct Sphenopsida. Fossil Sphenopsida are in this picture represented with two fossils from the order Equisetales, namely with a stem from *Archaeocalamites* from the Lower Carboniferous (*1*) and a stem with leaves from *Calamites* (*2*), another important coal-forming plant of the Carboniferous. At the *right* is a stem with leaves from *Sphenophyllum*, which belongs to the only other order of the Sphenopsida.

Grimaldi 2004). *Rhyniognatha hirsti* was clearly a chewing insect, but it is unclear whether its diet was spores and pollen, leaf and stem tissue, or other animals. The authors of this article take great pain to suggest from indirect evidence that this beast was a winged insect. Part of the excitement is certainly a Guinness book of records entry. If their interpretation is correct, then insect flight would have preceded that of pterosaurs, birds and bats by full 90, 170, and 270 million years, which are not small time spans. This hypothesis would at the same time challenge and concur with current models. The challenge is presented by the fact that insect wings are powered by thorax muscles, which are among the highest oxygen consumers we know in the animal kingdom. These high-metabolic costs need a good food source and an atmosphere rich in oxygen. However, the Devonian sported only about 15% oxygen, less than the current atmosphere. On the food side the early land animals formed a strange guild. Herbivores were rare and if present only active as microherbivores. The early arthropods had no answer to lignin. In addition plants lack excretion organs. Toxic waste products are therefore frequently stored in cell walls. The plant material was thus not only hard to digest, it was even poisonous. Those vegetative parts of plants that could be attacked were of low nutritional value. They lacked and still lack today sodium and aromatic amino acids. The early arthropods had

FIGURE 5.44. Pteridophyta V: Calamites. An artistic impression of a Carboniferous swamp with Calamites trees (class Sphenopsida, depicted at the *left*), which could reach 30 m in height, and true ferns of the class Filicopsida at the *right*; both plants are members of the Pteridophyta.

to wait until fungi and bacteria had made their attack on plant litter; the animals could only form the second wave of attack.

Detrivores and Taking to the Air

Most of the animals were detrivores (Figure 5.64) and many were predators. In fact, this has remained a characteristic of soil ecosystems until the present. Mites living on saprophytic fungi and collembolans are found in the lower layers of the soil food web. In the upper ranks, one finds today nematode-feeding mites and predaceous mites feeding on the latter (Neutel et al. 2002). When bacteria and fungi had done their job, the plant material was sufficiently detoxified and the nutritional value of the explant material was enhanced. True herbivory was not yet an option and this remained so for a while. Even in the Carboniferous fossil, evidence for insects chewing on living plant parts is weak (Figure 5.65). For example, paleobotanists looked for bite marks on leaves from *Neuropteris*: Only 4% showed evidence for chewing (Shear 1991). When looking today at trees during spring, some biologists will argue that this is still a respectable rate. However, evolution is a tough teacher (or "need makes inventors"). Spores, ovules, and seeds contain less toxin and they are nutritionally a much richer food source than the vegetative plant parts. It is only logical that the early insects specialized on this food source. However, there was a problem: The plants started to grow in the Devonian. Tracheophytes became shrubby plants ~1 m

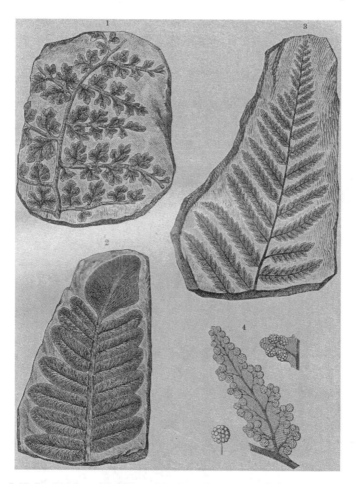

FIGURE 5.45. Pteridophyta VI: Filicopsida. Four fossil representatives from Filicopsida ferns of the Carboniferous: *Sphenopteris* (*1*), 2 *Neuropteris* (*2*), *Pecopteris* (*3*), and *Oligocarpia* (*4*), which resemble some living ferns.

tall. In the Carboniferous, the trend for arborescence was in full swing and plants were of sizable height. As the reproductive organs were passively lifted with the size increase, insect flight became a necessity. Now you understand why the discoverers of *Rhyniognatha* argue for a winged insect—it would just come in time to keep pace with the size evolution of plants. This is in evolutionary terms a healthy argument since other races can also be read from the fossil record. The predator–prey fight also took a new gear. Plant prey grew larger to avoid predation, but the insect predator participated in this size arms race, too. In some way, the competition got out of control and the largest insects that ever lived on the planet had evolved in the Carboniferous: *Meganeuron* with a wingspan that exceeded 60 cm. The winged insects of the Carboniferous had specialized mouthparts that allowed, for example, *Homaloneura* to tear apart the soft cones

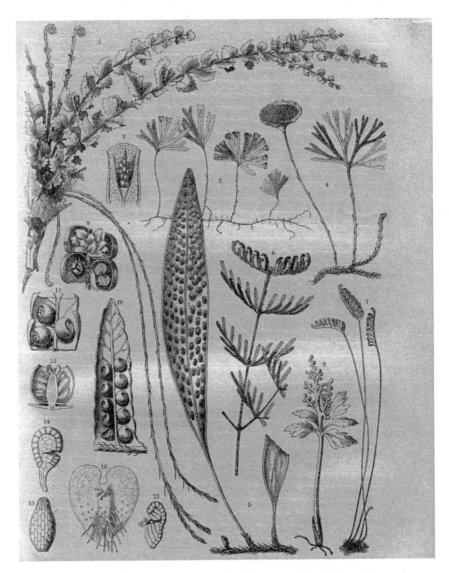

FIGURE 5.46. Pteridophyta VII: Filicopsida. The figure compiles different forms of extant ferns showing the widely distinct leave forms: (*1 Nephrolepis (1)*, *Trichomanes (2–3)*, *Rhipidopteris (4)*, *Polypodium (5)*, *Gleichenia (6)*, *Schizea (7)*, *Botrychium (8)*, *Cyathea (10)* and the sporangia (spore-producing organs) from *Gleichenia (9)*, *Cyathea (10—13)*, *Polypodium (14)*, and *Schizea (15)*. *16* shows a prothallium, the gametophyte of ferns.

of *Cordaites* trees. Others pierced ovules or fed by sucking with stylet-like mouthparts. Still others specialized on spores as demonstrated by fossils where the gut of the insects is entirely filled with spores. However, in the Carboniferous a long-term mutualistic insect–plant relationship started: pollination by insects

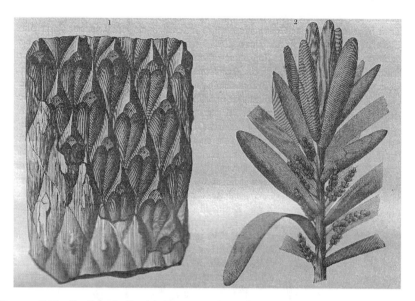

FIGURE 5.47. Conifers I. Cordaitales are perhaps the best-characterized order on the way to spermatophyta. They derive from Progymospermae in the Devonian, flourished with 30-m tall trees in the genus *Cordaites* in the Carboniferous, but became extinct in the Late Permian. In the popular cordaite–conifer hypothesis they gave rise through the intermediate of the primitive conifer family Lebachiaceae to the Coniferophytinae. The figure shows a stem with leaves and inflorescences from *Cordaites laevis*.

(Figures 5.66–5.68). The evidence is, however, indirect. The fern-seed pollen became so large that wind pollination became an unlikely option.

Insect Diversification in the Age of Angiosperms

Ants have a good fossil record: The oldest amber fossils are 100 million years old and come from the Early Cretaceous. However, these ambers contain both crown and stem groups of ants suggesting that the origin of ants substantially predates this time period. A large-scale molecular phylogeny of ants from nearly all 20 described subfamilies pointed to a shared common ancestor living 170 million years ago in the Middle Jurassic (Moreau et al. 2006). The authors asked the question why ants were so slow to diversify. Ants were rare in the Cretaceous and their march to ecological dominance began only in the Eocene 90 million years ago. They linked the dramatic diversification of ants to the rise of angiosperm-dominated forests. They suggested that the litter of angiosperm forests is more diverse than that of gymnosperm forests, providing more habitats. Ants could then exploit different food sources using predatory to a scavenger lifestyles. Still other ants lived on carbohydrate-rich honeydew secretions excreted through the anus of angiosperm sap–sucking insects like homoptera (aphids). These "trophobionts" receive for their nutritious excretions protection from predators

FIGURE 5.48. Conifers II: The evolutionary history of conifers began in the Carboniferous where descendants from the Cordaitales evolved what was lumped for convenience in the family Voltziaceae, here represented by *Voltzia heterophylla* from the Lower Triassic. They formed forests in the Triassic and Jurassic, but then became extinct. They gave rise to the Pinidae (pines), but not the Taxidae (yews).

and parasitoids by the tending ants. In addition, a third of tropical woody dicots produce extrafloral nectar or lipid-rich pearl bodies to attract ants that defend in turn "their" ant plants against herbivorous insects. Still other ants are "leaf foragers" to feed their fungal gardens, while foliovory (leaf eating as herbivores) has not evolved in ants. Some ant species glean food from the leaf lamina eating adherent fungal hyphae and spores. A recent analysis of food nitrogen sources

FIGURE 5.49. Conifers III: The order Pinales remained competitive against angiosperms in the cold climate of high altitudes and the Northern hemisphere, where they represent the dominant tree flora. The figure shows *Picea cembra*, typically growing at the limits of forests in high altitudes of the Alps, but also of Siberia.

in ants from tropical rainforest canopies revealed that plant-resource losses to insects greatly exceeded the previous estimates of 0.8 tons/ha/year. Already, the smaller former estimates were threefold greater than losses due to vertebrate herbivory (Davidson et al. 2003).

Ants were not the only insects that showed a radiation with the evolution of angiosperms. This was recently also demonstrated for Chrysomelidae or "leaf beetles" (Wilf et al. 2000; Figure 5.69). These animals lack a body-fossil

FIGURE 5.50. Conifers IV: Twigs with needles and ripe cones from the larch (*Larix europea*, *1*), the pine (*Pinus serotina*, *2*) and young (*3*, *4* cross section, *6* detail) and old (*5*) cones from *Cupressus sempervirens*. *Ginkgo biloba* (*7*) here shown with a leaf and immature seeds also belongs to the conifers, but forms a distinct class of mainly extinct plants (Ginkgoatae) from the class Pinatae.

record, but their feeding attack on rolled juvenile leaves of ginger and heliconias (moncot order Zingiberales) in the understories of Neotropical forests lead to so characteristic damage trails on the plants that their identification on fossil plants is diagnostic for the leaf beetles. The Zingiberales contain many compounds like tannins, phenols, alkaloids, and terpenes that are of potential defensive use against insect herbivores. Heliconia, a basal member of the Zingiberales group, harbors a high diversity of leaf beetles. At the same time, they conspicuously lack these defensive compounds. The fossil record of Zingiberales shows feeding attacks by leaf beetles near the time of the first appearance of the hosts for these leaf beetles in the Cretaceous. Insects and angiosperms showed thus a coupled diversification to numerical dominance in the animal and plant world such that their trophic association is a dominant feature of the evolving terrestrial ecosystems. Over half of all beetles are herbivorous and they fared well with this food source. Actually, when the British biologist Haldane was asked by a group of theologians what one can conclude as to the nature of the Creator from the study of His creation, he quipped, "An inordinate fondness for beetles" (Farell 1998). In fact, the insect order Coleoptera exceeds with respect to species

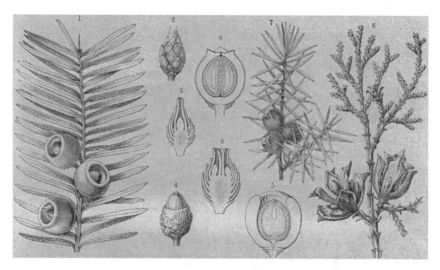

FIGURE 5.51. Conifers V: This figure shows further members of the cypress family, namely the oriental arborvitae (*Thuja orientalis*) with narrow, scale-like leaves pressed against the branchlets (*6*) and the juniper (*Juniperus communis*), which carries a cone encased in a fleshy fruit (*7, 8* cross section). The yew here depicted with *Taxus baccata* (*1*) belongs to a distinct subclass of the conifers (Taxidae) shows small female reproductive organs (*2, 3*), which form a cone (*4*) that is surrounded by a fleshy, red, cup-shaped aril (*5*). These are the only nontoxic parts of the plant in line with the function of fruits as a nutritional incentive for animals to disperse the enclosed seeds.

richness any other animal or plant group. The earliest beetles in the Permian were saprophages (eaters of dead organisms). Feeding on plants arose 50 million years later in the Triassic, but most of the beetles were still saprophages. At this time, angiosperms were nonexisting and the terrestrial plant world was represented by Bryophytes, Pteridophytes, Ginkgoales, Gnetales, Coniferales, Cycadales, and Bennettitales (Figures 5.40–5.57). Interestingly, the phylogenetic

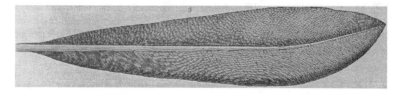

FIGURE 5.52. Cycadophytina I: Pteridospermae. Charles Darwin described the origin of angiosperms as "an abominable mystery" and it still remains so today. In older hypotheses the angiosperms (Magnoliophytina) and the Cycadophytina derive from the Progymnospermae via Pteridospermae (seed ferns), which flourished in the Carboniferous and Permian of the Palaeozoic, but died out in the Mesozoic. A rather modern-looking leaf with mesh-like venation is shown here for the seed fern *Glossopteris indica*, a characteristic fossil of the Gondwana flora.

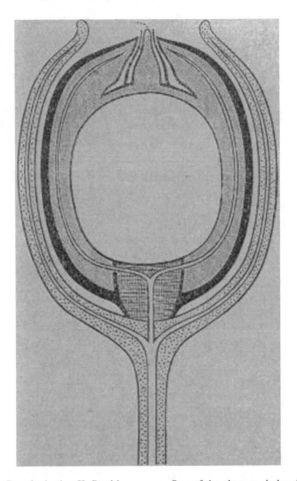

FIGURE 5.53. Cycadophytina II: Pteridospermae. One of the characteristics that permitted exploitation of the land was retention of the embryo within the maternal tissue. This trend culminated in the development of the seeds. The embryo is not only retained within the parent gametophyte, but the gametophyte in turn is encased in and protected by tissue from the sporophyte that produced it. Thus every seed includes tissues of three generations: the sporophyte plant, the microscopic gametophyte, and the second-generation sporophyte plant, which is the embryo. This development started with the seed ferns in the Devonian and the three tissues are reconstructed in the picture for the seed fern *Lyginodendron* from the Carboniferous.

tree of the phytophagous beetles shows as most basal branches conifer- and cycad-feeding beetle lineages. The larvae of these basal beetles feed on nutrient-rich reproductive structures, which apparently preceded foliage feeding (Farell 1998). When angiosperms arose, leaf mining and seed and root feeding evolved in beetles consistent with the plant–beetle coevolutionary model of Ehrlich and Raven.

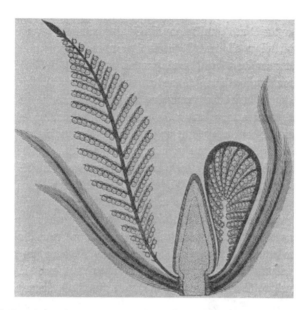

FIGURE 5.54. Cycadophytina III: Bennettitatae are an extinct class of plants that dominated with cycads the Middle Mesozoic times, which is also called the "Age of Cycads." The rise and fall of Bennettitatae paralleled that of the dinosaurs. In certain bennettites, reproductive organs were observed that resemble conspicuously "flowers" with a central cone-like female reproductive organ surrounded by whisks of male reproductive organs (*left bent, right extended*) in a ring-like arrangement surrounded by perianth-like leaves. Despite that strikingly similar structure, they are not related to the flowers of angiosperms. When the angiosperms rose to dominance in the Cretaceous, the Bennettitales lost the race and became quickly extinct leaving perhaps the Gnetatae.

Early Herbivorous Vertebrates

Pelycosaurs as First Vertebrate Herbivores

The follow-up of the tetrapod story, which we retold in one of the preceding chapters was hampered by Romer's Gap, a 30 million years period after *Ichtyostega*, which nearly lacked tetrapod fossils. This gap was in 2002 filled with a beautiful fossil named *Pederpes* (Clack 2002). Paleontologists are held to justify their name, this name is rather tricky since it can be read in two reading frames and in both it makes sense. One message the name conveys is *pes*, Latin for foot. It shows a modern foot with five digits in contrast to the earlier amphibians, which showed seven or eight digits. Also this animal does not look like an herbivore. The first proposal of potential herbivores is made by tetrapods in the transition from the Carboniferous to the Permian with edaphosaurs from the reptilian lineage and diadectids from the amphibians. Only later in the Permian unequivocal evidence for adaptations to feeding on vegetation was documented for pelycosaurs. Thus only relatively late animals found an access to the lush vegetation surrounding them. Until then this food source was, for

FIGURE 5.55. Cycadophytina IV: Cycas. Cycadatae are another branch that derived in parallel to the Bennettitatae from Pteridospermae. In the depicted *Cycas revoluta* the ovules develop on megasporophylls, which remain brown, but otherwise clearly resemble leaves. The ovules (the "buds" at the stem of the leaf in *1, 2* shows a cross section) are not surrounded by further tissue, Cycas is thus like the conifers a gymnosperm. The wind-borne pollen develops on other plants in cone-like microstroboli. Notably, at the time of pollination each ovule exudates a mucilaginous droplet and the pollen releases 0.3-mm large multiflagellate sperms, the largest in the plant and animal kingdoms. The plant thus recalls in its male reproductive behavior still its origin in the water, but has done the necessary adaptation to the conquest of the land.

vertebrates, untapped and became available only through the intermediate of fungi and bacteria. Animal detrivores funneled the primary productivity of plants into higher trophic levels. A very intensive life took place at the ground level where organic carbon was intensively traded. Gigantic myriapods testify the heat of the battle for food in the litter.

Lessons from Cretaceous Gut Contents: Titanosaurus Dined on Grass

Not all beasts of Zallinger's mural paintings in the Peabody museum of the Yale University are carnivores, there are also peaceful, although sometimes gigantic herbivores among them. They dine on conifers, cycas, and ferns. Early flowering plants, but no grass, are depicted. This is somewhat astonishing since grass provides today staple foods for much of humankind and its domesticated animals. Until last year there was actually not much reason for painters to change the

FIGURE 5.56. Cycadophytina V: Cycas. Today Cycadatae are represented only with a few living fossils, some of them resembling superficially palm trees.

picture since the fossil record of grasses (*Poaceae* or *Gramineae*) reached only back to 56 million years ago, the beginning of the Cenozoic and thus well after the demise of the dinosaurs. The dental features of the most prominent terrestrial plant eaters of the Late Cretaceous, titanosur sauropods, did not display specializations for grass eating, like grinding cheek teeth. Mammals show this specialization with hypsodont (high crowned) teeth in the Miocene. Actually, these teeth are necessary to digest such abrasive material like grass. The abrasiveness is a low-tech invention of grass against herbivory. Grass impregnates its structure with silica—this invention was apparently made by angiosperms 65 million years ago when they experienced a considerable herbivory pressure from herd forming dinosaurs. Phytoliths ("plant stones"), as these silica remains are called, are found today in basal angiosperms, monocots, and dicots as counterattack against insect and vertebrate herbivores. They induced an arms race leading to the modification of animal mouthparts to cope with these plant defenses. These phytoliths take very bizarre forms, they can be bilobate, cross-shaped, saddle-shaped and they are characteristic for the plant group (Piperno and Sues 2005). The exciting new finds are now coproliths associated with titanosaur skeletal remains that show phytoliths (Prasad et al. 2005). Phytoliths were found before in their fossilized dung and testified conifers and cycads in their diet in addition to bacterial colonies, fungal spores, and algal remains. Now unequivocal phytoliths of basal *Poaceae* were added to the list, which prolong the fossil record of grasses into the Late Cretaceous. They weren't prominent parts of the diet of these sauropods, but grasses extend thus definitively farther into the past than was thought before. In light of the new findings, another observation becomes clearer. A previously enigmatic gondwanatherian (in plain English:

FIGURE 5.57. Cycadophytina VI: Gnetatae. *Welwitschia mirabilis* is a surviving member of the plant order Gnetales, which probably derived from the extinct Bennettitatae and which share at the same time some floral characteristics with angiosperms. A single species exists in the Kalahari desert. It is anchored by a deep taproot in the ground and sports only two several-meters-long leaves growing from the basis. The plant in the foreground shows flowers carried on cones that grow in a ring covering the basis of the two leaves.

a mammal from the continent Gondwana) with highly hypsodont teeth is the early answer of mammals to the problem of making a living from grazing grass in the Cretaceous. When in this coevolutionary arms race grass used more and more silica for enforcing its structures, it started to monopolize the bioavailable silica to the detriment of other organisms that used silica for their casing like sponges. Sponges lost this battle with grass for the silica resource and the arrival of grasses broke the prominent role of sponges in the geological record.

A Bite of Plant Material by an Omnivore Like us

In the following, I will illustrate what we as omnivorous mammals can do with plant material, not that much in fact. You might have wondered why we cannot survive on leaves and grass like caterpillars and cows; in the end, we are called omnivores and have to deal with many different forms of foodstuff. Starving human populations have tried to eat many plant materials growing around them—to no avail. It turned out to be nondigestible, unyielding, even antinutritional or frankly toxic. Our digestion capacity is limited to relatively simple sugar compounds. For more complex sugar macromolecules, we have only learned to attack a relatively small number of bonds, notably those of the plant storage polysaccharide starch. For the somewhat more complex polysaccharides, we get some help from colonic bacteria. Next we will see that even professional herbivores like cattle need the assistance of microorganisms to come to grips with plant polysaccharides. As animals learned to deal with plants, plants had to

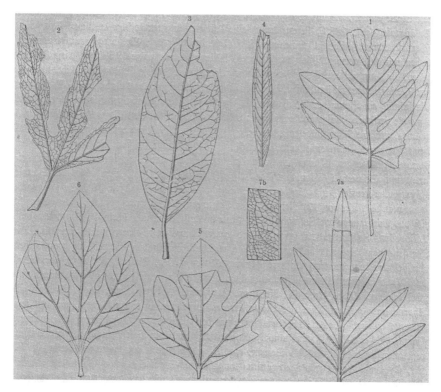

FIGURE 5.58. Angiosperms I: fossil leaves. The fossil record of angiosperms starts with the Cretaceous and presents already an astonishing differentiation with *Artocarpus* (*1*) (Hamamelididae) from Greenland and *Araliaephyllum* (*5*), and *Aceriphyllum* (*6*) from the Cretaceous Potomac formation. The Magnolia-like angiosperms are commonly believed to represent the oldest forms of angiosperms here represented with a leaf from the Lower Cretaceous in Portugal (*3*). The other leaves are attributed to *Aralia* (*2*), *Salix* (*4*) and *Sapindopsis* (*7, 7b magnified*).

design new strategies to thwart feeding attacks on them, which is the subject of the last two chapters in the herbivory context.

On Teeth and Enzymes

There is a great difference between carbohydrate and, for example, lipid digestion. While normal adults absorb about 95% of the dietary lipids, only part of the dietary carbohydrate is digestible. There are two reasons for this difference. One is size: Lipids come as only moderately sized molecules, while carbohydrates can be very large. The carbohydrates from the cell walls of plant material in our diet are molecules of enormous molecular weight and they are frequently part of tough plant tissues. In fact cellulose, the carbohydrate making up a substantial part of the cell wall of plants, was designed by evolution to confer mechanical strengths to plants. Tough or not, Nature has also endowed us with

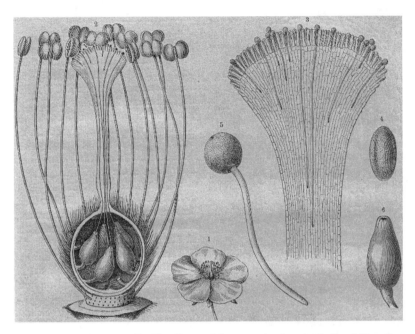

FIGURE 5.59. Angiosperms II: The flower. The most important distinguishing feature separating angiosperms (flowering plants) from gymnosperms is not the possession of flowers, but the fact that the ovules from angiosperms are enclosed in a carpel. The Greek words recall this fact: gymnosperms "marry nakedly," while angiosperms "marry under the blanket." The conditions are illustrated for the flower of *Helianthemum marifolium* (*1*). The flower is enlarged in (*2*) with petals and sepals removed to better visualize the many stamens separated into the anthers on a long filament, which produce the male gametophyte, the pollen (*4*). The stamens surround the single pistil, consisting of the stigma (enlarged in *3*), which is connected via a style to the five ovules (enlarged in *6*) encased in the carpel (ovary). When the pollen gets wetted on the stigma, it develops a tube (*5*) and grows down the style to reach the ovule, the female gametophyte, for fertilization.

tools to handle rough food. The toothed jaws of vertebrates can do wonders with many robust food materials—look how carnivores are cracking bones. However, digestion is more than just chewing food down to pieces that you can swallow. In that sense you can also eat straw. But you will not extract much energy from straw because we lack the digestive enzymes that can cut down the plant cell walls to their constituting monomers. At the enzymatic level, cellulose is a very sturdy material: It is glucose linked by β-1,4 bonds. We simply do not have enzymes able to attack this seemingly simple bond and such enzymes are indeed relatively rare in nature. Cows that make a living from eating cellulose rely for this digestive property on microbes in their rumen (more on that later). The reason is evolutionarily quite clear. Not only did plants need mechanical strength, but because they had to build this strength with biological material, they also had to resist attack by herbivores. It is thus not surprising that we have a lot of nondigestible

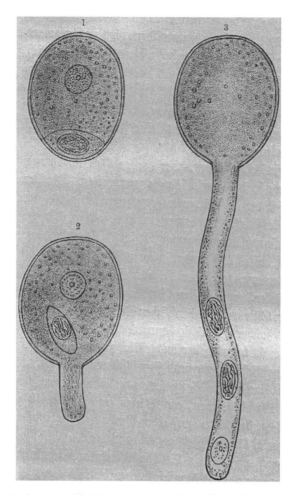

FIGURE 5.60. Angiosperms III: The male gametophyte. The pollen grain separates into two cells, the generative spindle-shaped cell and the vegetative tube cell with a round nucleus (*1*). The generative cell divides in the tube into the two sperm nuclei to prepare the characteristic double fertilization of angiosperms (*2*). At the tip of the tube is the nucleus of the vegetative cell (*3*).

carbohydrates, collectively called fiber, in our diet. There are substantial cultural and economical differences with the fiber content in the human diet. People from industrialized countries consume substantial amounts of so-called "refined" sugar in their diet and only small amounts of fiber. Lack of fiber in the Western diet has been linked to diseases that are more prevalent in industrial than in developing countries, like colon carcinoma and vascular diseases. Actually, part of the nondigestible carbohydrates becomes food to bacteria that grow in great number in our colon. Pectins, gums, and mucilages are digested by colonic bacteria. In contrast, cellulose and hemicellulose are only degraded to a small extent and lignin not at all and they are excreted unaltered with the feces.

Starch Digestion in Humans

Approximately 45–60% of the digestible carbohydrates in a European diet are in the form of starch, the storage form of glucose in plants. Amylose,

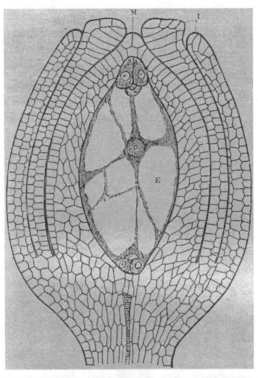

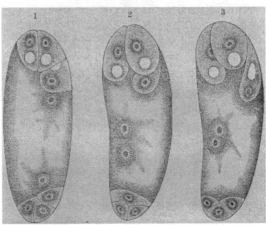

the nonbranched type of starch, consists of glucose residues in α-1,4 linkage. Amylopectin, the branched form of starch, has about one α-1,6 linkage per thirty α-1,4 linkages. These chemical linkages between glucose residues are much easier to crack than those in cellulose. Their digestion starts already in the mouth with the α-amylase secreted with saliva. Chewing starch in the mouth quickly gives you a sweet taste because of the release of oligosaccharides. However, the digestion of starch in the mouth is incomplete and the salivary α-amylase is inactivated in the stomach acidity. Starch digestion is resumed in the duodenum with a pancreatic α-amylase, which is 94% identical in its amino acid sequence to the salivary enzyme. This enzyme completes starch digestion in the lumen of the small intestine. Completes is perhaps too great a word: α-Amylase is an endoenzyme that cleaves internal, but not terminal α-1,4 linkages, nor does it cut α-1,6 linkages or α-1,4 linkages adjacent to α-1,6 linkages. The end result of starch digestion is therefore the disaccharide maltose, the trisaccharide maltotriose, and the limit dextrins around the branching points. The starch digestion is in fact finished by "membrane digestion" on the epithelia of the small intestine. These cells have a particular ultrastructure, they ride on the basal lamina which separates them from the lamina propria containing blood capillaries. The epithelial cell rises as a columnar structure ending in a brush border of so-called microvilli that enormously increases their absorptive surface. The glycocalyx covers the tips of the microvilli and consist

FIGURE 5.61. Angiosperms IV: The female gametophyte. (a): cut through the ovary, the central ovule is covered by two layers of integuments. The development of the female gametophyte starts with a single nucleus in the embryo sac cell. It forms a tetrad of four megaspores as a result of meiosis. Then a complicatedly orchestrated nuclear dance sets in (right). Three megaspores degenerate, one divides into the typical eight-nucleate, seven-celled female gametophyte ("embryo sac"). At the top near the microphyle (M) you see the synergid cells and slightly below the egg cell. The central cell shows two nuclei and opposite to the microphyle are the antipodal cells. As the pollen tube discharges its content into the female gametophyte, one sperm nucleus fuses with the egg nucleus forming the next-generation sporophyte, the new plant. The other pollen sperm nucleus fuses with the two polar nuclei in the central cell thus forming a triploid primary endosperm nucleus. This nucleus then divides further forming the endosperm tissue, which becomes filled with stored food like starches, oils, and proteins, which are recruited from the parental plant. The embryo grows thus from the food supplied by the parent as a form of accepted parasite. You can see that also angiosperms have their triumph of the egg story, where nutrition of the next generation became one crucial asset for the evolutionary success underlining again the close connection between food and sex. In legumes the embryo consumes the endosperm during its development, resulting in mature seeds with massive embryos and no sperm. A recent report on Amborella trichopoda shows a variant to the standard scheme of the nuclear divisions by displaying an extra synergid (Friedman 2006). As Amborella is the sole member of the most ancient extant angiosperm lineage, the variation testifies to an extensive experimentation phase during the early evolution of angiosperms, which occurred within a time span of 15 million years at about 130 million years ago.

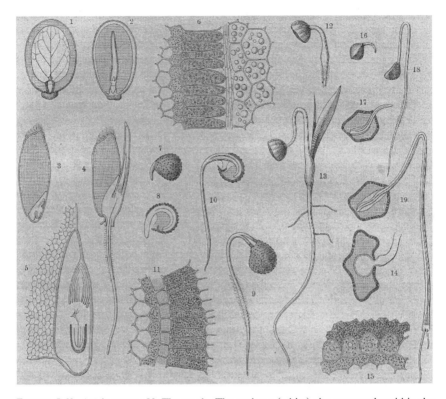

FIGURE 5.62. Angiosperms V: The seeds. The embryo (white) sits commonly within the endosperm (fine gray granulation) as shown for the *Ricinus* seed with cross sections (*1, 2*) along different planes, *Agrostemma* (*7–11*), *Tradescantia* (*12–15*) or the onion *Allium cepa* (*16–19*). The embryo of *Agrostemma githago* curls around the endosperm, while the embryo of the wheat (*Triticum vulgare*) sits apposed at one side of the large mass of the starchy endosperm (*3*). The bulk of the embryo consists of the fleshy body, the scutellum, which lies in contact with the endosperm (*4*) and serves as the digestive and absorptive structure during germination (*5, 6* show the absorptive cell layer of the scutellum in contact with the starch grain–filled endosperm cells). The food reserves in the endosperm powers the outgrowth of the stem and the leaves on one side and the root on the other side (*4*) and ensures nutrition before leaves and root become functional and can feed the new plant. The grains of grasses also feed now directly and indirectly a large part of the world population.

of a network of glycoproteins and membrane-bound digestive enzymes. One of them is sucrase–isomaltase. Limit dextrin is digested by isomaltase subunit to maltose and maltotriose. A second membrane-bound enzyme is glucoamylase, it cleaves maltose and maltotriose into the glucose monomers.

Disaccharides

The rest of the digestible carbohydrates in our diet consist of sucrose (table sugar, a disaccharide consisting of a glucose and fructose) and lactose. Sucrose

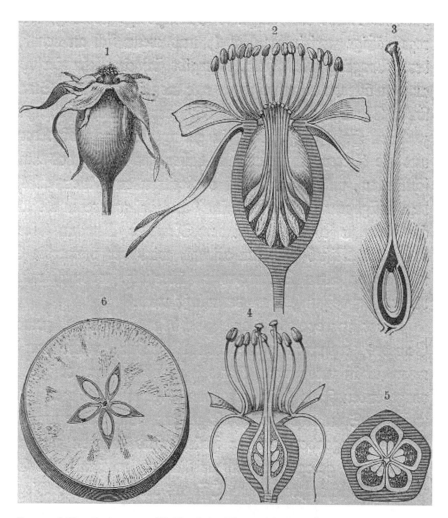

FIGURE 5.63. Angiosperms VI: The fruits. The seed is the reproductive body of angiosperms and gymnosperms, consisting of the undeveloped plant (embryo), the stored food
(endosperm) and a protective coat (testa). Seeds provide man with his most important foods
(e.g. wheat, maize, rice; beans, peas, peanuts, soybeans, nuts). Many plants surround their
seeds by an additional nutritious layer of different anatomical origin (e.g. olive, banana,
avocado, apple, orange, date palm) forming a fruit. Most fruits develop from a single pistil.
However, most pronounced in roses, additional flower parts like the stem axis and floral
tubes participate in fruit formation, resulting in an accessory fruit here represented by the
rose (*Rosa, 1–3*) and the apple (*Malus sylvestris, 4–6*). The fruit is an organ of dispersal
for the sperm attracting animals that eat the fruit, but cannot digest the protected seed,
which they excrete and disperse with their feces. This function was initially fulfilled by
reptiles, but was later largely taken over by the evolving mammals (here mainly primates,
including man, rodents, and bats) and birds. According to the senses of the dispersing
animals, fruits attract with intense colors (birds) or scents (mammals). Also invertebrates like ants play a prominent role in the dispersal of some plants via seeds and fruits.

FIGURE 5.64. Detrivores have an important function in the ecosystem and this lifestyle is widely distributed in insects. The figure shows carrion beetles (*Necrophorus vespillo*) which bury the dead mouse within minutes. They belong to the family Siphidae, order Coleoptera.

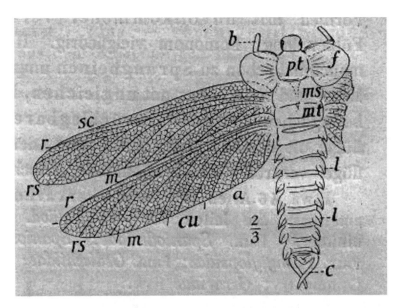

FIGURE 5.65. Paleodictyoptera- early flying insects are here represented by *Stenodictya* from the Carboniferous: pt: prothorax; f: paranotal lobe; forewing and hindwing (sc subcosta, r radius, rs sector radii, m medialis, cu cubitus) ms: mesothorax, mt: metathorax, l: abdomen, c: cerci. The paranotal lobe gave rise to wings in a hypothesis formulated by Müller in 1873. Currently the appendage hypothesis is more popular, which sees wing evolution starting from legs (Kukalova-Peck hypothesis 1983).

FIGURE 5.66. Pollination I. Pollination is the transfer of pollen grains from the stamen to the ovule-bearing organs. Pollination is another fascinating cross road between sex and food. Pollination by insects probably occurred in primitive seed plants. As flowering plants evolved in the Mesozoic, the first pollinators were probably beetles that had specialized on the reproductive organs from primitive plants. Beetles became pollen eaters, and pollination a chance event with spared pollen. Some visits of beetles can have devastating effects on the developing flower, like that of the beetle *Oxythyrea funesta* on the plant *Serratula lycopofolia*. The plant develops nectaria outside of the flower, which attracts the ant *Formica exsecta*, which in turn defends this food source against the landing of the beetle and spare the flower an unwelcome visit of a beetle. The figure alludes to the story of early insect herbivory on reproductive organs of primitive plants.

is the major transport form of sugar between the different plant tissues, while lactose is the sugar transferred with the milk to the breast-fed mammalian young. These two disaccharides make up about 30–40% of the digestible carbohydrates in our diet. The remaining digestible carbohydrates are the monosaccharides fructose and glucose, representing 5–10% of the total carbohydrates. Sucrose is split into its monomers glucose and fructose by the sucrase subunit of the sucrase–isomaltase membrane protein. Lactose is cleaved into its monosaccharides glucose and galactose by the membrane enzyme lactase. Uptake, not hydrolysis, is the rate-limiting step for sugar digestion except for lactase, which is a slow enzyme. The peak oligosaccharide activity is found in the proximal jejunum, much less is located in the duodenum and the ileum, none in the colon. As we have seen lactase activity is developmentally regulated, lactase is also more susceptible to enterocyte damage. Acute

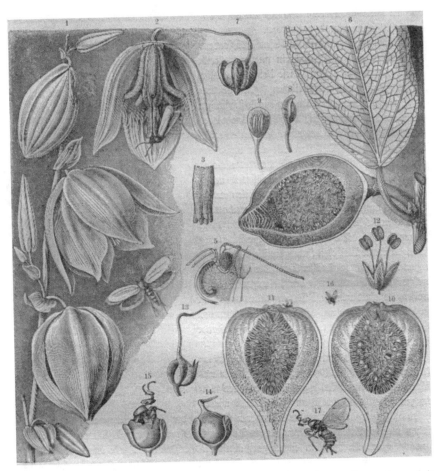

FIGURE 5.67. Pollination II. Probably in a next step other flowers began to exploit primitive gall-forming insects for pollination. The figure tells two fascinating stories in this field. In the *left* you can see the *Yucca filamentosa* visited by the moth *Pronuba yuccasella*. This moth takes more than a mouthful of pollen during a visit (center of the figure), which is carried to the next flower. The moth does not eat the pollen, but discharges it on the pistil and at the same time deposits its eggs into the flower. Equally remarkable is the story of the fig (*Ficus pumila*) at the *right* side. Short-styled female flowers are found next to a leaf (*top right*). They are penetrated by the fig wasp (*Blastophage grossorum*) (*11*), which deposit the egg that develops into a gall (*10*, with numerous small galls). The progeny insect leaves the gall (*15*) and gets over powdered with pollen when leaving the flower. During a visit to the next flower, the fig wasp deposits the pollen.

childhood diarrhea leads to temporary loss of lactase activity, which led in the past to the recommendation to withhold milk during a bout of diarrhea. Sucrose overfeeding can increase sucrase expression, this cannot be done with lactose on lactase.

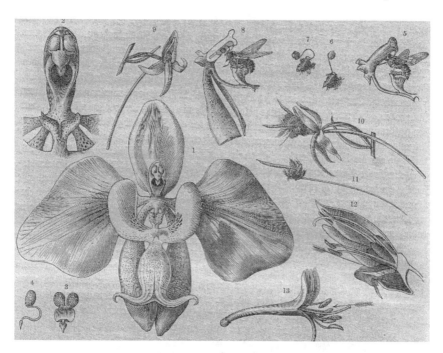

FIGURE 5.68. Pollination III. In more advanced flowers, nectar replaced pollen as reward for the pollinator. Flowers also developed guidance for the pollinator in the form of landing platforms. Intensive odor and bright color advertise the nectar to the pollinator. An impressive insect–flower coevolution took place and insects evolved sophisticated sucking mouthparts to forage on nectar. The picture shows such a seat for pollinating flies (5,8) in an attractive flower *Phalaenopsis schilleriana (1)*.

Transporters: Sugar–Sodium Copackers

Now the carbohydrates are in the form of monosacchrides and ready for transport across the plasma membrane of the intestinal epithelia. The SGLT1 transporter takes charge of glucose and galactose, GLUT5 cares for fructose. Both show 12 membrane-spanning helices, but SGLT1 has a higher molecular weight. SLGT1 has two structural requirements for its substrate: it must be a hexose in D configuration able to build the pyranose ring. This applies to glucose and galactose as dietary digestion products, but not to L-glucose (a product of carbohydrate digestion by some bacteria that produce mixtures of D- and L-glucose) nor to fructose, which builds as a ketose the furanose ring.

SGLT1 is a cotransporter. As the name implicates, it transports glucose together with another compound, in the case of SGLT1 a sodium ion. Cotransporters come in two types: symporter and antiporter, depending on the

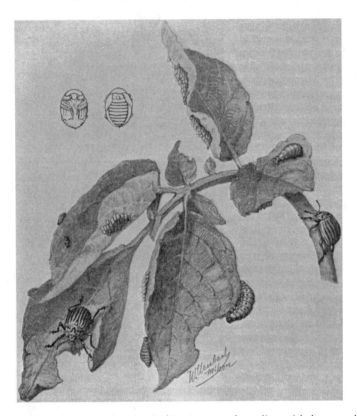

FIGURE 5.69. The Colorado potato beetle (*Leptinotarsa decemlineata*) belongs to the very species-rich family of Chrysomelidae (leaf beetle) of the order Coleoptera. It originally fed on buffalo bur (*Solanum rostratum*) in the Rocky Mountains, but with the westward moving farmers, who introduced potato (*Solanum tuberosum*), the beetle changed in the 1880s to a new host plant. Ten years later it was reported on the US east coast and arrived further ten years later in continental Europe. The picture shows egg masses, larvae, and adult beetles and on the *top left* a pupa.

transport direction of the two compounds. SGLT1 is a symporter: Sodium and glucose both travel from the lumen of the intestine into the epithelial cell. From an energetic viewpoint, SGLT1 is an interesting protein since it transports glucose against a concentration gradient. Remarkably, it can do this uphill transportation of glucose without using ATP. In fact, it derives the energy for the transport against the diffusion gradient by the cotransport of Na^+, which goes downhill. Glucose absorption is driven by the Na^+ electrochemical gradient. This gradient has two components: The tenfold higher Na^+ concentration outside when compared to the inside of the epithelial cell and the inside negative voltage across the apical cell membrane (–60 mV). Both driving forces can be multiplied: The chemical gradient "buys" a tenfold gradient, the electric gradient another tenfold glucose gradient allowing SGLT to achieve a 100-fold glucose concentration gradient across the membrane. However, this might not be enough since organisms might want to concentrate glucose even more, e.g., during urine production. Here the body recovers glucose from the primary urine filtrate. Avoiding the loss of useful calories is also an aspect of the quest for food. How is this achieved? The kidney SGLT form carries two Na^+ ions with each glucose molecule. As the concentration gradient enters the transport equation in squared form, glucose can now be concentrated 10,000-fold. This is the same underlying thermodynamic logic as we have seen before when the energy contained in the hydrolysis of a single ATP is not sufficient to allow an endergonic reaction to proceed. You simply couple the desired reaction to the hydrolysis of two or three ATPs.

Of course, there is a problem with SGLT-mediated glucose import: If Na^+ is constantly cotransported into the cell, you will loose the Na^+ concentration gradient across the apical intestinal membrane and thus the driving force for the transport. Hence, you have to drive intracellular Na^+ out again. This job is done by a primary active transport—the Na–K pump. This Na–K pump is energized by ATP hydrolysis and exchanges $3Na^+$ (outward) against $2K^+$ (inward). Now comes an important point. This Na–K pump is only localized in the basolateral membrane of the intestinal epithelial cell. By placing different transporters at the apical and basolateral membranes, epithelia can achieve direct transport of solutes across the epithelial layer, both in the absorption and in the secretion mode. In the intestinal cell, Na^+ is thus transported from the intestinal lumen across the epithelial cell into the interstitial space and from there into the blood. Glucose and fructose follow the same way. As glucose is concentrated in the intestinal cell, its export from the cell into the interstitial space can be done by facilitated diffusion. Consequently, this transport is mediated by GLUT2 located in the basolateral membrane.

Fructose

Fructose is transported into the intestinal epithelial cell across the apical (luminal) surface via the GLUT5 transporter. GLUT5 is an example for a uniporter-catalyzed transport. Helices 7, 8, and 11 contain aa that can form hydrogen bonds with the hydroxyl groups of the hexose. They are thought to form the alternatively inward and outward facing hexose-binding sites in the interior of the protein. Binding of the substrate to the outward-facing binding site triggers a conformational change in the uniporter, moving the bound hexose through the protein such that it is now bound to the inward-facing binding site. The sugar is then released to the inside of the cell. Finally, the transporter undergoes the reverse transformational change, inactivating the inward-facing hexose-binding site and regenerating the outward-facing site. The uniporter mediates facilitated diffusion, it only transports the fructose downhill of the concentration gradient.

Traffic Control

Accumulation of hexoses does not occur inside the cell since the intestinal epithelial cells contain the full complement of glycolytic enzymes and part of the absorbed hexoses is metabolized to lactate. Lactate diffuses into the blood and will be resynthesized to glucose in the liver cell via gluconeogenesis. Glucose concentration is maintained at 5 mM in the blood. This assures a sufficient concentration gradient for glucose import into the cells of the mammalian body, which express GLUT1. However, under conditions of starvation the blood glucose levels fall and in glucose-synthesizing liver cells the uniporter then transports glucose out of the cell. This is a useful reaction since it can now replenish the decreased blood glucose level by the liver cell that synthesizes glucose from lipid or protein degradation, which becomes visible as a loss of body fat and muscles.

A Bioreactor Fueled by Grass

Cellulase in Animals

Cellulose is one of the most abundant organic compound in the biosphere, it was estimated that the astronomical quantity of 10^{15} kg of cellulose is synthesized and degraded on earth each year. Cellulose is a nonbranched chain of glucose residues linked in β-1,4 linkages. We are surrounded by a major carbon and energy source, but cannot use it because we do not have the enzyme cleaving the β-1,4 linkages between the glucose residues in cellulose. *At first glance, this seems to be a terrible mistake of stepmother Nature that cries for remediation to ban hunger forever from human memory. However, we should think twice about that question: Evolution had long time periods to tinker with different solutions. There might be something really odd with cellulose that we do not possess a cellulase enzyme in our intestine.* Notably, we know of only two animals that express a cellulose-digesting enzyme: termites (Figure 5.70) and crayfish. Termites are well known for their capacity to digest what appears even more unpalatable than grass, namely wood. However, in both animals it is still unclear whether their cellulase contributes really to their nutrition.

The Consequence of the β-1,4 Glycosidic Bond

So what is the underlying problem in cellulose digestion that makes famines possible in human history? The answer is really trivial: It is the geometry of the glycosidic bond between the glucose residues in cellulose. More precisely, it is the difference between the α- and β-1,4 glycosidic bond between the glucose residues in starch/glycogen and cellulose, respectively. In both configurations, an oxygen atom links the C1 carbon from one glucose to the C4 carbon of the next glucose, hence a 1,4 bond. The α or β designation means that the hydroxyl group attached to the C1 carbon is below or above the plane of the six-member

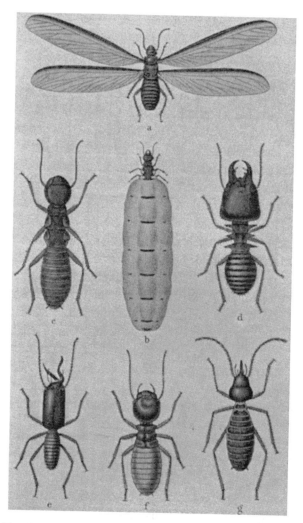

FIGURE 5.70. Termites: social insects of the order Isoptera. a young female from *Termes spinosius*; b queen from *T. gilvus*; c king from *Hodotermes ochraceus*; d–g soldiers from different species; f worker. Termites are mentioned in the book as one of the few truly cellulose-digesting animals.

pyranose ring of glucose. This difference has dramatic consequences. In the β-1,4 bond each glucose residue is related to the next one by a rotation of 180° and the oxygen in the pyranose ring of one glucose is hydrogen bonded to the 3-OH group of the adjacent glucose. The β configuration thus allows cellulose to form long straight chains, chains can align to fibers that are again linked by hydrogen bonds. This parallel alignment makes the fibrils rigid and insoluble and confers a substantial tensile strength to them—ideal for a polymer that has to fulfill a structural role. The α-1,4 bond, in contrast, forces the starch and glycogen-type glucose polymer into a hollow helix conformation. The structure is water-soluble and solubility is still further favored by the side chains. In fact, this is what you would expect for a storage molecule. In times of need, the animal or plant cell should have an easy access for enzymes to peel off glucose residues from this store of glucose. This is what we are doing already in our mouth: We produce the starch-digesting enzyme α-amylase in great quantities in our saliva. You can actually sense the result, if you keep a starch source in your mouth, it will soon develop a sweet taste. Nothing like this happens when you masticate grass.

Cellulolytic Bacteria

Actually, a mammalian herbivore that can extract energy from eating straw still lacks the capacity to digest cellulose. However, herbivores like ruminants teamed up with organisms that have learned to digest cellulose. The ability to digest cellulose is widely distributed among many genera in the domain of Eubacteria (but strangely absent in Archaea) and in the fungal groups of Eukarya. However, the system must have evolved relatively late, i.e., after the appearance of algae and land plants, since cellulose is only very rarely produced outside of this realm (a few bacteria, e.g., *Azetobacter*, within animals tunicates, a primitive chordata). Within eubacteria cellulolytic capacities developed in aerobic bacteria (*Actinomycetales* is a prominent order), as well as in anaerobic bacteria (e.g., *Clostridia* order). Physiologically, these bacteria fit into distinct environments: The anaerobic Gram-positive bacteria like *Ruminococcus* are found—as the name suggest—in a specialized fermentation chamber of the alimentary tract of ruminants where they start the digestion of cellulose (for a review see Krause et al. 2003). In contrast, the aerobic Gram-positive bacteria like *Cellulomonas* and the gliding bacterium *Cytophaga* are found in soil where their task is the degradation of plant litter falling on the ground. If one looks at the enzymatic systems elaborated by cellulolytic microbes, the complexity of the enzymatic task becomes immediately evident (for a review see Lynd et al. 2002). Microorganisms that use "noncomplexed" cellulase systems impress us by the sheer amount of different enzymes. This system has been studied intensively in the fungus *Trichoderma* used in industrial processing of cellulose. It displays five endoglucanases, two exoglucanases, and two β-glucosidases. Attack on cellulose starts with endoglucanases, enzymes that cut in the middle of the cellulose chain. However, they cannot attack the tightly aligned so-called crystalline parts of the cellulose complexes, they gain access to cellulose in the amorphous regions of the cellulose fibrils, regions of imperfect alignment of the cellulose chains. Why this organism has five different endoglucanases is not clear. Part of the answer may be that the laboratory experiments were conducted with purified cellulose while cellulose in nature frequently comes in association with hemicelluloses.

Hemicellulose and *Trichoderma*

Hemicellulose accompanies cellulose particularly in secondary cell walls and represents the second most important source of carbohydrates in nature. The backbone of hemicellulose (xyloglucan) is glucose in β-1,4 linkage. Actually, the core structure of cellulose and hemicellulose are identical although only cellulose is made at the plasma membrane. Hemicellulose is decorated with xylose and some other sugars. Hemicelluloses are only slightly branched and more easily degraded than cellulose in part because they do not form microcrystalline structures. When the endoglucanase EGI from the fungus *Trichoderma* shows a broad substrate specificity including a xylanase activity, it will assure the physical access to cellulose. The endoglucanase cuts increase the amorphous character of these regions, some chains peel off from the fibril and then become targets to exoglucanases, enzymes that digest cellulose from the ends in a processive way. *Trichoderma* has two of this kind: CBHI, which attacks cellulose from the reducing end (i.e., showing a free aldehyde group), and CBHII, which starts digestion from the nonreducing end (showing the hydroxyl group) of the cellulose chain cut by the endoglucanase. CBHI contains four surface loops that give rise to a tunnel with a length of 50 Å, the cellulose chain gets into this tunnel and processive cleavage of cellulose occurs. Also the 3D structure of two endoglucanases was resolved. Interestingly, it showed a surface groove and not a tunnel for cellulose binding and cleavage, probably reflecting the difficulty to get primary access to the intact cellulose chain even in the amorphous region. The 3D structure of the exoglucanase revealed that cellobiose (two linked glucose residues) is the major hydrolytic product as the cellulose chain passes through the tunnel (Divne et al. 1994, Rouvinen et al. 1990). Only very small amounts of cellodextrins (3–4 glucose residues) or of glucose monomers are released at the beginning of the reaction. Therefore, two β-glucosidases are still necessary to complete the digestion. Interestingly, these two enzymes are cell bound to keep the diffusion path of the released glucose small, which favors the absorption of glucose by the *Trichoderma* cell and discourages stealing of the glucose by organisms living in the same environment. However, the exoglucanases reduce the degree of polymerization in cellulose only very slowly. The fungus tries to compensate it by producing more of these enzymes, 80% of the total cellulases are these two exoglucanases. *Trichoderma* strains used in industrial cellulose degradation produce 0.33 g of cellulase protein per gram of utilizable carbohydrate (Lynd et al. 2002).

Cellulosome

Cellulose digestion is apparently a tedious business—this is also true for cellulolytic bacteria, which generally show slow growth on cellulose. This is apparently the price to pay for living on an assured carbon source. The price is paid in the form of a very specialized digestion apparatus built by many cellulolytic bacteria. They developed a real machine, a cellulosome, for cellulose digestion, which is very costly to construct. Cellulolytic bacteria are metabolic specialists; they are unable to use proteins or lipids as energy sources for their growth. No wonder that they are outcompeted by bacteria, which can use these alternative food sources. It thus appears logical that these so-called "complexed" cellulase systems are typically found in anaerobic environments like the rumen of herbivores where they live in consortia with high concentrations of other microorganisms. Stealing of your processed food product is under these conditions a big problem. The solution is to do all next to your cell wall; you organize a coordinated enzyme activity in a Henry Ford–like way, this time, however, as a disassembly line. Electron microscopy showed that cellulosomes are compact fist-like protuberances on the cell wall that open when attaching to cellulose. A *Clostridium* cellulosome has been studied in some detail (Lamed et al. 1994). It contains a large concatalytic scaffolding protein, CipA. On one side it is anchored to the cell wall via a type II dockerin/cohesin, on the other side an association of type I cohesins/dockerins link it to more than 20 different catalytic modules, including nine distinct endoglucanases, four exoglucanases, and five hemicellulases. The high efficiency of this megazyme has been attributed to the appropriate combination and spacing of enzymatic activities that can remove physical hindrances of other polysaccharides found in plant-derived crude cellulose. Some cellulosomes reach a molecular weight exceeding 10 MDa, but not all cellulosomes are so complex. We understand now that a significant metabolic effort has to be devoted to its synthesis. The justification for this effort is the high caloric value and natural abundance of cellulose in nature—a cellulolytic microbe is unlikely to starve in any habitat that carries plant life. The major cellulose degrading microbes in the rumen, *Ruminococcus albus* and *Fibrobacter succinogens*, have developed further adaptations like a glycocalyx envelope consisting of glycoproteins. This structure has multiple functions: It mediates attachment to the cellulose, protects against grazing by ciliates and bacteriophage attack in the rumen, and assures that not too many degradation products are lost by cross-feeding of nonadherent cellulolytic and noncellulolytic bacteria. The oligosaccharides are taken up and fermented via the Embden-Meyerhof pathway.

Ruminants: A Bit of Microbiology and Physiology

The digestive system of all mammalian herbivores shows specific adaptations to the fermentation of plant material, e.g., a large cecum (a postgastric fermentation chamber distal from the small intestine) and an increased large intestine. Ruminants went one step further by developing a huge dilatation of the esophagus, the rumen, and a restructuring of the stomach. A 500 kg bovine has a rumen containing 70 L of liquid. The feed consisted classically of grass, hay, and straw, but is today supplemented. In the traditional regime, the dry material is roughly half hemicellulose and half cellulose, 3% are lignins and <1% protein and lipids. The rumen receives in addition to the fodder daily about 60 L of saliva, which is buffered with bicarbonate. The turnover time differs for liquids and solids, but does not exceed 55 h for solids. This is sufficiently long for bacteria to proceed with cellulose degradation. Populations whose generation times are longer than the turnover time would simply be washed out of the system without leading to fermentation products. The rumen is the fermentation chamber of the cow. It contains a large number of microbes, both in absolute numbers (10^{10}–10^{11} bacteria, 10^4–10^6 ciliates, 10^2–10^4 fungi per gram) and in species composition. More than 200 prokaryotic species were identified, the population is diverse (remarkably, no single species makes more than 3% of the total community), but stable. There are cellulose-, hemicellulose-, starch-, and lactate-degrading bacteria, succinate-decarboxylating bacteria, and methanogenic archaea. Pyruvate produced in the Embden-Meyerhof pathway is then transformed into succinate (via net fixation of CO_2 by PEP carboxykinase), which is then decarboxylated to propionate. Decarboxylation of pyruvate leads also to acetate and in a side reaction to butyrate. These volatile fatty acids are partly taken up by the rumen mucosa and their absorption is finished in the omasum part of the stomach. The cow thus feeds on these fatty acids and not really on plant polysaccharides. The amount of fatty acids absorbed are substantial: These are daily about 3.7 kg acetate, 1.1 kg propionate, and 0.7 kg butyrate. They enter the bloodstream and are further oxidized by the host; they cover about 40% of the energy needs of the cow.

The reticulum part of the stomach sieves out the fibrous part of the rumen outflow, which is then regurgitated into the mouth for renewed chewing. The liquid part of the rumen is pumped into the omasum. The liquid contains now many ciliates and bacteria from the rumen fermentation chamber. These fermenting microorganisms become themselves food when the liquid flows then into the abomasum, which corresponds physiologically to a true stomach. Pepsin and HCl are secreted and digest the microorganisms and protists providing a nitrogen source for the cow. Lysozyme is secreted by the abomasum to assist in the digestion of the bacterial biomass produced in the rumen. Actually, 400 g of bacterial and 300 g of protist cells are added daily to the bovine diet. This is important since the feed is low in nitrogen. However, there is a dilemma: The microbes need first a nitrogen source to grow in the rumen. The bovine invented an elegant nitrogen supply. Urea is synthesized in the liver of mammals during amino acid degradation. Urea is in most mammals excreted with the urine as a waste product. Not so in ruminants: The urea produced in the liver reenters the alimentary tract via the saliva (ureohepatic cycle). Urea is decomposed in the rumen into ammonium ions, the preferred nitrogen source of bacteria, and CO_2.

Biotechnology Meets Complexity

In view of the economical importance of bovines as a source of milk, meat, and leather (wool in the case of sheep), substantial efforts were made to improve the fiber digestion in ruminants and thus animal productivity. This is a bold approach in view of the complexity of a symbiotic system linking the metabolism of a mammalian host and such a complex microbiota in the rumen. Genetically modified bacteria were developed under the assumption that the rumen microbiota does not produce the correct mixture of enzymes to maximize plant cell-wall degradation. This approach met technical difficulties; *Ruminococcus* and *Fibrobacter*, the most fibrolytic rumen bacteria, could not be genetically transformed. The introduction of a xylanase into *Butyrovibrio fibrisolvens* improved fiber digestibility in vitro, but the recombinant strains could not compete with *Ruminococcus* and did not persist in the rumen beyond 15 days. Collectively, all these attempts of introducing "ameliorated" strains failed, the inoculated microorganism usually disappeared from the rumen. This observation is an important lesson for the genetic engineer: Microbes do not exist in ecosystems of their own, but reproduce and persist as members of complex microbial communities. These communities generally show complex networks of metabolic cooperation between different microbes that cannot yet be simulated in the laboratory. An example is the transfer of cellodextrins from cellulolytic bacteria to noncellulolytic carbohydrate fermenting bacteria, which in turn produce ammonia and branched-chain fatty acids, which are consumed by the cellulolytic bacteria. We have here a simple case of cross-feeding. The relationships between microbes in the rumen are in reality even more complicated. Many bacteria produce inhibitors of other bacteria, they kill by secreting bacteriocins; bacteriophages were isolated from the rumen and ciliates feed on bacteria.

Plant Defense Against Herbivory

Cattle

Terrestrial plants have a problem: They are rooted into the ground. If a herbivore approaches, they cannot run away. This immobility of plants is so deeply rooted in our consciousness that moving woods sent terror waves to Macbeth (or a more welcome horror to the spectator of the *Lord of the Rings* movie). Plants cannot mount a fancy defense posture, but this does not mean that plants are defenseless against herbivorous predators. Plants have evolved many means to make them a difficult food source. Some defense measures are easily seen, e.g., thorns and spines (Figure 5.71). You feel pity for the hungry mouths of those mammals, which take a bite from them. Other plants look quite palatable, but they are carefully avoided by grazing cattle. If you hike across mountain pastures in the Alps, you will not miss *Veratrum album*, a 1 m high Liliaceae. Cattle eat it only once and the experience of the vertigo caused by its poison will deter them to repeat this unpleasant food experience. But there are many plants that do not sport spines and are not poisonous. How do they defend themselves against the many herbivores?

Ants

Actually, vertebrate herbivores are not the greatest threat to plants. If you take the tropical rainforest as richest manifestation of terrestrial plant life on earth, its ecologically dominant animals are ants. In fact, ants constitute about 20–40%

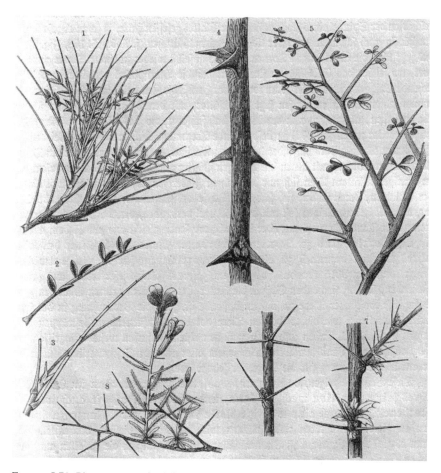

FIGURE 5.71. Plant weapons in defense mission are shown for the Fabaceae shrub *Astragalus tragacantha* (*1–3*) and the Fabaceae tree *Robinia pseudoacacia* (*4*). The figure displays also the defense structures against herbivores developed by *Cytisus spinosus* (*5*), *Berberis vulgaris* (*6, 7*) and *Vella spinosa* (*8*).

of the arthropod biomass in the rainforest canopies. This dominance of ants is for an ecologist surprising and is referred to in the literature as Tobin's paradox (Hunt 2003). The greatest animal biomass in a terrestrial ecosystem must be herbivores, situated at the second level of the trophic pyramid. The reason is based on sound principles of thermodynamics: Each transition from one to the next trophic level (primary producer–herbivore–carnivore) is paid by substantial reduction of cumulative biomass in the next higher trophic level because the rule of the thumb is that 10 g of food from the lower trophic level gives not more than 1 g of body mass in the next higher trophic level. However, ants do not feed (directly) on leaf tissue. The solution to this conundrum of the cryptic herbivores will be given in the next chapter.

Aphids

Plants are under assault by an armada of invertebrate herbivores. I will just mention three feeding guilds. There are the leaf-chewing insects like caterpillars and sap-sucking insects like aphids (Figure 5.72). Finally, there are nematodes and insects that attack from below the ground targeting the roots. The caterpillars, beetles, and locusts have a rather simple strategy: they have very efficient mouthpieces for chewing, they are very hungry and numerous, and at least at one stage in their life cycle airborne and thus very mobile. Potato beetles may defoliate an entire crop covering several acres within a day and locusts are known from Biblical times as a plague. Some pests actually grow with the victim: The European corn borer larvae attack the leaves of the young maize plants, while the mature animal becomes the dreaded stalk borer pest of adult maize.

In contrast, sap-sucking insects like aphids, leafhoppers, and thrips cause only minimal direct tissue damage because they have evolved a fine stylet with which they locate, penetrate, and drain sap from the phloem sieve cells of the plant's vascular system. This feeding stylet of aphids is such an efficient locator of the sieve elements that it represents still today the favorite needle of plant physiologists who want to sample the phloem fluid. For the plant this exploitation means a substantial loss of photosynthate, which is transported via the phloem from the sites of production to the sites of consumption. Heavy infestation can thus severely reduce the growth of the affected plant.

Nematodes

Nematodes are a major threat to plant roots. Also nematodes have evolved a sophisticated feeding stylet, which penetrates the cell wall of the root cell and contacts the plasma membrane of the cell without pushing through it. The nematode releases glandular secretions into the root cell, which increases their metabolic activity substantially. The cyst nematodes dissolve the plant cell walls and syncytial feeding structures consisting of up to 200 plant cells are formed. The root-knot nematodes induce aberrant series of mitosis without cell division resulting in giant cells. The nematode becomes sessile when it has found the feeding position and elaborates feeding tubes in the cytoplasm that connect to the end of the stylet. Strangely, the nematode builds a new feeding tube for each meal. The giant root cell becomes thus littered with hundreds of feeding tubes in its cytoplasm.

Fungi

All this sounds not very rosy. Not to mention that aphids are an excellent vector for the transmission of viral diseases. Also many fungi are lurking on or near the plant. The concept that fungi are saprophytes, meaning that they feed on dead organisms, can unfortunately not be generalized. Actually, fungi have a long history for predation of living photosynthetic cells. Fungal attack on the alga *Palaeonitella* was documented in a 400-million-year-old fossil (Taylor et al. 1992). That this fossil is a case for parasitism is revealed by distinct hyper-trophism of internodal cells of the alga, which represents still today the host

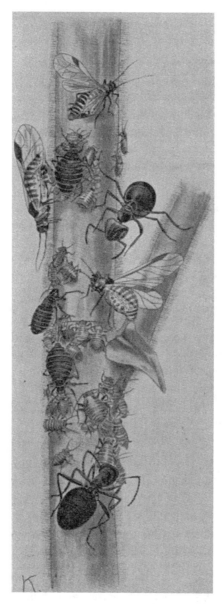

FIGURE 5.72. Aphids in different developmental stages sucking on a stem of a plant are tended by ants interested in the honeydew excretion of the ants. The ants will defend the aphids against the attack from *Aphidius* wasps.

response of two living species of the alga *Chara* to fungal attack. The host–parasite relationship between plants/algae and fungi has thus a very long evolutionary history (Figure 5.73). Interestingly, this fossil comes from the famous Rhynie Chert in Scotland, which yielded also the first undeniable higher land plants with *Rhynia* known to those visitors of natural history museums who also explore the botanical departments.

The Rules of the Game

Of course, plants are not passive spectators of these attacks—otherwise evolution would have wiped them out from the planet. Actually, a few plants living under nitrogen-poor conditions became active carnivores to cover their nutritional needs as predators of animals (Figures 5.74–5.76). However, here we want to explore diverse defence modes of plants. Research over the last years has provided us some fascinating insights into this silent war between herbivores and plants. In view of the immobile nature of plants, there is as expected a lot of chemical warfare. As is usual in the big game of evolution, all tricks are allowed and plants have mustered chemicals that range from fragrance to antinutrients, from pain killers to poisons. But beware, the same lack of rules apply for your enemy. The outcome is as usual an antagonistic coevolution. Those organisms that lost the race are only found in the annals of paleontology. There are many entries in this Book of Life and Death. *Only a very small fraction of life forms that existed once on earth still have living descendents. Most organisms had their time and were then discarded because they turned out not to be sufficiently adaptable to changing environments or the pressure of competitors and predators. Evolutionary biologists adopted a nice image for this everlasting evolutionary fight—the Red Queen. This is not a peculiar tropical ant queen, but a figure from Lewis Carroll's* "Through the Looking Glass," *in which Alice and the Red Queen have to run faster and faster only to stay at the same place relative to one another.*

Antinutrients

Let's start with an antinutrient. The plant might not be able to stop the chewing of the insect, but leaves or roots frequently contain protease inhibitors. Soybean trypsin inhibitor is likewise known to biochemists and nutritionists. The evolutionary purpose is clear: These plant peptides inhibit serine, cysteine, and aspartyl proteinases in the insect gut and interfere negatively with the proteolysis of the ingested food. Less amino acid is released and the growth and sometimes even the life of the herbivore is at risk. How efficient this system is becomes apparent when you compare the hornworm *Manduca*, a major pest of crops, grown on the wild-type tomato plant or the mutant *defenseless*, which is defective in the synthesis of the proteinase inhibitor as part of the systemic wound response (Howe et al. 1996). Compared to its growth on the mutant plant, you get only hornworm dwarfs on the wild-type tomato.

FIGURE 5.73. Fungal infections of cultivated plants: barley (1), wheat (2; the spores of various smuts are shown in 4–6), oat (3), rye (18; *Claviceps* fungus in different stages, 19–23, at the top is a perithecium, 22, the fruiting body of *Claviceps* that contains the ascospores), vine (16, and its fungus *Oidium*, 17) and potato (7, cut through the leaf, 8, and spores 9, 10). Rust is also found on grass (11, teleutospores of grass and cereal rust 12, 13) and shrubs like barberry (14; cut through the rust-infected barberry, 15).

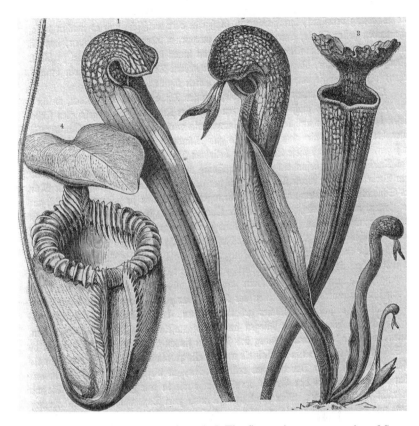

FIGURE 5.74. Plant weapons in attack mode I. The figure shows two species of Sarrace-niales that trap insects in urn-like leaves containing a digestive fluid (*1 Sarracenia variolaris, 3 Sarracenia laciniata*). These plants are true honey traps, where nectaria guide the insects down a gliding slope. At the end of the season, an 8-cm-high layer of insects is found in the 30-cm urns. The fun of biology is that one fly carrying the telling name *Sarcophaga sarraceniae* has learned to survive in this trap and can thus make a living for their larvae on the carcasses of the entrapped insects. *Darlingtonia california* (*2*), the cobra plant, follows the same strategy as does *Nepenthes villosa*. The latter has gone even a step farther by developing a very active digestion system at the bottom of the modified leaf: thousands of digestive glands per cm^2 secrete enzymes and acids. A structure analogous to an animal stomach has developed in plants. To make the analogy complete, teeth-like structures surround the entrance of the urn. The nitrogen-containing end products of digestion are absorbed by the plant through the walls of the pitcher.

Chemical Warfare

Plants produce a number of secondary metabolites that protect them against unfriendly takeover. Saponins like the steroidal glycoalkaloid avenacin belong to this class. As the name indicates, it is produced by oat plants (*Avena*), which become via this chemical defense very resistant against the fungal pathogen *Gaeumannomyces graminis*, which affects wheat and barley, but never oat.

FIGURE 5.75. Plant weapons in attack mode II. The plant order Nepenthales shows active insect predators. *Drosera* has numerous stalked glands (*1*) that secrete a clear sticky fluid. Many tentacles sit on the leaves of this plant (*4*), which progressively bend around an insect that landed on its leaves. The folding occurs on a time scale of minutes to hours (*2, 3*).

FIGURE 5.76. Plant weapons in attack mode III. Even quicker is *Dionaea*, another member of the Nepenthales order, commonly known as Venus fly trap, which works like a mouse trap. The plant is shown open in *1* depicting the spine-like teeth at the margin of the leaf and sensitive hairs in the trap (enlarged *3*). When an insect touches these hairs the trap snaps (*2* shows a cut through the closed trap). *Aldrovandia* (*4, 5*) is another genus of this plant order that preys on insects. Inserts 6 and 7 show gland cells. The closing of the trap is a very fast process and was recently reinvestigated with contemporary methodology (Forterre 2005).

This barrier is overcome by a fungal variety, which elaborates the detoxifying enzyme avenacinase. This enzyme removes the terminal glucose molecule from the saponin making it nontoxic. A number of other plants produce glucosinolates, sulfur-containing glucosides, that are as such nontoxic. They become biologically active only in response to tissue damage by the chewing insect. The enzyme myrosinase is released from its compartment and removes the glucose, which renders the compound unstable, leading to the release of toxic nitriles and thiocyanates. Cyanogenic glucosides are derived from tyrosine and can release hydrogen cyanide when the cell is disrupted. Again some insects, e.g., the butterfly *Heliconius*, can strike back and evolved enzymes that can deal with cyanogenic glucosides. The butterfly replaces the nitrile group by a thiol and preempts thus the release of cyanide and at the same time recovers a valuable nitrogen into the insect's own metabolism (Engler et al. 2000). The deterrent effect is better when it meets an unprepared herbivore like the flea beetle *Phylotreta*. As a specialist crucifer-feeder, it has never met cyanogenic glucosides since they are not produced by crucifers. Researchers have now transferred the biosynthetic pathway of one of these compounds, dhurrin, from its native host *Sorghum* to the model plant and crucifer *Arabidopsis*. The plant accepted this diversion of tyrosine into this new pathway and deterred the flea beetle larvae, only few animals started a leaf mine and many of them died. This type of research opens new possibilities for pest control in crop plants (Tattersall et al. 2001).

Salicylate and Jasmonate

Now to the pain killer salicylate, which is a member of the signal transduction pathway leading to resistance in plants. Systematic acquired resistance allows the plant to mount a robust, active, and broad spectrum defense throughout its whole body in response to a highly localized attack by a microbial pathogen. Necrosis induced by viral, fungal, or bacterial attack increases locally the concentration of salicylate. This leads to the activation of an as yet undefined phloem-mobile signal, which increases in distal parts of the plant the salicylate concentration and also that of volatile methyl-salicylate, which can signal to more distant sites. Other forms of attack induce different signals. For example, chewing insects and mechanical wounds induce the synthesis of jasmonate, a derivative of the membrane lipid linolenic acid. Its synthesis is mediated by a series of enzymes that are activated by the tissue damage. Jasmonate belongs to another signal cascade in which a gas (ethylene), a small peptide, proteolytically derived from prosystemin, electrical signals, and a volatile methyl jasmonate are involved. The salicylate and jasmonate pathways activate different sets of plant defense genes, but there is substantial communication between both pathways. Amplification of the signals leads to the synthesis of toxic compounds (allelochemicals). On the other hand, herbivores increased their production of enzymes that detoxify these allochemicals, their counterdefense system includes cytochrome P450, a microsomal enzyme system that we use for the transformation of pharmaceutical drugs.

The Reaction of the Insect Herbivore

For the herbivore there is a dangerous window between the encounter of the toxin and the induction of the detoxifying enzymes. The corn earworm *Helicoverpa zea* has found a solution. It intercepts the plant signal molecules jasmonate and salicylate. Both induce in the midgut and the fat body of the chewing insect the synthesis of cytochrome P450. When the plant lashes allochemicals at the insect, its detoxifying system is already in place. Its ability to eavesdrop on plant defense signals protects the herbivore (Li et al. 2002). Not enough with that, *Helicoverpa*'s saliva contains glucose oxidase as its principal enzyme, which counteracts the production of the toxic nicotine induced in tobacco plants by caterpillar feeding (Musser et al. 2002). The importance of jasmonate signaling was demonstrated by an antimessenger RNA approach to silence lipoxygenase, the first enzyme on the pathway to jasmonate. The hornworm *Manduca* gained weight faster on the silenced plant than on the wild type. The herbivore resistance could be restored by treating the silenced plant with jasmonate. Also the leafhopper *Empoasca* preferred the silenced plant over the wild type and caused much greater damage (Kessler et al. 2004). Jasmonate is not the only product of the linolenic acid oxidation pathway. Also volatile aldehydes are produced that play a major role in forming the aroma of many fruits and flowers. In fact, methyl jasmonate is known to the perfume industry as fragrant components of the essential oils of jasmine. Hydroperoxide lyase is such an enzyme leading to fragrant aldehydes with antimicrobial properties. Silencing of this enzyme in transgenic potato plants led to a twofold fecundity increase in sap-sucking aphids (Vancanneyt et al. 2001).

It is astonishing to realize that plant compounds which inspire poets to love songs like the fragrant of flowers should have so mundane biological functions as wound healing and herbivore defense. In the following chapter, we will realize that poets are not the only beings on earth that are attracted by volatiles from plants, poets share this property with such uninspiring animals as parasitoid wasps and entomophagic nematodes. However, the story is worth telling since it shows how ingenious plants can be in their defense against herbivores. Since plants cannot flee, they use volatile chemicals to cry for help. And they are heard or, better, sniffed.

The Enemy of My Enemy is My Friend

Parasitoids

Before discussing the role of plant volatiles in herbivore defense, we must first get acquainted with a group of insects that are liked by ecologists despite their rather rude life style. Parasitoids are insects, usually flies or wasps, whose immature stages feed on their hosts' bodies, usually other insects, and ultimately kill the host. The basis of their peculiar feeding mode is the egg-laying behavior of the adult female. They lay their eggs into other insects (Figure 5.77). The larval parasitoids develop inside the host individual, which is usually also a

FIGURE 5.77. An *Aphidius* wasp, a parasitoid wasp named after its victim, the aphid. The wasp belongs to the Hymenoptera, the aphid to the Homoptera order of insects. Aphids are plant sap-sucking insects that cause substantial damage to plants, not the least as transmitter of viral diseases.

preadult organism. Initially, the parasitoid does not cause much harm to its host. But then it starts to eat the host from the inside and normally kills it when the host goes into the pupal stage. What looks like a normal host pupa gives birth to a parasitoid instead of the host. You have certainly already seen these wasps with their egg-laying apparatus looking like a super-long sting (Figure 5.78, compare with another food prey specialization in Figure 5.79). You might find that this is a rather crude way of quest for food, but you might not expect them to represent about 10% of all species on earth. However, this figure should not surprise you since it reflects the great number of insect species, most of them are attacked by a parasitoid.

Finding the Victim

Now comes a problem: The parasitoid wasp does not eat the host, it uses it just as a cradle for its eggs. It needs to search its victims and foster mothers just for egg laying and this might not be an easy task with cryptic and evasive hosts. The problem is compounded by the selectivity of parasitoids for their host. Consumers are classified by ecologists as either monophagous (feeding on a single prey), oligophagous (having few preys) or polyphagous (having many prey types); the former are also called specialists while the latter is a generalist. While true predators generally have relatively broad diets, parasitoids are generally specialized, frequently even monophagous. How do they solve the problem of finding the appropriate host? The wasp uses chemical cues to locate the victims. It rubs the feces of the victim with its antennae for several minutes as if it has to memorize the odor and the animal shows subsequently host-seeking flight responses to these chemical attractants in a wind channel. Behavioral studies showed that the parasitoid wasp *Microplitis* uses two types of learned information for seeking the host. One is a nonvolatile contact chemical from the host and the other is a volatile chemical. Since researchers could teach the wasps to recognize vanilla odor, scientists suspected already nearly 20 years ago that the volatiles might be plant derived (de Moraes et al. 1998). This preference for perfumes should not surprise since adult parasitic wasps need nectar or some

Figure 5.78. A parasitoid wasp (*Rhammura filicauda*) from Africa sporting an extremely long ovipositor. Parasitoids deposit their eggs into other animals in which the young develop by eating up the host from the inside. This is a widely distributed group of animals, which has split into many species due to host-specialization. As parasitoids can also attack other parasitoids, Russian doll parasitism is known in this group.

other fragrant food source. Entomologists trained parasitoids to associate odors with food sources when feeding them on sucrose solutions and offering simultaneously vanilla or chocolate smell. If taste and smell sensation were temporally separated, no training effect was achieved. Females could also learn to associate

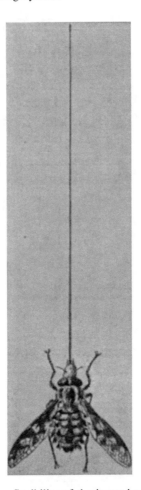

FIGURE 5.79. The evolutionary flexibility of the insect body plan is demonstrated with *Megistorrhynchus longirostris*, order Nemestrinae, from South Africa, showing an extraordinary long mouth part. Nemestrinae are relatives of Tabanidae, which are fierceful bloodsuckers. Tabanidae that attack animals with a thick skin develop large mouthparts, but the shown species uses its mouth part for nectar foraging. An even more exciting case is the hawkmoth (*Xanthopan morgani*) with a 22-cm long proboscis pollinating the orchid *Angraecum sesquipedale* with an extremely deep nectarium. Charles Darwin and Alfred Russel Wallace predicted the insect knowing only the plant 40 years before the insect was actually discovered.

their hosts for egg-laying with odors. They were also able to learn both odor associations, e.g., vanilla-to-hosts and chocolate-to-food. Interestingly, hungry females followed the odor they associated with food, while well-fed females followed the odor they learned together with the host (Lewis and Tumlinson 1988). When they are foraging for food or seeking a host, female parasitoids constantly monitor their environment with their olfactory system.

Terpenoids

The following data were published by scientists from the US Department of Agriculture and one could therefore suspect that they had a hindsight with their research (Turlings et al. 1990). They let larvae from the beet armyworm *Spodoptera exigua* chew on maize seedlings in a glass gas collector. The insect-infested, but not mechanically injured plants released a set of volatiles, mostly terpenoids. Terpenoids is a historical name derived from the German word for turpentine and refers to a vast class of chemicals that are formally (but not biochemically) derived from the condensation of isoprene (unsaturated isopentane) subunits. Chromatographic analysis revealed 11 terpenoids, with bergamotene as the most prominent compound. These volatiles attracted the parasitoid *Cotesia* in two-choice flight tunnel tests. Also synthetic terpenoids attracted the wasps. Interestingly, mechanically injured leaves on which the oral secretions of the beet armyworn were applied also became a source of volatiles that showed a comparable terpenoid composition and behavioral attraction for its specific parasitoid wasp.

Volicitins

The US entomologists then took a deeper look into the oral secretions of the caterpillar and identified a low molecular weight compound as the sole active principle and named it volicitin (i.e., volatile inciting). This new name hides a well-known class of compounds; volicitin is a linolenic acid product that is coupled to glutamine. In plants, the synthesis and release of volatiles seem to be under the control of jasmonic acid, which is also produced from linolenic acid via the octadecanoid signaling pathway. Volicitin could thus be a chance inducer of the plant defense system. But why should the caterpillar synthesize a compound that elicits plant chemicals directed against itself? It is most probable that volicitins are emulsifying agents necessary for membrane disruption— they resemble amphipathic detergents. Or does the plant here intercept a compound of the insect as an advance warning system? Volicitin activity is not diet related and also produced by caterpillars reared on an artificial diet, speaking in favor of an intrinsic insect function for this compound. Lewis and Tumlinson from USDA continued with their research and explored the next aspect of this fascinating system now consisting of three trophic levels: the plant, the herbivore, and the parasitoid. Sensing terpenoids is not enough for a busy parasitoid, to be an efficient olfactory signal the volatiles should contain chemical information on the attacking herbivore that can be read by the parasitoid.

Chemical Attraction

To test this possibility, they tested the volatiles released by tobacco, cotton, and maize in response to two closely related herbivores, *Heliothis virescence* (tobacco budworm) and *Helicoverpa zea* (maize earworm). In field trials, the tobacco budworm-specific parasitoid wasp *Cardiochiles nigriceps* distinguished the odor emanating from the tobacco plants infested by its host (De Moraes et al. 1998). The nonhost-induced volatiles were hardly more attractive to the wasp than undamaged control plants. The budworm-infested plants remained attractive even when the leaves plus the budworms were removed—the signal comes from the plant and not the victim for oviposition. The wasp could also distinguish its victim when it infested cotton. This is a remarkable result: It means that the wasp can distinguish between host and nonhost infestations on phylogentically distant plants that produce varying chemical blends. As a host-specific parasitoid, it locates its polyphageous host over a range of different habitats. In the past, insect physiologists have anticipated behavioral responses to be relatively hard-wired. The parasitoids convinced them that learning and context-dependent analysis of chemical cues confer substantial behavioral plasticity to wasps.

The next question is whether this seemingly efficient chemical attraction of wasps via the jasmonate pathway results also in an increased parasitism of the herbivore. This question is not trivial since the caterpillar comes under double attack. The jasmonate response attracts the wasp to its victim, but it induces also allochemicals in the plant that reduce the growth of the herbivore, which could negatively affect the reproduction success of the wasp. The net result of both processes was a twofold increase in parasite level in the herbivores (Thaler 1999). The volatiles released by the plant did not only reduce the growth rate and attracted parasitoids, it also attracted general predators that emptied the eggs and discouraged the oviposition rate of the herbivore. As a consequence, a tobacco plant confronted with three different leaf-feeding herbivores could reduce the number of herbivores by more than 90% when releasing volatiles (Kessler and Baldwin 2001). Elicitors of plant resistance may thus become useful in agriculture.

Rootworm Generalization

It is always important in biology to ask whether the reported observation represents an interesting case that adds just a new thread to the huge carpet of natural phenomena or a new general principle. Recent data obtained in maize confirm that we deal here with a new principle. The scene is now below the ground and this is now an interesting case of how far volatiles can carry a message within the soil as compared to the air. The herbivore is in this case Western corn rootworm *Diabrotica virgifera*, a voracious coleopteran pest that threatens maize production in central Europe. The beetle larvae feed on the roots of maize and a Swiss–German group of scientists asked the question whether the attacked plant releases a signal molecule that can attract a predator of the beetle larvae. The answer to their question came from experiments with a six-armed olfactometer (essentially a star-type arrangement of pots with plants connected via tubing to a central pot, all filled with soil) that allowed the researchers to give an organism that feeds on rootworms the choice between healthy and rootworm- infected maize plants. In addition, field trials were conducted (Rasmann et al. 2005). The plants released a compound that they identified as caryophyllene, this compound traveled rapidly through the sand and qualifies as an exceptionally suitable below ground signal. In addition, caryophyllene- either produced naturally or added to mechanically injured roots- attracted a predator of the rootworm, the entomopathogenic nematode *Heterorhabditis megidis*. The predators rushed to their prey at a velocity of 250 body lengths per day. Plants that released the below ground volatile showed 43% infected larvae compared to only 8% for a commercial US maize variety that did not produce caryophyllene. The commercial maize variety had apparently lost this signal chemical during the breeding program since the ancestor of modern maize, teosinte, still produced caryophyllene.

Herbivores: Patterns of Predation

Carnivores as Protectors Against Mammalian Grazers

Plants have developed a number of strategies to discourage mammalian herbivores. However, the best protectors of plants are probably still carnivorous mammals (Figures 5.80, 5.81). A 40-year study of mammalian herbivores in the Serengeti provided important insights into these predator–prey relationships, which poise the exploitation of the grassland by herbivores (Sinclair et al. 2003).

FIGURE 5.80. The spotted hyena (*Crocuta crocuta*) is a dog-like carnivore that depends for its food on the rest of the meals from more powerful carnivores; it also takes carrion and small animals.

FIGURE 5.81. Tapirs (*Tapirus*) are represented with four species of the family Tapiridae, order Perissodactyla. They live in the tropical forests in Malaysia and the New World. They eat plants, mainly leaves of trees, but they like also fallen fruits, especially the *Spondias* plum. They also eat soil to satisfy their salt needs. Their main enemies are the tiger in Asia, and the jaguar in America (*Leo onca*), the largest New World member of the cat family (Felidae). The jaguar is a solitary predator that preys on deer, tapirs, and capybaras.

This ecological study is a fascinating number game. The Serengeti is composed of open grassland and savannah. This ecosystem supports 28 species of ungulates and 10 species of carnivores that prey on the herbivores. Each predator specializes on a different prey: The lion preferentially attacks the large prey like wildebeest (*Connochaetes taurinus*) (Figure 5.82) and the zebra in the 170–250 kg weight range, which corresponds to its own weight of 150 kg. Lion can attack larger prey, but buffalo and giraffe comprise only 10% of its diet. However, still 40% of the lion's prey is lighter than 50 kg. The diet niche of the smaller carnivores (hyena, wild dog: upper limit 250 kg, leopard, cheetah: upper limit 100 kg) is nested within the prey range of the larger carnivores. The consequence is that if you are getting smaller as a prey, you have increasingly more enemies. Small prey like the Thomson's gazelle is confronted with an uncomfortable predator: prey ratio of 1:3. Small wonder that all recorded causes of death in this herbivore weight range is by predation. This rate falls with the weight of the herbivore: It is only 23 and 5% for buffalo and giraffe, respectively. The giant herbivores like the elephant (Figure 5.83), hippopotamus, and rhinoceros almost never suffer predation. Their mortality is mainly caused by food limitation. The threshold from a top-down control (predation) to a bottom-up (food) control is at 150 kg herbivore size. In the undisturbed situation, the Serengeti carries 1.1 predators per square kilometer (lions contribute 0.2 and hyenas 0.5 to this figure). The density of the Thomson gazelle is three animals per square

FIGURE 5.82. The wildebeest (*Connochaetes taurinus*), family Bovidae, order Artiodactyla, is a large mammalian grazer. Its impact on plants is kept in check by carnivores.

FIGURE 5.83. The African elephant (*Loxodonta africana*) is one of the two surviving species of the family Elephantidae, order Proboscidea. An adult African elephant weighs up to 7,500 kg and eats 300 kg of solid food per day. The food consists of grass, branches, and the fruits of various palm trees. The animal is too great for predation by any carnivore and its mortality is therefore only determined by shortage of food and water.

kilometer. In a natural experiment, the majority of the large predators was removed in part of the Serengeti. All herbivores below 60 kg weight were thus relieved from predation pressure. As expected, an increase in prey density was observed; the Thomson gazelle, for example, rose to 18 animals per square kilometer. When the predators immigrated from neighboring areas into the void niche the prey density quickly returned to the initial levels. Large herbivores like the giraffe were not predation limited; in fact, their density fell in the predator-free areas.

6
Eating Cultures

Choosing Food

To Eat or Not to Eat

We are omnivores. As the name indicates, we eat various foodstuffs. Other animals have opted for more restricted diets. Carnivores eat only meat in the large sense; herbivores eat plant material; detrivores eat decomposing organic material; many other "-vores" exist. Within each category, there are still substantial differences. Some animals are not very selective in a given food category, whereas others are specialized—in extreme cases, to a single food item. This has the advantage that you can optimize your search strategy and physiology to a single food item. This can of course sometimes be a risky decision when your food item gets under pressure. The panda and the koala are doomed should something happen to the bamboo or the eucalyptus tree, respectively. In contrast, if you were an anteater, one would guess that you are on the safer side in view of the fact how many ants populate the world. Living from a single organism has further inherent risks. Much prey does not like the idea to be eaten and plants frequently deploy chemical weapons to discourage herbivores. You might then be well advised to vary your food: By mixing different toxins, you can push each below a critical level. Other questions emerge: Does the target organism really offer all ingredients you need? Take a grazing cattle. The body composition of the cow differs substantially from that of grass. Will grass satisfy all nutrient requirements of a mammal? We have seen in the previous chapter that ruminants in a certain respect do not eat grass. For ecologists and ethologists, the "who eats whom?" is a central question, and many theoretical thinking went into this field. The optimal foraging theory, the marginal value theorem, the scalar expectancy theory, and Charnov's optimal diet model are important theoretical concepts in the field. Some achieve even mathematical descriptions of predator–prey relationships. Instead of presenting you equations, I will illustrate you general problems with examples. They are easier to grasp, hopefully entertaining, and they suffice to define the problem.

The Moose and Its Sodium Problem

My first example is the moose (*Alces alces*) (Figure 6.1), a herbivore of the northern hemisphere. G. E. Belovsky has developed some instructive concepts for the moose feeding on the shores of the Lake Superior (quoted from C. Barnard Animal Behaviour, Pearson 2004). Plants and animals have distinct physiologies. *It was said that our blood, which bathes all our body cells, still reflects in a certain respect the salty environment of the ocean.* We need salts like NaCl to maintain the ion concentration gradients across membranes that contribute to the electrical potential necessary for nerve and muscle activities. *Plants do not have nerves and muscles, and consequently, they do not really care about Na$^+$ ions.* If you are eating plants (moose browse the leaves of deciduous trees), you get nutritious food, but the leaves will be deficient in sodium ions. So foraging exclusively on leaves will not be an option for the moose, and if only for its sodium needs, the moose has to vary its food intake. If you have observed moose in nature, you will remember them standing in water. Moose crop aquatic plants in shallow lakes. The reason is simple: Aquatic plants are richer in sodium than terrestrial plants. With this strategy, the sodium problem has been settled, but not all problems have been solved. Aquatic plants contain less food energy than terrestrial plants. So you might propose that moose will have to eat more aquatic plants to get both energy and salt. But there is another constraint: the rumen capacity.

FIGURE 6.1. The moose (*Alces alces*) is the largest member of the deer family (Cervidae, order Artiodactyla).

Economical Decisions in Biology

Stated simply: Any animal has the gut it needs. It makes no sense to invest in a gut that is oversized for your needs. This would only be a waste of resources. As a first approximation, you can start with the idea that all organs are maintained to their needs. When the need does not persist any longer, the body economizes immediately on these investments. A classical example is domesticated animals. Under such conditions, the rancher puts them in an enclosure and mounts the guard against predators. Sharp senses then become a luxury function, and consequently the senses come quickly down in domesticated animals, when compared with their wild cousins. To address the problem with the moose, G. E. Belovsky developed an optimization approach known as linear programming. A rumen that provides the daily energy needs based on leave food is too small and the animal would quickly become deficient in salt. A rumen that satisfies the caloric need on the basis of aquatic plants would be too large and thus a waste of bodily infrastructure. The researcher plotted a type of phase diagram for the moose diet with intake of aquatic versus intake of terrestrial plants in gram. The food composition must be above the minimal energy level and above the minimal sodium level needed by the animal, and, at the same time, it must lie below the rumen capacity. The three lines intersect and form a triangle. All values on the surface area of the triangle fulfill all constraints and are thus possible solutions. When G. E. Belovsky looked what the moose actually took, he realized that they pushed for the maximal terrestrial plant intake that still satisfied the sodium needs. In other words, moose selected their diet to maximize the energy intake within the limits of the given rumen capacity.

Extension of the Issue

What are you doing if you are a mountain goat? Its plant diet is also deficient in sodium, but since goats frequently live in quite dry environments, aquatic plants are no realistic solution. Goats discovered that licking certain stones can help them to satisfy their sodium needs. Not only goats lick stones to meet their mineral requirements, children do it, too. The following story is a famous case in pediatrics. In the nineteenth century, poor British children did not receive enough milk to meet their calcium demands for bone growth. Since the mineral part of bones is calcium apatite, quite substantial amounts of dietary calcium are needed during the growth phase. In rural Britain, the classical wall painting of farm houses was made with chalk—children, without following any chemistry course, knew about that link and met their calcium needs by licking the walls. When the early wall paints were introduced which contained lead, severe intoxications came to the attention of a physician who linked it to the licking behavior. It is quite fascinating that the body can express its needs and direct its action in form of a complex behavior to satisfy its nutritional needs. This self-control does not work in all cases: I remember a medical case record from Zurich, in which a homeless person was hospitalized with strange symptoms until the physicians recognized their first case of scurvy that they had seen in their life. The patient lived exclusively from two food items, canned beans and biscuits, neither contained vitamin C. Apparently, he did not develop a craving for fresh food.

Nutrient-Specific Foraging

Animals frequently know their food needs very well. Nutrient-specific foraging was even reported in invertebrate predators (Mayntz et al. 2005). This observation is somewhat surprising since it was believed that the body composition of prey animals is nutritionally balanced for carnivores. These biologists used three polyphagous animals for their investigations. The first was a *Carabidae*, a highly mobile beetle (Figure 6.2). To test whether this ground rover actively selects prey according to nutrient content, they tested the food intake on locust powder mixed with either lipid or protein. They observed that the beetles preferred lipid locust powder when they were pretreated on a protein diet and a protein locust powder when pretreated on a lipid diet. In addition, when coming from a protein-rich diet, they consumed more food than when coming from a lipid pretreatment. Then the zoologist reared *Drosophila* flies on different diets, and they were surprised that they could vary the protein-to-lipid ratio in the animals over a relatively wide range, simply by offering a distinct diet. Researchers had anticipated that the body composition of the prey is relatively independent from diet. The next test animal was a wolf spider (Figure 6.3), a classical sit-and-wait predator. Again the predator tried to balance its food intake. When they came

FIGURE 6.2. Carabidae or ground beetles are one of the largest families in the insect order Coleoptera. They are fierce predators that inactivate their animal prey by biting them with their strong mouth parts, then they spit a digestive juice on the prey, which rapidly liquefies the prey. The beetles then take up the predigested juice from the prey. The figure shows from *left to right Carabus hortensis*, *Calomosa sycophanta*, and *Carabus auratus* with its larva under the stone.

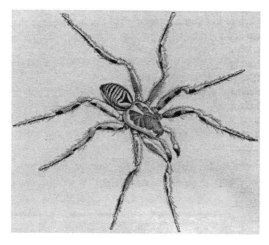

FIGURE 6.3. Wolf spider, here the tarantula of southern Europe (*Lycosa tarentula*), family Lycosidae, order Araneida. In the fourteenth century, the bite of the tarantula was linked to a dancing mania (Chorea saltatoria), which was a mere fright reaction from humans decimated by the Great Pestilence that swept across Europe in 1348 and not the effect from the poison of this spider.

from a diet on lipid-rich flies, they preferred protein-rich flies and vice versa. The next taster was a desert spider that builds durable webs. Prey items arrive spontaneously, and the spider cannot control the nutritional composition of the prey. However even these spiders control their diet composition by differential extraction of their prey. If the spiders came from a protein-poor diet, they left less nitrogen in the fly remnants after feeding than after coming from a protein-rich diet. These results now force theoretical ecologist to go back to model building and to incorporate the new observations into optimal foraging models that account not only for prey quantity but also for prey composition.

Leaves Chosen by Your Mother

In some animals, food selection is already made by the mother. Take the moth *Manduca sexta*. It lays its eggs mostly on solanaceous plants (potato, tomato). This is not an absolute rule; occasionally it lays eggs also on nonsolanaceous plants. The preference for the former plant as food source is clear: Larvae (also called tobacco hornworm) reared on solanaceous plants grow and develop much quicker on them than on alternative food sources. The larval stage is a critical step where the moth is exposed to more predators, pathogens, and parasitoids than the adult stage. So you best pass quickly through these stages. How does nature achieve fidelity of the larvae to the better food without compromising the capacity to exploit alternative food sources, if necessary? The larvae molt through different instar stages, and the researchers took fourth instar larvae reared on solanaceous foliage and put them on a solanaceous or a nonsolanaceous (cowpea) diet (del Campo et al. 2001). In the first case, all larvae started feeding on their known food, while only 22% of the solanaceous-reared larvae started feeding on the unusual food. In fact, 35% of the larvae on the new food source preferred to starve to death, which is a remarkable observation since the hornworm can complete their development on many nonhost plants. In contrast, hornworms reared on nonsolanaceous foliage did not care whether feeding continued on this food source or changed to a solanaceous food.

The Material Basis of the Decision

A small trick led to the identification of the mechanisms underlying this food choice process. The researchers soaked cowpea leave discs with an extract of potato leaves, and solanaceous-reared larvae preferred the treated leaves over the control leaves by an overwhelming vote of 96 to 4%. This biological test guided them through the purification procedures. The feeding stimulant was indioside D—a steroid substituted with a few sugar moieties. After the chemical analysis, they did the next step and explored the neurophysiological basis of the food recognition. The mouthpart of the larvae consists of a maxillary palp and a mandibule as mechanical feeding instruments. They are surrounded on the top by antennae and at the bottom by a structure, which looks like a tiny electrical plug, the sensilla styloconica. *There is no need to be afraid of these names: zoologists have to invent a word to name the observed structures for which no words exist in our everyday language. Scientific committees encourage Greco-Latin artificial words, which sound terribly scientific while in fact you could have also called it the "sensitive electric plug."* When the scientists made electrical recordings from these plugs/styloconia, the lateral sticks of this structure responded specifically to the application of the chemical cue indioside D (*here you have the same naming problem for something which has no name in our languages*), but only in the solanaceous-reared animals. Microsurgery closed the chain of arguments: When the sensilla sticks were cut on both sides of the mouth, the solanaceous-reared animals lost their preference and started readily on the cowpea diet. In this way, Nature accomplishes that the specialist herbivore keeps to its host plant. Nature must follow a different strategy when dealing with a generalist herbivore like the locust. Its taste receptors responses to amino acids, salts, and sugars which are found in many plants and not to a specific compound found only in a single plant family.

Koala Versus *Eucalyptus*: Choices for the Food Specialist

Even food specialists show preferences and clear-cut food choices. Take the case of the koala (*Phascolarctos cinereus*; Figure 6.4), which feeds exclusively on *Eucalyptus* (Figure 6.5) trees belonging to three species (Moore and Foley 2005). *Eucalyptus* trees differ in a number of characteristics. These trees vary in size—koalas prefer larger trees simply because they represent larger food patches. However, this comes at a price. Plants elaborate secondary metabolites to deter specifically herbivores. *Eucalyptus* trees do that via the synthesis of

FIGURE 6.4. The koala (*Phascolarctos cinereus*), the arboreal teddy bear of Australia, was attributed to the family Phalangeridae from the order Marsupialia.

FIGURE 6.5. The eucalyptus tree, which belongs to the order Myrtales (which includes the most important genera of tropical mangrove), family Myrtaceae, is here represented by its largest member *Eucalyptus regnans*. This tree is native to Australia and reaches more than 100 m of height and is thus the largest tree of the world. It reaches a height set by the physical limit of water transport in trees (Koch et al. 2004).

formylated phloroglucinol. Koalas, as specialized *Eucalyptus* "folio-vores," are adapted to these compounds and are tolerant to it as no other animal. And this is necessary because the larger trees are also producing higher concentration of this compound. However, the detoxification of these compounds also costs energy to the koala. It is thus only consequent that within a given size class, koalas visit trees with lower toxin concentrations. Avoiding intoxication is only one side, assuring a balanced nutritious diet is another side of their food choices, which the koalas have to take into consideration. *Eucalyptus* is extremely low in nitrogen content, but when the researchers conducted extensive leave chemistry, differences in nitrogen concentrations emerged. The strategy of the koalas is clear: Trees with high toxin concentrations are avoided, trees with medium toxin concentration were overvisited because they offered satisfying nitrogen sources, while trees producing low toxin and also low nitrogen concentrations were again avoided. Thus even in an extreme food specialist, food habitat differentiation occurs and sets in motion an arms race between defense reactions of the trees (by overexploitation koalas can cause *Eucalyptus* tree mortality) and counterdefenses by the specialist herbivore.

Food Choices at the Ecosystem Level

Choosing food is not only important for the nutritional decisions of individuals or a species, it also affects the structure of the entire ecosystems. If large numbers of species share the same ecological niche, it is commonly assumed that the coexistence of so many species is the consequence of finely divided food resources. This reasoning was applied to very different ecological niches ranging from coexisting insectivorous bats, herbivorous insects, and bacterial commensals in the human gut. Yet, biologists arrived to different conclusions when looking to herbivorous insects, coexisting in a tropical forest of New Guinea (Novotny et al. 2002). When they analyzed the data for 900 herbivorous insects on 50 plant species, they found that most insect species feed on several closely related plant species. They investigated two plant genera in detail: *Ficus* and *Psychotria*. Any two species in these genera shared 50% of their herbivorous species. This overlap in herbivore community decreased gradually with increasing phylogenetic distance between the plant hosts. Large and species-rich plant genera hosted more distinct leaf-chewing communities than small genera. This reflects simply the preference of the herbivores for a larger food resource basis. Monophagous insects, i.e., animals feeding on a single plant species, were rare in the tropical forest. On the average, a single host plant species was the food resource for 30 species of Coleoptera (beetles), 26 species of Lepidoptera (butterflies, moths), and 20 species of Orthoptera (locusts, crickets, grasshoppers). These observations are at variance with the notion that extremely specialized food interactions between herbivores and plants finely divide the plant food resource and permit the coexistence of numerous herbivorous species. Actually these ecological considerations had also an impact on other branches of zoology: It also decreased the estimate of arthropod diversity from 30 million to "only" five million species.

The Star-Nosed Mole has Studied the Optimal Foraging Theory

In the 1970s, a lot of ecological thinking dealt with the concepts as to how a predator makes its food choices. It was anticipated that such an important part of the life history of animals would be under natural selection and the animal's foraging activity should therefore maximize the rate of energy intake. Some of the work sounds rather theoretical, and there are many mathematical treatments of the problems. Yet, for the animal, optimal foraging is a crucial question of survival. If the animal spends too much time with prey of too low nutritious value, it will have a suboptimal food intake. You might say that it will have to spend more time on foraging to satisfy its caloric needs. However, many animals spend essentially the whole day with foraging—prolonging the search is not an option, they will starve. Even for an animal that is not searching food the whole day, the consequence of suboptimal foraging is hard; it will simply have less time for finding a mate, caring for its young, or defending its territory.

In Charnov's classical model, the rate of energy intake R can be described by a simple equation: $R = E/(T_s + T_h)$. In this equation, E stands for the energy content of a prey item, T_s is the time spent for searching the prey item, and T_h is the time needed to handle the food item. Well, if the mole *Condylura cristata* has understood the equation, it has several options. Simple arithmetic tells it that R is high if E is high. So, its first instinct will tell it to go for an earthworm instead of a cricket, not to speak of a small invertebrate—these yield 1100, 840, or only 10 kJ, respectively, as caloric value. However the mole has done its arithmetic course in the severe school of natural selection (Catania and Remple 2005). It learned there not to neglect the denominator. Keeping T_s or T_h or better both low will also boost R. T_s is influenced by choosing the right habitat—the right one is the richest with respect to the prey. As animals can move around, they have some freedom of choice. Very interesting observations were made with birds as to how they decide when the current feeding plot is still profitable and when they have to leave the plot. This is not a major problem for the mole: Its wetland habitat contains high densities of invertebrate prey. T_s can thus be kept low. Now, it must look for keeping T_h low. Going straight for the nutritious earthworm might not be the best choice if this prey necessitates a relatively long handling time. Less nutritious prey can be more profitable because it is the ratio E/T_h that matters. In addition, is the large prey sufficiently abundant to keep T_s low? Does the mole meet every 10 s an earthworm? You might argue that the mole takes what it gets: the earthworm and the small invertebrate. However, keeping T_s and T_h low means that you need a specialized sensorium and dentition to detect and eat the prey efficiently. Thus you have to make your choices. Selection worked on star-nosed mole (Figure 6.6) and made its choice: It goes for small prey. Its dentition is tweezers-like, which are optimized for precisely grasping small preys. They are greatly reduced when compared to the dentition in the eastern mole (*Scalopus aquaticus*). As its name indicates, the star-nosed mole has a bizarre nose—many times greater than the nose of the other moles and consequently more densely populated with sensory cells. The nose is specialized into a low-resolution periphery and a high-resolution

FIGURE 6.6. With the star-nosed mole (*Condylura cristata*), family Talpidae, order Insectivora, we explore the predictions of the optimal foraging theory. This North American mole frequently leaves its burrow and even swims.

central region. The optimization of the search and eating program is such that the zoologists needed a high-speed video system to film the foraging of this mole. The sequences are really quick: a 50 ms move forward followed by a 25 ms touch time. If no prey was detected, the next cycle of move and touch follows and so on. If a prey is detected, a decision must be made: to eat or not to eat. If the mole decides "no," it continues with the move–touch cycles; if it decided "yes," it will handle the food. For this handling, the mole takes mere 230 ms to achieve the capture and eating task. These extremely fast T times allow, under experimental conditions, the processing of 10 small food items in about 3 s. The energy intake or profitability reaches 40 kJ/s and is thus at the theoretical limit of the process. The system works really at the borderline of physiological capacities: Mammals need 12 ms to process tactile information in the somatosensory cortex and 5 ms for activating the appropriate muscle cells with a motor command. The 25 ms touch time does not leave much room for central processing and reflection on a decision. One could also formulate that the mole is "blind" because it could not be such an efficient forager with distance detection (sight, sound, or smell). A positive feedback from a tactile detection system means that the mole can directly open its mouth to catch the prey. Visual information processing is slow in comparison—human eye movement takes 200 ms and then you still have to reach the perceived prey.

To Lure with Food

If animals are under the dictate of optimal foraging, other organisms can exploit these constraints for their own purpose. I will illustrate this with an example from central Europe. When walking along riverbanks one can see there a nice, colorful plant that was not there twenty years ago. As one could suspect, it is one of the many examples of successful species invasion. The plant in question is *Impatiens glandulifera*. It is native to the Himalayas; it was introduced in Europe hundred years ago and started to spread from Czechoslovakia and has now reached the riverbanks of Austria and Germany. The native plants—like *Stachys*, *Lytrium*, and *Epilobium*, all nice and sizable flowers—are getting under pressure from this aggressive invader. Normally, invaders compete for food, water, light, and space. The Asian beauty has chosen a more insidious, but efficient strategy: It lures pollinators with food. In fact, the nectar sugar concentration of about 50% is quite normal when compared to the nectar from plants visited by bumblebees. The trick is, it produces more nectar (Chittka and Schurkens 2001). As science is about figures, the numbers are revealing: *I. glandulifera* secretes 0.5 mg nectar per flower and hour, while the other riverbank beauties achieve only 0.01–0.04 mg. Bumblebees had also "learned" their Charnov equation and realized that by visiting the Asian invader, they can reduce T_s—they get the same amount of nectar with fewer visits. Less time is spent in flying from one flower to the next, and since insect flight is a costly business, these economies count. The invader also invested strongly in advertisement using different senses. For the eyes, the flowers are big and intensively colored; for the nose, the Indian flower is strongly scented. The honey trap is prepared for the pollinator. The result was foreseeable: *I. glandulifera* experiences 2.5 visits per flower per 10-min observation period, which is four times higher than visits on *Stachys*—the highest nectar producer from the native flora. This is the investment part for the invader: To get into the new floral market, it paid a higher price. However, the reward was directly proportional to the investment. In mixed plots with *Impatiens*, the number of visits to the *Stachys* competitor was reduced by 50% and their seed set was reduced by 25%. Small wonder that the percentage of *I. glandulifera* along the riverbanks had increased over the last years. We should not blame this flower for exploiting Charnov's equation: Luring with food is a common strategy in animals. Male birds frequently present an attractive food item (or material for nest building) to the female partner in courtship. Have you ever thought about the evolutionary sense of your action when offering a flower bouquet to a lady?

Food Avoidance

There is some food that you better avoid to eat because it can make you sick. Some people experienced that food-associated illness left a lifelong aversion toward food items that were consumed just before the onset of the sickness. This phenomenon was now investigated in detail in a much simpler system—the nematode *Caenorhabditis elegans*. What to eat and what not is an important decision for *C. elegans*: In its natural habitat—the soil—this worm lives on bacteria. The soil is a complex microbial environment; the nematode will meet there more or less nutritious bacteria and also pathogenic bacteria. The pathogenic soil bacteria *Pseudomonas aeruginosa* and *Serratia marcescens* can multiply in the intestine of the nematode, and this can lead to the death of the animal after a few days. *C. elegans* protects itself from this danger by two strategies: innate immunity and food avoidance by aversive olfactory learning. Thereby they learn to associate a particular smell of the pathogenic bacterium with the visceral malaise and to avoid a second exposure. Since this nematode has a simple nervous system consisting of just 302 neurons, the identification of molecules, cells, and neuronal circuits associated with this behavior is greatly facilitated. As a first step, the researchers developed a behavioral test (Zhang et al. 2005). When they offered a choice of two bacteria, a nutritious *E. coli* and a pathogenic bacterium, to nematodes grown up on *E. coli*, the animals showed no particular avoidance to the pathogen; they were naïve. However, when they were already trained on plates containing a food and a pathogenic bacterium, the animals avoided the pathogen on the test plate. This observation speaks for associative learning, but it does not tell you whether they develop a positive association with the harmless or a negative association with the pathogenic bacterium. An olfactory maze assay, offering a more extended bacterial food choice, demonstrated both attractive and aversive components in the learning. They could dissociate both components, since a short time exposure of 4 h to the pathogen was sufficient to induce the aversive behavior, while this short experience could not induce attractive associations. Yet, the behavioral test does not lead to the molecular basis of this learning process. The researchers looked therefore into the literature and tentatively targeted serotonin as candidate signal molecule. Why serotonin? Neuroscientists knew that mood, food, and serotonin are connected. Food, especially carbohydrates like chocolates, was known to lift the mood in periods of stress—for example, in students preparing for examinations. Measurement of serotonin in the hypothalamus showed that serotonin levels rise in the anticipation of food and spike during a meal. Drugs that elevate serotonin levels in the brain are powerful appetite suppressants and were used to treat human obesity. Therefore, it was logical to look also into serotonin as a first guess for associative learning of food avoidance with *C. elegans*. Serotonin is synthesized in the body from the amino acid tryptophan via a sequential hydroxylation and decarboxylation reaction. The *C. elegans* genome contains a single tryptophan hydroxylase homologue, and a knockout mutant was very instructive (Sze et al. 2000). The mutant animals showed no serotonin accumulation but were viable. When presented with food, the wild type but not the mutant increased the pharyngeal pumping activity. The serotonin mutant had also effects on the metabolism-regulating neurosecretory axis. The mutant accumulated more fat and 15% of the mutants entered a nonfeeding, nonreproducing "dauer" (German for "enduring") state. Like cookies in children, the bacterial food upregulates the production, release, and response to serotonin in *C. elegans*. Serotonin signaling has been implicated in the control of mammalian feeding, metabolism, and body temperature regulation by the hypothalamus. The unifying feature of many of the serotonin responses in *C. elegans* is the coupling of food sensation to various motor and endocrine reactions. Food is a major reward in animal life, and we should thus not be surprised to see serotonin also linked to associative learning in lowly invertebrates.

The tryptophan hydroxylase mutant tph-1 in *C. elegans* was not impaired in general olfactory ability or the recognition of bacteria, but it was selectively unable to associate the physiological response to pathogenic bacteria ingestion with olfactory cues (Zhang et al. 2005). As so much is known about this simple invertebrate, the researchers could implicate a simple neuronal circuit with the associative learning. In response to bacterial infection, serotonin is produced in one specific neuron called ADF. Another neuron receives information about bacterial food from a sensory neuron and integrates at the same time the serotonin signal from ADF via a serotonin-gated ion channel leading to aversive learning. As no serotonin is emitted by the ADF neuron in the mutant animal, it has lost its capacity to avoid poisonous food.

Food Separating Species

Speciation Concepts

The rapidity of an event in biology is not an argument against gradual evolution in the context of the evolution theory. Evolutionary biologists count with the possibility of speciation occurring under our eyes at least when applying time spans used in human history. This "creation" of species under our eyes could be food for thought for creationists. Biologists have explanations for these events. It would be interesting to hear about alternative explanations from a creationist framework of thought. In science, the value of a hypothesis is commonly judged from the predictions you can make about the natural phenomenon under investigation that can be tested by observation or experimentation. As this book is about food and eating, I will illustrate the problem of speciation from the perspective of food choices by animals.

The crucial event for the origin of a new species is reproductive isolation. Reproductive isolation can be of two types. Two species are called prezygotic when they differ by courtship or mate choice or different breeding seasons. The species are called postzygotic when they interbreed but yield offspring of low viability or fertility. A commonly hypothesized situation for speciation is a new species evolving from its ancestor in geographical isolation. This process is called allopatric speciation, and the cause for isolation can be manifold (mountain range, river, disease outbreak, migration that separates an initially contiguous species). Almost all biologists accept that allopatric speciation occurs. However, the discussion between biologists starts when it concerns sympatric speciation, meaning when a new species evolves within the geographical range of its ancestor. Ernst Mayr, the great twentieth-century evolutionary biologist, who searched an extension of Charles Darwin's ideas into the biological discoveries of the twentieth century, contested in his 1942 classic *Systematics and the Origin of Species* (1999 reprint by Harvard University Press) that sympatric speciation can occur. By casting doubts, he did something very valuable in science: He stimulated others to look for evidence.

The Food Choice Shift from *Rhagoletis pomonella*

The 2004 edition of a brilliant textbook (Mark Ridley, *Evolution*, Blackwell 2004) quotes an example that comes close to sympatric speciation without fulfilling all tenets. The actor of this drama is *Rhagoletis pomonella*, a tephritid fly and pest of apples (Figure 6.7). The insect lays its eggs into apples, the maggot grows in the apple causing economical damage. As apples are under close

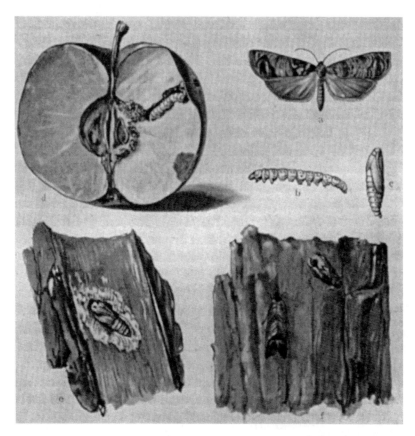

FIGURE 6.7. The codling moth belongs to the family Tortricidae, order Lepidoptera and is a major fruit pest in North America. The figure shows (a, f) the adult moth, (b) the caterpillar, (c) the pupa, (d) shows a caterpillar burrowing into the apple and (e) shows the caterpillar, after devouring the apple, emerging from the apple and pupating under a bark.

scrutiny in orchards, one can trace the first report on apple infestation with this fly to 1864 in North America. Before that date, the preferred host for oviposition was the hawthorn. In the meanwhile, this fly has included still further fruit plants (cherry, pears) as targets and did what biologists call a host shift. Females prefer to lay eggs into the fruits they grew up, and males tend to wait for mating on the fruit they grew up—the consequence is what biologists call assortative mating. *R. pomonella* from apple and hawthorn have, after about 140 generations, developed into two races that differ, for example, in their development time. Maggots in apple develop in 40 days, those in hawthorn in 60 days, and this in turn leads to a reproductive isolation because the adults are then sexually not active at the same time. However, when you put the two races together in the lab, they mate together indiscriminately, and they yield fertile offspring. In the biological species definition, no speciation has yet occurred. However, M. Ridley quoted convincing circumstantial evidence for sympatric speciation in phytophageous insects. There are about 750 fig wasp species, and each breeds on its own fig species. A similar case could be made with leaf miners—a parasitic gall wasp, whose larvae feed on plant tissue, lead to exotic transformation of plants (Figure 6.8).

The European Corn Borer

Ostrinia nubilalis feeds on about 200 weeds. When maize was introduced into Europe about 500 years ago, the insect included maize into its food plants. In France, two races of corn borers can be distinguished that differ in food plant utilization. The first is the "hop-mugwort E race" and the second is the "maize Z race." The races display host plant fidelity for oviposition, but mating does not occur primarily on the host plant. How do they distinguish each other? There are apparently E race- and Z race-specific sex pheromone blends that could lead to assortative mating. The researchers used a nice trick to explore the genetic differentiation of the two races. As mugwort (*Artemisia vulgaris*) is a C3 plant and maize a C4 plant, the two plants handle CO_2 differently and can thus be distinguished by stable ^{13}C carbon isotope analysis (Malausa et al. 2005). The animal tissue closely mirrors the isotope distribution of its food plant. The researchers caught 400 moths from 5 sites within 4-km distance and found, both with respect to wing and spermatophor analysis, a bimodal carbon isotope distribution without overlap. All but one E race females had mated with E males, and all but one Z race females had mated with Z males. Genetic analysis differentiated Z and E animals into two clearly distinct groups. Gene flow between these two races was thus reduced to <1% in the absence of spatial and temporal isolation (the flight periods overlap). Food choice preferences might be a major force here for the genesis of races, which get close to a sympatric speciation event.

Cichlid Fish in a Crater Lake

The case for sympatric speciation is coming close to proof with a recent work analyzing cichlid fish (Figure 6.9) in a small (<5 km), but deep (200 m) Crater Lake from Nicaragua (Barluenga et al. 2006). It is a homogeneous place containing an impoverished fauna due to its geographical isolation and recent origin (<23,000 years). Allopatric speciation is thus very unlikely. Yet, the lake contains two closely related cichlids, *Amphilophus citrinellus* and *A. zaliosus*. With respect to mitochondrial DNA sequences, they differ clearly from cichlids of the next lakes. The molecular data suggest that *A. zaliosus* has evolved from *A. citrinellus*. Despite that close similarity, the two species differ visibly even for nonzoologists. The two species mate assortatively. Stable heterospecific pairs can still form, but they do not lead to successful matings. This observation suggests prezygotic isolation by differences in courtship behavior. One might stop here and suspect sympatric speciation through sexual selection. This has actually been proposed to explain the outstanding species richness of cichlids in the East African Great Lakes, which contain more than 1,500 endemic species. In fact, the strong differences in body coloration could independently suggest sexual preferences as the underlying mechanism. However, the zoologists

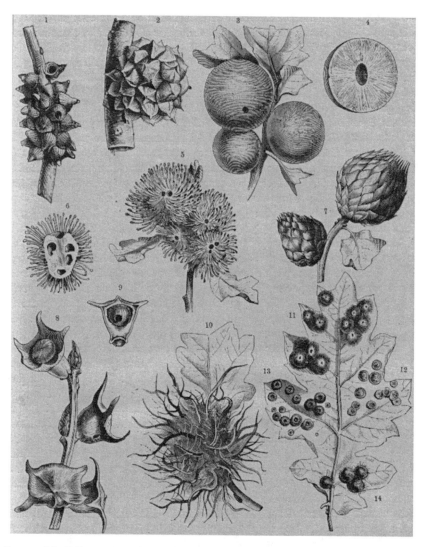

FIGURE 6.8. Gall wasps are insects from the family Cypnidae, order Hymenoptera. The female lays an egg through the long ovipositor into plant tissue, where a substance from the insect causes an abnormal outgrowth of the plant. The gall grows with the larva, which feeds on the plant tissue. The larva pupates within the gall. Oak is the host for many different gall wasp species giving rise to many diverse forms of galls as depicted in this figure. Interesting life stories can be told about some galls, e.g., the gall from *Amphilothrix sieboldi* (*top left*) produces a sweet secretion, which attracts ants that defend the gall against predators.

from a lacustrine University in Germany dived deeper into the Crater Lake and looked whether disruptive natural selection could explain the evolution of alternative habitat preferences in the cichlids of this small lake. They found that *A. citrinellus*, the form with a greater body height and the broader pharyngeal

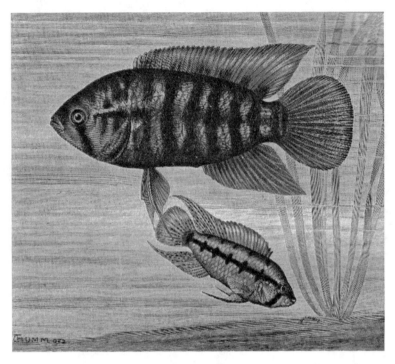

FIGURE 6.9. Cichlids are a species-rich fish family from the order Perciformes (here represented by the American *Cichlasoma facetum* at the *top* and *Heterogramma pleurotaenia* at the *bottom*). They are discussed as an example where food choice differences led to speciation.

jaw differed in habitat preference from the more slender *A. zaliosus*. The former is a benthic (bottom) forager, while the latter is living in the open water column (limnetic). Stomach content analysis indicated significant dietary differences. *A. citrinellus* feeds in decreasing order on the alga *Chara*, biofilm, insects, and other fish. *A. zaliosus* has a more restricted diet and biofilm dominated in its stomach content. These researchers suggested that sexual selection alone is unlikely to explain the speciation in cichlids from the African lakes if unaided by differentiation of the food space.

Behavior

Sharing Food and Other Goods: On Cheating and Altruism

Great Expectations with Capuchins

Biologists like to play with animals. Sometimes they find even an employer that pays them to play with animals and occasionally this playing provides deep insights into the human nature. Take the following scene where food choice becomes context dependent in a way that tells us that the "sense of fairness" is not an exclusively human trait. The plot is the following: The biologist gives capuchin monkeys (*Cebus apella*; Figure 6.10) token, which they exchange against food (in this example, a cucumber; Brosnan and de Waal 2003). They play the game. Then they introduced a second monkey, which also gets a cucumber against a token. No problem, both monkeys play the game. Then the second monkey gets for its token a grape, a higher appreciated food, or even worse they receive a grape without being obliged to return a token. The reaction of the first monkey can be violent: It might not only refuse to play the game, but it will throw the token or the cucumber out of the test chamber. Apparently, food takes another role—social context, fairness, and pride are now at stake. The animal refuses a food item, it would always have accepted when trading alone. However, now the human player has violated the fair trade social contract. Nonhumane primates apparently have a precise idea how food resources should

FIGURE 6.10. Capuchin monkeys, family Cebidae, are an example for the sense of justice in food sharing and for tooth wear of fruit-eating monkeys.

be divided in a social context. Interestingly, there was a strong gender difference: While female monkeys showed the described behavior, male capuchins were indifferent toward the interventions of the biologist, possibly suggesting an effect of hormones on the perception of social fairness.

To test the sense of fairness in capuchins, scientists constructed a special table, where food items could only be reached when both monkeys cooperated—they did this in 89% of the cases (de Waal and Berger 2000). In the next series of experiments, the effort of both animals was needed but only one animal was rewarded by food. The cooperation rate fell—as one could expect—to 39%. However, capuchins shared food spontaneously by dropping crumbs while the cooperating partner reached for the food. Significantly more pieces of food were shared after successful cooperation trials and helper capuchines pulled two to three times more often in cooperation trials if the preceding trial had been successful.

Deception with Chimps

In the view of biologists, humans are an unusually prosocial species—humans accept costs when helping congeners in one-way situations without expecting a direct reward. One might therefore expect a gradual increase in cooperative behavior and the concern for the welfare of others when moving from C. apella to higher nonhuman primates. For example, chimpanzees show cooperative hunting, food sharing, and coalitionary aggression (Figure 6.11). What about their sense of fairness when sharing food? To test that trait, anthropologists set out a simple but revealing test system (Silk et al. 2005). Chimps were trained to handle an apparatus to get a desired food item, and they were rarely making mistakes. When their own food was on stake, they chose the correct handle in 92% of the trials. To make the device instructive for the researcher, the apparatus had two options. The chimp could get the food only for itself or it could manipulate the system such that an unrelated, but long year cage mate would also get food. Importantly, the food came to the observing chimp without incurring an extra cost to the manipulating animal. The response was clear: In about half of the cases, the acting chimpanzee chose the food-sharing option, and the figure was exactly the same whether a second animal was present or not. To be sure, the outcome could have been worse. The acting chimpanzee could have followed an antisocial preference by avoiding food sharing in the presence of a second chimp. This was not the case. However, the chimpanzees were clearly indifferent to the welfare of the unrelated group member. The scientists assured that the first animal could see and hear the second, and the second understood the consequence of the action from the first animal as it displayed begging gestures. The sense of fairness has thus not evolved gradually in primates but is a derived property of the human species and might be linked to cultural learning and the evolution of moral thinking and the feeling of justice.

How strong this latter feeling is in humans was shown by Swiss researchers interested in the motivation of humans in economical interactions.

FIGURE 6.11. A juvenile chimpanzee (*Pan troglodytes*), man's closest relative, is explored for its tool use and culture development. The evolution of social behavior and food sharing is discussed with experiments conducted with chimpanzees.

Before switching to this interesting, although for biologists somewhat unconventional, reasoning, I should mention that German anthropologists detected a spontaneous inclination for altruistic helping in as young as 18-month-old children (Warneken and Tomasello 2006). The children helped adults unable to conduct a given action (e.g., open a door of a cabinet when the adult's hand were full). Notably, the children did help without hesitation—they reacted with an average latency of 5 s, needed no verbal request for help, eye contact, reward, or praise. These researchers from Leipzig were not convinced that altruism is a new evolutionary invention in humans. They argued that selfish reactions were provoked in the above-mentioned animal experiments because they involved food. Under natural conditions, chimpanzees have to compete over food, and the drive to acquire food for themselves might preclude their capacity to act on behalf of others. Therefore, they studied the reactions of young chimps in the same tasks as the infants. Notably, all chimps helped reliably in the tasks involving reaching. The animals needed a verbal encouragement to help and kept the object longer than infants before handing it over. In other tests, the chimps did not help, but the researchers noted that these were more demanding tasks. They concluded that children and chimpanzees are both willing to help but differ in their ability to interpret the other's need for help. The same group of Max Planck researchers conducted another revealing experiment with chimpanzees allowing insights into the evolution of cooperative behavior (Melis et al. 2006). In fact, according to their tests, their capacity for cooperation evolved already in the ancestor of chimpanzees and humans. I will skip the technically amusing details of this test. Briefly, a chimpanzee can try to solve a food-rewarded task either alone or in collaboration. Not surprisingly, they invited a collaborator significantly more often when they realized that the food-reaching task needed two actors compared to a situation where they could get the food in solo action. The fascinating story unfolds with the choice of the collaborator. When confronted with two potential collaborators for a task necessitating cooperation, the test animal chose a partner apparently on a random basis. However, chimpanzees differ in dexterity and insight into a given situation. There are good and poor collaborators. In cooperation with a poor collaborator, no food could be gained. The animals recalled this difference very well. In a repeat session, they regularly recruited the better collaborator. As this means better nutrition for both, the researchers argued for a strong selective pressure for good collaboration in animals that address food-seeking problems in teams. Actually, the strong human trend for collaboration between nonkin individuals might find here an evolutionary explanation.

Homo reciprocans

Sharing food became an issue for us because humans evolved in social groups. When man was hunting big game in the ice ages, the evolution of human cooperation was accelerated. Hunting animals as large as mammoth was a daring exercise and could only be achieved as teamwork. When the hunt was successful, a lot of meat was there to share. *Historians, biologists, social scientists, and economists have their own approaches how this sharing was settled. As a biologist, I reviewed mostly biological research literature, which treat humans somewhat as a naked ape, a usual professional deformation of biologists. Worse, I transferred conclusions from the biological context into the human society. To reverse this trend, I will now deliberately report on approaches from Swiss economists that have a bearing on the biological and philosophical "know yourself."* Their reasoning starts with the big game argument. Every member of the group benefits from the hunted meat and should participate at the hunt. Well, not all were physically fit to do so, restricting the hunting group to strong adult males. However, some were perhaps not inclined to do so. Hunting is dangerous, so isn't it more practical to keep a rear position during the hunt and fight for a front position during the sharing ceremony? What led people to participate in dangerous exercises? Biologists proposed the kin selection hypothesis where cooperation is the result of interactions between closely related individuals—the arguments of the selfish gene weighs less under this condition. Others pointed to reciprocity, a long-term tit-for-tat as the basis for this cooperation. Humans are like many animals very interested in a social position; part of the costly cooperation might derive from a social interest in reputation. You might also argue that unwritten laws developed quite early that punished free riders in such hunting operations. However, if you stay in a biological framework, punishment is costly for the punisher, an individual must identify the free rider and punish him running a personal risk in this conflict. The punishment hypothesis will only work when there are enough humans that are willing to do this as an altruistic punishment.

E. Fehr from an economy research group in Zurich and colleagues set out to look whether this behavior exists in human populations (Fehr and Gachter 2002). By lack of mammoths, he sent his Swiss students out to hunt for mammon. The details of their thoughtful money game do not interest here. In short, the students could invest in this game for the public good or cheat and reap a selfish benefit. Without punishment, it happened what happens everywhere. The students started with an average level of cooperation, and this level decreased sharply. Even smart students from Zurich are still the selfish naked apes. Then the game was repeated with punishment for the free rider. Yet, the punishment was costly to the person who did the punishment. The rules of the game were so that the student could not expect to meet the same players again, he could not count on reaping a future profit by educating his coplayer. As predicted by the authors, the level of cooperation increased to top levels. What are the mechanisms for this costly altruistic punishment? We are getting back to the capuchin monkeys—it is negative emotions, anger. The anger grew proportionally greater, the larger was the difference in the noninvestment of the cheater from the group norm. Remarkably, the cheater knew about these negative emotions and when asked about it, he anticipated even stronger averse reaction than actually occurred.

In fact, when biologists anticipate that the selfish gene theory comes from their field, they should read David Hume, who wrote in the eighteenth century: "every man ought to be supposed to be a knave and to have no other end, in all his actions, than his private interest." E. Fehr tells us that humans have a strong tendency for altruism if only in social punishment. This observation might explain some trends in human societies that are not easily understood from the selfish gene theory. It highlights why humans are so willing to punish those who violate social norms (Bowles and Gintis 2002).

On Human Altruism and Sanctions

As if he was feeling that his concept of the altruistic punishment as the fabric of human society will not flatter our self-perception, E. Fehr designed another economical game (Fehr and Rockenbach 2003). As food is not very motivating in an affluent society, real money was at stake, a quite obvious choice for an economist. In a clever game design, they played now investor and trustee. The investor had to trust the other person, but the trusted person was free to cheat. To be somewhat protected, the investor could impose in some games a fine for cheating trustees, but the investor could also voluntarily renounce on this punishment. They measured the back transfer of the trustee to the investor. Three major observations came out: (1) Overall, the money returned by the trustee was proportional to the money invested. (2) When a fine was imposed, less money was returned by the trustee than when no fine could be imposed. (3) When the investor renounced voluntarily on the fine, the highest amount of money was returned by the trustee. Apparently the trustee acknowledges being trusted. If it comes to the famous sticks and carrots, the carrots might be the better argument in human societies. Remarkably, the investor—once informed about the better feedback of the trusted trustee—did not change his behavior: The same percentage preferred to impose a fine for cheaters even when knowing that this reduces his return. Apparently, we are not simple automates that calculate the optimal result for our selfish interest—we have strong opinions on the moral legitimacy of sanctions even when they are against our selfish interest. Irrational human behavior that is against survival instinct like revenge or suicide bombing might have a root in this apparently deeply human feeling.

Empathy

Neurobiologist went to the neural basis of altruistic punishment (de Quervain et al. 2004) and empathy—the sharing of emotions, pain, and sensations with others (Singer et al. 2006). These neural responses were modulated by the perceived fairness of the other partner. In the experiments, the study subjects started a money game, where they could behave as fair or unfair players. In the next level, the subjects observed the fair or unfair players experiencing a painful punishment inflicted by the experimental system. The neuroscientist looked for the empathy reaction in the insular and cingulate cortex by neuroimaging methods. Less empathic activity was elicited in the brain of the observer when an unfair player was in pain. Then the neuroscientists looked into the reward center, the nucleus accumbens. People shared empathy with fair players but also liked the punishment of unfair players. Interestingly, men showed a stronger desire for revenge than women. There is thus a neural foundation for theories of social preferences and a predominant role for males in the maintenance of justice.

Kin-Altruism and "Food" Sharing

Of course, the field of food sharing has not been entirely left to economists and neurobiologists. The genetic basis for the evolution of social behavior has been studied by W. Hamilton in the 1960s, and now, also economists ask genetic questions ("how hungry is the selfish gene?"). The inclusive fitness model predicts that people favor those to whom they are most closely related. In humans, the model has explicit predictions for commonly observed behaviors like within-household violence, allocation of food, or childcare. Anthropologists have investigated food sharing in tribesmen from South America, and the effects of kin-altruism appeared to be modest. References for the above statements can be found in Bowles and Posel (2005). These authors had the idea to submit the predictions of the inclusive fitness model to a rigorous statistical test. They collected data on the remittances sent by South African migrant workers to their rural households of origin. They collected many further data that allowed an evaluation of the rural households with respect to the composition of the household and degree of relatedness with the sender of the remittance. Since the predictions of the model do not depend only on the degree of relatedness but on the age, sex, and health and other characteristics relevant to the so-called reproductive value of the beneficiary, they also collected these data. The model is rather complex: The sender should give more to his adolescent children instead of his infants (high infant mortality will prevent many from reaching reproductive age) or his parents (they have no reproductive activity). He should give more when at the end than at the beginning of his reproductive years. The authors accounted also for the foundation of secondary households at the place of migrant work. Overall, the effect of kinship was modest. Less than a third of the variation in the remittances could be explained by the applied kin-altruism model. The only observation sticking out was the wife's presence in the household, which resulted in 45% increase over the mean remittance. As the wife is genetically not related to the worker, the interpretation is ambiguous. There might be selfish-gene reasons (care of the migrant's children, expected future reproductive success with his wife) or altruistic motives toward nonkin, documented in behavioral experiments (Fehr and Fischbacher 2003).

Food Help and Social Queues

Kin selection for explaining eusocial animal behavior came also under criticism from zoologists working with Stenogastrinae hover wasps. These hymenopteran insects live in groups of up to 10 females. A single rank 1 female lays nearly all the eggs but rarely leaves the nest. Helper females collect food to feed the larvae of the dominant female. The problem is twofold: There is substantial variation in the help provided by the assisting females, and only 10% is explained by variation in relatedness. Kin selection is here apparently not the key to the understanding of help. The system is accessible to a quantitative analysis because there is a strict age ranking in the females with promotion of the next oldest female in the queue when the top female has died (Field et al. 2006). The help could be easily measured as the time spent outside of the nest. Notably, the researchers observed that the helpers worked harder when they were of lower rank. The helping effort dropped dramatically when a female was lifted from rank 3 to rank 2. They suggested an interesting explanation. Helpers will gain indirect fitness benefits through aiding natal nest-mates. If you are low on the social ladder, helping is your best option. However, the situation is different for a rank 2 female. She can hope for laying eggs herself and reaping the full genetic benefit. Yet, she must survive to achieve this position. Foraging is dangerous, you might meet predators and pathogens and that will negatively affect your survival probability. Staying home is then the safer bet. Furthermore, helpers should work less hard when the group of helpers increases because this is like rising in the rank—a large work force promises a larger survival of larvae should the individual become the first ranking female. The behavior of animals in experimentally manipulated groups where ranks were shifted and where the workforce was increased, concurred with the predictions of their model.

On Trust, the Amygdala and Oxytocin

Trust in other people has a clear physiological basis. Patients with bilateral damage to the amygdala showed marked effects. They rated people as trustworthy from their facial expressions, who were rated as untrustworthy by a control population. This defect was restricted to the visual interpretation of facial expressions and did not extend to verbal evaluation of people when short biographies of different characters were presented to them (Adolphs et al. 1998). E. Fehr added a new facet to his research when teaming up with clinical psychologists. They repeated the investor–trustee game, but half of the subjects were now treated with a nasal sniff of the peptide hormone oxytocin, best known for its physiological action in milk letdown and labor. This treatment significantly increased the investments of the donor when he was aware that he played with humans, but not when playing with a computer program (Kosfeld et al. 2005). Apparently, oxytocin rendered subjects more optimistic about the likelihood of a good outcome in social interactions by inhibiting the defense behaviors.

Evolution of Eating Strategies from First Principles

This chapter has made a wide swing around the quest for food topic, using approaches coming from different nonbiological disciplines. I would like to still add a pure mathematic analysis (Burtsev and Turchin 2006), which transgresses the constraints of game theory. The latter is characterized by a simple structure of the payoffs and only a small number of possible strategies. The model divides the world into patches that contain food resources or are empty. The eating agents can rest, eat the resource (but can store only a fixed maximal amount of energy), divide, move to the next patch (when the resource in the initial one is empty or consumed), or attack other agents in a neighboring patch. An agent that has exhausted its reserves dies. The number of possible behavioral strategies is extremely large, but in a simple run of the model only those emerged that were already determined by J. Maynard Smith using the theory of games in the 1970s. These were doves that never attack and flee when attacked; the hawks attacked other agents, and finally the bourgeois, which stays in the patch, attacks invaders, and does not care about neighbors. If the carrying capacity is large, the bourgeois strategy dominates. However, below a threshold, only doves and hawks compete. If the agents are now allowed to evolve cooperative strategies, new populations emerged: cooperative doves, ravens (predators recognize and do not attack in-group members), and starlings (which mob large intruding invaders). The researchers claim that their computer simulations have implications for the evolution of territoriality in animals (and private property in humans) but also admit limitations on their model. One strategy, well known in zoology, namely the cooperative attack (wolf strategy), did not evolve in their simulation. In a future extended model, they want to allow for horizontal ("cultural") and not only vertical ("genetic") transmission of traits to check whether this allows the evolution of human ultrasociality, i.e., cooperation between genetically unrelated individuals, so characteristic for humans in the animal kingdom.

Altruism or Personalized Nutrition in a Social Mammal

I will end the chapter with a peculiar eating culture in a eusocial animal. In eusocial organisms, there is division of labor between morphologically distinct castes. These species show communal breeding where "queens" are responsible for producing all the offspring, while workers forego reproduction and assist others to reproduce. Eusociality has interested biologists because it represents an extreme form of altruism in biology; it is best known from insects. In mammals only two strange species of African mole rats show this social structure (Figure 6.12). These are nearly hairless and sightless subterranean animals that disperse by extending their burrow system. In *Cryptomys* three types of animals were observed: the queen and the frequent workers, which perform more than 95% of the daily work performed by the colony, and finally there are the "lazy" types, which the zoologists called respectfully the "infrequent" worker (Scantlebury et al. 2006). In fact, the infrequent workers contribute between 25 and 40% of the individuals of the colony but perform less than 5% of the total work. The infrequent workers were heavier than the frequent workers and showed both higher amounts of fat and muscle mass. The likely reason for these two nutritional classes of workers became clear after heavy rainfall. Rainwater softened the soil and now allowed prospective forays by the infrequent workers. They now invest their fat reserves to tunnel through the earth in search for other colonies of mole rats. The heavier and lazier individuals are thus the reproductive form of the animals in the colony, and they need their fat stores to meet the high energy demands to find a partner without getting above ground where they would be helpless against predators.

Communicating on Food

Honeybee's Waggle Dance

One of the most complex systems in animal communication is honeybee's waggle dance. The dance resembles a miniaturized reenactment of the flight to the food source. The direction and distance to the food source are encoded by the elements of the dance. Karl von Frisch made his first observations on the

FIGURE 6.12. Mammals are defined by the possession of hairs, but the naked mole rat, a burrowing mammal from the plains and deserts of Africa, mostly lacks hairs. The figure shows a species from Somalia, *Fornarina phillipsi*, a close relative of the better known *Heterocephalus*. These animals nearly lack eyes and ears and dig tunnels by their front claws and incisors. They feed mainly on roots and bulbs of plants. This animal is also a eusocial mammal that has evolved nutritional castes.

FIGURE 6.13. Bees (Apidae) come in different forms. The most primitive are the Proapina (*1*) here represented by *Prosopis*, they forage only on flowers with easily accessible nectar. The next group is Prodilegina with collecting hairs (S) on the hindlegs, here represented by *Dasypoda* (*2*) and then Gastrilegina with a collecting brush under the belly, here depicted with *Megachile* (*3*). These are all solitary bees.

bee (Figures 6.13 and 6.14) language in 1921, but it took him decades before he could decode the information for a human observer. For example, the direction of the waggle run corresponds to the angle of heading to the food source relative to the current sun azimuth and is performed relative to a sensory reference. Sounds complicated? I can reassure you, not only for you: Even for the bees,

FIGURE 6.14. Social bees are bumblebees (*Bombinae*) and of course the honeybee, *Apis mellifica*, here represented by a male (*1*), a queen (*2*), and a worker (*3*). Their legs are magnified.

this behavior is rather difficult to learn, and the bee recruits might need more than an hour to decide what direction to take. The bees have to follow the successful forager-dancer to learn this information. They thereby also learn the odor of the food.

Bumblebees

Bumblebees have a simpler information system in place that is also fairly efficient. The forager returns to the nest and makes minutes-long excited runs across the nest, frequently bumping into other bumblebees, and it distributes the odor by buzzing the wings. The nest mates get aroused by this action and leave the hive in large numbers. In field experiments, they seek the odor of those flowers brought into the nest by the forager. Biologists believe that this is the evolutionary origin of the complex behavior in honeybees (Dornhaus and Chittka 1999). The rougher bumblebees are in many other aspects a useful reference for the more sophisticated behavior of the honeybee. In color-discriminating task (only one artificial flower offered sucrose, the others nothing), bumblebees showed a continuum of behavior. Some were making careful and correct decisions for the right rewarding color, but this took time. Other bumblebees opted for a quick and dirty strategy and visited any flower since the time to check a wrong source was anyway short. However, they could do better. When the landing on the wrong artificial flower was linked to a punishment (they were penalized with bitter quinine), also these looser bumblebees showed an increased discrimination capacity but at the expense of longer response times (Chittka et al. 2003).

Fitness Through Dancing?

Bees are more sophisticated in their food search behavior and must have an impressive cognitive capacity to work through all the information necessary for successful foraging. However, recently many biologists were rather critical about the basic tenets of the discoveries of K.v. Frisch. In one recent report, they asked whether honeybee colonies really achieve a higher ecological fitness through dancing. To test for that correlation, they disturbed the bees by alternatively offering diffuse light and orienting light. In one case, the dances were disoriented, and the bees could not transmit precise information on food location to the recruited bees. As an outcome parameter, they measured the mass of the hive serving as an indicator of foraging success. During the periods with oriented dancing, the colonies showed an overall greater amount of collected food, but this difference did not manifest in all seasons. In the autumn, no difference was associated with oriented dancing (both colony types lost weight); in the summer, both gained weight, but the oriented dancers gained somewhat more. Only in the winter (the study site was California), a marked effect was found, the colonies with the disoriented dancers lost weight, while the oriented dancers gained weight. Apparently, the waggle dance is ecologically important only when the food sources are hard to find or variable in richness (Sherman and Visscher 2002).

Precision of Communication

When these critical questions were asked, other followed. What is actually the precision of the conveyed information? If the coding of honeybees were really as precise as described by K.v. Frisch, it would be the most sophisticated nonprimate communication system in the animal kingdom. Doubts came up because recruits had to go through different dance sessions to get the message, some hesitated, still others did not find the food source at all. In a recent report, the biologists captured the recruited bees as they left the hive and attached them a harmonic transponder such that the researchers could track their paths. The scout bees had to convey the information that the artificial feeder was at right angle to the coordinates defined by the Sun and at 200-m distance. Overall, the recruited foragers performed very well. All but three of the electronically marked 19 bees arrived at the feeder, although some arrived just a few meters aside without finding the feeder and then returned without food to the hive. Despite some side wind, the bees kept very precisely the indicated angle and deviated only when trespassing the set 200-m mark. In these experiments, one could still argue that the bees used additional cues like the odor marks of the dancers for orientation. To exclude that possibility, the researchers displaced the hive and observed that the recruited bees followed as precisely the transmitted path information from the new release point as from the original start point (Riley et al. 2005). However, other researcher could trick out honeybees by sending them into a narrow tunnel for foraging. The scouts transmitted then a too long distance for food location to the recruited bees. Apparently, the bees calculate distance according to retinal image flow that they experience during the search. Their brain misreads the distance because the close tunnel walls increase the perceived optic flow (Esch et al. 2001). These experiments do not refute the precision navigation system of bees but reveal the physical parameters that the bees use for their distance counting.

Teaching

The ant *Temnothorax* has developed a way to teach its nest mates the way to a food source by a process called tandem running (Franks and Richardson 2006). The teaching leader ant goes ahead followed by the naïve learning ant. The teacher continues to run only when touched on its legs and abdomen by the antennae of the learning ant. The run is frequently interrupted by stops where the learner ant walks around probably memorizing landmarks—the teacher waits patiently. During the tandem run, both ants coordinate their speed: If the gap between the ants gets larger such that the physical contact is interrupted, the teacher slows down and the learner speeds up to remain in contact. Tandem running has no advantage for the teacher— actually it is fourfold slowed down in its foraging trip—but the learner reaches food in a significantly shorter time period. The gain is 30% acceleration in time. The exercise becomes cost-efficient for the nest because the behavior is "contagious." The former pupil, once it had detected the food source, runs back to the nest and becomes a teacher in turn. In this way, these ants propagate time-saving knowledge among foragers. Tandem running is thus superior to the strategy where workers from the same ant species carry nest mates to the food source. The carried ant does not become a teacher probably because it was transported with the head pointing rearward, and it was thus prevented from memorizing landmarks. Other ant species carry their nest mates with the head looking forward, and these carried ants become then carriers for the next run. Apparently, a big brain is not a requisite for teaching to occur in animals.

Decision Processes

Orientation by waggle dance plays a role in still another situation of the life history of the beehive—when the swarm leaves the old hive and searches for a new site. The swarm then sends the scouts out, which explore the environment searching a good site for a new hive in a promising food surrounding. The scouts dance again a waggle dance, but it follows different rules. The question is no longer to convey precise information on a food source, now competing places are proposed by different scouts. In house hunting, the number of waggle runs, which is initially proportional to the perceived quality of the site, declines with each successive dance. Other bees follow the dance of the scouts, some check it physically, and then the popularity of the different proposals is assessed by attrition. The dance with the most persistent scouts wins, and this is usually the highest quality site (Visscher 2003). This subject touches an interesting current research area about decision-making in animal groups on the move. This can be migrating grazing mammals in search of the most promising grassland, schools of fish, large groups of insects or birds. The group has now a difficult point to settle: How do uninformed individuals recognize those that are informed, and how do groups come to a collective decision (Couzin et al. 2005)? It is also a question of leadership since mathematical models showed that only a very small proportion of informed individuals are required to achieve great accuracy in the prediction as observed in the migrating beehive example. The mathematical model showed

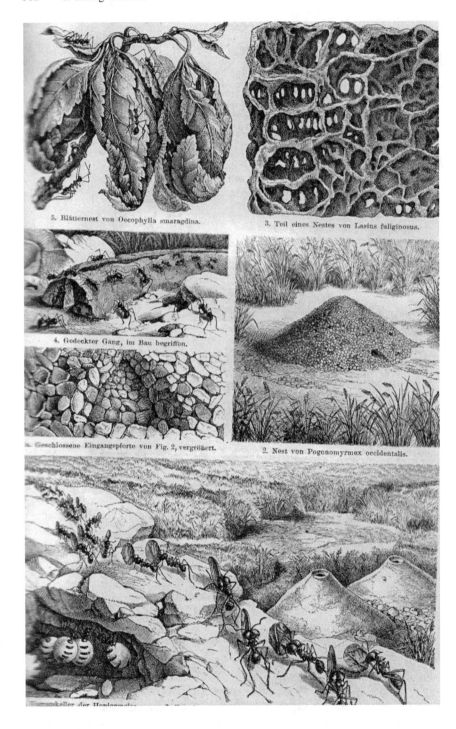

5. Blätternest von Oecophylla smaragdina.

3. Teil eines Nestes von Lasius fuliginosus.

4. Gedeckter Gang, im Bau begriffen.

a. Geschlossene Eingangspforte von Fig. 2, vergrößert.

2. Nest von Pogonomyrmex occidentalis.

Honigkeller der Honigameise.

that successful group foraging needs only limited cognitive ability. Biologists differentiate two basic decision modes: One is the despotic mode, where the decision is made by the most experienced group member. Again, mathematical treatment of the process tells us that this process only pays when the group size is small and the difference in information is large. In all other cases, they tell us that the second mode, the democratic mode, is more beneficial. The benefit derives from the property of the system to avoid extreme decisions, "democracy" in animals is a type of insurance system. It is not likely that each individual has the possibility to influence the decision process by a personal voting (Conradt and Roper 2003). The stage is mathematically prepared, and the ground is now free for the experimentalists to test the predictions.

Animal Technology

The Invention of Agriculture: Fungal Gardens of Ants

A Riddle

If you hear agriculture, your first idea might be the neolithical revolution by Homo sapiens about 10,000 years ago, but this figure is not tenable any longer— a better guess is 50,000,000 years ago. If you count, slightly astonished, the digits of this number, you will realize that this figure reaches back into the time just after the demise of the dinosaurs. Who has taken the step from hunting to targeted food production that merits the name of agriculture? If I wanted to push you on a wrong track, I would say the animals have the size of a cow and also consume daily an amount of leaves that the cow is getting as feed. If you try now to check vertebrates for this capacity, then you are definitely on a wrong track. Of course, I voluntarily misguided you: When I said the weight of a cow, I referred to the biomass of a single colony of leaf-cutter ants from the American tropics, which counts several millions of animals.

History of Leaf-Cutting Ants

The fungal gardens of these Central and South American ants became a showcase for an evolution-oriented ecological research that created new paradigms for the theory of symbiotic interactions (Figures 6.15 and 6.16). The

FIGURE 6.15. As social insects, ants build a variety of nests. In forests the common European ant *Formica rufra* constructs characteristic ant-heaps with many entrances (Figure 6.16, 8 shows a cut through the colony with brood chambers), while the agricultural Texan ant *Pogonomyrmex barbatus* builds a hillock with a central opening at the top (bottom, right), *P. occidentalis* covers the hillock with small stones and even closes the entrance (center, right). The nest of *Lasius fulginosius* is built from chewed wood (top, right), those of *Oecophylla smaragdina* are sewn from leaves (*top left*) using their silk-excreting larvae as sewing needles. Along their foraging ways, leaf cutter ants keep lanes free of grass (bottom, center), some of these lanes can even be covered (center, top left).

8. Stück des Nestes der Waldameise (Formica rufa) im Längsschnitt.

10 Interimsbauten von Tapinoma erraticum.

9. Blattlausställe von Lasius.

12. Kleiner Keu- 13. Großer Keu- 14. Ecitonilla 15. Thorictus Fo- 16. Xenoce- 17. Oecodoma cephalo- 18. Wanderameise
.enkäfer (Clavi- lenkäfer (Paus- claviventris. reli auf Myrmeco- phalus trilo- tus, Arbeiter.   (Anomma),
ger testaceus).   sus Favieri).   cystus.   bita.   Arbeiter.

inquiry into this fascinating system is also a good illustration as to how science is stepwise unfolding the secrets of nature. The leaf-cutters (Figure 6.17) are so characteristic when they come back from their foraging to the colony, each charged with a leaf and one after the other, that they attracted quite early the attention of human observers. Their earliest mentioning is in the Popul Vuh of the Mayan civilization in the nonscientific language of a creation myth. The first but wrong scientific conjecture was made by the British naturalist Bates. He proposed in 1863 that the leaves were used to protect the broods in the nest against the tropical rain. Ten years later, the engineer Thomas Belt got it right when he stated that the leaf-cutter ants use "leaves on which they grow a minute species of fungus, on which they feed. The ants are mushroom growers and eaters" (quoted from Schultz 1999). As frequently in science, the breakthrough discovery was done twice in the same year. The other person was a professional biologist, Fritz Müller, who described his discovery in a very early issue of the science journal *Nature*. After them several generations of zoologists have worked on ants, and entire books have been dedicated to ants, social insects, and gardening ants. Research on ants led also to the conception of an entire new branch in biology: sociobiology. I will not retrace this history, interesting as it is, but try to convince you that what these ants are doing is really an agricultural activity. After that outline, I will illustrate current research around this fascinating symbiosis system that got more and more trophic levels over the years.

Gardening Activity

First to the natural history of leaf-cutting ants (Currie 2001): When a foundress queen leaves the old nest, she takes care to carry a small clump of the fungus stored in a type of a mouth pocket on her nuptial flight. Once inseminated, the queen digs a small chamber, spits out the fungus and starts tending for it. She either uses her fecal fluids as manure or forages for substrate. Then she rears the first generation of workers, which then take over the task of tending the garden and foraging for the garden substrate. When they come back to the colony with a leaf, the ants lick the leaves to remove the wax layer that the plant has evolved to protect its leaves from fungal attack. Then follows a mastication, which cuts leaves to 2-mm pieces. This is supposed to remove the leaf-specific microbiota.

FIGURE 6.16. Various ants produce special breeding devices, e.g., *Tapinoma erraticum* build elongated structures along grasses and use them as brood chambers for their eggs (center, left). Still other ants like *Lasius flavius* construct shelters for their milking cows, the aphids (top, right; bottom, left). Ant-heaps are visited by beetles, some are tended by ants (center, left two animals), others are living on ants (center, fourth from left), still others display ant forms to mix into the colony (center, third from left). Some ants associate with plants, which they protect against leaf-cutter ants or herbivorous animals. On these "ant plants" the ants form air-borne nests. Some birds prefer ant plants also for their own nest-building activity, they probably profit from the protection conferred by the ants defending their trees (bottom, right).

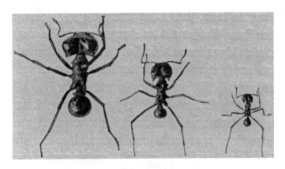

FIGURE 6.17. Leaf-cutting ants—here three differently sized workers from *Atta cephalotes*, family Formicidae, order Hymenoptera—do not eat leaves, but feed at home fungi with the plant material. They became thus inventors of gardening activities and had to solve many problems associated with biological food processing.

The resulting pulp is then enriched with a fecal droplet to provide some enzymes. The pulp is then posed on the top of the fungal garden, and it takes 6 weeks for the complete decomposition process. During that time, fresh material is continuously added on the top and the spent material is regularly removed from the bottom of the fungal garden. The workers take great care with the spent material, which is deposited in real refuse heaps at some distance from the colony. The ants have some gardening activity; they move around, eat parts of the fungus, and distribute the fungal enzymes with their fecal pellets over the culture. The ants prune the fungus, they open and close tunnels to the surface to achieve optimal humidity and temperature for the decomposition. The ants neither eat the leaves nor the fungus, but the gongylidia—special nutrient-rich hyphal tips that sprout out of the fungal biomass. Like a riddle in a fairy tail, the ants are thus the dominant herbivore of the Neotropics without eating a single leaf. The entomologist Wilson marveled that leaf-cutting ants have such "an efficient utilization of almost all forms of fresh vegetation that their invention can be properly called as one of the major breakthroughs in animal evolution." Individual colonies survive for 10 years. I think we can agree that this is true agriculture.

Diversity in Ants and the Associated Fungi

However, not all what glitters is gold—this description of early agriculture is overoptimistic. Recent research has revealed that life for leaf-cutter ants is not that easy. I will try to retrace the major steps of research in gardening ants over the last 10 years. Major efforts were invested in the molecular phylogeny of the fungus-growing ants. The *Attini* tribe is thought to have originated 50 million years ago, and the separation between the *Cyphomyrex* genus, which uses insect feces and corpses as gardening substrate, and *Trachymyrex* genus, which uses dead vegetative matter as food substrate, is older than 25 million years and backed by molecular data and animals included in amber (Hinkle et al. 1994). Only the higher attines (*Acromyrmex* and *Atta*) use fresh leaves and flowers of many plants as nutritional support for their gardens. The congruent evolutionary relationships of the higher ants and their fungal symbionts initially suggested cospeciation and asexual clonal propagation of the fungi for millions of years. This simple picture got cracks when a larger set of ants and their associated fungal gardens were investigated. The fungi are not of a monophyletic origin but came in three groups. Two belong to *Lepiotaceae*. Relatives of this group of mushrooms are also appreciated by human gourmets—a cousin is *Agaricus campestris* (the "champignon de Paris" of the French restaurants) and greater brother is *Macrolepiota procera*, a fine mushroom with a nut flavor. A third group of fungi from ant garden belongs to the *Apterostigma* group, which are distant relatives of the *Tricholoma* mushrooms (containing some comestible, but also some poisonous mushrooms). Particularly between the primitive ants and their fungi, numerous topological incongruency was detected on the respective phylogenetic trees suggesting horizontal transfer of fungi across the ant species and the secondary acquisition of fungi from a pool of free-living fungi (Chapela et al. 1994). Subsequent work showed that any single attine ant nest contains only a single fungal cultivar, whereas nests of the same ant species may contain distantly related cultivars even if they were only separated by a few centimeters. The same fungi might also be cultivated by distantly related ants (Mueller et al. 1998). These data refuted the long-standing model of fungal clonality spanning million of years, and this model was already questioned on theoretical grounds because asexuality should pose problems with the accumulation of deleterious mutations (Muller's ratchet argument). In fact, after accidental loss of the cultivar, ants are apparently forced to search replacements from neighboring colonies. Even if this "borrowed" fungus does not match their original clone, incompatibility is quickly overcome.

From Two to Three Players: *Escovopsis*

But what are these accidents that lead to the collapse of the fungal gardens? Fungus-growing ants maintain axenic (single species) "monocultures" of fungi, despite the continuous exposure of the substrate to microbes adapted to the plant material, which are passively carried with the leaves into the nest. Apparently, ants arrive to weed out their gardens from alien microbes. Humans have great difficulty to maintain monocultures of genetically homogeneous crops—parasites have frequently a devastating effect on these crops. How do ants succeed to keep their gardens pathogen-free? The answer is simple: They don't. When Cameron Currie and colleagues sampled thousands of fungal gardens from Panama, about 40% of the gardens yielded nonmutualistic fungi. Notably, the nonmutualists were dominated by a single fungus, *Escovopsis*. Nonmutualist is actually a euphemism, *Escovopsis* is a highly virulent pathogen of the fungal garden. After intentional infection, the fungal garden collapses within days. They are so radical that they leave no apparent microscopic evidence of the mutualistic fungi. *Escovopsis* is highly specialized to fungal garden and cannot be isolated from the environment. It is apparently horizontally transmitted from colony to colony, although it is not imported by the founding queen (Currie, Mueller et al. 1999). Phylogenetic studies demonstrated that *Escovopsis* is monophyletic and comes from a dubious parent company, which includes parasites of commercially cultivated mushrooms and the famous cereal fungal parasite *Claviceps purpurea*. At ancient levels, the phylogenies of the three organisms ant/mutualistic fungus/parasitic fungus are perfectly congruent, suggesting that what was initially described as an ant–fungus symbiosis is actually a tripartite coevolution. At recent evolution levels, frequent host switches were observed demonstrating traits of a fierce arms race (Currie et al. 2003).

A Fourth Player: *Streptomyces*

There is a German word saying the deeper you get into a forest, the more trees you are seeing. This sounds pretty trivial, but this means that you need to concentrate if you want to find your way out of this forest. This applies not only to forests of the brothers Grimm but even to the much smaller fungal garden. The Smithonian outpost at Panama added a new player to the garden party. To keep the symmetry, the turn is now at the helper side. Behind the curtain are *Streptomyces* bacteria. Actually, they do not stay behind the scene, they are found in a long-overlooked powdery whitish crust under the thorax of ants, in other ants they are found under the forelegs. While the place of association was variable, they were regularly associated with all groups of attine ants underlining their ecological importance. *Streptomyces* is actually a good choice for the ants, because it is the most notorious of all bacteria with respect to antibiotic production. In fact, most antibiotics developed for human pharmaceutical use are actinomycete (to which *Streptomyces* belongs) metabolites. *You see the humiliation of human pride after Copernicus is a never-ending story. First, I had to tell you that agriculture was not a human invention, and, now, leaf-cutter ants are also the inventors of a grass roots pharmaceutical industry.*

Streptomyces associated with attine ants are doing what they are expected to do. They do not produce antibiotics against saprophytic or entomopathogenic or other common fungi but show exclusive inhibitory activity against *Escovopis*. At the same time, it produces growth-promoting compounds for the mutualistic fungi (Currie, Scott et al. 1999). We have here the fourth partner in this ant symbiosis, which makes this system one of the most complex symbiotic associations discovered in nature.

If you have such a valuable antibiotic producer, ants have all interest to treat them carefully. This is indeed the case. The bacteria, which belong to a single genus, *Pseudonocardia*, are associated with all attine ant species examined and occur on each species in a given location on the cuticle. Females carry them away during their mating flights to transmit them to the offspring colonies. The location of the bacteria differs between ants: The paleo-attines, as the name indicates the oldest phylogenetic group of these gardening ants, have them under the forelegs. The "higher" attines have them under the propleural plates on the trunk. Cameron Currie and colleagues recently took a closer look on these propleura and carefully removed these filamentous bacteria. Below this filth, they detected crescent-shaped cavities. The ants have developed homes for their mutualistic bacteria to house them. Not enough with that, the cavities are underlain with an exocrine gland, and it seems that the ant feeds the bacteria by its glandular secretions. Actually, some worker ants have cavity openings essentially over the entire cuticle. A phylogenetic analysis demonstrated that the mutualistic association is of an early evolutionary origin.

As we live currently an antibiotic resistance crisis in medicine, this observation raises also the intriguing question as to why the parasitic fungus has over these long evolutionary periods not evolved a resistance against the antibiotic produced by the filamentous bacterium (Currie et al. 2006).

Why a Fungal Monoculture?

Recent research has added still further layers of complexity to the system. When ants weed their garden, they apparently do not remove only the parasitic fungi but also the mutualistic fungi that happen to contaminate their garden. Why do they do that? Wouldn't it be in the interest of the ants to abandon the monoculture to be more stable in case of parasite attack like humans are doing (Zhu et al. 2000)? *A possible answer to this paradox is that ant's weeding is actually an extended phenotype à la Dawkins of the symbiont, which does not like the idea to share the food resources with a competitor. This is a sound selfish instinct, which governs all of biology.* If garden fungi are brought together, they show incompatibility reactions, which are proportional to their genetic distance. All fungus-growing ants manure newly grown mycelia with their own feces. As the ants feed on the resident fungus, the latter can use the ants for spreading incompatibility factors via its digestion. In fact, these factors survive the gastric passage, and ants are now used as a mobile spraying unit for incompatibility factors whose molecular nature was not yet defined (Poulsen and Boomsma 2005).

On Inventions

Ants are clever animals, but a good idea in biology runs the risk to be copied. We should therefore not be surprised to hear about fungus-growing termites. This insect-fungus symbiosis has a single African origin. In contrast to the ant system, in termites no secondary domestication of other fungi than those belonging to the *Termitomyces* group has occurred, which do not exist any longer as free-living species (Aanen et al. 2002).

Some biologists think that ants are especially inventive animals and recent research showed that you can also quote them as the inventors of group hunting, which developed traps. This strategy is seen in Amazonian arboreal ant *Allomerus*. They live in leaf pouches of the ant-plant Hirtella. They cut hairs from the stem of the host plant and stitch them together by the mycelium of a purpose-grown fungus building a type of gallery with numerous holes. Workers hide within the gallery and wait for a large prey landing on the gallery in ambush position with their mandibles wide open. If a locust has landed, the first wave of ant aggressors grasps legs, wings, and antenna of the prey through these holes. They then stretch the prey on the gallery such that the next waves of ants can sting it to death (Dejean et al. 2005).

Ant Plants

I will finish our excursion into the fascinating world of ants with plants that use ants for their purpose (Figure 6.18). If you want to use somebody in biology,

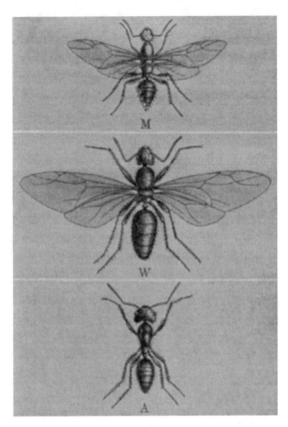

FIGURE 6.18. Class society in ants: male (M), female (W), and sexually not active worker (A) from the European ant *Camponotus ligniperda*. Its American cousin *C. femoratus* is known for the construction of ant gardens.

you need a shrewd strategy of exploitation or you must offer something for the service provided by a partner. Plants have learned to use insects and their offer is nectar produced in their flowers. With that free meal trick, they attract insect or bird pollinators that take over an important function in the propagation strategy of plants. Some plants also produce extrafloral nectar on vegetative parts, and myrmecophytes (in plain English "ant-plants") house and nourish specialized ant colonies on their leaves. In turn ants defend then their hosts against herbivores. The deal in this mutualistic relationship is fair and clear: food for defense. But there is a problem: What about cheater ants that take the nectar but do not show any inclination to defend the plant? In fact, there are enough nonsymbiotic ants foraging in the vegetation to spoil the tit-for-tat game. Researchers from the Max Planck Institute of Chemical Ecology in Jena found an interesting cue. Mutualistic ants preferred the nectar from "their" *Acacia* plants, while nonsymbiotic ants preferred the nectar of other plants. How was that achieved? Ant plants produce the disaccharide sucrose into the nectar as normal plants do, but they produce into the secretion also the enzyme invertase (also known as sucrase) that hydrolyses sucrose into glucose and fructose. Nonsymbiotic ants prefer sucrose over monosaccharides, and the converse is true for the symbiotic ants. This preference pattern forms an efficient filter against exploitation by cheater ants. However, why should the symbiotic ants prefer the monosaccharides? The reason became quickly clear: The symbionts, in contrast to the nonsymbionts, lack a sucrase activity in their intestines. As disaccharides cannot be absorbed by the intestinal epithelia, the symbiotic ants become tied to "their" host. This nutritional trick stabilizes the mutualistic relationship, and we have here a fascinating case of specific coevolution (Heil et al. 2005).

Tool Use and Caches in Crows

From New Caledonia to Old England: Corvus Betty

Traditionally, the manufacture and use of tools were seen as a specific attribute of humanity. Ethologists told us that we have to share this property with apes. Since apes are our cousins, we were not too much upset that our uniqueness was violated. However, then came bird watchers with surprising observation that question some of our prejudices on tool use, culture, and animal intelligence. The story started with G. Hunt, an eyewitness of tool use in the New Caledonian crow *Corvus moneduloides*. When this resourceful bird searches for prey in the holes of dead wood, it uses twigs with a hook. The tools were made on purpose and carefully carried during foraging. They placed the tool under their feet when on a perch and in the bill when flying. These tools are used many times, and they are made with a shrewd technique. The birds removed secondary twigs along the primary twig, which then becomes the hook. They spend several minutes to remove the leaves and the bark to get a neat hooked stick. These crows used another tool made from the tough leaves of the plant *Pandanus* forming a stepped-cut tool. This pointed, but sturdy leaf with rigid barbs along the uncut end was held longways in the bill when foraging for spiders, millipedes, and alike under detritus (Hunt 1996). The researcher returned to Grand Terre, the only island where wild crows were observed to manufacture tools of this great complexity. He observed a population-wide handedness (or should one better say footedness in birds) in toolmaking. This laterality in tool manufacture, which was independent of ecological factors, speaks in favor of the involvement of a neural program in this process. It was compared to the right-handedness in humans that may be a consequence of the evolution of language, which favors the left hemisphere (Hunt et al. 2001). Later observations confirmed that tool use was not a cultural trait linked to this island like in some monkeys. In the latter, the classical case is that of macaques from a tiny Japanese island, where one juvenile female invented 50 years ago the washing of sweet potatoes in the river to remove the dirt. Since then the potato washing has spread through the population and passed as a tradition to the next generations (de Waal 1999). That this process of toolmaking can be spontaneously learned was shown with a clever Caledonian crow called Betty. In her Oxford captivity, she learned to bend wires to retrieve food that she could not reach otherwise. Betty did the toolmaking in a minute and got the food within 2 min (Weir et al. 2002). The Oxford zoologists wanted then to settle whether toolmaking is a hard wired property of the bird's brain or whether it had to be learned from adult birds. For this purpose, they raised birds from the egg, half of them got teaching lessons in toolmaking from the zoologist, which they attended with apparent interest. The other half got no training. Food retrieval with hooked twigs was observed in both groups of animals at about 70 days of age. From Kew Botanical Garden, the researchers got also *Pandanus* leaves, which they used also as tools, although none fashioned the tools as their relatives from Grande Terre. While tool use is thus genetically acquired, the special tool form might still be influenced by social factors (Kenward et al. 2005).

Avian Tricks

Corvids are probably the most intelligent birds and are only rivaled by parrots and within mammals by monkeys and apes. In many respects, one can distinguish them as rule learners from rote learners like pigeons. They solve many tasks easily. One such task is the pulling of a piece of meat attached to a string and hanging from a perch. The raven has to pull the string with the beak, fix it in the new position with the feet, and pull again until it can reach the prey. It does not get mixed up when alternatives are offered like additional strings or stones fixed to the string or a piece of meat so heavy that he judges it useless to try it are offered (quoted from Emery and Clayton 2004).

  Corvids are exceptional but not the only case of tool use in birds. Herons float feathers as tools to attract fish, woodpeckers use small sticks, and the burrowing owl *Athene* waits silently next to its nest when it has placed mammalian dung around its burrow to entice its prey. Indeed, the dung beetle *Phanaeus* attracted by this bait makes a sizable part of its diet (Levey et al. 2004).

The Complicated World of Food Hoarding

Some corvids (jays (Figure 6.19), nutcrackers, and magpies) like squirrels hoard food in times of plenty, and they return to these caches in time of need, which can be after hours, days, or in the next winter. The scrub jay does it also in the laboratory, which allowed zoologists to explore its food retrieval strategies. Corvids managed the classical three memory Ws: *What* happened, *where*, and *when*? The tasks they solved required phenomenal memory capacities since they remembered the whereabouts of more than a thousand caches. Their home computer showed remarkable processing qualities. Captive jays were given wax moth larvae, their favorite but perishable food item or peanuts, a stable but less likened food. They hoarded this food but were allowed to retrieve it only after either a short- or long-waiting period. First they go for their favorite food, but they quickly learned that after several days of waiting the larvae became degraded and unpalatable. This experience made, they do not even try to search

FIGURE 6.19. The Eurasian jay (*Garrulus glandarius*), family Corvidae, order Passeriformes.

the larvae, and the next time, they go directly to the otherwise less-favored peanut. If the researchers replaced the larvae with a fresh one and they did not make the degradation experience, the jays continued to go for the worms because they did not make this negative experience. These experiments were the first conclusive behavioral evidence of episodic-like memory in animals (Clayton and Dickinson 1998). In the wild, the construction of caches is complicated by the social context. If jays are observed when hoarding food, they quickly come back and recache the food item. However, they do that only when they had previously made the experience of pilfering, either themselves as a pilferer or as a victim of food stealing. No such strategy was chosen when the jay could cache the food in private, i.e., unobserved (Emery and Clayton 2001). These experiments were the first demonstration that animals can remember the social context of past events and can adjust their present behavior to avoid potential detrimental consequences in the future. Different corvids differed in their memory capacities, e.g., jays remembered the caches of other birds for 2 days, while nutcrackers remembered only their own caches after this time period. We have here again an arms race between storers and pilferers, which leads to remarkable behavioral strategies, like waiting until a potential pilferer is distracted or lead the competitor arbitrarily away or even making false caches containing inedible food like a stone or nothing at all (quoted from Emery and Clayton 2004). A recent report added a further layer of complexity when Cambridge scientists discovered that food-caching scrub jays kept track of who was watching when (Dally et al. 2006). The scientists arranged for a situation where the storers cached in the presence of a dominant or a subordinate bird, or their partner or in private. Item recaching was greatest in the dominant condition, while partners were not perceived as a risk to cache safety.

Corvids gain increasingly a living within urban settlements, and sometimes use human civilization for food acquisition in astonishing ways. Corvids were previously observed to let nuts fall on rocks to get them opened. In a recent TV news item, I saw crows to drop nuts on streets in a Japanese city. They waited so that cars would crack the nut to recover the edible part. The news item showed that crows also used pedestrian crossing for this exercise to recover the nuts during the pedestrian's green period thereby avoiding getting accidentally under the cars.

On Stone Tools and Culture in Apes

What is Culture?

We associate our existence with culture, although when speaking of earlier hominids (species more closely related to humans than to chimpanzees) their tools are frequently referred to as industries to avoid the pretentious word of culture. This shy to use the term culture for animal behavior is probably again rooted in our anxiety to draw a firm line between us and the animals; such that we can maintain our Biblical God-like existence against the Darwinian Ape-like roots (Figure 6.20). One researcher brought this dilemma to the point when stating: The question of whether animals have culture is a bit like asking whether chicken can fly. There are certainly differences in the flying capacity of chicken and the albatros, but both can fly (de Waal 1999).

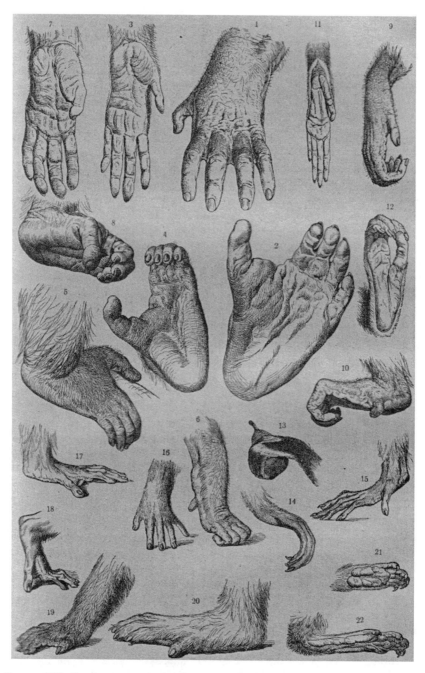

FIGURE 6.20. On the way to the handy man or the evolution of dexterity: Hands from the gorilla (*1, 2*), chimpanzee (*3–8*), orangoutang (*9, 10*), gibbon (*11–13*), the baboon (*19–20*) and the marmoset *Callithrix (21–22).*

Nutcracking Culture

One of the best-studied cultural practices of chimpanzees is nutcracking. Mothers share the nuts that they have just cracked with their youngsters, which thus immediately get the first lessons in nutcracking by onlooking. Nongenetic, demonstrated, and learned transmission of techniques is one of the definitions of culture. Observers estimated that it takes up to 5 years for a chimpanzee to learn an efficient way of cracking hard nuts like the *Panda* nut from West African rainforests. The reward is high: If the animal masters the technique, it can obtain more than 3,000 kcal of food per day. However, to get to the *Panda* seed, which sits in one of the hardest kernels, is not an easy task. The compression force to crack it open can only be achieved by using stone tools. The trick is to crack the kernel without smashing the seed. Management of this trick is essential if you want to consume a lot of nuts without loosing time to pick nut fragments from kernel remnants. Chimpanzees from a remote area in the Tai National Park in Cote d'Ivoire were observed to use stone hammers of igneous rocks weighing between 3 and 15 kg for this task (Mercader et al. 2002). They carry these stones for 100 m to 2 km from the source rock bed to the nutcracking place. Once at the place, they curate the hammers intensively and transport them from site to site. Notably, the scientists who explored this site were not zoologists or ethologists but anthropologists, and they took an unusual but interesting approach. They knew from their zoologist colleagues that the site was used between 1975 and 1996 (when the *Panda* tree died). Nevertheless, they applied formal archaeological techniques as if this was a human excavation site. They carefully documented 40 kg nutshells and 4 kg of broken stone. They concluded that the stones and the nutshells were associated by a behavior, that anvils were used and that the stones are limited to a zone around the anvils. In a conventional archeological site, the stones would be classified as hammer edges, flakes, and shatters. These stone tools were produced unintentionally by the chimpanzees but resembled the objects found in the Olduwan industry of the earliest hominin technologies. Stone tool use is not limited to chimpanzees. Also capuchin monkeys from dry forests in Brazil use stones to dig out tubers, roots, or insects; to crack seeds or to get to the inner pitch of thorny fruits. The researchers concluded that tool use became essential during the dry season with low food abundance; in their view, the nutritional need led to the invention (Moura and Lee 2004).

Chimp Cultures

But is stone use already a culture? Cultural anthropologists insist on the use of language for the definition of culture and make it thus an exclusive human trait. Biologists have more inclusive criteria and define culture as intergeneration transmission of behavior through social or observational learning. This definition covers what one could also call "traditions" like dialects in songbirds or potato washing in macaques. Actually, when chimpanzees from two western and four eastern African sites were compared based on accumulated 150 years of chimpanzee observation by field biologists, they listed their behavior into that shared by all, ecologically explained behavior and cultural variants (Whiten et al. 1999). To the cultural category belonged the nutcracking in western chimpanzees or, for example, distinct ant-eating habits. One ant-dip technique is done with a long wand in one hand stirring an ant nest, and the other hand is used to wipe off the ball of ants clawed into the wand and to lead it to the mouth. The other ant-dip "culture" is characterized by the use of a small stick, which collects a smaller number of ants and is directly transferred into the mouth. Chimpanzees are thus not so different from humans, in which differences between cultures are constituted by a multitude of variations in technology, social customs, and eating habits.

Cultural Conformism in Tool Use

The similarity in human and chimp culture goes even further. We tend to associate culture with creativity and novelty. One should not overlook that conformity to cultural norms is also an important ingredient of human and chimp culture. Conformity in tool use was demonstrated by psychologists working with captive chimps. Scientists have criticized that critical elements of chimpanzee behavior cannot be studied in captivity but only at the population level with wild animals. However, this limits the scientist to observational studies. To bridge the gap between the two types of research, scientists studied social learning where an animal teaches another animal a naturalistic foraging task. The psychologists educated a high-ranking female chimpanzee in the handling of an instrument that provided a desired food item, when correctly manipulated (Whiten et al. 2005). In fact, the food could be obtained in two alternative ways, either by lifting a level with a stick or by poking with a stick into an opening. The psychologists showed only one technique to the animal. When the instrument was introduced to naïve control animals, they manipulated it, but no animal succeeded to get the food item out. Then the researchers introduced the trained expert. The members of his group showed great interest in the food searching activity of the expert and most animals quickly learned the task—how to get the food. All animals practiced the mode that was taught to the expert. Then a follower individual discovered the second mode and practiced both techniques, some animals became then proficient in both techniques. Then the apparatus was removed. When reintroduced two months later, the chimps initially introduced into the poke or the lift technique, nearly exclusively practiced the first intro-duced technique despite the knowledge about the alternative technique. Despite the capacity to acquire particular local variants of a technique, most animals showed powerful conformist tendency to copy others. They discounted personal experience in favor of adopting perceived community norms. Apparently, the animals valued social bonds very high. Social norms contribute to the integration of groups. In macaques these norms are enforced by policing function of a small subset of respected individuals, which do conflict management. Removal of these individuals from the group destabilized the social niche and the group members disintegrated into smaller, less divers, less interacting subgroups when measured with respect to grooming, play, and contact-sitting. Policing thus not only controls conflict but also plays a critical role in social learning and cultural traditions (Flack et al. 2006).

Sex Differences

I want to mention another similarity between the chimp and human cultures, which concerns sex differences in learning. Scientists observed chimpanzee mothers in Tanzania teaching their offspring the local technique of fishing termites with a stick (Lonsdorf et al. 2004). The striking observation was that the daughters learned the termite fishing technique from their mothers nearly 2 years earlier than sons. Daughters were also more proficient as measured by the number of termites gathered per dip. The reason became also clear: Daughters observed the mothers more carefully than the sons—which spent more time on playing—and imitated carefully the stick technique, while the sons were relatively careless in that respect. The biologists were quite impressed by this difference since similar sex differences in social learning were found in human children.

Orangoutang Cultures and We

The classical question is again whether chimpanzees are special or whether their case can be generalized. A subsequent study documented similar geographical variation in orangoutang (Figure 6.21) behavior from Sumatra and Borneo (van Schaik et al. 2003). The list of cultural variants in behavior again included eating habits like using leaf gloves to handle spiny fruits, tool use to poke holes into the nests of social insects, and breaking hollow twigs to suck ants out of their nests. These two studies push back the origin of manual skills and material culture to about 14 million years ago when the ancestor of orangoutangs, chimpanzees, and humans lived. Human culture differs clearly in having symbolic elements, stressing the cognitive elements. This study addressed another question, which can perhaps in the human context not be addressed due to the complexity and the contentiousness of the issue. The researchers asked what factors correlated positively with the inventiveness of orangoutangs. *Many languages keep proverbs that address this question. In German one word is "need is the best inventor" or stated negatively "a full stomach does not like studies." Is there a basis to this statement? The researchers formulated the question more scientifically and asked whether higher rates of innovation are linked to marginal ecological conditions ("necessity"). When they used percent feeding time of orangoutangs on tree cambium as an index of food scarcity, they obtained no support for the necessity hypothesis. Other words link creativity with playful exploration (free time; the poet F. Schiller stated "humans are only then really humans when they play"). The idea of our purpose-free fundamental research culture is partially based on this idea. The inventiveness was significantly linked with playtime, but the two factors were correlated negatively. Interestingly, the*

FIGURE 6.21. Juvenile orangoutang (Malay: man of the wood; Latin: *Pongo pygmaeus*), large man-like ape from Borneo and Sumatra, family Pongidae.

relationship between the time spent in association and inventiveness were significantly correlated but only for nondependant animals. This is actually an index to opportunities for social learning.

Human's Progress?

The Diet of Australopithecus

Hominid Successions

The term hominid refers to the human clade subsequent to the divergence from our common ancestor with chimpanzees. Until quite recently, a hominid specimen attributed to the genus *Ardipithecus* from the 5.8 My old Middle Awash formation in Ethiopia fitted best with the suggested divergence of the human/chimpanzees lineages dated to 6.5 My by molecular methods (Haile-Selassi 2001). Spectacular fossil finds have now pushed back this time horizon. The current hominid family tree starts with *Sahelanthropus tchadensis* at about 7 My ago (Brunet et al. 2002). This specimen is currently the closest position in the hominid clade to the last common ancestor of chimpanzees and humans. Actually, the similarity with apes were so marked that some researchers proposed the genus name *Sahelpithecus* instead of *Sahelanthropus* (Wolpoff et al. 2002). The original authors insisted on the belonging to the hominid lineage with a virtual reconstruction of the deformed cranium of the specimen by computer tomography, which even suggested bipedalism in this earliest hominid (Zollikofer et al. 2005). The hominid tree continues with *Ardipithecus ramidus*, which bridges the time period from 5.8 to 4.2 My. At 4 My, the first *Australopithecus* is distinguished with *A. anamensis*, then comes the famous "Lucy", the 1-m high *A. afarensis* women. This branch ends perhaps with *A. africanus* (the "Taung child") at the 2.5 My time point. If one accepts this chronology and taxonomy, one could argue that until that evolution stage, no stone industry had developed in these early hominids since the oldest findings of the Olduvai stone culture goes only back to 2.5 My ago.

The "Gracile" *Australopithecus*

The aforementioned australopithecines are summarized as "gracile" (lightly built) types, characterized by a size of up to 1–1.5 m and a weight between 25 and 50 kg (there was a considerable dimorphism between the sexes; a rule of thumb in primates tells that intellectual progress was linked to the disappearance of this sex difference). Brain size was 400–500 cc, which is comparable to that of chimpanzees when corrected for body mass. Possibly, its intellectual capacity did not surpass that of extant chimps, which could explain why these hominids never learned to manufacture tools. However, australopithecines were bipedal as demonstrated by the famous footprints of Laetoli (Day and Wickens 1980). Their habitat was relatively open consisting of grassland intermingled with trees.

From the critical time period around 4 My ago where we place the origin of *Australopithecus*, we possess now fossils from a woodland context in Ethiopia as judged from the associated vertebrate assemblage (White et al. 2006). This contrasts with the aquatic, grassland, and bush land with gallery forest context described australopithecines found 1,000 km to the south in Kenya (Leakey et al. 1995). In comparison with *Ardipithecus*, the oldest australopithecines had crossed the threshold of megadontia (big teeth). The craniodental features of the fossils display hypertrophy, evolved under the selection for intensified mastication of the food. This trend keeps on over the next two million years and is only violated by the genus *Homo*, but only subsequent to the appearance of stone tools, which apparently relieved the pressure on teeth during eating.

Diet Deduced from Tooth Size and Shape

Intensive efforts were made to deduce the food eaten by australopithecines from the fossil remains (Teaford and Ungar 2000). Information was garnered from tooth size, tooth shape, enamel structure, dental microwear, and jaw biomechanics displayed by the fossils. Australopithecines showed small incisors, but large flat molars and the postcanine teeth became larger and larger. Tooth size analysis suggests seed, leaf, and berry eating. The tooth shape analysis can provide further hints. Comparative analysis in primates demonstrated that tough, difficult to fracture food is sheared between the edges of sharp crests. In contrast, hard, brittle food, which is difficult to penetrate, is crushed between planar surfaces. Insect exoskeletons and leaves are best manipulated by concave crested teeth. Fruit eaters use flatter cusped teeth. According to the tooth shape analysis, australopithecines were capable of processing buds, flowers, shoots, nuts, and soft fruit, but not coated seeds or leaves. Notably, they were dentally not prepared to eat meat. However, after some decomposition meat becomes less tough. Via a scavenger lifestyle, some meat might thus have found entrance into the diet of australopithecines as indirectly suggested by isotope evidence (Sponheimer and Lee-thorp 1999). *Australopithecus* was probably a scavenger and ate what remained from the prey made by larger animal predators. This interpretation was deduced from dental scratches on the bones of the prey first by the animal predator and only later by *Australopithecus*.

Tooth Wear

The thick enamel of australopithecines suggests abrasive food in the diet. The scanning microscope can reveal tiny pits and scratches on the tooth enamel ("microwear patterns"), diagnostic of the diet. Grass feeding leads to linear scratches on teeth; leaves in contrast polish the teeth; bone eating results in tiny pits in the enamel. Unfortunately, the established methods of studying microwear are error-prone and therefore anthropologists looked for more reliable methods combining scanning confocal microscopy with fractal analysis (Scott et al. 2005). They first calibrated their method with the monkeys *C. apella*, which eats fruit flesh and hard, brittle seeds, and *Alouatta palliata*, which consumes leaves and other tough food. With these data they analyzed the microwears on teeth of *Australopithecus africanus* and *Paranthropus robustus*. They deduced tough food for the former and a diet composed of more hard and brittle food for the latter. However, they warned to oversimplify this pattern into a dichotomous food preference because it may only reflect fall-back food choices consumed during some critical periods of seasonal food shortage.

In other studies, australopithecine teeth showed evidence for a fruit-containing diet like in chimpanzees and not for the tough plant material ingested by orangoutangs. *Homo erectus*, in contrast, showed teeth with signs of heavy wear and tear with pits, striation, and polish suggesting a nonspecialized omnivore. Dental microwear showed at the same time gorilla-like fine wear striae and baboon-like pits indicating a wide range of diets.

Trends for Dental Robustness

The australopithecines finally showed thick mandibular bones capable of developing a substantial torsion and bite force with a high occlusal load. These hominids had the mastication properties to eat fibrous, coarse food. When looking for time trends, *A. anamensis* adapted to hard abrasive food, *A. afarensis* developed the increased mandibular robustness, and *A. africanus* the increase in postcanine tooth size. Overall the comparison of *Australopithecus* with *Ardipithecus* suggested a dietary shift in face of climatic variability. Hard abrasive food became apparently increasingly important in the Pliocene. For the paleontologists, this change signals an ecological breakthrough involving a niche and food expansion with an intensified exploitation of more open African Pliocene habitats.

From Hominid Stone Tools to the Control of Fire

The Oldest Industrial Complex: Olduvai

The Olduwan stone-tool industry was named after the 1.8-My-old stone artifacts after the Olduvai Gorge in Tanzania, but these are not the oldest pieces of this form. This culture can be traced back to 2.5 My ago from Gona in Ethiopia (Semaw et al. 1997) or to 2.3 My in Turkana, Kenya (Roche et al. 1999). Human technology had a humble beginning here: It only needed to select a suitable pebble from a streambed, to strike it with another stone to produce sharp-edged flakes or crude choppers. The stone cores show evidence of pitting and bruising. The working edges are very sharp. Late Pliocene hominids had already mastered the basics of stone-tool manufacture, although only multipurpose tools like hammerstones suitable for pounding activities were made. Knapping stones, understanding of the fracture mechanics was already well understood. Refits of flakes to the original stone showed how they decided to strike the original pebble.

The tools cannot be dated directly, their age must be inferred from the archae-ological stratum that contained the artifacts. The faunal remains include many mammals and within them mainly grazing species like bovids, suids, equids ("cows," "swines," "horses"), fish, reptiles (shell fragments of large tortoise) and ostrich-egg fragments. While this demonstrates perhaps some hominid collection strategy, no evidence of man-made action was found on the animal remains. The most striking aspect of this stone industry is that it did not change over a long time period, spanning 2.5—1.5 My ago; in the literature, this is called the technological stasis of the Oldowan industrial complex. It is tempting to link this era to the life span of a specific hominid lineage. Hominids living before 2.5 My were unable to this technological "achievements," while new populations emerging at 1.5 My ago climbed to new heights of stone technology due to higher brain capacities and thus technical intelligence.

The Maker of the Olduvai Tools

But what face do we fix to the Oldowan industry? Here, one must probably directly confess our current ignorance. Human remains regularly make headlines in newspapers and fill also the columns of research journals like *Nature* and *Science*. However, basically the populations of early hominids were small, the likelihood of fossilization was low, therefore the fossil record is still very fragmentary. Actually, the number of skeletal remains is not so bad, but we frequently do not know whether this lineage died out or led to modern humans. The case is vividly demonstrated with the maker of the Olduwan industry. The maker was variably referred to as *Homo rudolfensis* (a name now mostly rejected by anthropologists) or *Australopithecus rudolfensis*, while most researchers prefer a new genus name with *Kenyanthropus rudolfensis*.

From the "gracile" australopithecines, the actors of the previous section, the makers of the Olduvai tools were differentiated by paleontologists as "robust" australopithecines. These were heavily built hominids with small brains and large jaws and big chewing teeth. These hominids are currently placed in at least two different genera *Australopithecus* (controversially also called *Homo* or to avoid conflict *Kenyanthropus*) *rudolfensis* and *Paranthropus* with two lines *P. robustus* and *P. bosei*. From these hominids, we possess as oldest remains the fairly complete skull from Lake Turkana dated at 2.5 My ago. These robust hominid types are found in eastern Africa between 2.3 and 1.3 My ago. The robust stature is commonly attributed to their diet that required heavy chewing and possibly consisted of low quality vegetable food, such as roots and nuts. Morphologically, *Paranthropus* showed initially a rapid dental evolution, which was then followed by a long stasis. The apparent conformity between a consistent dietary adaptation, dental, and technological stasis makes a powerful circumstantial case for linking *Paranthropus* with the Olduvai industry (Wood 1997).

The genus *Homo* is currently thought to start with *Homo erectus/ergaster* and not with *Homo habilis*, which is currently delegated to the *Australopithecus* lineage. However, from its brain size (500–650 cc) and its extension in time, which fits at least part of the Oluwan industry period, *Homo* (alias *Australopithecus*) *habilis* could also still fit the ticket. This would justify the older name of this hominid as the "handy man." As in many cases of hominid evolution, we must conclude that the jury is still out and we do not yet know in what hands to lay the Olduvai stones.

The Face Behind the Acheulean Technology

As abrupt as the Olduvai technology made its appearance, so suddenly was the striking change to the Acheulean stone technology at about 1.6 My (Figures 6.22 and 6.23). This industry impressed as bifaces, teardrop-shaped hand axes that were worked around all of the margins. They persisted until about 150 Ky ago and became the most popular "Swiss Army knife" of the Paleolithic. The earliest finds are from Konso in Ethiopia where roughly made biface hand axes and picks of up to 27-cm length were manufactured (Asfaw et al. 1992). The flakes are untrimmed. In contrast to the Olduvai stone tools, the Acheulean tools are not only closely associated to large mammal bones but the bones showed hominid-induced modifications like percussion pits, flake scars, and cut marks. Butchery is their apparent task, while the food use of the Olduvai stone tools is less clear, but might be linked to resistant plant material. Again in contrast to the previous period, the Acheulean culture provided already in the Konso sediment a hominid mandible that allowed the identification of the carrier of this culture. It is *H. erectus*. African specimens are sometimes called *H. ergaster*, and *H. erectus* is reserved to fossils found outside of Africa. To circumvent this ambiguity some anthropologists use the clumsy name *H. erectus/ergaster*. The Nariokotome boy described by Walker and Leakey is the most complete early hominid ever found.

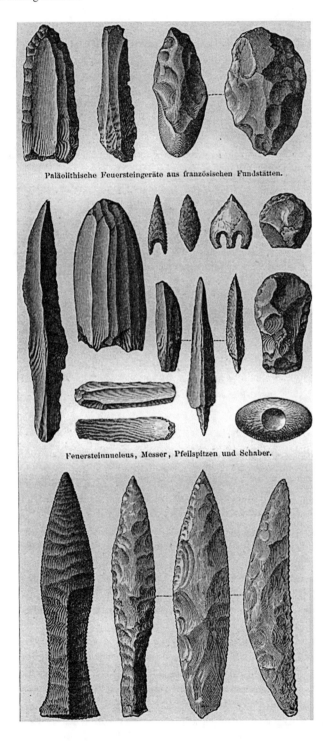

Paläolithische Feuersteingeräte aus französischen Fundstätten.

Feuersteinnucleus, Messer, Pfeilspitzen und Schaber.

It dates from 1.6 My ago and comes from Lake Turkana in Kenya. It is a male, young adolescent 1.5-m tall, which would have reached modern proportions as an adult (1.8 m and 68 kg), but showed a significantly smaller brain size (900 cc vs. 1,500 cc for modern humans). Its large body size would have provided heat and dehydration tolerance; together with its advanced tool kit it would have contributed to the success of this early species of our genus *Homo*. It lived in many ecological settings and saw a wide geographical distribution: *H. erectus* was found in Indonesia and China more than hundred years ago and became famous as Java Man and Peking Man, respectively. *H. erectus* was also found with a mandible in Dmanisi, Georgia, dated to 1.6 and 1.8 My ago (Gabunia and Vekua 1995). It shared a number of similarities with both African and Chinese representatives. A clear-cut specimen of *H. erectus* dated to 1 My ago was also found in Awash, Ethiopia making the distinction of an African *H. ergaster* from an Eurasian *H. erectus* doubtful (Asfaw et al. 2002).

Butchery

In Ethiopia *H. erectus* lived in an open grassland habitat with water margins as demonstrated by equid, bovine, and also hippo fossils. Butchery of these large mammals is evident. The different African and Eurasian specimens represent probably demes, i.e., communities below the species level. Many aspects of *H. erectus* remain still elusive: Its origin is undefined and the direction of its dispersal is not clear. Did he emigrate from Africa to Eurasia, stood in the Caucasus at the gates of Europe without intruding this continent (was the climate too cold or the carnivores too big?) and taking route to Asia? Or was he an immigrant from Asia into Africa? The stone tools are traditionally differentiated into a western and eastern Acheulean separated by the Movius Line extending from the Caucasus to the Bay of Bengal. However, at 800 Ky ago, the eastern and western industries were not so different to justify speaking of different cultures associated with different species (Yamei et al. 2000; Goren-Inbar et al. 2000). The lithic findings both at Gesher Benot in Israel and in Bose, China, reflect adroit technical skills and in-depth planning abilities suggesting complex hominid behavior at the beginning of human globalization.

Old Age Food Care

A beautifully preserved nearly complete skull from the same layer in Dmanisi yielding the *H. erectus* mandible demonstrated an astonishing insight into the complexity of our genus 1.7 My ago. It shows an edentate (toothless) skull, all the

FIGURE 6.22. The figures illustrate the evolution of the stone tools used by humans through the different periods of the Stone Age. The Stone Age is commonly divided into a Paleolithic period, where further subperiods can be distinguished according to the stone technology. The periods are named according to French archeological sites as Acheuléen, Moustérien, and Magdalénien in this temporal order.

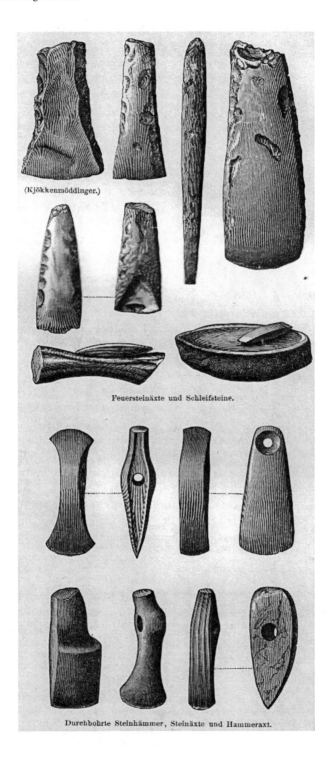

(Kjökkenmöddinger.)

Feuersteinäxte und Schleifsteine.

Durchbohrte Steinhämmer, Steinäxte und Hammeraxt.

maxillary and most of mandibular teeth were lost due to aging or a pathological process (Lordkipanidze et al. 2005). The remarkable observation is that this occurred well before death as the mandibular bone showed substantial resorption. The skull was associated with stone artifacts and animal bones showing cut and percussion marks. Meat consumption was probably the nutritional basis for this population living at this high latitude. The survival of a toothless individual suggests that it fed on soft tissue like bone marrow, brain tissue, and soft plants or depended for the mastication on the help of other individuals from his clan.

Fire Use

Mythology and fairy tales do not place the events in a historical time frame. Hänsel and Gretel from the Grimm brother would probably loose a lot of their attraction to a broader audience, when a cultural anthropologist would place the event to a famine that occurred 1625 in Marburg, Germany (this is purely fictive). However, scientists are a curious blend of people and cannot resist to search for indications of a place and a time even when reading fairy tales. To them it might come as a satisfaction that a famous act of the Greek mythology can now be located and dated. Prometheus feels pity with the earthlings that live under miserable conditions, and he steals the fire from the jealous Greek gods who do not want to share it with humans. There is definitively something special with the Holy Land: Israeli archaeologists found the earliest evidence of hominid control of fire. The place is again the above-mentioned Gesher Benot at the Dead Sea associated with the Acheulean stone industry, and the time is about 800,000 years ago. The site shows evidence for localized burned flint and wood (wild olive), heated to 500 °C, which they interpret as hearths. Alternative explanations were discarded: peat fire by lack of substrate, wildfire following lightning by the localized nature of the heating and the sparing of adjacent wood, and finally underground fire as burning roots because it would not have developed the 500 °C heat observed in the purported hearths (Goren-Inbar et al. 2004). If confirmed such an early use of fire would have led relatively early to dramatic changes in hominid behavior with respect to food preparation, defense against wild animals, and social interaction (light and heat in the night). Other archaeologists found indirect evidence for the controlled use of fire at the much younger Peking Man site dated to 500,000 years ago but noted the absence of ash residues and hearth features (Weiner et al. 1998).

FIGURE 6.23. The stone artefacts at the top are from a Danish culture known as the *Kjökkenmöddinger*. These people lived through the period described as Mesolithic. This was a difficult period when compared to the period of the great hunters. From their diet they left mainly heaps of oyster and clamshells, but few big games. At the lower part of the figure are stone tools from the Neolithic.

On Cooking and a Varied Diet

Homo erectus is also the inventor of cooking. This habit did not only contribute substantially to food safety but it also reduced the advantage of big teeth. Cooked food requires much less cutting, tearing, and grinding than raw food. One of the distinguishing features of *H. sapiens* from *H. erectus* is the further diminution in the size of the teeth. Extant hunters like Kalahari Bushmen or Australian Aborigines do not live from the meat of big game alone. Snakes, birds and their eggs, locusts, scorpions, centipedes, tortoise, mice, hedgehogs, fish, crustaceans, and gastropods figure on their menu plan. These animal morsels were complemented by plant delicacies. The latter included fleshy leaves, fruits, nuts, roots, and seeds like hackberry. *This type of variable diet was probably not far from that of early hunters. When considering the effect of diet on human health, one should keep in mind that we were selected by evolution for this type of food. Members of early and poor agricultural societies have a monotonous diet in comparison, and our food has diverged substantially from that of our ancestors.*

Palaeofaeces

For example, paleofecal deposits from a place in Texas that showed archaeological evidence for 10,000 years of intermittent occupation by prehistoric hunter-gatherers were recently investigated by DNA technology (Poinar et al. 2001). These 2000-year-old samples identified—in the plant diet—agave, yucca, sunflower, hackberry, acorn, edible nightshades, legumes were detected by amplification of a chloroplast gene, and cactus was identified microscopically. The animal diet consisted, according to DNA analysis, of sheep, goat, pronghorn antelope, and cottontail rabbit. Microscopy identified small mammals (packrat, squirrel) and fish as part of the diet. The authors pointed to two important implications of the data. One is more technical: Archaeologists emphasize the role of small mammals in hunter-gatherer diets. The sampling methods underestimate systematically the contribution of large animals. They were butchered at the kill site and only the pure meat was carried back to the occupation site, leaving thus only few large animal remains. The other is nutritional: An individual stool sample contained food derived from four animal and four plant sources. These humans were clearly omnivores. In addition, this is a remarkably rich diet. Cave hunter-gatherers had a more varied and nutritionally sound diet than humans dependent on early agriculture.

Hunters and Gatherers: The Origin of Grandmother's Recipe

The Reinsdorf Spear

Archaeology is literally finding the needle in the haystack, but this improbability of findings does not discourage the followers of this science. Apparently, it spurns only the age-old instinct of the hunter and gatherer in us. And then there are always these stories of the absolute lucky characters that struck gold. Here, I am not alluding to Heinrich Schliemann who decorated his wife with the gold treasure of Priamos at Troy. I think here of H. Thieme, a curator in Germany, who fights an impossible battle against a 50-m-high shovel-excavator that digs through a brown coal mine. His task is to stop the engine should it unearth an interesting archeological find, in its tons per minute progress. What he found made his fellow archaeologists speechless (Dennell 1997). In the Reinsdorf Interglacial dated to 400,000 years ago, he found notched pieces of wood for holding stone tools and 1-m-long sticks for stunning small prey or a stabbing spear for killing an already wounded or corned animal at short distance. But the best is still to come: He found three complete wooden spears, 2-m long, made out of very hard wood, carefully manufactured from a 30-year-old spruce tree. The spear has the aerodynamic form of a modern javelin and is certainly destined for big game hunting (Thieme 1997). Associated with these finds were a thousand of large mammalian bones, some of them with cut marks from butchery. The bones included those of straight-tusked elephants, rhinoceros, red deer, bears, and horses. The site also yielded a hearth. According to this site meat from hunting may have provided a great share to the diet of the hunters. Although this picture fits well to that widely distributed in laymen, professional archaeologists were wary with the issue of big game hunting in the Lower and even the Middle Paleolithic. They saw hominids more as scavengers competing with hyenas for animal prey, killed by big carnivores. The evidence were tooth marks from hominids on animal bones overlaying and not underlaying those of bigger carnivores, which according to this evidence killed the prey and only the leftover was the meager meal of hominids. Evidence for purposeful hunting was too fragmented to convince archaeologists and big game hunting was delayed until the arrival of fully modern humans at about 40,000 years ago. The spears from Schöningen changed this judgment: They had the same proportions, with the center of gravity a third of the way from the sharp end, and the tip of the spear came from the base of the tree trunk where the wood is the hardest. The spears are not the work of a 15 min culture; substantial planning activity was necessary to achieve these hunting weapons. The finds added to other evidence.

Homo heidelbergensis as Hunter

Although with an age of 120,000 years much younger, another spear from Lehringen in Germany was found where you would expect it: between the ribs of a straight-tusked elephant. A tip of a likely spear was found in Clacton, England, from a comparable period as that of Schöningen and a 500,000-year-old scapula from a rhinoceros unearthed in Boxgrove, England, showed a circular hole pointing to wounding with a spear. Boxgrove delivered also the hunter; to be precise, it was the tibia from the earliest Englishman. The size and robustness of the leg bone left no doubt about the male sex, the hunter's weight was calculated to 80 kg. He had the strength to throw the Schöningen spear. The stratum containing the tibia was associated with typical flint bifaces of the Acheulean industry, but the tibia was attributed to another species of the genus *Homo: H. heidelbergensis* (Roberts et al. 1994). This species has a larger brain than *H. erectus* (1,200 cc vs. 900 cc) and has a wide distribution over Europe. The type species is the famous mandible from Mauer 1 near Heidelberg, Germany. Other finds are from Greece (Petralona cranium), Ethiopia (Bodo cranium), and even footprints from volcanic ash in Italy (Mietto et al. 2003). Most anthropologists think that *H. heidelbergensis* originated from *H. erectus* in Africa at about 1 My ago and gave rise to more recent *Homo* species, including *H. sapiens*. Taken together the data mean that Europe was well colonized with *H. heidelbergensis* at 500,000 years ago, while evidence for an earlier human occupation of Europe by for example *H. erectus* is lacking (Gamble 1994).

The Gatherer

The traditional view sees the tribes with adult males as specialized hunters and women as gatherers. Although direct evidence for this scenario is lacking, it would fit classical prejudices and probably natural inclinations. Extant male humans have a greater aggressiveness and a better 3-D orientation in the environment predisposing them for hunting. On the other side, one sees women caring for children, gathering fruits, mushrooms, tubers, roots, grains, and small animal prey. Probably, cooking also became part of their task in that time period. We still see this heritage in psychological tests of modern women who easily outcompete their male counterparts if the parallel execution of multiple tasks is requested. Food preparation and not only food foraging became part of the everyday activity. We see this trend when we look to a hominid tree that depicts the hominid fossils on 2-D graph with an age and postcanine teeth size as axes. It has a definitive Y-shape. At the stem is the hypothetical chimp-like, forest-dweller, arboreal and fruit-eating ancestor of the hominid tree. The australopithecine trunk is then bifurcating into the *Paranthropus* branch with big teeth and the *Homo* branch with small teeth. The small teeth of the *Homo* branch does not necessarily indicate meat eating, but vegetable material was perhaps prepared to make it more palatable to smaller teeth.

Anthropologists on Grandmothers

In a *Nature* editorial titled "We Are What We Ate" anthropologists discussed the value of observations in a hunter and gatherer population in Tanzania (Wood and Brooks 1999). What struck the anthropologists was the role of the postmenopausal grandmothers. They cared for themselves, and their grand-children by using foraged tubers. Thus their daughters are released to gather food and to become pregnant again. *Hunter-gatherer societies must follow the animals imposing a substantial mobility on the clan. The spacing of birth is therefore anticipated to be 4 years that a mother has only to carry a single child during their migration. Infanticide was supposed to be frequent as excessive child caring could overstretch young women. Here physically fit grandmother could play an important, and some would say decisive, role in the development of modern humans. Via selection, this cooperation of grandmothers could lead to important changes ranging from longer lifespan, an extended growth phase with delayed maturity giving more time to social learning, and also an increased adult body mass. The fertility of the clan would profit from the cooperation of the grandmothers (the "grandmother hypothesis"). But what is the evidence for this hypothesis?*

Life Histories in Mammals

First, some facts: Human children, unlike other primates including chimpanzees, are unable to feed themselves when they reach the weaning age. The food we eat is too difficult that a young child could handle it. Assistance in food provisioning, food preparation, and even feeding will be important. If you see assistance in animals you see more frequently premenopausal individuals, which help their parents to breed.

Mammals fall along a continuum of life histories. At one end are species were maturation is quick, fertility is high, and adults die young (e.g., mice). At the other end, maturity is delayed, reproduction is slow, and adults usually live long enough to grow old (e.g., elephants). Interestingly, chimpanzees and woman have comparable life histories; both show female fertility peaks before 30 years of age and virtually none after 45 years. However, then the curves dissociate: chimpanzee survival rates fall along with fertility; in the wild less than 3% of the adults are older than 45. In contrast, even in the hunter-gatherer population a third of the women are over 45 years (Hawkes 2004). *Grandmothers could be useful and the toothless Georgian H. erectus individual demonstrates that its clan shared this belief. Formulating a hypothesis and proving evidence for it are, however, two different things.*

Epidemiologists on Vital Statistics

Recently, a cooperation between biologists and epidemiologists set out to test the hypothesis with demographic data from a multigenerational individual database, involving 3,000 women from traditional farming societies in Finland and Canada in the preindustrial age. Indeed, the evaluation showed that after extensive control for confounding factors, women gained two extra grandchildren for every decade they lived after age of 50 years. Both sons and daughters experienced this granny effect; they show greater fecundity and raise more children to adulthood and start reproducing at a younger age. This grandmother effect disappeared when the grandmother lived more than 20 km away (Lahdenperä et al. 2004). These data have enormous implications for current life history discussions. It shows not only that grandmothers contribute to the survival of the tribe by increasing the fertility of their children and by decreasing childhood mortality; the implications go farther: Senescence and lifespan can now be interpreted as a selected evolutionary trait. Notably, the study showed that the hazard of death in the grandmother increased sharply when their daughters passed through menopause. Now they can slip into the role of the grandmother and the grand-grandmother becomes dispensable. Notably, the data showed that the grandmother effect showed some attrition when their age increased beyond 60 years, which is intuitively plausible since their physical fitness will decrease. *There is some food for thought in these data. One might postulate that the current trend of delaying childbearing in European societies could lead to a further increase in lifespan. Then for the male readership, they could ponder the question why females outlive male in most industrial societies. Is the stronger sex selected for physical robustness during active hunting adulthood, but since we do not contribute so much to childcare, men are from an evolutionary viewpoint dispensable at an earlier age? A complicating factor is certainly that male fertility declines with age, but does not know an andropause. How does the preference of many women for older man and men for younger women influence evolutionary life histories?*

Grandmother's Recipe: Another Hypothesis

There is another hunter-gatherer effect that might still exist in our subconsciousness and that influences our food preferences. In the food industry, marketing groups have two contradictory strategies to sell their products. One is the label "New," which is perceived as an attractive argument to buy a better food; the other, although less frequently used, is "Grandmother's recipe." I think we have here a basic dilemma of human nutrition, which is explained by the grandmother hypothesis in the gatherer societies. Let's do the theoretical experiment to hunt with early humans. We have seen that mammals were the favorite game—the only question was to find them and to kill them. We heard of toxic animals that discourage the appetite of their predators with poisons. I am not aware that mammals use this strategy: they run, they fight, they hide, perhaps they stink, but once culled there is practically no risk associated with the consumption of the meat (parasites and infectious agents put aside). Hunters did not need much toxicology knowledge in their business. The situation is very different for the gathering women. Plants use the chemical club extensively in their fight against herbivores. This might be just antinutrients or bitter compounds to dampen the appetite, but many vegetables do not shy away from frankly killing the predator with potent toxins. Making a distinction and knowing plants, their nutritional and pharmacological properties, became essential for the survival of the clan. And here we are back with the contradictory marketing strategy of the food industry. If you are too conservative in your food selection (a German proverb says "The farmer eats only what he knows"), you avoid intoxication, but in a clan, which lives below its nutritional needs, survival depends also on the finding of new, better, and more nutritious food items. The farmer has no problem, he has sown what he is eating and he weeds out of his field what does not belong there. The situation was different for the women gatherer. She had to choose what Mother Nature was offering her. Worse, since hunting dictates the migration behind the game, they get in constantly changing places. Geographical landmarks became useless for remembering nutritious plant material; they had to memorize the plant. While only a handful of games existed, the plants get into the thousands. Small errors like confusion of A. campestris, *the champignon of Paris, with a poisonous* Amanita *mushroom could literally wipe out the family. Women had therefore to learn to distinguish and the best teacher would be her own mother, who has one generation more experience. However, to what extent can you trust the judgment of your mother? Here, humans apparently chose a simple, but efficient criterion: Ask the oldest women of the clan. She has the largest botanical knowledge and she has never made a fatal error, otherwise she would simply not have reached old age.*

Extensions

All traditional societies follow this scheme (old men have aspirated to this position perhaps with the transition from matrimonial to patrimonial family structures, but this characteristically did not extend to eating and cooking, except for the chefs in Western restaurants). Before the industrialization of food production, any food item was a potential risk and substantial knowledge was necessary to prepare a healthy meal. It is to these basic instincts of survival that the advertisement "Grandmother's recipe" alludes. Men distrusted women throughout history for poisoning them, and they knew that women did know a lot about plants and their secret powers (not only toxins, but also hallucinogens and even medicinal properties). The borderline between an old women knowing herbs and their healing power to witches that could in their imagination wipe out entire clans and societies was small and delicate. I suspect also that the resistance of substantial parts of the European population against the perceived and, frequently, rather theoretical risks of genetically modified food strikes an ancient emotional string of the gatherer societies. You had to eat, but you were never really sure whether you would see the next morning when using a new food item in your diet. All scientific arguments of a young male scientist in a white lab coat will not placate the deeply emotional concerns of the consumers (which include many individuals who studied science), they probably need a grandmother addressing their concerns after a lifetime consumption of GMO (genetically modified organism) food.

On Neanderthals and Cannibalism

His Place on the Tree

In the previous section, I took the liberty to anticipate relatively humane thinking, but we lack clear evidence whether we can do that. With respect to our evolutionary ancestry, most anthropologists see us as the follower of H. heidelbergensis. *Thus genes from the Schöningen hunter should be in us. However, we are not the only offspring of this tree: Long before* H. sapiens *walked on earth another* Homo *species deviated from the* H. heidelbergensis *branch, namely* H. neanderthalensis. *This is actually the first early human where the art of ancient DNA analysis could complement the fossil finds. The museum guarding the historical remains of the first Neanderthal find offered a bone piece on the altar of molecular biology, and they were rewarded. The scientists succeeded to amplify some mitochondrial DNA out of it* (Krings et al. 1997). More recently mitochondrial DNA of even better quality was recovered from a Neanderthal child in the Caucasus who was dated to 29,000 years BP. Now even a Neanderthal genome sequencing project is on its way (Green et al. 2006). The extinction of the Neanderthal line is currently fixed at 28,000 years BP; the child is thus one of the last survivors of this lineage. The German and the Caucasian Neanderthal subjects were separated not only by 2,500 km geographical distance but also by about 200,000 years from their common ancestor. The last common ancestor of *H. sapiens* and *H. neanderthalensis* lived according to the mitochondrial DNA argument perhaps 700,000 years ago (Ovchinnikov et al. 2000). The split of *H. sapiens* from the ancestor branch (being it *H. heidelbergensis* or other offshoots of *H. ergaster* like *H. rhodensis* is controversial, see Stringer 2003) is also well documented now. The Herto find in Ethiopia gives a minimum fossil age to our ancestors of 160,000 years (White et al. 2003), while the Kibish find, also from Ethiopia, would push the earliest well-dated anatomically modern humans to nearly 200,000 years BP (McDougall et al. 2005). In contrast, the expansion of our species did not occur before 60,000 years ago. There is thus clear fossil and archaeological evidence that *H. sapiens* and *H. neanderthalensis* are contemporary. How much contact they had is not clear.

Cold and Food Adaptations

In Spain, the late Neanderthal and *H. sapiens* occupied different territories separated by the Ebro River. Archaeologists did not find evidence that behavioral and technological transfer occurred between both populations (Mellars 1998). The relatively short, stocky bodies of the Neanderthals were interpreted as a cold adaptation to the periglacial north of Europe. Yet, most of the Neanderthals lived in the temperate areas of southern Eurasia. Their large noses and faces were probably due to the pressures of heavy chewing of their diet and the use of jaws as tools and not for the warming of cold air streams in nasal passage as was also proposed. In France, both populations occurred side by side: Neanderthals are associated with the "Mousterian" stone culture, which is developed from the Levallois technique. These tools were shaped by removing flakes from a core, followed by removal of one final flake, which formed the tool itself. Fireplaces are regularly found, and hunting remained the basis foraging activity. The relatively cold climate restricted fruits probably to a few berries. Parallel you find the "Aurignacian" tools associated with *H. sapiens*. In addition to stone tools, you find numerous bone tools, many are pointed and might have served as spear tips.

Emancipation

The French "Châtelperronian" culture led to an archaeological emancipation of the Neanderthals (Bahn 1998). Bone tools were found and a wealth of ornaments were found. Animal teeth, ivory beads, and bones with perforations suggested necklaces. These Neanderthals had objects created for visual display on the body, we have to anticipate that these ornaments communicated some meaning; Neanderthals thus elaborated and transmitted autonomous codes. Anatomical investigations showed that the hand of the Neanderthals was capable of substantial dexterity (Niewoehner et al. 2003), other investigation of the larynx suggested the use of speech. Suddenly, the shaggy, subhuman brutes transformed into cultured human beings. Tool use became more sophisticated. Levallois flakes of the Mousterian level dated to 40,000 years age showed a black substance on their surface. Chemical analysis identified it as remnants of a hafting material used by Middle Paleolithic people to glue handles on their tools. The raw bitumen glue was heated to extreme temperature before application (Boëda et al. 1996). We do not know with certainty whether we can attribute these complex multicomponent tools to the Neanderthals. However, the initial simple idea of archaic hominids making simple tools that were replaced by anatomically modern hominids that made complex tools cannot any longer be maintained.

Dark Sides: Cannibalism?

The appreciation of the Neanderthals has lived cycles of extremes. If you enter the archaeological museum in Neanderthal near Düsseldorf, Germany, you see a pretty Neanderthal woman receiving you in the entrance hall. She looks as if she would have appreciated the Arcy necklace. But beware: Despite all speculation on sexual attraction between human species in science fiction movies, anatomical hybrids between both human forms have not yet been clearly identified. Some anthropologists speak of the Neanderthals as a subspecies of H. sapiens. However, the postulated morphological continuity between both forms could not be demonstrated, even the late Neanderthals showed very distinct anatomical features, which suggest a reproductive barrier between both populations (Hublin et al. 1996).

The dark side of the Neanderthal man was a subhumane brute with a club in his fist as he was represented near the old museum in Düsseldorf. Archaeology has experienced rather dramatic swings in the assessment of the Neanderthal culture. The dark side started with archaeological findings in the 1930s that were interpreted as cannibalism in Neanderthals. The early cannibalism signs on the Monte Circeo skull turned out to be a misinterpretation; they were caused by gnawing hyena. Somewhat better is the evidence from a finding in Valence, France, where Neanderthal bones showed cut marks from a flint tool and fresh bone breakage (Defleur et al. 1993). This looks like quite good evidence, but we should not precipitate a conclusion; ethnographic studies showed mortuary practices involving bone defleshing and secondary burial of bones, but we are not aware of sophisticated burial rites in Neanderthals.

American Cannibalism

Cannibalism in humans is a very contentious subject that touches psychological barriers (Diamond 2000). We should not blame the Neanderthals; cannibalism was widespread in modern humans, too, and this as recent as in the aftermath of the Russian Revolution. There it was a widespread phenomenon that was not only explained by starvation as documented in an excellent historical account (O. Figes *A People's Tragedy*). Many opponents have refused to accept the evidence for cannibalism, but archaeologists have gone very far to prove the case (Marlar et al. 2000). Excavation at a prehistoric Puebloan site in Colorado that was precipitously abandoned for unknown reason showed the bodies of people of both sexes, disarticulated, defleshed, and apparently cooked. The cutting stone tools showed molecular evidence for blood and the cooking pots demonstrated traces of myoglobin of human origin. A coprolite (ancient fecal deposit) of human origin was found near one hearth. It contained no starch granules from a vegetarian diet but again showed human myoglobin. This observation does not show how common cannibalism was on the territory of the USA.

Modern Cannibalism: From Mourning Rites to Kuru

Cannibalism in contrast never became a widespread eating habit, and there might be good reasons for it, which have nothing to do with religious or ethical feelings. The case is best illustrated by a ritual cannibalism practiced by modern humans in the Fore region of Papua New Guinea. Endocannibalism is, in that region, a rite of mourning and respect for the dead kinsmen. The exploration of this rite was actually not done by ethnologists but by neurologists of the National Institutes of Health in Bethesda, especially D.C. Gajdusek, who lived himself through glory and misery like few of his fellow scientists. In his report, women opened the dead with their bare hands when they prepared the tissue for consumption. Horrible as it sounds, they scooped the liquefied brain by hand in bamboo cylinders. What followed brought the tribes to the brink of extinction and changed the male/female ratio in the Fore region to 3:1. The reason became tragically clear a few years after the start of epidemiological investigations. Children of both sexes crowded around the women during these rites, while men stood apart. Women and children apparently got infected by a mysterious agent that defied for a long time a molecular definition. Yet, epidemiological observations gave no indication for its contagion. No disease was ever seen in foreigners living in this area, while many children from the Fore tribes developed, sometimes after decades, the illness when living for years abroad. More than 2,500 cases were reported since the medical investigations started in the late 1950s. Kuru, as the disease was called, has been disappearing gradually during the last 40 years since the rites were suspended. The remaining cases showed a conspicuous rise in age of onset that pointed to an exposure at these cannibalistic rites.

The Symptoms

Even if the incubation period could be very long (which earned the illness the qualifier a "slow virus disease" in its early days), when the symptoms of Kuru set in, death within a year or so was the inevitable outcome. The subjects had self-diagnosed unsteadiness of stance and gait, they searched support of a stick for walking, tremor and ataxia set in and the patients needed the help of tribesmen to walk. Kuru means shivering, which is a cardinal symptom, in this period. Then follows a sedentary phase characterized by emotional instability with outburst of pathological laughter and gradual loss of speech, while dementia was not observed. In the third stage, sitting becomes impossible; the patients become incontinent, dysphagia leads to starvation and then to death within 9 months. The pathology is restricted to the brain, showing a spongiform encephalopathy and amyloid-containing plaques. The laboratory findings are unlike to any other infectious disease, no immune or inflammatory response is induced, but the brain material of the victim contains high titers of infectious material that could after much testing be transmitted to laboratory animals. The mysterious agent belongs to an entirely new class of infectious proteins (prions, more on it later in the BSE section). At the moment it is sufficient to conclude that cannibalism is evolutionary a dangerous strategy since it involves dietary exposure to congeners – thus no species barrier provides protection against disease transmission.

The Hobbit: Wanderer Between the Worlds

Are we Orphans?

Humans are fascinated by the idea to establish a contact with other intelligences on a different planet. Martians visited at least in science fiction films our home planet. The hope for life on Mars or on the Jupiter satellite Europe has not yet faded, but even the most optimistic biologists do not expect more than microbial life forms at these places. Astrophysicists are used to look for planets in other solar systems and designed the SETI program emitted into the cosmos. Until now we have not received an answer to our messages. Perhaps the probability of extraterrestrial life has increased after the observation of planets circling around other suns. After the demise of the Neanderthals, H. sapiens saw itself as a cosmic orphan. Then a team composed of Australian and Indonesian archaeologists took the scientific and the lay world by surprise: They found another member of Homo family whose demise was dated as recent as 18,000 years ago. We were not alone, the Cro Magnon modern man, the painter of the caves of Lascaux and Altamira, who is so close to our feelings when we look to his artistic remains, could entertain a conservation with two other species of humans if he had some sense of tourism and cross-species language capacities, if the third contemporary of H. sapiens was capable of speech.

Homo florensis

It needed traveling to the Indonesian island of Flores, but this should not have presented a major problem; 40,000 years ago, a spectacular expansion of the modern humans started which led over the next few ten thousands of years to a colonization of the entire globe. This was not a that remarkable feat—*H. erectus*, the likely ancestor, was already a daring explorer and did not shy away to cross the street of Sunda as a seafarer as demonstrated by stone tools dated to 900,000 years ago. The lack of a land bridge was also suggested by an impoverished fauna that consisted of giant reptiles (the Komodo dragon), giant rats, tortoise, and a pygmy elephant (*Stegodon*; Morwood et al. 1998). This wide radiation of early humans calls for a reappraisal of the intellectual capacity of *H. erectus*. However, no skeleton remains accompany the stone tools. This is the more unfortunate because the descendant of this early invasion is a particularly strange member of the human branch. The archaeologists called it *Homo florensis*, but it was quickly dubbed the Hobbit even in scientific journals. The cranium and skeleton from a likely female individual suggests a human being of a mere 1-m size and less than 28-kg weight and a spectacular low brain volume of 380 cc (Brown et al. 2004). All figures are lower than the estimates for australopithecines, which existed several millions of years earlier. Anthropologists made a virtual endocast of the skull. The investigation of the original is delicate: The skull was not fossilized and had the consistency of wet papier mâché. It was slightly damaged during the excavation and redamaged by an Indonesian scientist that could not believe the finding (*a worthy story of Pride and Prejudice on its own*). This study excluded a microencephalic individual, and the statistical analysis of the brain form, using principle component analysis, puts it closest to *H. erectus*, the likely first colonizer of Flores (Falk et al. 2005).

Size Reduction and Diet

A human pygmy was also excluded: African pygmies are taller than 1.4 m and have brain volumes well above 1,000 cc. Explanations of the small body size of African pygmies are based on thermoregulatory advantages of small size for life in hot and humid forests either as enhanced evaporative cooling or reduced internal heat production. The reduction in size is achieved by reduced production of the insulin-like growth factor in puberty when brains are already fully outgrown. The most likely explanation for small size of *H. florensis* is insular dwarfism. The Hobbit is most likely the product of a long period of evolution from probably normal-sized *H. erectus*, who arrived on this island 900,000 years ago. On this relatively small island, small body size was adaptive. In the absence of agriculture, tropical rainforests offer a very limited supply of calories for hominids. The scientists argued that selection should favor smaller individuals with reduced energy requirements. The impoverished fauna means also lack of predators, the Komodo dragon being the sole danger. Like in domesticated animals that loose the predator pressure, the sensory system is reduced, which might result in the observed dramatic reduction in brain size. The bones found in the cave point to fish, snakes, frogs, tortoise, birds, and rodents as food sources (Morwood et al. 2004). The only big game of this human species was neonates from the pygmy elephant *Stegodon*, which disappeared from this island 12,000 years ago. We know that *H. sapiens* arrived in this region by island hopping, some 55,000–35,000 years BP. Apparently, both human species coexisted thousands of years. We ignore whether there were contacts between both species and whether *H. sapiens* could communicate with a hominid of such small brain size. The small brain size also raises an important question with respect to the stone tools that accompanied the *H. florensis* skeletons. The stone tools associated with the cave are simple, but they show clear signs of designed elaboration. Some anthropologists doubted that a small-bodied hominin with a brain size of about 400 ccm and thus comparable to that of an *Australopithecus* could have mastered stone tool use. The question was now settled positively by the description of stone tools from a much older cave in Flores dated to 800,000 years BP (Brumm et al. 2006). The tools from both caves are in the same technological tradition suggesting *H. florensis* as a stone-tool maker, which apparently challenges our preconceptions on the necessary brain size for such technological activities.

Late Pleistocene Megafauna Extinction: An Early Blitzkrieg?

An Australian "Blitzkrieg"?

Was H. sapiens *responsible for the disappearance of* H. florensis? H. sapiens *has a bad reputation on the fifth continent.* The Cenozoic is commonly divided into the Tertiary (Figures 6.24 and 6.25) and the Quaternary Period. The latter is divided into the Pleistocene and the Holocene. In the Pleistocene (Figure 6.26), Australia sported 24 genera of megafauna: giant marsupials like the 3-m high, short-faced kangaroo *Procoptodon*, marsupial equivalents of rhinoceros and leopards, 1-ton carnivorous lizards like a too-big Komodo dragon, giant crocodiles, the giant bird *Genyornis*. In the Late Pleistocene, 23 of the 24 genera of megafauna disappeared leaving behind only the *Macropus* kangaroos (Diamond 2001; Figure 6.27). Two theories were proposed to explain this collapse of the megafauna. One is the overkill theory, where human hunters drove the megafauna to extinction. There is even a "blitzkrieg" version of this scenario, where the newly arriving humans quickly culled the large herbivores out of existence and thereby killed indirectly also the large carnivores that lost their food source. Despite all courage attributed to the early humans, it is judged unlikely that humans attacked the 1-ton lizards looking like a blown-up version of the extant Komodo dragon. The alternative hypothesis is extinction due to climate and vegetation change. The dilemma can only be solved by precise data of human arrival and the kinetics of megafauna disappearance.

Timing

The arrival of humans in Australia can relatively precisely be dated to 56 ± 4 thousand years ago. All Australian land mammals, reptiles, and birds weighing more than 100 kg experienced a continent-wide extinction at 46 ± 2 thousand years ago with remarkable synchrony in eastern and western Australia. It occurred at least 20,000 years before the Last Glacial Maximum and also before the aridity increased dramatically in Australia. The data are consistent with a human role in extinction, but the blitzkrieg version can be discarded since it took 10,000 years after human arrival to take effect (Roberts Flannery et al. 2001).

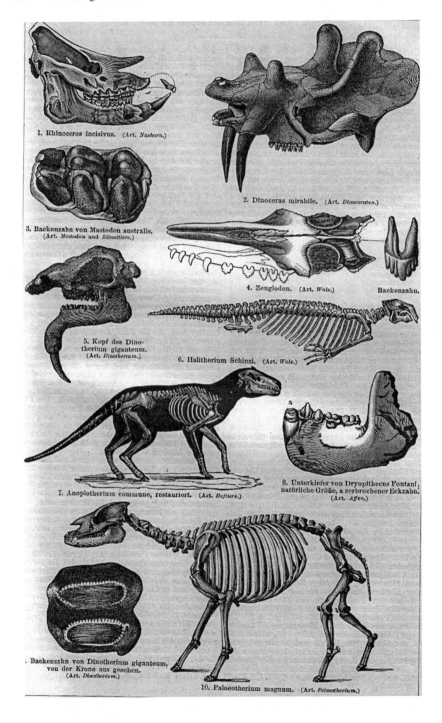

1. Rhinoceros incisivus. (Art. *Nashorn*.)

2. Dinoceras mirabile. (Art. *Dinoceraten*.)

3. Backenzahn von Mastodon australis.
(Art. *Mastodon* und *Rüsseltiere*.)

4. Zeuglodon. (Art. *Wale*.)

Backenzahn.

5. Kopf des Dino-
therium giganteum.
(Art. *Dinotherium*.)

6. Halitherium Schinzi. (Art. *Wale*.)

7. Anoplotherium commune, restauriert. (Art. *Huftiere*.)

8. Unterkiefer von Dryopithecus Fontani,
natürliche Größe, a zerbrochener Eckzahn.
(Art. *Affen*.)

. Backenzahn von Dinotherium giganteum,
von der Krone aus gesehen.
(Art. *Dinotherium*.)

10. Palaeotherium magnum. (Art. *Palaeotherium*.)

Humans therefore first triggered ecosystem disruption, as a result of which the megafauna became extinct. Other biologists observed an interesting trend: The likelihood of extinction was actually not correlated with the body size but with slow reproduction rate. In their words, it was an extinction of the "bradyfauna" ("*brady*," Greek for "slow") and not of the megafauna ("*mega*," Greek for "great"). Exceptions to this rule were arboreal and nocturnal animals, inhabitants of dense forests and animals living in high altitudes and high latitudes. These are all regions where humans had less access and where the predation pressure was consequently lower (Cardillo and Lister 2002).

Recent statistical analysis showed that threatened and nonthreatened terrestrial mammals differed in mean body weight (1.4 vs. 0.14 kg; Cardillo et al. 2005). When taking 3 kg as a cutoff, regression models showed no intrinsic biological traits that correlated with extinction risk in smaller animals, while such traits were found in larger animals (population density, neonatal weight, and litters per year). Larger animals are more affected by forest fragmentation and predation pressure. For example, in neotropical forests, subsistence hunters' prey preference increases abruptly for mammals above 6.5 kg.

Smoking Gun in Australia

Researchers tried to decipher the events around the ecosystem collapse that occurred with human arrival in Australia by analyzing eggshells and teeth from herbivorous animals (Miller et al. 2005). Plants that use the C3 or C4 pathway of carbon fixation leave a distinct isotopic composition, which can still be detected in the minerals of the fossils. They thus reconstructed the diet of the extant emu (*Dromaius novaehollandiae*) and the extinct giant flightless bird *Genyornis newtoni* over the critical time period. Before human arrival, the emu ate a wide range of food sources and utilized abundant nutritious grasslands constituted by C4 plants in wet years and relied more on shrubs and trees, i.e., C3 plants, in drier years. After the arrival of humans, emus shifted to C3 dietary sources. *Genyornis* consumed a more restricted diet and relied heavily on C4 plants. The emu can in contrast tolerate a pure C3 plant diet. Also teeth of the herbivorous marsupial wombat told the same story. The same emu eggshell observation was made in three places characterized by three different climates,

FIGURE 6.24. Tertiary 1. The Tertiary Period opens the Cenozoic era; it began 65 million years ago. Tertiary life forms are distinctly modern. The most remarkable development was the diversification of mammals. Exaggerated skull forms like in *Dinotherium* (*5, 9*) and *Dinoceras* (*2*) strike the eyes. Molars from elephant-like Mastodons (*3*) were found and early rhinoceros fossils (*1*) were described. Cuvier made his famous investigations describing an early hoofed animal, *Anoplotherium* (*7*), and *Paleotherium* (*10*). Monkeys are represented with the famous jaw of the ape *Dryopithecus* (*8*), a tree-dwelling fruit eater with some relationship to the ancestry of humans. Other mammals returned from the land back into the water as documented by *Zeuglodon* (*4*), an ancestor of whales and *Halitherium* (*6*).

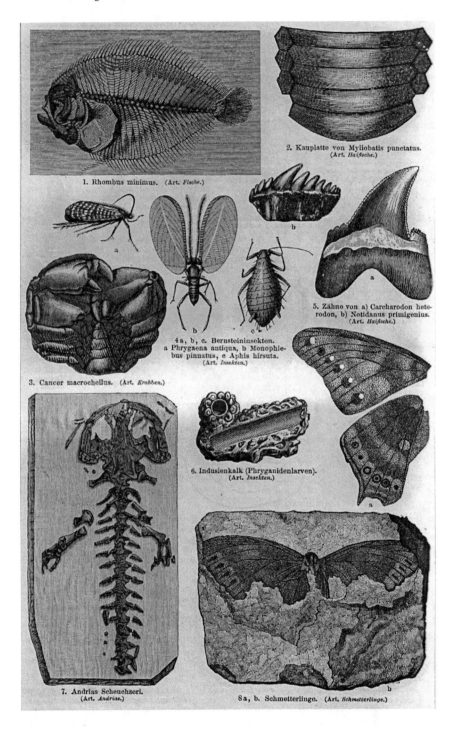

1. Rhombus minimus. (Art. *Fische.*)

2. Kauplatte von Myliobatis punctatus.
(Art. *Haifische.*)

4a, b, c. Bernsteininsekten.
a Phrygaena antiqua, b Monophlebus pinnatus, c Aphis hirsuta.
(Art. *Insekten.*)

5. Zähne von a) Carcharodon heterodon, b) Notidanus primigenius.
(Art. *Haifische.*)

3. Cancer macrocheilus. (Art. *Krabben.*)

6. Indusienkalk (Phryganidenlarven).
(Art. *Insekten.*)

7. Andrias Scheuchzeri.
(Art. *Andrias.*)

8a, b. Schmetterlinge. (Art. *Schmetterlinge.*)

making climate factors an unlikely explanation. The authors favored fire use by humans as a cause. They speculated that humans might have hunted along the fire front, cleared passageways, or promoted the growth of preferred fire-resistant plants. Thus a drought-adapted tree/shrub/grassland mosaic was transformed into modern chenopod/desert scrub. According to this interpretation, dietary specialization and not feeding strategy (browsing vs. grazing) was the predictor of extinction.

Alaska Paints a Different Picture

A US arctic biologist used the excellent preservation state of large mammalian species in the permafrost from Alaska to explore the critical time period around the human arrival to explore the effect of humans on the megafauna with radio-carbon dating of the animal remains (Guthrie 2006). Humans became visible to archaeologists at about 12,000 BP. This time period coincides precisely with the demise of the horse *Equus ferus*. However, the researcher does not believe that humans were the cause for the extinction of the horse. He proposes several arguments against this interpretation. The horse shrank in size before extinction pointing to a causal nutritional factor for its disappearance. The mammoth coexisted with humans for about one millennium, before it got extinct. Furthermore, the bison, sparse before the arrival of humans, expands around the time of the human arrival and remains in place despite being traditional a favorite human hunting target. The wapiti (*Cervus canadensis*) and the moose arrive with humans and stay there. This is apparently not the Australian blitzkrieg scenario. Looking for alternative ecological explanations, the biologist investigated the pollen record and deduced for the time period, before human arrival, a very cold dry mammoth steppe with xerophyllic grasses, sedges, and sages. This period is followed by a warmer and wetter transition period—the grasses explode and edible woody plants like *Salix* become abundant. This climate and vegetation change attracts humans and wapiti from northern Eurasia. Shortly after the climate changes to taiga and tundra vegetation, the dwarf birch (*Betula*), highly defended against herbivory by its toxin content, becomes the dominant plant in the pollen record. Three thousand years later the willow (*Populus*) follows, whose leaves have the highest nutrient content of northern plants. Wapitis and bisons

---

FIGURE 6.25. Tertiary 2. Other vertebrates are also documented in the fossil record, like modern-looking bony fishes (*Rhombus, 1*), sharks are represented with many teeth (5) and rays with dental plates (2). A historically particularly interesting fossil is that of a giant salamander (7). Scheuchzer described this fossil in 1732 as "Homo diluvi testis"—as a fossil evidence for humans that drowned during the deluge described in the Biblical record. Invertebrates like fragile butterflies are also documented in the Tertiary (8). Of outstanding beauty are many insects trapped in amber. They are frequently preserved in great detail (4). Insert 3 shows a crab fossil and insert 6 shows a characteristic chalk from the Tertiary formed by the larvae of caddisflies (Phryganeidae).

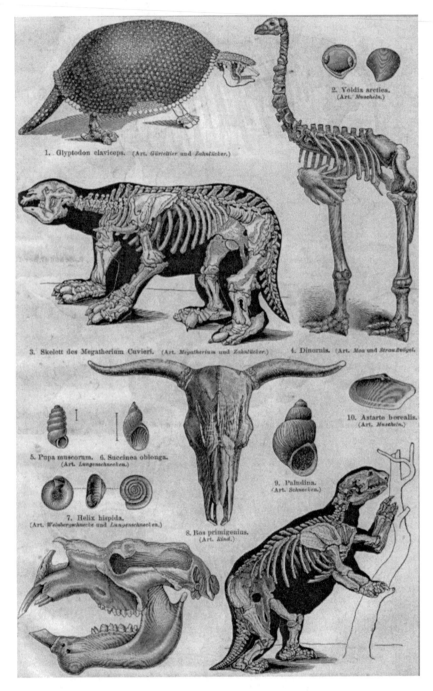

FIGURE 6.26 I.

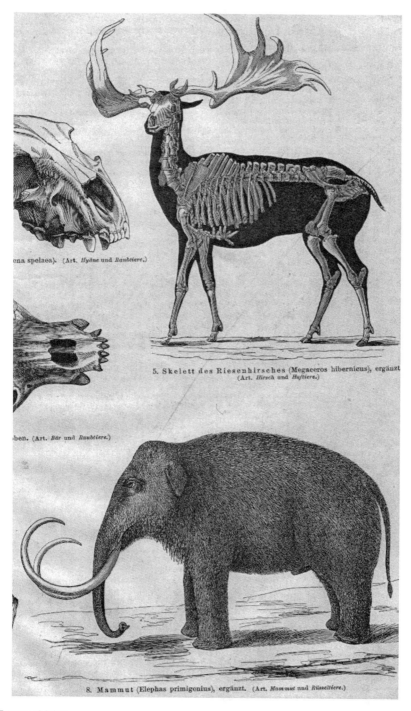

ena spelaea). (Art. *Hyäne* und *Raubtiere.*)

5. Skelett des Riesenhirsches (Megaceros hibernicus), ergänzt
(Art. *Hirsch* und *Huftiere.*)

ben. (Art. *Bär* und *Raubtiere.*)

8. Mammut (Elephas primigenius), ergänzt. (Art. *Mammut* und *Rüsseltiere.*)

FIGURE 6.26 II.

can make heavy use of this new food. Caecalids, animals like the horse, elephant, and rhinos—characterized by a large hindgut diverticulum (cecum)—loose their competitive advantage. Under poor vegetation conditions, they can tolerate high volumes of poor quality forage due to the higher food throughput through their gut. Climate change, not human hunters, are behind the fauna changes in Alaska.

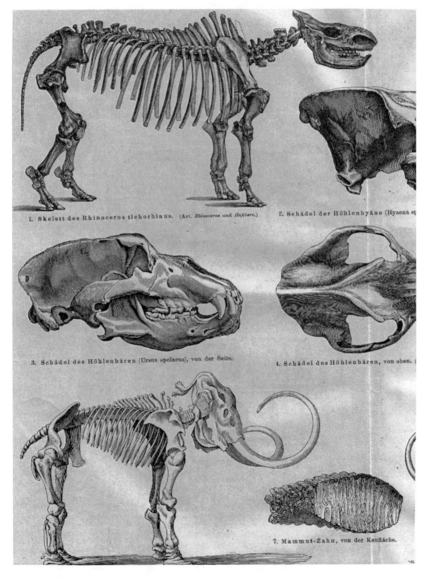

1. Skelett des Rhinoceros tichorhinus. (Art. *Rhinoceros* und *Haftiere*.)   2. Schädel der Höhlenhyäne (Hyaena s)

3. Schädel des Höhlenbären (Ursus spelaeus), von der Seite.   4. Schädel des Höhlenbären, von oben. (

7. Mammut-Zahn, von der Kaufläche.

FIGURE 6.26 III.

The Irish Elk

Also in other regions, megafauna extinction was observed, but it showed distinct kinetics and causes: Sabre-toothed tigers, mastodons, woolly mammoths, and the Irish elk survived until the end of the Pleistocene epoch, while the mammoths roamed their last exiles still at the time of the Egyptian pharaohs. With respect to the causes, let's take the example of the Irish elk (*Megaloceros giganteus*; Figure 6.26; Pastor and Moen 2004; Stuart et al. 2004). Female elks are probably to be blamed for the demise of the species. They selected males with large antlers, their instinct has told them that this is a sign of fitness for males if they find sufficient food to build and shed these antlers each year. This task should not be underestimated from the nutritional side. The animal's huge antlers weighed 40 kg and spanned 3.5 m. It was calculated that the antlers contained 8 kg calcium and 4 kg phosphate, which is a tremendous burden of mineral intake for the bulls. Furthermore, the extreme antlers prevented them the access into dense forests. All went well as long as the Irish elk could forage in its native willow and birch shrub habitat, existing 20,000 years ago in central and Western Europe. Then the climate cooled in the Last Glacial Maximum around 20–12,000 years ago. The vegetation in Europe changed to short-stature and unproductive tundra. Open grass-shrub plant communities persisted only near the Urals, the exile of the elks. After the great cold, the Irish elk came back to the British islands and Denmark, during the Late Glacial Interstadial. The rapid warming resulted in a productive

FIGURE 6.26. Cenozoic era, Quaternary Period. The Pleistocene (Diluvium) started one million years ago and preceded the most recent geological period, the Holocene. The figures document some giant animal forms: within birds the emu-like *Dinornis* (I, *4*) from New Zealand dwarfed its nineteenth-century describer Owen; within marsupials the Australian *Diprotodon* reached the size of a rhinoceros (I, bottom left); *Megatherium* (I, *3*) the largest ground sloth, an edentate mammal, reached the size of an elephant. As suggested by the reconstruction of *Mylodon* (I, bottom right), another ground sloth, this giant fed on leaves from trees and bushes. Also from South America comes *Glyptodon* (I, *1*), a giant armadillo-like mammal. Its heavy armour suggests harsh predation pressure. It probably fed on plants, carrion, and insects. Leading fossils of the Northern German Diluvian loess are the snails Pupa, Uccinea, Helix and Paludina (*1, 5, 6, 7, 9*). Giant forms were also described for the Pleistocene rhinoceros (III, *1*), the giant Irish elk *Megaceros* (II, *5*), and the mammoth (III, bottom left). The specimens of the mammoth preserved in the permafrost soil showed a heavy fur as an adaptation to global cooling during the Ice Ages (II, *8*). Caves yielded a rich fauna like the cave bear *Ursus speleus* (III, *3*). Also the precursor of the domesticated cattle *Bos primigenius* is documented on this figure (I, *8*). The Irish elk (*Megalocerus giganteus*, II, *5*) shows increased growth of the antlers induced by sexual selection. The female's sexual preferences drove this species into extinction when the food basis of the animal changed with the changing climate. The mammoth (*Mammuthus primigenius*) is an extinct genus of elephants, which was still depicted in the cave paintings of the hunter–gatherers from early Europe.

FIGURE 6.27. Red kangaroo (*Macropus rufus*), one of the few large marsupials surviving the arrival of humans in Australia.

grass and sedge vegetation. At about 10,000 years, the temperatures dropped again, the open steppe-tundra returned, and the animals were again forced out of Europe back to the Urals, where their fate became sealed at 7,000 years ago. Western populations tried to survive by trading reduced sexual display organs against nutritional relieve (Gonzalez et al. 2000). The final blow was probably given by the Mesolithic hunters.

The Woolly Mammoth

A different extinction history was reconstructed for the woolly mammoth *Mammuthus primigenius* (Figure 6.26) (Lister 1993; Stuart et al. 2004; Guthrie 2004). The mammoth is special for extinct animals. You can admire nice specimen displaying trunk and fur in Russian museums because they were exquisitely preserved in the permafrost soil of Northern Siberia. The mammoth is, with the cave bear, a prominent example for paleogenomics. Its genome is about to be reconstructed from frozen samples. Actually, it is a metagenome project since only half of the DNA (but this is a remarkably high amount for ancient DNA samples) is elephant DNA (Poinar et al. 2006). The analysis of the mitochondrial DNA allowed placing the mammoth closer to the Asian than to the African elephant, which separated 6 Ma ago. The mammoth split from the Asian line 0.5 Ma later (Krause et al. 2006). The nonelephant half of the "mammoth" DNA is also interesting for this book since it allows a glimpse into the ancient gut microbiota and the decomposers of the mammoth carcass. Numerically, bacteria dominated the non-mammoth DNA with proteobacteria, firmicutes, actinobacteria, bacteroidetes, and chlorobi bacteria, in this decreasing order. Archaea were found with lesser frequency as were fungal taxa like *Ashyba*, *Aspergillus*, *Neurospora* and *Magnaporthe*. The last rank was filled with soil-inhabiting eukaryotes like *Dictyosteliida* and *Entamoeba*. Although the researchers clearly targeted the ancient mammoth DNA, the associated microbial DNA might be a bonus for researchers interested in the organisms following sequentially in the mineralization of dead mammalian bodies. After this short excursion into paleogenomics, back to the extinction history of the woolly mammoth, which differs from that of the Irish elk. The mammoth's presence in Europe does not show the large gaps as the giant elk; its natural habitat apparently persisted through this time, and it was well adapted to the cold. Warming of the climate replaced the previous rich, mosaic vegetation of herbs and grasses (the steppe-tundra) by a less diverse and less productive boggy tundra, mixed with coniferous forests. The mammoths left Europe to seek refuges in Northern Siberia and on islands in the Arctic Sea. On the Wrangel island, they persisted until 3,500 years ago. Paleo-Eskimos arrived only 3,000 years ago and are thus not a likely cause of their extinction. Food limitations and lack of predation caused the development of dwarf mammoths. Island dwarfism is a common phenomenon and produced 1-m small elephants (*Elephas falconeri*) on Malta. The extinction of mammoths on small islands in the Bering Street was linked to habitat destruction by sea level rises and inbreeding depression by too small populations. Hunting was probably not involved.

Mesolithic Marasm

Large mammals were well adapted to the arid and windy conditions with a treeless, short grass–sedge–sage sward. Mammoth, bison, and reindeer found their different optimal diets and habitats. In North America the transition to the Holocene brought dramatic changes. Landscape changes included the creation of lakes, bogs, shrub tundra, forests, soil paludification, low-nutrient soils, and plants highly defended against herbivory. Horses experienced a rapid body size decline before they went extinct (Guthrie 2003). The decline reflected on one side a restricted access to optimal food sources, but on the other side also increasing competition with other herbivores. Zoologists observed a general replacement of caecalid species that use a large cecum gut-diverticulum in food processing (woolly mammoth, woolly rhino, and horses) by ruminants (bison, moose, and reindeer).

Reindeers, the favorite game animal of the European paleolithic hunters, followed the retreating glaciers to maintain their preferred habitat. With the evasion of their preferred meat source, the hunter-gatherers entered a severe crisis, which is by some historians called the "Mesolithic marasm." The open landscape ideal for hunting disappeared; the retreating glaciers left many lakes and bogs; forests reclaimed large areas. People were plagued by myriads of mosquitoes, and we have in addition ample evidence for strife between the tribes. In Denmark the former hunters lived mainly of mussels, and the diet was sometimes so small that you also find the bones of the earliest domesticated dogs on the refuse heaps. *The cultural heights of the hunters who created 15,000 years ago the wonderful paintings in the caves of Altamira in Spain and Lascaux in France were forgotten. The Mesolithic art is primitive in comparison, the tribes were poor, fighting for their nutritional survival and weakened by frequent feuds with neighboring groups. The time was ripe for a next major invention in the quest for food, which changed not only the foraging strategy of humans but also the face of the planet. Over the last thousand years, humans became the world's greatest evolutionary force linked to mass extinction and habitat destruction that was before only seen with climate changes.*

The Spread of Early Agriculture

A Transition Period

In the words of a prominent scientist, plant and animal domestication is the most important development in the past 13,000 years of human history (Diamond 2002). A surprising observation is that the step from a hunter-gatherer to a farmer society was done at about the same time (i.e., 9,000–4,000 years ago) in several regions of the world: The Near East and China in Asia; the Sahel, West Africa and Ethiopia in Africa; the Andes, Amazonian, Mesoamerica, and the Eastern US in America; New Guinea in Oceania. There is a clear east–west alignment in the Old World imposed by the ease of spread between comparable climate zones along similar latitudes. Only in the New World do we see a north–south gradient dictated by the form of the continent. Europeans tend to look to the Fertile Crescent in the Near East as the cradle of agriculture because here their crops and animals were first domesticated and because this region soon also became the birthplace of their civilization. *Some historians see these links between agriculture and culture as a necessity. However, neither the farmers from New Guinea nor those from the Eastern USA developed a higher civilization.*

Agriculture relieved humans from the obligation to follow the hunted game animals on their migratory paths. They could settle down, found villages, and develop new tools like pottery that was too fragile or forges that were too heavy that hunters could carry them on their migrations. Hunter societies could only grow slowly because over-hunting of their game would destroy their food basis. In contrast, farmers could produce a surplus of food: They could nourish more mouths, and their wives could raise more infants due to the sedentary life style. The storable food sources freed some people from food production, and they became craftsmen, inventors, soldiers, bureaucrats, and nobles. The moving tribe that had to split when it became too big contrasted with the village that could grow, and fuse with other villages to become the many "kingdoms" mentioned in the Bible.

Hardship

One should refrain to paint a too rosy picture of this transition time. Actually, many memories of this historical process were kept in the Biblical report. With a little fantasy, you still recognize the good old time of the heydays from the "Altamira" hunter society in the Paradise. Humans were expelled from this hunter Paradise, where everything was for free; you just had to stretch out your hand. You might want to interpret the archangel who guarded the entry to the paradise as the climate change that occurred after the last glaciation. Adam is condemned to agricultural activity, and he perceives this as a punishment. Early farming was probably a lot of hardship. Indeed, the fieldwork is expressively cursed in the Old Testament. Surely, the regular work on the field was much less fun than following the old human instinct of hunting and gathering, which is still so strong in modern man as everybody can tell from self-observation. Deadly conflicts as reported between the first sons of man ("Adam"), suggesting splits between those that tamed animals and became pastoralists (Abel) and those domesticating the plants and working on the fields (Cain). The former still kept more traditions of the hunter society, but Abel was slain by Cain, disproving from the beginning the concept of the peaceful vegetarian that conquered the world by the superiority of his inventions. The transition to agriculture was really a dangerous, if not murderous period. Medical epidemiologists tell us that many infectious diseases were not efficiently transmitted in the sparsely populated hunter societies. The crowding of humans in villages and their close living with their domesticated animals created new possibilities for the transmission of infectious diseases. It is probably no chance observation that humans share more serologically related microbes with cattle than with lemurs. We should also not keep a too idealistic idea about food abundance in the early farming society. The efficiency of the early crops was still low, the basic inventions in animal husbandry had still to be made. The diet was surely less varied than in the hunter-gatherer society. In short, the transition was characterized by more work, worse nutritional condition as demonstrated by lower adult stature and heavier disease burden.

Anthropologists noted poor dental health, but amazingly also an early neolithic tradition of dentistry in farmers from Pakistan (Coppa et al. 2006). Holes were apparently quite skillfully drilled into the enamel on occlusal surfaces of teeth, and the patients continued to chew on these tooth surfaces. The protodentists probably used a bow-drill tipped with a flint head. This technique was practiced for 1,500 years but abandoned at 6,500 BP. The reason why it was given up is not clear since the population continued to suffer from poor tooth health.

A Point of No Return?

Why should this step be done? There are several answers. First, the early farmers could not foresee the consequence of their action. Another problem was that there was no way back into the hunter society. The big herds had disappeared from Europe, and the growing population needed more and more food, which could anyway not be provided by hunting. After the first growth, there was no way back except in the form of a collapse of the civilization. In the Classic Maya period at about 750 AD, the Yucatan lowlands counted perhaps up to 10 million inhabitants. The Maya experienced after the Terminal Classic period a demographic disaster as profound as few in human history. Multiyear droughts ruined the food basis of this society, and the densely populated urban centers were abandoned permanently (Haug et al. 2003). How vulnerable even modern societies are with respect to their agricultural basis is illustrated by the Dust Bowl event in the USA.

The Expansion

However, once established, the farmer society developed a dynamic that pushed aside all hunter-gatherer societies. For example, European agriculture originated in the Near East about 9,000 years ago. The Neolithic revolution reached in Europe almost all areas suitable for agriculture, and the precise contour waves of its progress can be projected on the European map in 500 years intervals starting from east to west Anatolia, reaching the Aegean basin from where three different trajectories can be deciphered. A southern route goes westward over Italy, Southern France to the Iberian Peninsula. The propagation was quick with a spread of 10 km per year suggesting a maritime transport. The neolithic populations can easily be identified by their fully terrestrial diet, while the Mesolithic populations showed a 50% marine component in their diet. In Portugal, a stepwise replacement of late Mesolithic by early neolithic settlements occurred between 6,000 and 5,000 years ago (Zilhao 2001). A second wave goes from the Aegean world into what is today Romania where it split into a western wave progressing via Germany to Ireland and another wave via Ukraina, Russia, and Scandinavia (Sokal et al. 1991). However, it was not only the new agricultural technique that spread, but at the same time the Indo-European language family replaced all other dialects in Europe (and as the name suggests also to the east as far into Asia as India), with the remarkable exception of the Basque language.

The Ukranian Horse Versus the Anatolian Cattle

As we will see below, the domestication of the major European crops originated in a small region of eastern Anatolia. Anatolia is also the home of one, if not the oldest Indo-European language, the Hittite. Was the Proto-Indoeuropean language given as part of the cultural package that included agricultural techniques as well as domesticated crops and animals (Diamond and Bellwood 2003)? Linguists have tried to deduce the home of the Proto-Indoeuropean by searching their shared words. The most widely shared vocabulary concerns the horse associated with transport (wheel, chariot) and riding, then comes cattle herding and milking, followed by words describing many other domesticated animals (pig, sheep and wool, goat, dog, goose, duck, bee). Interestingly, plant crops and trees do not belong to this linguistic package. This is the dictionary of a pastoral society, not that of a sedentary, agrarian crop-oriented population. Was horse the motor of this dispersal of the early agricultural life style? The horse had multiple functions: They were eaten by the hunters, used for working on the field and also for riding. Riding became, however, only possible with the invention of the bit. Archaeologists argued that the use of the bit should leave a trace on the teeth of the horses, and they searched for the earliest signs of horse teeth with signs of wear. This observation should lead them to the propagator of the Proto-Indoeuropean language. The oldest finds of horses with used teeth were preserved by the Ukrainian Sredny Stog culture dated to 4000 BC (Diamond 1991). This archaeological evidence fits remarkably well with the linguistic analysis of N.S. Trubetzkoy, who placed the home of this language family to the north of the Black Sea. It fits also with the superior role that the horseman played in military-driven colonization throughout history as still demonstrated much later by the conquest of Latin America through the Spaniards. Does such a warrior model speak in favor of a stronger demic diffusion than in the case of the spread of Anatolian cattle culture? Human geneticists are undecided, they speak currently of a 22% Neolithic versus a 30% Paleolithic contribution to the current European gene pool, and they distinguish with classical markers a southeast to northwest gradient. Y chromosome and mitochondrial DNA data also show gradients, but they are difficult to compose in a single picture. *As in the analysis of the central metabolism, one can borrow here the term "palimpsest" from archaeologists, a document that was overwritten in later time but which still allows to read partly the original text. Too many later genetic intrusions have occurred in the long history of Europe to expect a clear message.* The situation is clearer in other regions of the world. Genetic evidence for the demic diffusion model of agriculture could be provided for India (Cordaux et al. 2004). Demic diffusion of the Han languages in China could also be demonstrated.

Domestication

The Garden of Eden: Domestication of Crops

The Garden of Eden at Karacadag?

Let's now review the paleobotanical evidence, which allows to localize the Garden of Eden or at least the East of Eden. This claim sounds astonishing, but the evidence is quite convincing. Near East neolithic agriculture was based on three cereals (einkorn wheat, emmer wheat, and barley (Figures 6.28 and 6.29)), four pulses (lentil, pea, chickpea, and bitter vetch), and a fiber crop (flax). If you map the native distribution area of these plants, you define a core area between the upper Tigris and Euphrates rivers as the center for the innovation. Apparently, the cool, dry climate of the Younger Dryas (about 9000–8000 BC named after the Alpine plant *Dryas octopetala*) triggered the end of the nomadic life style and the beginning of farming. A number of archaeological data support this conclusion: Excavations indicate a wealthy society with plenty of food that used stone sickles and "stepped" stone quern for grinding cereal crops. Glossed flint sickle blades from the Jordan valley suggested that wild emmer wheat was harvested as early as 7800 BC (Lev-Yadun et al. 2000). Wheat became the major cereal in Europe and countries colonized by Europe. The wild einkorn wheat *Triticum monococcum* subspecies *boeticum* is the wild relative of

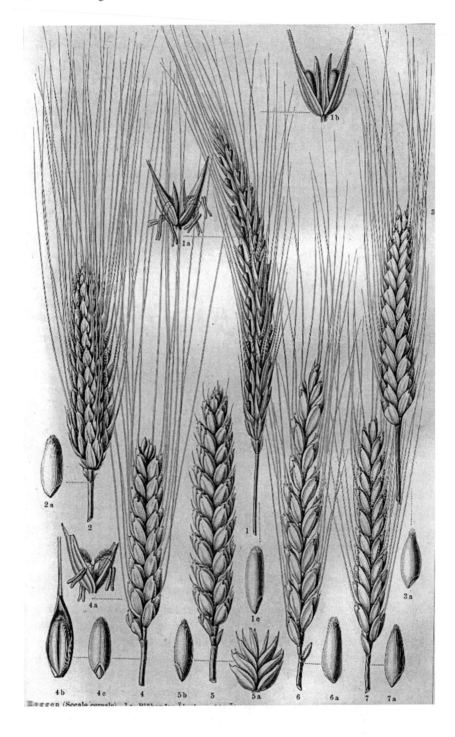

the domesticated einkorn wheat *T. monococcum* subspecies *monococcum*. Plant geneticists characterized 1,300 lines of einkorn, which were localized to 5 km with respect to their origin, for agronomical and taxonomical traits (Salamini et al. 2002). The genetic distance within this group was evaluated using multiple DNA markers. Phylogenetic tree analysis showed that one group originating from the volcanic Karacadag mountain in southeast Turkey is distinctly separated from the remaining groups, while all cultivated "einkorns" are closely related among themselves. This observation raises the possibility that the Karacadag lines are the closest relatives of the wild progenitor that gave rise to cultivated einkorn about 10,000 years ago.

The Selection Process

The morphological traits of the Karacadag lines show, however, a wild cereal with low seed weight and a very brittle rachis. The early farmers selected for three basic characters when dealing with wheat. For an optimal nutritional value, they selected first for larger seeds. But this was not enough; in the wild forms, the spikelets of the ear fall apart at ripening through fragmentation of the rachis ("shattering"). The early farmers therefore selected for plants with a tough rachis. This is lethal for a wild form since it would effectively prevent dispersal of the seeds but an essential property for harvesting.

As shattering is such a visible marker of domestication, agricultural archaeologists working in the Near East could date the domestication process (Tanno and Willcox 2006). They investigated many charred wheat spikelets and dated the sites. Wild cereals with dehiscent ears shatter at maturity into spikelets identified by smooth scars. Domesticated cereals do not shatter, but separate when threshed producing jagged scars. A site dated to 10,000 years BP yielded no indication for domesticated wheat. This mutant form represented 30% of the spikelets between 9,300 and 8,500 and rose to 60% by 7,500 years BP. The mutant took thus more than a millennium to become established. The archaeologists suspected that the early farmers probably harvested before the spikelets fell to avoid loss of grains—the work of selection was thus slowed down. Also selection for larger grain size was slow, the size was maintained between 9,500 and 6,500 years BP.

The selection for reduced shattering was also crucial for rice domestication. Plant geneticists working with the domesticated rice *Oryza sativa* and *O. nivara*, a closely related wild relative, identified a genetic locus where one allele made that

◄────────────────────────────

Figure 6.28. Wheat and rye. 1-Rye (*Secale cereale*) with the flower (*1a*) and grain (*1c*). The other depicted plants are various varieties of wheat: *Triticum turgidum* (*2*) was cultivated in England, the diploid *T. monococcum* (*3*) and the tetraploid *T. dicoccum* (*7*) was popular in Germany, but have largely disappeared from the fields; the tetraploid *T. durum* (*4*) is used to make pasta. The most important variety is the hexaploid *T. aestivum* = *T. vulgare* (*5*), which is used to make bread. *T. spelta* (*6*) is still relatively popular in southern Germany.

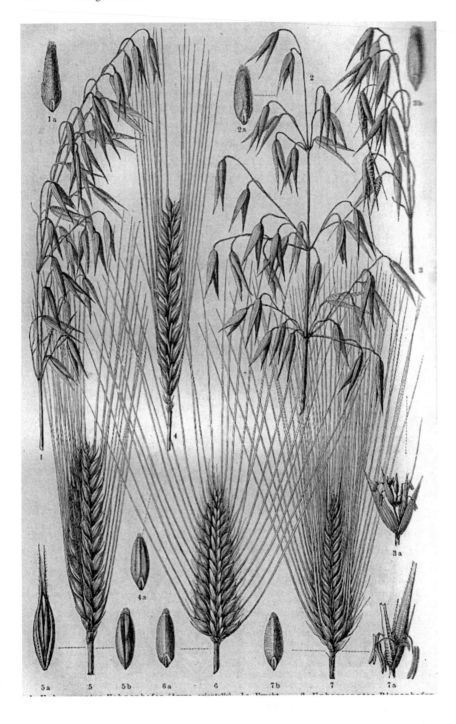

the plant shed all mature grains when hand-tapped, while another allele allowed only partial shedding under vigorous hand shaking. Many of the agronomically important characteristics of domesticated plants are determined by quantitative trait loci (QTL). These traits are typically affected by more than one gene, as well as by the environment. There can also be interactions between QTLs ("epistasis"), which further complicates analysis. The rice geneticists were lucky: They could link the reduced shattering to a single gene expressed in the pedicel junction, where mature grains separate from the mother plant (Li et al. 2006). This is a logical finding. The falling of old leaves and of ripe fruits is crucial for proper plant function and regulated in an abscission zone. In *O. nivara* a nicely marked abscission zone consisting of a one-cell layer is histologically clearly visible, while in *O. oryzae* this layer is discontinuous and even lacking near vascular bundles. Domestication of cereals was a delicate balance. The abscission zone should be sufficiently stabilized to prevent spontaneous shattering after grain maturation. However, the abscission zone should not be too much suppressed to avoid problems with threshing. Yet, there are still other problems with domestication that had to be solved.

In the wild form the leaf-like structures that protect the seed ("glumes") are attached tightly to the seed ("hulled" genotype). The farmers selected forms that released the seed during threshing ("free-threshing" genotype). The first domesticated einkorn was a diploid species. This form was abandoned in the Bronze Age and replaced by a tetraploid form. This form was accidentally crossed with the wild cereal *Aegilops tauschii*, yielding a hexaploid hybrid, which was most suitable for baking and became thus the world's leading crop. This hexaploid bread wheat (*Triticum vulgare*) and the tetraploid hard wheat (*T. durum*) represent the final steps of *Triticum* domestication. The genetic analysis of the mutant genes in wheat is complicated by the high degree of ploidy and the large genome of this crop (Salamini et al. 2002).

Still Earlier Steps in the Jordan Valley

Some plants left their imprints in the Bible: in the Old Testament the primogenitur of a hunter (Esau) is sold against a lentil meal from a farmer/pastoralist (Jacob). The choice of the lentil as the plant with the closest amino acid profile to animal

---

FIGURE 6.29. Barley and oats. At the *top*, you see three varieties of oats: *Avena orientalis* (*1, 3*) and *A. sativa* (*2*), only the latter is economically important. Oats are mainly used a livestock feed, but they have also a place in breakfast cereals and cookies. At the *bottom*, different varieties of barley are depicted: *Hordeum distichum* (*4*), *H. vulgare* (*5*), *H. zeocriton* (*6*), and *H. hexastichum* (*7*). Half of the barley is used as livestock feed, the rest goes in human consumption; 10% is used in beer production. The history of wheat (*T. dicoccum*) and barley cultivation reaches back to the neolithical revolution. Barley was the main bread cereal of the land of the Bible, Greece, and Rome and remained in Europe the prominent bread cereal until the sixteenth century. Rye and oats were introduced only in the Bronze and Iron Age.

proteins is probably an intended allusion. In the New Testament the fig tree (*Ficus carica*) plays a prominent role in parables. We should thus not be surprised when Israeli agro-archaeologists recently came up with good evidence for the domestication of figs in a pre-pottery Neolithic village near Gilgal in the Jordan valley (Kislev et al. 2006). The finds consisted of carbonized figs and drupelets that were stored in a site dated to 11,400 years ago and thus fully a thousand years before previous early estimates for the Neolithic Revolution. A major player in this plot is the parasitoid fig wasp *Blastophaga* that inserts her ovipositor into the style of the female flower. This infestation prevents the dropping of the fruit and allows maturation of the fruit. Actually, villagers picked up a mutant that evolved parthenocarpy—the development of a fruit without pollination and fertilization. The fruit became soft, sweet and edible, while the wildtype was nonpalatable. Without human intervention, the mutant would have been lost due to its sterility—embryoless fruits do not produce new trees. However, figs were an easy early domestication because cuttings of the sterile fig developed roots more easily than any other fruit tree and could thus be propagated vegetatively. This easy growing of fig roots should not surprise since the *Ficus* genus is known for its abundant growing of air roots (Figure 6.30). The genus comprises a number of species (e.g., *F. benjaminica*) that literally strangulate trees on which they grow by descending a large mass of air roots. In Gilgal the figs were stored

FIGURE 6.30. Ficus trees with characteristic oversized roots. The strong development of roots allows parasitic ways of life to several members of the *Ficus* genus. However, the ease of growing roots from a cut branch of *Ficus* is also at the basis of the vegetative propagation of *Ficus carica* which was the key to the early domestication of the fig in the Jordan valley.

with wild barley and wild oat suggesting a subsistence strategy consisting of a gathering of annual plants from wild stands together with early domestication.

The ancient world revered the fig and its tree, it was associated with the cult of Bacchus; figs became the food at the public table of the Spartans; figs were the major source of the food for the slaves from antiquity. In public display the fig tree gave shadow to the founder twins of Rome in front of their wolf cage and the tree became a symbol of prosperity.

Surviving stores of lentils from the same period showed that wild lentils (*Lens orientalis*), the progenitor of *L. culinaris*, were gathered in Syria. Its domestication was initially impossible due to a 90% seed dormancy. Since wild lentils yield only 10 seeds per plant, its cultivation was impossible. Israeli archaeologists dated a hoard of a million lentils found in the Biblical city of Nazareth at 8,800 years ago. As this store of lentils was contaminated with seeds from a weed of present-day lentil fields, namely *Galium tricornutum*, the lentil dormancy mutant was found and the lentil was cultivated by this time. A large granary with wild barley (*Hordeum spontaneum*), the progenitor of *H. vulgare* contaminated with wild oat (*Avena sterilis*) was also found in Gilgal, the site of the fig findings. Clearly, the inhabitants of this village sowed a number of wild crops and were engaged in predomistication cultivation at this early time (Weiss et al. 2006).

Tomato Fruit Size

In contrast to the slow increase in wheat grain size, rapid evolution of fruit size has accompanied the domestication of most fruit-bearing crop. The precursor of the domesticated tomato (*Lycopersicon esculentum*) probably weighted only a few grams. Fruits are seed dispersal organs of plants, they must thus be sufficiently large to house enough seeds, but small enough to be commensurate with the size of the animals, which disperse the seeds. On most continents these are small rodents and birds setting a clear upper limit for the fruit size. A notable exception is New Zealand, whose isolated geographical island position prevented the colonization with small mammals. Here giant, 7-cm-long, flightless grasshoppers (*Deinacrida rugosa*) take at least for some plants the role of small rodents as seed disperser. Their seeds pass the intestinal tract of the insects intact and show even a higher germination rate than undigested control seed (Duthie et al. 2006). A tomato fruit larger than a 1 cm would be unsuitable for seed dispersal. However, tomatoes have a substantial unused genetic capacity: Modern tomatoes can weigh as much a 1 kg. US plant geneticists identified a QLT, which contributed about 30% of the fruit size (Frary et al. 2000). They narrowed their analysis down to a single gene whose highest expression level was in carpels, the structure which ultimately develops into fruit. Carpels of large-fruited tomatoes were already heavier at flowering. The gene controlled the number of carpel cells. The mutation was located in the upstream promoter region of the gene and the geneticists suspected a negative regulator of cell division as the gene function, which concurs with similarities to the *ras* oncogen.

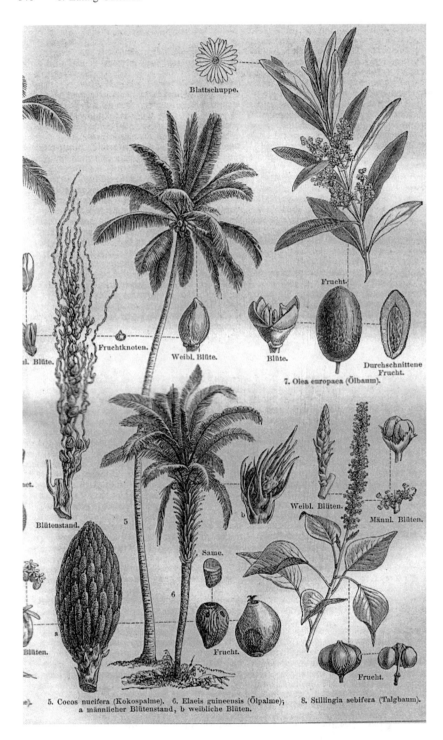

Blattschuppe.

Fruchtknoten.

Weibl. Blüte.

Blüte.

Frucht.

Durchschnittene
Frucht.

7. Olea europaea (Ölbaum).

al. Blüte.

net.

Blütenstand.

Blüten.

5

6

Same.

Frucht.

b

Weibl. Blüten.

Männl. Blüten.

Frucht.

5. Cocos nucifera (Kokospalme). 6. Elaeis guineensis (Ölpalme);   8. Stillingia sebifera (Talgbaum).
a männlicher Blütenstand, b weibliche Blüten.

Ethiopia

Not everything was invented in the Fertile Crescent (Figure 6.31). Important mutants were selected in other cradles of agriculture like Ethiopia. Resistance to fungal pathogens is an important trait for a cereal to become an important crop. Barley (*Hordeum vulgare*) has played an important role in Old World agriculture. The majority of the cultivated elite varieties of barley are resistant to the widespread powdery mildew fungus. Plant geneticists could now show that the resistance is the consequence of a loss of gene function. The seven-membrane Mlo protein is not any longer expressed because a complex tandem repeat array in the 5' regulatory sequence leads to an aberrant transcript. Since this gene product is essential for the infection process, the mutant shows a strong resistance phenotype. In line with this hypothesis, excision of the repeat induced in the laboratory led to the restoration of the Mlo function and restored also the sensitivity to the fungus (Piffanelli et al. 2004). Interestingly, this mutation was detected in landraces from traditional farmers in Ethiopia during the 1930s and might reflect the breeding efforts of their ancestors to achieve pathogen-resisting barley.

Maize in Mesoamerica

We get a genetically even clearer picture when changing to maize (Figure 6.32), a crop that was domesticated in Mesoamerica. Archaeological research in the humid lowlands at the Gulf Coast of Mexico provided the earliest record of maize cultivation. Large pollen typical of domesticated maize (*Zea mays*) appeared about 5000 BC. A single manihot pollen characteristic for domesticated manioc was sighted at 4600 BC, while sunflower seeds and cotton were only spotted 2500 BC when farming had already expanded (Pope et al. 2001).

Already in 1939 the prominent geneticist George Beadle argued that maize could have been selected as a new variant of teosinte. He speculated that maize was only separated by five major mutations from the wild plant

FIGURE 6.31. Fat- and oil-providing plants. At the *left* you see the coconut (*Cocos nucifera*, 5), which yields cocobutter, which is rich in saturated fatty acids and free of cholesterol. It is used for margarine, ice cream, and chocolate. This palm tree (up to 30 m tall) was domesticated 3,000 years ago in Malaysia and was exported from there over the entire Pacific and Indian Ocean. Next to it is the oil palm (*Elaeis guineensis*, 6) from West Africa. With a harvest of 6 tons palm oil per hectare of cultivated land, it provides the highest yield for any oil-producing plant and is after soybean the second important source of vegetable oil. The two palms are shown with their male (**a**) and female (**c**) flowers. The fruit of the olive (*Olea europea*, 7) contains 40% oil and has thus the highest caloric content of all known fruits (200 kcal/100 g). It was cultivated in the Near East and became a characteristic tree for the Mediterranean basin from antiquity to now. From Biblical times onwards it was associated with many values, the olive twig became a symbol of conciliation and peace. *Stillingia sebifera* (8) is a Euphorbiaceae from China, its seeds are surrounded by a fat layer.

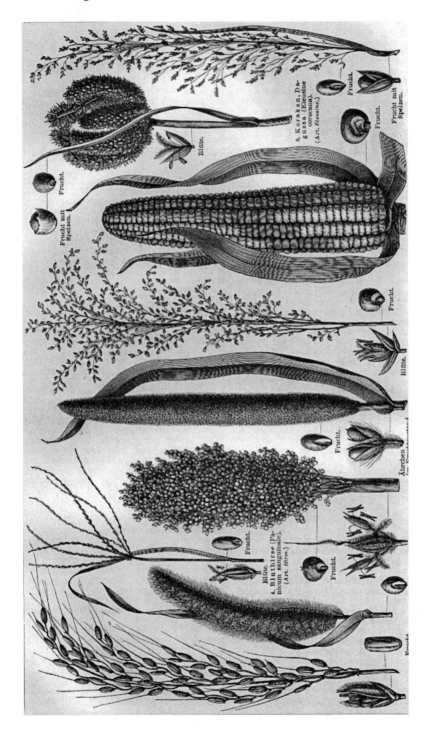

(Martienssen 1997). One of these major mutations was indeed identified in 1997 and called *teosinte branched 1*, (in short *tb1*; Doebley et al. 1997). It affects the apical dominance of the plant. Whereas teosinte plants bear many long axillar branches that are tipped by male inflorescences (tassels), maize plants are upright and not branched and show only few, short axillar branches tipped by female ears. The growth of the ears is controlled by *tb1* and the *terminal ear 1* genes. The *tb1* gene was cloned and offered a surprise: it did not differ from the gene sequence of teosinte. The protein showed similarity to the *cycloidea* gene product from snapdragon, which decides whether a radially or a bilaterally symmetrical flower is built. However, the maize *tb1* gene is expressed at twice the mRNA level of the teosinte allele. This observation suggests that gene regulation changes underlie the transition from teosinte to maize. This was proven subsequently.

The plant breeder selects seeds from the preferred form and culls out the undesired forms during a series of crossing experiments. The trait of interest increases in frequency and becomes thus finally fixed. One should thus expect that the genetic diversity is drastically decreased in the crop when compared to the wild type. This was, however, not the case for the protein-coding part of the *tb1* gene. This explains why maize is still today such a variable crop. However, the regulatory 5' noncoding part of the *tb1* gene showed evidence for a strong selection (Wang et al. 1999). The geneticists deduced that it took only a few hundred years to fix the *tb1* mutation and that Balsas teosinte from the southwestern Mexico was the ancestor of maize.

Plant geneticists suspect, however, that maize domestication was a complex process that acted on a large number of traits and genes. Important morphological differences between teosinte and maize are currently attributed to genetic changes in five chromosomal regions. Recently, US scientists added another gene to this list (Vollbrecht et al. 2005). It is called *ramosa1* and it is a transcription factor, which controls inflorescence architecture by imposing a short branch identity. The mutant was detected a century ago on the field of a farmer. The plant showed branched tassels and ears and looked so different that it was initially described as a new species, *Zea ramosa*.

The Southward Spread of Maize

The rise of historical states in ancient South America like that of the Wari or Inca was maize-dependent. However, the arrival of maize from Mesoamerica

◄——

FIGURE 6.32. The figure shows extra-European cereals, from *left to right*: rice (*Oryza sativa* with flower and grain, *1*) and various forms of millet—the foxtail millet *Setaria italica* (*2*), Durra *Sorghum* (*5*), pearl millet *Pennisetum* (*6*) and the broomcorn millet *Panicum miliaceum* (*3*), *Panicum sanguinale* (*4*). At the *right* side you see the finger millet *Eleusine coracana* (*8*) and the teff *Eragrostis abyssinica* (*9*). Millets are important staple food in Asia and western Africa. During the Middle Ages millet provided major grains in Europe. America's contribution to the cereals is maize or simply corn (*Zea mays, 7*).

into the Andes was poorly documented. Recent archaeological evidence pushed the beginning of Andean maize cultivation back to 4000 BP and that at an impressive altitude of 3,600 m above sea level in Peru (Perry et al. 2006). At this level two agricultural zones meet: In the temperate *qheshwa* zona between 2,300 and 3,600 m, maize is produced under intensive irrigation, while in the cool *echadero* zone from 3,600 to 4,000 m, tuber cultivation is practiced. The investigation of starch granules and phytoliths identified mainly maize, half of them with evidence of onsite processing into flour by grinding. Interestingly, the agricultural archaeologists found also remains of arrowroot (*Maranta*), which is a typical plant of the lowland tropical zone that cannot be grown above 1,000-m altitude. The scientists interpret this as evidence for the deliberate transport of plant foods and thus human traffic between the tropical forest and the highland.

Agricultural Archeology

An outpost of the Smithsonian Institute in Panama showed that simple microscopic methods could contribute to the elucidation of the history of early plant domestication in humid tropical areas that are hostile to the preservation of organic remains. In one study, they used phytoliths, microscopic siliceous remains of plants. In a prehistoric site from coastal Ecuador, they found phytoliths with forms and sizes characteristic for domesticated *Cucurbita* (squash and gourd), which produced oil- and protein-rich seeds appreciated by the latest foragers and earliest farmers in America (Piperno and Stothert 2003). The remains were dated to 9,000 years ago. Since stone tools occasionally contained still identified remains from blood, the researchers argued that milling stones might preserve in crevices starch grains that can be identified and dated and thus allow to retrace part of the domestication history of roots and tubers that are otherwise only poorly documented in the archaeological record. In fact, on milling stones from a shelter in Panama dated between 7,000 and 5,000 years ago, they found starch grains (Figure 6.33) identifiable as manioc (*Manihot esculaenta*), yams (*Dioscorea*) and arrowroot (*Maranta*) mixed with maize starch (Piperno et al. 2000).

A Filled Cooking Pot

In fortunate situations we can have a direct look into the food bowl and are not obliged to rely on microscopic analysis of the food item. A nice archeological paper (Lu et al. 2005) describes a Late Neolithic dish, which might also provide a strong argument on the question who invented noodles. Spontaneously, you might associate Italians with this popular staple food, but also Arabs and Chinese claimed to have invented them first. Some data point to pasta that was already prepared by the Romans, giving noodles the respectable age of 2,000 years. The abovementioned bowl was filled with what looks like noodles. It was found in a 4,000-year-old Neolithic settlement near the Yellow River in China. To make the Chinese claim to the invention of noodles clear, the Chinese authors noted a resemblance between these 0.3-cm thin, 50-cm long yellow noodles

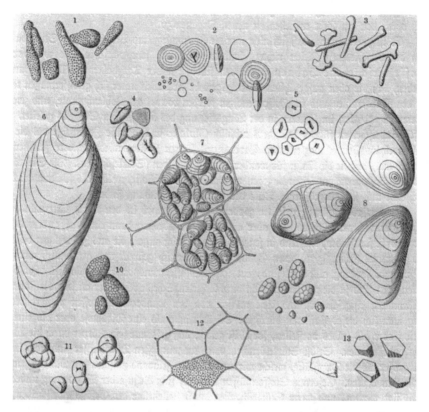

FIGURE 6.33. Agricultural archaeologists deduce the nature of domesticated plants from starch remains associated with stone tools on. This is possible since different plants show distinct forms of starch grains as demonstrated here with some examples (*1* Agrostemma; *2* wheat; *3* Euphorbia; *4* beans; *5* maize; *7, 8* potato; *9* oat; *12* rice; *13* millet).

with modern La-Mian noodles from China. Bad luck for the Roman noodle's claim to fame and antiquity. The authors provide us some further details on these proto-noodles. Modern pasta and noodles are made from durum and bread wheat. Phytolith and starch grain analysis, however, identified modern millet as the basis of this early food item, which would also explain the very delicate character of these noodles.

New Guinea

Also the remote New Guinea belongs to the cradles of agriculture. As the early farmers collected plants from their environment, each center has its own local character. In excavations of the Kuk Swamp, archaeologists found stone tools dated to 7,000 BP. They contained raphids (calcium oxalate crystals) and phytoliths of two starch-containing plants. They were identified as remains from taro (*Colocasia esculenta*) and banana (*Musa*, for other tropical fruits

FIGURE 6.34. Tropical fruits. From *left to right, top*: guava (*Psidium guajava, 3*) from Central America, one of the richest sources of vitamin C; persimmon (*Diospyros kaki, 2*) from China and Japan is rich in vitamin A; an "apple" from Malaysia (jambu, *4*); the custard apple (*Annona, 1*). *Bottom*: sapodilla (*Achras zapota, 6*) from Central America—a traditional food plant of the Mayas and Aztecs, initially a basis for the chewing gum; the avocado (*Persea americana, 7*) is an unusual fruit since it was appreciated by the Aztecs more for its lipid content (linoleic acid) than its sweetness; cashew nut (*Anacardium occidentale, 8*) from Brazil; mango (*Mangifera indica, 9*) was cultivated 4,000 years ago in Indomalaysia; mangosteen (*Garcinia mangostana, 10*) from Malaysia is rated as one of the most delicious tropical fruits.

see Figure 6.34), the most important food staples in New Guinea before the European introduction of sweet potato. At the same time mountain rainforests were cleared by fire to grassland. Soon followed wetland agriculture with mounding (phase 2) and ditched cultivation (phase 3; 4,000 years ago; Denham et al. 2003). However, in other respects New Guinea did not follow the familiar aspects of agricultural societies in other world regions. New Guinea represents one of the most colored patchwork of languages, no demic diffusion was observed and no social stratification and civilization developed in New Guinea. The invention of agriculture is thus not a one-way into a technical culture.

Taming the Beast

Conditions for Breeding

Despite substantial variability in plant and animal species that were domesticated in the different world regions, it remains striking that a relatively small set of species got a place in agriculture. Humans have tried in vain to domesticate many animals, a number of examples are documented in the written documents of ancient Egypt that will amuse a modern farmer. A more recent famous fiasco is the unsuccessful taming of the zebra by European settlers in South Africa. The zebra turned out to be incurable vicious and endowed with a much better peripheral vision than horses. They escape the lasso even of experienced rodeo cowboys. There is thus a minimal list of requirements the animal breeder expects to be fulfilled (Diamond 2002): The diet should be easily provided, the growth rate should be quick, no long birth spacing, no nasty disposition (aggression toward the breeder), possibility to breed in captivity, and no panic in enclosures, and the animal should follow the leader dominance of the breeder.

Goat

Notably, we have to return to the Fertile Crescent when we want to retrace the early history of animal domestication. The initial domestication of goats (*Capra hircus*) could be traced to 10,000 years ago in the Zagros mountain of western Iran. The researchers deduced their claim from studying the distribution of animal bones associated with human activities. First they established a bimodal distribution curve between the sexes and then searched for the selective culling of young males and delayed slaughter of females as indication of herding. In hunted populations, adult males were well represented, then the first shift was seen when herding started first within the geographical distribution range of goats in the mountain, followed by dispersal of goat herding into the lowlands (Zeder and Hesse 2000).

Cattle

The limited geographic ranges of the wild progenitors of many of the primary European domesticated species excluded a central European origin for them and pointed again to eastern Anatolia or the Near East as their homeland. This is not the case for the cattle, humans' favorite beast of burden. Central Europe knows the wild ox (*Bos primigenius*) or aurochs, which became extinct in the seventeenth century when the last animal was killed in Poland. Scientists collected 400 samples from Europe (including four extinct British wild oxen from museums), Africa, and the Near East, and determined mitochondrial DNA sequences. They argued that genetic loci from a center of origin are expected to retain more ancestral variation, while lineage pruning through successive colonization should lead to a reduction of genetic variability in derived populations. Phylogenetic tree building set first as expected the zebu (*Bos indicus*) apart from the European ox (*Bos taurus*); they are separated by about 100,000 years. It should be noted that the distinction of both oxen as different species is not justified from the biological species definition. Both taxa are completely interfertile. The aurochs from the museum clusters near the *B. taurus* branch. The African cattle diversity is clustered around a haplotype that is absent from the European samples and occurs only with low frequency in the Near East. The haplotype analysis traced the origin of the European cattle to east Anatolia and the Near East (Troy et al. 2001). What about the situation of cattle domestication in Africa? Livestock research institutes located in Africa set out to sample cattle from 50 populations of 23 African countries and investigated them for 15 microsatellite markers (Hanotte et al. 2002). *They got some advance help from painters. Rock artists showed that the earliest African cattle was the humpless* B. taurus, *the humped cattle was first depicted in Egyptian tomb paintings of the XII dynasty. As suggested by the painters, the geneticists got a mixed message. The zebu influence was the highest at the horn of Africa and suggested import of the zebu by Arabian contact. The zebu was actually domesticated in the Indus valley 6,000 years ago. Arabs brought the zebu together with the domesticated chicken and the camel. These contacts were no one way paths since India received cereals domesticated in Africa like sorghum and the finger millet. However, the initial expansion was by* B. taurus *most likely derived from a single region of origin and it reached the southern part by traveling an eastern route. The genetic structure of the African cattle population was markedly influenced by the rinderpest epidemics in the late nineteenth century and by the fact that it experienced population bottlenecks due to attrition from tropical livestock diseases.* The statistical analysis of the genetic data pointed to at least three centers of domestication contributing to the current genetic make-up of cattle in Africa.

Horse

An even more complicated genetic structure was revealed when about 200 domesticated horses were investigated for mitochondrial DNA sequences. In the analysis the modern horse lineages coalesce only 300,000 years BP, much too early when compared to the domestication of horses, which took place according to archaeological evidence 6,000 years ago. Numerous matrilines (mitochondria are only inherited from the mother) must have been incorporated into the gene pool of the domestic horse. The genetic data suggest that horses were initially captured over a wide geographical area and used for nutrition and transport. When the supply of wild horses dwindled because of overexploitation or environmental changes, captive breeding started in many different regions. In the history of horse use, the technique of capturing, taming, and rearing of wild animals was transferred between the populations and not the domesticated breeds (Vila et al. 2001).

Pig

Archaeological record of pig bones suggested that pigs were first domesticated 9,000 years ago in the Near East. We have here a case as in cattle where central Europe has an own wild boar population. Did the domesticated pig come in an early agricultural import package with the cattle and sheep from the Near East? Mitochondrial sequences from 700 wild and domesticated pigs revealed that the European wild boar is the principal source of modern European domesticated pigs. The sequence data placed the origin of the wild boar in Southeast Asia from where they dispersed across Eurasia. There were also multiple centers of pig domestication across Eurasia (Larson et al. 2005).

Domestication of Moulds: Aspergillus

Microorganisms are frequently overlooked when recalling the history of domestication despite the fact that they played important roles for food and feed production. We mentioned already the important cultural role of lactic acid bacteria in milk fermentation, practiced in dairy countries. A similar cultural role was played by *Aspergillus* moulds for the fermentation of soy sauce, miso, and sake in Asia and especially in Japan. *Aspergilllus* (Figures 6.35 and 6.36) belongs to the Deuteromycetes (Fungi imperfecti) and produces sexual ascospores or asexual conidiospores on a highly characteristic structure called an aspergillum. Its recruitment into food production also involved substantial taming. In the *Aspergillus* genus you find one of the most useful food fermenting microorganism, *Aspergillus oryzae*, which is the source of industrial enzymes and organic acids, but also *A. flavus*, a saprophyte growing on crops like corn, peanuts, cottonseed, and tree nuts. However, it is not just a food spoilage organism: *A. flavus* produces aflatoxin B1, which is one of the most potent liver toxins. In fact, a strong epidemiological link between dietary exposure to aflatoxin and hepatocellular carcinoma was demonstrated in populations from China and sub-Saharan Africa. Scientists showed that this was not a trivial domestication process. *A. oryzae* was used for thousands of years in oriental food manufacturing and carries its scientific name from its common association with rice plants. In old Japanese reports, the seed culture was derived from rice smut, where *A. oryzae* is found in association with a fungal plant pathogen. Also the analysis of the genome sequences from *A. oryzae* demonstrated that it was itself once a toxic mould. It contains all the 25 genes that *A. flavus* needs for the biosynthesis of aflatoxin (Machida et al. 2005). However, in *A. oryzae* these genes are not expressed, possibly due to a regulatory mutation. *A. oryzae* shows a 37-Mb genome encoding 12,000 genes and is thus 8 Mb larger

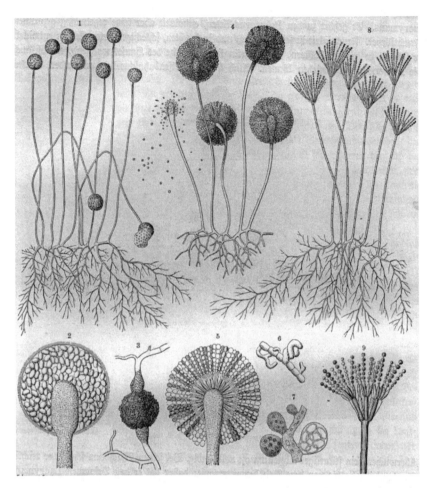

FIGURE 6.35. Different fungi. *1–3 Mucor mucedo* (order Zygomycetales, class Phycomycetes, 2 sporangium), *4, 5, 7 Aspergillus niger* and *6, 8, 9 Penicillium crustaceum* (order Plectascales, class Ascomycetes). *2, 5, 9* sporangia with conidia. These organisms are commonly called moulds. Mould is not a systematic name, only the indication for superficially growing fungal hyphae. Moulds play an important role in food processing, both in the positive and in the negative sense.

than the genome of the pathogenic and allergenic fungus *Aspergillus fumigatus*. The latter mould is a prolific conidia producer, which might explain its allergic properties. However, none of the nine major allergens is a spore surface protein. *A. fumigatus* is an opportunistic pathogen in immunocompromised patients, but the genome sequence showed few cues to the underlying pathogenic mechanisms. The authors observed many genes that are upregulated at temperatures higher than in the environment (Nierman et al. 2005). However, this does not reflect an adaptation to the warm-blooded host. *A. fumigatus* is quite thermotolerant (up to 70° C) because it is frequently isolated from compost. Its primary niche is rotting vegetable.

FIGURE 6.36. Hundred years later: molds seen with the scanning electron microscope. Sporangia with conidia from the molds *Aspergillus restrictus* (*top*) and *Syncephalastrum racemosum*.

Alignment of these two and a third *Aspergillus* genome revealed the presence of similarly organized genome segments (syntenic regions) and *A. oryzae*-specific genome segments in a mosaic pattern. The latter gene segments are enriched for genes involved in metabolism. *A. oryzae* produces copious amounts of industrially useful enzymes like amylases and proteinases. The reason is clear: Within the koji culture, *A. oryzae* grows on the surface of steamed rice or ground soybean, where it has to digest starch and proteins to get the sugars and amino acids it needs. In fact, *A. oryzae* encodes 135 secreted proteinase genes, more than 1% of the total genome is dedicated to this single task. Even more impressive is the expansion of genes involved in secondary metabolism. However, in contrast to yeast, no duplication of the *A. oryzae* was diagnosed; the *A. oryzae*-specific DNA segments are most likely the result of horizontal gene transfer.

Fishery

Contemporary Fishery Problems

Sustainable Fishery?

The introduction of a new top predator, humans, into the food chain raised a number of problems. I will illustrate the problem with some figures taken from a report that questioned whether the current rate of global fishery is sustainable

(Pauly and Christensen 1995). If you subdivide the global primary productivity in dry weight of annually created biomass, about 60% is produced by terrestrial ecosystems. Humans take a disproportional part of this primary productivity: estimates are between 35 and 40%, which is used directly (food, fiber) or indirectly (animal feed) or foregone (urban sprawl). The authors calculated that 8% of the world's aquatic primary productivity is required to sustain the current rate of fishery. This figure was based on the world fishery statistics from 1988 to 1991 (annually 95 million tons of fish), augmented by the discard of bycatch (27 million tons), and a recalculated average rate of energy transfer of 10% from one trophic level to the other. The latter figure indicates that 1 kg biomass of prey at trophic level x creates 100 g biomass in its direct predator at trophic level $x+1$. The food chain is efficient and short in the ocean: for example in the open ocean, the catch (major species are tunas, bonitos) is on average at trophic level 4, in upwelling systems (major species: anchovy, sardine) it is lower (2.8), and intermediate at nontropical shelves (major species: cods, hakes, and haddocks; trophic level: 3.5). Upwelling and shelves provide the major catches, and in these ecosystems, about 30% of the primary productivity is needed to sustain the current rate of fishery. The authors concluded that the prospect of increasing these catches is dim. They contradicted also earlier conclusions that "human influence on the lowest trophic levels in the ocean outside of severely polluted areas is minimal and human exploitation of marine resources therefore seems insufficient." The discussion over the last 10 years since this report appeared has amply verified the bleak prospect for some areas of fishery. What are possible consequences? Do we have to catch the prey of tuna (mainly small pelagic fish), i.e., do we have to go down the trophic level with our fishery? Or do we need better management of the available stocks? Or do we even need a complement of agriculture for the oceans?

In Cod We Trust

The Collapse Which Could be Forseen

Fishing is the catching of aquatic wildlife and was likened to the hunting of bison (Pauly et al. 2002). The comparison was taken by the authors on purpose to point to two facts: Modern men returned to the food supply practices of the gone hunter-and-gatherer society and when used to feed a large population and when using modern technology, the prey population as big as it might appear will finally collapse. The bisons of the Great Plains were hunted down with the help of the gun, and high-tech fleets netted the cod out of the ocean. *I like reading old biology books, one has the illusion to be a bit cleverer than our ancestors and one can ask whether the bright minds of their time could foresee future events. One of my favorite books is the late-nineteenth-century zoology book series known in Germany as the "Brehm" from which already Darwin quoted extensively in his book The Descent of Man. The author of the article on the North Atlantic cod reported with much surprise on the unbelievable abundance of cod off Norway, which nourished an entire industry in Northern Europe.*

With the same awe, eighteenth-century zoologists would have reported on the bison herds. Both populations have collapsed. However, the author of the Brehm was already asking himself whether this exploitation rate of the cod would be sustainable without knowing what late-twentieth-century hunting technology we would lash on the cod. The problematic strategy of the "harvesting without sowing" was already apparent to biologists around 1900.

History of Fishing

Fishing is associated with *H. sapiens* nearly from its biological beginnings. A finely worked bone tool industry containing barbed bone points was found at an archaeological site in Zaire, which was dated to 90,000 years BP (Yellen et al. 1995). The site was characterized by abundant catfish remains. Fish exceeding 2 m in length was apparently harpooned in shallow water during spawning and thus provided a major meat input for the hunters. Very similar tools were found in this region about 50,000 years later pointing to a remarkable conservation of a fishing culture. The long-term exploitation of this meat source was possible since the population apparently consisted only of nuclear family units. Fishery researchers distinguish three periods of human impact on marine ecosystems. The aboriginal use refers to subsistence exploitation of near shore ecosystems by human cultures using simple watercraft and fish extraction technology. Colonial use comprised a systematic exploitation of shelf seas by foreign mercantile powers (Jackson et al. 2001). The last period was called the global pattern of maritime resource consumption and is characterized by a rapid worldwide depletion of predatory fish communities, both in the open ocean and the continental shelves (Myers and Worm 2003).

Signs of Depletion

The first step to an analysis providing a scientific basis for future restoration efforts is an exact quantification of the extent of fish depletion. Data were therefore compiled from standardized research trawl surveys over the last 40 years. The large demersal (bottom-dwelling) fish such as cod, flatfishes, skates, and rays all showed a sharp decline in different geographical areas. Data for the open ocean were obtained from the Japanese fishing fleet that uses pelagic longlines as the most widespread fishing gear. Longlines contain many hooks in a single line and thus represent a long baited transect through the water. Most world regions still yielded between 8 and 10 catches per 100 hooks in the 1960s, but these values dropped to 2 in the 1980s with decreasing trend. Formerly productive areas had to be abandoned and higher catches were only found at the periphery of exploited areas. Calculations showed an initial decline of 16% per year and the global ocean has lost more than 90% of the large predator fish species. The researchers suspected that most fish managers might not be aware of the true magnitude of the fish decline since the majority of the decline occurred in the first years of exploitation. For example, during the first 5 years of industrialized trawl fishing, the Golf of Thailand lost 60% of the large demersal fish. The researchers warned furthermore that reliance on recent data might be misleading since the data provide only minimum estimates for the unexploited communities, and referred to this as the "missing baseline" problem.

Detailed investigation off Newfoundland where Canada introduced severe restrictions on cod fishing painted a bleak picture for some large, once widely distributed demersal fishes. Skates (*Raja*) occurred in the St Pierre Bank with 0.6 million individuals in the 1950s, this dropped to 0.2 millions in the 1960s to a mere 500 individuals in the 1970s, and in the last 20 years none has been caught (Casey and Myers 1998; Figure 6.37). This 1-m-large elasmobranch (cartilageous-fish-like sharks) has few natural enemies but gets accidentally into the nets of fishers. Consequently, it once sported a leisurely life history. It grows slowly, mature at about 11 years, and lays only about 50 eggs per year.

The Overlooked Biology

Bony fishes like the cod invest heavily in many eggs and could thus have a fabulous fecundity and appear therefore better equipped to cope with the problem of overfishing. However, fishing is a very selective process going for the larger animals and this dramatically compromises the egg production. A ripe female red snapper (*Lutjanus campechanus*) measures 61 cm, weighs 12.5 kg, and produces 9,300,000 eggs. You need 212 younger females of 42-cm length and 1.1-kg weight to produce the same amount of eggs (Pauly et al. 2002). Going for the larger animals as is done with the current extraction technologies means that you

FIGURE 6.37. Skates (*left: Raja clavata; right: R. batis*), skates feed on mollusks, crustaceans, and fishes, dropping down on the prey from above. We speak about skate depletion as a sign of over-fishing the oceans.

target the Achilles' heel even of prodigiously fertile bony fish. Cod can live for many years and only reaches sexual maturity by the age of 4 years. However, cod is fully exploited by the year of two. Substantial mortality sets in before many fish had a chance to reproduce. Currently, only about 4% of 1-year-old cod survive to maturity. The result is clear, even when formulated in fishery jargon: low survival of recruits into the spawning stock. Or more bluntly as stated in the title of the report: *Potential Collapse of the North Sea Cod Stock*" (Cook et al. 1997) and this came just on the heels of the Canadian cod demise.

A Bit of Politics

The Canadian government reacted and ordered in 2003 an end to all cod fishery off Newfoundland and Labrador, while the EU only tighten the cod quota. The reason for the distinct reaction is differences in the extent of cod decline in both regions (99% off Canada and "only" 90% in the North Sea) and a stronger fishery lobby in Europe (Schiermeier 2003). At the moment there is a discussion between scientists and fishery managers about the burden of proof for the effect of fishery on the marine ecosystem (Dayton 1998). Scientists argue that traditional fishery management aims to optimize the catch of commercially important species, but eventually cause the collapse of the targeted species itself. They complain that those profiting from the public resources are not required to prove that their actions cause no damage. Much higher margins of safety are requested from industries running nuclear power plants or selling drugs for human use. They argue that thousands of square kilometers of benthic (sea bottom) habitat and invertebrate communities have been obliterated by trawling. Other scientists argue that the sea bottom is literally ploughed by trawling. Still others found this analogy inappropriate and likened trawling to clear cutting forests in the course of hunting deer (Pauly et al. 2002). Some fishery scientists would put a rather long list of well-known fishes on the extinction threat list, while fishery managers claim that the declines are still statistically blurred by natural variability in abundance.

Recovery?

To shed more light on these statistics, more data evaluation is needed. When 90 marine fish stocks that experienced a 15-year decline were followed 40% continued with the decline, 50% showed some recovery, and only 7% had fully recovered. The latter belonged all to clupeids (herring and relatives). The recovery of clupeids was attributed to their younger maturation age and the high selectivity of the clupeid catch technology (mid-water trawls on schools identified acoustically). Gadeid (cod and relatives) did not recover (Hutchings 2000). The latter observation disappointed and surprised fishery researchers. The Canadian government had already closed the directed fishing for the northern cod in 1992, but even after a decade-long moratorium, its population size remained at a historical low. What went wrong? Or are our models inadequate? Scientists draw attention to the fact that the life history of cod changed even before overt evidence for population decline became evident (Olsen et al. 2004). They had observed that cod continually shifted toward maturation at earlier ages and smaller sizes. Actually this is exactly what life-history theory predicts: Increased mortality at a given age and size of maturity selects for an earlier age of maturity. In parallel with the drop in annual survival that cod experienced over the last 20 years, age at 50% maturity decreased from 6 to 5 years. Ecologists have here a theoretical explanation: With reduced biomass, the remaining fish experience greater resource availability and could thus grow quicker. However, the cod collapse period was characterized by poor conditions for growth and the early maturing females were also smaller.

Darwinian Selection

The analysis actually supported another hypothesis, namely that fishery pressure selected for an early-maturing genotype. The concern is now great that current fishery practices have induced a strong evolution in cods, perhaps also by immigration of genotypes that fixed the population at lower stock levels. *This would not be the first case that human intervention has selected for an unwanted phenotype; think on the use of antibiotics to fight bacterial infections, which resulted perversely, but logically in a growing population of antibiotic-resistant germs.* In fact, there are experimental data with fish that we should not ignore, namely the Darwinian consequences of selective harvest by fishery (Conover and Munch 2002). Fishing mortality is highly selective, regulatory measures impose minimal sizes for the catches. Exploited stocks therefore typically show truncated size and decreased age distribution. This pressure could easily favor genotypes with slower growth and earlier age at maturity. US scientists sampled silversides (*Menidia menidia*) from a large gene pool of fish. They took 1,100 juveniles and imposed 10% mortality by harvesting the largest, in another experiment the smallest, and in a third experiment random 10% of the individuals over multiple generations. Interestingly, the small-harvested line showed much higher total harvest and mean weight of the individuals than the large-harvested line. The random-line took an intermediate position. Clearly, the selection on adult size caused the evolution of a range of traits (e.g., egg size, larval growth rate). The genetic changes might be irreversible, challenging the hopes for the recovery of severely battered fish stocks. These researchers called for a rethinking of our current regulation (all fish below a given size range are protected) and suggested a contrary protection policy, namely that all fish above a given size are protected. With respect to current fish extraction technology, this might not be an easy task. The definition of marine reserves could there be a more practicable solution. Fish would be under no size selection pressure and could diversify outside of artificial fishery constraints. Such reserves were propagated both as conservation as well as fishery management tools. Available data indicate that reserves lead to rapid increase in biomass, abundance, average size of organisms, and increased species diversity. In the Caribbean, the combined biomass in the reserve tripled in 3 years. Notably the biomass doubled also in adjacent fishing areas. Apparently, increased egg-output supplied juveniles and large-sized adults into the neighboring fishing grounds (Roberts, Bohnsack et al. 2001).

Trophic Interactions

We now get to trophic cascades, which we encountered in the previous chapter with the introduction of a top predator, the fox. Here we see cascading effects by the fishery-induced removal of a top predator, the cod. Parallel to the cod decline, the Canadian shelf saw an increase in pelagic (water column) fish and shrimps and crabs, which were the preferred prey of cod (Frank et al. 2005). Their commercial landing now increased substantially reflecting a population relieved from predation. Interestingly, seal biomass also increased in this area. Seals got now unlimited access to the pelagic fish prey. However, the disappearance of cod had also effects at lower levels of the food chain. Zooplankton larger than 2 mm diminished in numbers. They are the food basis for pelagic fish and shrimps and crabs. In contrast, smaller zooplankton and phytoplankton increased numerically. The takeout of the cod restructured the entire food web and the data should remind you the story of the otter and the kelp, which I told you before. We have here a general ecological principle emerging from different food web conditions. It is not clear whether this new equilibrium can get fixed or whether cod can again gain ground. As a recovery of the Canadian cod was not yet observed, one might suspect that the reestablishment of the old equilibrium will take much longer than suspected, if it occurs at all.

Fishing Down the Food Web

Interestingly, the combined shrimp and crab landings from the aforementioned study area now far exceed that of the groundfish fishery it replaced. This illustrates a phenomenon, which became known in the scientific fishery literature as "Fishing down the marine food web" (Pauly et al. 1998). Landings from global fisheries have shifted in the last half-century from large piscivorous fishes toward smaller invertebrates and planktivorous fishes. When the catches for this time period are plotted as a function of their trophic level, the scientists observed a decrease from 3.3 to 3.1 in marine areas and from 3.0 to 2.8 for freshwater areas. This trend was seen over many regions of the world, but a number of exceptions were observed. Part of the variation was due to the high temporal variation in anchovy landings in the South Pacific or the sharp decline in Antarctic catches, which are now dominated by krill. Ecological food web theory wants that catches of organisms from lower trophic level should exceed those of organisms at higher trophic level as observed in the previous study. However, when applied to larger geographical areas, the highest landings are not associated with the lowest trophic levels. Curious backward-bending features were observed when both parameters were plotted against each other. This discrepancy with theory probably indicates that we have a very incomplete picture of the complexity of marine food webs. Anyway, fishing down the food web is not a good alternative: As the authors stated, zooplankton is not going to reach our dinner plates in the foreseeable future. Actually, there are also theoretical limits to this practice.

Effect of Overfishing

Overfishing in a lower trophical level will only decrease the predation pressure in the next lower trophical level. These populations will increase in size and become so dense that they are much more susceptible to infectious diseases (Jackson et al. 2001). Indeed, there are some indications for emerging marine diseases as a consequence of both climatic changes and anthropogenic factors (Harvell et al. 1999). The final specter is a "microbialization" of the global coastal ocean. Overfishing was linked to population explosions of microbes and areas like the Chesapeake Bay and the Baltic Sea are now bacterially dominated ecosystems with a trophical structure totally different from what they showed only a century ago (Jackson et al. 2001). *Microbiologists are anyway convinced that microbes are the silent rulers of the world. We should think twice whether we should act in a way that we see us confronted more or less alone against the microbes without any intermediate fauna. This is likely a battle, which we cannot win. Are "no take zones" the only remedy against the decline of fishery or should we conduct fishery not with a hunter-gatherer, but a farmer mentality? In other words is aquaculture a sustainable alternative?*

Aquaculture

Farmed Fish

Is aquaculture the transition from the hunting to the farming type of fish production? This question cannot be answered in a single sentence and as usual in science a differentiated picture must be painted (reviewed in Naylor et al. 2000). First a few numbers as orientation: Over the last decade, the weight and value of farmed fish and shellfish has doubled. Currently about a quarter of all fish consumed by humans is farmed fish. Asia accounts for 90% of the global aquaculture production: China takes here with two-thirds, the lion's share. With respect to the trophic level, very different types of species are farmed: At one end are giant clams, which obtain most of their energy from symbiotic algae; followed by filter feeders like mussels; then come herbivorous fish like the carp; at the other end are carnivorous fish like salmon. Carp is mainly produced in China for low-income households and its production does not need extra feed input (Figure 6.38). The situation is different for salmon (Figure 6.39) farmed in Norway and catfish farmed in the USA. Both need compound feed in the form of fish biomass. Salmon farming can thus not relieve the fish extraction stress from the oceans since the fishmeal and fish oil are derived from ocean catches of small pelagic fish. There are other ecological problems associated, for example, with salmon farming: The intensive fish farming leads to wastewater

FIGURE 6.38. The carp (*Cyprinus carpio*) illustrates why it makes a difference to use herbivorous instead of piscivorous fish in aquaculture.

laden with uneaten feed and fish feces. The high fish concentration favors the transmission of infectious diseases. The latter problem is also relevant for the ocean population since escape of farmed salmon is a common problem. In the Northern Atlantic, as much as 40% of the salmon caught on sea are escapees from fish farms. This contribution now sensitively alters the genetic makeup of the wild population.

Shifting the Problem

It is not easy to replace fishmeal and fish oil by plant-derived material. Fish convert carbohydrates to energy inefficiently, and plant material is deficient in two essential amino acids (lysine and methionine) and some fatty acids. In contrast, farmed carp is mostly reared without the use of compound feed and is grown in inland ponds. Carp production is thus independent from the marine ecosystem. However, other aquaculture systems depend on wild seeds to stock the aquaculture ponds instead of relying on hatchery-reared finfish or shellfish larvae. Other aquaculture systems like shrimp (Figure 6.40) ponds have transformed large areas of mangroves (Figure 6.41) and coastal wetlands. Earlier, they provided food and shelter to many juvenile fish that was caught in the past as adults in offshore fishery. It was calculated that about 400 g of fish is lost by habitat conversion for every kilogram of shrimp produced in Thailand. To return to our initial question, it was concluded that if the growing aquaculture is to contribute fish in a sustainable way, it must reduce wild fish inputs in feed and adopt ecologically sound management practices. Otherwise it will only cause a shift of the problem of sustainability of fishery from one to the next ecological system.

FIGURE 6.39. Salmon (*Salmo salar*) and salmon trout (*S. trutta*), typical piscivorous fishes are shown here.

The Lesson of the Lake Victoria

History of a Lake

The region around Lake Victoria became a disaster zone because of the compound impact of tribal feud, genocide and wars, and the AIDS epidemic. This was not always so: It was once a fertile contact zone between diverse African agricultural traditions (Phillipson 1986). Cereals like millet and sorghum came from the north and met bananas, yams, and sweet potatoes. Bananas came to Africa via the Indian subcontinent. Harpoon-fishing settlements were dated as early as 11,000 years ago in this region. Like in Europe the change from a hunter-gatherers to a food-producing economy was accompanied by the spread of a language, here that of the Bantu languages into southern Africa. Lake Victoria has nearly twice the area of Switzerland and is the home of a rich and diverse fish population. The lake harbors an estimated 500 endemic species of cichlid fishes, which are herbivores and detritus feeders. Mitochondrial DNA sequences

FIGURE 6.40. Shrimps are together with crabs and lobsters members of the order Decapoda, subphylum Crustacea, here represented with *Palaemon serratus* (*top*) and *Crangon vulgaris* (*bottom*). They are important in trophic interactions in the sea.

identified them as members of a genetically closely related superflock (Meyer et al. 1990). The degree of sequence variation was lower than in the human species and pointed to a recent origin of this superflock. Teleosts (bony fishes) represent with about 25,000 species half of all living vertebrate species, but such a diversification into 500 species within a single lake is astonishing. Initially it was suspected that the diversification occurred within the lake, but more recent sequencing work revealed that the Lake Victoria flock was several times seeded from cichlids living in the lake Kivu (Verheyen et al. 2003). This lake harbors much less fish species (26, of which 15 are cichlids), but the lake is much deeper (it belongs to the East African rift system) and is thus geologically much more stable. Lake Victoria is relatively shallow and experienced a nearly total dryness as recently as 15,000 years ago. This dramatic event has not sterilized the lake. However, another event in the recent historical past has upset the biological equilibrium of the lake. Fishery biologists fear that this manmade event will not only endanger the biological diversity in the lake but also lead to the destruction of the economical basis for the population living around the lake, as vividly depicted in a prize-winning documentary film ("Darwin's Nightmare"). There

FIGURE 6.41. *Rhizophora conjugata* forms mangrove, i.e., dense thickets of shrubs and trees along tidal estuaries characterized by exposed, supporting roots. Respiratory roots are common to many species through which air enters into the plant aquaculture.

are fundamentally different methods between journalists and scientists when they report on events. Scientists essentially make experiments in the laboratory or observe events in the field and report their data in peer-reviewed scientific journals. In addition, they meet at conferences where they discuss data and exchange interpretations of the observations. However, they generally do not travel to the spot. So I cannot comment on the political and economical implications of the events documented in the film. I will limit the story to data, which were published in scientific journals and they are dramatic enough.

Species Introduction

In the public discussion, there is a misconception such that natural and artificial introduction of new life forms are distinguished. The natural introductions are considered as safe, while genetically modified organisms containing a few well-characterized alien genes capture media attention. Biologists would argue that the true dangers lurk in the incautious introduction of natural species into new environments. Virologists also distrust certain agricultural practices like the proximity of pig and poultry rearing in Asia, which favors the mixing of avian and mammalian influenza genes, raising the specter of an imminent human influenza pandemic. We should not forget that Nature is not only our loving mother, but also the greatest bioterrorist.

The Nile Perch

In the 1950s, the proposal to introduce the Nile perch, *Lates nilotus*, into Lake Victoria was made. Ecologists warned against this introduction because it is a large piscivorous fish that grows to a size larger than humans. This introduction was against ecological commonsense since short food chains are ecologically the most efficient and the yields of the predator can never be as great as that of its prey (Barel et al. 1985). In 1960 the Nile perch was introduced and the drama took its way. The local fishery was based on the catch of cichlids. Trawl fishery had already some effect on this fish stock, which represented the food basis for the local fishery. Now cichlids experienced another stress, the heavy predation by the much larger perch. The explosion of the perch—it already represented, in several lake areas, 80% of the biomass—is clearly not sustainable and will be followed by the decline of the predator when the prey population comes down. *The local population did not like the oily flesh of the perch. It cannot be sun-dried as the cichlids and must be smoked, which led to deforestation around the lake. Important sociological consequences were observed. The small fishers lost their cichlid catches, but they could not switch to perch catching since it needed stronger nets and better boats. The processing of the fish filets is now done in modern fish factories in Mwanza associated with airstrips, where the fish is directly imported to Europe. Wealth was created only for a small part of the population, and the rest fell in even deeper poverty and now live from the discards of the Nile perch processing in the factories. Ecologists complain that the introduction of the Nile perch is irreversible and that the biomass- and species-rich fish population of the Lake Victoria is perhaps gone forever.*

Risk Perception

This tragic introduction of an alien fish reminds us that fishery technology is sometimes more advanced than the ecological knowledge about the consequences of such actions. We should thus think twice before we change the environment by deliberate introduction of new species. This practice caused a lot of human tragedy, but the human mind has apparently an inclination to deal more with perceived theoretical risks (I would argue that GMO belong to them) than with real dangers.

On Fishery, Bushmeat, and SARS

Fishing off Ghana: Economics and Ecosystem Resources

Science journals report relatively rarely on economical research questions. Yet, we live only in one reality, and it is only the human mind that divides the exploration of our interior and our environment into different branches of knowledge. Especially macroecological research questions cannot be separated from economical decisions of the human society. The modern democratically governed societies are pluralistic and they follow sometimes contradictory strategies reflecting distinct interests of different pressure groups. To illustrate this point in our quest for food survey, I have chosen a recent research article in Science (Brashares et al. 2004). One camp is represented by a new branch in biology called Conservation Biology, which deals with the dramatically accelerated decline of species loss due to human transformation of the planet. Many biologists argue that mankind and its economical activity has become a major evolutionary force on a multitude of organisms (the "terminator," in the movie language). The species richness and at the same time the ecological vulnerability of the tropical rainforest made this zone a focus of conservation efforts. "Green" groups encourage us to buy or adopt parcels of the tropical rain forest to decrease the decline of biological diversity. At the other side, the European Union follows a subsidizing policy that goes sometimes straight against this goal. For example, industrialized countries possess—at least in the view of many ecological scientists—a too large and too industrialized fishing fleet. This industry is increasingly subsidized, EU financial support increased from $6 mio in 1981 to $350 mio in 2001. Owing to the decrease of catches in northern seas, we now see increasingly foreign fishing fleets off West Africa, the lion's share coming from the EU. What were the consequences? At a simple level, the fish stocks declined and may face imminent collapse. Furthermore, the technically superior European fleets outcompeted the local fishery with the consequence that the fish supply to the local markets declined and prices increased. In Ghana the supply of fish varied between 230,000 and 480,000 tons a year and showed up to 24% changes between consecutive years. As fish is the primary source of animal protein in the diet from West Africans, there are only two consequences: Either the population goes into temporary protein malnutrition or it seeks alternative animal protein sources.

The Shift to Bushmeat

The latter link was investigated in the *Science* report by the conservation biologists. They observed that monthly bushmeat sales in local markets and the estimated hunting intensity in wildlife reserves were negatively correlated with regional fish supply. Bushmeat selling and the counts of hunters reported by wildlife rangers were furthermore closely related to annual rates of wildlife decline. Over the last 30 years, the biomass of 41 mammals in the reserves declined by 76% and a number of species became locally extinct. The researchers tested this association of observations. The correlation decreased with increasing distance of the reserve from the coast, speaking for a causal connection between the observations. Apparently, we see here a transfer of harvest pressure from aquatic to terrestrial resources. *Thus by subsidizing fish industry, the EU endangers the wildlife survival in West Africa. One might argue—as in the case of the Lake Victoria—that some ecological consequences of economical decisions could not be foreseen and they became only evident after substantial research efforts. A posteriori, one gets the impression that some of the consequences could have been anticipated by pure logical reasoning. As a biologist, I wonder whether governments should have more professional ecologists on decision boards to explore the potential ecological consequences of political decisions. Leaving this job to emotionally and politically motivated environmentalists led, to my opinion, to a concentration on minor issues that deal more with sentimental concerns of citizens of wealthy societies than with ecological disasters of the developing world.*

Food Safety

Food industry spends a lot of energy and money on food quality and safety. Due to these efforts, cases of food poisoning that can be tracked to industrial food sources came substantially down over the last century. However, fresh food that is consumed raw represents still an important concern. Oysters are an illustration, and the Centers of Disease Control in Atlanta traced, as another example, large numbers of gastroenteritis in the USA to the import of raspberries from Central America. Irrigation of the plants from the bottom with contaminated water was not a problem for the fruits. In this specific case, contaminated water was used to dilute pesticides that were sprayed from the top with airplanes and could thus contaminate the fruits. *Bushmeat belongs to a different category for food microbiologists. Virologist will remember the discussions in the early phase of the AIDS epidemic, where exposure of humans to apes and monkeys was discussed for the origin of the new human virus. Nowhere is the contact between wildlife and humans so close as in the bushmeat handling, which is a multibillion-dollar trade in the tropics. During handling and butchery of the wild animal, exposure to the blood of the animal is inevitable and creates an opportunity for viral transmission. We have somewhat forgotten that the quest for food is a dangerous business not the least because our animal prey can infect us.*

SARS

This risk was recently demonstrated again by the SARS epidemic. Severe acute respiratory syndrome, as it reads in full, emerged as a new human disease associated with pneumonia in November 2002 in the Guangdong Province of China. The early phase remained largely unrecognized or underreported. The disease attracted attention when a major outbreak occurred in a hospital of Guangzhou and a hotel in the nearby Hong Kong in February 2003. Epidemiologists call this a superspreader event where single patients infected as many as 300 other individuals (Dye and Gay 2003). Outside of these events, which mark the middle phase of the epidemic (Chinese Consortium 2004), the transmission dynamics was, with 2.7 secondary infections per case, less dramatic such that public health interventions could finally cope with the epidemic (Riley et al. 2003), leading to the decline of the case numbers in the third, late phase. However, at that time the disease had already spread to 25 countries around the world with epicenters as far away as Canada. The epidemic ended in July 2003, the nightmare of a pandemic did not become a reality. Despite all public disarray, the international and the Chinese research community, assisted by the WHO as a fireworker, organized a relatively structured approach to the disease. A contributing factor was certainly the early warning by avian influenza infections in Hong Kong, which led to fatalities in humans and heightened the alert of virologist for the possible emergence of a devastating influenza epidemic from China.

Coronaviruses

You might now ask what SARS has to do with food production: a lot in fact. The connection became clear when laboratories in the USA, Canada, Germany, and Hong Kong isolated and then sequenced a coronavirus as the causative agent of this epidemic (e.g., Rota et al. 2003, Marra et al. 2003). These isolations followed directly on the heel of the epidemic and the various reports were published in the first half of 2003. The agent turned out to be a known virus. It belonged to the coronavirus group, which are large, enveloped, positive-strand RNA viruses—in virological parlance, this means that the viral genome encodes the information for the viral proteins. Coronaviruses cause respiratory and enteric diseases in humans and animals. Human coronaviruses were up to that epidemic only, associated with mild upper respiratory tract infections, but some animal coronavirus like TGE cause deadly enteric infections in swine. Coronaviruses contain the largest genomes of any RNA viruses, and the SARS isolates showed genome lengths that clustered around 29,750 nucleotides. The genome organization resembled closely that of the known coronaviruses, but its sequence forms a distinct group that is not closely related to any of the previously characterized coronaviruses.

The Epidemiologists

Under such conditions, the task is commonly passed to the epidemiologist to provide hints to the potential source of a new virus. A review of the early patient data by a WHO fact-finding mission in April 2003 revealed to an epidemiologist that 9 of the 23 early patients worked in the food industry. Also people working in the vicinity of food markets and workers in specialty food restaurants were overrepresented (Normile and Enserink 2003). These data were later substantiated by serological surveys: 13% of 508 animal traders whose blood was sampled during the outbreak in May 2003 tested positive for IgG antibodies in the quickly developed SARS virus immunoassay. Control groups showed only 1–3% prevalence rates. Notably, traders that handled the masked palm civet, a distant relative of the house cat, were the most likely to show SARS-specific antibodies (Enserink and Normile 2003). This is not an unlikely finding: Civets are regularly eaten in China. In fact, they are preferred for their tasty meat and are handled as a delicacy by a growing sector of wealthy consumers. In China, a lot of folklore surrounds food items, civets are for example fabled to strengthen the body against winter chills. The demand for wildlife cuisine in China is thus high, and illegal poaching and husbandry of wildlife is widespread. Many families in the rural area make a living by providing this wildlife to cities (Liu 2003).

The Virus Hunters

Now the race was open for the virus hunters. This is unfortunately a respectable business for professional virologists. One needs in addition to technical skills a detective nose for this business. Guided by the epidemiological data, Yi Guan et al. (2003) went into live animal markets where they borrowed animals from vendors. None of them was found to be ill, but PCR diagnosis tools showed that from the many sampled species four of the six palm civets scored positive, the two negative animals yielded a live virus isolation from nasal secretions. They were sequenced, and it was 99.8% identical to the human isolates and differed mainly by a 29-nt insertion, upstream of the structural N gene. Interestingly, the earliest human SARS virus isolates still contained this 29-nt segment but later isolates lost this segment possibly as an adaptation to human-to-human virus transmission. The researchers cautioned that their isolation of the SARS virus from civets might not have identified the true animal reservoir of the virus. Civets might have got infected in the markets and much larger investigations in feral animals are needed to settle the question. In fact, also a raccoon dog from the investigated market yielded a closely related virus.

   Indeed, experimental infection of civets with human SARS virus resulted in overt clinical disease, which is not expected for a viral reservoir where asymptomatic infection should dominate. Chinese virus hunters then went after bats. Bats are a good bet since flying mammals can be dynamic transmitters of disease. Bats are known to do that with rabies and more recently with Hendra and Nipah viruses, two other emerging zoonotic viruses. Furthermore, with 4,800 described species, bats represent 20% of all mammalian species. *Finally, the Chinese virologists knew that bat meat is considered a delicacy by Chinese gourmets and another strong link is the use of bat feces in traditional Chinese medicine.* Equipped with that knowledge, they screened nasopharyngeal and anal swabs of bats, rodents, and monkeys in the Hong Kong area. They detected a coronavirus sequence related to the SARS virus in 40% of the anal swabs from the insectivorous Chinese horseshoe bats (*Rhinolophus sinicus*) by using PCR technology (Lau et al. 2005). None of the positive bats showed clinical symptoms, but many showed an antibody response and, interestingly, high serum titers correlated with low anal virus excretion. A parallel study from mainland China confirmed the findings (Li et al. 2005). Both studies showed closely related sequences for this coronavirus, much closer related to SARS than to another recently isolated bat coronavirus. *Rhinolophus* roosts in caves and feeds on moths and beetles. However, the cave-dwelling fruit bat *Rousettus leschenaulti* also showed serological evidence for coronavirus infection. These fruit bats (Figure 6.42) were found by the virus detectives in the markets of southern China. The current hypothesis imagines that they were the asymptomatic source for virus spillover to susceptible animals exposed in the markets like the civet. The spread of the virus to susceptible animals might have provided the necessary amplification to achieve intrusion into the human population. Notably, the greatest genetic variation existed between the bat and human coronavirus in the S1 domain of the S protein, which is involved in receptor binding of the virus. Coronaviruses

FIGURE 6.42. The dog-faced bat (*Rousettus*), a member of the Megachiroptera; flying foxes forage mostly on fruits and flowers.

apparently have the genetic diversity and flexibility to create variant S proteins that might enable them to cross the species barrier.

Transmission Dynamics

The SARS epidemic did not spread worldwide not because of our superior scientific knowledge, but because we were lucky that its transmission dynamics could be countered by age-old containment measures, which might not be the case for other respiratory infections. However, the havoc that this limited epidemic, involving about 7,000 cases and 700 deaths, spelled on the world economy can give us a foretaste on a new influenza pandemic that might lurk somewhere in China. Also in this case, agricultural practices might be at the basis of the problem. In the case of influenza, it is the proximity of poultry, pigs, and the human producer/consumer that provide an appropriate combination of viral incubation vessels to initiate a new influenza pandemic. The prospects and stakes are clearly formulated in scientific journals (see the series of reports on the threat of an avian flu epidemic in the May 26, 2005 issue of *Nature*; Aldhous and Tomlin 2005). *International air traffic will assure the quick transport of human patients in the incubation period and have already realized the One-World concept for infectious diseases. The bottom line of these reports was that we are not necessarily much better prepared to such an event as in the case of the 1918 Spanish flu, and the WHO estimates that such an avian influenza pandemic could again claim millions of human lives worldwide.*

Again a Bit of Politics

Governments and the public are relatively complacent about these prospects and deal with—in the view of many biologists—minor issues of food safety like the GMOs in the food chain at a time when another – and this time a real-biological bomb is ticking. Political economy is the art of shifting limited resources to the most urgent public needs. This means that research money spent on the health impact of electric powerlines (apparently a non-issue according to leading medical journals) and the safety of GMOs in food (probably of greater psychological than technical concern), cannot be spent on virological food safety and agricultural food production issues. The SARS crisis showed that the WHO was already overstretched with its resources when confronting a numerically still limited epidemic. Money problems at the WHO are perennial, and the WHO is definitively understaffed to face a major influenza epidemic, for example. The discussion in this chapter centered on SARS not because it is the only case. Viruses coming from tradi-tional Asian animal husbandry condition are manifold, and similar warnings could be drawn from other emerging viral infections like Nipah virus that affected the Thai pig farmers. We should thus not see us as the ultimate predator on earth, where only food limitation puts a break on our population growth. We should realize that we are also the food for many organisms, or in more benign cases, the involuntary feeders of many organisms. This subject will be the content of the next chapter in our quest for food survey.

7
We as Food and Feeders

Prey of Microbes

A Lion's Share?

Becoming Food at Death

As we are with respect to our body a physical part of the creation, we represent of course a source of food for other organisms. Since we have experienced a spectacular population growth over the last 10,000 years or so, other organisms will not overlook the naked ape as food. Psychologically, we have lost the impression of representing an integral part of the food chain. However, there are a number of ways for us to become food for other organisms. If our body is buried after death, a sequence of detrivores take care of it. People throughout time were aware of that, and the paintings of the late medieval time in Europe frightened their observers with the specter of decomposing human bodies. In the western hemisphere, we have learned to blot out the idea that death is part of our life as for any other organism and we leave this impression to the specialists of forensic medicine. On the other hand we have got used to the idea that we are the top predators in the biosphere, that we have lost the memory that we can fall victim to predators, large and small.

Attack by Carnivores

Large carnivores still take their toll. In Tanzania lions (Figure 7.1) killed more than 560 people over the last 15 years, and the trend is increasing (Packer et al. 2005). Lions pull people out of bed, attack nursing mothers, and catch playing children. A typical scenario is an attack on a farmer sleeping in a makeshift hut on his fields to protect the crops from nocturnal raids by bush pigs. The rising trend is easily explained: humans intrude into areas where lions live and since they eliminate the natural prey of these large carnivores (kudu, zebra, hartebeest, and the like), lions take humans as a surrogate prey. Similar cases could be told for the Asian tiger, but overall only a negligible part of the world population becomes prey to carnivores.

1. **Tiger** (Felis Tigris). ¹/₁₂. (Art. *Tiger.*)

FIGURE 7.1. The tiger (*Leo tigris, top*) and the lion are the largest representatives of the cat family (Felidae). The tiger probably evolved in Northern Europe, but moved from there to Siberia and Asia. Individual tigers are known as man-eaters, but normally tigers hunt deer and wild hog. The lion (*Leo leo, bottom*), the king of the animals in the folklore, hunts many animals, ranging in size from insects to giraffes, but its preferred prey is antelope-sized. An old lion might become a man-eater when unable to hunt quicker prey.

Microbes Live on us and from us

So where do we become food ourselves? In affluent societies, more people die of cancer, cardiovascular diseases, or simply of old age than of infectious diseases. This was not always so; just 100 years ago, infectious diseases were a major killer and this remained so in many parts of the world, which we euphemistically call developing countries. If you look at infectious diseases with the eyes of a biologist, you see microorganisms, which make a living from us. In benign cases—and this is the majority—they feed on us without doing too much harm to us. In more serious cases they literally eat us up. It was said that humans fight with insects for supremacy on the globe and it is not yet decided who will win the war. The battle with mosquitoes will be illustrated. Personally, I would not like to bet on either of these two combatants. My personal bet is different: in the beginning of time, microbes were the rulers, and most likely, they will be the dominant form of life at the end of time. Whether mammals will survive the second half of the evolutionary path, which is still ahead of us (if we accept the life expectancy of our sun given by astrophysicists), remains doubtful to me, while bacteria will certainly survive. As this question is idle speculation, I will now explore with you how we become substrate for the growth of microorganisms.

The Haunted Hunter...

Simian Foamy Virus

In a previous section, it was reported that the decline of fish in the local West African market motivated people to search for alternative protein sources in the nearby bush. This bushmeat hunting not only had negative consequences on the population size of animal wildlife, but might be the basis of one of the largest medical challenges currently facing humankind. A recent study in the *Lancet* illustrates this point (Wolfe et al. 2004). Epidemiologists asked 1,800 people from nine rural villages in Cameroon in equatorial Africa; 61% reported direct exposure to primate blood primarily through hunting and butchery. Notably, 16% of the exposed people showed serological evidence of a past infection with simian foamy virus. In 1% the viral exposure was confirmed by a more rigorous test (western blot) and viral RNA was detected in the subjects. As foamy viruses are largely species specific, the sequence analysis suggested three independent cross-species infections. When these sequences were projected on a large number of simian foamy viruses, transmission to humans from three different primates could be identified by phylogenetic analysis (*Cercopithecus*, *Mandrillus*, and *Gorilla*). All the virus-positive persons confirmed contact with animals either through butchery and eating of monkey, chimpanzee or gorilla hunted by guns, bows, or wires or through a pet monkey. Simian foamy viruses do not cause a disease in the natural host nor showed any of the virus-positive human signs of disease in this or previous studies with zoo workers who experienced simian foamy virus infections. Although the simian-to-human transmission is relatively high (about 2%), no human-to-human transmission was yet reported.

The Nature of Virus Infections

Why should we then worry about these data if it is such an innocuous virus? First, we should recall that probably the majority of all viral infections cause only slight or no symptoms in the infected individual. That viruses can cause disease is surely correct, but most viruses likely go unnoticed because only very few virologists take the pain to isolate viruses from healthy subjects or animals since it will be hard to motivate grant organizations to pay for such studies. The equation of virus = disease thus clearly suffers from a strong observation bias. However, there is a caveat to this reassuring statement about the benign nature of most viruses. Even if the virus is, through millions of years of coevolution, well adapted to its exploited animal and thus harmless to its natural host, it might be maladapted to an occasional heterologous host, where it can cause serious disease. And there is another disturbing point in this study. Simian foamy virus is a retrovirus like human T-lymphotropic viruses, which are associated with lymphoma and leukemia in humans. This infection is most prevalent in tropical forest regions of equatorial Africa, which suggests likewise a zoonotic origin.

AIDS as a Zoonosis?

Zoonosis is a human disease caused by the transmission of an animal virus. The simian origin of human T-lymphotropic viruses is now well accepted (Vandamme et al. 1998). To put the impact of zoonosis in the right perspective, there is also good evidence that AIDS is likewise a zoonosis (Hahn et al. 2000). The case is relatively clear for HIV-2: the human virus and that of monkeys (specifically that of sooty mangabeys) share a specific accessory protein and are phylogenetically closely related; sooty mangabeys are numerous in west African countries where HIV-2 infection is endemic. The animal host is infected with HIV-2-like virus at substantial frequency. Sooty mangabeys are frequently hunted for food, and orphans are kept as pets. Cumulatively, this evidence represents a smoking gun. A pandemic as in the case of HIV-1 did not occur because HIV-2 apparently lacks the capacity for efficient human-to-human spread outside of western Africa. The global human disease became a reality with the introduction of HIV-1 into the human population. In fact, it turned out to be much more difficult to trace the origin of HIV-1. Part of the problem is that HIV-1 is not a single virus but comprises three distinct virus groups termed M, N, and O. The predominant M group consists of more than 10 clades denoted as subtypes A to K. They represent at least three separate transmission events. Sequencing and phylogenetic analysis suggested a chimpanzee origin for HIV-1. This is not an unlikely origin since chimpanzee meat figures in African bushmeat markets. It is currently believed that the HIV-1 group M pandemic arose as a consequence of a single virus transmission event from a chimpanzee followed by a "starbust" radiation of numerous viral lineages in the new human host.

Origin and Spread of AIDS

An understanding of the origin of the AIDS pandemic is of substantial public health interest, but the epidemiological situation leading to the spread of the infection is complex and not yet clearly settled. Several events were tentatively associated with the origin of the AIDS epidemic. An interesting scenario is the association with commercial logging of tropical forests. This activity necessitates an intrusion into the bush and thus assures close contact with primates. The logging is accompanied by a network of other economical activities ranging from local commercial bushmeat trade over road construction to the arrival of sex workers. In combination, this setting could have paved the way for the spread of the original zoonotic infection. However, one should mention that alternative scenarios were proposed with respect to the origin of the AIDS pandemic. An especially dire hypothesis independent of the human quest for food and sex was the link with the polio vaccination campaign. According to this hypothesis, the polio vaccine might have been contaminated with the AIDS virus during the production of the vaccine on kidney cell cultures from chimpanzees held in tropical Africa. This scenario is, however, unlikely for two reasons. Old stocks of poliovirus vaccine used in Africa failed to provide evidence for HIV contamination and the timing of the diversification of the pandemic M group HIV-1 in the human population points to an introduction of the virus into the human population that predates the polio vaccination efforts in Africa. One elaborate computer study of full-length gene sequences of many M group isolates pointed to 1930 as the most likely date for this transmission event (Korber et al. 2000), a full 30 years before the start of polio vaccination in Africa. Furthermore, stored historical serum samples revealed a seropositive individual just before the onset of the polio vaccination campaign.

The Danger of Bushmeat

The bushmeat link is therefore a more logical hypothesis for the origin of AIDS. What makes bushmeat more dangerous than other meat used as a human food source? Humans have hunted for meat for ten thousands of years, and one should expect an adaptation of the human population to meat consumption and the associated problems of virus exposure. However, there are two peculiar problems with this form of bushmeat. The first problem is the fact that according to the archeological record, monkeys and apes did not figure in the human food list—any unusual wild animal as meat source represents a potential source for the transmission of potentially harmful viruses because humans had not adapted to live with the specific viruses associated with this particular meat source. Second, viruses are generally considered to be species specific and the likelihood of trans-species infections decreases with the phylogenetic distance separating the two animal host species. A primate virus is thus more likely to infect a human being than viruses infecting more distantly related mammals. This argument also explains why we are not victims to viral diseases transmitted from fish despite the fact that fishery is also a form of hunting of wild animals.

... And the Risks of Animal Farming

Infections of the Early Farmer

On the basis of these arguments, one would conclude that farming of domesticated animals is a safer source of animal meat for human consumption. At first glance this argument seems straightforward. However, its logic is not entirely watertight. On one side, animal husbandry is a relatively recent human occupation, and just 10,000 years ago, wild or half-wild animals were brought into close contact with humans. Physical proximity to an animal is a requisite for the transmission of an animal virus to us and vice versa. One can thus expect a period of extensive viral exchange between the pastoralist and its herd followed by a period of mutual adaptation to the respective heterologous viruses. Prominent biological anthropologists anticipate therefore that the early relationship between domesticated animals and humans during the Neolithical Revolution was not really a honeymoon. They postulate a lot of hardship from cross-species viral and bacterial infections during this phase. The greater stability of the food basis in the early farming societies when compared to the deteriorating hunting conditions during the Mesolithic was paid by the introduction of new viral diseases into the human population. We lack direct evidence for this process, but indirect data support this model.

Measles Virus

Take measles, a major killer of children on a global scale. Measles is a paradoxical viral infection. Humans are the only known host of measles, and historical records allowed tracing back measles for at least 2,000 years into human history. However, measles cannot be a genuine human infection. Classical epidemiological experiences on islands demonstrated that measles infection dies out on small islands; under this condition, measles propagation needs an external influx of humans (as occurred with the British troops during World War II on the Faeroer islands) to ignite a new wave of measles infection. To persist in the host species, measles needs to regularly encounter newly susceptible individuals who are contributed only by newborns since once infected with measles, it leaves a life-long immunity in humans. Only populations that exceed a quarter of a million individuals fulfill the requirement for measles transmission. Measles could thus only spread in the human population when agriculture provided in historical times the nutritional basis for supporting such a population size. Interestingly, the closest relative of measles is the rinderpest virus infecting bovids (Sharp 2002). *Cattle are thus the likely origin of the human measles virus. It does not need a lot of fantasy to imagine the effect of newly introduced morbilliviruses into the human population during the Neolithic Revolution. Some anthropologists suspect that the rapid replacement of the Mesolithic hunter societies by the early farmer societies was caused by the collapse of the hunters when they met the farmers excreting measles. There is indirect evidence that this process of rapid replacement of one society by another occurred once again 500 years ago, namely during the colonization of the New World by the Spaniards. As mentioned in a previous chapter, the literal melting of huge Indian armies in face of the small Spanish troops was probably caused or at least influenced by the effect of measles infections unknown to the autochthonous population. Notably, the American population did not domesticate cattle and thus lacked the experience with the rinderpest virus* (reviewed in the lively book from J. Diamond "Guns, Germs and Steel" 1997).

Influenza Virus

One should not conclude from these two historical examples that animal farming is now a risk-free business for the human population with respect to viral infections. Dangers lurk in peculiar agricultural and market practices even today. The most threatening example is that of influenza viruses. Influenza A viruses possess a segmented single-stranded RNA genome, each encoding one of the eight viral proteins. Influenza A viruses infect humans, swine, horses, seals and a large variety of birds; the viruses are antigenically quite complex. Importantly, aquatic birds are the source of all influenza viruses in other species. Older studies on wild ducks in Canada demonstrated a number of remarkable observations: up to 20% of juvenile birds are infected with influenza virus when the birds congregate prior to migration, while none of the birds showed any symptoms of infection. This is commonly observed in influenza infection of wild birds. High viral loads are sent into the environment with the feces of the ducks leading to important but transient viral contamination of lakes.

Not all influenza infections are so benign. The catastrophic "Spanish Influenza" pandemic, which killed between 20 and 40 million people worldwide—more than all casualties in World War I—could be traced by sequence analysis to a virus from swine. This is a true detective story where virus hunters went through embedded historical tissue samples from soldiers who died in 1918 in military hospitals and to graves in the permafrost region. Since a swine epidemic occurred at the same time, the virus was probably transmitted from swine to human before efficient human-to-human transmission occurred. Notably, pigs serve as host for the replication of both avian and human influenza viruses. In a popular hypothesis, pigs are seen as a "mixing vessel" where genetic reassortants between avian and mammalian/human influenza viruses are created. Pigs may thus play a pivotal role in the generation and transmission of avian influenza viruses to humans.

Resurrection of the Flu

Exciting progress was made recently when the Spanish flu virus was literally raised out of its formalin (Taubenberger et al. 1997) and permafrost (Reid et al. 1999) grave. Initially this historical virus isolate led an *in silico* existence in the database. The analysis of its sequence corrected the older interpretation quoted in the preceding paragraph (Taubenberger et al. 2005). The polymerase sequences differed from the avian consensus sequence at only a small number of sites, making it more likely that it was derived from an avian source shortly before the pandemic. Analysis of the now-completed genome of the 1918 influenza isolate suggested that this virus was not a reassortant virus like the later 1957 and 1968 strains, but entirely an avian-like virus that had adapted to humans. Notably, a number of changes observed in the 1918 isolate were also observed in the currently circulating, highly pathogenic H5N1 avian virus. The next step was even more breathtaking. Under high containment conditions, approved by NIH and CDC, researchers (under the protection of antiviral prophylaxis) used reverse genetics to unearth the 1918 virus not as a computer event, but as a real-life existence that could be propagated in cell culture as any other replication-competent influenza virus (Tumpey et al. 2005). The researchers wanted to grab the deadly secrets from the virus, and deadly it was. Mice started to lose weight and died a mere 3 days after infection. The damage was pronounced in the lungs, but the pathology remained restricted to the respiratory tract. The virus had apparently learned other dirty tricks: for example, it could be propagated in cell culture without the help of trypsin. Ordinary influenza A viruses need trypsin for the proteolytic cleavage of their hemagglutinin. In mice, virus titers of the 1918 isolate skyrocketed in the lung; likewise the lethal doses dropped by at least a factor of hundred in comparison with reference influenza viruses. In addition, the 1918 virus was lethal for fertile chicken eggs, a pathogenic feature of avian H1N1 viruses.

Why China?

Epidemiologists have always wondered why China is the source of new pandemic influenza strains. When new pandemic strains replaced the predominant H1N1 subtype (H refers to the hemagglutinin and N to a neuraminidase, the two major external proteins of the virus, and the numbers to different serotypes), like in the 1957 Asian flu with an H2N2 subtype or the 1977 Hong Kong H3N2 subtype, avian genes contributed to the new pandemic strains. Virologists suspected that the access of pigs to ponds with ducks as practised in traditional family animal rearing in China might explain the geographical predilection for the emergence of new influenza virus subtypes in this region.

The Avian Flu

The "*Hannibal ante portas*" is still a clear warning in our days: avian influenza strains are clearly knocking at our doors. This became evident in 1997 in Hong Kong when 18 confirmed cases and six deaths were reported that could be traced to the avian H5N1 subtypes (de Jong et al. 1997). The H5N1 virus was nonpathogenic in duck, but highly pathogenic in chicken and humans. It was prevalent in chicken from live bird markets in Hong Kong, but failed to transmit efficiently from human to human. A potential catastrophe was prevented by a massive culling of 1.5 million chicken within 3 days and the cleaning of markets and the separation of markets for live chicken and live aquatic birds. Later the authorities banned ducks, geese, and quails from the markets and imposed vaccination with an inactivated H5N1 vaccine for all domestic and imported poultry. These measures were successful and no H5N1 viruses were isolated from domestic poultry or humans in Hong Kong since 2004 (Webster and Hulse 2005). However, they have not stamped out H5N1 influenza—since December 2004, 41 people have been infected with H5N1 in Vietnam, where both commercial poultry farming and backyard poultry are expanding. The strains are highly virulent for humans and caused at least 16 deaths (Tran et al. 2004). H5N1 is now endemic in ducks and highly virulent forms were now reported in wild waterfowl such as geese at the Qinghai Lake, where large number of dead geese were counted (Chen et al. 2005).

Geese showed not only diarrhea, but also neurological signs such as tremor and opisthotonus (a retroflection of the head). Brain lesions and pancreatic necrosis were seen by the veterinary pathologists. Virus was isolated from the viscera, the brain, and also from the oropharynx and the cloaca of sick and dead animals. The latter two routes of virus excretion are directly relevant for the transmission of the virus. Infected gulls (*Larus*) were also detected at this lake, which is significant since gulls eat carrion and might thus get infected from dead geese (Liu et al. 2005; Figure 7.2).

The importance of this observation, which was disputed by the Chinese government (Butler 2005), is given by the fact that this lake in western China is an important aggregation and breeding ground before the animals migrate southward to Myanmar or over the Himalayas to India. Some Asian geese take even a migratory route over Europe. Because infected birds excrete influenza viruses with their feces, the droppings of migratory birds have a great potential to spread the disease. As cats and tigers turned out to be susceptible to these avian viruses, the potential of transspecies infection is clearly given. The situation is compounded by the dynamic nature of the H5N1 genotypes in eastern Asia. Since its isolation in 1999 from a goose in Guandong, China, numerous reassortants have been observed over the last 5 years; many muster genes from up to four different viral sources via "antigenic shift." Also the individual genes showed a high rate of amino acid substitutions ("antigenic drift"); the virus is thus clearly not in a stasis, but actively evolving. In contrast to the "starbust" feature of the phylogenetic tree from HIV-1 M subtypes, the H5N1 gene sequences show a successive replacement of older by newer types as if the virus goes repetitively through genetic bottlenecks in its adaptation to new hosts and ecological settings (Guan et al. 2004).

Many industrialized countries are now preparing for the potentiality of an H5N1 epidemic. Prominent veterinary virologists such as Albert Osterhaus from Rotterdam University are ringing the alarm bell that migratory birds might spread influenza virus on their path between the northern summer and the southern winter quarters. Casualties with H5N1 in wild waterfowls are now also seen in Europe, where swans seem to be especially afflicted. Migratory birds in the winter residence of Nigeria showed individuals infected with H5N1. Likewise, eastern Turkey experienced an outbreak of H5N1 flu in poultry, and persons in close contact with infected chicken (Figure 7.3) contracted the disease

FIGURE 7.2. Herring gull (*Larus argentatus*), family Laridae, order Charadriiformes, in the foreground.

and some died of it. The WHO has listed 152 laboratory-confirmed cases of human H5N1 infection over the last 2 years with a fatality rate of 50%. So far, most cases still involve direct transmission from poultry to humans. However, the first probable cases of human-to-human transmission have in the meanwhile

FIGURE 7.3. Chicken (*Gallus gallus*) represent today a major source of animal proteins to human nutrition by providing meat and eggs, both in industrialized and developing countries. The figure shows the likely ancestor of the domesticated chicken, *G. gallus* ferruginous, from Bankiva, Malaysia. Chicken were initially not raised for meat or egg production, but held for cockfighting. The food use of chicken was propagated by chicken raisers only at about 1900 and became important after 1920.

been reported (Ungchusak et al. 2005). If this mode of transmission will become established, the specter of a pandemic flu outbreak in the human population becomes possible. Are we prepared?

On Models and Superspreaders

Epidemiologists made computer calculations to simulate the impact of different control measures on an epidemic taking Thailand as a model case (Ferguson et al. 2005). The critical parameter is the basic reproduction number R_0, a transmissibility parameter. It is defined as the average number of secondary cases generated by a typical primary case in an entirely susceptible population. If $R_0 > 1$, an infection will spread, with higher values it will explode, whereas with $R_0 < 1$, the chain of transmission will die out. To get R_0 below 1, public health has three principle measures. First, one can reduce the contact rate in the population, e.g., by closing schools or airports. Second, one can reduce the infectiousness of the infected person, e.g., through drug treatment or classically by quarantine. Third, one can reduce the susceptibility of uninfected individuals by antiviral prophylaxis or vaccination. You might wonder why I detail the infection process here, but do not forget we have here an especially insidious predator (the virus)–prey (us) pair, where the predator multiplies after each successful predation event. The simulation showed that with $R_0 = 1.5$, the infection remained for 30 days around the seeding location of human-to-human transmission, swept over the country in 90 days, became thereafter international, and was over in the model country at day 200, when 33% of the population became infected. With $R_0 = 1.8$, the infection rate is 50%. The authors for the Thailand simulation found that a combination of geographically targeted prophylaxis and social distance measures could wipe out the epidemic if cases are rapidly identified and 3 million courses of antiviral drugs are stockpiled. The drug now plays the role of an antinutrient or a chemical armor against predation. How tight is this strategy? In Europe several governments have started to stockpile drugs such as oseltamivir for the prophylaxis of health personnel and for the treatment of patients. This drug is a neuraminidase inhibitor and impedes the spreading of the virus through the body of the infected person. With antimicrobial drugs, the development of drug resistance is a problem. In that respect, a case report from Vietnam is alerting (Le et al. 2005). A 14-year-old girl without contact with poultry cared for her 21-year-old brother, who suffered from an H5N1 infection. The girl received oseltamivir prophylactically, but nevertheless contracted the virus apparently from her brother (some viral clones were sequence identical). She was then treated with oseltamivir and recovered. However, virologists isolated a drug-resistant H5N1 virus from her that showed a characteristic neuraminidase mutation that is known to confer resistance to this drug. The simulation depended on the value anticipated for R_0. However, the model also varies with different distributions of the R_0 value in an infected population. For sexually transmitted and vector-borne disease, epidemiologists used the "20/80% rule," according to which 20% of the cases cause 80% of the transmissions. When evaluating the transmission data of the relatively well-investigated SARS epidemic, scientists identified individuals with exceptionally high R_0 values, who were called superspreaders. This skew in the R_0 distribution led to entirely new transmission dynamics for epidemics (Lloyd-Smith et al. 2005).

Avian Flu Vaccines

Finally, a word on vaccines. Currently H5N1 vaccine production depends on a supply of embryonated eggs to produce inactivated subvirion vaccines (Webby et al. 2004). The virus produced on two eggs is needed to immunize one person; this is much cheaper than producing flu vaccine on cell culture. However, in the case of a pandemic with avian flu, chicken might be the critical transmitter and the authorities will call for massive destruction of chicken flocks. In view of the hundred millions of doses needed, there might not be enough laying hens around to satisfy the demand for embryonated eggs. Furthermore, H5N1 is quickly lethal for the embryonic egg—it might kill the egg without producing a high harvest of progeny virus. Ironically, rescue might come from another virus. Scientists from the CDC in Atlanta introduced the H5 hemagglutinin into a replication-deficient human adenovirus vector. It produced both humoral and cell-mediated immune responses against various H5N1 strains in mice and protected the animals against lethal challenge with flu virus (Hoelscher et al. 2006).

We are not Alone

The take-home message is clear. Despite decades of active research into influenza viruses, we are still at the mercy of this dynamic virus that seems to look for breaches in the species barrier. A recent example is the transmission of H3N8 equine influenza virus to dogs. Veterinarians became aware of this problem after virological investigations in an outbreak of severe respiratory disease with high mortality affecting greyhounds from various US racing grounds. The virus isolated from the affected dogs was according to genome sequence analysis a clear-cut equine virus. The viral hemagglutinin showed adaptation to the new host by changing a few critical amino acid positions. Even more disturbing was the observation that according to serological evidence, the equine virus transgressor had sneaked into the pet dog population (Crawford et al. 2005). Apparently, the viral empire strikes vigorously back, and we are getting cornered by influenza viruses coming from different fronts.

Natural is not Necessarily Healthy

We are well advised to keep an open eye on these avian influenza viruses since they could become a threat to human health comparable to the AIDS pandemic with millions of deaths. With safer sex measures and blood screening, the AIDS pandemic can at least theoretically be contained in the industrialized world; a pandemic influenza will be much more contagious and thus more disruptive for the world economy as already demonstrated by the smaller SARS epidemic. *It is important that a wider public realizes that these widely known viral infections are intimately linked to backyard animal rearing combined with the dangers of live animal markets. In Europe there are at present strong currents with the consumer to appreciate food from natural sources in the neighborhood that are minimally processed over products from established agricultural practices that are processed by the food industry. Many biologists think that these consumers commit an important error when equating "natural" products with healthy products. They are probably influenced by romantic feelings that were so prevalent in different epochs of European art and thinking. During those periods, ideals of the antiquity and Mother Nature were painted that never existed except in the heads of artists and philosophers. Also the Christian faith with its belief in a caring father and the heavenly commandment to subject nature to our order might have influenced this benign perception of nature. Even if I used in several passages of the book the metaphor of "Mother Nature," we should not anticipate a caring principle behind this name. As stressed in another context, the "Mother Nature" picture corresponds more to the Hindu trinity of gods combining at the same time the creation of the new (Brahma), the maintenance of the existing (Vishnu), and also the destruction of the old (Shiva). Romantic westerners forget all too easily that Shiva is part of our "Mother Nature" concept. Prominent veterinary virologists such as Albert Osterhaus suspect that free-range hens on commercial European poultry farms could play a crucial role in the transmission of avian H5N1 influenza infection from migratory birds to the human population.*

One gets the impression that these production facilities are maintained for a romantic feeling about a species-adequate animal husbandry or a sense of guilt toward fellow creatures, which we maintain as a food source. Europeans are there in a dilemma since the Christian belief did not formulate special commands toward animals. Ecology-oriented biologists claim more rights for wild animal life frequently for a purely egoistic reason, namely a concern for the future of the human civilization, but are less expressive with domesticated animal rights. The public knows subconsciously that a Rousseau-type "retour à la nature" will not lead domesticated animals and agriculture back into a terrestrial paradise. The Old Testament vision of the prophet where the lion lies with the sheep sets the stage for eschatological hopes for a world freed from the quest for food, but at the same time, the public knows that such a world cannot be populated with animals selected by evolution on the planet earth.

Problems of Food Safety: BSE

The Underused Cattle

We mentioned prion disease as a danger of cannibalism in remote New Guinea. Prion disease has now shaken the confidence in public health systems in Europe. It is difficult to refrain from political undertones and to restrict the discussion to basic scientific arguments when recounting the BSE crisis in Europe. I will try a neutral account. In the opening chapter, I mentioned that the European culture and all overseas cultures built on its ideals are deeply rooted on a cattle cult. Dairy and beef industries are important parts of the economic life and make a major part of the dietary identity of its people. However, only few parts of cattle actually go into the human food chain. Human imagination found other uses: less appreciated organs go into pet food, hides into the leather industry, bones into gelatin production for the film industry and as one of the most important binders into the food industry. This current practice has not created any major problems. Yet, still substantial parts of the body of the cattle remain unused, and technologists searched for alternative use of waste material like horns and hoofs and offal. If all this is crushed to its elemental biochemical constituents, the resulting "bone meal" should be a valuable addition to the feed of agronomically important animals. Apparently, no biologists trained in evolutionary and ecological thinking participated in the development of these rendering industries and the technologists overlooked that cattle not only were turned into carnivores when this type of material entered their feed, but actually became cannibals. As strict herbivores, cattle's only carnivore activity is when the dam eats its own placenta. Unfortunately, the lessons of the ritual cannibalism in the Fore tribe were not taken as a warning. Veterinary epidemiologists now paint the same scheme as in kuru. Perhaps one rare individual developed a spontaneous form of this disease, which would have ended with the death of that animal. By the feeding of bone meal, the agent was introduced into the food chain. In cattle the infected animal had a chance to enter into the rendering system and the infectivity could be amplified.

A Short History of BSE

Epidemiologists have postulated that other factors added to this scenario to explain why the BSE epidemic occurred only in the 1990s. One interesting, but not proven argument is the energy crisis which forced the low-profit rendering industry to change their technology to decrease the price of the rendering process. This change could have allowed the infectious agent to slip into the product. Whatever the exact epidemiological explanation, the fact is that the number of infected animals skyrocketed in Great Britain. The distribution of cases— *or its reporting practice*—was patchy in continental Europe. Then the British government reacted with a ban on the feeding of bone meal, made BSE a notifiable disease, explored the distribution of the infectious agent in the body of clinically affected animals, and declared several organs like the brain unfit for human consumption. The best scientific minds of British epidemiology and biomathematicians as well as veterinary and human neuropathologists addressed the problem and came up with discoveries that impressed the scientific observers in the food industry, which had never before been confronted with such an enigmatic agent in their food safety evaluation systems. Overall, the research in BSE is a scientific success story and—although with some delay—the BSE cases came dramatically down, verifying the correctness of the scientific assumptions.

Communication Problems

However, from the communication side the BSE crisis was a disaster that destroyed the confidence in the public health systems. Personally, I think that the grandmother argument was neglected. Scientists and politicians alike under-estimated the deep emotional fears of humans toward food and the risks of food poisoning from the hunter-gatherer phase of human evolution, which repre-sents the overwhelming part of human evolution. Only recently have humans started with agriculture where the risk of poisonous plants has been dramatically reduced. However, the safest food can still be spoiled by microbial overgrowth, therefore the threat of food poisoning remained real over most of the historical time and experienced a dramatic decrease only very recently with the industrial-ization of food processing. The lay public literally used their "gut feeling" and not their brain feeling to address this safety issue. Furthermore, they got in the wake of the BSE crisis another crisis, which confirmed the bleak presage that BSE is potentially only the top of an iceberg.

How Safe is Meat?

I will illustrate the problem of food safety with three recent publications asking for the presence of prions in skeletal muscle of infected animals. German scien-tists used transgenic mice expressing bovine prion protein (Buschmann and Groschup 2005). These mice turned out to be good sentinels for prion detection as they were 10 times more sensitive toward bovine prion infectivity than cattle. When they searched different anatomical sites in BSE-afflicted cows for infec-tivity, they found it only in the central and peripheral nervous system and not in lymphatic tissue. The only exception was the Peyer's patches of the distal ileum, which is most likely the site of entry for BSE infectivity. Apparently, upon oral exposure to the infectious agent the BSE infectivity spreads centripetally to the central nervous system via the enteric nervous system and peripheral nerves. Amniotic fluid and colostrum lacked prion infectivity suggesting that neither intrauterine transmission nor milk feeding is a mode for vertical transmission of the disease from the dam to the calf. With a single exception, which might represent an experimental error, no muscle samples from BSE cows transmitted the disease to transgenic mice. However, the presence of prions in muscle meat cannot be dismissed. In fact, in North American mule deer, kept in captivity for meat production, a prion disease was described. Animals afflicted with this chronic wasting disease demonstrated regularly prion infectivity in muscle tissue that could be transmitted to mice. The longer and more variable incubation period in the inoculated mice compared to those exposed to brain samples suggested lower infectivity titers in the muscle than in the brain (Angers et al. 2006). While these data could suggest some caution toward mule deer meat consumption, scientists knew that sheep afflicted with a prion disease called scrapie show a widespread anatomical distribution of infectivity. In naturally infected sheep, the PrP^{Sc} was detected in muscles several months before clinical disease onset (Andreoletti et al. 2004). The titer was 5,000-fold lower than in brain, and the muscle infectivity was concentrated over spindles, highly innervated structures that ensure muscle proprioception. As scrapie is a sheep disease known in Britain for more than two centuries, sheep-derived PrP^{Sc} has in appreciable quantities entered the human food chain. Nevertheless, on the basis of epidemiological data dietary exposure to scrapie-infected sheep is currently considered nonhazardous to humans. However, one should not trust this reassuring evidence from sheep too much as there is good epidemiological evidence that BSE has entered the human population as discussed next.

The New Threat

The new disease is called in scientific slang the vCJD epidemic. The abbreviation stands for variant Creutzfeldt–Jacob disease, which most likely represents the intrusion of the BSE agent into the human population. Again, on the scientific side it represents a success story. Only few countries outside of Britain would have noticed this new disease—only scrupulous screening of death certificates and enormous neuropathological dedication led to the definition of this disease in young adults. Numerically, the toll of this disease is not yet heavy and happily the rate of new case identification is decreasing. However, the jury is still out because the long incubation period of the transmissible spongiform encephalopathies does not yet allow the conclusion that we have already seen most of the food-mediated transfer of BSE into the human population. Actually, the greater medical concern nowadays is that BSE crept subclinically into the human population (millions of people were exposed to BSE-tainted food products, but only a few dozen developed vCJD until now) and can now be further transmitted by medical "cannibalism" (blood transfusion, organ transplantation). There is precedence to iatrogenic transmission of CJD when stunted children were treated with growth hormone that was not produced by genetic engineering, but was isolated from the pituitary gland of a large number of human cadavers. Unfortunately, one dead donor was incubating CJD and infected a sizable number of French children.

An Unusual Pathogen: The Prion Hypothesis

Kuru is the last, apocalyptic chapter of virology textbooks like "Field's Virology," but it fits this classification as a virus only with the original Latin meaning of virus as a poison, not the modern definition of a viral particle. Despite substantial effort and heated discussions at scientific conferences that bordered on political or religious strife, no genetic material in the form of nucleic acids could be associated with the infectious agent. This lack of a genome could also well explain the disturbing resistance of the agent toward chemical and physical actions. Standard medical sterilization procedures were insufficient to destroy the infectivity. To come to grip with a model for this mysterious disease, S. Prusiner developed over the years the prion hypothesis, which is nothing less than a new class of infectious agent consisting of a rogue protein with an aberrant 3-D conformation called PrP^{Sc}. PrP^{Sc} imparts its misshapen conformation to its normal cellular counterpart, the PrP^{C} protein. PrP^{C} is thus misfolded under the influence of PrP^{Sc} and can then transfer the newly acquired pathological conformation to further normal PrP^{C} proteins. In this way PrP^{Sc} can "replicate" and become an infectious agent.

Opponents maintained, for example, a virino hypothesis, i.e., an agent containing its own nucleic acid enveloped in host-encoded proteins. In a recent editorial, prion disease researchers compared this discussion with a key paper from 1840 by Joseph Henle (Zou and Ganbetti 2005). Henle proposed at this early time that infectious diseases are caused by "contagia animate"; we would today say microbes, and not by miasma, poisoned air. Actually, the name malaria still in use today reflects this old and outdated "bad air" hypothesis. Now biologists are striving to fulfill Koch's postulates for prions. Some progress along this line was achieved recently when a truncated fragment of mouse PrP^{C} protein could be misfolded in vitro into β sheet-rich fibrils. After intracerebral inoculation, these fibrils could cause disease, albeit only after a long incubation period and in transgenic mice highly overexpressing the truncated PrP^{C} protein (Legname et al. 2004). Now, scientists mixed an excess of PrP^{C} with small amounts of PrP^{Sc} and conversion of PrP^{C} was observed. The trick was that they now used sonication that disrupted the new fibrils and they added the result of the first round of amplification into a new test tube containing fresh PrP^{C}. Like in a type of PCR, this cycle of incubation, PrP^{C} to PrP^{Sc} conversion, sonication, and transfer to a fresh tube was repeated many times. The researchers calculated that the initial brain inoculum contained perhaps 10^{11} molecules of PrP^{Sc}, but the series of amplifications implicated a dilution of 10^{-40}. It is thus physically impossible that the initial material is transferred into the last tube. Nevertheless, the last tube led to a scrapie disease in wild-type hamsters that was identical in all respects to that produced by the initial infectious brain material (Castilla et al. 2005).

Critics might still maintain that some type of RNA might have been amplified along with PrP^{Sc}, but this hypothesis lacks any experimental evidence and also appears somewhat farfetched.

Going for our Blood

Real-life Draculas

On Vampires: Real and Imagined

As a scientist you might wonder about the boom of horror and disaster films, which fill the box offices today. Some people want to see it as a sign of degeneration of our civilization, which begs its end by anticipating its annihilation. However, as a scientist you should differentiate your judgment. Some films (e.g., Armageddon, Outbreak) have such a clear scientific core message that you can easily quote articles from scientific journals that deal with the same subject. Other films seem to appeal only to our lower instincts. Yet for a biologist, appealing to instincts is not necessarily negative. Some films call on what one could characterize as a collective memory of humankind. Take as example perhaps the oldest and most successful subject of horror films. The plot is in Transylvania—already the naming of this real province in Romania gives strange feelings, a mixture of fright and delight typical for these types of films. And as in fairy tales there is a real-life nucleus to the story. Briefly, a wicked Romanian count transforms every night into a bat-like blood-sucking creature, which rests in its grave during daytime. It can only be put to eternal rest when a wooden peg is pierced through its heart. Prominent in the fight against the evil is a scientist, actually a zoologist specialized in bats and who uses garlic as a vampire repellent. Actually you can identify historical and real-life aspects in this story. A zoologist will think of blood-sucking bats of the family Phyllostomidae (also called "vampires" by zoologists). Already Alexander von Humboldt reported on blood-sucking bats during his travels through South America, following accounts from physicians accompanying Spanish soldiers in the New World. Horses were the main victims: bats sucked blood from superficial vessels in the skin, which could still bleed the next morning. The bats punched out a small area of the skin—the wounds were only harmful when flies subsequently laid their eggs into the wound. Humans suffered mainly from painless and generally harmless bites on the toes and the nose. Blood-sucking by bats is only prominent in Latin America and so rare in Europe that it is an unlikely source for the Dracula legend. In fact, Romanians still remember a cruel king with a Latinized name of Dracula. He impaled Turkish prisoners of war, hence perhaps the origin of the wooden dagger through the heart. During the period of the reign of Habsburg Empress Maria-Theresia, rumors of vampires reached Vienna and the Empress sent her court physician to Romania—the core for the legend of the professor fighting against the vampires. Actually, he reported that the suspected cadavers were vividly colored with a reddish skin and when the peg was pierced into the body by frightened fellow villagers, an intensively red liquid oozed out giving the impression of liquid blood. Modern-day pathologists suspect that these corpses suffered from a bacterial putrefaction process where the red color came from a bacterial pigment. The piercing with the wooden dagger would have caused a splash of the bacterial culture, which could have infected the bystanders of this rite, who would have suffered similar putrefaction after death from a bacterial infection.

Mosquito Biting

Tragically, Dracula is still amongst us and when the night falls he sets out to get his blood meal and to spread death in the world. However, it is neither a devil nor a bat, but a little fly or in Spanish a "mosquito" (Figure 7.4). Their mouth parts seem to be tailor-made for blood sucking. The proboscis consists of two stylets that slide against one another when piercing the skin in search for a blood vessel. When not directly successful, the mosquito can retreat the needle and change the angle for injection to have a second try in its quest for blood. Inside the proboscis are two hollow tubes. Through one of them the animal injects its saliva, through the other it draws the blood of the victim. This double-barrel high-tech surgical device uses a lot of pharmacology during its action. To prevent blood clotting, which would clog the feeding tube, the insect injects with its saliva an anticoagulant. The mosquito also injects antiinflammatory chemicals. You might hear the insect or see it when it is sitting on your skin, but you will not feel it during the blood transfusion. Even this elegant feeding device needs more than a minute for a blood meal. During that process you will not experience a pain reaction, which is important for the animal to escape unharmed. Once away, an intense irritation will set in, which leaves you an itching memory. At that moment the culprit is already away. However, it has not gone far—it might sit on the wall or the ceiling of your room. This should not surprise us: the mosquito has taken a substantial amount of blood, which increased its weight by a factor of four (Budiansky 2002). Small wonder that rest and digestion of the protein-rich blood meal and absorption of the nutrients are now the priority.

Blood Meal

The mosquito surrounds the stolen blood by a peritrophic matrix consisting of protein and chitin. Proteases secreted by the insect gut then penetrate this matrix and liberate smaller hydrolysis products from the blood. Microvilli-bound enzymes finish the digestion before absorption by the midgut cells. The nutrients travel to the fat body, roughly corresponding to a mixture of the liver and adipose tissue of our body, where egg proteins and lipids are synthesized. These synthesis products then go via the hemolymph to the insect ovary, where they are used for egg development, which takes 3 days until oviposition. During this time, the female mosquito takes no food (Holt et al. 2002). After egg

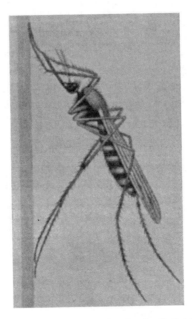

FIGURE 7.4. *Anopheles*, a genus of the mosquito family Culicidae, order Diptera, where the female is a blood-sucking animal and can transmit malaria.

laying, a new victim is searched for a new blood meal since only the energy-rich blood can sustain this demanding reproductive activity. This becomes clear when looking at *Culex pipiens* populations (the mosquito known to Europeans and North Americans) sporting distinct feeding habits. Subpopulations that do not feed on blood lay only about 60 eggs once in their lifetime. Blood-seeking subpopulations in contrast can produce 400 eggs per blood meal and this happens repeatedly during their life (Budiansky 2002). Only egg-laying females suck blood. Early after metamorphosis into an adult mosquito, the animals take sugar meals from plant nectar to maintain basal metabolism and to power the flight. Flying is important to find a mate and the victim for the first blood sucking. As producing sperms is a low-energy business, male mosquitoes remain lifelong vegetarians.

Chemical Cues

Mosquitoes come in many forms and are not limited to specific climate zones. In fact during the short summer period, Artic zones become so densely populated by mosquitoes that dead caribous lacking blood were described. Some mosquitoes hunt during daytime and use visual cues; some even chase their victims. However, most mosquitoes hunt in the night and have thus to rely on other senses to find their prey. Mosquitoes differ in their searching profile, which determines their host specificity. Commonly, they are attracted by a 37 °C skin temperature, moisture, CO_2, and specific odors. The odor preference was tested using olfactometers. The notorious *Anopheles gambiae* mosquito bites its human victims on the feet and ankles. It is a well-known observation that the human foot odor bears a remarkable resemblance to a few strong-smelling cheese specialties. This parallel is not fortuitous: *Brevibacteria* are found in the moist clefts between the toes as well as in strong-smelling cheese types. Currently, devices that emit such odors to detract mosquitoes from their human victims are in development (Enserink 2002).

The *Anopheles* genome

The design of antimosquito measures is now helped by the sequence of the *A. gambiae* genome (Holt et al. 2002). The size of the genome of this insect is 278 Mb and is substantially larger than that of the fruit fly *Drosophila melanogaster*, which shows only a 122-Mb genome size. However, both genomes show a comparable number of about 13,000 genes. The difference in size is thus largely due to variation in intergenic DNA. Half of the genes are orthologs and share an average sequence identity of 56% (Zdobnov et al. 2002). This percentage corresponds to the genetic distance separating humans from the pufferfish. This is a surprising observation since the two dipterans (the zoological term for these two-winged insects) diverged only 250 My ago, while the two vertebrates separated from a common ancestor 450 My ago. These two insects apparently diverged on the fast lane. Inspection of the *Anopheles* genome identified 276 G protein-coupled receptor genes that play roles in many pathways affecting nearly all aspects of the mosquito's life cycle (Hill et al. 2002). Especially prominent are chemosensory receptors: nearly 160 genes were classified as either odorant or gustatory receptor genes. Thus about 1.5% of the proteome is dedicated to these chemical senses. Impressive as this percentage appears, it is not unusual. Animals have to deploy substantial care for finding food and deciding on its suitability for eating. Mosquitoes are thus not an exception in nature; *D. melanogaster*, which lives on rotting fruits, shows a comparable percentage of chemosensory genes. The nematode *Caenorhabditis elegans* uses even 6% of its proteome for chemical sensing. However, the DNA blueprint is essentially static information, which allows only limited inferences on the involvement of the different genes in the various life processes. This DNA sequence information can now be made dynamic by mRNA expression analysis. When this is done for blood-fed and nonblood-fed *Anopheles*, researchers found about 100 upregulated and 70 downregulated genes. Prominent under the upregulated genes were protein digestion and protein and lipid synthesis genes. As expected, muscle-related genes and genes encoding enzymes for sugar meal digestion were downregulated (Holt et al. 2002).

The Mosquito as a Vector

Bioinformatic analysis also identified 242 genes in the *Anopheles* genome involved in innate immunity. Adaptive immunity does not occur in insects and is in fact limited to chordates. This figure compares to 185 corresponding genes in *Drosophila*. The excess of the mosquito genes is especially marked in the recognition category of innate immunity genes (Christophides et al. 2002). This expansion of immunity genes probably reflects the different lifestyles of both dipterans. Here we touch an important point of the mosquito's feeding strategy, which has crucial medical implications. When feeding, mosquitoes become exposed to human pathogens, and some of them actually use the mosquito as a vector. If the mosquito wants to survive on its nutritious, but potentially dangerous blood meal, it must find an answer against getting infected by the many blood-borne pathogens. The pathogen has of course no interest to kill the host, it only searches for a lift from one to the next human victim as its transmission relies on cycling between the vertebrate and the insect host. There are many mosquito-borne diseases, but the numerically most prominent and the most hideous is malaria. In fact, *A. gambiae* carries the deadliest of the malarial parasites, namely *Plasmodium falciparum*.

Hitchhiking the Blood Sucker

Plasmodium: Life Cycle

Malaria is endemic in more than 100 countries girdling the equatorial zone of the world. It accounts for 300 million cases and 1 million deaths each year, with most deaths occurring in African children. Unfortunately, this is not the only mosquito-borne disease. Mosquitoes transmit dengue, yellow fever, and Japanese encephalitis virus and even parasitic worms causing lymphatic filariasis. The latest addition to this list is the West Nile virus in the United States—industrial countries are thus not spared. The *Plasmodium* parasite (Figure 7.5) leads a complicated life that puzzled scientists for over 100 years since its role in malaria was established. Already the naming of the different stages is complicated (Wirth 2002). When injected by the mosquito the *Plasmodium* is in the sporozoite stage, which travels to the liver. There it goes through several stages to become merozoites that infect the red blood cell. In this cell the *Plasmodium* develops into a so-called trophozoite. As the name indicates this is the major feeding stage of the *Plasmodium*. Actually all clinical signs of malaria such as fever, chills, anemia, and cerebral malaria can be explained by the altered state of the infected red blood cell. The trophozoite then goes again through a merozoite stage. *Plasmodium* is released from disrupted red blood cells only to infect new red blood cells. Finally, up to 10% of the erythrocytes of the patient is infected with the malarial parasite. If the *Plasmodium* does not develop further, it would be locked in the infected patient. Therefore, the merozoites in some red blood cells still take another turn: they develop into gametocytes. With the next blood meal mosquitoes ingest these cells. Within the gut of the insect the gametocytes develop into male and female gametes (reproductive cells), which fuse to a zygote. This becomes an ookinete, the strange name means "moving egg" and describes the traveling of this cell through the gut wall. It transforms then into an

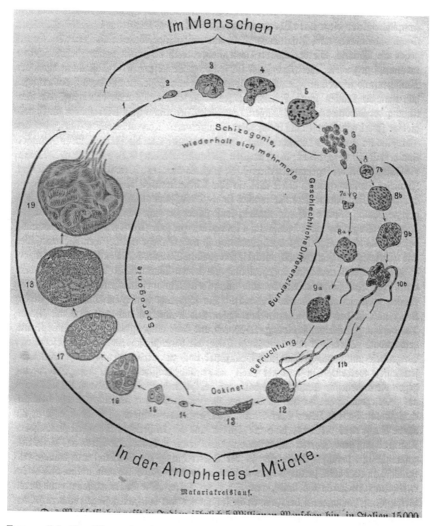

FIGURE 7.5. The life cycle of *Plasmodium vivax* starts in humans with the injection of the sporozoite (*1*) from the saliva of the mosquito. The sporozoites enter liver cells, where they mature into tissue schizonts (*2–5*), which rupture to release merozoites (*6*). Once within the bloodstream, merozoites invade red blood cells and either mature to erythrocyte parasites (ring trophozoite → schizont → merozoites, not depicted) or they differentiate into sexual forms: male gametocytes (*7b–11b*) and female gametocytes (*7a–11a*). However, the protists proceed with the following sexual stages only in the next mosquito vector. Within the mosquito midgut, the haploid male gametocyte loses its flagellum to become a male gamete (*11b*), and fertilizes a haploid female gamete (*9a*) to produce a diploid zygote (*12*). The diploid zygote then transforms into an ookinete (*13*), which invades the gut of the mosquito to become an oocyst (*14*). It then undergoes a meiotic reduction division to produce the haploid sporozoites (*15–18*) that migrate to the salivary gland (*19*) to complete the infection cycle. The sporozoite then infects the next human victim.

oocyst, which finally bursts releasing many sporozoites that travel to the salivary gland of the biting insect.

Fungi as a Weapon

The cycle in the insect takes about two weeks to complete. This is a relatively long period and thus a potential Achilles' heel of the parasite. If something happens to the mosquito vector during this period, the parasite is stopped. Therefore, even slow-killing agents can have a great effect on malarial transmission. One of the most recent antimalaria approaches targets just this window of opportunity. Researchers sprayed two entomopathogenic fungi on surfaces commonly used by mosquitoes in the postfeeding rest period. Mortality was observed only after a week, but reached 90% before the ingested gametocytes could develop into newly infectious sporozoites. The sporulating fungi were quicker and had grown out of the mosquito cadavers in 70% of the infected animals. In addition the appetite of the infected mosquitoes for blood was lost even before that (Blanford et al. 2005). Importantly, these fungi can infect and kill the mosquito without being ingested. Contact with the mosquito just for the 1-day rest period of the mosquito after a blood meal was sufficient to infect the vector. Even more importantly, infections of the mosquitoes with the fungi were observed under real-life conditions during a field study in Tanzania (Scholte et al. 2005).

Herbal Medicines?

Alternatives in the fight against the insect vector are chemical repellents either applied to bed nets or directly on the skin. Some chemicals showed a dose-dependent protection from insect bites with standardized mosquitoes. Many consumers in industrialized countries are now weary about chemicals and prefer plant extracts believing that this is natural (although this is ultimately of course also chemical). And indeed soybean oil had a moderately good repellent effect, rehabilitating the good old folklore of using garlic against count Dracula. *However, not all plants will do the job: Citronella extracts, widely used against mosquito bites in industrialized countries, turned out to be ineffi-cient in controlled tests (Fradin and Day 2002). The ecology-minded consumer should therefore think twice about soft medicine approaches against vector-borne diseases. Clinical tests with herbal extracts have more negative outcomes than just Citronella plants. Echinacea plants recently joined this list when their extract turned out to be of no effect against rhinovirus infection (Turner et al. 2005).*

Plasmodium Genomics

Now biologists have a great aid in their fight against malaria: the 23-Mb-large genome of *P. falciparum* has been determined (Gardner et al. 2002). As genomics is a much more powerful tool when it can be done on a comparative basis, the sequence of the rodent malarial parasite *Plasmodium yoelii* (Carlton et al.

2002) adds substantially to the antimalaria approaches, not least because it puts the analysis of the mouse malaria model on a firmer footing. *P. falciparum* musters 5,300 genes, slightly less than the baker's yeast, which shows 5,800 genes. If you look at the many lives of the parasite, it is a quite remarkable feat to achieve this with so few genes. However, you should not forget: yeasts are not so primitive, and it needs only a little more than twice as many genes to make a mosquito. Surprisingly, humans are made with less than 5–10 times the amount of genes you find in a baker's yeast, although at the level of gene products the factor is 15. It seems that nature constantly humiliates human pride. After Copernicus' and Darwin's revelations, we are taught modesty again by geneticists. However, there are also positive aspects to these humbling number games. The genomes of man, mosquito, and malarial parasite are now available; the tools are there to attack the killer by systematic approaches. Most of these genomics and postgenomics approaches are concerned with targets for drugs or vaccine development. As this book is written under an "eating" heading, I will illustrate the power of these approaches for the metabolism of the parasite.

Plasmodium Metabolism

About 700 enzymes were annotated in the *P. falciparum* genome, about 14% of all predicted proteins. This is a small percentage when compared to the 25–33% of enzymes in the proteome from bacteria and archaea. Apparently as a parasite, plasmodia dedicate fewer genes to enzymatic activity. In the erythrocyte, *P. falciparum* relies principally on anaerobic glycolysis for energy production. NAD^+ is regenerated by conversion of pyruvate to lactate. A complete enzyme set for the TCA cycle was identified, but it is not used for energy production via the respiratory chain as critical components of the ATP synthase are lacking. The TCA cycle probably functions as a supplier of intermediates for other pathways. The genome sequence predicts that the parasite can synthesize pyrimidines de novo, while it is incapable of de novo purine synthesis, which must therefore be imported. Likewise the parasite lacks the enzymes for amino acid biosynthesis. Amino acids are obtained by salvage from the host and by globin digestion. The trophozoite uses hemoglobin from the erythrocyte cytoplasm as an important food source. In the food vacuole the globin is hydrolyzed to small peptides, which then provide the amino acids needed by the parasite. The parasite synthesizes its own heme de novo. An important metabolic activity is the degradation of the heme moiety of the hemoglobin into hemazoin—large amounts of heme are toxic to the parasite.

Drug Targets

This heme degradation pathway is actually the target of classical drugs against malaria, namely quinine and especially chloroquine. The sobering fact is that despite all research on malaria, the impact of malaria in Africa has been increasing over the last decade mainly because of the slow, but constant advance of chloroquine-resistant malaria (Arrow et al. 2005). Fortunately, there is a Chinese herbal drug emerging that is currently the treatment of choice:

artemisinin from the plant *Artemisia annua*. This is now the showcase where a drug is cultivated by farmers on the field. Notably, the hunger of the parasite for hemoglobin is the target for this new drug: artemisinin also interferes with the degradation pathway of the heme.

Plasmodium Proteomics

However, there are clear limitations to the use of genomics in malaria research: only 35% of the predicted proteins have an identifiable function. Alternative approaches are thus needed to get into the secrets of this elusive pathogen. Large-scale mass spectrometry allowed a detailed proteome analysis. One study identified 1,300 (Lasonder et al. 2002), another even 2,400 proteins (Florens et al. 2002). The proteins were associated with different developmental stages, which allowed a first attribution of many hypothetical proteins to different stages. A complication was the observation of a substantial overlap of proteins found in different stages. For example, the antigenically variant proteins of the *var* and *rif* genes are not only found on the surface of infected erythrocytes, but were unexpectedly also expressed in sporozoites (Florens et al. 2002). These proteins are exported to the surface of the infected erythrocyte, where they mediate adherence of the infected red blood cells to endothelial receptors on the host blood vessels. These proteins represent important virulence factors, but owing to their location, they are also very exposed to the attack of the immune system. If the infected cell is destroyed before the parasite has completed its life cycle in the red blood cell, the infection chain will be interrupted. Immune evasion is thus an important task for the parasite. It achieves this task by having 59 *var* and 149 *rif* genes that can be expressed alternatively. The crucial role of the *var* genes is highlighted by their chromosomal position. In 24 of the 28 chromosome ends, the *var* genes are the first transcriptional unit. This telomeric organization is consistent with the exchange between chromosome ends as a means of immune evasion (Gardner et al. 2002).

Transcriptome

The next step providing new insights was the mRNA transcription studies on microarrays (Le Roch et al. 2003). Notably, 88% of the predicted genes were expressed in at least one stage of the life cycle. Half of the expressed genes were constitutively expressed identifying them as housekeeping proteins. Others were expressed in a cell-cycle-dependent way, which led to the definition of 15 expression clusters allowing tentative attributions of functions by the guilty through association principle. Interestingly, from *var* genes found at the chromosome ends (subtelomeric *var*) all but one showed low, constitutive expression and only one of them showed differential expression. Some *var* genes are located toward the centromere ("middle"part) of the chromosomes and they showed even more highly regulated expression (40- and 150-fold increases over basal level in sporozoite and trophozoite, respectively).

Organizing the Membranes

Apparently, you have to work hard to get a free meal from our erythrocytes. Some of the problems are self-created. The parasite contains many intracellular membrane systems and is also surrounded by a plasma membrane like any eukaryotic cell. In addition, within the erythrocyte it sits in a so-called parasitophorous vacuole, which is bound by a membrane; all are then wrapped with the erythrocyte membrane. This Russian doll construction of membrane systems gives you protection and creates new metabolic compartments to conduct parasite-specific metabolism. However, these multiple membranes have to be crossed by the substrates and end products of metabolism necessitating many protein transporters. Researchers had identified multiple tags at the N terminus of proteins destined for export. As a first ticket, a signal sequence recruited the protein into the secretory pathway within the parasite—a well-known process from other organisms. However, when this signal was cleaved, a second ticket became visible in some proteins that got them still further, the vacuolar transport signal. The barcode destined for further transport by the translocon in the vacuolar membrane was now deciphered (Marti et al. 2004; Hiller et al. 2004). As we live in the era of the -omes (genomes, transcriptome, proteome), researchers dubbed the ensemble of these exported proteins the malaria secretome (Przyborski and Lanzer 2004). The sheer size of the malaria secretome surprised even experienced malaria researchers: 320–400 proteins showed this translocon ticket. In fact, the malarial parasite rebuilds its home and directs the erythrocytes to its own nutritional and developmental needs.

Nutrient Import

As I am here concerned with the quest for food, I will focus in the finishing paragraph on one of these nutrient transporters. To put it into perspective: it takes a mere 48 h for the parasite to grow from an infecting merozoite to a huge trophozoite that gives rise to 20–40 new parasites. To fuel this enormous growth, the parasite needs nutrients, but the erythrocyte was not built to satisfy this appetite. The nutrients available in the uninfected erythrocyte simply could not sustain this need for food molecules (Kirk 2000). Hence the parasite had to target transporters also to the erythrocyte membrane. Not surprisingly, trophozoite-infected erythrocytes exhibited a 150-fold higher conductance than uninfected cells. The current was mainly carried by anions and abruptly abolished by channel blockers. Patch-clamp techniques identified a small ion channel present in about 1,000 copies per infected cell, which transports anions, sugars, purines, amino acids, and organic cations (Desai et al. 2000). This channel is probably also used to export metabolic waste like the enormous amount of lactate produced by the parasite.

From Cucania to Schlaraffia

With this transporter in place the parasite has unlimited access to small nutrients carried in the serum of its victim: the table is ready and as long as the host survives this hard ride, the parasite lives in Schlaraffia. This imaginative country became popular with a fairy tale by Bechstein, a follower of the Grimm brothers: it is the country where milk and honey flows, sausages grow instead of fences around the houses, grilled pigeons are flying into the mouth of the gourmands. Here laziness is a virtue and diligence a vice. The story does not lack the fountain of youth where you can exchange your aging better half against a young one. The name of the land derives from the older German "slur" (lazy fellow) and "affe" (monkey, here a stupid person who believes the story). Of course, you must ask a mute for the way. Schlaraffia is not in Germany, clerics of the Dark Ages fabled of the land Cucania, the Latin poet Lucian knew it, and the promised land of Moses carries traits of Schlaraffia. You can categorize this story as one of the endless myths on the lost paradise. For a biologist the story is nevertheless interesting: it symbolizes the revolt of man against the world as it is. It is a reverted world. Natural laws are reversed: organisms are growing younger and humans are freed from the laws of thermodynamics imposing the quest for food. To be precise and interestingly for this story, it eliminates only the quest part for food, not the food itself. Eating as such is seen as a divine pleasure, only in this dream it comes cost free. The French are the most consequent Schlaraffians with their proverb: qui dort dîne *(who sleeps, eats). Recall that the ancients imagined their gods still eating nectar and ambrosia. Scholars do not agree on what Homer meant by ambrosia except that it was a food needed to confirm the immortality of the Greek gods and has a strong fragrant character. Even the God of the Old Testament smells with satisfaction the food, which is burned in his honor.*

Let's get back to biology: the dream of the promised land is only of short duration for the malarial parasite—the host is not well, loses appetite, burns food calories excessively in fever bouts, and might die eventually. If the parasite does not want to die with its victim it must get to the next host. You have heard that this is not an easy road for plasmodia, necessitating the transition through mosquitoes. Did other organisms find the land where milk and honey flows?

Going for our Gut

The Land Where Milk and Honey Flows

Complexity of the Gut Microbiota

It is not only man dreaming of Schlaraffia, a land that is surrounded by a wall of gruel, and to get there you have to eat your way through. Many organisms have this dream and when considered superficially, the microbial inhabitants of our gut come close to this goal. The gut microbiota is the unseen majority sitting in us. It has been calculated that our commensal microbial flora exceeds our body cells by a factor of ten with respect to cell numbers. On a cell basis we are thus 90% bacteria and only 10% human. Of course, bacterial cells are much smaller than human cells, but they still contribute the appreciable amount of 1.5 kg to our body weight. Most bacteria reside in the alimentary tract, the highest numbers are reached in the colon with an excess of 10^{11} bacteria per ml of colonic material. This population also exhibits a high turnover when considering that up to a half of the 120 g of feces, which we produce per day, is bacterial biomass, live or dead. Living in the gut of somebody else is a priori the closest you can get to Schlaraffia in biology. You get the food from what your host is eating. From what we learned in our survey of the quest for food, one should distrust this idyll and suspect that life is not as easy as one thinks.

Complementary Approaches

To every place where a table with food is ready, consumers will arrive. There is probably no place on the earth which fulfills the basic physicochemical requirements for life that is not colonized by organisms. If one organism has discovered a food source, more or less related fellow organisms are not far. This also applies to the gut. Researchers around D. Relman from Stanford University recently examined the intestinal "microbiome" of just three healthy adult subjects (Eckburg et al. 2005). Many studies like this were conducted before; what sets this study apart is the fact that they did not restrict their search to fecal material, but extended their sampling to six different anatomical sites of the colon. In addition, they sequenced more than 10,000 ribosomal DNA samples and obtained bacterial and archaeal sequences in the ratio of 10:1. In this sequence set they identified 400 bacterial phylotypes; the majority were novel sequences from organisms that cannot yet be cultivated. Most of the inferred organisms were members of the Firmicutes (e.g., *Clostridium*) and Bacteroidetes phyla. Proteobacteria to which the famous gut bacterium *E. coli* belongs yielded only few sequences, which was not so surprising as *E. coli* represents only 0.1% of the bacteria in the colon (Figure 7.6). So if there is Schlaraffia, it is so densely populated by so diverse prokaryotes that the free meal idea becomes a pure illusion. This dense and varied bacterial population is also a challenge for the researcher. *Efforts to cultivate these bacteria or to describe their genetic variety as a function of the host species, host genetics, age, sex, geography, health, and disease becomes a Herculean task, if not a chapter from the myth of Sisyphus.* Research on this task uses three different approaches:

1. Despite the relatively hopeless situation, many research papers describe just the number of different species by culture-dependent and culture-independent methods.
2. Other researchers put their hope into the so-called second Human Genome Project where the composition of the gut microbiota will be described by a systematic comparative metagenomics approach (Relman and Falkow 2001). With the rapid development of sequencing technologies and the ameliorating computational capacities to assemble prokaryotic genomes from shotgun sequencing, this approach does not lack attraction. It could reveal to us 2–4 million genes that are associated with the aggregated genome of the about

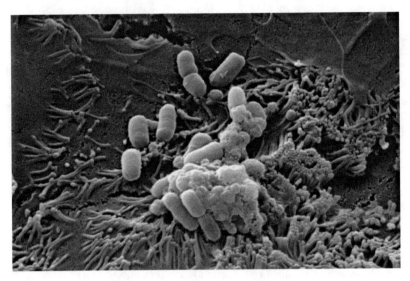

FIGURE 7.6. *Escherichia coli* adhering to the apical side of a differentiated intestinal cell line showing microvilli. The borders between three eukaryotic cells are visible allowing a size comparison between higher cells and bacterial cells.

1,000 bacterial species that make up the microbiome and which indirectly add to our metabolic capacities. However, it is not a trivial task to reconstruct a biological reality from *in silico* data.

3. Therefore, one could quote here a third school perhaps best represented by J. Gordon's group in St Louis, who address the complexity, at first glance, by two paradoxical approaches. At one side they use a variety of experimental techniques combining the available -omics techniques with physiological experiments in animals conducted with the approaches of many different biological disciplines. On the other side, they simplified the experimental system dramatically by working in most of their studies with germ-free mice that were monoinoculated with a single bacterial species.

Control the Access

One might also approach the complexity of host-microbe interaction in the gut by applying first principle arguments. You might argue that the host does not appreciate the idea too much to share its food with uninvited guests. It will in all probability restrict the growth of its commensals. The control will not be easy. Anatomically, the gut lumen is, despite its difficult access for the physician and experimental researcher, not an internal part of the body—it is an external world. Even worse: it has to be filled regularly with food items that contain microbes. If you look at the human situation with the eyes of a bioengineer you might try to control the access of microbes to the gut in the first place.

This is what human evolution actually did: relatively proximal in the alimentary tract it placed a highly acidic stomach. In a cyclic pattern, triggered by the meals, the pH of the gastric juice changes between highly acid conditions of pH values near 1 to values approaching relative neutrality (pH of 5). In parallel with this pH cycles the bacterial content in the stomach juice changes between 10^2 and 10^4 colonies per ml. Thus only low titers of bacteria enter the small intestine.

Developing Priorities

Next another engineering request is encountered: your own digestion priority should be put over the digestion by the commensals assuming that you have the enzymatic equipment to deal with the specific food item. For us, proteins are not too difficult to digest, neither are fats. However, our Achilles' heel is the digestion of polysaccharides of plant origin. The major sites of nutrient absorption in humans are in decreasing order the duodenum, the jejunum, and the ileum. A bioengineer would thus design a bacteria-free zone over this region to avoid competition for nutrients, which we can digest and absorb. This is exactly what is observed: bacterial counts of the mucosa and lumen of the jejunum rarely exceed 10^4 per ml. Here I have to specify that great geographical variations complicate the picture. Human populations from developing countries have a more substantial and complex microbiota in the jejunum: viable counts reach up to 10^8 per g for the luminal content and 10^6 for the epithelia-associated fraction. In contrast to the contaminated small bowel syndrome known from patients in industrialized countries, who show steatorrhoea (increased fecal fat) and vitamin B12 deficiency, a high bacterial load in the small intestine is widely distributed in Asian populations. However, while being common, it might not be without consequence: the jejunal mucosa in people from the developing world is frequently flat and shows leaf-like villi. It is currently unknown whether this condition impairs the absorption of nutrients and contributes to the undernutrition in these regions. A complicating factor in populations from developing countries might already be the high prevalence of subjects producing only small amounts of hydrochloric acid in the stomach. This impaired acid production makes these people not only more susceptible to cholera (a subject producing normal acid levels needs 10^4 *Vibrio cholerae* to get infected, while the infectious dose is lower for low acid producers). It also has a lower capacity to sterilize the food material reaching the small intestine.

In the context of our armchair speculation along teleological principles we could postulate that the human host made a deal with the bacteria. It provides them a home in the colon, a region with a naturally flat mucosa endowed with only marginal absorptive capacity except for short-chain fatty acids. The latter is produced by the anaerobic polysaccharide-digesting commensal flora. These short-chain fatty acids can be absorbed and utilized by the host. In addition, it leaves them that part of the food which the human host cannot handle due to limitations of its enzymatic apparatus.

Nutrition of *Bacteroides*

From the diet, 8–18 g nonstarch polysaccharide, 8–40 g starch, 2–8 g oligosaccharides, 2–10 g unabsorbed sugars, 10–15 g protein and peptides, and 6–8 g
fats reach the colon each day. This is a decent food filling for a fermentation
vessel. As we will soon see, host components can be added to the food list of
the gut bacteria. The colon is thus a great prospect for saccharolytic bacteria
that have the necessary enzymatic capacities to deal with complex polysaccharides. You might take a glimpse into the contract signed between humans and
Bacteroides thetaiotaomicron by reading the genome of the latter. *B. thetaiotaomicron* (it has even for microbiological standards a tongue-twisting name) is
the numerically dominant gut microbe. It owns with 6.3 Mb a relatively large
genome for bacteria. Fittingly, it encodes the greatest amount of polysaccharide-
digesting enzymes of all sequenced prokaryotes. It comprises 172 glycosylhy-
drolases, which cover with their cleavage specificity a broad range of glycosidic bonds encountered in polysaccharides of plant origin (Xu and Gordon
2003; Xu, Bjursell et al. 2003). The next prolific carbohydrate digester of the
human intestine is *Bifidobacterium* (Figure 7.7), the dominant gut microbiota
of the breast-fed baby. The sequenced *B. longum* strain (Schell et al. 2002)
contains 39 glycosylhydrolase genes, which is still remarkable when compared
to the mere 18 glycosylhydrolase genes in *E. coli*. *B. thetaiotaomicron* has
also outer membrane proteins, which bind starch to the bacterial cell surface.
The bound starch is then hydrolyzed by an outer-membrane-bound α-amylase.
The membrane location of these enzymes minimizes the diffusion of digested
products and thus the cross-feeding of competitors. The oligosaccharide is
then taken up by a porin into the periplasmic space where another α-amylase
creates together with an α-glucosidase glucose monomers. Glucose is broken
down in the cytoplasm to pyruvate via glycolysis; fermentation processes then
create short-chain fatty acids (SCFA: acetate, propionate, butyrate in the ratio
70:20:10). In a European diet, about 50 g of carbohydrate is typically fermented
per day, yielding 0.5 mol of SCFA with a total energy value of 150 kcal—the
energy equivalent of one extra yogurt pot (Hooper et al. 2002). As SCFA are
directly taken up by the colonic epithelium, they thus significantly contribute
to the energy supply in man (McNeil 1984). The cooperation contract is thus
honored from both sides and one could speak of a relation of mutual benefit,
a symbiosis.

Has *Bacteroides* "Domesticated" Humans?

In a series of fascinating papers, the Gordon lab has extended this host–microbial
relationship in the intestine to unexpected areas. The researchers colonized
germ-free mice with *B. thetaiotaomicron* and followed the global transcriptional responses of the intestine to bacterial colonization using microarray
technique (Hooper et al. 2001). The expressions of activators for digestive
enzymes (colipase), transporters of glucose into the gut epithelium, and
immunoglobulin transport out of the gut epithelial cell were increased, while

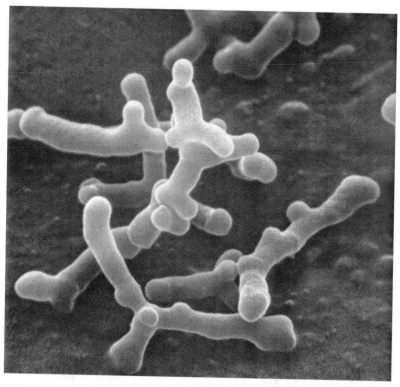

FIGURE 7.7. *Bifidobacteria* are nonmotile, nonsporing rods of varied shapes that frequently display V-shaped terminal clubs. They belong to the high GC content Gram-positive bacteria and are a pioneer colonizer of the human intestinal tract particularly in breast-fed babies. They lack a pathogenic potential and are thus the classical commensal flora of the gut not only of warm-blooded vertebrates, but also of some insects.

lactase expression decreased. Spectacular was the upregulation of a protein involved in epithelial barrier function. Notable was also the increased expression of angiogenin-3, which plays a role in the growth of blood vessels. Indeed, it was observed that germ-free mice did not complete the normal postnatal gut development. The mice had arrested capillary network development. The colonization with *B. thetaiotaomicron* induced this developmental program and the maturation of the intestinal capillary system was completed within 10 days (Stappenbeck et al. 2002). *B. thetaiotaomicron* also induced the production of antimicrobial peptides from Paneth cells in the bottom of the crypts from the gut epithelium. In vitro activity was demonstrated against competing bacteria including pathogens. Not enough with that, the introduction of a microbial flora to germ-free mice produced 60% increase in body fat content despite reduced food intake. A circulating lipoprotein lipase inhibitor is suppressed by *Bacteroides* leading to a greater deposition of triglycerides in fat cells. The gut microbiota could thus affect the energy harvest from the diet and energy storage in the body (Bäckhed et al. 2004). The researchers extended their investigations recently to mutant mice strains that were genetically disposed to obesity (*ob/ob* mice), which showed concomitantly a shift in the microbiota from Bacteroidetes to Firmicutes (Ley et al. 2005). The murine host and its gut microbiota thus showed unexpected interactions and cross-talks speaking for a finely tuned coevolution between these partners over millions of years.

Double Strategy: Fucose Versus Glycan

I will illustrate this relationship by mentioning a mouse enzyme that contributes food to *B. thetaiotaomicron* in the gut. The enzyme in question is α 1,2-fucosyltransferase, whose expression in the small intestinal epithelium is induced by *B. thetaiotaomicron*. This enzyme adds a terminal fucose pentose sugar to glycoconjugates in the gut mucosa (Bry et al. 1996). Where is the benefit for the gut microbiota here? The answer is straightforward: the mammalian host sacrifices a substantial amount of biological material to its gut microbiota. This host-derived bacterial food comprises 5–8 g secreted digestive enzymes, 3 g IgA antibodies, 2–3 g mucins and 20–30 g desquamated epithelial cells per day. This is a substantial extra amount of food if you learned to use it. The genome sequence of *B. thetaiotaomicron* is a "read my lips" recipe: *B. thetaiotaomicron* has a fucose-utilization cluster, which enables it to live from fucose residues cleaved off from host glycoconjugates. The *fuc* operon is preceded by a repressor gene. FucR blocks the transcription of this operon unless fucose is present, which prevents the inhibitory action of the repressor. When the fucose concentration decreases and *B. thetaiotaomicron* loses this valuable carbon source, another function of fucose becomes apparent. Fucose is a corepressor of another locus called *csp* for *control of signal production*. This signal molecule, whose molecular nature is not yet defined, will induce the murine α 1,2-fucosyltransferase. The gut bacterium convinces its host to fucosylate its glycoconjugates to provide food for its commensal (Hooper et al. 1999). In this way *B. thetaiotaomicron* has the flexibility to change from diet-derived polysaccharide to a host-derived glycan foraging. This double strategy makes the gut commensal relatively independent of variations in the dietary food supply. When the diet is the food source, *B. thetaiotaomicron* assembles on food particles and outer-membrane polysaccharide-binding proteins and glycoside hydrolases are induced. The metabolic reconstruction of RNA microarray data is consistent with the delivery of the hexoses mannose, galactose, and glucose to the glycolytic pathway, and the pentoses arabinose and xylose are funneled into the pentose phosphate pathway. If mice were deprived of sugars in the diet, *B. thetaiotaomicron* changes to an endogenous (= host-derived) source of glycans in the cecum. In accordance with the different chemical structure of this food source, sialidase, hexosaminidases, mucin-desulfating sulfatases, and chondroitin lyases are induced (Sonnenburg et al. 2005).

Value and Limitations of the Reductionist Approach

A number of studies illustrate now the human–*Bacteroides* symbiosis, where the human host provides shelter and food and the bacterium offers contributions to the postnatal gut development, host physiology, and nutrition. This view culminates in a review entitled "Honor Thy Symbiont" (Xu and Gordon 2003). We should not forget, however, that the data were obtained with the reductionist approach. This approach has a long tradition in biology and is one of the most powerful approaches in structuring biological complexity. Yet it comes at a price, namely that it describes only a 1-D picture of the complexity surrounding us. While being in the tradition of Max Delbrueck, one should not forget that current biological model systems reduce the biological complexity to systems like phage lambda interacting with *E. coli* K-12 in a test tube or on an agar plate. The consequence of the reductionist approach in the case of phage lambda was that on the one hand it became one of the best-studied biological systems, but on the other hand we do not have a single study that describes the interaction of a coliphage with *E. coli* in its natural niche, the gut, in any molecular detail. This lack of data now renders phage therapy approaches of *E. coli* diarrhea with orally applied phages rather difficult (Figure 7.8).

If I now transfer this argument to the gut microbiota, we see currently a strong trend for the development of probiotic, i.e., health-promoting, bacteria. Interestingly, the majority of the approaches were conducted with gut bacteria belonging to the genera *Lactobacillus* and *Bifidobacterium*. Some like the *Lactobacillus rhamnosus* strain GG showed beneficial health effects in a number of carefully controlled clinical trials. However, the genome of this strain has not yet been published, consequently -omics techniques were not yet applied to this

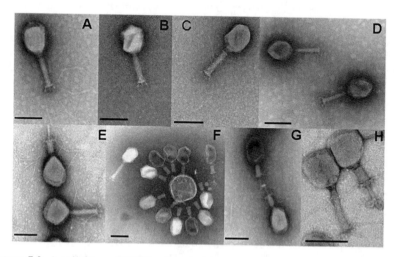

FIGURE 7.8. A collection of T4-like phages that target pathogenic *E. coli* strains, which are developed in the laboratory of the author to test the phage therapy concept against *E. coli* diarrhea as proposed by the codiscoverer of phages. *Panel F* shows T4 phages in various stages of injection of their DNA genome into a remnant of an *E. coli* cell.

strain and its interaction with the host and competing gut microbes has not yet been described to any greater molecular extent. We can thus only relatively vaguely define the probiotic properties of such strains.

Therefore, we definitively need both fields to reach a common ground, meaning that the *Bacteroides* work reaches out for more realistic experimental studies (not only strain combinations, but also probiotic clinical trials), whereas the *Lactobacillus* and *Bifidobacterium* work extends from black box approaches in clinical trials to mechanistic and molecular approaches to elucidate the basis of probiotic properties.

The Thin Line Between Symbiont and Pathogen

Bacteroides fragilis

So far we got the impression that *B. thetaiotaomicron* is a versatile gut bacterium that has grown to intestinal dominance because it learned in its evolution a number of lessons which led to a superior adaptation in its niche. Mice and *B. thetaiotaomicron* exchange nutrients for mutual benefit. In fact, bacterial signals have become important cues for the maturation of the intestine and important regulators of host physiology. All these observations point to a symbiotic relationship (both partners benefit) and not just a commensal relationship (neither is harmed). This adaptive success is reflected by the fact that the genus *Bacteroides* accounts for about 30% of the fecal isolates. We have seen the Greek god Proteus several times in the current book and he lifts his head again. However, even if *B. thetaiotaomicron* numerically dominates this fraction, it is not the only *Bacteroides* species in our gut. The cell numbers of *B. fragilis* is 10- to 100-fold lower than those of *B. thetaiotaomicron*. *These bacteria are probably not brothers, but more distant cousins. The differences between two bacterial species can be deceivingly large. Even within the confines of a single bacterial species, different isolates can differ by as much as 15% in gene content (e.g., Salama et al. 2000 for the stomach pathogen* Helicobacter pylori*), which is enormous when considering that the application of this microbial standard would require to place all primates into a single species. Humans and chimpanzees are, for example, 98.7% identical in their genomic DNA sequence (Enard et al. 2002). This broad bacterial species definition practised by current taxonomists also explains why we have, at the moment of writing, less than 8,000 recognized bacterial species (http://www.bacterio.cict.fr).*

Capsular Polysaccharide Variation

When analyzing *B. fragilis* a different picture emerged. *B. fragilis* is covered with multiple polysaccharides. The structure of two polysaccharides, PSA (polysaccharide A) and PSB, was elucidated and consists of repeating units containing a terminal α 1,2-fucose moiety. The bacterium proteome contains an enzyme that handles fucose, the biochemical details are somewhat complicated, but the essential feature is that this bacterial enzyme looks like a fusion of two mammalian enzymes handling fucose. Not surprisingly, *B. fragilis* can take up fucose from the medium and incorporates it with the help of this hybrid enzyme into capsular polysaccharides and glycoproteins (Coyne et al. 2005). Glycoproteins are rare in bacteria. The authors of this article concluded that the bacterium used a host-like pathway for bacterial surface fucosylation to serve as a molecular disguise for *B. fragilis* against the immune system of the host. *B. thetaiotaomicron* also possesses this gene, but fucose is apparently more used as food than for molecular mimicry as also reflected by the lower number of capsular polysaccharide genes. It thus seems that both species derive from a common ancestor, which split into two ecological lines despite the fact that both bacteria still sit in the same niche. However, the mammalian gut has many niches. Somewhat simplified *B. thetaiotaomicron* occupies the lumen, whereas *B. fragilis* targets the mucosal surface. This specialization within the same ecological setting necessitated genetic adaptations, which can still nicely be read from the blueprint of the two bacterial genomes (Kuwahara et al. 2004).

Two Different Lifestyles

Both bacterial species cope well in their environment essentially by deploying two strategies: first, the exceptional capability to use a wide range of dietary polysaccharides due to gene amplification and second, the capacity to create variable surface antigens with capsular polysaccharides. *B. thetaiotaomicron* excels in the first strategy, while *B. fragilis* is a master in the second. The absolute number of capsular polysaccharide clusters is not so different between both species: nine versus seven. The genomes differ substantially in genome size: 6.2 vs. 5.2 Mb. While *B. fragilis* is smaller, it excels in flexibility. The Comstock lab in Boston documented its genomic dynamics. They raised antibodies against the wild-type strain expressing the different capsular polysaccharides. Then they created mutants, which inactivated the expression of a single polysaccharide cluster. When the antiserum against the wild-type strain was now absorbed with the mutant strain, it pulled out all antibodies against the shared polysaccharides, and only antibodies against the mutated polysaccharide remained in solution. In this way the investigators obtained monospecific sera. To their surprise these sera labeled, in electron microscopy and flow cytometry, only a portion of any given culture. Positive and negative colonies split again into both phenotypes defining a reversible on–off phenotype. Genetic analysis revealed the reversible inversion of a DNA segment ahead of the polysaccharide synthesis cluster. This operation brings the promoter in either the on or the off position (Krinos et al. 2001). The assembly process of the *B. fragilis* genome was actually complicated by the extensive amount of DNA inversions that occurred even in the bacterial colony from which the DNA was extracted for sequencing (Cerdeno-Tarraga et al. 2005). The DNA inversions control variable gene expression and the genome researchers from the Sanger Center distinguished several types: On one side, there is recombination across inverted repeats that flank the promoter. They also identified 30 different enzymes, which resembled site-specific DNA recombinases ("invertases"). One enzyme, the Mpi recombinase, apparently plays the master regulator for the switches ahead of the capsular polysaccharide clusters (Coyne et al. 2003). *B. thetaiotaomicron* has also these genetic elements, but it seems to use it less frequently than *B. fragilis*. On the other side, *B. fragilis* also has intergenic shufflons, where gene segments coding for protein domains are exchanged, creating proteins with new biochemical properties.

Pathogenic and Probiotic Properties

This phase variation is an essential part of the success for the colonization of gut epithelia by *B. fragilis*. We pay for this evolutionary success: *B. fragilis* is the major anaerobic Gram-negative bacterium isolated from abscesses, soft tissue infections, and bacteremia that arise by contamination from the gut. A number of comparisons are instructive here: *B. fragilis* accounts for between 4 and 13% of the normal human fecal microbiota, but is present in 63–80% of human *Bacteroides* infections; *B. thetaiotaomicron* accounts for between 15 and 29% of the fecal microbiota, but is associated with only 13–17% of infection cases (Cerdeno-Tarraga et al. 2005). Except for two hemolysin-like genes, the two strains do not differ in virulence-related genes if one does not account for the supplementary capsular polysaccharide genes and their regulation. B. thetaio-taomicron *has thus clinically too much residual potential for pathogenicity to qualify as a health-promoting, probiotic strain. The food industry screens for members of the natural human microbiota that have health-promoting properties. Lead strains come from the genera* Lactobacillus *and* Bifidobacterium. *The credential of the first is its long safe use in food fermentation and of the second is its dominance in the gut microbiota of the breast-fed baby. The idea of probiotic bacteria is not new and goes back to Metchnikoff, a pioneer medical researcher who propagated the idea to modify our gut microbiota by alimentary means to achieve health benefits.* That this idea is not just part of a "green" health folklore is demonstrated by clinical tests that objectively demonstrated measurable health effects after oral application of a gut-derived *Lactobacillus* strain (Sarker et al. 2005). Changing the composition of our gut microbiota might be an innovative approach when addressing nutritional problems such as the current obesity epidemic (Ley et al. 2005).

Antibiotic use in Animal Rearing

In the agricultural context, this approach is already a long-established, although currently hotly debated, practice. To shift the ratio of the gut bacteria by different diets might be difficult when realizing the nutritional flexibility of *B. thetaio-taomicron* (Sonnenburg et al. 2005). In animal husbandry, antibiotics are added to the feed. This intervention significantly increases the weight gain of the animals. *As the quantities of antibiotics used are in the subtherapeutical dose range, it seems unlikely that the extra weight increase is gained through a reduced incidence of infectious disease. This idea is not far-fetched because in developing countries nutritionists documented that repeated episodes of infectious diseases, especially of diarrhea, cause a measurable flattening of the weight increase curve of children. Malnutrition in these countries is the combined effect of insuf-ficient food intake and diarrhea. The alternative interpretation for the effect of antibiotics in animal feed is a shift in the composition of the gut microbiota with a concomitant increase in the efficiency of conversion of fuel in the diet to body mass. If this hypothesis should be verified it would question the symbiotic character of the established gut microbiota as it would be responsible for a drain of food energy. In contrast to a substantial literature, which investigates the effect of antibiotics feeding on the relentless rise of antibiotics-resistance genes in bacterial pathogens, not much is known on the nutritional effects of antibiotics (Gustafson and Bowen 1997). Surprisingly, the jury is still out in this eminently important practical question.*

Immune Response to Commensals

Even if the opinion of researchers on the role of specific bacteria in the gut micro-biota still remains split, there is an objective mean to assess their role. You simply ask the immune system of the host. Robert Koch has established microbial rules to establish a link between a bacterium and a given disease. Chance isolation of pathogenic bacteria from healthy subjects is commonplace in diarrhea epidemi-ology. A common corollary in infectious disease medicine is to ask whether the patient responded to the isolated bacterium immunologically. Again: the host defenses against the very heavy load of intestinal commensal bacteria are poorly understood. This is surprising as >70% of the daily produced antibodies drain down the gut (Macpherson and Uhr 2004) and are quickly lost with the feces. Why should the body dedicate so much synthetic activity to secretory antibodies that have such a short half-life compared to the circulating antibodies, which have half-lives measured in months? The answer of an evolutionary biologist would probably be this: because they fulfill an important role to protect the body against microbial attack coming from the gut. This seems plausible, but what are the target organisms? The secretory immune system is not particularly efficient against enteric pathogens—repeated episodes of diarrhea traced to the same enteropathogenic species are well known. Also the researchers trying to develop vaccines against common gut pathogens like E. coli *and* V. cholerae *suffered a lot from the short-lasting immune response to these antigens. Apparently, the gut immune system does not have a good immunological memory.* The gap in this knowledge was recently filled with important data from the University of Zurich (Macpherson et al. 2000). The researchers reported that—in contrast to conventional views—a large proportion of the intestinal IgA is directed against cell wall antigens and proteins of commensals. These antibodies are specifically induced in response to the presence of commensals like *Enterobacter cloacae*, the predominant commensal in the mice of the Zurich lab. Notably the antibody response is independent of T cells and the authors speculated that their data shed light on a primitive, specific, antibody-dependent immune system that developed in the gut as a defense line against the gut microbes lurking "ante portas." The system is quite sophisticated: intestinal dendritic cells located in the mucosa take up and retain small numbers of live commensals. These cells are apparently

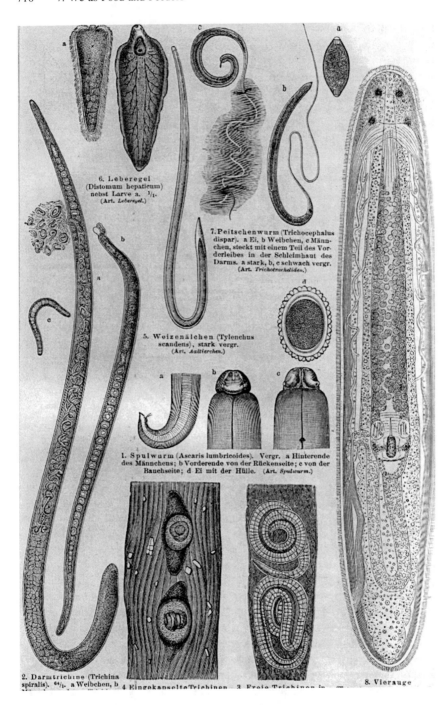

6. Leberegel (Distomum hepaticum) nebst Larve a. ¹/₁. (Art. *Leberegel*.)

7. Peitschenwurm (Trichocephalus dispar). a Ei, b Weibchen, c Männchen, steckt mit einem Teil des Vorderleibes in der Schleimhaut des Darms. a stark, b, c schwach vergr. (Art. *Trichotrachelides*.)

5. Weizenälchen (Tylenchus scandens), stark vergr. (Art. *Aaltierchen*.)

1. Spulwurm (Ascaris lumbricoides). Vergr. a Hinterende des Männchens; b Vorderende von der Rückenseite; c von der Bauchseite; d Ei mit der Hülle. (Art. *Spulwurm*.)

2. Darmtrichine (Trichina spiralis). ⁶⁴/₁. a Weibchen, b   4 Eingekapselte Trichinen.   3 Freie Trichinen in   8. Vierauge

on a spying mission to induce the selective IgA response needed to protect the mucosa against penetration by commensals (Macpherson and Uhr 2004).

Intestinal Desquamation and Peristalsis

In this context, I have another subject that puzzled me. What is actually the evolutionary reason for the high rate of renewal of the intestinal epithelia? Approximately, every 3 days the entire intestinal mucosal epithelia are replaced. As the gut was designed by evolution for surface amplification, the number of cells desquamated is enormous and represents a substantial metabolic load for the body. Is it imaginable that the desquamation is a defense process against gut bacteria that succeeded to get a foothold on the epithelia? Before they could do harm, they are removed by the exfoliation of the villi? This suspicion was recently confirmed by the reaction of mice experimentally infected with the nematode (Figure 7.9) *Trichuris trichura*, the 4-cm-long whipworm. Worldwide, about 800 million people are infected with this parasite, which sticks with its anterior whip to the superficial mucosa of the cecum and ascending colon, each worm sucking 0.005 ml blood per day. This amount is so small that infections remain normally asymptomatic. Only heavily infected people suffer mild anemia and bloody diarrhea. However, one should not underestimate the worm burden: the life expectancy of the adult worm is 1 year and a single female produces 5,000 eggs per day. The eggs are transmitted to new hosts by fecal contamination. In countries with low hygiene level, 75% of school children are thus infected with this worm. It is thus clearly better not to have these uninvited guests. Mice strains differ in their susceptibility to the worm and it was revealing to compare susceptible with resistant strains. In the susceptible mice, a significant crypt hyperplasia was observed, while the resistant mice showed a greater loss of cells from this proliferative compartment. In the latter, the cells moved at a much faster rate along the crypt–villus axis. The crypt contains a pool of

---

FIGURE 7.9. Parasitic worms. Four Nematoda worms (Phylum Aschelminthes, round-worms): *1. Ascaris lumbricoides* (*a*, rear; *b* and *c*, front-end views of the worm). The durable egg (*d*) is ingested with contaminated vegetables and fruits, the animal hatches in the intestine, travels from there into the lungs, then back into the mouth and arrives after swallowing in the feces to begin a new life cycle. *2. Trichinella spiralis* causes a food-borne disease when uncooked muscle meat containing encysted worms is eaten (*a*, female; *b*, male worm; *c*, young larvae). The worm reproduces in the intestine and the larvae then invade the muscle fibres where they grow; the center, bottom shows free larvae in the muscle (right) and encysted larvae (left). *Panel 7* shows a closely related, common, but relatively harmless human parasite *Trichocephalus dispar* (*a*, egg; *b*, female; *c*, male invading with its body the gut mucosa). *Anguina tritici* (*5*) is a plant parasite; the worm infects wheat, the grain becomes filled with worm larvae. The fluke *Fasciola hepatica* (*6*, class Trematoda) belongs to the worm phylum Plathelminthes (flatworm). It causes the liver rot disease in sheep. The adult animal shows at the top a muscular sucker with which it attaches to the host. *Panel 8* shows a representative of the Nemertini class of flatworms, the depicted animal is a freshwater inhabitant.

stem cells, which divide and give rise to enterocytes, enteroendocrine cells, and goblet cells that migrate in vertical bands to the top of the villi. The enterocyte differentiates during this journey into a mature absorptive cell, the goblet cells secrete the mucus, and the enteroendocrines sense the presence of the nutrients and respond with the release of hormones. These cells reach within 5 days the exclusion zone at the tip of the villus, from which they are shed into the lumen at a rate of 10^{11} cells per day in humans. When exposed to the whipworm, the resistant mice increase the cellular turnover and drive the "epithelial escalator" to expel *Trichuris* (Cliffe et al. 2005). As one could expect, the interplay between worm and host is more complicated because the escalator is in addition under immune control. The worm has apparently learned to induce one cytokine that favors crypt cell hyperplasia creating a neat niche and another that slows the escalator. On the other hand, the goblet cells secrete a substance that disorients the chemosensory apparatus of the nematode, leading to the loss of its foothold (Artis et al. 2004).

Peristalsis is partly explained by the need to push the nutrient bolus through the intestine, while at the same time it serves to flush the gut lumen. Luminal bacteria, which do not divide more rapidly than the rate of peristalsis, will not increase in number. Some enteropathogenic bacteria have a high replication rate in the small intestine as indicated by their high numbers in the stool (e.g., 10^8 cfu = colony forming units V. cholerae/ml cholera stool). The increased peristalsis observed in diarrhea patients might be an adaptive response. The host increases the flushing to get rid of the pathogen by expulsion. Therefore medication to decrease the peristalsis (e.g., loperamide) is regarded by some clinicians as counterproductive. If you consider further that—as mentioned before—the majority of the antibodies synthesized by the body are secreted into the gut, it could appear that the body builds a high firewall against unfriendly intrusions by its gut microbiota. Even gut bacteria that enjoy a GRAS (generally regarded as safe) status like Lactobacilli *are not a rare finding in blood samples sent to the clinical microbiologist. Even harmless gut bacteria—once in the circulation—might have a residual pathogenic potential. In the case of lactobacilli, subjects at risk of developing pathologies with lactobacilli are those with defects on the heart valves.*

Critical Mutations that Lead to Loss of Control over the Gut Microbiota

The analysis of a mutation in mice underlined the importance of secretory IgA as the front line defense in the gut mucosa. The mutation concerns an RNA editing enzyme called AID, a cytidine deaminase. The biochemical action of this enzyme is of lesser interest in our context. However, if the gene is defective, class switching from IgM to IgA-producing cells does not take place in the gut lamina propria, and somatic hypermutation of the IgA antibodies does not occur either. Therefore these mice secrete great amounts of IgM into the gut (Fagarasan et al. 2001). This immune defect has important consequences: the mice lose the control over the anaerobic microbial flora in the small intestine, which expands 100-fold. This increased antigenic load leads to a hyperplasia of isolated lymphoid follicles in the gut, which is probably induced secondarily by the increased antigen load in the small intestine. In fact, antibiotic treatment of the AID$^{-/-}$ mice abolished the exaggerated immune reaction toward the gut microbiota (Fagarasan et al. 2002). When using culture-independent techniques such as 16S rRNA sequencing, these researchers discovered that the lion's share of the increased microbiota was contributed by a group of bacteria, which they called SFB (segmented filamentous bacteria). These bacteria replaced the lactobacilli found in young and clostridia and *Bacteroides* found in older AID$^{+/+}$ mice (Suzuki et al. 2004). SFB strongly attach to the gut epithelium, which was not prevented by the normal levels of secretion of antimicrobial proteins by the Paneth cells at the bottom of the crypts. When primed IgA B cells from normal mice were introduced into the circulation of the mutant mice, the SFB retreated to the lower segments of the intestine. Apparently, the local immune system controls where it wants to have the gut commensals.

Bacteroides and the Immune System

If we look at all these data: are the gut microbes friends or foes? What is the purpose of the immune response to these commensals? Does it want to restrict the microbes to the lower parts of the gut to limit nutritional competition? Or is its main purpose to prevent inappropriate immune activation of the gut mucosa by adhering microbes potentially leading to inflammation? Recent work showed that the gut microbiota is important for the development of the immune system. It was, for example, demonstrated that monocolonization of germ-free mice with *B. fragilis* is sufficient to correct immunological defects that occur in the absence of a bacterial microflora in the gut of axenic animals. The researchers identified polysaccharide A (PSA) as the bacterial component that directs the maturation of the developing immune system. T-helper cell imbalances and lympoid organogenesis were corrected. Mechanistically, PSA activates CD4$^+$ T cells via dendritic cells and elicits appropriate cytokine production (Mazmanian 2005).

To get more clarity on the role of *Bacteroides* it would be interesting to investigate the immune response against this gut bacterium. Researchers working with streptococci have already realized that there is anyway a thin line between commensals and pathogens (Gilmore and Ferretti 2003). The play with the variable expression of the capsular polysaccharides in *B. fragilis* is a well-known immune evasion strategy in pathogens. It is probably significant that the mucosa-associated *B. fragilis* plays these cards more actively than the more lumen-restricted *B. thetaiotaomicron*. Probably, *Bacteroides* is a bit of all: symbiont, commensal, and pathogen. With respect to the latter attribution: enterotoxigenic *B. fragilis* is a small, but significant contributor to diarrheal disease in developing countries (Pathela et al. 2005).

Coprophagy

We still lack a clear answer to many questions concerning the role of the gut microbiota for humans. Here I just want to mention a question that puzzled me and which might be relevant in this context. If the stool is composed in its majority of bacteria, it would represent a valuable source of relatively easy to digest proteins in times of starvation. Yet except in a recent Russian film from northern Finland (Kukuschka) I have never come across reports that coprophagy was practised during famines. Humans have a strong aversion against their excrements, although this seems to develop only in early childhood and is thus not inborn. The gut is not a major organ for the elimination of metabolic waste when compared to urine. Only relatively few biochemical waste is excreted by the feces (e.g., the heme ring without the iron atom). The feces should thus not be too toxic from a chemical side. In support of this hypothesis, coprophagy, the eating of one's own feces, is well known to laboratory biologists as mice and rabbits practice it in captivity. Primates are instructive in this respect. Primates have two principle strategies to place gut microbes. Microbial fermentation can occur in the enlarged foregut (stomach) as in langurs (*Presbytis*) and colobus monkeys (*Colobus*). They have large stomachs with many pouch-like sacs. Cecum-fermenting animals like the gorilla and the sportive lemur (*Lepilemur*) have simple stomachs, but a large, divided cecum. In the latter group food is digested in the stomach before microbial fermentation takes place. Simple carbohydrates from fruit and nectar are digested and absorbed in the proximal parts of the intestine, while the complex polysaccharides are handled in the distal parts by the gut microbes. The advantage of this strategy is that plant material is not well broken down in the upper parts where highly efficient absorptive epithelia are found—this means that plant toxins pass into the lower part of the gut and are thus mainly handled by the gut microflora. The microbes detoxify the plant material and spare the liver this work and the chemical insults. If toxins are freed unmodified, then a remedy is that the colon epithelium is less absorptive and the toxic load to the body is still kept low. The price you pay for this strategy is that your exploitation of the nutrients in the food is reduced. Gorillas have a simple answer to this dilemma, they ingest their own feces as observed in the wild. There is thus a second passage of food material through the intestine after detoxification of the plant material. Why do humans not practice coprophagy even in times of hunger? One could imagine that the microbial content of our feces poses a risk when recycled via a second round of ingestion. There might be a trade-off between diarrhea risk and extra nutrition. Cooking could, however, shift the infection risk ratio. I will stop this issue here, which will be as emotional as cannibalism — because of lack of data. In the scientific literature human coprophagy is exclusively reported in psychiatric patients.

Janus Faces: The Case of Vibrio cholerae

Human Categories do not Fit a Complex Reality

Bacteria do not have an obligation to fit into human categories like symbiont, commensal, or facultative pathogen. These categories were developed by scientists to mentally order their observations. The diversity in Nature perplexes our minds more than it unsettles the organisms. I will illustrate the multifunctional ecological role of microbes with V. cholerae *and its phages. In fact, extreme genomic diversity was described in coastal bacterioplankton populations (Thompson et al. 2005) and diversity might be a guarantee for biological stability. Evolution is the ultimate opportunist: why shouldn't an organism profit from opportunities that are offered to it? It does not care about good and evil—categories, which got a sense in the biological world only after the arrival of self-conscious human beings, at best a few million years ago. Nature is autonomous and not obliged to respect our scientific terminology frequently coined with an eye on the impact of microbes on human life. Indeed, there might be an uninterrupted continuum from symbionts to pathogens and some organisms have several roles in the natural order, which obliges them to carry Janus faces when seen with our eyes. Janus is that double-faced Roman god who was in Roman liturgy invoked as the first of any gods and thus a good patron for biology.*

Vibrio cholerae in the Environment

I will illustrate the Janus analogy with recent research papers dealing superficially with another gut bacterium, *V. cholerae*, the cause of cholera. This bacterium is such a dreaded pathogen in human history that its categorization in our human system of bacterial utility seems to be an easy task: it is a culprit. However, while not being innocent, it could plead guilty only by association with bad company. Slightly simplified, the following story can be told for *V. cholerae*: this bacterium can be found in many coastal waters in the tropics and subtropics. Its major habitat seems to be brackish waters and estuarine systems (Colwell 1996). This observation fits with the epidemiology of cholera—the disease originated in coastal areas such as the Bay of Bengal and in South America. The reason for this niche preference is probably the nutritional basis of this marine organism. *V. cholerae* produces a mucinase, an enzyme that degrades mucin. These compounds are encountered in the gut, hence its association with the human gut, but in the natural environment *V. cholerae* finds mucin-like substances in algae (*Volvox*) or in the mucilaginous sheath of the cyanobacterium *Anabaena*. *V. cholerae* also elaborates a chitinase and plays an important role in the remineralization of chitin in the ocean. The carapace of copepods is an important source for marine chitin. We should thus not be surprised that *V. cholerae* is associated with these organisms that already took center stage in other parts of our book. Cells of *V. cholerae* also colonize the gut of copepods (for the original references see: Lipp et al. 2002). After cellulose, chitin is the most abundant polymer on earth. Arthropods shed it with each molt and with their fecal pellets. Small wonder that *V. cholerae* growing on chitin induces the expression of a regulon consisting of 41 genes involved in chitin colonization, digestion, transport, and assimilation. Growth on chitin also induces in *V. cholerae* a type IV pilus that is in other bacteria associated with DNA uptake. Indeed, chitin thus induces natural competence in *V. cholerae* (Meibom et al. 2005). This observation has two facets. One is nutritional: extracellular DNA is present in tens of micrograms per liter of seawater; this concentration is three to four orders of magnitude higher in sediments and biofilms. DNA at that concentration becomes an interesting source of carbon, energy, nitrogen, and phosphorus (Bartlett and Azam 2005). DNA, which is taken up by transformation, can in addition promote DNA repair and genetic diversity. This process could explain the mosaic structure of the *V. cholerae* genome and the presence of the phage-encoded cholera toxin gene in cells that lack the receptor for the phages carrying this gene.

    V. cholerae was also found associated with egg masses of the nonbiting midge *Chironomus*, which lays its eggs into fresh water ponds. The eggs form a row, are folded in loops, and are embedded in a thick gelatinous casing. The gelatin is consumed by *V. cholerae* resulting in bacterial growth, while the released insect eggs mostly failed to hatch (Broza and Halpern 2001). So for *Chironomus* and *Anabaena, V. cholerae* is a predator; for a copepod, it is a commensal. Rita Colewell speculated that the cholera toxin CT, which is responsible for the profuse diarrhea observed in cholera patients, helps in the osmoregulation of the copepod by facilitating the efflux of Na^+ out of the gut when CT binds the CT receptor. The cholera toxin was therefore possibly designed for a different, perhaps even mutualistic, purpose before it became a potent toxin for humans. Part of the beneficial effect of CT can be still observed in humans: when given in small doses CT is one of the most potent adjuvant for eliciting a secretory immune response to mucosal antigens (Hajishengallis et al. 1995).

Context-dependent Toxins

This Janus-face of bacterial virulence factors is even better documented for a marine Vibrio, *V. fischeri*. This Vibrio releases a tracheal cytotoxin, which acts in synergy with lipopolysaccharide (LPS) to trigger tissue development in its mutualistic symbiosis with the squid *Euprymna scolopes*. The details of this process are of minor interest here: they concern the development of a light-emitting organ in the squid where the luminous bacterium offers its help. The toxin induces an inflow of squid hemocytes, followed by apoptosis (controlled cell death) induced by LPS resulting in epithelial regression. The toxin plays an important role as morphogen for this squid organ (Koropatnick et al. 2004). *Bordetella pertussis*, the cause of whooping cough, and *Neisseria gonorrhoea* elaborate a similar compound with comparable effects: the loss of ciliated cells from mammalian respiratory epithelia and the fallopian tube epithelia, respectively. *However, this is destructive pathology, not destructive morphogenesis. It is hard to tell whether this compound is a mutualistic morphogen or a toxin. The answer is context-dependent, a typical Janus face. Evolution uses its inventions for many purposes. One could argue that humans as evolutionary newcomers simply got in the way of microbes and suffer collateral damage as a mere evolutionary accident. However, humans became so abundant on the globe that microbes would miss an important ecological niche if they would not exploit this possibility. Actually, microbes do not weigh their chances and choose on their survival strategies. The evolution game selects simply those organisms that succeeded to multiply in the most efficient way, whatever the logic of their replication strategy.*

Epidemic Cholera Strains

V. cholerae has another fascinating Janus face. Epidemic cholera strains belong to just two O-serotypes determined by the chemical structure of the LPS decorating the bacterium: O1 and O139, the latter appeared only in 1992 (Faruque et al. 2003). Why are just two serotypes pathogenic from the about 200 O-serotypes described in environmental *V. cholerae* strains? The answer to this paradox is the host range of a filamentous phage CTXΦ, which carries the cholera toxin genes *ctxA,B* (Waldor and Mekalanos 1996). It resembles in its genetic organization the classical filamentous *E. coli* phage M13, the famous early cloning vector. This toxin-encoding phage recognizes a receptor on the susceptible *V. cholerae* cell that is called the toxin-coregulated pilus (TCP). This structure is a fiber of the polymerized pilin protein (TcpA). TCP functions as an important virulence factor and is the main intestinal colonization factor of the pathogen. This virulence factor/phage receptor is not encoded by the core part of the *V. cholerae* chromosome. Actually, here I must speak of chromosomes in plural as quite unusually for the sequenced bacteria, *V. cholerae*'s genome consists of a large and a small chromosome of about 3 and 1 Mb in size (Heidelberg et al. 2000). The region encoding the TcpA protein belongs to a 40-kb-long mobile DNA element that was initially described as another filamentous phage, with TcpA being the major coat protein of this purported second filamentous phage (Karaolis et al. 1999). These data gave rise to wide-ranging models of how the cooperation between two phages created a pathogen. The phage nature of the genetic element was later questioned and it is probably better described as a pathogenicity island with phage-like properties. The current data suggest a model of a benign ancestor *V. cholerae* strain O1, which acquired TCP by lateral gene transfer and was then infected by the temperate phage CTXΦ, which itself had somewhere acquired the *ctxA,B* genes, possibly by an incorrect excision event from a heterologous host. This sequence of events created the epidemic *V. cholerae* O1 El Tor strain. Epidemic *V. cholerae* strains show an unusual low degree of genetic diversity in microarray analysis with only 1% differences in gene content, suggesting a recently diverged lineage (Dziejman et al. 2002). The newly emerged epidemic *V. cholerae* serotype O139 belongs to this lineage and has experienced a replacement of the O1- by an O139-specific LPS gene cluster, resulting in a serotype switch (Faruque et al. 2003).

Cooperation with a Phage

If I remain in anthropocentric view, *V. cholerae* cannot just point to the CTXΦ phage as the culprit. It was not simply invaded by a parasitic mobile DNA element. There is a lot of evidence which points to a close collaboration between phage and bacterium in the pathogenic process. Just to mention some recent research literature: The bacterium takes care to integrate the phage as a prophage, in some strains both in the large and the small chromosome to keep it safe. The phage lacks its own integrase; this task is fulfilled by two host-encoded recombinases (Huber and Waldor 2002). Furthermore, the integrated prophage alone could not export the toxin out of the intact cell. The secretion of the cholera toxin as well as the liberation of the phage for the dissemination of this key virulence factor is mediated by a host-encoded *e*xtracellular *p*rotein *s*ecretion (*eps*) type II secretion system (Davis et al. 2000). Indeed the key virulence factors TCP and CT are under the control of master regulators, which are encoded in the ancestral *Vibrio* chromosome. ToxR is regarded as the central virulence regulator and a set of about 60 genes, termed the *toxR* regulon (involved in colonization, toxin production, and bacterial survival in the host) are coregulated by ToxR in response to external stimuli (Bina et al. 2003).

Adaptation to Human Gut

V. cholerae shows a remarkable adaptation during passage in the human host by changing its transcription profile. In the stool of cholera patients, the bacterium exhibited high expression levels of genes required for nutrient acquisition and motility when compared to *V. cholerae* grown in the laboratory. Notably *V. cholerae* excreted by the patients showed a 700-fold higher infectivity than broth-grown bacteria. The bacteria thus adapted their transcription pattern for optimal transmission of the pathogen, thus laying the ground for epidemic spread (Merrell et al. 2002). Oddly, the key virulence genes like TCP and CT were not expressed, while a previously overlooked small gene was highly induced. Is this the program for reentry of *V. cholerae* into the environment before infecting the next host? Other researchers argued that bacteria recovered from stools are in a different physiological state than bacteria grown in the upper intestine where replication and pathogenesis occur. To clarify this point they reevaluated the transcription profile of *V. cholerae* in the rabbit ileal loop model (Xu, Dziejman and Mekalanos 2003). Analysis of the transcripts suggests that the in vivo expression pattern represents a response to the stress associated with nutrient limitation, scarcity of iron and oxygen.

The Difficult Relationship of Bacteria with Phages

In the quest for the "human food", the phage and the bacterium play hand in hand (Figure 7.10). This is not that surprising because the success of the bacterial lysogen is also the success of the prophage as seen from the selfish DNA model (Canchaya et al. 2003b). The phage gets a free lift for its genome into niches of the biosphere, which it would not reach on its own. The bacterium can explore the sequence space carried by a mobile DNA element for genes that increase its survival capacity. Both gene systems cooperate to the detriment of the human host or any host for which the lysogenic bacterium is pathogenic. As both phage and bacterial DNA share the same chromosome during lysogeny, they also have at least temporarily the same destiny. As the prophage can again resume a new replication cycle, which can kill the host, the bacterium plays with

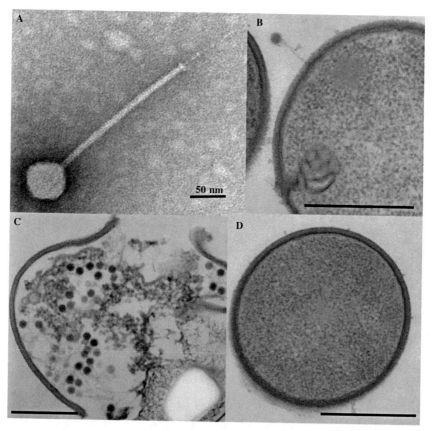

FIGURE 7.10. *Streptococcus thermophilus* phage Sfi21 resembles morphologically the famous temperate *E. coli* phage lambda (**a**). It adsorbs to the cell wall of this yogurt-fermenting strain and injects its DNA into the bacterial prey (**b**), intracellular phage replication then leads to the lysis of the cell and release of progeny phage, which can then infect new cells. This amplification of phage can quickly kill the bacteria in the fermentation vat, leading to the loss of the product. Phage contamination is a major problem in many industrial food fermentation processes. Sometimes, the phage opts for a different replication strategy by integrating its DNA into the bacterial chromosome—it becomes then a parasite at a molecular level or a selfish DNA.

fire and the cooperation aspects are still complicated by a parallel arms race between the virus and the host bacterium. To allude to the title of the section, the V. cholerae *phage has a clear Janus face for the bacterium, but only a grim face for us as the host of the pathogenic bacterium.*

Cholera Cycles

Is this really so? Two recent reports by two old hands in cholera research added a surprising twist to the natural history of cholera, which might be Janus' laughing cholera face for us. Rita Colwell explored the links between the cycles of infectious diseases and global climate change. In this context she has developed cholera as a paradigm of climate impact on human health (Colwell 1996). The analysis of time series of cholera cases in Bangladesh associated cholera dynamics with the frequency of El Nino-Southern Oscillation events describing temperature anomalies in the ocean (Pascual et al. 2000). However, a review of the literature on many infectious diseases found no unequivocal examples of natural changes in disease severity resulting from directional climate warming per se (Harvell et al. 2002). The cyclic nature of cholera in the Ganges delta region is, however, uncontested. Epidemics usually occur twice a year: a major one after the monsoon and a minor one during the spring.

Phages as Drivers of Cholera Cycles?

J. Mekalanos from Harvard University and B. Nair from the diarrhea hospital in Dhaka, Bangladesh, made fascinating observations that could provide an alternative explanation for these cycles. They found that water samples contained either epidemic *V. cholerae* or O1- and O139-specific phages, but rarely both. The time curve of their appearance showed a characteristic cycling as if the accumulation of cholera phages in the environmental water ended the epidemic. As lytic cholera phages are efficient predators of their host bacterium, the observation offers even a potential causal relationship. Most interestingly, phages that were closely related to those viruses which attacked the epidemic strains were released from lysogenic *V. cholerae* strains found in the environment (Faruque et al. 2005a,b). During the early phase of the epidemic, phage titers in the environment were low. As the epidemic progressed, an increasing number of patients excreted phage with their stools. At the end of the epidemic, up to 100% of the patients' stools tested positive for phages and titers could reach up to 10^8 phage particles per ml of watery stool. Phage might thus be responsible for the collapse of the epidemic. We are now back to old concepts. John Snow linked contaminated Thames River water pumped into the city of London to the observed cholera cases in 1855. When clean water was provided, the cases ceased in London. For cost reasons this environmental engineering approach is not practical in developing countries.

Phage Therapy

Therefore the British army and the WHO used in the 1940s—apparently with some success—an idea developed by Felix d'Hérelle at the beginning of the twentieth century. He proposed to use phages as therapeutic and prophylactic agents against bacterial infections. This idea led to phage preparations by American pharmaceutical companies like Elli Lily in the 1930s. The German and the Red Army used phages in World War II against dysentery and wound infections and US labs conducted promising research in that field until the 1940s when the development of antibiotics replaced most scientific efforts in phage therapy. The idea to use phages was nearly forgotten were it not for a few labs in the Soviet Union maintaining this tradition to our days. Currently, there is a growing interest to use the hunger of phages for bacteria as the basis for medicine (Brüssow 2005). The hypothesis that the wax and wane of the cholera epidemics is possibly triggered by cyclic activation of phages from lysogenic cholera strains might be of practical application. Notably, a British medical officer in India added in the 1940s cholera phages to wells, apparently with some success against cholera epidemics. *This would be a nice illustration of the Janus faces of cholera phages, which bring virulence factors into* V. cholerae *making it a potent human pathogen, but which also provide the little helpers that man can use in his fight against this disease. Biology is perhaps always Janus-faced, it is not dominated by the "either–or" as the human mind wants reality, but by the ambiguous "as well…as" so disliked by human categorization efforts. Evolution does not think in alternative options, it considers all options for an organism. From the viewpoint of a phage the "either–or" makes no sense. What the cholera phages must follow is the optimal amplification of their genomes if they want to stay in place. Under some conditions it will be more appropriate to associate with a bacterium and to contribute virulence factors to further this cohabitation, while under other conditions it will be more profitable to consider all susceptible bacteria as a prey. Perhaps the best strategy for the phage is to get to the crest of the epidemic wave passively as a prophage and then to use this top position to run the maximal havoc under the bacteria by replicating as a lytic phage. What we see as a Janus face is only our view because we get on and from the hook with these two viral replication strategies, while phages further their progress with both.*

From Gut to Blood: The Battle for Iron

Iron Thievery: From Enterobactin to Lipocalin

A pathogenic bacterium that wants to grow in the human body must overcome the formidable barrier presented by transferrin, which keeps the free-iron concentration in the blood as low as 10^{-26} M. This concentration is too low to sustain bacterial growth. Similarly, many body secretions contain lactoferrin, namely tears, semen, and human milk, hence its name. Like transferrin it binds two ferric ions and prevents the growth of bacteria. Iron-lactoferrin must also have some bacteriocidal effect as bacteria do not grow after exposure to it even when iron is provided that exceeds the iron binding capacity of lactoferrin. Many aspects in biology are the result of an evolutionary arms race. This is also the case for iron. An *E. coli* pathogen that causes a bacteremia must cope with the low blood iron concentration. The answer is clear-cut: if transferrin has a binding constant for the Fe^{3+} complex of 10^{24} at the physiological pH 7, enterobactin counters with the astronomical Fe^{3+} binding constant of 10^{52}. Checkmate for the host? Biology would be a poor game if the host would now give up. In fact, if we had not designed countermeasures, our evolutionary survival might have been at stake. Actually, the rescue in this iron fight comes not only from our sophisticated adaptive immune system, but from an old trick also used by bacteria: iron thievery (Flo et al. 2004). After peritoneal infection of mice with *E. coli*, the synthesis of lipocalin is part of the acute phase response. The protein is especially prominent where it is most needed: in the liver, which filters everything that comes directly from the gut, and in the spleen, which clears bacteria that escaped the screening in the liver. Mice, in which the lipocalin 2 gene was knocked out, succumb quickly to *E. coli* infection, while wild-type mice survive the same challenge unharmed. The addition of a recombinant lipocalin expressed from a plasmid to the acute phase serum of the knockout mice restored the bacteriostatic effect. This is nearly a fulfillment of Koch's postulates for a genetic experiment. The key to the mechanism of lipocalin's action is that it binds enterochelin specifically with a high affinity (10^{-10} M). It steals the siderophore from *E. coli* and starves thereby the bacterium for iron. The effect is quite specific despite the fact that *Salmonella*, *Brucella*, and *Corynebacterium diphtheriae* are also inhibited—they all elaborate an enterochelin-like siderophore. *Staphylococcus aureus*, however, is not affected by lipocalin 2 because it uses a different iron uptake system. Ferrichrome is not bound by lipocalin 2. Ferrichrome, experimentally added to the system, therefore allows the growth of *E. coli* in the presence of lipocalin 2, both in vitro and in vivo.

Getting Independent from Iron

It now becomes clear why *E. coli* uses so many different iron uptake systems in parallel. Other bacteria, like lactic acid bacteria, do not require iron in their metabolism; they can thus grow under extreme iron shortage conditions. Interestingly, lactobacilli make up a nonnegligible part of isolates from blood samples sent to the clinical laboratory for diagnostic purpose. As they lack pathogenic potential, their presence in the blood is surprising and might be explained by their transgression from the gut (where they are a prominent commensal) and their iron sufficiency.

An Adaptive Value of Anemia?

The existence of an iron battle line between bacteria and humans became already apparent to English physicians in the nineteenth century when they tried to treat anemia. Anemia is characterized by a diminished erythrocyte count and reduced hemoglobin concentration. It has three principle causes: enhanced blood loss, and increased destruction or decreased synthesis of erythrocytes. Decreased synthesis is frequently due to insufficient availability of iron. That this shortage could be alleviated by oral iron supplementation was already clear to nineteenth-century physicians. However, in few patients they saw something rather odd: first an amelioration of the anemia — a pale anemic woman got rosy cheeks again. However, the physician's delight was of short duration. The recovery phase was followed by a rapid decline and the patient died of tuberculosis. The medical background for this case report is clear from our previous paragraph. In cases of iron-shortage anemia, mycobacteria are apparently kept in check by the lack of available body iron. Once supplemented, mycobacteria could resume their growth, which led to the demise of the patient.

Hemolysins

As the pathogen–host interaction is one of the most dynamic aspects of coevolution in biology, bacterial pathogens have developed a number of strategies to actively deal with the problem of iron shortage in the blood. Iron is found in high concentrations in red blood cells, bound to the O_2 transporter protein hemoglobin. How to get to these resources? Many pathogenic bacteria solved this problem with the help of a simple device called a hemolysin. The name already indicates how they do the trick: they simply lyse red blood cells, which leads to the release of the iron-containing blood pigment. The task is not that easy: Enterohemorrhagic *E. coli* (EHEC) dedicate four genes to its synthesis, the *hylCABD* operon. Its genetic footprints testify that it is a recent acquisition of pathogenic *E. coli*. Its GC base content is much lower than the *E. coli* average and it is associated with other virulence genes. It forms part of a pathogenicity island, a piece of mobile, horizontally acquired DNA. HlyA or α-hemolysin belongs to a family of membrane-targeted toxins also found in other pathogens like *B. pertussis*, the cause of the whooping cough. Hemolysin is synthesized as an inactive protoxin and has to mature by the action of HylC. The latter is an acyl transferase, which adds a fatty acid stolen from the fatty acid biosynthetic pathway. HylA has a distinct domain structure: first a hydrophobic pore-forming domain, then at the center two HylC-binding regions FAI and FAII (referring to the two *f*atty *a*cids), then a repeat region (RTX), and finally a C-terminal secretion signal. The lipidation of HylA is absolutely required for its pore-building properties in the target cell. However, HylA must first get out of the bacterial cell. This export process is achieved by the next two proteins encoded by the *hyl* operon, HylB and HylD, which assemble in the bacterial inner membrane to form a translocase complex. The toxin export requires hydrolysis of ATP mediated by HylB and contact with the translocase complex. TolC from the outer membrane is then recruited by the transfer complex. TolC is a homotrimer forming a hollow tapered cylinder. One part consists of an α-helical tunnel, which crosses the periplasmic space between the inner and the outer membranes, and a β barrel that spans the outer membrane (Koronakis et al. 2000). In concert with the HylB, D translocase complex, TolC mediates the gated exit of HylA out of the cell. The repeat region of HylA binds first Ca^{2+} and then the glycoprotein glycophorin on the membrane of the red blood cell. An oligomerization of HylA and insertion into the erythrocyte membrane follows. This leads to a ring-like structure visible in electron microscopy with a 2.5 nm inner transmembrane pore. The alteration of the membrane permeability causes lysis and death of the erythrocyte, which provides iron to the bacterial cell. Actually, the lytic action is not limited to red blood cells, but extends to immune cells and epithelial cells. The latter is important since the EHEC strain O157 is noninvasive and remains restricted to the gut. Does HylA target the ferritin stores in the mucosal epithelia? Without being of obvious use to the food pathogen *E. coli* O157, one of the most feared consequences, the hemolytic uremia syndrome, is in fact linked to the lysis of erythrocytes in the circulation.

Shiga Toxin

Why do I tell you this facet of the quest for iron at some length? Actually, *E. coli* O157 is a good example to demonstrate the dynamic nature of the battle for iron. This class of pathogenic *E. coli* has appeared in humans only very recently. The first cases were described in 1982. It got its alternative name STEC (Shiga-toxin producing *E. coli*) from another toxin that it elaborates, the Shiga-like toxin. The experimental character of the O157 *E. coli* toxins and their involvement in the quest for iron become even clearer with the Shiga-like toxins. They come in two forms, Stx1 and Stx2. Stx1 differs from the better known Shiga toxin of *Shigella dysenteriae* by a single amino acid replacement. They belong to the AB-type toxins and are composed of an enzymatically active A subunit which is surrounded by a pentameric B subunit ring (Stein et al. 1992; Fraser et al. 1994). To be fully active the A subunit must be cleaved at the disulfide loop connecting the A1 and A2 fragments. The activation is done by the protease elastase from the intestinal mucus (Kokai-Kun et al. 2000). The B subunit mediates the binding reaction to neutral glycosphingolipids, namely Gb3. Interestingly, in the intestinal Caco2 cells, Gb3 expression is induced by butyrate together with the differentiation of villus cell differentiation markers. Butyrate is produced by the normal resident enteric flora at high concentrations in the human colon, the site of STEC infection (Jacewicz et al. 1995). After binding, the Shiga toxin is internalized by receptor-mediated endocytosis at clathrin-coated pits. Translocation occurs from the endoplasmic reticulum to the cytosol where the A subunit functions as a glycohydrolase. It cleaves a specific adenine from the 28 S rRNA and inhibits thereby the binding of aminoacyl-tRNA to the ribosome. At the end, an irreversible block of protein synthesis results. The toxin action on the intestinal epithelial cells of the colon leads to a submucosa congestion and hemorrhage. The stool shows blood and pus, which led to the alternative name EHEC for these pathogens, enterohemorrhagic *E. coli*. Now the likely evolutionary sense of the toxin becomes evident. Again, it could be iron stealing by the bacterium from the decaying erythrocytes in the intestine. This suspicion is reinforced when looking into the transcriptional control of the *stx* genes. A functional promotor was identified directly upstream of the *stx1* gene. The activity of this promoter is regulated by the environmental iron concentration via a mechanism involving the iron-dependent Fur transcriptional repressor, which is thought to bind to a site near the promoter (Calderwood and Mekalanos 1987). Only when the iron concentration in the bacterial cell is low and iron becomes growth limiting, the expression of the Shiga toxin is induced. We see here that bacteria have added a new gear into the arm's race for iron acquisition. This is not an isolated case. Another classical AB toxin is diphtheria toxin elaborated by *C. diphtheriae*. Its A subunit is an ADP-ribosyltransferase, which covalently modifies the elongation factor-2, thereby inhibiting chain elongation during protein synthesis (Holmes 2000). The expression of the diphtheria toxin is regulated via the DtxR transcriptional regulator that binds in an iron-dependent way to the operators of many bacterial genes. Notably, Fur and DtxR are the master regulators of large bacterial iron regulons (genes scattered through the genome, but submitted to a joint expression control). You realize that a major theme of bacterial pathogenicity is the quest for iron.

An Emerging Pathogen: E. coli O157

I promised you an insight into the experimental character of bacterial evolution. Until now the iron acquisition mechanisms of *E. coli* seem rather cute. However, a deeper look into the system reveals that the system is not yet poised. Actually, STEC has despite its name no export mechanism for its toxin and thus suffers a drawback not known to HylA. There are other indications that the Shiga toxin is a newcomer to STEC. *stx1* and *stx2* are encoded on prophages; they are thus not part of the bacterial genome, but came into bacteria via the integration of the genome from temperate phages. Numerous bacterial pathogens owe at least some virulence factors to such integrated viral genomes (Figure 7.11) (Brüssow et al. 2004). The export mechanism is also provided by the phage genome. Directly downstream of the *stx* genes, you find the phage lysis cassette consisting of a holin and a lysin. The holin builds a pore into the inner membrane of *E. coli* through which the lysin can exit from the cell. The lysin attacks the cell wall and the weakening or digestion of the cell wall leads to an osmotic explosion of the cell if a major part of the cytoplasmic content had not already passed through the holin pores. With that cell lysis the Shiga toxin is also released. *The problem is: the bacterial cell, which produced the toxin, is dead and by any definition a dead cell cannot any longer profit from the iron supplied by the intestinal hemorrhage induced by the Shiga toxin. The situation is not totally unknown to bacteria. As the infecting bacteria are clonally related, the arguments of the selfish gene do not apply to the interaction between these bacteria. If a minority commits suicide, the majority of their genetically identical brethren can profit from the iron rain into the intestine.*

Colicins

A comparable situation was already described for colicins. Colicins are proteins encoded on the plasmids of *E. coli* that are exported and kill closely related, but not identical bacteria. The idea is a chemical club that is used to wipe out related bacteria that are by definition the most potent competitors for the nutritional resources in their ecological niche. Identical bacteria also carry on their plasmid an immunity function that protects the cell against the toxic effect of the colicin (colicin E1 permealizes the cytoplasmic membrane, E2 and E3 degrade DNA and rRNA, respectively). The selfish gene argument is thus satisfied under this condition. The medically widely applied antibiotics are also part of these chemical clubs used by microbes to clear their environment from nutritional competitors. Bacteria were thus under the selective

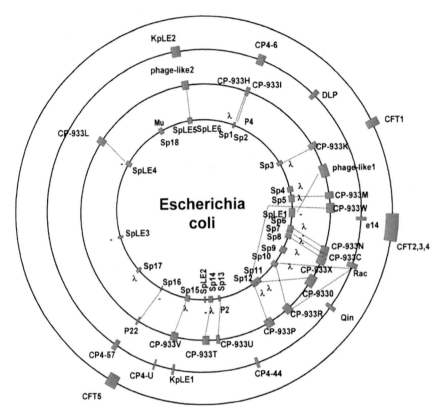

FIGURE 7.11. Molecular parasites: viruses integrate their genome into the chromosomes of both prokaryotes and eukaryotes. The figure shows the circular genome maps of four different about 4-Mb large *E. coli* genomes, the black rectangles represent integrated prophage genomes. The food pathogen *E. coli* O157 strains are literally littered with prophage DNA. The majority of them are not any longer infectious and have suffered various degrees of genetic degeneration. Our own genome also carries many proviruses and remnants from viral integration events.

force of antibiotics long before the compounds were used in medicine. The corresponding immunity factors, the antibiotic resistance genes, were also already at hand, which explains why some antibiotics have only a short lifetime after introduction in medical practice. In the case of colicins, only some bacteria produce them and die for the benefit of their congeners. STEC might have accepted death of some cells by prophage induction to the iron benefit of the remainder.

Recruiting Bystander Cells

STEC cells have "discovered" another way around the iron problem and the suicidal toxin production. As the toxins are encoded on a phage, another solution is at hand. A few STEC induce the prophage, which leads to the production of infectious phages from the lysed cell. This phage can then infect bystanders like the commensal E. coli in the colon of the infected host. These cells then undergo lytic infections, which lead to an amplification of Shiga toxin production without harming the STEC cells. They are actually "immune" to the action of the released phage via the immunity repressor, which is constitutively expressed from the resident prophage in STEC. While still somewhat speculative, this model of STEC infection is backed by a number of recent biological observations (Gamage et al. 2003) and can explain a number of otherwise paradoxical clinical observations (like the detrimental effect of antibiotics in STEC patients).

Pathogenicity as Nonadaptation to a New Host?

In fact, STEC might be a good example for a newly evolved pathogen. Pathogens making a hard ride with their host are frequently considered a newcomer in that system. The systemic effects of a noninvasive STEC like the kidney failure leading to dialysis in pediatric patients might be an evolutionary unwanted unbalanced side effect of an intestinal pathogen. This conclusion is not only backed by the recent emergence of EHEC in humans. The molecular archaeology of O157 also suggested the sequential acquisition of the *rfb* genes (encoding the O-serotype determinants), phage stx2 and then phage stx1 within perhaps 100 years. In accordance with the theory, O157 *E. coli* strains are not pathogenic in the intestine of cattle—their natural host and reservoir—and they have arrived in humans only quite recently, probably mediated by some changes in food consumption or processing.

Viruses Going for Gut or Genome

Portrait of a Killer Virus

Perceived and Real Killers

Many viruses feed on us, we call this phenomenon infectious disease, but from the viewpoint of viruses we are simply the substrate for their growth, providing nutrients and a good deal of biosynthetic activity. In this section I deal with rotavirus. You might raise your eyebrows as you have not yet heard from this killer, but this is a question of media attention. There is a strong bias between perceived threats and real killers (Glass 2004). We are frightened by West Nile virus, Ebola, and SARS headlines in the newspapers because they remind us that we are not really on the top of the food pyramid threatened only by our own suicidal instincts. These viruses remind us that we are perhaps not directly food, but at least a suitable substrate for our viral predators. Ebola— how dreadful the disease might be—isn't an efficient killer. The real killers are other viruses: influenza that kills each year 37,000 persons in the United States and had and still has the potential of great pandemics as demonstrated by the 1918 Spanish Flu, which killed more people than World War I. Rotavirus (RV), although less well known, kills yearly about 500,000 children and this on a regular basis. As these deaths occur mainly in the developing world and go along with a disease that we are used to see as a nuisance and not a real threat, namely diarrhea, its death toll remains unnoticed. However, children in the industrialized world also become growth substrate for RV. As they are quickly hospitalized and generally well nourished, the death rate is very low.

Rotavirus Epidemiology

To put this quest for food by a single virus in perspective: every child, irrespective of its place of birth, experiences during its first years of life a few episodes of RV infections, the first episode tends to be symptomatic, it causes a diarrhea frequently associated with vomiting. *In this respect, RVs are part of what epidemiologists call—somewhat floppily—a democratic infection.* It is very hard to find people that lack serological evidence of RV exposure: you have to go to a tribe living in a remote area of Amazonia or you must follow a missionary on isolated islands of Oceania, who apparently seeded an RV epidemic on all islands he visited. Approximately 1 in 10 children in the industrialized world sees a physician for this diarrhea; 1 in 100 gets hospitalized. In fact, during the RV winter season, about 5% of all pediatric hospitalizations are due to RV diarrhea. On a purely quantitative basis, we are the feeding basis for a massive RV replication. The situation is worse in the developing countries where RV meets children already weakened by malnutrition. In many parts of the poorer corners of the world children experience from 5 to 10 episodes of diarrhea per year. Thus a substantial part of the world children population spends up to 15% of their time on diarrhea. This condition acerbates malnutrition, and a negative spiral of growth retardation sets in. Of course, not all of these infections are RV-induced. Many bacteria and parasites actually target our gut as a feeding substrate. However, RV infections in children from developing countries tend to be clinically more severe than non-RV diarrhea. There is not much consolation in the fact that we are not the only target of RV, we share this destiny with the young of many mammals and birds.

Transmission

Let's look a bit into the feeding strategy of RV. Gastroenteritis agents generally have a fecal–oral infection route. The diarrhea agent is explosively discharged with the stool and evolutionary biologists interpret this as an adaptation to efficient dispersal. The fecal microbe then contaminates water or food sources and is thus recycled via the oral infection way back into the intestine of another subject where the next round of replication resumes, thus maintaining the agent in the population. As the drinking water in the industrialized world is chlorinated and the food for children is frequently produced industrially, it is not evident why RV infections are so prevalent in the northern hemisphere. Here epidemiologists tell us that RV behaves like a respiratory infection; in the United States, they even appear to be blown with the winds if one looks to the kinetics of the annual RV epidemics starting on the west coast in late fall, crossing the continent, and running out on the East coast in late spring. This peculiar pattern was interpreted by the fact that small doses of RV might be sufficient to get infected, while about 10^4 vibros are needed to contract cholera.

Histopathology

Once in the intestine, RV infects the mature enterocytes covering the villi of the intestinal mucosa. RV-induced diarrhea occurs early after the infection before significant histopathology is evident. In experimentally infected piglets a patchy viral antigen distribution is seen in the enterocytes with a decreasing intensity from the duodenum over the jejunum to the ileum. At 12 hours post infection, an early profuse secretory diarrhea is the probable consequence of the production of a viral enterotoxin, the nonstructural protein NSP4 (Ball et al.1996). The elaboration of a toxin is an absolute rarity in viral infections. The pathological effects of viral infections are normally the consequence of the cytopathic effects of the viral multiplication that leads to the death of the infected cell. In rare cases viral pathology is also the consequence of a derailed immune reaction against the invader. The cytopathic effect of RV infection becomes clear at 24 to 72 hours post infection. The viral antigen is widespread in the mucosa and viral titers are high in the intestinal lumen. The villi become atrophied and blunted. Signs of inflammation are seen in the mucosa, the epithelial cells are vacuolated, and the lamina propria is infiltrated by mononuclear cells.

Pathophysiology

The loss of the differentiated enterocytes leads to maladsorption. The lactase activity is lost and the impeded absorption of lactose leads to a further osmotic drag into the intestinal lumen, which aggravates the secretory diarrhea. In that phase, children show various degrees of dehydration ranging from sunken eyes and fontanella to loss of turgor in the skin. If uncorrected, the water and electrolyte loss can lead to acidosis, neurological disturbances, and death. It is absolutely critical that the children are at that moment rehydrated. The World Health Organization advocates an oral rehydration solution consisting of glucose and sodium chloride. The glucose is a nutrient, but its main function in the solution is to lead to a concomitant import of sodium into the intestinal mucosa. The trick is actually that glucose transport from the gut is, in the normal physiological situation, powered against the higher glucose concentration within the enterocytes by the cotransport of sodium downhill the concentration gradient (sodium is higher in the lumen than in the enterocyte). Other sodium transporters are frequently nonfunctional in diarrhea patients; sodium without glucose would thus not be absorbed. In severe cases the glucose–sodium solution must be instilled intravenously. With these measures death can be prevented and the patient recovers within days.

Self-Limited Infection

The acute infection and diarrheal disease normally resolves within 7 days and the intestinal mucosa again looks normal. This *restitutio ad integrum* is explained by the great regeneration power of the intestinal mucosa. Immature enterocytes are constantly born in the crypts of the mucosa and the differentiating enterocytes travel from the crypt to the top of the villi from which they slough off in a physiological growth cycle. In this way, the entire intestinal mucosa is renewed in every human being within days. Were it not for the water and electrolyte disturbances, RV infections would only be a rather benign gut disease. Not all viral infections of the gut are that respectful. Some enteric coronaviruses (the family to which SARS belongs) target the regenerative crypt cells for viral replication. When these cells die as a consequence from viral infection, no new enterocytes are regenerated and the intestinal villi atrophy permanently. The absorption is permanently interrupted and death is the logical consequence. Fortunately, this is not a human disease, but occurs only in swine (transmissible gastroenteritis virus, TGE).

Target Specificity

Many viruses are very selective for their target cell. RV exclusively uses the enterocytes as support for its propagation. Only in experimental animals with combined immunodeficiency, liver infections by RV were reported. How does RV target suitable cells? The virus is not only delicately poised towards its target cell, the enterocyte, it also targets apparently a developmental regulated viral receptor, as RV infection is a disease of the young animal and RV is, under natural conditions, very species specific. Only few human individuals were infected by bovine RV in the field despite the fact that humans can be experimentally infected with bovine RV (this was actually the first-generation RV vaccine; Roberts 2004). Only a single case of an avian RV possibly infecting a calf was reported (Brüssow et al.1992).

Rotavirus Genome

To get an answer, we must look into the molecular organization of RV. Its genome appears like a messenger from another world. Actually, its total genome size consists of about 19,000 bp divided into 11 segments of double-stranded RNA ranging from 3,300 to 660 bp. Each segment codes for a single protein except segment 11, which codes for two nonstructural proteins via frameshifting. There are noncoding regions at both ends of these simple one-gene chromosomes that have to fulfill a number of tasks. They contain a minimal promoter and transcriptional enhancers, signals for replication and translational enhancers.

(a)

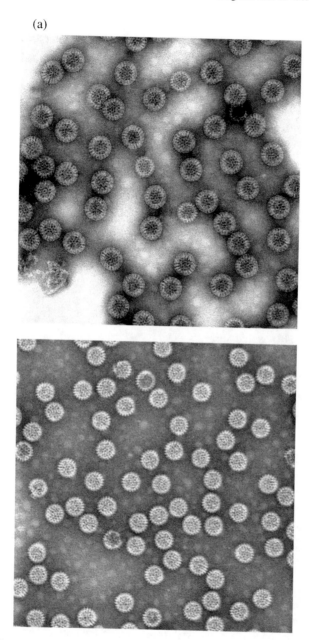

FIGURE 7.12. Double-layered rotavirus particles containing the double-stranded RNA genome (bottom) and triple-layered rotavirus particles lacking the RNA genome ("empty capsids") (top) as seen by negative stain electron microscopy. Both particle types are noninfectious, the first because it lacks the viral antireceptor, the second because it lacks the genetic information.

Translation

The mRNAs transcribed from them are capped, but not polyadenylated and need the help of the viral NSP3 (non structural protein 3) for translation. NSP3 associates with the poly(A)-binding protein of the translation initiation complex of the ribosome, which leads to the replacement of the cellular mRNA by RV mRNA. In this way, shortly after infection of the cell, the ribosomes synthesize RV proteins nearly exclusively. *The pirate has taken control of the boarded ship.*

RNA Replication

The viral proteins NSP2, 5, and 6 are associated with the so-called viroplasms, large virus core–building factories with nearly a crystal structure where RNA replication in the nascent virion cores takes place. NSP2 has been investigated in some detail (Jayaram et al. 2002). NSP2 oligomerizes into octamers, two stacked rings of four proteins leaving a central pore of 35 Å, not unlike a doughnut. The authors suggest that this NSP2 complex functions as a molecular motor during genome replication and packaging. Viral RNA binds to the external groove, the viral polymerase in the central pore.

The Enzymatically Active Viral Particle

Different virus-like particles could be reconstituted with recombinant RV proteins. Particles assembled from VP2 resembled viral cores and those assembled from VP1, 2, 3, and 6 resembled double-layered empty viral particles also observed in infected cells. The core of two related viruses, namely reovirus (Reinisch et al. 2000) and bluetongue virus (Grimes et al. 1998), was solved by X-ray analysis and the double-layered RV particle was analyzed by cryo-electronmicroscopy, and radial-density profiles were computed. This allowed a location of the different proteins in the RV particle (Prasad et al. 1996). The center is built by a dense convolute of RNA strands. In the better-resolved bluetongue virus, the RNA helices are tightly coiled, reminiscent of the dense DNA packaging observed in the head of phage particles. Only two empty volumes were observed—the likely place of the transcription complexes in bluetongue virus. A thin shell of VP2 is observed surrounding the central RNA convolute. It is overlaid by a more robust VP6 shell and underlaid at specific symmetry positions by a flower-like structure consisting of VP1, the RNA polymerase, and VP3, the mRNA capping enzyme. The VP2 and VP6 shells leave a hole over the flower cups through which the mRNA is extruded (Reinisch et al. 2000). *Why all this complicated structure and the associated enzymes? The answer is simple: the invader has to first make cautious steps that the cell does not detect the presence of a pirate on the ship. The regular crew of the cellular ship has a simple cue: they screen the ship for double-stranded RNA and when they detect this sign of an unfriendly takeover a vigorous defense response is set in motion.* RV and its relatives therefore avoid presenting free dsRNA at any step of their multiplication cycle. dsRNA is always hidden in a protein shell.

Transcription takes place from the uncoated viral particle directly after entry into the cell. Uncoating means, however, that only the outer protein shell consisting of VP4 and VP7 is blown away, induced by the low Ca^{2+} concentration in the cytoplasm. The double-layered particle thus starts with transcription and the viral mRNA is capped in exactly the same way as cellular mRNA. The cell has no cue to detect it as foreign RNA. In later steps, the mRNA also serves as a template to synthesize the complementary strand. Before replication starts, the mRNA is complexed with proteins in the viroplasm—again no dsRNA becomes visible to the cell. *In this way the cell is overrun before it actually takes notice of the unfriendly takeover.*

The Infectious Viral Particle

Now RV needs the final brush-up, the outer layer with the proteins that allow the targeting of the next prey when the newly made RVs are released from the dying cell. This process takes an interesting detour. NSP4, which we mentioned already as the first viral enterotoxin, has still another job: it is a glycoprotein that inserts into the membrane of the endoplasmic reticulum, attracts the nascent double-layered particle, and leads to the transient build-up of an intracellular membrane-enveloped particle before the outer shell proteins VP4 and VP7 take their final position in the infectious triple-layered particle. VP7 assembles as a third shell surrounding the mature virion. This layer is interrupted by spikes, built by VP4 that project beyond the VP7 shell. As VP7 and VP4 are the most exposed proteins they become the target of the immune system leading to neutralizing antibodies and cytotoxic T lymphocytes. As they are under immune selection, it is not surprising that RV differentiated in more than a dozen G serotypes (VP7 antigens) and an equal number of P serotypes (VP4 antigens).

Viral Spikes and Cell Entry

The X-ray structure and the structural rearrangements of VP4 relevant for the membrane penetration have been recently resolved (Dormitzer et al. 2004). Three distinct conformations correspond to steps of RV entry into a new target cell. When released from the previous cell, VP4 is noncleaved. In the intestine, trypsin provides the first cue: this pancreatic enzyme signals the appropriate ecological niche and triggers a first structural change by the tryptic cleavage of VP4. It is critical for any virus to make sure that it is really on the target. If the penetration process is initiated on a nonpermissive cell, the infectivity is lost. RV apparently uses a cautious strategy: first it searches with the viral spike protein the primary receptors, which are sialic acids in the context of gangliosides of the cellular membrane (Delorme et al. 2001). The following steps are only known in their outline: when the appropriate ganglioside is bound, the spike ejects the cleaved tip only held by noncovalent bonds. This process unmasks a hydrophobic region in the spike that inserts into the host membrane. The viral spike protein then interacts with the secondary receptors, a specific integrin. Then RV rolls apparently to still another tertiary receptor consisting of a complex of three membrane proteins. *This multistep entry of RV into the cell has been likened to the steps of a minuet dance of court balls where a dancer moves in small steps along a line of contradancers sharing frequent kisses* (Lopez and Arias 2004). After these confirmations of the right target cell, the spike protein folds back and creates a breach through which the RV enters the next cell. Notably, this entry process of a nonenveloped virus closely resembles that determined for enveloped viruses (Modis et al. 2004; Gibbons et al. 2004) unifying principles are thus also found in the biological world of viruses.

Bluetongue Variation

An interesting variation of this entry scheme is provided by the bluetongue virus, a cousin of the RV. It causes a hemorrhagic fever in sheep characterized by prolonged viremia, which is ecologically necessary as the disease has to be transmitted by blood-sucking insects. To avoid the onslaught of the immune system in the blood, the bluetongue virus (the name describes a characteristic clinical feature in sheep) hides out in erythrocytes. As it has to replicate in both an invertebrate and a mammalian host, it is confronted with the problem of how to build two recognition systems. Bluetongue virus has solved this problem elegantly: the triple-layered particle contains, as in the corresponding RV particle, the recognition proteins for the mammalian cell receptor of this virus, while the double-layered particle from bluetongue virus (which is noninfectious in RV) is infectious for the insect cell.

A Remarkable Design...

We have now seen what quest for food means for the RV. RV is for the standard of RNA viruses already a complex virus. RNA viruses have to solve a few biochemical problems with the constraint that RNA probably does not allow the use of really large genomes. One additional problem is the need to contribute the enzymes for their replication as such enzymes are unknown to the cell. The virus has to think on its camouflage and then an efficient takeover of the cells in order to build a stable transport vehicle for its own genome. Finally, they need a carefully poised protein apparatus to find and then breach into the next susceptible cell. I wonder whether molecular biologists would not despair if they are asked to achieve all this with a mere 18,500 bp of genetic information when starting from scratch. The common argument is that of R. Dawkins "Blind Watchmaker," evolution had very long time periods to tinker around to achieve the solution that we see now. How long is actually a contentious issue in virology. Animal virologists are used to phylogenetic trees, but no fossil record or molecular clock exists for viruses.

... Coming from Where?

How old are dsRNA viruses? Are they living fossils from the early genomes of the RNA world? Is RNA double-strandedness a solution to confer some error

*reduction in information storage? Except for the proteins involved in the camou-
flage business, no obvious links exist to cellular proteins, making it not very
probable that RV evolved from cellular genes. The exotic chemical structure of
their genes makes this hypothesis from the start unlikely. On the other hand,
RV has an extended family tree. RVs are known from mammals, birds, and
fish. The family* Reoviridae *is found not only in vertebrates, but also in insects
(orbiviruses, coltiviruses) and plants (phytoreoviruses). More distant relatives
are even found in prokaryotes as demonstrated by the* Pseudomonas *phage φ 6
(Karrasch et al. 1995). However, the phage differs in important details from
RV. It contains an internal lipid membrane (but recall the NSP4-associated
transitory membrane of RV) and only three dsRNA segments that each encodes
4–5 proteins.*

A Glimpse into the World of Retroelements

HIV: A Newly Emerged Old Virus

Evolutionary biologists state that old host–parasite relationships are characterized by a rather benign nutritional interaction. The parasite is best served by
a surviving host because it continues to provide a good growth substrate. If you
meet deadly human microbes, you might expect to see newcomers into the host–
parasite relationship. Such a great killer of humans is HIV. At first glance HIV
is not a likely reference when you want to explore old relationships reaching
back into the evolutionary past. The earliest isolate of HIV dates from 1959 in
Kinshasa. This sounds terribly recent and fits with two other observations. The
AIDS epidemic is so dramatic that it could not have been overlooked in the
past. AIDS is thus most likely a new disease in the medical record. Another
element fitting with the recent origin of HIV is its mutation rate, which has been
estimated to 10^{-2} substitutions per site per year. This fast rate is the product of
the high mutation rate of its reverse transcriptase and the rapid replication rate
in human victims (Sharp 2002).

The Ritual of Reverse Transcription and Provirus Integration

However, other characteristics point to an ancient origin of HIV. Its genome
is RNA, actually two identical (as far as HIV enzymes can achieve high
fidelity) strands of single-stranded RNA, which are linked by noncovalent bonds.
Furthermore, virions contain $tRNA_{Lys}$ bound to the genome RNA, where it
serves as a primer. With the help of its *pol* gene the viral genome shuttles
between RNA and DNA forms. Integration of the DNA provirus into the human
chromosome is an obligatory step in the replication cycle of HIV. The course
of reverse transcription is complex and involves a lot of molecular acrobatics.
A single-stranded viral RNA genome is copied into a double-stranded viral DNA
genome. The viral integrase then carries the linear DNA copy into the nucleus,
nibbles away two nucleotides from both ends, creating 3'-recessed ends. Then

follows a staggered cut in the chromosomal DNA, strand transfer, and repair of the cuts. This is a complicated molecular dance, which is with small variants faithfully executed by the various retroviruses. *This frozen ritual has the smack of a genetic element that stems from the transition period of the RNA into the DNA world.*

A "Complex" Retrovirus...

Like other retroviruses HIV shows the usual gene constellation *LTR–gag–pol–env–LTR*(large terminal repeat-group specific antigens–polymerase–envelope). Retroviruses with this genome organization are widely distributed in vertebrates. HIV-1 and the other infectious human retroviruses belong to the "complex" retroviruses, which are characterized by a set of six regulatory genes surrounding the *env* gene. These extra genes were new to retrovirologists and they made life very hard to scientists and prevented up to now the development of an efficient vaccine despite enormous research efforts. *To put this dilemma in perspective: the HIV-1 genome is only about 9,400-nt long, while far more than 10,000 research papers have been written on this virus since its discovery as the cause of AIDS in 1984. More than one paper per nucleotide or if you look at AIDS conferences, more than one researcher per nucleotide. It is astonishing that despite this human intellectual investment, the virus still slips through our hands. The grim perspective is that in many parts of the world the epidemic will run full course and might only come to a standstill when it has selected a genetically resistant or—in the more optimistic version—educated human population.*

... Targeting Our Defense System

AIDS has another lesson for us. HIV targets for its replication the CD4+ T-helper/inducer subset of the lymphocytes as its food or growth substrate. As these cells play a crucial role in the immune defense of the body against microbial invaders, the victim loses the most important line of defense and becomes helpless to fight HIV. Notably, AIDS patients suffer and finally die from attacks by other microbes if not supported by substantial and costly medical interventions. Strikingly, many of the coinfecting microbes were pretty unknown to clinical microbiologists before the AIDS epidemic. This observation shows that a major duty of our immune system is to fence off microorganisms that want to prey on us. Microbial predators of all kinds surround us, but we do not realize it because of our immunological armor.

Ancient Proviruses

Killing a host population is only one option for a rogue virus. Other retroviruses opted for an even more hideous strategy: they succeed in infecting germ cells and in integrating their provirus stably into the host chromosome. This integration can lead to the vertical transmission of the provirus via germ cells. Sequences that resemble endogenous retroviruses represent about 0.1% of the human genomic DNA. These are quite sizable numbers. HERV-H for *h*uman *e*ndogenous *r*etro*v*irus integrated into the tRNA *h*istidine gene comes in about 1,000 copies per genome, followed by HERV-K and HERV-E with 100 copies in the $tRNA_{Lys}$ and $tRNA_{Glu}$ genes, respectively. They represent "fossil" infections and insertion into the germ line that occurred in the evolutionary past. None of them belongs to the "complex" retroviruses, possibly underlining the more recent origin of the latter. There might be a subtle interplay between exogenous and endogenous lifestyles. As long as a "new" exogenous retrovirus meets a virgin susceptible population, it has an essentially unlimited supply of target host individuals. Its spread will be horizontal transmission of an exogenous infectious retrovirus. After a while viral interference will develop, the target cell will already carry a provirus from a related family, which will prevent the superinfection. The possibilities of horizontal spread become limiting and the retrovirus will be forced into vertical spread through genome colonization in germ cells. The hit-and-run lifestyle of the exogenous virus is replaced by an intimate, long-term association between the provirus and the host.

Evolutionary Trade-offs

Provirus genes that have a negative effect on the host or are neutral with respect to selection will be lost. This hypothesis actually neatly explains what happens with endogenous proviruses: ancient proviruses tend to be defective, and accumulate stop codons and deletions. If a retrovirus ORF has a beneficial function, it might be maintained. Syncytin is involved in human placenta morphogenesis. It represents a captured *env* gene from an endogenous retrovirus (Mi et al. 2000). An HERV-E insertion provides an LTR promoter for pleiotrophin expression in the placenta (Schulte et al. 1996). Another HERV-E is inserted upstream of a pancreatic amylase gene and provides a salivary gland specific enhancer that allows amylase secretion already in the mouth, thus conferring a sweet taste to a cereal-rich diet for humans (Ting et al. 1992).

Provirus–Prophage Analogy?

These observations with proviruses show striking parallels with the situation of prophages in bacteria. However, beyond the fact that retroviruses and temperate phages can integrate their genomes into the host chromosome and leave its replication to the diligence of the cell, no other obvious links exist between both viral systems. For example, the prophage and provirus integrase do not share sequence similarities and their genome organization is totally unrelated.

Gypsy

What is the evolutionary reach of retroviruses beyond vertebrates? Retroviruses with the usual gene constellation *LTR–gag–pol–env–LTR* were also found in *Drosophila*, where they were described as Gypsy element. Gypsy is transmitted via the germ line as a provirus and is under the control of the host gene *flamenco*. Putative endogenous retroviruses are common in insects, but Gypsy in *Drosophila* is the only invertebrate retroelement for which infectivity has been shown (Kim et al. 1994). With gypsy we got on a slippery road. Retroviruses are only the tip of an iceberg. Gypsy is also classified as an LTR retrotransposon. The LINE or Alu and SINE sequences making up so much of the human genome sequence (see below) are classified as non-LTR retrotransposons and retrotranscripts, respectively. Even bacteria contain retroelements, where they are called retrons. The list is by no means exhaustive (for a review see Bushman, Lateral DNA Transfer: Mechanisms and Consequences, Cold Spring Harbor Laboratory Press, New York, 2002). The function of many of these elements is unclear. Some might in their current state only represent purely parasitic DNA.

The Sense of Life

We touch here an eminent philosophical question of modern biology that also matters in a quest for food survey. Classical philosophers and founders of religion have come up with theories about the goal of life. In our robust anthropocentric (a biologist would say egocentric) view, most hypotheses deal with the goal of human life. Many theories link these goals to ethical commitments or our relationship with God or the transcendent. As natural scientists, biologists should not have a strong opinion on the transcendental, whatever their private opinion on the subject. At an individual level any opinion is allowed, even necessary that allows you to find your personal place in the cosmos during your life time. You can not wait that future centuries of scientific research or philosophical thinking do provide you an answer. Two conditions for a fruitful discussion between science and religious belief are necessary. As a natural scientist you should accept that the transcendent is outside of the framework of your science. Natural scientists start with the working hypothesis that Nature follows only physico-chemical laws. The relative explicatory success of this approach as demonstrated by the current scientific research is, however, by no means a philosophically valid proof that the transcendent does not exist. Scientists should – already from a methodological viewpoint - refrain from strong opinions on religious beliefs. Religious people should, on the other hand, likewise refrain from meddling with the philosophical basis of natural sciences as a working hypothesis to explain the natural world around us. Religions are not a method for an alternative explanation for the natural world around us. It is not the role of Christian faith to dispute the evidence for fossils, geology and the evolution of species. True religious beliefs should provide us with indications how the world should be organized in face of the transcendent, which reaches into our life as a personal or collective experience. Religions should not be against reason, but beyond reason. Where human reason can meaningfully reach, we should let it run its course. Where science cannot any longer count, observe and measure and where we need an answer in our life time, religions might provide us (including the scientist) an answer. The founder of the Christian faith saw these two faces of the world when telling people to give God what belongs to God and to give Cesar what belongs to Cesar. High profile scientists and theologians should meet to define the borderline between science and religious beliefs (see H. Küng Der Anfang aller Dinge-Naturwissenschaft und Religion, Piper 2005).

Back to biology and reasoning concerning the sense of life within the limits of its framework. Ever since Darwin, biologists they raised their voices that human life is but a small contribution to life. Life must have a "sense" in the absence of ethical and religious terms. The latter terms depend on the presence of a self-conscious intelligent mind or—to use the formulation of the Bible—a mind which knows to distinguish the good and the evil. In the history of life, such minds are a very recent development, which hardly looks back more than 100,000 years. To formulate it agnostically without hurting religious feelings: what was the sense of life left in the geological record and represented by the current life forms before the arrival of human consciousness? Biologists came up with some answers. To come to grip with this complex question, let's concentrate on some key arguments in this debate.

From Thomas to Darwin

Thomas of Aquino was not really a biologist, but this great Christian thinker of the thirteenth century had a sound interest in the natural world surrounding him. He counted the instinct to preserve the human race as one of the natural inclinations of our species. The preservation of human life became one of the prime goals of human laws and the task of the society was to guarantee this right of physical integrity for the individual. Species conservation was a main goal and this gave us the right to eat the sheep. Species were perceived as immutable until Darwin. He finally perceived that species developed, split into new species, which competed for a place under the sun. The picture got very dynamic and the extant species were seen as the descendent of a long series of extinct species. Although Darwin knew about the power of breeding for the species issue, he had no notion of genetics because the Austrian monk Gregor Mendel missed to send him a letter describing the experiments he did with peas in the garden of his monastery. Due to this handicap, Darwin's reasoning on the physical basis of heredity had to remain vague.

The Impact of Genetics ...

The marriage of evolution and genetics had to wait and took place only in the 1930s when the geneticist R.A. Fisher prepared in his book The Genetical Theory of Natural Selection *a synthesis of the Darwinian thinking on evolution with population genetics. Now the individual took the place of the species as the actor in evolution. The next step was taken in the 1960s by W.D. Hamilton and G.C. Williams, who emphasized the role of the gene in evolution. This development is only a logical consequence of the population genetics set out in the 1930s. The gene took center stage in the writings of J. Maynard Smith and R. Dawkins in the 1970s.*

... and Molecular Biology

The discussion in biology has thus definitively taken a reductionist slope from life → species → individual → gene. Small wonder then that molecular biologists like L. Orgel and F. Crick still went one step further to the basis of modern biology when they formulated the selfish DNA hypothesis in the early 1980s (Orgel and Crick 1980; Orgel et al. 1980). This hypothesis rests on the observation that in higher animals and plants, genes are sparsely distributed on the chromosomes while in prokaryotic genomes between 85 and 95% of the DNA is actually encoding proteins. This percentage is only 1.5% in the human genome. Many biologists looked at the noncoding repetitive DNA as "junk."

From Junk DNA ...

This interpretation is not farfetched: much of this DNA is indeed derived from mobile elements, 45% of the human DNA would fall into the category of mobile DNA or their action. LINE elements comprise about 20% of the human genome. LINE is an acronym for *L*ong *IN*terspersed DNA sequence *E*lement. They are typically several kilobases long and are found in thousands of copies per genome. Their generic structure is quite revealing: a 5' untranslated region (UTR) is followed by an orf that encodes Gag-like functions. In some insect elements this Gag still detectably resembles nucleocapsid proteins from retroviruses. Then follows a second orf combining an endonuclease, a reverse transcriptase, and an RNaseH function. However, no sequence similarity with retrovirus RT is observed. RT enzymes are one of the most divergent sets of enzymes known, suggesting that RT is among the evolutionary oldest enzymes. This interpretation is not incompatible with its role it probably had to play in catalyzing the transition from the RNA into the DNA world. The LINE ends with a 3' UTR and a polyA tail. Many LINE are truncated and rearranged. Another repetitive DNA is called SINE, which stands for *S*hort *IN*terspersed DNA sequence *E*lement. They range in length from 130 to 300 bp and occur as isolated dispersed DNA elements throughout the genome. The most frequent representative of SINE is the Alu sequence, so called because it contains a single AluI restriction site. This 300-bp sequence shows sequence relatedness to the 7SL RNA. Alu sequences occur in more than a million copies in the human genome and thus add another 11% of "junk" DNA to the human genome. Proviruses, DNA transposons, and pseudogenes, derived from the action of the retrotransposon machinery on cellular mRNA, add to the rest of the mobile DNA in the human genome. Of course, all this "junk" DNA reflects our current insight into our genome organization. We have seen that nature is an opportunist, even the "junk" DNA might have taken some function in our genome. For example, in many eukaryotes the centromeres of the chromosomes contain abundant mobile DNA elements, which may have found a useful role as architectural elements. In fruit flies, telomeres of the chromosomes are formed by LINE-like DNA.

... to Selfish DNA

However, the mainstream interpretation is still that of Orgel and Crick, where selfish DNA seeks to achieve Darwinian success by expanding its copy number, while it does not contribute a selective advantage for its host. They profit from the permissiveness of higher eukaryotes with respect to the genome size. No such selfish DNA could develop in prokaryotes, which strive to an economy in their genome size, which rarely goes beyond 10 Mb.

An informational molecule wants world power? What is the take-home message from these considerations? First, it confirms that there is a grain of truth in the word of some researchers who declared that "humans are descended from viruses as well as from apes." Second, I think this seems to mean that the quest for food is not a hot pursuit in its own sake. Having a metabolism and eating is not a biological goal in itself, it is only a means to another goal. And this ultimate goal is to remain on the scene as an informational macromolecule, where the nucleic acid sequence spells out your identity. J. Monod said once that the ultimate aim of a bacterium is to become two. One could paraphrase that the ultimate aim of life is defined by the zeal of a nucleic acid molecule containing a specific sequence to fill the biosphere. For a philosopher this might seem a minimalist goal, but for many biologists this hypothesis has a strong explicatory power. In his book The Selfish Gene *R. Dawkins showed in an eloquent way how far this basic idea reaches out from the inner realm of biology into the understanding of human behavior.*

There is no inherent unidirectional trend in evolution to develop ever more complicated life forms with ever more sophisticated eating strategies. It is obvious that our world is biologically more complex than the Archaean Sea or what it was during the Cambrian Revolution. However, we are still surrounded with reductionist organisms. They achieve their goals with minimal means. Cheaters in the quest for food come in many forms. The LINEs and SINEs in our chromosomes that get replicated at our cost without spending time in the quest for food are only one extreme, but at the same time the most successful form.

8
An Agro(-Eco)nomical Outlook: Feeding the Billions

Malthus: Doomsday Versus Science and Technology?

No World Without Hunger

In a previous section of this book we have seen that each of us easily feeds trillions of guest workers in the gut. However, these are tiny bacteria and consequently their appetite is small. But what is the prospect for the human society to feed 10 billions of human consumers on the planet? Opinions on this subject were split. Two thousand years ago and with a much smaller world population, the Man of Galilee was rather sceptical on this subject; he told his disciples that poor people will be a constant company to human society. He did not foresee a society without poverty and hunger on earth. He was feeding the 5,000, not the billions. On a purely historical basis he was right—there is hardly any period in human history where the poor and the hungry were unknown. Feeding people was not the main concern of philosophers and thus the discussion got only heated with an anonymous pamphlet published in 1798. The title reads, "An Essay on the Principles of Population, as it Affects the Future Improvement of Society." The title sounds rather benign, but the reactions to this publication were discordant. The contemporaneous poet-philosopher Samuel Coleridge spoke of "the monstrous practical sophism of Malthus." A generation later Karl Marx spoke of Malthus as "a shameless sycophant of the ruling classes." Closer to our days, the economist John Maynard Keynes described the essay as "a work of youthful genius." Malthus' pamphlet also inspired biology. Charles Darwin confessed that this reading was a crucial moment for the development of his evolution theory. Also the other co-discoverer of the evolution theory, Alfred Russell Wallace, quoted Malthus' essay as perhaps the most important book that influenced his thinking.

Malthus Thesis

So what are the basic tenets of this essay? Malthus noted "the power of population is indefinitely greater than the power in the earth to produce subsistence for man." He argued that the population shows a geometric growth, while food production grows only arithmetically. Then, he continued, "this implies a strong and constantly operating check on population from the difficulty of subsistence. This difficulty must fall somewhere and must necessarily felt by a large portion of mankind" (quotations from Short 1998). The Biblical mission "Be fruitful and multiply" will thus lead to a crisis. Once again, as a biologist I admire the insight of the religious poetry in the book of Genesis. This word is the quintessence of biology, more precisely the first half of the biological credo— it's other half being Darwin's "Survival of the Fittest". However, the prophet forget (says a biologist) to tell us that this order was given to all organisms — not just humans. We took this order literally and became over a few thousand years the dominant evolutionary force, but possibly also the nemesis for higher life forms on our planet. But back to Malthus. The conclusions drawn from his basic idea were dividing people. The Malthusian pessimism was taken as legitimacy for the 1834 New Poor Law Act in Britain, which did away with the relief for the destitute. Friedrich Engels, editor of the Communist Manifesto, *saw in this pamphlet an illustration of Thomas Hobbes' theory of the* "bellum omnium contra omnes" *(a war of all against all) and the foundation of the bourgeois economic theory of competition (Puntis and Kirpalani 1998). Darwin's mind was well prepared for this essay because he already appreciated the struggle for existence, which he saw everywhere in his observations of animals and plants. Darwin's followers like Ernst Haeckel in Germany then made the step from Darwin's theory into the human society, paving the way for social Darwinism.*

Condorcet

Today many scientists claim that Malthus was a false prophet. The opponents use different arguments. Some argue that Malthus has taken the basic idea from the Marquis of Condorcet. He was with equal intensity a mathematician, philosopher, and politician. Despite his tragic fate (in fact already the French Revolution was devouring its children), he had an optimistic view of the world, typical for the period of Enlightenment. He saw the implications of population growth and food production, but he predicted that the threat of widespread famine could be prevented by universal education, sexual equality, and the progress of reason and science. The combination of these factors would lead to a fall of the birth rate and continued improvement of the human fate. The question who is the right prophet and who is the false one is not just a choice between Malthus and Condorcet, it continues to our days (Raven 2002).

Reasons for Pessimism

If one looks to scientific publications one gets the impression that whether a scientist is Malthusian or not depends somewhat on the working field. Agronomists are overall more optimistic than ecologists to feed the world. Physicians tend to be more pessimistic than molecular biologists probably because the former are more directly confronted with human misery in the developing countries. Medical writers noted that the twentieth century was not instilling a lot of optimism. They counted more than 100 million people killed in war and genocidal atrocities and another 100 million who perished from famine and diseases during that time period. Some even postulated a connection between war and a demographic trap. They see ethnic cleansing as clearing farmland of its previous occupants in order to enable aggressors to sustain expanding populations (Last 1993). This was the argument of Nazi Germany ("People without space") for heading eastward. The same argument was used in the War of Yugoslavia and in the genocide of Rwanda, to name a few.

Entrapment

Medical epidemiologists introduced the idea of "entrapment." They defined a population as trapped if it exceeded the carrying capacity of its own ecosystem and has nothing to trade to obtain food and cannot migrate into other ecosystems to remedy the situation. The consequences are dire: people migrate in misery to some urban slum (forced migration), they hope to be supported indefinitely by food aid from industrialized countries or they die. This poses a severe dilemma for aid agencies as they have to chose between measures against childhood mortality or community health when both cannot be achieved with the available means in such populations (King and Elliott 1993). In many cases ecologically entrapped people go for violent conflicts as documented in a study by the University of Toronto. Environmental change led to conflicts for farmland (e.g., Bangladesh–Assam; The Philippines) or water resources (Senegal–Mauritania; Israel–Palestine) (Homer-Dixon et al. 1993). I remember a couple of comments from the *Lancet* during the Rwanda war where the writers stated that the people already lived beyond the ecological carrying capacity of the region. Not the least reason for the hostility and the genocide, which followed, were the distinct agricultural traditions separating Hutus and Tutsis, the Biblical conflict between crop farmers and cattle ranchers.

Food Aid

Even outside of war zones, famines are frequent events in sub-Saharan Africa. At the peak of the 2003 food crisis in southern Africa more than 15 million people needed humanitarian assistance for survival. The very fact that food aid can be delivered means that hunger has a peculiar note in our world. As stated by Amartya Sen, famine means today lack of access to food rather than lack of food per se. This applies to the global level, but can also be said for individual countries like India. Despite that "optimistic" note, the fact remains that according to a census done for the UN Millennium program 0.8 billion people—about 14% of the world population—are chronically or acutely malnourished (Sanchez and Swaminathan 2005a).

Distribution of Hunger

The distribution of hunger shows regional centers: most cases are found in South and Southeast Asia and sub-Saharan Africa, where the hunger prevalence is as high as 30%. Yet there are distinct differences: in South Asia food insecurity is primarily due to poor distribution and lack of purchasing power. The situation in Africa in contrast is characterized by low per capita food production. The reasons are manifold, but at the top is the depletion of soil fertility followed by problems of weeds, pests, and disease. Over decades small-scale farmers have, with their harvests, removed large quantities of nutrients from their soils. Owing to lack of education, capital, or resources, not enough manure or fertilizers were added to replenish the soils. It was calculated that annually 1 ha of average African soil loses 22 kg nitrogen, 2.5 kg phosphorus, and 15 kg potassium (Sanchez 2002). To remedy the situation $4 billion worth fertilizer would be needed. The actual costs would be even higher as you would have to add substantial transport costs to get them to the African fields. The UN Millennium Project, which wants to cut world hunger in half by 2015, gave a dire message: hunger in Africa is a vicious link between unhealthy people and unhealthy soils (Sanchez and Swaminathan 2005b).

Reasons for Optimism

Molecular biologists see the world differently. A lead article titled "Malthus Foiled Again and Again" for a special issue of *Nature* on "Food and the Future", Trewavas (2002) pointed to a number of important figures: currently one person can be fed by food grown on $2,000\,m^2$, this figure was $20,000\,m^2$ in Malthus's time. After Malthus came Justus Liebig and the fertilizer, the development of the Haber–Bosch process, plant breeding in 1900, the mechanization of agriculture, the Green Revolution of the 1950s, and today biotechnology. These are not small innovations: the global output of fixed nitrogen from the chemical industry exceeds that of the biological nitrogen cycle. Every time people predicted imminent disaster, an innovation opened new opportunities. However, even if you are an optimist, the challenge ahead is enormous.

Population and Harvest Growth

The outcome will crucially depend on the growth kinetics of the world population. Optimistic demographers see the twenty-first century as a crisis period with respect to food security: the population will grow to about 9–10 billion by 2050 and then growth will level off. However, the stabilization of the population size will not mean immediate relief. Again in the optimistic scenario the world population will get richer and as a consequence the meat consumption will increase. As cattle has to be fed with cereals and as about ten kilograms of cereal have to be fed to grow 1 kg of meat, the cereal yields will have to grow substantially, probably it will have to double to keep pace. In the past, harvest increases could be achieved by plowing up new land. This will not be possible any longer—half of the good soil is already being used for agriculture. The rest is covered by forests, which we should not touch for various reasons (e.g., biodiversity as a basis for ecosystem services, Loreau et al. 2001). The crop increases must therefore come from maximizing yields with the available land.

Sustainability

There is another constraint: we should aim to ensure that the exploitation is done in a sustainable form. If we want to preserve the land as a capital for the future and if we want to live from the interest without touching the capital, we must minimize the damage to the ecosystem. Ecosystems are providing us more than food: they give us clothes from plant fiber, fuel, and construction material for houses. Forests protect us against avalanches and flooding, they moderate climate, store atmospheric CO_2, regenerate fertile soils, and purify water. Many researchers are convinced that agriculture can meet the food needs of 8–10 billions of people. However, it is not yet clear how we will cope with the associated ecological problems like fertilizer and pesticide draining into aquatic habitats. A number of priorities can already be defined (Tilman et al. 2002). We will have to increase the efficiency of nutrient use by a type of "precision agriculture" (Cassman 1999; Matson et al. 1998). Fertilizers are only added when and where they are needed. This needs substantial education of farmers and investments in soil testing. Slow-release nutrients are needed, which do not drain away easily, and fertilizers need multiple and closer applications to the root system. Likewise we need to increase the efficiency of water use: 40% of the crop production comes from 16% of land that is irrigated. Yield increases will only be realistic with higher water investment, but pumping in excess of water recharge capacities was seen not only in China, India, and Bangladesh, but also in the United States. Alternatively, crops with increased drought tolerance could increase yields in water-limited environments (Charles 2001). Another priority must be the restoration of soil fertility with classical measures like crop rotation, reduced tillage, the use of cover crops, fallow periods, manure and fertilizer application (Drinkwater et al. 1998). Next come measures against disease and pest control. This is especially important when considering that 60% of human food is provided by just three cereals, namely wheat, rice, and corn. Over the last millennia it was not only humans covering the globe, but also these three once rare weedy species that came to world dominance. Resistance against diseases and pests is, in biology, always a transitory property. The pathogens and pests evolve with their hosts and will seek breaches into the plant defense systems. As a final layer in the current agricultural system comes an equally impressive number of farm animals belonging to a small number of species, which take a big share of the resources on our planet. Rangeland occupies about a fifth of the world's land surface and supports 3.3 billions of cattle, sheep, and goat (Raven 2002). They are likewise subject to specific pathogens. In fact, in ecological terms the ruminant–grassland connection is a highly efficient converter of low-quality forage into high-protein human food like dairy products and beef and a provider of manure, which can close the nitrogen cycle to a certain extent. Science and technology seems to suggest that we can meet the challenge—we can feed the poor (Huang et al. 2002)—but at what price? We should not forget that the next 50 years of agricultural practices will shape, perhaps irreversibly, the surface of the planet (Tilman et al. 2002). Humans became the world's greatest evolutionary force (Palumbi 2001). It is our responsibility to use this power respectfully if we don't want to spoil the earth for future human generations and many other life forms on earth.

From the Green Revolution to Organic Farming

Rationale and Impact of the Green Revolution

The development of high-yielding crop varieties for developing countries that began in the late 1950s became known as the Green Revolution (Evenson and Gollin 2003). Its historical core was a cooperation between US plant breeders and a maize improvement institute in Mexico and later a rice research institute in the Philippines. The breeders could thus rely on rich stocks of genetic resources. Using breeding techniques they incorporated dwarfing genes into crops that led to the development of shorter varieties characterized by stiffer straws. The energy which normally would go into straw and leaves went now into grain production. The new varieties also responded better to fertilizers. The breeders developed first a basic plant type, a high-yielding semi-dwarf, which was later further modified according to local needs. The hybrids were quickly accepted by the farmers, but the impact was variable in different parts of the world. Asia profited largely from the Green Revolution: over the last 30 years food production tripled, the rural poor decreased by half, malnutrition dropped from 30 to 18%, and the prices of cereal crops came down by 70%. Comparable success was reported for Latin America. This success story was not observed in Africa where crop yields increased by only 28%. There are several reasons for this failure. One was the lack of elite germ plasma for African crops like cassava or tropical beans. Research into plants that can be cultivated under semiarid conditions like sorghum, millet, and barley were only initiated in the 1980s. While in Asia yield growth accounted for almost all increases in food production, this was not achieved in Africa. Here increases were only realized by extending the areas under cultivation.

The Dark Side of the Green Revolution

On a quantitative basis the extent of this Green Revolution was impressive: food production approximately paralleled the population growth and grain prices are now near historical lows; yet the prospects for a continued revolution are mixed. Some researchers are already calling for a greening of the Green Revolution (Tilman 1998). The adjective "green" had in the 1960s not yet political connotations; "green" then did not confer to an ecological orientation. The Green Revolution was in fact a high-tech agriculture including the use of fertilizers, herbicides, pesticides, irrigation, and a substantial mechanization. On the ecological side there was a price to pay: the use of fertilizers led to the release of greenhouse gases, the genetic diversity of the cultivated crops was lost, groundwater was contaminated with pesticides and herbicides, the fertilizers were drained into surrounding freshwater and coastal marine areas leading to eutrophication. In fact, during the Green Revolution the global use of nitrogen fertilizers increased 7-fold, that of phosphorus 3.5-fold. A further threefold increase for both fertilizers is expected until 2050. You have to replace the nitrogen which you remove with the crops to the soil. The consequence of no nitrogen addition to the soil has now become evident in many rural areas of Africa where the soils are more and more depleted of nutrients. In this region, many rural activities resembled more "mining" approaches than agriculture. In the long run, fertile soils are essential for a sustainable agriculture. During the Green Revolution agricultural yields have doubled, but at the same time one third of the global arable land has been lost to erosion (Pimentel et al. 1995). The consequences are clear: as agricultural land is degraded and abandoned, more forests are cut and converted to agricultural land. However, tropical forests soils are not adapted to crop production and the new land will soon be exhausted—a vicious cycle sets in.

Erosion

The cause of soil degradation has a name: erosion. The basic force behind erosion is the kinetic energy of raindrops falling on the soil or wind blowing over the soil. The raindrop splash transports soil particles into the air where they become air bound and can be blown away over many miles. Erosion rate is greater at slopes due to the force of running water. Of the world's agricultural land, one-third is used for crop production the remaining two thirds are pastures for livestock grazing. The cropland suffers more from erosion because the tilled land lacks a plant cover. In China and Bangladesh crop residues are not left on the land, but are sometimes collected with the roots and burned for fuel. About 80% of the agricultural land suffers at least from moderate erosion, half of the pastures are overgrazed. Erosion takes away between 15 (USA, Europe) and 40 tons/ha/year (Asia). This has important consequences: topsoil depth in the United States was estimated to be 23 cm in 1776 and is now down to 15 cm.

Soil Quality

Soil rebuilds only slowly: something between 200 and 1,000 years are needed to give 2.5 cm topsoil. The topsoil determines a number of properties that are crucial for plant growth. Soil retains and stores water, which is an important property when considering that 1 ha of maize (7,000 kg biomass) transpires 4×10^6 l of water. Organic matter in the soil increases soil porosity, provides nutrients that enhance the growth of roots and soil biota. A hectare of good soil contains about

Gemeiner Regenwurm, Lumbricus herculeus San. Natürliche Größe.

FIGURE 8.1. Earthworms, here represented with its most common species *Lumbricus terrestris*, subclass Oligochaeta from the phylum Annelida, occur in practically all soils that provide moisture and food to them. Earthworms eat decaying organisms and in doing so they ingest large amounts of soil, which corresponds to their own weight per day. They live normally near the soil surface, but can tunnel further down thereby aerating and draining the soil. They carry decaying material into their burrows; organic material becomes thus available to plant roots.

1,000 kg of earthworms (Figure 8.1; already Charles Darwin investigated the importance of this animal for soil fertility), 1,000 kg of arthropods (including many predators of plant pests) and 150 kg each of protozoa and algae. To that you add 1,700 kg of bacteria and 2,700 kg of fungi, which recycle the nutrients in the soil. The sheer quantity—and the astounding genetic diversity—of the soil microbes underlines their crucial role for this ecosystem. Microbiologists will admit that their knowledge of microbes from fresh water, wastewater, and ocean water is much more developed than that of soil. The soil and its microbiology is certainly a New Frontier. Fertile topsoil contains 100 tons of organic matter per hectare. One ton of topsoil typically contains 1–6 kg of nitrogen which can fall to 0.1–0.5 kg in severely eroded soil. In such soils, plant growth is stunted and crop yield declines.

Loss of Nitrogen in Soil

The loss of soil organic matter occurs with any conversion of natural ecosystems into permanent agricultural land. This process has been intensively studied in various areas. In temperate zones, the losses are most rapid during the first 25 years of cultivation, leading to a reduction of soil carbon and nitrogen by 50% when compared to the preagricultural amounts. In tropical soils these losses may occur within 5 years (Matson et al. 1997). After that period, the C and N content become stabilized at this value and even intensive addition of fertilizers cannot replenish this stock. Reduced tillage led to small, but significant increase from 53 to 61% of the preagricultural level. It was estimated that only 40–60% of the added nitrogen is used by plants. The rest is left in the soil. In a process called nitrification the microbiota oxidize NH_4^+ under aerobic conditions to NO_3^- (with some by-products $NO > N_2O$). In principle, microbes can close the N cycle under anaerobic conditions when they reduce NO_3^- to N_2, N_2O, or NO in a process called denitrification. If the loop is not closed or the biochemical activities are overwhelmed by too large pulses of mobile forms of nitrogen, NO_3^- is lost and leaches in the groundwater or flows away with rivers. US rivers now show 3- to 10-fold higher nitrate levels than in 1900. Nitrate in the drinking water is a public health concern. In fresh water, nitrate additions can cause eutrophication, which leads to fish loss and blooms of toxic algae in stratified water via oxygen consumption. Fertilizer use also causes the release of NO_x gases leading to damaging ozone levels in the fields. Phosphorus is also lost from fertilizers leading likewise to eutrophication. In addition, substantial amounts of pesticides do not reach their target, they are carried away and then kill nontarget organisms or represent public health concerns because of their hormone mimetic and immunosuppressive side activities. *In summary, not all what is called Green Revolution is really green in our current eco-political understanding.*

The Pennsylvania Experience in Organic Farming

Agronomists with scientific motivation led a number of surprisingly "green" projects testing tenets of organic farming and the results were encouraging. One is a 15-year study with three distinct maize/soybean agro-ecosystems, which was conducted in Pennsylvania (Drinkwater et al. 1998). The test included a conventional system with maize/soybean rotation, mineral nitrogen fertilizer for maize plantation, and pesticide use as needed. It was compared to a system that cultivated crop biomass of legumes and grasses, which were then fed to beef cattle. Manure was returned on the field as nitrogen source for maize. In the third system, maize received the nitrogen directly from legumes through incorporation of leguminous biomass before maize plantation. Ten-year average for maize yields was nearly identical in the three compared systems (7,100–7,170 kg/ha). Quantitatively the overall nitrogen additions were similar. However, there were significant differences in soil nitrogen storage: the conventional system lost nitrogen, the manure system gained nitrogen, the legume system remained constant. The systems also differed in nitrate leaching: the conventional system lost twice as much as the other two regimes. The CO_2 emissions from the two organic systems were lower than from the conventional system, which also needed a 50% higher energy input. Is the way to sustainable agriculture opened and a really green revolution ahead?

The Mexican Experience

The greatly increased use of fertilizer nitrogen did not only increase nitrate runoff, but also released N-containing gases into the atmosphere. Fertilized agriculture is the main source of the greenhouse gas N_2O. Agronomists from Stanford teamed up with Mexicans to investigate the need of fertilizer in a wheat system in one of Mexico's major breadbaskets (Matson et al. 1998). They compared conventional fertilized cultures with those receiving less nitrogen, or nitrogen later in the season or less nitrogen at a later time. All reduced regimes led to reduced losses of nitrogen to the environment without affecting yield and grain quality. As the fertilizer costs equaled that of planting, irrigation, and harvest, reduced fertilization means not only less emission of nitrogen, but also real cost savings.

Organic Apples from Washington

Promising data for organic farming were also obtained by US horticulture specialists. They investigated the yield, quality, and profitability of three apple production systems. A conventional, an organic, and a mixed ("integrated") system were compared. In the organic system pesticides were not used and instead of fertilizers the agronomists used compost and mulch, animal and green manure. Again the yields were comparable. Soil rating was higher in the organic and integrated system than in the conventional system. Tree growth was identical. The organic apples were considered sweeter, and the integrated had a better flavor than the conventional apples. Energy efficiency was also greater in the organic system. This success on many fronts came as a surprise. A news article in *Nature* already asked: "Organic—is it the future of farming?" (Macilwain 2004).

The Swiss Way

A subsequent study conducted in Switzerland provided a more differentiated picture, but it still inspired the journalist from *Science* to the heading: "Organic farms reap many benefits" (Stokstad 2002). Two conventional systems (chemical fertilizer and manure) were compared with two organic systems that used only manure and plant extracts to control pests (Mäder et al. 2002). The organic system delivered 90% of the winter wheat grain harvest and the grassland yields ranged from 70 to 100% of the conventional exploitation. Only the potato harvests were with about 60% substantially lower than the yield under conventional conditions. Overall, the mean crop yields were 20% lower in the organic system, but they were achieved with 50% less nutrient (N, P, K) input. The energy efficiency of organic farming was thus very favorable. Cost-efficient production is only one aspect in agriculture. The maintenance of soil fertility is equally important for a sustainable land use. Here organic farming excelled in several respects: organic plots showed a greater soil stability, which led to higher microbial biomass and biodiversity and associated biochemical activities. The soil micro-biota from the organic plots decomposed more plant material and showed higher abundance of earthworms, carabids, and spiders and demonstrated more root colonization with mycorrhiza than conventional farming. The effect of organic farming thus extends to several trophic levels and is characterized by efficient resource utilization. In an interesting letter exchange in *Science* (Goklany 2002) the relevance of the observations was questioned by another observation that a 20% yield reduction would be fine in well-fed Europeans, but unacceptable for the hungry population. This would—the critics stated—translate into more hunger or more new land use for agriculture. The Swiss authors countered that the current practice forced to abandon one third of the arable land because of loss of soil fertility. Sustainability of the agricultural system must thus remain a priority if we want to feed not only the next two generations, but many human generations to come.

From Biodiversity to the Wood Wide Web: On Rice and Grassland Productivity

This section starts with a deceivingly simple experiment which has great practical implications for agronomy, but at the same time substantial theoretical implications for the scientific discussion of biodiversity in ecology (Zhu et al. 2000). The scene is China and the crop is rice. Both the place and the plant are significant. The economical development of China is confronted with the problem of the population explosion because with a population exceeding 1 billion people, they represent every sixth human being on the globe.

Ehrlich Versus Borlaugh

Rice, wheat, and maize are the cereals that feed 60% of the people on the globe. To appreciate the paper, we need a bit of history (Mann 1997). The prospect for the communist Chinese society to feed its population was rather bleak. Back in 1969, Paul Ehrlich, author of the influential book *The Population Bomb*, forecasted that a horde of famished Chinese would invade Russia. This Malthusian prediction fortunately did not become true. It illustrates the chronic difficulty when even the brightest minds in science project ecological scenarios for the real world. The defusing of the bomb in China has many reasons: First and foremost the Green Revolution, then particularly in China the one-child population policy, and finally the liberalization of economical life within a communist political framework. The work of the plant geneticist and the Nobel Peace Prize winner Norman Borlaugh initiated the Green Revolution. His work goes back to 1943 when he tried to breed high-yielding wheat varieties that resisted stem rust, a major fungal disease in Latin America. With funds from the Rockefeller Foundation and the Mexican Ministry of Agriculture, he obtained both goals: high yield and rust resistance, but he overdid it. The plant was so top-heavy with grains that it fell. However, he and his colleagues did not give up. They looked into strain collections for varieties that had short stout stalks. Such a dwarf variety was found in Japan. This was, at first look, an impossible variety to work with: it was frequently sterile, grains were unusable, and it succumbed to rust. However, 7 years later Borlaugh and his colleagues could combine the best of both worlds and got a plant that increased the harvest from 0.8 to 8 tons/ha. The success was so impressive that in 1960, US Foundations together with the International Rice Research Institute in the Philippines tried the same for rice. They succeeded by using the same strategy crossing high-yield, disease-resistant varieties and dwarf varieties of rice.

The Limits of Growth

In the 1990s the world grain production fell below the FAO and World Bank projections. The reasons are complex. To some the slippage in global grain productivity has a political background, mainly the collapse of the Soviet Union. Borlaugh is less concerned about the slowing of the Green Revolution; he argues that it still has to spread into areas which it never reached and which need it most, i.e., Africa. However, the agricultural economist L. Brown from the Worldwatch Institute fears that the Green Revolution has lost momentum. Agronomists are less worried, but remind that what was achieved by plant breeding has reached a ceiling. The harvest index, the weight of grain per weight of the plant, was 0.25 at the beginning of the century and it is now at 0.5. This can perhaps still be increased to 0.65, but soon you get to the physical limits. A plant cannot be all grain, it needs leaves, stalks, and roots to grow. Like in organic farming, we might need more biology instead of agronomical technology.

Rice and the Fungus

And this is where the abovementioned research report sets in. The success of the breeding approach has led to widespread monocultures. If you reduce the number of species and the number of genetic varieties within a species in agronomy, you create enormous possibilities for plant pathogens. Those that can grow on the breed have a wide field for spreading (Figure 8.2). An important rice pathogen especially in the cool, wet climate of the Yunnan province in China is the blast disease induced by the fungus *Magnaporthe grisea*. It causes necrosis on leaves and panicles. Like many fungal pathogens *M. grisea* interacts with plants by what is called gene-for-gene basis (Baker et al. 1997). Genetic variety from the plant side is thus the best remedy against this type of pathogens. Already Darwin knew about the advantages in growing different types of plants when targeting higher yields. The Chinese agronomists observed that farmers applied empirically the recipe of Darwin. They were interested in growing a glutinous (sticky) rice variety that achieved higher market values than the conventional hybrid rice of the region. However, the sticky variant grew taller, had lesser yields, and was very susceptible to the fungus. What they did was to sow a row of glutinous rice between four rows of hybrid rice. The latter was not only less susceptible to this fungus in general, but the two rice varieties did not share *M. grisea* variants able to grow on both plants. The agronomists evaluated this empirical technique scientifically and the success was striking. In monoculture the glutinous variety suffered blast disease on 20% of the panicles, in mixed culture it was only 1%. Even in the hybrid majority the extent of blast disease decreased from 2.3 to 1%. In the mixture the glutinous variety produced 18% of the harvest although it was planted only at a 9% level. As the farmers harvest the rice by hand, the higher valued rice variety could be collected separately resulting in a substantial economical gain due to its higher market price. The hybrid harvest remained at the level of its monoculture, which was with 10 Mg/ha the highest in the world. The researchers observed that the new planting practice affected directly

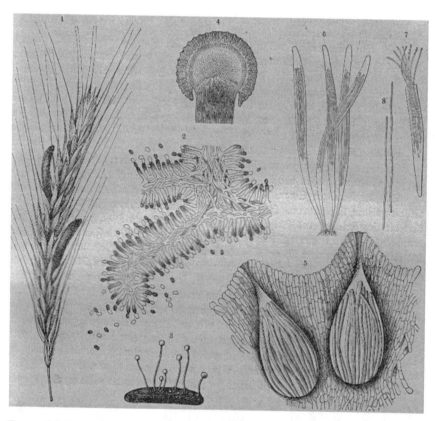

FIGURE 8.2. Fungal infections decrease the yield of many crops or make it unfit for human consumption due to their toxic content, like ergot produced by *Claviceps purpurea*, which belongs to the Pyrenomycetidae group of the fungal class Ascomycetes. The fungus infects the pistil of grasses, most commonly of rye (*Secale cereale*). The infection leads to a secretion of honey dew, which attracts insects that transmit the conidia (*2*) of the fungus to uninfected plants. Once the fungus has used up the pistil tissue, it hardens into a sclerotium (*1*). It falls down and develops fruiting bodies (*4*), which harbor perithecia (*5*). They contain the long asci, releasing thread-like spores, which can then reinfect new pistils of grasses. The use of fungicides has strongly reduced ergots in cereal fields, but the heavy use of chemicals in intensive agriculture came at a price.

the genetic structure of the pathogen population. The monoculture fields were dominated by one or a few fungus strains. Fields with mixtures showed complex pathogen populations without a dominant strain. Due to the visible success, farmers quickly adopted the new planting technique to the whole province and fungicidal sprays could be abandoned.

The Cedar Creek plots

The Chinese experiment could not provide mechanisms that explained the results. However, the results are trustworthy because they build on an old tradition in biology starting with Darwin's observation in the *Origin of Species* from 1859, the "diversity begets stability" hypothesis formulated by the ecologist Charles Elton in the late 1950s, and the mathematical treatment of ecosystem dynamics by Robert May in the 1970s (Kareiva 1996). The experimental verification started only about 10 years ago with a famous experiment in the grassland (Figure 8.3) of Minnesota, the Cedar Creek area. In later versions, David Tilman and colleagues planted 147 plots each 3 m by 3 m with seeds of 1, 2, 4, 6, 8, 12, or 24 species drawn from a pool of species native to the North American grassland. For those who have seen the fields, the experiment offers in addition to statistical rigor an esthetically appealing appearance. The first experiments lived through a period of severe drought, the most severe Minnesota has seen over 50 years. This allowed the researchers to measure resistance to drought by calculating the plant community biomass change over the stress period. The greatest dependence showed plots with nine or fewer species. They measured another parameter also: resilience, i.e., the rate of return to preexisting conditions after pertur-bation. Again the same observation was made: the species-poor plots showed the greatest difficulty to return to the status quo ante (Tilman and Downing 1994). In a follow-up study they verified that plant cover and light penetration were increasing with species richness. The agronomists observed that soil mineral nitrogen, the main limiting nutrient in terrestrial systems, was utilized more completely when there was a greater diversity of species as demonstrated by

FIGURE 8.3. An artistic view of the steppe in Ukrania with characteristic prairie plants; *top*, from *right to left*: *Stipa pennata, Jurinea mollis, Syrenia angustifolia, Salvia austriaca, Dianthus polymorphus*, and *Astralgus virgatus*; *bottom*, from *right to left*: *Festuca vaginata, Astralgus onobrychis, Iris variegata, Achillea ochroleuca, Astralgus exscapus*, and *Ranunculus pedatus*.

measuring nitrate concentrations in and below the rooting zone. These observations suggested a lower leaching loss of nitrate from the ecosystem. That resource utilization was significantly greater at higher plant diversity was also observed in native undisturbed grasslands (Tilman et al. 1996). In a more recent report from the same group, the researchers first demonstrated in a 7-year experiment that 16-species plots attained 2.7 times greater biomass than a monoculture (Tilman et al. 2001). The result was comparable whether biomass was measured as aboveground living biomass important for grassland used as cattle pastures or as total biomass measured as above- plus belowground biomass.

Sampling Effect Versus Niche Complementarity?

These and similar results from other groups have generated substantial debate with respect to the explanation of these results. Competing hypotheses were "sampling effects" and "niche complementarity." In the first scenario there are a few high-performing species (e.g., presence of legumes on a low-N soil) and the chances to include these plants increase with species richness in the plot. In the second scenario better results come from interspecies differences in resource requirements. These plants can exploit spatial and temporal resources and habitat differences. There might also be positive interactions between coexisting species. These two hypotheses lead to different predictions with respect to transient and persistent effects and the performance of the higher-diversity plots to the best-performing monoculture. The authors concluded that their observations, especially those of overyielding species combinations, are best explained by the niche complementarity hypothesis and suggested that even with the wisest choices, monocultures will be less productive than species combinations. What is true for a single site in ecological research cannot directly be extrapolated to other geographical areas. Therefore, the confirmation of the major conclusion from Minnesota in European grasslands was an important event (Hector et al. 1999). In the European study many climate zones were included ranging from cool Sweden to hot Greece, maritime Ireland to continental Germany. Detailed analyses showed that despite the clear overall trend, not all individual sites showed the dependence of the aboveground biomass to species richness. Differences were even observed between two sites from the same country: one site in the UK showed the expected species-richness-dependent biomass relationship, but the second did not. Regression analysis showed that in European grasslands one plant, the nitrogen-fixing clover *Trifolium pratense*, had a major effect on productivity and could thus cause sampling effects.

No Insurance in Switzerland . . .

Of note, a detailed analysis of the data from the Swiss site of this European BIODEPTH project went against common wisdom in ecological research. The scientists observed that species diversity actually decreased the stability of ecosystem functioning. They simulated drought perturbation in a Cedar creek-like sowing experiment using a transparent polycarbonate roof that could be raised above individual plots, which prevented watering by rain. Surprisingly the species-rich systems showed a greater reduction in biomass production under perturbation than the species-poor plots and this on the background of a system that showed clear productivity increases with species richness under unperturbed conditions (Pfisterer 2002). These observations went into the face of the insurance hypothesis in ecology. However, this hypothesis impressed more by its inherent logic than by a large body of experimental evidence. Logic suggests that having a variety of species insures an ecosystem against environmental perturbations. Some plant species might be more resistant toward drought, others to flood, still others to fire, insect, or vertebrate herbivores. Possessing many species with different genetic potentials buffers against irreversible losses in case of major damage. In an editorial on this article the reviewer noted that the most likely explanation of the diversity–productivity link, namely niche complementarity among species in a diverse plot, could be its downfall when faced by perturbation (Naeem 2002). Too many species are interconnected and the fall of one might drag down interconnected species.

... But in Mongolia

The editorial also criticized that many data in the field of ecology were created by poking (small experiments) and peeking (observational studies). However, the combinatorial studies are also not without pitfalls. The Swiss BIODEPTH study was also small. Small wonder that its conclusions were disputed by a large observational study of two types of grasslands in the Inner Mongolia, a rhizome grass-dominated and a bunchgrass-dominated community (Bai et al. 2004). These grasslands were observed over 24 years and the researchers looked for the dominant climate factors and then biological factors correlated with aboveground biomass productivity. The only significant climate variable having an impact was January-to-July precipitation. The observation that this steppe ecosystem achieved high species richness, productivity, and ecosystem stability simultaneously in the late succession stage of its development looked like a clear return to the insurance hypothesis. Also the latest evaluation of the Cedar Creek field experiment by David Tilman et al. (2006) pointed to a clear positive correlation between biodiversity and ecosystem stability. However, individual species stability was inversely correlated with species numbers sown into the fields. As this experiment is now running more than 10 years, the system can be tested over substantial variation of precipitation and daily high temperatures in the growing seasons. Stepwise regression analysis showed no correlation with group composition of C3 and C4 grasses, legumes, weed species, or soil fertility. The only significant factor was root biomass, which is quite logic. Roots are the perenniating structures of herbaceous perennial species that provide a large store of nutrients to buffer growth against environmental perturbance. The authors concluded that biodiversity might be a reliable approach to assure the supply of grasses for livestock fodder and biofuel in the future.

Digging Deeper into the Wood Wide Web

The published combinatorial studies, elegant as they are, suffer another drawback. Most of them remained restricted to one trophic level, that of the primary producers. Extensions into food webs including herbivores or decomposers are necessary to get more realistic answers for real ecosystems. Some work has been conducted in that direction. Most attention was paid to microbe–plant interaction of either the positive type (mycorrhizal symbiosis) or the negative type (plant pathogens). Arbuscular mycorrhizal fungi (AMF) make mutualist associations with the roots of 80% of all terrestrial plant species and thus become functional extensions of the plant roots, increasing the uptake of nutrients, mainly of phosphorus. Not only the aboveground plants are divers, also the AMF show species diversity. To assess whether AMF affected plant diversity, botanists grew combinations of 11 plant species with 4 AMF species in the greenhouse. Eight plants depended totally on AMF and biomass varied significantly among treatments with different AMF species (van der Heijden et al. 1998). The authors then manipulated in 70 macrocosmos field experiments the number of AMF species added while maintaining the same 15 plant species in each experiment. The aboveground plant diversity increased with belowground AMF diversity. Further analysis revealed increased hyphae length with more AMF species added as well as larger shoots and root biomass. Plant phosphorus content increased, while soil phosphorus decreased with AMF species number. The richness of AMF species thus led to an increased hyphal foraging capacity. These data are interesting as previous work had demonstrated that AMF connect the roots of different plants in the forest directing a flow of nutrients between different tree species (Simard et al. 1997). This system was dubbed a "wood wide web." Agricultural practice might plow up this web because the genetic diversity of AMF in woodland was much larger than that of neighboring farm land, which is dominated by a single species not found in the woodland (Helgason et al. 1998). A fascinating report linked rarity or invasiveness of plants in an ecosystem to feedback with the soil biota. Rare plants suffered a relative decrease in growth on "home" soil in which pathogens had a chance to accumulate. In contrast, invasive plants profited from beneficial interaction with AMF (Klironomos 2002) without paying the price of being submitted to suppression by soil pathogens in the "non-home" soil while still suffering suppression on their "home" soil (Packer and Clay 2000).

The Rice Blast Fungus: A Threat to World Food Security?

In a previous section we discussed perceived and real killers of humankind. The public might be less aware that there are other infectious agents that could literally kill millions of people without infecting a single human being. To solve this riddle, I am speaking of the cereal killer (Talbot 2003) *Magnaporthe grisea*, the causative agent of rice blast, which we already met in a previous section. As we got acquainted with viral, bacterial, and protozoal pathogens in our survey, we still need a fungal pathogen to complete the picture. The threat is real, and annually rice quantities sufficient to feed 60 million people are destroyed by this pathogen alone. From the Yunnan experience, one might draw the premature conclusion that this organism is not an important foe. However, we should not underestimate it—all recent molecular data point to a formidable enemy.

Life Cycle

Infections of rice plants occur when the fungal spores land on the leaf of rice. Plants have evolved to deal with invasion and the intact leaf surface cannot be assailed by many pathogens. This barrier is not a problem for *M. grisea*. After adhesion of the spore to the leaf surface via a glue released from the tip of the spore, the spore germinates and creates a terminal swelling called an appressorium. The latter is heavily pigmented with melanin and causes the rust-like appearance on the infected plants. After adhesion, the appressorium shows a frantic biochemical activity: it accumulates glycerol derived from glycolytic intermediates and degradation of stored fatty acids. The accumulation of glycerol leads to a concentration exceeding 3 M within 8 h. This corresponds to a turgor pressure, which is equivalent to 40 times the pressure in a car tire—the highest ever observed in an organism (de Jong et al. 1997). This pressure of 8 MPa ruptures the leaf cuticle, the waxy layer that protects the plant leaf. The germinating spore then sends hyphae into the plant tissue, which invade the plant. Young rice plants die from this infection, older plants lose the grains. After spreading in the plant, the fungus regains the leaf surface, sporulates, and spreads via air to uninfected plants.

This does not sound very reassuring and in fact, fungicides are only of moderate efficiency against this pathogen and plant breeders cannot develop resistance systems quickly enough to keep this fungal pathogen under control.

Two Modes of Infection

To make matters worse: plant pathologists from Norwich found recently that *M. grisea* is more versatile than previously believed. Taxonomical analysis led to a close association of this rice pathogen with a fungal wheat pathogen, *Gaeumannomyces graminis*, which causes the "take-all root rot" disease of a number of agronomic cereal species. Wheat ears show the so-called whitehead symptoms. *G. graminis* is a soil-borne pathogen that attacks the root and from there it spreads systemically. Normally, a fungal pathogen uses either the aerial or the root infection route, but the close taxonomical affinity of *M. grisea* with a soil pathogen motivated the researchers to inoculate the roots of rice with *M. grisea*. Strikingly, this pathogen, which is the paradigm for a foliar pathogen, has all the features of a classical root pathogen (Sesma 2004). Genetic experiments demonstrated that these are really two different infection modes mastered by a single pathogen. A melanin-deficient *M. grisea* mutant was unable to penetrate barley leaves, but could still infect barley roots. As melanin deficiency was known to interfere with the glycerol accumulation process in the appressorium, the defect for leaf infection is easily understood. What alternative infection mechanism is used in root infection? The lead was provided by the soil-borne wilt fungus *Fusarium*, which requires a mitochondrial carrier protein for colonization. The genome sequence from *M. grisea* showed a related gene. Genetic knockout mutants proved the link: the mutant could still adhere to the root, but cannot spread any longer through the vascular tissue. Above the ground, infection of the leaves was, in contrast, not affected by this mutation.

Gene-for-Gene Immunity

The plant–pathogen interactions are governed by the same rules irrespective of whether aboveground or belowground infections occur. Pathogenic microorganisms in plants are known to secrete proteins directly into host plant cells to disturb host cell signaling or to interfere with the plant innate immune system. As this is a classical case of host–pathogen coevolution, the plant has in turn evolved proteins that specifically recognize the effector proteins from the pathogen. One speaks of a gene-for-gene system of immunity: one gene in the host causes resistance to a pathogen effector protein encoded by a single pathogen gene. For example, a *M. grisea* strain virulent on the fully susceptible cultivar Nipponbare was avirulent on leaves of the rice cultivar CO39. The same pathogenic strain also got stuck in root infections of the cultivar CO39: it could only colonize the outermost root cells.

Disturbingly, *M. grisea* was already isolated from diseased wheat in Brazil (Valent 2004). It is thus important to keep an open eye on this pathogen, which targets major staple food. If it would escape control by the resistance mechanisms of the currently used rice and wheat cultivars, the consequences for the human food basis would be dramatic. The consequence of the Irish potato famine can still be read from the telephone book of the Boston area.

Genomics

The analysis of the *M. grisea* genome allows first insights into the genetic basis of fungal disease in plants (Dean et al. 2005). With a 40-Mb genome encoding 11,000 genes it does not excel the genome of a nonpathogenic filamentous fungus. *Neurospora crassa*, which like *M. grisea* belongs to the same pyrenomycete group of fungi, encodes 10,000 genes. *N. crassa* is the bread mould and it is a saprotroph, i.e., it feeds on dead or decaying matter and can thus serve as a nonpathogenic reference (Galagan et al. 2003). Gene density is high in both fungi with 1 gene per 3.5 kb. To propagate on plants, *M. grisea* must undergo a coordinated series of morphological and physiological changes. A relatively large group of genes are G-protein-coupled receptors which sense and integrate environmental signals. The genome revealed 61 new receptors, some of them are upregulated in specific stages of the infection process like the appressorium formation. Then there are virulence-associated signaling pathways, one is of generic importance for any fungal plant pathogen. In addition, a large part of the genome encodes secreted proteins, which include cutinases that attack the waxy polymer of the leaf cuticle. It also encodes carbohydrate-binding proteins that seem to prevent the action of plant-encoded chitinases on the fungal cell wall. Also a number of secondary metabolic pathways were described. It is somewhat disturbing that the nonpathogenic filamentous fungus *N. crassa* encodes likewise enzymes for secondary metabolites (nonribosomal peptide synthetases, polyketide synthetases, enzymes involved in diterpene metabolism) and even genes involved in plant pathogenesis. This similarity prevents the use of comparative genomics for the identification of important virulence genes in *M. grisea*.

Variability and Recombination

Another disturbing factor emerging from the genetic analysis is the high degree of variability not only between the two filamentous fungi (which differ from each other as much as humans from amphibians), but also between variants of *M. grisea* capable of infecting formerly resistant rice strains. The researchers suspected that the high percentage of 10% repeat DNA in the genome might be responsible for this variability. The repeat DNA belongs to mobile selfish DNA such as retrotransposons and DNA transposons. *N. crassa* has a genetic defense mechanism in place that inactivates transposons by inducing targeted mutations in repeat elements called repeat-induced point mutation or RIP, a system only known from fungi. In addition, transposon gene silencing is achieved by DNA methylation of cytosines specifically in the repeats (Selker et al. 2003). In contrast, *M. grisea* has been unable to stop the proliferation of DNA transposons. One element occurs in over 100 apparently intact copies that display 99.4% DNA sequence identity. By transposition these elements can inactivate pathogen effector genes and thereby allow an escape of the fungus from plant resistance mechanisms that are specifically based on the recognition of these fungal pathogenicity proteins. Furthermore, these repeats can be the target for extensive DNA genome rearrangements. As a result, the plant is confronted with an extremely dynamic fungal genome. On the other side, it is not clear how the fungal pathogen can tolerate such extensive genetic changes. During the sexual stages meiotic recombination could lead to a recombination catastrophe. In view of the possible economical and humanitarian effect of an uncontrolled infection of such an important staple crop, we are well advised to push the research on this fungal plant pathogen. Research on plant pathogens received much less grant money than research on human pathogens. However, for the human victim it does not matter whether he dies from a human virus or from the lack of food because a fungus has killed his staple food.

The Story of the Rice Seedling Blight

Another fungal rice infection has recently revealed its secrets. The fungus is *Rhizopus* and the disease it causes is the rice seedling blight. The fungus secretes a toxin, rhizoxin, which binds to rice tubulin and thus interferes with mitosis. This toxin kills the rice root cells and the fungus makes a dinner from the dead roots. There are fascinating side stories with this toxin, not the least that its antimitotic activity is currently evaluated as an anticancer drug. This medical interest led to a structural analysis of this toxin, which belongs to a chemical class called macrocyclic polyketides. The dilemma was that the fungus genome does not contain the enzymes to synthesize it. The genes were finally found, but they were of bacterial origin (Partida-Martinez and Hertweck 2005). Horizontal gene transfer was excluded—the fungal cell contained a live bacterial endosymbiont that could be classified into the Bulkholderia group, known to contain medical and veterinary pathogens. The nearest relative is *Pseudomonas* also well known to plant pathologists. In a remarkable demonstration of Koch's principles, researchers from Jena cured the fungus from the bacterium with antibiotic treatment. The fungus lost its toxic function. They then succeeded to grow the bacterium in bacteriological culture, where it continued to produce the toxin. Finally, they reintroduced the bacterium into the fungus, which restored the toxic property of the fungus. In this remarkable symbiosis, the fungus provides its plant cell digestion capacity, the bacterium its toxic killer substance and both profit from the nutrients released by the decaying plant root. Such symbiotic relationships are not so rare. In a previous chapter we learnt that sponges are a rich source of antitumor substances; in at least one case such a substance is produced by a bacterial symbiont of a marine sponge (Piel et al. 2004). The same principle is worked out by insects or tunicates (Piel 2002). These associations are not necessarily to the detriment of the plant: bacteria also occur as symbionts of mycorrhizal fungi. However, the common denominator in both bacterial–fungal associations is food-sharing with or without the consent of the plant.

Sowing Golden Rice in the Field?

The Carotene-Producing Golden Rice

A frequently asked question is whether genetically modified (GM) crops can solve the pressing nutritional problems of developing countries. To my judgment this question currently cannot be clearly answered because it touches different layers of problems, which need distinct analyses. I will try to illustrate the problem with a concrete example. At the first level are the gene technology problems, which might still be the most tractable. Take the golden rice developed at the ETH in Zurich: it was genetically engineered to allow the production of β-carotene in the seeds (Ye et al. 2000). To achieve its synthesis from precursors, which are present in the rice seed, the biosynthetic apparatus of the rice endosperm needed a complementation with two new enzymes: first a phytoene synthase that condensed two C20 isoprene derivatives (geranylgeranyl diphosphate) to the C40 compound phytoene. The researchers used here a gene from daffodil (*Narcissus pseudonarcissus*). However, there was another block in the flow of these precursors into β-carotene. Phytoene had to undergo several oxidations by desaturases to yield lycopene. This biosynthetic activity is provided by a gene from the soil bacterium *Erwinia uredovora*. Endowed with these two foreign genes the golden rice starts to synthesize β-carotene. As β-carotene is colored, the rice changes from white to golden. β-carotene is a provitamin A carotenoid and can thus be used to combat vitamin A deficiency, which is a widespread problem in rice-eating countries. Vitamin A deficiency affects over 250 million people and causes in children night blindness and a depressed immune reaction, which makes them more susceptible to measles infections and diarrheal diseases. The public health impact of vitamin A complementation could thus be enormous. However, there are technological hurdles. The maximal carotenoid level achieved in transgenic rice was $0.8 \mu g$ β-carotene per g dry weight of rice. Children would have to eat an unrealistic amount of rice to satisfy their recommended daily allowance of $300 \mu g$ vitamin A in the 1–3 years age range. This problem was recently solved by replacing the daffodil phytoene synthase in Golden Rice 1 with the corresponding maize gene (Paine et al. 2005). The ameliorated Golden Rice 2 now has a much higher substrate flow through this pathway and produces $37 \mu g$ β-carotene per g dry weight of rice. Sixty grams of rice is a typical daily child's portion and Golden Rice 2 could in some estimates satisfy about 40% of the daily need of vitamin A (Grusak 2005).

Absorption and Perception Problems

However, there are still problems: a dietary oil intake is needed to achieve the absorption of the lipophilic β-carotene. The amount of oil needed is unknown and unfortunately the diet of the poorest, i.e., those most in need of golden rice, is often low in dietary oil. In this example we shift from genetic engineering problems, which were just a few months ago settled, to physiology problems in nutrition. Not enough with that, we meet now logistical and economical constraints of distributing oil for a better absorption of β-carotene. *In addition to the economical constraints, you have still to add psychological barriers even for a not-for-profit product, developed by a public institution of a neutral country to fix an important public health problem. It is often stated that the well-fed European consumer is the most reluctant to GM food because he or she can allow being choosy and must not follow the imperative demands of the gut. A standard argument is that only when real health benefits can be communicated to the consumer, the acceptance of GM food will increase as if we would eat with our conscious brain parts. I don't think that this argument is correct. Even people in need of vitamin A are choosy and their problem is the color of the golden rice. Asian people are used to the white color of their rice dish and they are not eager to change to colored rice, whatever the health reason behind. You have to remember that rice is the main food intake of these populations. They have thus strong emotional links with their food. Imagine an Irishman in a pub who gets a Guinness beer, which is for some nutritional reason, let's say because it is cardioprotective, of golden color. Nutritionists have lived this dilemma already in the past. You could cope with the vitamin A deficiency by simple nutritional interventions. You simply send out food chemists who screen the local plants for species rich in vitamin A. You collect or cultivate these plants and add them in small amounts to the diet. In fact, nutritionists tried this approach with Asian children. The Asian children with marginal vitamin A levels did what children from all over the globe will do, they carefully sorted out the green additions from their meal. The children have certainly not yet heard of the GMO discussion, they reproduce possibly only an old inherited (?) aversion against nonwhite food that was once (?) of survival value against food intoxication.*

Where Market Forces Become Inefficient

Green biotechnology makes still other arousal reactions. In the industrialized western societies this research is mainly conducted in the private sector. Not surprisingly, the research strategies of these companies go for cash crops cultivated in temperate climate zones of the world. Many Europeans deplore the privatization of top class agricultural research to companies and they have a deep distrust of their activities. They mention a truism that a private company will not look for the public global good, but for profit. However, a private company cannot produce for consumers that lack the purchase power to buy the product. Similar problems exist for the pharmaceutical industry. Public health specialists discuss here models where international organizations order large amounts of drugs from private companies against diseases like malaria, which offer the companies a reasonable profit margin. This creates a market incentive for these companies. The WHO or other aid agencies then sell these drugs to the local distribution channels in Africa, which allows them some profit while still remaining affordable to the African patient. Of course the taxpayer of the industrialized countries would then have to pay the bill for the price difference in the form of a solidarity pact with the poorer countries. The only alternative is that the public sector takes care of this applied research and development, which market forces cannot serve by lack of demand expressed in money terms. The third possibility is to abandon the people in need, who lack the purchase power. Whether this truly represents an alternative is questionable because it would undermine ethical values on which human societies are or should be based.

GM Cotton in the United States and India

Can the current products of private companies be of interest to poor small-scale farmers in developing countries? Currently, about 99% of the global GM crop acreage is devoted to the cultivation of insect-resistant (Figure 8.4) and herbicide-tolerant plants. The use of insect-resistant maize is profitable for the US farmer because it results in gains due to pesticide savings. The environment gets less exposed to highly hazardous chemicals like organophosphates, carbamates, or pyrethroids belonging to defined toxicity classes. Yield advantages for insect-resistant cotton in the United States are less than 10% in comparison with control cotton. The reason is that the pest pressure is relatively weak for cotton grown in the United States. The situation is different for India (Qaim and Zilberman 2003). The US company Montsanto developed Bt cotton. This GM plant contains a gene from the bacterium *Bacillus thuringiensis*. Its expression confers a high degree of resistance to several ballworms that are major pests of cotton in India. The gene was introduced by an Indian partner into Indian cotton hybrids and approved in 2002. Field trials for biosafety evaluations were conducted in 2001 on nearly 400

FIGURE 8.4. Common African cotton pests, *left* (from *top to bottom*): *Gelechia*, *Zonocerus*, *Dysdercus*, *right*: *Heliothis*, *Phonoctonus*, *Calidea*, *Alcides*.

farms. Bt cotton was compared to isogenic cotton lacking the Bt gene and another local non-GM control. The latter two breeds needed an average of 3.6 sprays against the ballworm. Bt cotton required only 0.6 sprays. The effect was specific as no benefit was seen with respect to spraying against sucking pests. Most notable for the Indian farmer were the substantial gains in yield. Both cotton controls yielded 830 kg/ha, while the Bt cotton gave 1,500 kg/ha. Over a 4-year observation period, 60% harvest increases were measured. Indian small-scale farmers are frequently in debt and cannot afford to buy chemicals when their crop is threatened. Bt cotton is thus of substantial interest to them. *However, the relationship between local governments and biotech companies has not been easy, complaints come from both sides.*

GM Cotton in China

Therefore the governments from some countries strive to develop a strong biotech home basis. China is currently developing the largest plant biotechnology capacity outside of North America (Huang et al. 2002). The biotechnology research is state-funded and adds to the already most successful agricultural research system in the developing world. In the Chinese agronomical sector more than 70,000 scientists are employed. In 2002 China's Office of Genetic Engineering had already approved more than 250 cases of GM crops, animals, or recombinant microorganisms. The response especially from small farmers in China was quick: from 1997 to 2000 the area sown with Bt cotton increased from 2,000 to 700,000 ha. By 2000 already 20% of all sown cotton were Bt varieties. The greatest benefit to farmers is the reduced cost of pesticides calculated at $760 per hectare. The use of pesticides, which also affected the health of the farmers, could be reduced by 80%. Only 5% of the Bt cotton farmers in comparison with 22% of the control cotton farmers reported signs of poisoning with the highly toxic pesticides.

Hesitant Extensions

Extensive programs also exist for other agricultural crops. Transgenic rice resistant to major pests (stem borer, planthopper, bacterial leaf blight) has passed environmental release tests. In contrast to developed countries where about half of the field trials are for herbicide resistance, China's field trials are 90% dominated by those that target insect and disease resistance in crop plants. *Despite their potential for hunger reduction, GM food crops are not even used in countries like China that showed an aggressive commercialization of Bt cotton and heavy investments in GM food crops. This decision shows a substantial psychological barrier for GM food even in developing countries.* The first farm-level preproduction trials with GM rice were just evaluated in China (Huang et al. 2005). Two Chinese-created rice strains, one containing a Bt gene and another a cowpea trypsin inhibitor gene (Tu et al. 2000), were compared to conventional rice in several villages. Yield increase in the GM rice was modest and ranged from 6 to 9%. However, the pesticide use was reduced by an impressive 80% in the GM rice fields from 3.7 to 0.5 applications resulting in 17 kg less pesticide use per ha. As pesticide application caused a number of health problems in the farmer, the sowing of GM rice improved health in a measurable way.

Herbicide-resistance GM crops

About 10% of the global crop production is lost through weed infestation. The weed is pushed back by about 100 different chemical forms of herbicides causing costs of $ 10 billions per year. However, many herbicides do not distinguish weeds from crop. Here GM plants can help. Take the example of glyphosate, an "environmental friendly" herbicide because it is readily degraded to nontoxic compounds in the soil. The crop plant receives the gene from a glyphosate-resistant *E. coli* strain, which replaces the plant enzyme that is inhibited by the herbicide. The transgenic crop plant can now grow in the presence of the herbicide glyphosate while the weed dies. Now consider the case of a herbicide-resistant crop plant; take one of the first GM crops, *Brassica napus*, the oilseed rape or canola. It is widely grown in North America and its release was planned in Australia.

Pollen Distribution

If you grow such a plant, it will produce pollen, which is essentially a lightweight transport form of the plant genome designed for dispersal. GM canola has a number of weeds as cousins, but there are also farmers that might want to grow non-GM canola in the neighborhood. So what is the risk of involuntary dispersal of the herbicide-resistance gene in the environment? An Australian team of scientists set out to test just this by collecting seeds from 63 control canola fields growing next to a GM canola field. Overall, 48 million canola individuals were tested for herbicide resistance; 23 fields showed no pollen dispersal. In the fields yielding resistant seedlings the frequency of resistant plants was mostly below 0.05%. The flow of the genes was restricted to 3 km from the source. Edge effects, i.e., differences between the front and the rear end of the contacting fields, were not observed, but the genetic variety of the canola in the adjacent test field was an important determinant for the receptivity (Rieger et al. 2002).

On Sense and Sensibility

Should we now be worried about this result? As I said in the beginning of this section the question can from the scientific side not be answered objectively. We enter here again the game of "sense and sensibility." Consider eating an exotic fruit like a papaya. I doubt that you will ask some ecological information where this fruit is growing or that you check at an NCBI site whether the papaya genome was sequenced and whether it might contain genes that you would not like to ingest. In fact, you will probably not consider the possibility that you will incorporate the fruit's gene and that you get increasingly "Papayan" when declaring it your favorite fruit.

Gene Transfer?

However, as a sceptical scientist you should be worried because this point has not been investigated to any greater detail. The gene conferring resistance to the abovementioned herbicide glyphosate in GM soya was only recently investigated in humans suffering from an open access to the intestine due to an underlying medical problem. The resistance gene fared badly and did not survive gastrointestinal passage and did not show up in the intestinal mucosa (Netherwood et al. 2004). This is the sense argument: one would not expect that eating represents an efficient way for stable incorporation of foreign DNA into the human DNA. *The draft of the human DNA sequence showed a number of putative bacterial-like genes: however, later research demonstrated that this was bacterial DNA contaminating the human DNA during cloning and sequencing, not a genuine part of the human genome. I am not aware that anything like apple DNA showed up in the human genome. Yet this does not exclude the possibility that under rare conditions DNA from the gut content is not taken up by the mammalian host. Experiments suggested that the DNA from a bacteriophage of the gut bacterium E. coli can be taken up in small pieces by the mouse gut. With very low frequency this DNA finds its way into the genomic DNA of the liver cells of these mice. This was not a specific property of the phage DNA, a plasmid DNA did it, too (Schubbert et al. 1997). To my knowledge these data have not been repeated by another laboratory.*

Sensitivity Against Sense

Let's admit for pure discussion purpose that it occurs. Does it matter? Are we not anyway constantly bombarded by viruses that seek access to our genome—our genome sequence is the best witness for how successful viruses were in this respect. Without too much exaggeration one can say that the essence of biology is gene transfer, vertical gene transfer via sexuality, but also horizontal gene transfer. F. Bushman starts his excellent book "Lateral Gene Transfer" with the breathtaking opening: "Given the very high concentrations of bacteria and viruses in seawater, and the tremendous volume of water in the ocean, it follows that gene transfer between organisms takes place about 20 million billion times per second in the ocean." Some of these transfers actually matter as pathogens like *V. cholerae* were created in this way. *However, do the single defined genes in the GM crops have the potential to cause harm? Or do we cultivate only our "sensitivity" instead of our "sense" by maintaining an outmoded biological philosophy of created, immutable species exempt of gene transfer processes? In my opinion, we need a lot more "sense" especially the old-fashioned "British common sense" in this discussion.*

The Case of the Mexican Pollen

A *Nature* paper from 2001 reported the presence of transgenic DNA constructs in native landraces grown in remote mountains in Mexico. The authors went even further by reporting a high frequency of transgene insertions into a diversity of genomic contexts of the maize plants. In a way this is the complement to the mouse-phage/plasmid DNA experiments. They noted that the result is more striking as a moratorium on the planting of transgenic maize had been imposed in 1998 in Mexico (Quist and Chapela 2001). The paper was a bombshell in the international discussion. Science is a long argument with thesis and antithesis. The inverse PCR methodology used by these authors soon came under fire and *Nature* stated later that it would not have published the paper if the criticism had cropped up while the paper was under review (Marris 2005). A repeat survey testing 150,000 Mexican maize seeds from 2003 and 2004 was done and found no transgenes at all (Ortiz-Garcia et al. 2005). One report might suffer technical problems (probably the first), but it is also possible (but less likely) that both are right and the transgenes were lost over the four seasons separating the two samplings. Yet this would mean that the initially reported integrations into the genome were highly unstable and readily lost. On an evolutionary time scale this is anyway the fate of most laterally transferred DNA. *However, under the line this is again a case where big conclusions were based on small data. One should look twice before getting nervous, but when looking into the public discussion the impression is that some like it hot and want to get nervous as a type of personal philosophy even before looking twice. Here again philosophy matters for everyday life to settle the old conflict between gut feeling versus brain analysis. Or we need more of the Latin word "sine studio et ira" (without zeal and anger) for our actions and feelings.*

A Story Without End?

A Cross-road

It was the best of times, it was the worst of times, it was the period where humans rose to dominance over the Earth's ecosystem, and it was the final period of rapid agricultural growth. It was the age of wisdom, where more scientists worked than in all periods of human history before combined. It was the age of foolishness where wars and genocides claimed in a single century more than 100 million human lives. It was an epoch of belief than human reason could lead us to economic management of our environment. It was the epoch of incredulity into the rational nature of humanity, which explored the carriage capacity of the planet by unsustainable exploitation of the ecosystem services. It was the season of Light—never before had so many human beings lived in unseen prosperity in the western world, that physicians face too much of food intake as a major scourge in public health, it was the season of Darkness, where so many people were undernourished and in poverty as never before in human history when looking to the absolute numbers. Whether we had everything before us, whether we had nothing before us, spring of hope and winter of despair as the next signpost, we do not know. Ecological predictions are as difficult as economical predictions. When these paradoxes were written down in 1870 by Charles Dickens (exceptionally not Charles Darwin) he projected them back in time to the period of the French Revolution; with equal justification he could have projected them into the future at the time of the end of the Green Revolution. Whether we like it or not, we live in a period of crisis, in the sense of the Greek theater where the plot turns to the good or to the bad side and both options are possible. There are numerous objective signs for this perception.

Human Resource Management

If you take a number of scientific reports over the last years the message is always the same, even the titles are interchangeable. One report reads "Human appropriation of renewable fresh water" (Postel et al. 1996). It offers a difficult-to-read account packed with numbers that are intended to provide an overall budget of fresh water use in our planet. The numbers are not so precise as they appear because the methods to approximate them are inherently imprecise. However, the take-home message is clear: *Homo sapiens* is using something like a quarter of this most crucial support system of terrestrial life. The next report reads "Human appropriation of photosynthesis products" (Rojstaczer et al. 2001). The authors try an approximation of the terrestrial net primary production, which represents the net energy (production minus respiration) created by carbon fixation on land. It provides a long table packed with numbers and the precise figures cannot hide the great uncertainties that affected the determination of many values. They then asked the question how much of these products are used by humans and again their best estimate comes close to the one-quarter value, which we met for the fresh water use. Of course, the imprecise contributing figures give great error bars for the final calculation, which range from 10 to 55% human use of photosynthesis. The lower and the upper ranges of the estimates determine whether we speak of a substantial use of the primary productivity by a single species or an alarming situation. Our grip on the marine ecosystem seems less complete: "only" 8% of the primary production is used by humans. However, this figure underestimates our capacities. In areas which can easily be reached, like the continental shelf system, our share is again up to 24–35% (Pauly and Christensen 1995). The idea to harvest without sowing was too attractive to humanity and is vividly expressed by the fact that about 60% of the world population lives within 100 km from the coast.

Humans as the Dominant Evolutionary Force

Our domination of the Earth's ecosystems can be expressed in many other figures, from land transformation to profound alterations of biogeochemical cycles (Vitousek et al. 1997). Human activities have driven an estimated one third of the bird species to extinction, while 18% of the mammals are threatened by extinction. We are also the cause of an impressive loss of genetic variation and are homogenizing biodiversity by introducing alien species in many ecosystems. Humans are now the world's greatest evolutionary force: we impose strong selections on bacteria, pests, and weeds by using huge amounts of antibiotics, pesticides, and herbicides (Palumbi 2001). And we have not yet reached the height of our impact: the need to feed 9 billion people in 2050 will enormously strain the carriage capacity of our planet: the forecasts of the agriculturally driven global environmental changes (Tilman et al. 2001) are such that we cannot expect to get through this period with the means of the Green Revolution without a major loss of biodiversity.

Biodiversity

The concern of conservation biology is not about the fancy objects of some crazy birdwatchers. We should not forget that the most striking feature on earth is the existence of life. For the moment we do not know whether life is a cosmic orphan. The most striking feature of life is its diversity. We do not know whether this breathtaking variety is just a caprice of Mother Nature, who likes to impress with her exuberant creativity. We must admit that we do not yet understand the reason why Nature issues so many tickets for different organisms inhabiting the same niche—we saw it when discussing gut commensal or grassland ecology. Many biologists suspect that biodiversity has something to do with sustainability, stability, and resilience of ecosystems (Tilman 2000). We are thus about to reduce biodiversity—and once extinct these organisms are gone for ever—before we even have a primitive understanding of its meaning. Many organisms will be gone before scientists even had a chance to detect and describe them. Perhaps we cannot run a sustainable ecosystem, which provides all ecosystem services on which human survival depends, without biological diversity—we saw several agricultural examples in our survey. If you deal with a complex system that you do not understand, caution might be a good advisor.

Esthetics...

Our attitude toward the ecosystems on earth has also to do with esthetics in the philosophical sense. There is a surprising unity in the perception of beauty: flowers that evolved forms, colors, and scents for pollinators are also perceived as beautiful by humans. What female birds prefer on the plumage of their male mates, also appeals to our eyes. Beauty is not a priceless extra for those who have satisfied all other basic instincts of life. A life without the beauty of the surrounding nature in all its diversity might not be worth living.

... And Ethics

Our position in the global ecosystem also has a lot to do with ethics. Ethics anticipates reason a free will, something that has for the moment only evolved in human beings. We should have the courage to realize that some commandments venerated in the Judeo-Christian world like "be fruitful and multiply" and "fill the earth and conquer it" are not divine commands, but simply the biological rules of the selfish gene. Thanks to our superior intellectual capacity that has no pair in the biological world, we grew to world power and are now the master of our fellow creatures. As we are part of the creation and as we depend on the creation for our own survival, we have to develop a responsibility for life and not only for our own species. By developing this responsibility we will probably also best assure the survival of our species, so it will also be a type of self-respect when we respect biodiversity. I do not like the term "bioethic" as if a human ethic, an animal ethic, and so forth existed. Biology has no ethic because most organisms lack the capacity for ethical reactions. We as humans have this capacity and we are obliged to use this capacity. The *"sapere aude"* (dare to think) of Immanuel Kant does not only encourage us to use our reason to orient us in the world. This reason should also tell us that a number of ethical commitments of the major religious beliefs are important rules, but they need to be extended. As the Christian ethics went from "Love thy neighbor" to "Love the foreigner," we should now include the whole creation into that ethic. This has nothing to do with the love for your pet animal, which is fine as long as you do not exaggerate it; this has also nothing to do with joining a group of activists against animal experiments in the laboratory or against the slaughtering of animals for human food. I do not think that you can individualize your ethical commitment toward an individual animal. You would then logically have to condemn the feeding mode of any carnivore. However, I do think that we have an ethical obligation not to destroy ecosystems in our quest for food. I read a nice word play that describes this idea in its shortest form: "The Value of Nature and the Nature of Value" (Daily et al. 2000). The article is more on economics than ethics as it calls for a real market price for ecosystem services, which are at the moment taken as free. A realistic price for extracting water or O_2 or adding CO_2 or nitrate, ammonium or phosphorus into the global cycle must be added to the calculation of economical goods if we want to realize what economical practices are sustainable. We definitely need a new valuation system that guides us in our decision making, otherwise higher life forms derived from mammals or birds or whatever animal might not enjoy a bright future in the few billions of years that the nuclear reactor in the all-nourishing sun has still reserved for biological evolution on our planet.

References

Aanen, D. K., P. Eggleton, C. Rouland-Lefevre, T. Guldberg-Froslev, S. Rosendahl, and J. J. Boomsma. 2002. The evolution of fungus-growing termites and their mutualistic fungal symbionts. Proc. Natl. Acad. Sci. U. S. A **99**:14887–14892.

Abrahams, J. P., A. G. Leslie, R. Lutter, and J. E. Walker. 1994. Structure at 2.8 A resolution of F1-ATPase from bovine heart mitochondria. Nature **370**:621–628.

Abram, N. J., M. K. Gagan, M. T. McCulloch, J. Chappell, and W. S. Hantoro. 2003. Coral reef death during the 1997 Indian Ocean Dipole linked to Indonesian wildfires. Science **301**:952–955.

Adolphs, R., D. Tranel, and A. R. Damasio. 1998. The human amygdala in social judgment. Nature **393**:470–474.

Aguzzi, A. and M. Heikenwalder. 2003. Prion diseases: Cannibals and garbage piles. Nature **423**:127–129.

Ahlberg, P. E. and J. A. Clack. 2006. Palaeontology: a firm step from water to land. Nature **440**:747–749.

Ahlberg, P. E. and A. R. Milner. 1994. The origin and early diversification of tetrapods. Nature **368**:507–514.

Ahlberg, P. E., J. A. Clack and H. Blom. 2005. The axial skeleton of the Devonian tetrapod Ichthyostega. Nature **437**:137–140.

Aldhous, P. and S. Tomlin. 2005. Avian flu special: avian flu: are we ready? Nature **435**:399.

Alexander, R. M. N. 1996. Tyrannosaurus on the run. Nature **379**:121.

Almaas, E., B. Kovacs, T. Vicsek, Z. N. Oltvai, and A. L. Barabasi. 2004. Global organization of metabolic fluxes in the bacterium *Escherichia coli*. Nature **427**: 839–843.

Anbar, A. D. and A. H. Knoll. 2002. Proterozoic ocean chemistry and evolution: a bioinorganic bridge? Science **297**:1137–1142.

Anderson, C. L., E. U. Canning, and B. Okamura. 1998. A triploblast origin for Myxozoa? Nature **392**:346–347.

Anderson, D. M. and B. A. Keafer. 1987. An endogenous annual clock in the toxic marine dinoflagellate *Gonyaulax tamarensis*. Nature **325**:616–617.

Anderson, R. T., F. H. Chapelle, and D. R. Lovley. 1998. Evidence against hydrogen-based microbial ecosystems in basalt aquifers. Science **281**:976–977.

Andersson, I., S. Knight, G. Schneider, Y. Lindqvist, T. Lundqvist, C.-I. Bränden, G. H. Lorimer. 1989. Crystal structure of the active site of ribulose-bisphosphate carboxylase Nature **337**:229–234.

Andersson, S. G., A. Zomorodipour, J. O. Andersson, T. Sicheritz-Ponten, U. C. Alsmark, R. M. Podowski, A. K. Naslund, A. S. Eriksson, H. H. Winkler, and C. G. Kurland. 1998. The genome sequence of *Rickettsia prowazekii* and the origin of mitochondria. Nature **396**:133–140.

Andreoletti, O., S. Simon, C. Lacroux, N. Morel, G. Tabouret, A. Chabert, S. Lugan, F. Corbiere, P. Ferre, G. Foucras, H. Laude, F. Eychenne, J. Grassi, and F. Schelcher. 2004. PrPSc accumulation in myocytes from sheep incubating natural scrapie. Nat. Med. **10**:591–593.

Angers, R. C., S. R. Browning, T. S. Seward, C. J. Sigurdson, M. W. Miller, E. A. Hoover, and G. C. Telling. 2006. Prions in skeletal muscles of deer with chronic wasting disease. Science **311**:1117.

Apesteguia, S. and H. Zaher. 2006. A Cretaceous terrestrial snake with robust hindlimbs and a sacrum. Nature **440**:1037–1040.

Aristegui, J., C. M. Duarte, S. Agusti, M. Doval, X. A. Alvarez-Salgado, and D. A. Hansell. 2002. Dissolved organic carbon support of respiration in the dark ocean. Science **298**:1967.

Arita, M. 2004. The metabolic world of *Escherichia coli* is not small. Proc. Natl. Acad. Sci. U. S. A **101**:1543–1547.

Arlettaz, R., G. Jones, and P. A. Racey. 2001. Effect of acoustic clutter on prey detection by bats. Nature **414**:742–745.

Armbrust, E. V., J. A. Berges, C. Bowler, B. R. Green, D. Martinez, N. H. Putnam, S. Zhou, and D. S. Rokhsar. 2004. The genome of the diatom *Thalassiosira* pseudonana: ecology, evolution, and metabolism. metabolism. Science **306**:79–86.

Arnold, K. and K. Zuberbuhler. 2006. Language evolution: semantic combinations in primate calls. Nature **441**:303.

Arrigo, K. R. 2005. Marine microorganisms and global nutrient cycles. Nature **437**: 349–355.

Arrow, K. J., H. Gelband, and D. T. Jamison. 2005. Making antimalarial agents available in Africa. N. Engl. J. Med. **353**:333–335.

Artis, D., M. L. Wang, S. A. Keilbaugh, W. He, M. Brenes, G. P. Swain, P. A. Knight, D. D. Donaldson, M. A. Lazar, H. R. Miller, G. A. Schad, P. Scott, and G. D. Wu. 2004. RELMbeta/FIZZ2 is a goblet cell-specific immune-effector molecule in the gastrointestinal tract. Proc. Natl. Acad. Sci. U. S. A **101**:13596–13600.

Asfaw, B., Y. Beyene, G. Suwa, R. C. Walter, T. D. White, G. WoldeGabriel, and T. Yemane. 1992. The earliest Acheulean from Konso-Gardula. Nature **360**:732–735.

Asfaw, B., W. H. Gilbert, Y. Beyene, W. K. Hart, P. R. Renne, G. WoldeGabriel, E. S. Vrba, and T. D. White. 2002. Remains of *Homo erectus* from Bouri, Middle Awash, Ethiopia. Nature **416**:317–320.

Ashida, H., Y. Saito, C. Kojima, K. Kobayashi, N. Ogasawara, and A. Yokota. 2003. A functional link between RuBisCO-like protein of Bacillus and photosynthetic RuBisCO. Science **302**:286–290.

Au, W. W. and K. J. Benoit-Bird. 2003. Automatic gain control in the echolocation system of dolphins. Nature **423**:861–863.

Babcock, G. T. and M. Wikstrom. 1992. Oxygen activation and the conservation of energy in cell respiration. Nature **356**:301–309.

Backhed, F., H. Ding, T. Wang, L. V. Hooper, G. Y. Koh, A. Nagy, C. F. Semenkovich, and J. I. Gordon. 2004. The gut microbiota as an environmental factor that regulates fat storage. Proc. Natl. Acad. Sci. U. S. A **101**:15718–15723.

Bahatyrova, S., R. N. Frese, C. A. Siebert, J. D. Olsen, K. O. Van Der Werf, R. Van Grondelle, R. A. Niederman, P. A. Bullough, C. Otto, and C. N. Hunter. 2004. The native architecture of a photosynthetic membrane. Nature **430**:1058–1062.

Bahn, P. G. 1998. Neanderthals emancipated. Nature **394**:719, 721.

Bai, Y., X. Han, J. Wu, Z. Chen, and L. Li. 2004. Ecosystem stability and compensatory effects in the Inner Mongolia grassland. Nature **431**:181–184.

Bailey, J. V., S. B. Joye, K. M: Kalanetra, B. E. Flood, and F. A. Corsetti. 2007. Evidence of giant sulphur bacteria in Neoproterozoic phosphorites. Nature **445**:198–201.

Baker, A. C. 2001. Ecosystems: Reef corals bleach to survive change. Nature **411**: 765–766.

Baker, B., P. Zambryski, B. Staskawicz, and S. P. Dinesh-Kumar. 1997. Signaling in plant-microbe interactions. Science **276**:726–733.

Ball, J. M., P. Tian, C. Q. Zeng, A. P. Morris, and M. K. Estes. 1996. Age-dependent diarrhea induced by a rotaviral nonstructural glycoprotein. Science **272**:101–104.

Balter, M. 2004. Search for the Indo-Europeans. Science **303**:1323.

Balter, M. 2005. Evolution. Ancient DNA yields clues to the puzzle of European origins. Science **310**:964–965.

Ban, N., P. Nissen, J. Hansen, P. B. Moore, and T. A. Steitz. 2000. The complete atomic structure of the large ribosomal subunit at 2.4 A resolution. Science **289**:905–920.

Barasch, J. and K. Mori. 2004. Cell biology: iron thievery. Nature **432**:811–813.

Barbeau, K., E. L. Rue, K. W. Bruland, and A. Butler. 2001. Photochemical cycling of iron in the surface ocean mediated by microbial iron(III)-binding ligands. Nature **413**:409–413.

Bardgett, R. D., R. S. Smith, R. S. Shiel, S. Peacock, J. M. Simkin, H. Quirk, and P. J. Hobbs. 2006. Parasitic plants indirectly regulate below-ground properties in grassland ecosystems. Nature **439**:969–972.

Barel, C.D.N., R. Dorit, P.H. Greenwood, G. Fryer, N. Hughes, P.B.N. Jackson, H. Kawanabe, R.H. Lowe-McConnell, M. Nagoshi, A.J. Ribbink, E. Trewavas, F. Witte, and K. Yamaoka. 1985. Destruction of fisheries in Africa's lakes. Nature **315**:19–20.

Barluenga, M., K. N. Stolting, W. Salzburger, M. Muschick, and A. Meyer. 2006. Sympatric speciation in Nicaraguan crater lake cichlid fish. Nature **439**:719–723.

Bartlett, D. H. and F. Azam. 2005. Microbiology. Chitin, cholera, and competence. Science **310**:1775–1777.

Bates, T. S., R. J. Charlson, and R. H. Gammon. 1987. Evidence for the climatic role of marine biogenic sulphur. Nature **329**:319–321.

Battistuzzi, F. U., A. Feijao, and S. B. Hedges. 2004. A genomic timescale of prokaryote evolution: insights into the origin of methanogenesis, phototrophy, and the colonization of land. BMC Evolut. Biol. **4**:44.

Baumiller, T. K. and F. J. Gahn. 2004. Testing predator-driven evolution with Paleozoic crinoid arm regeneration. Science **305**:1453–1455.

Baysal, B. E., R. E. Ferrell, J. E. Willett-Brozick, E. C. Lawrence, D. Myssiorek, A. Bosch, M. A. van der, P. E. Taschner, W. S. Rubinstein, E. N. Myers, C. W. Richard, III, C. J. Cornelisse, P. Devilee, and B. Devlin . 2000. Mutations in SDHD, a mitochondrial complex II gene, in hereditary paraganglioma. Science **287**:848–851.

Beatty, J. T., J. Overmann, M. T. Lince, A. K. Manske, A. S. Lang, R. E. Blankenship, C. L. Van Dover, T. A. Martinson, and F. G. Plumley. 2005. An obligately photosynthetic bacterial anaerobe from a deep-sea hydrothermal vent. Proc. Natl. Acad. Sci. U. S. A. **102**:9306–9310.

Beaugrand, G., P. C. Reid, F. Ibanez, J. A. Lindley, and M. Edwards. 2002. Reorganization of North Atlantic marine copepod biodiversity and climate. Science **296**:1692–1694.

Beaugrand, G., K. M. Brander, L. J. Alistair, S. Souissi, and P. C. Reid. 2003. Plankton effect on cod recruitment in the North Sea. Nature **426**:661–664.

Becker, D., M. Selbach, C. Rollenhagen, M. Ballmaier, T. F. Meyer, M. Mann, and D. Bumann. 2006. Robust Salmonella metabolism limits possibilities for new antimicrobials. Nature **440**:303–307.

Becks, L., F. M. Hilker, H. Malchow, K. Jurgens, and H. Arndt. 2005. Experimental demonstration of chaos in a microbial food web. Nature **435**:1226–1229.

Behrenfeld, M. J., A. J. Bale, Z. S. Kolber, J. Aiken, and P. G. Falkowski. 1996. Confirmation of iron limitation of phytoplankton photosynthesis in the equatorial Pacific Ocean. Nature **383**: 508–511.

Beja, O., L. Aravind, E. V. Koonin, M. T. Suzuki, A. Hadd, L. P. Nguyen, S. B. Jovanovich, C. M. Gates, R. A. Feldman, J. L. Spudich, E. N. Spudich, and E. F. DeLong. 2000. Bacterial rhodopsin: evidence for a new type of phototrophy in the sea. Science **289**:1902–1906.

Beja, O., E. N. Spudich, J. L. Spudich, M. Leclerc, and E. F. DeLong. 2001. Proteorhodopsin phototrophy in the ocean. Nature **411**:786–789.

Beja-Pereira, A., G. Luikart, P. R. England, and G. Erhardt. 2003. Gene-culture coevolution between cattle milk protein genes and human lactase genes. Nat. Genet. **35**: 311–313.

Bekker, A., H. D. Holland, P. L. Wang, D. Rumble, III, H. J. Stein, J. L. Hannah, L. L. Coetzee, and N. J. Beukes. 2004. Dating the rise of atmospheric oxygen. Nature **427**:117–120.

Belevich, I., M. I. Verkhovsky, and M. Wikstrom. 2006. Proton-coupled electron transfer drives the proton pump of cytochrome c oxidase. Nature **440**:829–832.

Bengston, S., and Z. Yue. 1992. Predatorial borings in Late Precambrian mineralized exoskeletons. Science **257**:367–369.

Bengtson, S., and Y. Zhao. 1992. Predatorial borings in Late Precambrian mineralized exoskeletons. Science **257**:367–369.

Ben Shahar, Y., A. Robichon, M. B. Sokolowski, and G. E. Robinson. 2002. Influence of gene action across different time scales on behavior. Science **296**:741–744.

Benner, S. A., A. D. Ellington, and A. Tauer. 1989. Modern metabolism as a palimpsest of the RNA world. Proc. Natl. Acad. Sci. U. S. A **86**:7054–7058.

Benson, S. D., J. K. Bamford, D. H. Bamford, and R. M. Burnett. 2004. Does common architecture reveal a viral lineage spanning all three domains of life? Mol. Cell **16**: 673–685.

Benson, S. D., J. K. Bamford, D. H. Bamford, and R. M. Burnett. 1999. Viral evolution revealed by bacteriophage PRD1 and human adenovirus coat protein structures. Cell **98**:825–833.

Beres, S. B., G. L. Sylva, K. D. Barbian, B. Lei, and J. M. Musser. 2002. Genome sequence of a serotype M3 strain of group A Streptococcus: Phage-encoded toxins, the high-virulence phenotype, and clone emergence. PNAS **99**:10078–10083.

Bergh, O., K. Y. Borsheim, G. Bratbak, and M. Heldal. 1989. High abundance of viruses found in aquatic environments. Nature **340**:467–468.

Bersaglieri, T., P. C. Sabeti, N. Patterson, T. Vanderploeg, S. F. Schaffner, J. A. Drake, M. Rhodes, D. E. Reich, and J. N. Hirschhorn. 2004. Genetic signatures of strong recent positive selection at the lactase gene. Am. J. Hum. Genet. **74**:1111–1120.

Bettstetter, M., X. Peng, R. A. Garrett, and D. Prangishvili. 2003. AFV1, a novel virus infecting hyperthermophilic archaea of the genus acidianus. Virology 315:68–79.

Bibby, T. S., J. Nield, and J. Barber. 2001a. Iron deficiency induces the formation of an antenna ring around trimeric photosystem I in cyanobacteria. Nature 412:743–745.

Bibby, T. S., J. Nield, F. Partensky, and J. Barber. 2001b. Oxyphotobacteria. Antenna ring around photosystem I. Nature 413:590.

Bibby, T. S., I. Mary, J. Nield, F. Partensky, and J. Barber. 2003. Low-light-adapted *Prochlorococcus* species possess specific antennae for each photosystem. Nature 424:1051–1054.

Bibikov, S. I., L. A. Barnes, Y. Gitin, and J. S. Parkinson. 2000. Domain organization and flavin adenine dinucleotide-binding determinants in the aerotaxis signal transducer Aer of *Escherichia coli*. Proc. Natl. Acad. Sci. U. S. A 97:5830–5835.

Bina, J., J. Zhu, M. Dziejman, S. Faruque, S. Calderwood, and J. Mekalanos. 2003. ToxR regulon of *Vibrio cholerae* and its expression in vibrios shed by cholera patients. Proc. Natl. Acad. Sci. U. S. A 100:2801–2806.

Bishop, J. K., T. J. Wood, R. E. Davis, and J. T. Sherman. 2004. Robotic observations of enhanced carbon biomass and export at 55 degrees during SOFeX. Science 304:417–420.

Blanford, S., B. H. Chan, N. Jenkins, D. Sim, R. J. Turner, A. F. Read, and M. B. Thomas. 2005. Fungal pathogen reduces potential for malaria transmission. Science 308:1638–1641.

Blank, R. J., and R. K. Trench. 1985. Speciation and symbiotic dinoflagellates. Science 229:656–658.

Block, B. A., H. Dewar, S. B. Blackwell, T. D. Williams, E. D. Prince, C. J. Farwell, A. Boustany, S. L. Teo, A. Seitz, A. Walli, and D. Fudge. 2001. Migratory movements, depth preferences, and thermal biology of Atlantic bluefin tuna. Science 293:1310–1314.

Block, B. A., S. L. Teo, A. Walli, A. Boustany, M. J. Stokesbury, C. J. Farwell, K. C. Weng, H. Dewar, and T. D. Williams. 2005. Electronic tagging and population structure of Atlantic bluefin tuna. Nature 434:1121–1127.

Block, S. M. 1997. Real engines of creation. Nature 386:217–219.

Bodelier, P. L., P. Roslev, T. Henckel, and P. Frenzel. 2000. Stimulation by ammonium-based fertilizers of methane oxidation in soil around rice roots. Nature 403:421–424.

Boëda, E., J. Connan, D. Dessort, S. Muhesen, N. Mercier, H. Valladas, and N. Tisnérat. 1996. Bitumen as a hafting material on Middle Palaeolithic artefacts. Nature 380:336–338.

Boekema, E. J., A. Hifney, A. E. Yakushevska, M. Piotrowski, W. Keegstra, S. Berry, K. P. Michel, E. K. Pistorius, and J. Kruip. 2001. A giant chlorophyll-protein complex induced by iron deficiency in cyanobacteria. Nature 412:745–748.

Boetius, A., K. Ravenschlag, C. J. Schubert, D. Rickert, F. Widdel, A. Gieseke, R. Amann, B. B. Jorgensen, U. Witte, and O. Pfannkuche. 2000. A marine microbial consortium apparently mediating anaerobic oxidation of methane. Nature 407:623–626.

Boisvert, C. A. 2005. The pelvic fin and girdle of Panderichthys and the origin of tetrapod locomotion. Nature 438 :1145–1147.

Boklage, C. E. 1997. Sex ratio unaffected by parental age gap. Nature 390:243.

Bordone, L. and L. Guarente. 2005. Calorie restriction, SIRT1 and metabolism: understanding longevity. Nat. Rev. Mol. Cell Biol. 6:298–305.

Bowles, S. and H. Gintis. 2002. Homo reciprocans. Nature 415:125–128.

Bowles, S. and D. Posel. 2005. Genetic relatedness predicts South African migrant workers' remittances to their families. Nature **434**:380–383.

Boxma, B., R. M. de Graaf, G. W. van der Staay, T. A. van Alen, G. Ricard, T. Gabaldon, A. H. van Hoek, Moon-van der Staay SY, W. J. Koopman, J. J. van Hellemond, A. G. Tielens, T. Friedrich, M. Veenhuis, M. A. Huynen, and J. H. Hackstein. 2005. An anaerobic mitochondrion that produces hydrogen. Nature **434**:74–79.

Boyd, P. W., A. J. Watson, C. S. Law, E. R. Abraham, and J. Zeldis. 2000. A mesoscale phytoplankton bloom in the polar Southern Ocean stimulated by iron fertilization. Nature **407**:695–702.

Boyd, P. W., C. S. Law, C. S. Wong, Y. Nojiri, A. Tsuda, M. Levasseur, S. Takeda, and T. Yoshimura. 2004. The decline and fate of an iron-induced subarctic phytoplankton bloom. Nature **428**:549–553.

Boyer, P.D. 1993. The binding change mechanism for ATP synthase–some probabilities and possibilities. Biochim. Biophys. Acta. **1140**:215–250.

Brainerd, E. L. 2001. Caught in the crossflow. Nature **412**:387–388.

Brashares, J. S., P. Arcese, M. K. Sam, P. B. Coppolillo, A. R. Sinclair, and A. Balmford. 2004. Bushmeat hunting, wildlife declines, and fish supply in West Africa. Science **306**:1180–1183.

Brasier, M. and J. Antcliffe. 2004. Paleobiology. Decoding the Ediacaran enigma. Science **305**:1115–1117.

Brasier, M. D., O. R. Green, A. P. Jephcoat, A. K. Kleppe, M. J. Van Kranendonk, J. F. Lindsay, A. Steele, and N. V. Grassineau. 2002. Questioning the evidence for Earth's oldest fossils. Nature **416**:76–81.

Bray, D. 1995. Protein molecules as computational elements in living cells. Nature **376**:307–312.

Brazeau, M. D. and P. E. Ahlberg. 2006. Tetrapod-like middle ear architecture in a Devonian fish. Nature **439**:318–321.

Breitbart, M., P. Salamon, B. Andresen, J. M. Mahaffy, A. M. Segall, D. Mead, F. Azam, and F. Rohwer. 2002. Genomic analysis of uncultured marine viral communities. Proc. Natl. Acad. Sci. U. S. A **99**:14250–14255.

Bricelj, V. M., L. Connell, K. Konoki, S. P. Macquarrie, T. Scheuer, W. A. Catterall, and V. L. Trainer. 2005. Sodium channel mutation leading to saxitoxin resistance in clams increases risk of PSP. Nature **434**:763–767.

Bricelj, V. M., L. Connell, K. Konoki, S. P. Macquarrie, T. Scheuer, W. A. Catterall, and V. L. Trainer. 2005. Sodium channel mutation leading to saxitoxin resistance in clams increases risk of PSP. Nature **434**:763–767.

Brierley, A. S., P. G. Fernandes, M. A. Brandon, F. Armstrong, N. W. Millard, S. D. McPhail, P. Stevenson, M. Pebody, J. Perrett, M. Squires, D. G. Bone, and G. Griffiths. 2002. Antarctic krill under sea ice: elevated abundance in a narrow band just south of ice edge. Science **295**:1890–1892.

Britten, R. J. and D. E. Kohne. 1968. Repeated sequences in DNA. Hundreds of thousands of copies of DNA sequences have been incorporated into the genomes of higher organisms. Science **161**:529–540.

Brochier, C. and H. Philippe. 2002. Phylogeny: a non-hyperthermophilic ancestor for bacteria. Nature **417**:244.

Brocks, J. J., G. A. Logan, R. Buick, and R. E. Summons. 1999. Archean molecular fossils and the early rise of eukaryotes. Science **285**:1033–1036.

Brosnan, S. F. and F. B. de Waal. 2003. Monkeys reject unequal pay. Nature **425**:297–299.

Brown, P., T. Sutikna, M. J. Morwood, R. P. Soejono, Jatmiko, E. W. Saptomo, and R. A. Due. 2004. A new small-bodied hominin from the Late Pleistocene of Flores, Indonesia. Nature **431**:1055–1061.

Broza, M. and M. Halpern. 2001. Pathogen reservoirs. Chironomid egg masses and *Vibrio cholerae*. Nature **412**:40.

Brumm, A., A. Fachroel, G. D. van den Bergh, M. J. Morwood, M. W. Moore, I. Kurniawan, D. R. Hobbs, and R. Fullagar. 2006. Early stone technology on Flores and its implications for *Homo floresiensis*. Nature **441**:624–628.

Brunet, M., F. Guy, D. Pilbeam, H. T. Mackaye, A. Likius, D. Ahounta, A. Beauvilain, and C. Zollikofer. 2002. A new hominid from the Upper Miocene of Chad, Central Africa. Nature **418**:145–151.

Brüssow, H. and R. W. Hendrix. 2002. Phage genomics: small is beautiful. Cell **108**: 13–16.

Brüssow, H. 2005. Phage therapy: the *Escherichia coli* experience. Microbiology **151**:2133–2140.

Brüssow, H., C. Canchaya, and W. D. Hardt. 2004. Phages and the evolution of bacterial pathogens: from genomic rearrangements to lysogenic conversion. Microbiol. Mol. Biol. Rev. **68**:560–602, table.

Brüssow, H., O. Nakagomi, G. Gerna, and W. Eichhorn. 1992. Isolation of an avianlike group A rotavirus from a calf with diarrhea. J. Clin. Microbiol. **30**:67–73.

Bry, L., P. G. Falk, T. Midtvedt, and J. I. Gordon. 1996. A model of host-microbial interactions in an open mammalian ecosystem. Science **273**:1380–1383.

Bryant, D. A. 2003. The beauty in small things revealed. Proc. Natl. Acad. Sci. U. S. A **100**:9647–9649.

Bshary, R., and A.S. Grutter. 2006. Image scoring and cooperation in a cleaner fish mutualism. Nature **441**:975–978.

Budiansky, S. 2002. Creatures of our own making. Science **298**:80–86.

Buesseler, K. O., J. E. Andrews, S. M. Pike, and M. A. Charette. 2004. The effects of iron fertilization on carbon sequestration in the Southern Ocean. Science **304**:414–417.

Buhl, J., D.J.T. Sumpter, I.D. Couzin, J.J. Hale, E. Despland, E.R. Miller, and S. J. Simpson. 2006. From disorder to order in marching locusts. Science **312**:1402–1406.

Bult, C. J., O. White, G. J. Olsen, L. Zhou, R. D. Fleischmann, G. G. Sutton, J. A. Blake, and J. C. Venter. 1996. Complete genome sequence of the methanogenic archaeon, Methanococcus jannaschii. Science **273**:1058–1073.

Burkholder, J. M., E. J. Noga, C. H. Hobbs, H. B. Glasgow, Jr., and S. A. Smith. 1992. New 'phantom' dinoflagellate is the causative agent of major estuarine fish kills. Nature **358**:407–410.

Burton, R. A., S. M. Wilson, M. Hrmova, A. J. Harvey, N. J. Shirley, A. Medhurst, B. A. Stone, E. J. Newbigin, A. Bacic, and G. B. Fincher . 2006. Cellulose synthase-like CslF genes mediate the synthesis of cell wall (1,3;1,4)-beta-D-glucans. Science **311**:1940–1942.

Burtsev, M. and P. Turchin. 2006. Evolution of cooperative strategies from first principles. Nature **440**:1041–1044.

Buschmann, A. and M. H. Groschup. 2005. Highly bovine spongiform encephalopathy-sensitive transgenic mice confirm the essential restriction of infectivity to the nervous system in clinically diseased cattle. J. Infect. Dis. **192**:934–942.

Butcher, S. J., T. Dokland, P. M. Ojala, D. H. Bamford, and S. D. Fuller. 1997. Intermediates in the assembly pathway of the double-stranded RNA virus phi6. EMBO J. **16**:4477–4487.

Butler, D. 2005. Bird flu: crossing borders. Nature **436**:310–311.

Buttner, M., D. L. Xie, H. Nelson, W. Pinther, G. Hauska, and N. Nelson. 1992. Photosynthetic reaction center genes in green sulfur bacteria and in photosystem 1 are related. Proc. Natl. Acad. Sci. U. S. A **89**:8135–8139.

Calderwood, S. B. and J. J. Mekalanos. 1987. Iron regulation of Shiga-like toxin expression in *Escherichia coli* is mediated by the fur locus. J. Bacteriol. **169**: 4759–4764.

Caldwell, M. W. and M. S. Y. Lee. 1997. A snake with legs from the marine Cretaceous of the Middle East. Nature **386**:705–709.

Canchaya, C., G. Fournous, and H. Brüssow. 2004. The impact of prophages on bacterial chromosomes. Mol. Microbiol. **53**:9–18.

Canchaya, C., G. Fournous, S. Chibani-Chennoufi, M. L. Dillmann, and H. Brüssow. 2003a. Phage as agents of lateral gene transfer. Curr. Opin. Microbiol. **6**:417–424.

Canchaya, C., C. Proux, G. Fournous, A. Bruttin, and H. Brüssow. 2003b. Prophage genomics. Microbiol. Mol. Biol. Rev. **67**:238–276.

Canfield, D. E. 1998. A new model for Proterozoic ocean chemistry. Nature **396**:450.

Cardillo, M., G. M. Mace, K. E. Jones, J. Bielby, O. R. Bininda-Emonds, W. Sechrest, C. D. Orme, and A. Purvis. 2005. Multiple causes of high extinction risk in large mammal species. Science **309** :1239–1241.

Cardillo, M. and A. Lister. 2002. Death in the slow lane. Nature **419**:440–441.

Carlton, J. M., S. V. Angiuoli, B. B. Suh, T. W. Kooij, M. Pertea, J. C. Silva, M. D. Ermolaeva, and D. J. Carucci. 2002. Genome sequence and comparative analysis of the model rodent malaria parasite *Plasmodium yoelii* yoelii. Nature **419**:512–519.

Carroll, R. L. 2005. Palaeontology: between water and land. Nature **437**:38–39.

Casey, J. M. and R. A. Myers. 1998. Near extinction of a large, widely distributed fish. Science **281**:690–692.

Cassman, K. G. 1999. Ecological intensification of cereal production systems: yield potential, soil quality, and precision agriculture. Proc. Natl. Acad. Sci. U. S. A **96**: 5952–5959.

Castilla, J., P. Saá, C. Hetz, and C. Soto. 2005. In Vitro Generation of Infectious Scrapie Prions. Cell **121**:195–206 .

Castresana, J. and M. Saraste. 1995. Evolution of energetic metabolism: the respiration-early hypothesis. Trends Biochem. Sci. **20**:443–448.

Catania, K. C. and F. E. Remple. 2005. Asymptotic prey profitability drives star-nosed moles to the foraging speed limit. Nature **433**:519–522.

Cavalier-Smith, T. 1987. Eukaryotes with no mitochondria. Nature **326**:332–333.

Cerdeno-Tarraga, A. M., S. Patrick, L. C. Crossman, G. Blakely, V. Abratt, N. Lennard, and J. Parkhill. 2005. Extensive DNA inversions in the *B. fragilis* genome control variable gene expression. Science **307**:1463–1465.

Chang, D. E., D. J. Smalley, D. L. Tucker, M. P. Leatham, W. E. Norris, S. J. Stevenson, A. B. Anderson, J. E. Grissom, D. C. Laux, P. S. Cohen, and T. Conway. 2004. Carbon nutrition of *Escherichia coli* in the mouse intestine. Proc. Natl. Acad. Sci. U. S. A **101**:7427–7432.

Chapela, I.H., S.A. Rehner, T.R. Schultz, and U.G. Mueller. 1994. Evolutionary history of the symbiosis between fungus-growing ants and their fungi. Science **266**:1691–1694.

Chapelle, F. H., K. O'Neill, P. M. Bradley, B. A. Methe, S. A. Ciufo, L. L. Knobel, and D. R. Lovley. 2002. A hydrogen-based subsurface microbial community dominated by methanogens. Nature **415**:312–315.

Chappellaz, J., T. Bluniert, D. Raynaud, J. M. Barnola, J. Schwander, and B. Stauffert. 1993. Synchronous changes in atmospheric CH4 and Greenland climate between 40 and 8 kyr BP. Nature **366**:443–445.

Charles, D. 2001. Agbiotech. Seeds of discontent. Science **294**:772–775.

Charlson, R. J., J. E. Lovelock, M. O. Andreae, and S. G. Warren. 1987. Oceanic phytoplankton, atmospheric sulphur, cloud albedo and climate. Nature **326**:655–661.

Chavez, F. P., J. Ryan, S. E. Lluch-Cota, and C. M. Niquen . 2003. From anchovies to sardines and back: multidecadal change in the Pacific Ocean. Science **299**:217–221.

Chen, J.-Y., J. Dzik, G. D. Edgecombe, L. Ramsköld, and G.-Q. Zhou. 1995. A possible Early Cambrian chordate. Nature **377**: 720–722.

Chen, H., G. J. Smith, S. Y. Zhang, K. Qin, J. Wang, K. S. Li, R. G. Webster, J. S. Peiris, and Y. Guan. 2005. Avian flu: H5N1 virus outbreak in migratory waterfowl. Nature **436**:191–192.

Chen, I. A., R. W. Roberts, and J. W. Szostak. 2004. The emergence of competition between model protocells. Science **305**:1474–1476.

Chiappe, L. M., A. R. Davidson, M. C. McKenna, P. Altangerel, and M. Novacek. 1994. A Theropod Dinosaur Embryo and the Affnities of the Flaming Cliffs Dinosaur Eggs. Science **266**:779–782.

Chikhi, L., R. A. Nichols, G. Barbujani, and M. A. Beaumont. 2002. Y genetic data support the Neolithic demic diffusion model. Proc. Natl. Acad. Sci. U. S. A **99**: 11008–11013.

Chin, K., T. T. Tokaryk, G. M. Erickson, L. C. Calk. 1998. A king-sized theropod coprolite. Nature **393**:680.

Chinese SARS Molecular Epidemiology Consortium. 2004. Molecular evolution of the SARS coronavirus during the course of the SARS epidemic in China. Science **303**:1666–1669.

Chisholm, S. W. 2000. Stirring times in the Southern Ocean. Nature **407**:685–687.

Chittka, L. and S. Schurkens. 2001. Successful invasion of a floral market. Nature **411**:653.

Chittka, L., A. G. Dyer, F. Bock, and A. Dornhaus. 2003. Psychophysics: bees trade off foraging speed for accuracy. Nature **424**:388.

Christophides, G. K., E. Zdobnov, C. Barillas-Mury, E. Birney, S. Blandin, C. Blass, P. T. Brey, and F. C. Kafatos. 2002. Immunity-related genes and gene families in Anopheles gambiae. Science **298**:159–165.

Cifelli, R. L. and B. M. Davis. 2003. Marsupial Origins. Science **302**:1899–1900.

Clack, J. A. 2002. An early tetrapod from 'Romer's Gap'. Nature **418**:72–76.

Clark, L. L., E. D. Ingall, and R. Brenner. 1998. Marine phosphorus is selectively remineralized. Nature **393**:426.

Clausen, J. and W. Junge. 2004. Detection of an intermediate of photosynthetic water oxidation. Nature **430**:480–483.

Clayton, N. S. and A. Dickinson. 1998. Episodic-like memory during cache recovery by scrub jays. Nature **395**:272–274.

Cliffe, L. J., N. E. Humphreys, T. E. Lane, C. S. Potten, C. Booth, and R. K. Grencis. 2005. Accelerated intestinal epithelial cell turnover: a new mechanism of parasite expulsion. Science **308**:1463–1465.

Coale, K. H., K. S. Johnson, F. P. Chavez, K. O. Buesseler, R. T. Barber, M. A. Brzezinski, and Z. I. Johnson. 2004. Southern Ocean iron enrichment experiment: carbon cycling in high- and low-Si waters. Science **304**:408–414.

Coale, K. H., K. S. Johnson, S. E. Fitzwater, R. M.l Gordon, S. Tanner, F. P. Chavez, L. Ferioli, C. Sakamoto, P. Rogers, and F. Millero. 1996. A massive phytoplankton bloom induced by an ecosystem-scale iron fertilization experiment in the equatorial Pacific Ocean. Nature **383**:495–501.

Coates, M. I. and J. A. Clack. 1990. Polydactyly in the earliest known tetrapod limbs. Nature **347**:66–69.

Coates, M. I. and J. A. Clack. 1991. Fish-like gills and breathing in the earliest known tetrapod. Nature **352**:234–236.

Coates, J. C. and M. de Bono. 2002. Antagonistic pathways in neurons exposed to body fluid regulate social feeding in *Caenorhabditis elegans*. Nature **419**:925–929.

Cody, G. D., N. Z. Boctor, T. R. Filley, R. M. Hazen, J. H. Scott, A. Sharma, and H. S. Yoder, Jr. 2000. Primordial carbonylated iron-sulfur compounds and the synthesis of pyruvate. Science **289**:1337–1340.

Cohn, M. J. 2002. Evolutionary biology: lamprey Hox genes and the origin of jaws. Nature **416**:386–387.

Coleman, M. L., M. B. Sullivan, A. C. Martiny, C. Steglich, K. Barry, E. F. DeLong, and S. W. Chisholm. 2006. Genomic islands and the ecology and evolution of Prochlorococcus. Science **311**:1768–1770.

Colwell, R. R. 1996. Global climate and infectious disease: the cholera paradigm. Science **274**:2025–2031.

Conover, D. O. and S. B. Munch. 2002. Sustaining fisheries yields over evolutionary time scales. Science **297**:94–96.

Conradt, L. and T. J. Roper. 2003. Group decision-making in animals. Nature **421**: 155–158.

Conway Morris, S. 1993. The fossil record and the early evolution of the Metazoa. Nature **361**:219–225.

Conway Morris, S. 2000. The Cambrian "explosion": slow-fuse or megatonnage? Proc. Natl. Acad. Sci. U. S. A **97**:4426–4429.

Cook, R. M., A. Sinclair, and G. Stefánsson. 1997. Potential collapse of North Sea cod stocks. Nature **385**:521–522.

Cooper, D. J., A. J. Watson, and P. D. Nightingale. 1996. Large decrease in ocean-surface $CO2$ fugacity in response to in situ iron fertilization. Nature **383**:511–513.

Coppa, A., L. Bondioli, A. Cucina, D. W. Frayer, C. Jarrige, J. F. Jarrige, G. Quivron, M. Rossi, M. Vidale, and R. Macchiarelli. 2006. Palaeontology: early Neolithic tradition of dentistry. Nature **440**:755–756.

Cordaux, R., E. Deepa, H. Vishwanathan, and M. Stoneking. 2004. Genetic evidence for the demic diffusion of agriculture to India. Science **304**:1125.

Cornish-Bowden, A., and M. L. Cárdenas. 2002. Systems biology: Metabolic balance sheets. Nature **420**:129–130.

Cote, F., D. Levesque, and J. P. Perreault. 2001. Natural 2',5'-phosphodiester bonds found at the ligation sites of peach latent mosaic viroid. J. Virol. **75**:19–25.

Cote, I. M. and K. L. Cheney. 2005. Animal mimicry: choosing when to be a cleaner-fish mimic. Nature **433**:211–212.

Couzin, I. D., J. Krause, N. R. Franks, and S. A. Levin. 2005. Effective leadership and decision-making in animal groups on the move. Nature **433**:513–516.

Covert, M. W., E. M. Knight, J. L. Reed, M. J. Herrgard, and B. O. Palsson. 2004. Integrating high-throughput and computational data elucidates bacterial networks. Nature **429**:92–96.

Coyne, M. J., K. G. Weinacht, C. M. Krinos, and L. E. Comstock. 2003. Mpi recombinase globally modulates the surface architecture of a human commensal bacterium. Proc. Natl. Acad. Sci. U. S. A **100**:10446–10451.

Coyne, M. J., B. Reinap, M. M. Lee, and L. E. Comstock. 2005. Human symbionts use a host-like pathway for surface fucosylation. Science 307:1778–1781.

Crawford, P. C., E. J. Dubovi, W. L. Castleman, I. Stephenson, E. P. Gibbs, L. Chen, and R. O. Donis. 2005. Transmission of equine influenza virus to dogs. Science 310:482–485.

Crick, F. H. 1968. The origin of the genetic code. J. Mol. Biol. 38:367–379.

Croll, D. A., J. L. Maron, J. A. Estes, E. M. Danner, and G. V. Byrd. 2005. Introduced predators transform subarctic islands from grassland to tundra. Science 307:1959–1961.

Cross, R. L. 2004. Molecular motors: turning the ATP motor. Nature 427:407–408.

Csete, M. E. and J. C. Doyle. 2002. Reverse engineering of biological complexity. Science 295:1664–1669.

Currie, C. R. 2001. A community of ants, fungi, and bacteria: a multilateral approach to studying symbiosis. Annu. Rev. Microbiol. 55:357–380.

Currie, C. R., U. G. Mueller, and D. Malloch. 1999. The agricultural pathology of ant fungus gardens. Proc. Natl. Acad. Sci. U. S. A 96:7998–8002.

Currie, C. R., M. Poulsen, J. Mendenhall, J. J. Boomsma, and J. Billen. 2006. Coevolved crypts and exocrine glands support mutualistic bacteria in fungus-growing ants. Science 311:81–83.

Currie, C. R., J. A. Scott, R. C. Summerbell, and D. Malloch. 1999. Fungus-growing ants use antibiotic-producing bacteria to control garden parasites. Nature 398:701–704.

Currie, C. R., B. Wong, A. E. Stuart, T. R. Schultz, S. A. Rehner, U. G. Mueller, G. H. Sung, J. W. Spatafora, and N. A. Straus. 2003. Ancient tripartite coevolution in the attine ant-microbe symbiosis. Science 299:386–388.

Suttle,C.A., A.M. Chan, and M.T. Cottrell. 1990. Infection of phytoplankton by viruses and reduction of primary productivity. Nature 347:467–469.

Dacey, J.W.H., and S.G. Wakeham. 1986. Oceanic dimethylsulfide: production during zooplankton grazing on phytoplankton. Science 233: 1314–1316.

Dacks, J. B. and W. F. Doolittle. 2001. Reconstructing/deconstructing the earliest eukaryotes: how comparative genomics can help. Cell 107:419–425.

Daeschler, E. B., N. H. Shubin, and F. A. Jenkins, Jr. 2006. A Devonian tetrapod-like fish and the evolution of the tetrapod body plan. Nature 440:757–763.

Daffonchio, D., S. Borin, T. Brusa, L. Brusetti, P. W. van der Wielen, H. Bolhuis, M. M. Yakimov, and S. Hoog. 2006. Stratified prokaryote network in the oxic-anoxic transition of a deep-sea halocline. Nature 440:203–207.

Daily, G. C., T. Soderqvist, S. Aniyar, K. Arrow, P. Dasgupta, P. R. Ehrlich, C. Folke, A. Jansson, B. Jansson, N. Kautsky, S. Levin, J. Lubchenco, K. G. Maler, D. Simpson, D. Starrett, D. Tilman, and B. Walker. 2000. Ecology. The value of nature and the nature of value. Science 289:395–396.

Dally, J.M., N. J. Emery, and N.S. Clayton. 2006. Food-caching western scrub-jays keep track of who was watching when. Science 312:1662–1665.

Dalton, R. 2002. Squaring up over ancient life. Nature 417:782–784.

Damste, J. S., G. Muyzer, B. Abbas, S. W. Rampen, G. Masse, W. G. Allard, S. T. Belt, J. M. Robert, S. J. Rowland, J. M. Moldowan, S. M. Barbanti, F. J. Fago, P. Denisevich, J. Dahl, L. A. Trindade, and S. Schouten. 2004. The rise of the rhizosolenid diatoms. Science 304:584–587.

Danial, N. N., C. F. Gramm, L. Scorrano, C. Y. Zhang, S. Krauss, A. M. Ranger, S. R. Datta, M. E. Greenberg, L. J. Licklider, B. B. Lowell, S. P. Gygi, and S. J. Korsmeyer. 2003. BAD and glucokinase reside in a mitochondrial complex that integrates glycolysis and apoptosis. Nature 424:952–956.

Daros, J. A. and R. Flores. 2004. Arabidopsis thaliana has the enzymatic machinery for replicating representative viroid species of the family Pospiviroidae. Proc. Natl. Acad. Sci. U. S. A **101**:6792–6797.

Darst, C. R. and M. E. Cummings. 2006. Predator learning favours mimicry of a less-toxic model in poison frogs. Nature **440**:208–211.

Davidson, D. W., S. C. Cook, R. R. Snelling, and T. H. Chua. 2003. Explaining the abundance of ants in lowland tropical rainforest canopies. Science **300**:969–972.

Davis, B. M., E. H. Lawson, M. Sandkvist, A. Ali, S. Sozhamannan, and M. K. Waldor. 2000. Convergence of the secretory pathways for cholera toxin and the filamentous phage, CTXphi. Science **288**:333–335.

Davis, R. W., L. A. Fuiman, T. M. Williams, S. O. Collier, W. P. Hagey, S. B. Kanatous, S. Kohin, and M. Horning. 1999. Hunting behavior of a marine mammal beneath the antarctic fast Ice. Science **283**:993–996.

Day, M. H., and E. H. Wickens. 1980. Laetoli Pliocene hominid footprints and bipedalism. Nature **286**:385–387.

Dayton, P. K. 1998. Reversal of the burden of proof in fisheries. Science **279**: 821–822.

de Baar, H. J. W., J. T. M. D. Jong, D. C. E. Bakker, B. M. Löscher, C. Veth, U. Bathmann, and V. Smetacek. 1995. Importance of iron for plankton blooms and carbon dioxide drawdown in the Southern Ocean. Nature **373**:412–415.

de Bono, M., D. M. Tobin, M. W. Davis, L. Avery, and C. I. Bargmann. 2002. Social feeding in *Caenorhabditis elegans* is induced by neurons that detect aversive stimuli. Nature **419**:899–903.

De Duve, C. 1991. Blueprint for a Cell: The Nature and Origin of Life, Neil Patterson Book.

de Duve, C. and S. L. Miller. 1991. Two-dimensional life? Proc. Natl. Acad. Sci. U. S. A **88**:10014–10017.

de Jong, J. C., E. C. Claas, A. D. Osterhaus, R. G. Webster, and W. L. Lim. 1997. A pandemic warning? Nature **389**:554.

De Marais, D. J. 2000. Evolution. When did photosynthesis emerge on Earth? Science **289**:1703–1705.

de Quervain, D. J., U. Fischbacher, V. Treyer, M. Schellhammer, U. Schnyder, A. Buck, and E. Fehr. 2004. The neural basis of altruistic punishment. Science **305**: 1254–1258.

de Waal, F. B. 1999. Cultural primatology comes of age. Nature **399**:635–636.

de Waal, F. B. and M. L. Berger. 2000. Payment for labour in monkeys. Nature **404**:563.

Dean, R. A., N. J. Talbot, D. J. Ebbole, M. L. Farman, T. K. Mitchell, M. J. Orbach and Bruce W. Birren. 2005. The genome sequence of the rice blast fungus *Magnaporthe grisea*. Nature **434**:980–986.

Deckert, G., P. V. Warren, T. Gaasterland, W. G. Young, A. L. Lenox, D. E. Graham, R. Overbeek, M. A. Snead, M. Keller, M. Aujay, R. Huber, R. A. Feldman, J. M. Short, G. J. Olsen, and R. V. Swanson. 1998. The complete genome of the hyperthermophilic bacterium *Aquifex aeolicus*. Nature **392**:353–358.

Deecke, V. B., P. J. Slater, and J. K. Ford. 2002. Selective habituation shapes acoustic predator recognition in harbour seals. Nature **420**:171–173.

Defleur, A., O. Dutour, H. Valladas, and B. Vandermeersch. 1993. Cannibals among the Neanderthals? Nature **362**:214.

Dehnhardt, G., B. Mauck, and H. Bleckmann. 1998. Seal whiskers detect water movements. Nature **394**:235–236.

Dehnhardt, G., B. Mauck, W. Hanke, and H. Bleckmann. 2001. Hydrodynamic trail-following in harbor seals (*Phoca vitulina*). Science **293**:102–104.

Deisenhofer, J., O. Epp, K. Miki, R. Huber, and H. Michel. 1985. Structure of the protein subunits in the photosynthetic reaction centre of *Rhodopseudomonas viridis* at 3Å resolution. Nature **318**:618–624.

Dejean, A., P. J. Solano, J. Ayroles, B. Corbara, and J. Orivel. 2005. Insect behaviour: arboreal ants build traps to capture prey. Nature **434**:973.

del Campo, M. L., C. I. Miles, F. C. Schroeder, C. Mueller, R. Booker, and J. A. Renwick. 2001. Host recognition by the tobacco hornworm is mediated by a host plant compound. Nature **411**:186–189.

del Giorgio, P. A., J. J. Cole, and A. Cimbleris. 1997. Respiration rates in bacteria exceed phytoplankton production in unproductive aquatic systems. Nature **385**:148–151.

del Giorgio, P. A. and C. M. Duarte. 2002. Respiration in the open ocean. Nature **420**:379–384.

Dell'Anno, A. and R. Danovaro. 2005. Extracellular DNA plays a key role in deep-sea ecosystem functioning. Science **309**:2179.

Delorme, C., H. Brüssow, J. Sidoti, N. Roche, K. A. Karlsson, J. R. Neeser, and S. Teneberg. 2001. Glycosphingolipid binding specificities of rotavirus: identification of a sialic acid-binding epitope. J Virol. **75**:2276–2287.

C., H. Brüssow, J. Sidoti, N. Roche, K.A. Karlsson, J.R. Neeser, and S. Teneberg. 2001. Glycosphingolipid binding specificities of rotavirus: identification of a sialic acid-binding epitope. J Virol. **75**:2276–2287.

DeLong, E. F., C. M. Preston, T. Mincer, V. Rich, S. J. Hallam, N. U. Frigaard, A. Martinez, M. B. Sullivan, R. Edwards, B. R. Brito, S. W. Chisholm, and D. M. Karl. 2006. Community genomics among stratified microbial assemblages in the ocean's interior. Science **311**:496–503.

DeLong, E. F. 2000. Resolving a methane mystery. Nature **407**:577, 579.

Delsuc, F., H. Brinkmann, D. Chourrout, and H. Philippe. 2006. Tunicates and not cephalochordates are the closest living relatives of vertebrates. Nature **439**: 965–968.

De Moraes, C. M., W. J. Lewis, P. W. Paré, H. T. Alborn, and J. H. Tumlinson. 1998. Herbivore-infested plants selectively attract parasitoids. Nature **393**:570–573.

Denham, T. P., S. G. Haberle, C. Lentfer, R. Fullagar, J. Field, M. Therin, N. Porch, and B. Winsborough. 2003. Origins of agriculture at Kuk Swamp in the highlands of New Guinea. Science **301**:189–193.

Dennell, R. 1997. The world's oldest spears. Nature **385**:767–768.

Depew, M. J., T. Lufkin, and J. L. Rubenstein. 2002. Specification of jaw subdivisions by Dlx genes. Science **298**:381–385.

Derry, L. A. 2006. Atmospheric science. Fungi, weathering, and the emergence of animals. Science **311**:1386–1387.

Desai, S. A., S. M. Bezrukov, and J. Zimmerberg. 2000. A voltage-dependent channel involved in nutrient uptake by red blood cells infected with the malaria parasite. Nature **406**:1001–1005.

D'Hondt, S., S. Rutherford, and A. J. Spivack. 2002. Metabolic activity of subsurface life in deep-sea sediments. Science **295**:2067–2070.

D'Hondt, S., B. B. Jorgensen, D. J. Miller, A. Batzke, R. Blake, B. A. Cragg, H. Cypionka, and J. L. Acosta. 2004. Distributions of microbial activities in deep subseafloor sediments. Science **306**:2216–2221.

Di Stefano, M., G. Veneto, S. Malservisi, L. Cecchetti, L. Minguzzi, A. Strocchi, and G. R. Corazza. 2002. Lactose malabsorption and intolerance and peak bone mass. Gastroenterology **122**:1793–1799.

Diamond, J. and P. Bellwood. 2003. Farmers and their languages: the first expansions. Science **300**:597–603.

Diamond, J. M. 1991. The earliest horsemen. Nature **350**:275–276.

Diamond, J. M. 2000. Talk of cannibalism. Nature **407**:25–26.

Diamond, J. M. 2001. Palaeontology. Australia's last giants. Nature **411**:755–757.

Diamond, J. 2002. Evolution, consequences and future of plant and animal domestication. Nature **418**:700–707.

Dismukes, G. C., V. V. Klimov, S. V. Baranov, Y. N. Kozlov, J. DasGupta, and A. Tyryshkin. 2001. The origin of atmospheric oxygen on Earth: the innovation of oxygenic photosynthesis. Proc. Natl. Acad. Sci. U. S. A **98**:2170–2175.

Divne, C., J. Stahlberg, T. Reinikainen, L. Ruohonen, G. Pettersson, J. K. Knowles, T. T. Teeri, and T. A. Jones. 1994. The three-dimensional crystal structure of the catalytic core of cellobiohydrolase I from *Trichoderma reesei*. Science **265**:524–528.

Dobson, C. M., G. B. Ellison, A. F. Tuck, and V. Vaida, V. 2000. Atmospheric aerosols as prebiotic chemical reactors. Proc. Natl. Acad. Sci. U. S. A **97**:11864–11868.

Dodd, A. N., N. Salathia, A. Hall, E. Kevei, R. Toth, F. Nagy, J. M. Hibberd, A. J. Millar, and A. A. Webb. 2005. Plant circadian clocks increase photosynthesis, growth, survival, and competitive advantage. Science **309**:630–633.

Doebley, J., A. Stec, and L. Hubbard. 1997. The evolution of apical dominance in maize. Nature **386**:485–488.

Doolittle, W. F. 1999. Phylogenetic classification and the universal tree. Science **284**:2124–2129.

Dormitzer, P. R., E. B. Nason, B. V. Prasad, and S. C. Harrison. 2004. Structural rearrangements in the membrane penetration protein of a non-enveloped virus. Nature **430**:1053–1058.

Dornhaus, A. and L. Chittka. 1999. Insect behaviour: Evolutionary origins of bee dances. Nature **401**:38–38

Doudna, J. A. and T. R. Cech. 2002. The chemical repertoire of natural ribozymes. Nature **418**:222–228.

Douglas, S., S. Zauner, M. Fraunholz, M. Beaton, S. Penny, L. T. Deng, X. Wu, M. Reith, T. Cavalier-Smith, and U. G. Maier. 2001. The highly reduced genome of an enslaved algal nucleus. Nature **410**:1091–1096.

Downward, J. 2003. Cell biology: metabolism meets death. Nature **424**:896–897.

Drinkwater, L. E., P. Wagoner, and M. Sarrantonio. 1998. Legume-based cropping systems have reduced carbon and nitrogen losses. Nature **396**:262–265.

Dufresne, A., M. Salanoubat, F. Partensky, F. Artiguenave, I. M. Axmann, V. Barbe, S. Duprat, and W. R. Hess. 2003. Genome sequence of the cyanobacterium *Prochlorococcus marinus* SS120, a nearly minimal oxyphototrophic genome. Proc. Natl. Acad. Sci. U. S. A **100**:10020–10025.

Duthie, C., G. Gibbs, and K. C. Burns. 2006. Seed dispersal by weta. Science **311**:1575.

Dyall, S. D., M. T. Brown, and P. J. Johnson. 2004a. Ancient invasions: from endosymbionts to organelles. Science **304**:253–257.

Dyall, S. D., W. Yan, M. G. Delgadillo-Correa, A. Lunceford, J. A. Loo, C. F. Clarke, and P. J. Johnson. 2004b. Non-mitochondrial complex I proteins in a hydrogenosomal oxidoreductase complex. Nature **431**:1103–1107.

Dye, C. and N. Gay. 2003. Epidemiology. Modeling the SARS epidemic. Science **300**:1884–1885.

Dyhrman, S. T., P. D. Chappell, S. T. Haley, J. W. Moffett, E. D. Orchard, J. B. Waterbury, and E. A. Webb. 2006. Phosphonate utilization by the globally important marine diazotroph Trichodesmium. Nature **439**:68–71.

Dziejman, M., E. Balon, D. Boyd, C. M. Fraser, J. F. Heidelberg, and J. J. Mekalanos. 2002. Comparative genomic analysis of *Vibrio cholerae*: genes that correlate with cholera endemic and pandemic disease. Proc. Natl. Acad. Sci. U. S. A **99**: 1556–1561.

Eckburg, P. B., E. M. Bik, C. N. Bernstein, E. Purdom, L. Dethlefsen, M. Sargent, S. R. Gill, K. E. Nelson, and D. A. Relman. 2005. Diversity of the human intestinal microbial flora. Science **308** :1635–1638.

Edman, K., P. Nollert, A. Royant, H. Belrhali, E. Pebay-Peyroula, J. Hajdu, R. Neutze, and E. M. Landau. 1999. High-resolution X-ray structure of an early intermediate in the bacteriorhodopsin photocycle. Nature **401**:822–826.

Edwards, J. S., R. U. Ibarra, and B. O. Palsson. 2001. In silico predictions of *Escherichia coli* metabolic capabilities are consistent with experimental data. Nat. Biotechnol. **19**:125–130.

Edwards, J. S. and B. O. Palsson. 2000. The *Escherichia coli* MG1655 in silico metabolic genotype: its definition, characteristics, and capabilities. Proc. Natl. Acad. Sci. U. S. A **97**:5528–5533.

Egholm, M., O. Buchardt, L. Christensen, C. Behrens, S. M. Freier, D. A. Driver, R. H. Berg, S. K. Kim, B. Norden, and P. E. Nielsen. 1993. PNA hybridizes to complementary oligonucleotides obeying the Watson-Crick hydrogen-bonding rules. Nature **365**:566–568.

Einsle, O., F. A. Tezcan, S. L. Andrade, B. Schmid, M. Yoshida, J. B. Howard, and D. C. Rees. 2002. Nitrogenase MoFe-protein at 1.16 A resolution: a central ligand in the FeMo-cofactor. Science **297**:1696–1700.

Elena, S. F., J. Dopazo, R. Flores, T. O. Diener, and A. Moya. 1991. Phylogeny of viroids, viroidlike satellite RNAs, and the viroidlike domain of hepatitis delta virus RNA. Proc. Natl. Acad. Sci. U. S. A **88**:5631–5634.

Elston, T., H. Wang, and G. Oster. 1998. Energy transduction in ATP synthase. Nature **391**:510–513.

Embley, T. M. and W. Martin. 2006. Eukaryotic evolution, changes and challenges. Nature **440**:623–630.

Emery, N. J. and N. S. Clayton. 2004. The mentality of crows: convergent evolution of intelligence in corvids and apes. Science **306**:1903–1907.

Emery, N. J. and N. S. Clayton. 2001. Effects of experience and social context on prospective caching strategies by scrub jays. Nature **414**:443–446.

Enard, W., P. Khaitovich, J. Klose, S. Zollner, F. Heissig, P. Giavalisco, K. Nieselt-Struwe, E. Muchmore, A. Varki, R. Ravid, G. M. Doxiadis, R. E. Bontrop, and S. Paabo. 2002. Intra- and interspecific variation in primate gene expression patterns. Science **296**:340–343.

Enattah, N.S., T. Sahi, E. Savilahti, J.D. Terwilliger, L. Peltonen, and I. Järvelä. 2002. Identification of a variant associated with adult-type hypolactasia. Nature Genetics **30**:233–237.

Engel, M. S. and D. A. Grimaldi. 2004. New light shed on the oldest insect. Nature **427**:627–630.

Engelberg-Kulka, H. and R. Hazan. 2003. Microbiology. Cannibals defy starvation and avoid sporulation. Science **301**:467–468.

Engler, H. S., K. C. Spencer, and L. E. Gilbert. 2000. Preventing cyanide release from leaves. Nature **406**:144–145.

Enserink, M. 2002. What mosquitoes want: secrets of host attraction. Science **298**:90–92.

Enserink, M. and D. Normile. 2003. Infectious diseases. Search for SARS origins stalls. Science **302**:766–767.

Erickson, G. M. 2001. The bite of Allosaurus. Nature **409**:987–988.

Erickson, G. M., S. D. Van Kirk, J. Su, M. E. Levenston, W. E. Caler, and D. R. Carter. 1996. Bite-force estimation for *Tyrannosaurus rex* from tooth-marked bones. Nature **382**:706–708.

Ermler, U., W. Grabarse, S. Shima, M. Goubeaud, and R. K. Thauer. 1997. Crystal structure of methyl-coenzyme M reductase: the key enzyme of biological methane formation. Science **278**:1457–1462.

Esch, H. E., S. Zhang, M. V. Srinivasan, and J. Tautz. 2001. Honeybee dances communicate distances measured by optic flow. Nature **411**:581–583.

Estes, J. A., M. T. Tinker, T. M. Williams, and D. F. Doak . 1998. Killer whale predation on sea otters linking oceanic and nearshore ecosystems. Science **282**:473–476.

Evenson, R. E. and D. Gollin. 2003. Assessing the impact of the green revolution, 1960 to 2000. Science **300**:758–762.

Fagarasan, S., M. Muramatsu, K. Suzuki, H. Nagaoka, H. Hiai, and T. Honjo. 2002. Critical roles of activation-induced cytidine deaminase in the homeostasis of gut flora. Science **298**:1424–1427.

Fagarasan, S., K. Kinoshita, M. Muramatsu, K. Ikuta, and T. Honjo. 2001. In situ class switching and differentiation to IgA-producing cells in the gut lamina propria. Nature **413**:639–643.

Falk, D., C. Hildebolt, K. Smith, M. J. Morwood, T. Sutikna, P. Brown, Jatmiko, E. W. Saptomo, B. Brunsden, and F. Prior. 2005. The Brain of LB1, *Homo floresiensis*. Science **308**:242–245.

Falkowski, P. G., M. E. Katz, A. H. Knoll, A. Quigg, J. A. Raven, O. Schofield, and F. J. Taylor. 2004. The evolution of modern eukaryotic phytoplankton. Science **305**:354–360.

Falkowski, P. G., M. E. Katz, A. J. Milligan, K. Fennel, B. S. Cramer, M. P. Aubry, R. A. Berner, M. J. Novacek, and W. M. Zapol. 2005. The rise of oxygen over the past 205 million years and the evolution of large placental mammals. Science **309**:2202–2204.

Falkowski, P. G. 2006. Evolution. Tracing oxygen's imprint on earth's metabolic evolution. Science **311**:1724–1725.

Farrell, B. D. 1998. "Inordinate Fondness" explained: why are there So many beetles? Science **281**:555–559.

Faruque, S. M., D. A. Sack, R. B. Sack, R. R. Colwell, Y. Takeda, and G. B. Nair. 2003. Emergence and evolution of *Vibrio cholerae* O139. Proc. Natl. Acad. Sci. U. S. A **100**:1304–1309.

Faruque, S. M., M. J. Islam, Q. S. Ahmad, A. S. Faruque, D. A. Sack, G. B. Nair, and J. J. Mekalanos. 2005. Self-limiting nature of seasonal cholera epidemics: Role of host-mediated amplification of phage. Proc. Natl. Acad. Sci. U. S. A **102**:6119–6124.

Faruque, S. M., I. B. Naser, M. J. Islam, A. S. Faruque, A. N. Ghosh, G. B. Nair, D. A. Sack, and J. J. Mekalanos. 2005a. Seasonal epidemics of cholera inversely correlate with the prevalence of environmental cholera phages. Proc. Natl. Acad. Sci. U. S. A **102**:1702–1707.

Faxen, K., G. Gilderson, P. Adelroth, and P. Brzezinski. 2005b. A mechanistic principle for proton pumping by cytochrome c oxidase. Nature **437**:286–289.

Fedonkin, M. A. and B. M. Waggoner. 1997. The Late Precambrian fossil Kimberella is a mollusc-like bilaterian organism. Nature **388**:868–871.

Fehr, E. and S. Gachter. 2002. Altruistic punishment in humans. Nature **415**:137–140.

Fehr, E. and B. Rockenbach. 2003. Detrimental effects of sanctions on human altruism. Nature **422**:137–140.

Fehr, E. and U. Fischbacher. 2003. The nature of human altruism. Nature **425**:785–791.

Feng, A. S., P. M. Narins, C. H. Xu, W. Y. Lin, Z. L. Yu, Q. Qiu, Z. M. Xu, and J. X. Shen. 2006. Ultrasonic communication in frogs. Nature **440**:333–336.

Fenton, B. and J. Ratcliffe. 2004. Animal behaviour: eavesdropping on bats. Nature **429**:612–613.

Ferguson, N. M., D. A. Cummings, S. Cauchemez, C. Fraser, S. Riley, A. Meeyai, S. Iamsirithaworn, and D. S. Burke. 2005. Strategies for containing an emerging influenza pandemic in Southeast Asia. Nature **437**:209–214.

Ferre-D'Amare, A. R., K. Zhou, and J. A. Doudna. 1998. Crystal structure of a hepatitis delta virus ribozyme. Nature **395**:567–574.

Ferreira, K. N., T. M. Iverson, K. Maghlaoui, J. Barber, and S. Iwata. 2004. Architecture of the photosynthetic oxygen-evolving center. Science **303**:1831–1838.

Ferris, J. P., A. R. Hill, Jr., R. Liu, and L. E. Orgel. 1996. Synthesis of long prebiotic oligomers on mineral surfaces. Nature **381**:59–61.

Ferris, M. J., and Palenik. 1998. Niche adaptation in ocean cyanobacteria. Nature **396**: 226–228.

Fiegna, F., Y. T. Yu, S. V. Kadam, and G. J. Velicer. 2006. Evolution of an obligate social cheater to a superior cooperator. Nature **441**:310–314.

Field, C. B., M. J. Behrenfeld, J. T. Randerson, and P. Falkowski. 1998. Primary production of the biosphere: integrating terrestrial and oceanic components. Science **281**:237–240.

Field, J., A. Cronin, and C. Bridge. 2006. Future fitness and helping in social queues. Nature **441**:214–217.

Finn, M. W. and F. R. Tabita. 2003. Synthesis of catalytically active form III ribulose 1,5-bisphosphate carboxylase/oxygenase in archaea. J. Bacteriol. **185**:3049–3059.

Fire, A., S. Xu, M. K. Montgomery, S. A. Kostas, S. E. Driver, and C. C. Mello. 1998. Potent and specific genetic interference by double-stranded RNA in *Caenorhabditis elegans*. Nature **391**:806–811.

Fischer, E. and U. Sauer. 2003. A novel metabolic cycle catalyzes glucose oxidation and anaplerosis in hungry *Escherichia coli*. J. Biol. Chem. **278**:46446–46451.

Flack, J. C., M. Girvan, F. B. de Waal, and D. C. Krakauer . 2006. Policing stabilizes construction of social niches in primates. Nature **439**:426–429.

Flannery, T. F., M. Archer, T. H. Rich, and R. Jones. 1995. A new family of monotremes from the Creataceous of Australia. Nature **377**:418–420.

Flatz, G. and H. W. Rotthauwe. 1973. Lactose nutrition and natural selection. Lancet **2**:76–77.

Flo, T. H., K. D. Smith, S. Sato, D. J. Rodriguez, M. A. Holmes, R. K. Strong, S. Akira, and A. Aderem. 2004. Lipocalin 2 mediates an innate immune response to bacterial infection by sequestrating iron. Nature **432**:917–921.

Florens, L., M. P. Washburn, J. D. Raine, R. M. Anthony, M. Grainger, J. D. Haynes, J. K. Moch, and D. J. Carucci. 2002. A proteomic view of the *Plasmodium falciparum* life cycle. Nature **419**:520–526.

Forterre, Y., Jan M. Skotheim, Jacques Dumais, and L. Mahadevan. 2005. How the Venus flytrap snaps. Nature **433**:421–425.

Forey, P. and P. Janvier. 1993. Agnathans and the origin of jawed vertebrates. Nature **361**:129–134.

Foster, R. G. 2005. Neurobiology: bright blue times. Nature **433**:698–699.

Fox, R. C. and C. S. Scott. 2005. First evidence of a venom delivery apparatus in extinct mammals. Nature **435**:1091–1093.

Fradin, M. S. and J. F. Day. 2002. Comparative efficacy of insect repellents against mosquito bites. N. Engl. J. Med. **347**:13–18.

Frank, K. T., B. Petrie, J. S. Choi, and W. C. Leggett. 2005. Trophic cascades in a formerly cod-dominated ecosystem. Science **308**:1621–1623.

Franks, N. R. and T. Richardson. 2006. Teaching in tandem-running ants. Nature **439**:153.

Frary, A., T. C. Nesbitt, S. Grandillo, E. Knaap, B. Cong, J. Liu, J. Meller, R. Elber, K. B. Alpert, and S. D. Tanksley. 2000. fw2.2: a quantitative trait locus key to the evolution of tomato fruit size. Science **289**:85–88.

Fraser, M. E., M. M. Chernaia, Y. V. Kozlov, and M. N. James. 1994. Crystal structure of the holotoxin from *Shigella dysenteriae* at 2.5 A resolution. Nat. Struct. Biol. **1**:59–64.

Freiberg, C., R. Fellay, A. Bairoch, W. J. Broughton, A. Rosenthal, and X. Perret. 1997. Molecular basis of symbiosis between Rhizobium and legumes. Nature **387**: 394–401.

Friedman, W. E. 2006. Embryological evidence for developmental lability during early angiosperm evolution. Nature **441**:337–340.

Frigaard, N. U., A. Martinez, T. J. Mincer, and E. F. DeLong. 2006. Proteorhodopsin lateral gene transfer between marine planktonic Bacteria and Archaea. Nature **439**: 847–850.

Fry, B. G., N. Vidal, J. A. Norman, F. J. Vonk, H. Scheib, S. F. Ramjan, S. Kuruppu, K. Fung, S. B. Hedges, M. K. Richardson, W. C. Hodgson, V. Ignjatovic, R. Summer-hayes, and E. Kochva. 2006. Early evolution of the venom system in lizards and snakes. Nature **439**:584–588.

Fuhrer, T., E. Fischer, and U. Sauer. 2005. Experimental identification and quantification of glucose metabolism in seven bacterial species. J. Bacteriol. **187**:1581–1590.

Fuhrman, J. 2003. Genome sequences from the sea. Nature **424**:1001–1002.

Fuhrman, J. A. 1999. Marine viruses and their biogeochemical and ecological effects. Nature **399**:541–548.

Fuhrman, J. A. and D. G. Capone. 2001. Nifty nanoplankton. Nature **412**:593–594.

Gabunia, L. and A. Vekua. 1995. A Plio-Pleistocene hominid from Dmanisi, East Georgia, Caucasus. Nature **373**:509–512.

Gage, D. A., D. Rhodes, K. D. Nolte, W. A. Hicks, T. Leustek, A. J. Cooper, and A. D. Hanson. 1997. A new route for synthesis of dimethylsulphoniopropionate in marine algae. Nature **387**:891–894.

Galagan, J. E., S. E. Calvo, K. A. Borkovich, E. U. Selker, N. D. Read, D. Jaffe, and B. Birren. 2003. The genome sequence of the filamentous fungus *Neurospora crassa*. Nature **422**:859–868.

Gamage, S. D., J. E. Strasser, C. L. Chalk, and A. A. Weiss. 2003. Nonpathogenic *Escherichia coli* can contribute to the production of Shiga toxin. Infect. Immun. **71**:3107–3115.

Gamble, C. 1994. Palaeoanthropology. Time for Boxgrove man. Nature **369**:275–276.

Garcia-Pichel, F. and O. Pringault. 2001. Microbiology. Cyanobacteria track water in desert soils. Nature **413**:380–381.

Garcia-Ruiz, J. M., S. T. Hyde, A. M. Carnerup, A. G. Christy, M. J. Van Kranendonk, and N. J. Welham. 2003. Self-assembled silica-carbonate structures and detection of ancient microfossils. Science **302**:1194–1197.

Gardner, M. J., N. Hall, E. Fung, O. White, M. Berriman, R. W. Hyman, J. M. Carlton, A. Pain, and B. Barrell. 2002. Genome sequence of the human malaria parasite *Plasmodium falciparum*. Nature **419**:498–511.

Gardner, T. A., I. M. Cote, J. A. Gill, A. Grant, and A. R. Watkinson. 2003. Long-term region-wide declines in Caribbean corals. Science **301**:958–960.

Gatenby, R. A. and R. J. Gillies. 2004. Why do cancers have high aerobic glycolysis? Nat. Rev. Cancer **4**:891–899.

Gattuso, J. P. and R. W. Buddemeier. 2000. Ocean biogeochemistry. Calcification and CO_2. Nature **407**:311, 313.

Gauthier-Clerc, M., Y. Le Maho, Y. Clerquin, S. Drault. 2000. Penguin fathers preserve food for their chicks. Nature **408**:928–929.

Geffeney, S., E. D. Brodie, Jr., P. C. Ruben, and E. D. Brodie, III. 2002. Mechanisms of adaptation in a predator-prey arms race: TTX-resistant sodium channels. Science **297**:1336–1339.

Geffeney, S. L., E. Fujimoto, E. D. Brodie, III, E. D. Brodie, Jr., and P. C. Ruben. 2005. Evolutionary diversification of TTX-resistant sodium channels in a predator-prey interaction. Nature **434**:759–763.

Genin, A., J. S. Jaffe, R. Reef, C. Richter, and P. J. Franks. 2005. Swimming against the flow: a mechanism of zooplankton aggregation. Science **308**:860–862.

Gerdes, S. Y., M. D. Scholle, J. W. Campbell, G. Balazsi, E. Ravasz, M. D. Daugherty, A. L. Somera, and A. L. Osterman. 2003. Experimental determination and system level analysis of essential genes in *Escherichia coli* MG1655. J. Bacteriol. **185**:5673–5684.

Gewin, V. 2006. Genomics: discovery in the dirt. Nature **439**:384–386.

Giaever, G., A. M. Chu, L. Ni, C. Connelly, L. Riles, S. Veronneau, S. Dow, A. Lucau-Danila, K. Anderson, and M. Johnston. 2002. Functional profiling of the *Saccharomyces cerevisiae* genome. Nature **418**:387–391.

Gibbons, D. L., M. C. Vaney, A. Roussel, A. Vigouroux, B. Reilly, J. Lepault, M. Kielian, and F. A. Rey. 2004. Conformational change and protein-protein interactions of the fusion protein of Semliki Forest virus. Nature **427**:320–325.

Gilbert, M. J., C. R. Thornton, G. E. Wakley, and N. J. Talbot. 2006. A P-type ATPase required for rice blast disease and induction of host resistance. Nature **440**:535–539.

Gilbert, W. 1986. Origin of life: The RNA world, Nature **319**:618–618.

Gilmore, M. S. and J. J. Ferretti. 2003. Microbiology. The thin line between gut commensal and pathogen. Science **299**:1999–2002.

Giordano, M, J. Beardall, and J.A. Raven. 2005. CO_2 concentrating mechanisms in algae: mechanisms, environmental modulation, and evolution. Annu. Rev. Plant Biol. **56**:99–131.

Giorgio, M., E. Migliaccio, F. Orsini, D. Paolucci, M. Moroni, C. Contursi, G. Pelliccia, L. Luzi, S. Minucci, M. Marcaccio, P. Pinton, R. Rizzuto, P. Bernardi, F. Paolucci, and P. G. Pelicci. 2005. Electron transfer between cytochrome c and p66Shc generates reactive oxygen species that trigger mitochondrial apoptosis. Cell **122**:221–233.

Giovannoni, S. J., H. J. Tripp, S. Givan, M. Podar, K. L. Vergin, D. Baptista, L. Bibbs, J. Eads, T. H. Richardson, M. Noordewier, M. S. Rappe, J. M. Short, J. C. Carrington, and E. J. Mathur. 2005. Genome streamlining in a cosmopolitan oceanic bacterium. Science **309**:1242–1245.

Giovannoni, S. J., L. Bibbs, J. C. Cho, M. D. Stapels, R. Desiderio, K. L. Vergin, M. S. Rappe, S. Laney, L. J. Wilhelm, H. J. Tripp, E. J. Mathur, and D. F. Barofsky. 2005. Proteorhodopsin in the ubiquitous marine bacterium SAR11. Nature **438**:82–85.

Giraud, E., J. Fardoux, N. Fourrier, L. Hannibal, B. Genty, P. Bouyer, B. Dreyfus, and A. Vermeglio. 2002. Bacteriophytochrome controls photosystem synthesis in anoxygenic bacteria. Nature **417**:202–205.

Glass, R. I. 2004. Perceived threats and real killers. Science **304**:927.

Gliwicz, M. Z. 1986. Predation and the evolution of vertical migration in zooplankton. Nature **320**:746–748.

Goklany, I. M. 2002. The ins and outs of organic farming. Science **298**:1889–1890.

Gold, T. 1992. The deep, hot biosphere. Proc. Natl. Acad. Sci. U. S. A **89**:6045–6049.

Gomez-Gutierrez, J., W. T. Peterson, A. De Robertis, and R. D. Brodeur. 2003. Mass mortality of krill caused by parasitoid ciliates. Science **301**:339.

Gonzalez-Pastor, J. E., E. C. Hobbs, and R. Losick. 2003. Cannibalism by sporulating bacteria. Science **301**:510–513.

Gonzalez, S., A. C. Kitchener, and A. M. Lister. 2000. Survival of the Irish elk into the Holocene. Nature **405**:753–754.

Gordeliy, V. I., J. Labahn, R. Moukhametzianov, R. Efremov, J. Granzin, R. Schlesinger, G. Buldt, T. Savopol, A. J. Scheidig, J. P. Klare, and M. Engelhard. 2002. Molecular basis of transmembrane signalling by sensory rhodopsin II-transducer complex. Nature **419**:484–487.

Goren-Inbar, N., C. S. Feibel, K. L. Verosub, Y. Melamed, M. E. Kislev, E. Tchernov, and I. Saragusti. 2000. Pleistocene milestones on the out-of-Africa corridor at Gesher Benot Ya'aqov, israel. Science **289**:944–947.

Goren-Inbar, N., N. Alperson, M. E. Kislev, O. Simchoni, Y. Melamed, A. Ben Nun, and E. Werker. 2004. Evidence of hominin control of fire at Gesher Benot Ya'aqov, Israel. Science **304**:725–727.

Grand, R. J., R. K. Montgomery, D. K. Chitkara, and J. N. Hirschhorn. 2003. Changing genes; losing lactase. Gut **52** :617–619.

Grassly, N. C., C. Fraser, and G. P. Garnett. 2005. Host immunity and synchronized epidemics of syphilis across the United States. Nature **433**:417–421.

Gray, J. M., D. S. Karow, H. Lu, A. J. Chang, J. S. Chang, R. E. Ellis, M. A. Marletta, and C. I. Bargmann. 2004. Oxygen sensation and social feeding mediated by a C. elegans guanylate cyclase homologue. Nature **430**:317–322.

Gray, M. W., G. Burger, and B. F. Lang. 1999. Mitochondrial evolution. Science **283**:1476–1481.

Gray, M. W. 1998. Rickettsia, typhus and the mitochondrial connection. Nature **396**: 109–110.

Gray, R. D. and Q. D. Atkinson. 2003. Language-tree divergence times support the Anatolian theory of Indo-European origin. Nature **426**:435–439.

Green, R. E., J. Krause, S. E. Ptak, A. W. Briggs, M. T. Ronan, J. F. Simons, L. Du, M. Engholm, J. M. Rothberg, M. Paunovic, and S. Pääbo. 2006. Analysis of one million base pairs of Neanderthal DNA. Nature **444**:330–336.

Grenfell, B. and O. Bjornstad. 2005. Sexually transmitted diseases: epidemic cycling and immunity. Nature **433**:366–367.

Grenfell, B. T., O. N. Bjornstad, and J. Kappey. 2001. Travelling waves and spatial hierarchies in measles epidemics. Nature **414**:716–723.

Grimes, J. M., J. N. Burroughs, P. Gouet, J. M. Diprose, R. Malby, S. Zientara, P. P. Mertens, and D. I. Stuart. 1998. The atomic structure of the bluetongue virus core. Nature **395**:470–478.

Grotzinger, J. P., and D. H. Rothman. 1996. An abiotic model for stromatolite morphogenesis. Nature **383**:423–42.

Grubb, T. C. 1972. Smell and Foraging in Shearwaters and Petrels. Nature **237**:404–405.

Grusak, M. A. 2005. Golden Rice gets a boost from maize. Nat. Biotechnol. **23**: 429–430.

Grutter, A. S. 1999. Cleaner fish really do clean. Nature **398**:672–673.

Gu, Y., Z. H. Zhou, D. B. McCarthy, L. J. Reed, and J. K. Stoops. 2003a. 3D electron microscopy reveals the variable deposition and protein dynamics of the peripheral pyruvate dehydrogenase component about the core. Proc. Natl. Acad. Sci. U. S. A **100**:7015–7020.

Gu, Z., L. M. Steinmetz, X. Gu, C. Scharfe, R. W. Davis, and W. H. Li. 2003b. Role of duplicate genes in genetic robustness against null mutations. Nature **421**: 63–66.

Guan, Y., B. J. Zheng, Y. Q. He, X. L. Liu, Z. X. Zhuang, C. L. Cheung, S. W. Luo, P. H. Li, L. J. Zhang, Y. J. Guan, K. M. Butt, K. L. Wong, K. W. Chan, W. Lim, K. F. Shortridge, K. Y. Yuen, J. S. Peiris, and L. L. Poon. 2003. Isolation and characterization of viruses related to the SARS coronavirus from animals in southern China. Science **302**:276–278.

Guan Y., L. L. Poon, C. Y. Cheung, T. M. Ellis, W. Lim, A. S. Lipatov, K. H. Chan, K. M. Sturm-Ramirez, C. L. Cheung, Y. H. Leung, K. Y. Yuen, R. G. Webster, J. S. Peiris. 2004. H5N1 influenza: a protean pandemic threat. Proc. Natl. Acad. Sci. U. S. A. **101**:8156–8161.

Guerrero, R., C. Pedros-Alio, I. Esteve, J. Mas, D. Chase, and L. Margulis. 1986. Predatory prokaryotes: predation and primary consumption evolved in bacteria. Proc. Natl. Acad. Sci. U. S. A **83**:2138–2142.

Guimera, R. and L. A. Nunes Amaral. 2005. Functional cartography of complex metabolic networks. Nature **433**:895–900.

Gupta, R., Z. He, and S. Luan. 2002. Functional relationship of cytochrome c_6 and plastocyanin in Arabidopsis. Nature **417**:567–571.

Gustafson, D. E., Jr., D. K. Stoecker, M. D. Johnson, W. F. Van Heukelem, and K. Sneider. 2000. Cryptophyte algae are robbed of their organelles by the marine ciliate *Mesodinium rubrum*. Nature **405**:1049–1052.

Gustafson, R. H. and R. E. Bowen. 1997. Antibiotic use in animal agriculture. J. Appl. Microbiol. **83**:531–541.

Guthrie, R. D. 2003. Rapid body size decline in Alaskan Pleistocene horses before extinction. Nature **426**:169–171.

Guthrie, R. D. 2004. Radiocarbon evidence of mid-Holocene mammoths stranded on an Alaskan Bering Sea island. Nature **429**:746–749.

Guthrie, R. D. 2006. New carbon dates link climatic change with human colonization and Pleistocene extinctions. Nature **441**:207–209.

Haag, A. 2005. Marine biology: whale fall. Nature **433**:566–567.

Haak, W., P. Forster, B. Bramanti, S. Matsumura, G. Brandt, M. Tanzer, R. Villems, C. Renfrew, D. Gronenborn, K. W. Alt, and J. Burger. 2005. Ancient DNA from the first European farmers in 7500–year-old Neolithic sites. Science **310**:1016–1018.

Hagmann, M. 2002. Gunter Wachtershauser profile. Between a rock and a hard place. Science **295**:2006–2007.

Hahn, B. H., G. M. Shaw, K. M. De Cock, and P. M. Sharp. 2000. AIDS as a zoonosis: scientific and public health implications. Science **287**:607–614.

Haile-Selassie, Y. 2001. Late Miocene hominids from the Middle Awash, Ethiopia. Nature **412**:178–181.

Hajishengallis, G., S. K. Hollingshead, T. Koga, and M. W. Russell. 1995. Mucosal immunization with a bacterial protein antigen genetically coupled to cholera toxin A2/B subunits. J. Immunol. **154**:4322–4332.

Hallam, S. J., N. Putnam, C. M. Preston, J. C. Detter, D. Rokhsar, P. M. Richardson, and E. F. DeLong. 2004. Reverse methanogenesis: testing the hypothesis with environmental genomics. Science **305**:1457–1462.

Halpern, B. S., K. Cottenie, and B. R. Broitman. 2006. Strong top-down control in southern California kelp forest ecosystems. Science **312**:1230–1232.

Hamm, C. E., R. Merkel, O. Springer, P. Jurkojc, C. Maier, K. Prechtel, and V. Smetacek. 2003. Architecture and material properties of diatom shells provide effective mechanical protection. Nature **421**:841–843.

Hanczyc, M. M., S. M. Fujikawa, and J. W. Szostak. 2003. Experimental models of primitive cellular compartments: encapsulation, growth, and division. Science **302**: 618–622.

Hannon, G. J. 2002. RNA interference. Nature **418**:244–251.

Hanotte, O., D. G. Bradley, J. W. Ochieng, Y. Verjee, E. W. Hill, and J. E. Rege. 2002. African pastoralism: genetic imprints of origins and migrations. Science **296**:336–339.

Häring, M., G. Vestergaard, R. Rachel, L. Chen, R. A. Garrett, and D. Prangishvili. 2005. Independent virus development outside a host. Nature **436**:1101–1102.

Harvell, C. D., K. Kim, J. M. Burkholder, R. R. Colwell, P. R. Epstein, D. J. Grimes, E. E. Hofmann, E. K. Lipp, A. D. Osterhaus, R. M. Overstreet, J. W. Porter, G. W. Smith, and G. R. Vasta. 1999. Emerging marine diseases–climate links and anthropogenic factors. Science **285**:1505–1510.

Harvell, C. D., C. E. Mitchell, J. R. Ward, S. Altizer, A. P. Dobson, R. S. Ostfeld, and M. D. Samuel. 2002. Climate warming and disease risks for terrestrial and marine biota. Science **296**:2158–2162.

Haug, G. H., D. Gunther, L. C. Peterson, D. M. Sigman, K. A. Hughen, and B. Aeschlimann. 2003. Climate and the collapse of Maya civilization. Science **299**:1731–1735.

Haumann, M., P. Liebisch, C. Muller, M. Barra, M. Grabolle, and H. Dau. 2005. Photosynthetic O2 formation tracked by time-resolved x-ray experiments. Science **310**: 1019–1021.

Hawkes, K. 2004. Human longevity: the grandmother effect. Nature **428**:128–129.

Hays, G. C. 1995. Zooplankton avoidance activity. Nature **376**:650–650.

Hayes, J. M. 1996. The earliest memories of life on Earth. Nature **384**:21–22.

Hayward, T. L. 1993.The rise and fall of Rhizosolenia. Nature **363**:675–676.

Heckman, D. S., D. M. Geiser, B. R. Eidell, R. L. Stauffer, N. L. Kardos, and S. B. Hedges. 2001. Molecular evidence for the early colonization of land by fungi and plants. Science **293**:1129–1133.

Hector, A., B. Schmid, C. Beierkuhnlein, M. C. Caldeira, M. Diemer, P. G. Dimitrakopoulos, J. A. Finn, and D. J. Read. 1999. Plant diversity and productivity experiments in European grasslands. Science **286**:1123–1127.

Hederstedt, L. 2003. Structural biology. Complex II is complex too. Science **299**:671–672.

Heidelberg, J. F., J. A. Eisen, W. C. Nelson, R. A. Clayton, M. L. Gwinn, R. J. Dodson, D. H. Haft, and C. M. Fraser. 2000. DNA sequence of both chromosomes of the cholera pathogen *Vibrio cholerae*. Nature **406**:477–483.

Heil, M., J. Rattke, and W. Boland. 2005. Postsecretory hydrolysis of nectar sucrose and specialization in ant/plant mutualism. Science **308**:560–563.

Helgason, T., T. J. Daniell, R. Husband, A. H. Fitter, and J. P. Young. 1998. Ploughing up the wood-wide web? Nature **394**:431.

Hendrix, R. W. 2004. Hot new virus, deep connections. Proc. Natl. Acad. Sci. U. S. A **101**:7495–7496.

Hendrix, R. W., J. G. Lawrence, G. F. Hatfull, and S. Casjens. 2000. The origins and ongoing evolution of viruses. Trends Microbiol. **8**:504–508.

Hendrix, R. W., M. C. Smith, R. N. Burns, M. E. Ford, and G. F. Hatfull. 1999. Evolutionary relationships among diverse bacteriophages and prophages: all the world's a phage. Proc. Natl. Acad. Sci. U. S. A **96**:2192–2197.

Hennes, K. P. and M. Simon. 1995. Significance of Bacteriophages for Controlling Bacterioplankton Growth in a Mesotrophic Lake. Appl. Environ. Microbiol. **61**: 333–340.

Henze, K. and W. Martin. 2003. Evolutionary biology: essence of mitochondria. Nature **426**:127–128.

Herek, J. L., W. Wohlleben, R. J. Cogdell, D. Zeidler, and M. Motzkus. 2002. Quantum control of energy flow in light harvesting. Nature **417**:533–535.

Hess, W. R., F. Partensky, G. W. van der Staay, J. M. Garcia-Fernandez, T. Borner, and D. Vaulot. 1996. Coexistence of phycoerythrin and a chlorophyll a/b antenna in a marine prokaryote. Proc. Natl. Acad. Sci. U. S. A **93**:11126–11130.

Hesselbo, S. P., D. R. Grocke, H. C. Jenkyns, C. J. Bjerrum, P. Farrimond, H. S. Morgans Bell, and O. R. Green. 2000. Massive dissociation of gas hydrate during a Jurassic oceanic anoxic event. Nature **406**:392–395.

Hessler, A. M., D. R. Lowe, R. L. Jones, and D. K. Bird. 2004. A lower limit for atmospheric carbon dioxide levels 3.2 billion years ago. Nature **428**:736–738.

Hetz, S. K. and T. J. Bradley. 2005. Insects breathe discontinuously to avoid oxygen toxicity. Nature **433**:516–519.

Hill, C. A., A. N. Fox, R. J. Pitts, L. B. Kent, P. L. Tan, M. A. Chrystal, A. Cravchik, F. H. Collins, H. M. Robertson, and L. J. Zwiebel . 2002. G protein-coupled receptors in *Anopheles gambiae*. Science **298**:176–178.

Hiller, N. L., S. Bhattacharjee, C. van Ooij, K. Liolios, T. Harrison, C. Lopez-Estrano, and K. Haldar. 2004. A host-targeting signal in virulence proteins reveals a secretome in malarial infection. Science **306**:1934–1937.

Hinchliffe, P. and L. A. Sazanov. 2005. Organization of iron-sulfur clusters in respiratory complex I. Science **309** :771–774.

Hinkle, G., J. K. Wetterer, T. R. Schultz, and M. L. Sogin . 1994. Phylogeny of the attine ant fungi based on analysis of small subunit ribosomal RNA gene sequences. Science **266**:1695–1697.

Hinrichs, K. U., J. M. Hayes, S. P. Sylva, P. G. Brewer, and E. F. DeLong. 1999. Methane-consuming archaebacteria in marine sediments. Nature **398**:802–805.

Hoegh-Guldberg, O., R. J. Jones, S. Ward, and W. K. Loh. 2002. Communication arising. Is coral bleaching really adaptive? Nature **415**:601–602.

Hoehler, T. M., B. M. Bebout, and D. J. Des Marais. 2001. The role of microbial mats in the production of reduced gases on the early Earth. Nature **412**:324–327.

Hoelscher, M. A., S. Garg, D. S. Bangari, J. A. Belser, I. Lu, I. Stephenson, R. A. Bright, J. M. Katz, S. K. Mittal, and S. Sambhara. 2006. Development of adenoviral-vector-based pandemic influenza vaccine against antigenically distinct human H5N1 strains in mice. Lancet **367**:475–481.

Hogan, K. B., J. S. Hoffman, and A. M. Thompson. 1991. Methane on the greenhouse agenda. Nature **354**:181–182.

Hoganson, C. W. and G. T. Babcock. 1997. A metalloradical mechanism for the generation of oxygen from water in photosynthesis. Science **277**:1953–1956.

Holmes, R. K. 2000. Biology and molecular epidemiology of diphtheria toxin and the tox gene. J. Infect. Dis. **181 Suppl** 1:S156–S167.

Holstein, T. and P. Tardent. 1984. An ultrahigh-speed analysis of exocytosis: nematocyst discharge. Science **223**:830–833.

Holstein, T. W., M. Benoit, G. V. Herder, G. Wanner, C. N. David, and H. E. Gaub. 1994. Fibrous Mini-Collagens in Hydra Nematocysts. Science **265**:402–404.

Holt, R. A., G. M. Subramanian, A. Halpern, G. G. Sutton, R. Charlab, D. R. Nusskern, P. Wincker, and S. L. Hoffman. 2002. The genome sequence of the malaria mosquito *Anopheles gambiae*. Science **298**:129–149.

Homer-Dixon, T. F., J. H. Boutwell, and G. W. Rathjens. 1993. Environmental Change and Violent Conflict: Growing Scarcities of Renewable Resources Can Contribute Instability and Civil Strife. Scientific American **268**:38–45.

Hooper, L. V., T. Midtvedt, and J. I. Gordon. 2002. How host-microbial interactions shape the nutrient environment of the mammalian intestine. Annu. Rev. Nutr. **22**:283–307.

Hooper, L. V., M. H. Wong, A. Thelin, L. Hansson, P. G. Falk, and J. I. Gordon. 2001. Molecular analysis of commensal host-microbial relationships in the intestine. Science **291**:881–884.

Hooper, L. V., J. Xu, P. G. Falk, T. Midtvedt, and J. I. Gordon. 1999. A molecular sensor that allows a gut commensal to control its nutrient foundation in a competitive ecosystem. Proc. Natl. Acad. Sci. U. S. A **96**:9833–9838.

Hoppe, H. G., K. Gocke, R. Koppe, and C. Begler. 2002. Bacterial growth and primary production along a north-south transect of the Atlantic Ocean. Nature **416**:168–171.

Horner, J. R. and D. B. Weishampel. 1998. A comparative embryological study of two ornithischian dinosaurs. Nature **332**:256–257.

Hou, S., R. W. Larsen, D. Boudko, C. W. Riley, E. Karatan, M. Zimmer, G. W. Ordal, and M. Alam. 2000. Myoglobin-like aerotaxis transducers in Archaea and Bacteria. Nature **403**:540–544.

Howe, G. A., J. Lightner, J. Browse, and C. A. Ryan. 1996. An octadecanoid pathway mutant (JL5) of tomato is compromised in signaling for defense against insect attack. Plant Cell **8**:2067–2077.

Hrdy, I., R. P. Hirt, P. Dolezal, L. Bardonova, P. G. Foster, J. Tachezy, and T. M. Embley. 2004. Trichomonas hydrogenosomes contain the NADH dehydrogenase module of mitochondrial complex I. Nature **432**:618–622.

Hu, Y., Y. Wang, Z. Luo, and C. Li. 1997. A new symmetrodont mammal from China and its implications for mammalian evolution. Nature **390**:137–142.

Hu, Y., J. Meng, Y. Wang and C. Li. 2005. Large Mesozoic mammals fed on young dinosaurs. Nature **433**:149–152.

Huang, C. Y., M. A. Ayliffe, and J. N. Timmis. 2003. Direct measurement of the transfer rate of chloroplast DNA into the nucleus. Nature **422**:72–76.

Huang, J., S. Rozelle, C. Pray, and Q. Wang. 2002. Plant biotechnology in China. Science **295**:674–676.

Huang, J., R. Hu, S. Rozelle, and C. Pray. 2005. Insect-resistant GM rice in farmers' fields: assessing productivity and health effects in China. Science **308**:688–690.

Huang, J., C. Pray, and S. Rozelle. 2002. Enhancing the crops to feed the poor. Nature **418**:678–684.

Huber, C. and G. Wachtershauser. 1997. Activated acetic acid by carbon fixation on (Fe,Ni)S under primordial conditions. Science **276**:245–247.

Huber, C. and G. Wachtershauser. 1998. Peptides by activation of amino acids with CO on (Ni,Fe)S surfaces: implications for the origin of life. Science **281**:670–672.

Huber, C., W. Eisenreich, S. Hecht, and G. Wachtershauser. 2003. A possible primordial peptide cycle. Science **301**:938–940.

Huber, K. E. and M. K. Waldor. 2002. Filamentous phage integration requires the host recombinases XerC and XerD. Nature **417**:656–659.

Hublin, J. J., F. Spoor, M. Braun, F. Zonneveld, and S. Condemi. 1996. A late Neanderthal associated with Upper Palaeolithic artefacts. Nature **381**:224–226.

Huey, R. B. and W. J. Moody. 2002. Neuroscience and evolution. Snake sodium channels resist TTX arrest. Science **297**:1289–1290.

Huey, B. and P. D. Ward. 2005. Hypoxia, Global Warming, and Terrestrial Late Permian Extinctions. Science **308**:398–401.

Hughes, T. P., A. H. Baird, D. R. Bellwood, M. Card, S. R. Connolly, C. Folke, R. Grosberg, O. Hoegh-Guldberg, J. B. Jackson, J. Kleypas, J. M. Lough, P. Marshall, M. Nystrom, S. R. Palumbi, J. M. Pandolfi, B. Rosen, and J. Roughgarden. 2003. Climate change, human impacts, and the resilience of coral reefs. Science **301**:929–933.

Hunt, G. R. 1996. Manufacture and use of hook-tools by New Caledonian crows. Nature **379**:249–251.

Hunt, G. R., M. C. Corballis, and R. D. Gray. 2001. Animal behaviour: Laterality in tool manufacture by crows. Nature **414**:707.

Hunt, J. H. 2003. Ecology. Cryptic herbivores of the rainforest canopy. Science **300**: 916–917.

Hutchings, J. A. 2000. Collapse and recovery of marine fishes. Nature **406**:882–885.

Hutchinson, J. R. and M. Garcia. 2002. Tyrannosaurus was not a fast runner. Nature **415**:1018–1021.

Hutchison, C. A., S. N. Peterson, S. R. Gill, R. T. Cline, O. White, C. M. Fraser, H. O. Smith, and J. C. Venter. 1999. Global transposon mutagenesis and a minimal Mycoplasma genome. Science **286**:2165–2169.

Hutchison, V. H., H. G. Dowling, and A. Vinegar. 1966. Thermoregulation in a brooding female Indian python, *Python molurus bivittatus*. Science **151**:694–696.

Ianora, A., A. Miralto, S. A. Poulet, Y. Carotenuto, I. Buttino, G. Romano, R. Casotti, G. Pohnert, T. Wichard, L. Colucci-D'Amato, G. Terrazzano, and V. Smetacek. 2004. Aldehyde suppression of copepod recruitment in blooms of a ubiquitous planktonic diatom. Nature **429**:403–407.

Ibarra, R. U., J. S. Edwards, and B. O. Palsson. 2002. *Escherichia coli* K-12 undergoes adaptive evolution to achieve in silico predicted optimal growth. Nature **420**:186–189.

Irigoien, X., R. P. Harris, H. M. Verheye, P. Joly, J. Runge, M. Starr, D. Pond, R. Campbell, R. Shreeve, P. Ward, A. N. Smith, H. G. Dam, W. Peterson, V. Tirelli, M. Koski, T. Smith, D. Harbour, and R. Davidson. 2002. Copepod hatching success in marine ecosystems with high diatom concentrations. Nature **419**: 387–389.

Ishii, N., M. Fujii, P. S. Hartman, M. Tsuda, K. Yasuda, N. Senoo-Matsuda, S. Yanase, D. Ayusawa, and K. Suzuki. 1998. A mutation in succinate dehydrogenase cytochrome b causes oxidative stress and ageing in nematodes. Nature **394**:694–697.

Itoh, H., A. Takahashi, K. Adachi, H. Noji, R. Yasuda, M. Yoshida, and K. Kinosita. 2004. Mechanically driven ATP synthesis by F1-ATPase. Nature **427**:465–468.

Iverson, T. M., C. Luna-Chavez, G. Cecchini, and D. C. Rees. 1999. Structure of the *Escherichia coli* fumarate reductase respiratory complex. Science **284**:1961–1966.

Iwata, S., J. W. Lee, K. Okada, J. K. Lee, M. Iwata, B. Rasmussen, T. A. Link, S. Ramaswamy, and B. K. Jap. 1998. Complete structure of the 11-subunit bovine mitochondrial cytochrome bc1 complex. Science **281**:64–71.

Iwata, S., C. Ostermeier, B. Ludwig, and H. Michel. 1995. Structure at 2.8 A resolution of cytochrome c oxidase from Paracoccus denitrificans. Nature **376**:660–669.

Jacewicz, M. S., D. W. Acheson, M. Mobassaleh, A. Donohue-Rolfe, K. A. Balasubramanian, and G. T. Keusch. 1995. Maturational regulation of globotriaosylceramide, the Shiga-like toxin 1 receptor, in cultured human gut epithelial cells. J. Clin. Invest **96**:1328–1335.

Jackson, J. B., M. X. Kirby, W. H. Berger, K. A. Bjorndal, L. W. Botsford, B. J. Bourque, R. and R. R. Warner. 2001. Historical overfishing and the recent collapse of coastal ecosystems. Science **293**:629–637.

Jaenicke, R. 2005. Abundance of cellular material and proteins in the atmosphere. Science **308**:73.

Janssens, I. A., A. Freibauer, P. Ciais, P. Smith, G. J. Nabuurs, G. Folberth, B. Schlamadinger, R. W. Hutjes, R. Ceulemans, E. D. Schulze, R. Valentini, and A. J. Dolman. 2003. Europe's terrestrial biosphere absorbs 7 to 12% of European anthropogenic CO_2 emissions. Science **300**:1538–1542.

Janvier, P. 1998. Forerunners of four legs. Nature **395**:748–749.

Jarvela, I., E. N. Sabri, J. Kokkonen, T. Varilo, E. Savilahti, and L. Peltonen. 1998. Assignment of the locus for congenital lactase deficiency to 2q21, in the vicinity of but separate from the lactase-phlorizin hydrolase gene. Am. J. Hum. Genet. **63**:1078–1085.

Javaux, E. J., A. H. Knoll, and M. R. Walter. 2001. Morphological and ecological complexity in early eukaryotic ecosystems. Nature **412**:66–69.

Jayaram, H., Z. Taraporewala, J. T. Patton, and B. V. Prasad. 2002. Rotavirus protein involved in genome replication and packaging exhibits a HIT-like fold. Nature **417**:311–315.

Jeong, H., B. Tombor, R. Albert, Z. N. Oltvai, and A. L. Barabasi. 2000. The large-scale organization of metabolic networks. Nature **407**:651–654.

Jermiin, L. S., L. Poladian, and M. A. Charleston. 2005. Evolution. Is the "Big Bang" in animal evolution real? Science **310**:1910–1911.

Ji, Q., Z. X. Luo, S. A. Ji, and Z. Luo. 1999. A Chinese triconodont mammal and mosaic evolution of the mammalian skeleton. Nature **398**:283–284.

Ji, Q., Z. X. Luo, C. X. Yuan, J. R. Wible, J. P. Zhang, and J. A. Georgi. 2002. The earliest known eutherian mammal. Nature **416**:816–822.

Ji, Q., Z. X. Luo, C. X. Yuan, and A. R. Tabrum. 2006. A swimming mammaliaform from the Middle Jurassic and ecomorphological diversification of early mammals. Science **311**:1123–1127.

Johnson, Z. I., E. R. Zinser, A. Coe, N. P. McNulty, E. M. Woodward, and S. W. Chisholm. 2006. Niche partitioning among Prochlorococcus ecotypes along ocean-scale environmental gradients. Science **311**:1737–1740.

Johnston, A. M., J. A. Raven, J. Beardall, and R. C. Leegood. 2001. Carbon fixation. Photosynthesis in a marine diatom. Nature **412**:40–41.

Jones, R. H. and K. J. Flynn. 2005. Nutritional status and diet composition affect the value of diatoms as copepod prey. Science **307**:1457–1459.

Jordan, P., P. Fromme, H. T. Witt, O. Klukas, W. Saenger, and N. Krauss. 2001. Three-dimensional structure of cyanobacterial photosystem I at 2.5 A resolution. Nature **411**:909–917.

Jorgensen, B. B. 2001. Biogeochemistry. Space for hydrogen. Nature **412**:286–7, 289.

Joyce, G. F. 2002. The antiquity of RNA-based evolution. Nature **418**:214–221.

Junge, W. and N. Nelson. 2005. Structural biology. Nature's rotary electromotors. Science **308**:642–644.

Kaiser, J. 2004. American Society of Human Genetics meeting. New prostate cancer genetic link. Science **306**:1285.

Kálmán, L., R. LoBrutto, J. P. Allen, and J. C. Williams. 1999. Modified reaction centres oxidize tyrosine in reactions that mirror photosystem II. Nature **402**:696–699.

Kamath, R. S., A. G. Fraser, Y. Dong, G. Poulin, R. Durbin, M. Gotta, A. Kanapin, N. Le Bot, S. Moreno, M. Sohrmann, D. P. Welchman, P. Zipperlen, and J. Ahringer. 2003. Systematic functional analysis of the *Caenorhabditis elegans* genome using RNAi. Nature **421**:231–237.

Kanamaru, S., P. G. Leiman, V. A. Kostyuchenko, P. R. Chipman,V. V. Mesyanzhinov, F. Arisaka, M. G. Rossmann. 2002. Structure of the cell-puncturing device of bacteriophage T4. Nature **415**:553–557.

Karaolis, D. K., S. Somara, D. R. Maneval, Jr., J. A. Johnson, and J. B. Kaper. 1999. A bacteriophage encoding a pathogenicity island, a type-IV pilus and a phage receptor in cholera bacteria. Nature **399**:375–379.

Kareiva, P. 1996. Diversity and sustainability on the prairie. Nature **379**:673–674.

Karner, M. B., E. F. DeLong, and D. M. Karl. 2001. Archaeal dominance in the mesopelagic zone of the Pacific Ocean. Nature **409**:507–510.

Karrasch, S., P. A. Bullough, and R. Ghosh. 1995. The 8.5 A projection map of the light-harvesting complex I from *Rhodospirillum rubrum* reveals a ring composed of 16 subunits. EMBO J. **14**:631–638.

Kasting, J. F. 1993. Earth's early atmosphere. Science **259**:920–926.

Kasting, J. F. and J. L. Siefert. 2002. Life and the evolution of Earth's atmosphere. Science **296**:1066–1068.

Katinka, M. D., S. Duprat, E. Cornillot, G. Metenier, F. Thomarat, G. Prensier, V. Barbe, and C. P. Vivares. 2001. Genome sequence and gene compaction of the eukaryote parasite *Encephalitozoon cuniculi*. Nature **414**:450–453.

Kaufman, A. J. and S. Xiao. 2003. High CO_2 levels in the Proterozoic atmosphere estimated from analyses of individual microfossils. Nature **425**:279–282.

Ke, A., K. Zhou, F. Ding, J. H. Cate, and J. A. Doudna. 2004. A conformational switch controls hepatitis delta virus ribozyme catalysis. Nature **429**:201–205.

Keefe, A. D., S. L. Miller, G. McDonald, and J. Bada. 1995. Investigation of the prebiotic synthesis of amino acids and RNA bases from CO_2 using FeS/H_2S as a reducing agent. Proc. Natl. Acad. Sci. U. S. A **92**:11904–11906.

Keegstra, K. and J. Walton. 2006. Plant science. Beta-glucans–brewer's bane, dietician's delight. Science **311**:1872–1873.

Keeling, P. J. 2001. Parasites go the full monty. Nature **414**:401–402.

Keller, M., E. Blöchl, G. Wächtershäuser, K. O. Stetter. 1994. Formation of amide bonds without a condensation agent and implications for origin of life. Nature **368**:836–838.

Kellis, M., B. W. Birren, and E. S. Lander. 2004. Proof and evolutionary analysis of ancient genome duplication in the yeast *Saccharomyces cerevisiae*. Nature **428**: 617–624.

Kemp, A. E. S., and J. G. Baldauf. 1993. Vast Neogene laminated diatom mat deposits from the eastern equatorial Pacific Ocean. Nature **362**:141–144.

Kemp, A. E. S., R. B. Pearce, I.u Koizumi, J. Pike, and S. J. Rance. 1999. The role of mat-forming diatoms in the formation of Mediterranean sapropels. Nature **398**:57–61.

Kennedy, M., M. Droser, L. M. Mayer, D. Pevear, and D. Mrofka. 2006. Late Precambrian oxygenation; inception of the clay mineral factory. Science 311:1446–1449.

Kenward, B., A. A. Weir, C. Rutz, and A. Kacelnik. 2005. Behavioural ecology: tool manufacture by naive juvenile crows. Nature 433:121.

Keppler, F., J. T. Hamilton, M. Brass, and T. Rockmann. 2006. Methane emissions from terrestrial plants under aerobic conditions. Nature 439:187–191.

Kerfeld, C. A., M. R. Sawaya, S. Tanaka, C. V. Nguyen, M. Phillips, M. Beeby, and T. O. Yeates. 2005. Protein structures forming the shell of primitive bacterial organelles. Science 309:936–938.

Kerr, R. A. 2004. Climate change. Three degrees of consensus. Science 305:932–934.

Kerr, R. A. 2005. Earth science. The story of O_2. Science 308:1730–1732.

Kessler, A. and I. T. Baldwin. 2001. Defensive function of herbivore-induced plant volatile emissions in nature. Science 291:2141–2144.

Kessler, A., R. Halitschke, and I. T. Baldwin. 2004. Silencing the jasmonate cascade: induced plant defenses and insect populations. Science 305:665–668.

Kim, A., C. Terzian, P. Santamaria, A. Pelisson, N. Purd'homme, and A. Bucheton. 1994. Retroviruses in invertebrates: the gypsy retrotransposon is apparently an infectious retrovirus of Drosophila melanogaster. Proc. Natl. Acad. Sci. U. S. A 91:1285–1289.

Kim, J. and D. C. Rees. 1992. Structural models for the metal centers in the nitrogenase molybdenum-iron protein. Science 257:1677–1682.

King, C. H., E. B. Shotts, Jr., R. E. Wooley, and K. G. Porter. 1988. Survival of coliforms and bacterial pathogens within protozoa during chlorination. Appl. Environ. Microbiol. 54:3023–3033.

King, M. and C. Elliott. 1993. Legitimate double-think. Lancet 341:669–672.

King, N. and S. B. Carroll. 2001. A receptor tyrosine kinase from choanoflagellates: molecular insights into early animal evolution. Proc. Natl. Acad. Sci. U. S. A 98: 15032–15037.

Kingston, T. and S. J. Rossiter. 2004. Harmonic-hopping in Wallacea's bats. Nature 429:654–657.

Kirk, K. 2000. Malaria. Channelling nutrients. Nature 406:949, 951.

Kislev, M.E., A. Hartmann, and O.Bar-Yosef. 2006. Early domesticated fig in the Jordan valley. Science 312:1372–1374.

Klausmeier, C. A., E. Litchman, T. Daufresne, and S. A. Levin. 2004. Optimal nitrogen-to-phosphorus stoichiometry of phytoplankton. Nature 429:171–174.

Klironomos, J. N. 2002. Feedback with soil biota contributes to plant rarity and invasiveness in communities. Nature 417:67–70.

Knight, J. 2004. Giardia: not so special, after all? Nature 429:236–237.

Knight, S., G. Schneider, Y. Lindqvist, T. Lundqvist, and C.-I. Brändén. 1989. Crystal structure of the active site of ribulose-bisphosphate carboxylase. Nature 337:229–234.

Knight, T. M., M. W. McCoy, J. M. Chase, K. A. McCoy, and R. D. Holt. 2005. Trophic cascades across ecosystems. Nature 437:880–883.

Kobayashi, T., T. Saito, and H. Ohtani. 2001. Real-time spectroscopy of transition states in bacteriorhodopsin during retinal isomerization. Nature 414:531–534.

Kobayashi, K., S. D. Ehrlich, A. Albertini, G. Amati, K. K. Andersen, and N. Ogasawara. 2003. Essential Bacillus subtilis genes. Proc. Natl. Acad. Sci. U. S. A. 100:4678–4683.

Koch, G. W., S. C. Sillett, G. M. Jennings, S. D. Davis. 2004. The limits to tree height. Nature 428:851–854.

Koentges, G. and T. Matsuoka. 2002. Evolution. Jaws of the fates. Science 298:371–373.

Kokai-Kun, J. F., A. R. Melton-Celsa, and A. D. O'Brien. 2000. Elastase in intestinal mucus enhances the cytotoxicity of Shiga toxin type 2d. J. Biol. Chem. **275**:3713–3721.

Kolber, Z. S., F. G. Plumley, A. S. Lang, J. T. Beatty, R. E. Blankenship, C. L. VanDover, C. Vetriani, M. Koblizek, C. Rathgeber, and P. G. Falkowski. 2001. Contribution of aerobic photoheterotrophic bacteria to the carbon cycle in the ocean. Science **292**:2492–2495.

Kolber, Z. S., C. L. Van Dover, R. A. Niederman, and P. G. Falkowski. 2000. Bacterial photosynthesis in surface waters of the open ocean. Nature **407**:177–179.

Konneke, M., A. E. Bernhard, l. T. de, Jr., C. B. Walker, J. B. Waterbury, and D. A. Stahl. 2005. Isolation of an autotrophic ammonia-oxidizing marine archaeon. Nature **437**: 543–546.

Korber, B., M. Muldoon, J. Theiler, F. Gao, R. Gupta, A. Lapedes, B. H. Hahn, S. Wolinsky, and T. Bhattacharya. 2000. Timing the ancestor of the HIV-1 pandemic strains. Science **288**:1789–1796.

Koronakis, V., A. Sharff, E. Koronakis, B. Luisi, and C. Hughes. 2000. Crystal structure of the bacterial membrane protein TolC central to multidrug efflux and protein export. Nature **405**:914–919.

Koropatnick, T. A., J. T. Engle, M. A. Apicella, E. V. Stabb, W. E. Goldman, and M. J. McFall-Ngai. 2004. Microbial factor-mediated development in a host-bacterial mutualism. Science **306**:1186–1188.

Kosfeld, M., M. Heinrichs, P. J. Zak, U. Fischbacher, and E. Fehr. 2005. Oxytocin increases trust in humans. Nature **435**:673–676.

Krause, D. O., S. E. Denman, R. I. Mackie, M. Morrison, A. L. Rae, G. T. Attwood, C. S. McSweeney. 2003. Opportunities to improve fiber degradation in the rumen: microbiology, ecology, and genomics. FEMS Microbiology Reviews **27**:663–693.

Krause, J., P. H. Dear, J. L. Pollack, M. Slatkin, H. Spriggs, I. Barnes, A. M. Lister, I. Ebersberger, S. Paabo, and M. Hofreiter. 2006. Multiplex amplification of the mammoth mitochondrial genome and the evolution of Elephantidae. Nature **439**: 724–727.

Krings, M., A. Stone, R. W. Schmitz, H. Krainitzki, M. Stoneking, and S. Paabo. 1997. Neandertal DNA sequences and the origin of modern humans. Cell **90**:19–30.

Krinos, C. M., M. J. Coyne, K. G. Weinacht, A. O. Tzianabos, D. L. Kasper, and L. E. Comstock. 2001. Extensive surface diversity of a commensal microorganism by multiple DNA inversions. Nature **414**:555–558.

Kuhl, M., M. Chen, P. J. Ralph, U. Schreiber, and A. W. Larkum. 2005. Ecology: a niche for cyanobacteria containing chlorophyll d. Nature **433**:820.

Kuhlbrandt, W. 2000. Bacteriorhodopsin–the movie. Nature **406**:569–570.

Kuhlbrandt, W. 2003. Structural biology: dual approach to a light problem. Nature **426**:399–400.

Kujoth, G. C., A. Hiona, T. D. Pugh, S. Someya, K. Panzer, S. E. Wohlgemuth, T. Hofer, A. Y. Seo, R. Sullivan, W. A. Jobling, J. D. Morrow, H. Van Remmen, J. M. Sedivy, T. Yamasoba, M. Tanokura, R. Weindruch, C. Leeuwenburgh, and T. A. Prolla. 2005. Mitochondrial DNA mutations, oxidative stress, and apoptosis in mammalian aging. Science **309**:481–484.

Kuma, A., M. Hatano, M. Matsui, A. Yamamoto, H. Nakaya, T. Yoshimori, Y. Ohsumi, T. Tokuhisa, and N. Mizushima. 2004. The role of autophagy during the early neonatal starvation period. Nature **432**:1032–1036.

Kuokkanen, M., N. S. Enattah, A. Oksanen, E. Savilahti, A. Orpana, and I. Jarvela. 2003. Transcriptional regulation of the lactase-phlorizin hydrolase gene by polymorphisms associated with adult-type hypolactasia. Gut **52**:647–652.

Kupfer, A., H. Muller, M. M. Antoniazzi, C. Jared, H. Greven, R. A. Nussbaum, and M. Wilkinson. 2006. Parental investment by skin feeding in a caecilian amphibian. Nature **440**:926–929.

Kurisu, G., H. Zhang, J. L. Smith, and W. A. Cramer. 2003. Structure of the cytochrome b_6f complex of oxygenic photosynthesis: tuning the cavity. Science **302**:1009–1014.

Kuwahara, T., A. Yamashita, H. Hirakawa, H. Nakayama, H. Toh, N. Okada, S. Kuhara, M. Hattori, T. Hayashi, and Y. Ohnishi. 2004. Genomic analysis of *Bacteroides fragilis* reveals extensive DNA inversions regulating cell surface adaptation. Proc. Natl. Acad. Sci. U. S. A **101**:14919–14924.

La Roche, J., G. W. van der Staay, F. Partensky, A. Ducret, R. Aebersold, R. Li, S. S. Golden, R. G. Hiller, P. M. Wrench, A. W. Larkum, and B. R. Green. 1996. Independent evolution of the prochlorophyte and green plant chlorophyll a/b light-harvesting proteins. Proc. Natl. Acad. Sci. U. S. A **93**:15244–15248.

La Scola, B., S. Audic, C. Robert, L. Jungang, L. de, X, M. Drancourt, R. Birtles, J. M. Claverie, and D. Raoult. 2003. A giant virus in amoebae. Science **299**:2033.

Lahdenpera, M., V. Lummaa, S. Helle, M. Tremblay, and A. F. Russell. 2004. Fitness benefits of prolonged post-reproductive lifespan in women. Nature **428**:178–181.

Lamed, R., J. Tormo, A. J. Chirino, E. Morag, and E. A. Bayer. 1994. Crystallization and preliminary X-ray analysis of the major cellulose-binding domain of the cellulosome from Clostridium thermocellum. J. Mol. Biol. **244**:236–237.

Lancaster, C. R., A. Kroger, M. Auer, and H. Michel. 1999. Structure of fumarate reductase from *Wolinella succinogenes* at 2.2 A resolution. Nature **402**:377–385.

Lang, B. F., G. Burger, C. J. O'Kelly, R. Cedergren, G. B. Golding, C. Lemieux, D. Sankoff, M. Turmel, and M. W. Gray. 1997. An ancestral mitochondrial DNA resembling a eubacterial genome in miniature. Nature **387**:493–497.

Larimer, F. W., P. Chain, L. Hauser, J. Lamerdin, S. Malfatti, L. Do, M. L. Land, D. A. Pelletier, and C. S. Harwood. 2004. Complete genome sequence of the metabolically versatile photosynthetic bacterium *Rhodopseudomonas palustris*. Nat. Biotechnol. **22**:55–61.

La Roche, J., P. W. Boyd, R. M. L. McKay, and R.J. Geider. 1996. Flavodoxin as an in situ marker for iron stress in phytoplankton. Nature **382**:802–805.

Larsen, P. L. and C. F. Clarke. 2002. Extension of life-span in *Caenorhabditis elegans* by a diet lacking coenzyme Q. Science **295**:120–123.

Larson, G., K. Dobney, U. Albarella, M. Fang, E. Matisoo-Smith, J. Robins, S. Lowden, H. Finlayson, T. Brand, E. Willerslev, P. Rowley-Conwy, L. Andersson, and A. Cooper. 2005. Worldwide phylogeography of wild boar reveals multiple centers of pig domestication. Science **307**:1618–1621.

Lasonder, E., Y. Ishihama, J. S. Andersen, A. M. Vermunt, A. Pain, R. W. Sauerwein, W. M. Eling, N. Hall, A. P. Waters, H. G. Stunnenberg, and M. Mann. 2002. Analysis of the *Plasmodium falciparum* proteome by high-accuracy mass spectrometry. Nature **419**:537–542.

Last, J. M. 1993. War and the demographic trap. Lancet **342**:508–509.

Lau, S. K., P. C. Woo, K. S. Li, Y. Huang, H. W. Tsoi, B. H. Wong, S. S. Wong, S. Y. Leung, K. H. Chan, and K. Y. Yuen. 2005. Severe acute respiratory syndrome coronavirus-like virus in Chinese horseshoe bats. Proc. Natl. Acad. Sci. U. S. A **102**:14040–14045.

Lazcano, A. and S. L. Miller. 1996. The origin and early evolution of life: prebiotic chemistry, the pre-RNA world, and time. Cell **85**:793–798.

Le Roch, K. G., Y. Zhou, P. L. Blair, M. Grainger, J. K. Moch, J. D. Haynes, L. De, V, A. A. Holder, S. Batalov, D. J. Carucci, and E. A. Winzeler. 2003. Discovery of gene function by expression profiling of the malaria parasite life cycle. Science **301**:1503–1508.

Le, Q. M., M. Kiso, K. Someya, Y. T. Sakai, T. H. Nguyen, K. H. Nguyen, N. D. Pham, H. H. Ngyen, S. Yamada, Y. Muramoto, T. Horimoto, A. Takada, H. Goto, T. Suzuki, Y. Suzuki, and Y. Kawaoka. 2005. Avian flu: isolation of drug-resistant H5N1 virus. Nature **437**:1108.

Leakey, M. G., C. S. Feibel, I. McDougall, and A. Walker. 1995. New four-million-year-old hominid species from Kanapoi and Allia Bay, Kenya. Nature **376**:565–571.

Lee, S. Y., Z. Wang, C. K. Lin, C. H. Contag, L. C. Olds, A. D. Cooper, and E. Sibley. 2002. Regulation of intestine-specific spatiotemporal expression by the rat lactase promoter. J. Biol. Chem. **277**:13099–13105.

Legname, G., I. V. Baskakov, H.-O. B. Nguyen, D. Riesner, F. E. Cohen, S. J. DeArmond and S. B. Prusiner. 2004. Synthetic Mammalian Prions. Science **305**:673–676 .

Leigh, G. J. 2003. Chemistry. So that's how it's done–maybe. Science **301**:55–56.

Leiman, P. G., P. R. Chipman, V. A. Kostyuchenko, V. V. Mesyanzhinov, and M. G. Rossmann. 2004. Three-dimensional rearrangement of proteins in the tail of bacteriophage T4 on infection of its host. Cell **118**:419–429.

Leman, L., L. Orgel, and M. R. Ghadiri. 2004. Carbonyl sulfide-mediated prebiotic formation of peptides. Science **306**:283–286.

Lemieux, C., C. Otis, and M. Turmel. 2000. Ancestral chloroplast genome in *Mesostigma viride* reveals an early branch of green plant evolution. Nature **403**:649–652.

Leopold, A. S., M. Erwin, J. Oh, and B. Browning. 1976. Phytoestrogens: adverse effects on reproduction in California quail. Science **191**:98–100.

Lesser, M. P., C. H. Mazel, M. Y. Gorbunov, and P. G. Falkowski. 2004. Discovery of symbiotic nitrogen-fixing cyanobacteria in corals. Science **305**:997–1000.

Lev-Yadun, S., A. Gopher, and S. Abbo. 2000. Archaeology. The cradle of agriculture. Science **288**:1602–1603.

Levey, D. J., R. S. Duncan, and C. F. Levins. 2004. Animal behaviour: use of dung as a tool by burrowing owls. Nature **431**:39.

Lewis, C. L. and M. A. Coffroth. 2004. The acquisition of exogenous algal symbionts by an octocoral after bleaching. Science **304**:1490–1492.

Lewis, S., T. N. Sherratt, K. C. Hamer, and S. Wanless. 2001. Evidence of intra-specific competition for food in a pelagic seabird. Nature **412**:816–819.

Lewis, W. J., and K. Takasu. 1990. Use of learned odours by a parasitic wasp in accordance with host and food needs. Nature **348**: 635–636.

Lewis, W. J., and J. H. Tumlinson. 1988. Host detection by chemically mediated associative learning in a parasitic wasp. Nature **331**:257–259.

Ley, R. E., F. Backhed, P. Turnbaugh, C. A. Lozupone, R. D. Knight, and J. I. Gordon. 2005. Obesity alters gut microbial ecology. Proc. Natl. Acad. Sci. U. S. A **102**: 11070–11075.

Li, C., A. Zhou, and T. Sang. 2006. Rice domestication by reducing shattering. Science **311**:1936–1939.

Li, C. W., J. Y. Chen, and T. E. Hua. 1998. Precambrian sponges with cellular structures. Science **279**:879–882.

Li, K. S., Y. Guan, J. Wang, G. J. Smith, K. M. Xu, L. Duan, A. P. Rahardjo, P. Putha-vathana, C. Buranathai, and J. S. Peiris. 2004. Genesis of a highly pathogenic and potentially pandemic H5N1 influenza virus in eastern Asia. Nature **430**:209–213.

Li, W., Z. Shi, M. Yu, W. Ren, C. Smith, J. H. Epstein, H. Wang, G. Crameri, Z. Hu, H. Zhang, J. Zhang, J. McEachern, H. Field, P. Daszak, B. T. Eaton, S. Zhang, and L. F. Wang. 2005. Bats are natural reservoirs of SARS-like coronaviruses. Science 310:676–679.

Li, X., M. A. Schuler, and M. R. Berenbaum. 2002. Jasmonate and salicylate induce expression of herbivore cytochrome P450 genes. Nature 419:712–715.

Lichtenegger, H. C., T. Schoberl, M. H. Bartl, H. Waite, and G. D. Stucky. 2002. High abrasion resistance with sparse mineralization: copper biomineral in worm jaws. Science 298:389–392.

Lin, S. J., M. Kaeberlein, A. A. Andalis, L. A. Sturtz, P. A. Defossez, V. C. Culotta, G. R. Fink, and L. Guarente. 2002. Calorie restriction extends *Saccharomyces cerevisiae* lifespan by increasing respiration. Nature 418:344–348.

Lindell, D., M. B. Sullivan, Z. I. Johnson, A. C. Tolonen, F. Rohwer, and S. W. Chisholm. 2004. Transfer of photosynthesis genes to and from *Prochlorococcus* viruses. Proc. Natl. Acad. Sci. U. S. A 101:11013–11018.

Lindell, D., J. D. Jaffe, Z. I. Johnson, G. M. Church, and S. W. Chisholm. 2005. Photosynthesis genes in marine viruses yield proteins during host infection. Nature 438:86–89.

Lipp, E. K., A. Huq, and R. R. Colwell. 2002. Effects of global climate on infectious disease: the cholera model. Clin. Microbiol. Rev. 15:757–770.

Lister, M. 1993. Mammoths in miniature. Nature 362: 288–289.

Little, A. F., M. J. van Oppen, and B. L. Willis. 2004. Flexibility in algal endosymbioses shapes growth in reef corals. Science 304:1492–1494.

Liu, J. 2003. SARS, wildlife, and human health. Science 302:53.

Liu, J., H. Xiao, F. Lei, Q. Zhu, K. Qin, X. W. Zhang, X. L. Zhang, D. Zhao, G. Wang, Y. Feng, J. Ma, W. Liu, J. Wang, and G. F. Gao. 2005. Highly pathogenic H5N1 influenza virus infection in migratory birds. Science 309:1206.

Lloyd-Smith, J. O., S. J. Schreiber, P. E. Kopp, and W. M. Getz. 2005. Superspreading and the effect of individual variation on disease emergence. Nature 438:355–359.

Loeb, V., V. Siegel, O. Holm-Hansen, R. Hewitt, W. Fraser, W. Trivelpiece, and S. Trivelpiece. 1997. Effects of sea-ice extent and krill or salp dominance on the Antarctic food web. Nature 387:897–900.

Loftus, B., I. Anderson, R. Davies, U. C. Alsmark, J. Samuelson, P. Amedeo, P. Roncaglia, M. Berriman, and N. Hall. 2005. The genome of the protist parasite *Entamoeba histolytica*. Nature 433:865–868.

Longo, V. D. and C. E. Finch. 2003. Evolutionary medicine: from dwarf model systems to healthy centenarians? Science 299:1342–1346.

Lonsdorf, E. V., L. E. Eberly, and A. E. Pusey. 2004. Sex differences in learning in chimpanzees. Nature 428:715–716.

Lopez-Garcia, P. and D. Moreira. 1999. Metabolic symbiosis at the origin of eukaryotes. Trends Biochem. Sci. 24:88–93.

Lopez, S. and C. F. Arias. 2004. Multistep entry of rotavirus into cells: a Versaillesque dance. Trends Microbiol. 12:271–278.

Lordkipanidze, D., A. Vekua, R. Ferring, G. P. Rightmire, J. Agusti, G. Kiladze, A. Mouskhelishvili, M. Nioradze, M. S. Ponce de Leon, M. Tappen, and C. P. Zollikofer. 2005. Anthropology: the earliest toothless hominin skull. Nature 434:717–718.

Loreau, M., S. Naeem, P. Inchausti, J. Bengtsson, J. P. Grime, A. Hector, D. U. Hooper, M. A. Huston, D. Raffaelli, B. Schmid, D. Tilman, and D. A. Wardle. 2001. Biodiversity and ecosystem functioning: current knowledge and future challenges. Science **294**: 804–808.

Lowe, D. C. 2006. Global change: a green source of surprise. Nature **439**:148–149.

Lowell, B. B. and G. I. Shulman. 2005. Mitochondrial dysfunction and type 2 diabetes. Science **307**:384–387.

Lu, H., R. A. Forbes, and A. Verma. 2002. Hypoxia-inducible factor 1 activation by aerobic glycolysis implicates the Warburg effect in carcinogenesis. J. Biol. Chem. **277**:23111–23115.

Lu, H., X. Yang, M. Ye, K. B. Liu, Z. Xia, X. Ren, L. Cai, N. Wu, and T. S. Liu. 2005. Culinary archaeology: Millet noodles in Late Neolithic China. Nature **437**:967–968.

Lu, Y. and R. Conrad. 2005. In situ stable isotope probing of methanogenic archaea in the rice rhizosphere. Science **309**:1088–1090.

Lundqvist, T. and G. Schneider. 1991. Crystal structure of activated ribulose-1, 5-bisphosphate carboxylase complexed with its substrate, ribulose-1,5-bisphosphate. J. Biol. Chem. **266**:12604–12611.

Lunzer, M., S. P. Miller, R. Felsheim, and A. M. Dean. 2005. The biochemical architecture of an ancient adaptive landscape. Science **310**:499–501.

Luo, Z. X., A. W. Crompton, and A. L. Sun. 2001. A new mammaliaform from the early Jurassic and evolution of mammalian characteristics. Science **292**:1535–1540.

Luo, Z. X., Q. Ji, J. R. Wible, and C. X. Yuan. 2003. An Early Cretaceous tribosphenic mammal and metatherian evolution. Science **302**:1934–1940.

Luo, Z. X. and J. R. Wible. 2005. A Late Jurassic digging mammal and early mammalian diversification. Science **308**:103–107.

Luria, S. E. and M. Delbrück. 1943. Mutations of bacteria from virus sensitivity to virus resistance. Genetics **28**:491–511.

Lutzoni, F., M. Pagel, and V. Reeb. 2001. Major fungal lineages are derived from lichen symbiotic ancestors. Nature **411**:937–940.

Lynd, L. R., P. J. Weimer, W. H. van Zyl, and I. S. Pretorius. 2002. Microbial cellulose utilization: fundamentals and biotechnology. Microbiol. Mol. Biol. Rev. **66**: 506–77, table.

Lyons, T. W. 2004. Geochemistry: warm debate on early climate. Nature **429**:359–360.

Ma, H. W. and A. P. Zeng. 2003. The connectivity structure, giant strong component and centrality of metabolic networks. Bioinformatics. **19**:1423–1430.

Machida, M., K. Asai, M. Sano, T. Tanaka, T. Kumagai, G. Terai, K. Kusumoto, T. Arima, O. Akita, and H. Kikuchi. 2005. Genome sequencing and analysis of *Aspergillus oryzae*. Nature **438**:1157–1161.

Macilwain, C. 2004. Organic: is it the future of farming? Nature **428**:792–793.

MacLean, R. C., and I. Gudelj. 2006. Resource competition and social conflict in experimental populations of yeast. Nature **441**: 498–501.

Macpherson, A. J., D. Gatto, E. Sainsbury, G. R. Harriman, H. Hengartner, and R. M. Zinkernagel. 2000. A primitive T cell-independent mechanism of intestinal mucosal IgA responses to commensal bacteria. Science **288**:2222–2226.

Macpherson, A. J. and T. Uhr. 2004. Induction of protective IgA by intestinal dendritic cells carrying commensal bacteria. Science **303**:1662–1665.

Maden, B. E. 1995. No soup for starters? Autotrophy and the origins of metabolism. Trends Biochem. Sci. **20**:337–341.

Mader, P., A. Fliessbach, D. Dubois, L. Gunst, P. Fried, and U. Niggli. 2002. Soil fertility and biodiversity in organic farming. Science **296**:1694–1697.

Maher, K. A. and D. J. Stevenson. 1988. Impact frustration of the origin of life. Nature **331**:612–614.

Malausa, T., M. T. Bethenod, A. Bontemps, D. Bourguet, J. M. Cornuet, and S. Ponsard. 2005. Assortative mating in sympatric host races of the European corn borer. Science **308**:258–260.

Maldonado, M., C. Carmona, M. J. Uriz, and A. Cruzado. 1999. Decline in Mesozoic reef-building sponges explained by silicon limitation. Nature **401**:785–788.

Maliga, P. 2003. Plant biology: Mobile plastid genes. Nature **422**:31–32.

Mandal, M., B. Boese, J. E. Barrick, W. C. Winkler, and R. R. Breaker. 2003. Riboswitches control fundamental biochemical pathways in Bacillus subtilis and other bacteria. Cell **113**:577–586.

Mann, C. 1997. Reseeding the green revolution. Science **277**:1038–1043.

Mann, D. A., Z. Lu, and A. N. Popper. 1997. A clupeid fish can detect ultrasound. Nature **389**:341–341.

Mann, N. H., M. R. Clokie, A. Millard, A. Cook, W. H. Wilson, P. J. Wheatley, A. Letarov, and H. M. Krisch. 2005. The genome of S-PM2, a "photosynthetic" T4-type bacteriophage that infects marine Synechococcus strains. J. Bacteriol. **187**: 3188–3200.

Mann, N. H., A. Cook, A. Millard, S. Bailey, and M. Clokie . 2003. Marine ecosystems: bacterial photosynthesis genes in a virus. Nature **424**:741.

Margulis, L., M. F. Dolan, and R. Guerrero. 2000. The chimeric eukaryote: origin of the nucleus from the karyomastigont in amitochondriate protists. Proc. Natl. Acad. Sci. U. S. A **97**:6954–6959.

Marlar, R. A., B. L. Leonard, B. R. Billman, P. M. Lambert, and J. E. Marlar. 2000. Biochemical evidence of cannibalism at a prehistoric Puebloan site in southwestern Colorado. Nature **407**:74–78.

Marra, M. A., S. J. Jones, C. R. Astell, R. A. Holt, A. Brooks-Wilson, Y. S. Butterfield, J. Khattra, and R. L. Roper. 2003. The Genome sequence of the SARS-associated coronavirus. Science **300**:1399–1404.

Marris, E. 2005. Four years on, no transgenes found in Mexican maize. Nature **436**:760.

Marti, M., R. T. Good, M. Rug, E. Knuepfer, and A. F. Cowman. 2004. Targeting malaria virulence and remodeling proteins to the host erythrocyte. Science **306**:1930–1933.

Martienssen, R. 1997. The origin of maize branches out. Nature **386**:443, 445.

Martin, J. H., K. H. Coale, K. S. Johnson, S. E. Fitzwater, R. M. Gordon, S. J. Tanner, C. N. Hunter, V. A. Elrod, J. L. Nowicki, T. L. Coley, R. T. Barber, and S. Lindley. 1994. Testing the iron hypothesis in ecosystems of the equatorial Pacific Ocean. Nature **371**:123–129.

Martin, J. H., and S. E. Fitzwater. 1988. Iron deficiency limits phytoplankton growth in the north-east Pacific subarctic. Nature **331**:341–343.

Martin, T. 2006. Paleontology. Early mammalian evolutionary experiments. Science **311**:1109–1110.

Martin, W. and T. M. Embley. 2004. Evolutionary biology: Early evolution comes full circle. Nature **431**:134–137.

Martin, W. and M. Muller. 1998. The hydrogen hypothesis for the first eukaryote. Nature **392**:37–41.

Martini, A. M., J. M. Budai, L. M. Walter, and M. Schoell. 1996. Microbial generation of economic accumulations of methane within a shallow organic-rich shale. Nature **383**:155–158.

Mason S. D., R. A. Howlett, M. J. Kim, I. M. Olfert, M. C. Hogan, W. McNulty, R. P. Hickey, P. D. Wagner, C. R. Kahn, F. J. Giordano, and R. S. Johnson. 2004. Loss of skeletal muscle HIF-1α results in altered exercise Endurance, PLoS Biology 2: No. 10

Masoro, E. J., B. P. Yu, and H. A. Bertrand. 1982. Action of food restriction in delaying the aging process. Proc. Natl. Acad. Sci. U. S. A 79:4239–4241.

Matson, P. A., R. Naylor, and I. Ortiz-Monasterio, I. 1998. Integration of environmental, agronomic, and economic aspects of fertilizer management. Science 280:112–115.

Matson, P. A., W. J. Parton, A. G. Power, and M. J. Swift. 1997. Agricultural iIntensification and ecosystem properties. Science 277:504–509.

Matsumura, K. 1995. Tetrodotoxin as a pheromone. Nature 378:563–564.

Matsuzaki, M., O. Misumi, I. Shin, S. Maruyama, M. Takahara, S. Y. Miyagishima, T. Mori, K. Nishida, and T. Kuroiwa. 2004. Genome sequence of the ultrasmall unicellular red alga Cyanidioschyzon merolae 10D. Nature 428:653–657.

Matz, C. and S. Kjelleberg. 2005. Off the hook–how bacteria survive protozoan grazing. Trends Microbiol. 13:302–307.

Matz, C., P. Deines, J. Boenigk, H. Arndt, L. Eberl, S. Kjelleberg, and K. Jurgens. 2004. Impact of violacein-producing bacteria on survival and feeding of bacterivorous nanoflagellates. Appl. Environ. Microbiol. 70:1593–1599.

Matz, C. and K. Jurgens. 2005. High motility reduces grazing mortality of planktonic bacteria. Appl. Environ. Microbiol. 71:921–929.

Mayntz, D., D. Raubenheimer, M. Salomon, S. Toft, and S. J. Simpson. 2005. Nutrient-specific foraging in invertebrate predators. Science 307:111–113.

Mazmanian, S., C. Liu, A. Tzianabos, and D. Kasper. An Immunomodulatory Molecule of Symbiotic Bacteria Directs Maturation of the Host Immune System. Cell 122:107–118.

McCauley, E., R. M. Nisbet, W. W. Murdoch, A. M. de Roos, W. S. C. Gurney, 1999. Large-amplitude cycles of Daphnia and its algal prey in enriched environments. Nature 402:653–656.

McDermott, G., S. M. Prince, A. A. Freer, A. M. Hawthornthwaite-Lawless, M. Z. Papiz, R. J. Cogdell, and N. W. Isaacs. 1995. Crystal structure of an integral membrane light-harvesting complex from photosynthetic bacteria. Nature 374:517–521.

McDougall, I., F. H. Brown, and J. G. Fleagle. 2005. Stratigraphic placement and age of modern humans from Kibish, Ethiopia. Nature 433:733–736.

McFadden, G. 1999. Chloroplasts. Ever decreasing circles. Nature 400:119–120.

McHenry, C. R., A. G. Cook, and S. Wroe. 2005. Bottom-feeding plesiosaurs. Science 310:75.

McKay, D. S., E. K. Gibson, Jr., K. L. Thomas-Keprta, H. Vali, C. S. Romanek, S. J. Clemett, X. D. Chillier, C. R. Maechling, and R. N. Zare. 1996. Search for past life on Mars: possible relic biogenic activity in martian meteorite ALH84001. Science 273:924–930.

McNeil, B. and L. Harvey. 2005. Energy well spent on a prokaryotic genome. Nat. Biotechnol. 23:186–187.

McNeil, N. I. 1984. The contribution of the large intestine to energy supplies in man. Am. J. Clin. Nutr. 39:338–342.

Meehl, G. A. and C. Tebaldi. 2004. More intense, more frequent, and longer lasting heat waves in the 21st century. Science 305:994–997.

Meibom, K. L., M. Blokesch, N. A. Dolganov, C. Y. Wu, and G. K. Schoolnik. 2005. Chitin induces natural competence in Vibrio cholerae. Science 310:1824–1827.

Meier, T., P. Polzer, K. Diederichs, W. Welte, and P. Dimroth. 2005. Structure of the rotor ring of F-Type Na+-ATPase from Ilyobacter tartaricus. Science **308**:659–662.

Melis, A. P., B. Hare, and M. Tomasello. 2006. Chimpanzees recruit the best collaborators. Science **311**:1297–1300.

Mellars, P. 1998. The fate of the Neanderthals. Nature **395**:539–540.

Mellars, P. 2006. A new radiocarbon revolution and the dispersal of modern humans in Eurasia. Nature **439**:931–935.

Melov, S., J. Ravenscroft, S. Malik, M. S. Gill, D. W. Walker, P. E. Clayton, D. C. Wallace, B. Malfroy, S. R. Doctrow, and G. J. Lithgow. 2000. Extension of life-span with superoxide dismutase/catalase mimetics. Science **289**:1567–1569.

Melyan, Z., E. E. Tarttelin, J. Bellingham, R. J. Lucas, and M. W. Hankins. 2005. Addition of human melanopsin renders mammalian cells photoresponsive. Nature **433**: 741–745.

Mercader, J., M. Panger, and C. Boesch. 2002. Excavation of a chimpanzee stone tool site in the African rainforest. Science **296**:1452–1455.

Merrell, D. S., S. M. Butler, F. Qadri, N. A. Dolganov, A. Alam, M. B. Cohen, S. B. Calderwood, G. K. Schoolnik, and A. Camilli. 2002. Host-induced epidemic spread of the cholera bacterium. Nature **417**:642–645.

Messner, K. R. and J. A. Imlay. 2002. Mechanism of superoxide and hydrogen peroxide formation by fumarate reductase, succinate dehydrogenase, and aspartate oxidase. J. Biol. Chem. **277**:42563–42571.

Meyer, A., T. D. Kocher, P. Basasibwaki, and A. C. Wilson. 1990. Monophyletic origin of Lake Victoria cichlid fishes suggested by mitochondrial DNA sequences. Nature **347**:550–553.

Meyer, J. L., E.T. Schultz, and G. S. Helfman. 1983. Fish schools: An asset to corals. Science, **220**:1047–1049.

Mi, S., X. Lee, X. Li, G. M. Veldman, H. Finnerty, L. Racie, E. LaVallie, X. Y. Tang, P. Edouard, S. Howes, J. C. Keith, Jr., and J. M. McCoy. 2000. Syncytin is a captive retroviral envelope protein involved in human placental morphogenesis. Nature **403**:785–789.

Michel, H. 1998. The mechanism of proton pumping by cytochrome c oxidasex127e comments]. Proc. Natl. Acad. Sci. U. S. A **95**:12819–12824.

Mietto, P., M. Avanzini, and G. Rolandi. 2003. Palaeontology: Human footprints in Pleistocene volcanic ash. Nature **422**:133.

Millard, A., M. R. Clokie, D. A. Shub, and N. H. Mann. 2004. Genetic organization of the psbAD region in phages infecting marine Synechococcus strains. Proc. Natl. Acad. Sci. U. S. A **101**:11007–11012.

Miller, G. H., M. L. Fogel, J. W. Magee, M. K. Gagan, S. J. Clarke, and B. J. Johnson. 2005. Ecosystem collapse in Pleistocene Australia and a human role in megafaunal extinction. Science **309**:287–290.

Miller, S. R., S. Augustine, T. L. Olson, R. E. Blankenship, J. Selker, and A. M. Wood. 2005. Discovery of a free-living chlorophyll d-producing cyanobacterium with a hybrid proteobacterial/cyanobacterial small-subunit rRNA gene. Proc. Natl. Acad. Sci. U. S. A **102**:850–855.

Mills, M. M., C. Ridame, M. Davey, J. La Roche, and R. J. Geider. 2004. Iron and phosphorus co-limit nitrogen fixation in the eastern tropical North Atlantic. Nature **429**:292–294.

Miralto, G., Barone, G. Romano, S. A. Poulet, A. Ianora, G. L. Russo, I. Buttino, G. Mazzarella, M. Laabir, M. Cabrini, and M. G. Giacobbe. 1999. The insidious effect of diatoms on copepod reproduction. Nature **402**:173–176.

Mironov, A. S., I. Gusarov, R. Rafikov, L. E. Lopez, K. Shatalin, R. A. Kreneva, D. A. Perumov, and E. Nudler. 2002. Sensing small molecules by nascent RNA: a mechanism to control transcription in bacteria. Cell **111**:747–756.

Mitchell, P. 1961. Coupling of phosphorylation to electron and hydrogen transfer by a chemiosmotic type of mechanism. Nature **191**:144–148.

Miyakawa, S., H. Yamanashi, K. Kobayashi, H. J. Cleaves, and S. L. Miller. 2002. Prebiotic synthesis from CO atmospheres: implications for the origins of life. Proc. Natl. Acad. Sci. U. S. A **99**:14628–14631.

Miyashita, H., H. Ikemoto, N. Kurano, K. Adachi, M. Chihara, and S. Miyachi. 1996. Chlorophyll d as a major pigment. Nature **383**:402–402.

Modis, Y., S. Ogata, D. Clements, and S. C. Harrison. 2004. Structure of the dengue virus envelope protein after membrane fusion. Nature **427**:313–319.

Mojzsis, S. J., G. Arrhenius, K. D. McKeegan, T. M. Harrison, A. P. Nutman, and C. R. Friend. 1996. Evidence for life on Earth before 3,800 million years ago. Nature **384**:55–59.

Mojzsis, S. J. 2003. Global change: probing early atmospheres. Nature **425**:249–250.

Moldowan, J. M. and N. M. Talyzina. 1998. Biogeochemical evidence for dinoflagellate ancestors in the early cambrian. Science **281**:1168–1170.

Molina-Heredia, F. P., J. Wastl, J. A. Navarro, D. S. Bendall, M. Hervas, C. J. Howe, and M. A. De La Rosa. 2003. Photosynthesis: a new function for an old cytochrome? Nature **424**:33–34.

Montoya, J. P., C. M. Holl, J. P. Zehr, A. Hansen, T. A. Villareal, and D. G. Capone. 2004. High rates of N_2 fixation by unicellular diazotrophs in the oligotrophic Pacific Ocean. Nature **430**:1027–1032.

Moore, B. D. and W. J. Foley. 2005. Tree use by koalas in a chemically complex landscape. Nature **435**:488–490.

Moore, L. R., G. Rocap, and S. W. Chisholm. 1998. Physiology and molecular phylogeny of coexisting *Prochlorococcus* ecotypes. Nature **393**:464–467.

Moore, P. B. and T. A. Steitz. 2002. The involvement of RNA in ribosome function. Nature **418**:229–235.

Moran, M. A., A. Buchan, J. M. Gonzalez, J. F. Heidelberg, W. B. Whitman, R. P. Kiene, J. R. Henriksen, and N. Ward. 2004. Genome sequence of *Silicibacter pomeroyi* reveals adaptations to the marine environment. Nature **432**:910–913.

Moreau, C. S., C. D. Bell, R. Vila, S. B. Archibald, and N. E. Pierce. 2006. Phylogeny of the ants: diversification in the age of angiosperms. Science **312**:101–104.

Moreira, D., H. Le Guyader, and H. Philippe. 2000. The origin of red algae and the evolution of chloroplasts. Nature **405**:69–72.

Morel, F. M. M., J. R. Reinfelder, S. B. Roberts, C. P. Chamberlain, J. G. Lee , and D. Yee. 1994. Zinc and carbon co-limitation of marine phytoplankton. Nature **369**:740–742.

Morowitz, H. J., J. D. Kostelnik, J. Yang, and G. D. Cody. 2000. The origin of intermediary metabolism. Proc. Natl. Acad. Sci. U. S. A **97**:7704–7708.

Morris, R. M., M. S. Rappe, S. A. Connon, K. L. Vergin, W. A. Siebold, C. A. Carlson, and S. J. Giovannoni. 2002. SAR11 clade dominates ocean surface bacterioplankton communities. Nature **420**:806–810.

Morse, D., P. Salois, P. Markovic, and J. W. Hastings. 1995. A nuclear-encoded form II RuBisCO in dinoflagellates. Science **268**:1622–1624.

Morwood, M. J., R. P. Soejono, R. G. Roberts, T. Sutikna, C. S. Turney, K. E. Westaway, W. J. Rink, J. X. Zhao, G. D. van den Bergh, R. A. Due, D. R. Hobbs, M. W. Moore, M. I. Bird, and L. K. Fifield. 2004. Archaeology and age of a new hominin from Flores in eastern Indonesia. Nature **431**:1087–1091.

Morwood, M. J., P. B. O'Sullivan, F. Aziz, and A. Raza. 1998. Fission-track ages of stone tools and fossils on the east Indonesian island of Flores. Nature **392**:173–176.

Moura, A. C. and P. C. Lee. 2004. Capuchin stone tool use in Caatinga dry forest. Science **306**:1909.

Mueller, U. G., S. A. Rehner, and T. R. Schultz. 1998. The evolution of agriculture in ants. Science **281**:2034–2038.

Murakami, A., H. Miyashita, M. Iseki, K. Adachi, and M. Mimuro. 2004. Chlorophyll d in an epiphytic cyanobacterium of red algae. Science **303**:1633.

Murata, T., I. Yamato, Y. Kakinuma, A. G. Leslie, and J. E. Walker. 2005. Structure of the rotor of the V-Type Na^+-ATPase from *Enterococcus hirae*. Science **308**:654–659.

Musser, R. O., S. M. Hum-Musser, H. Eichenseer, M. Peiffer, G. Ervin, J. B. Murphy, and G. W. Felton. 2002. Herbivory: caterpillar saliva beats plant defences. Nature **416**:599–600.

Muth, G. W., L. Ortoleva-Donnelly, and S. A. Strobel. 2000. A single adenosine with a neutral pK_a in the ribosomal peptidyl transferase center. Science **289**:947–950.

Myers, R. A. and B. Worm. 2003. Rapid worldwide depletion of predatory fish communities. Nature **423**:280–283.

Naeem, S. 2002. Biodiversity: biodiversity equals instability? Nature **416**:23–24.

Nagoshi, E., C. Saini, C. Bauer, T. Laroche, F. Naef, and U. Schibler. 2004. Circadian gene expression in individual fibroblasts: cell-autonomous and self-sustained oscillators pass time to daughter cells. Cell **119**:693–705.

Namslauer, A., A. S. Pawate, R. B. Gennis, and P. Brzezinski. 2003. Redox-coupled proton translocation in biological systems: proton shuttling in cytochrome c oxidase. Proc. Natl. Acad. Sci. U. S. A **100**:15543–15547.

Nandhagopal, N., A. A. Simpson, J. R. Gurnon, X. Yan, T. S. Baker, M. V. Graves, J. L. Van Etten, and M. G. Rossmann. 2002. The structure and evolution of the major capsid protein of a large, lipid-containing DNA virus. Proc. Natl. Acad. Sci. U. S. A **99**:14758–14763.

Narbonne, G. M. 2004. Modular construction of early Ediacaran complex life forms. Science **305**:1141–1144.

Nath, D. 2004. Cell biology: pathogen propulsion. Nature **427**:407.

Naylor, R. L., R. J. Goldburg, J. H. Primavera, N. Kautsky, M. C. Beveridge, J. Clay, C. Folke, J. Lubchenco, H. Mooney, and M. Troell. 2000. Effect of aquaculture on world fish supplies. Nature **405**:1017–1024.

Nealson, K. H. 2005. Hydrogen and energy flow as "sensed" by molecular genetics. Proc. Natl. Acad. Sci. U. S. A **102**:3889–3890.

NEEL, J. V. 1962. Diabetes mellitus: a "thrifty" genotype rendered detrimental by "progress"? Am. J. Hum. Genet. **14**:353–362.

Nelson, K. E., R. A. Clayton, S. R. Gill, M. L. Gwinn, R. J. Dodson, D. H. Haft, E. K. Hickey, J. D. Peterson, and C. M. Fraser. 1999. Evidence for lateral gene transfer between Archaea and bacteria from genome sequence of *Thermotoga maritima*. Nature **399**:323–329.

Nelson, W. A., E. McCauley, and F. J. Wrona. 2005. Stage-structured cycles promote genetic diversity in a predator-prey system of Daphnia and algae. Nature **433**:413–417.

Netherwood, T., S. M. Martin-Orue, A. G. O'Donnell, S. Gockling, J. Graham, J. C. Mathers, and H. J. Gilbert. 2004. Assessing the survival of transgenic plant DNA in the human gastrointestinal tract. Nat. Biotechnol. **22**:204–209.

Neutel, A. M., J. A. Heesterbeek, and P. C. De Ruiter. 2002. Stability in real food webs: weak links in long loops. Science **296**:1120–1123.

Nevitt, G. A., R. R. Veit, and P. Kareiva. 1995. Dimethyl sulphide as a foraging cue for Antarctic Procellariiform seabirds. Nature **376**:680–682.

Newcomb, W. W., R. M. Juhas, D. R. Thomsen, F. L. Homa, A. D. Burch, S. K. Weller, and J. C. Brown. 2001. The UL6 gene product forms the portal for entry of DNA into the herpes simplex virus capsid. J. Virol. **75**:10923–10932.

Nielsen, E. E., M. M. Hansen, C. Schmidt, D. Meldrup, and P. Gronkjaer. 2001. Fisheries. Population of origin of Atlantic cod. Nature **413**:272.

Nierman, W. C., A. Pain, M. J. Anderson, J. R. Wortman, H. S. Kim, J. Arroyo, M. Berriman, K. Abe, and D. W. Denning. 2005. Genomic sequence of the pathogenic and allergenic filamentous fungus *Aspergillus fumigatus*. Nature **438**:1151–1156.

Niewoehner, W. A., A. Bergstrom, D. Eichele, M. Zuroff, and J. T. Clark. 2003. Digital analysis: Manual dexterity in Neanderthals. Nature **422**:395.

Nisbet, E. G., J. R. Cann, C. Lee, and V. Dover. 1995. Origins of photosynthesis. Nature **373**:479–480.

Nisbet, E. G. and N. H. Sleep. 2001. The habitat and nature of early life. Nature **409**: 1083–1091.

Nishimura, Y., T. Yoshinari, K. Naruse, T. Yamada, K. Sumi, H. Mitani, T. Higashiyama, and T. Kuroiwa. 2006. Active digestion of sperm mitochondrial DNA in single living sperm revealed by optical tweezers. Proc. Natl. Acad. Sci. U. S. A **103**: 1382–1387.

Nissen, P., J. Hansen, N. Ban, P. B. Moore, and T. A. Steitz. 2000. The structural basis of ribosome activity in peptide bond synthesis. Science **289**:920–930.

Noji, H., R. Yasuda, M. Yoshida, and K. Kinosita, Jr. 1997. Direct observation of the rotation of F_1-ATPase. Nature **386**:299–302.

Noller, H. F., V. Hoffarth, and L. Zimniak. 1992. Unusual resistance of peptidyl transferase to protein extraction procedures. Science **256**:1416–1419.

Norell, M. A., J. M. Clark, L. M. Chiappe, and D. Dashzeveg. 1995. A nesting dinosaur. Nature **378**:774–776.

Norell, M. A., J. M. Clark, D. Demberelyin, B. Rhinchen, L. M. Chiappe, A. R Davidson, M. C. McKenna, P. Altangeral, and M. J. Novacek. 1994. A theropod dinosaur embryo and the affinities of the Flaming Cliffs dinosaur eggs. Science **266**:779–782.

Normile, D. and M. Enserink. 2003. SARS in China. Tracking the roots of a killer. Science **301**:297–299.

Novacek, M. J. 1985. Evidence for echolocation in the oldest known bats. Nature **315**: 140–141.

Novotny, V., Y. Basset, S. E. Miller, G. D. Weiblen, B. Bremer, L. Cizek, and P. Drozd. 2002. Low host specificity of herbivorous insects in a tropical forest. Nature **416**: 841–844.

Nussbaumer, A. D., C. R. Fisher, and M. Bright. 2006. Horizontal endosymbiont transmission in hydrothermal vent tubeworms. Nature **441**:345–348.

O'Dowd, C. D., M. C. Facchini, F. Cavalli, D. Ceburnis, M. Mircea, S. Decesari, S. Fuzzi, Y. J. Yoon, and J. P. Putaud. 2004. Biogenically driven organic contribution to marine aerosol. Nature **431**:676–680.

Ohman, M. D. and H. J. Hirche. 2001. Density-dependent mortality in an oceanic copepod population. Nature **412**:638–641.

Ohmoto, H., Y. Watanabe, and K. Kumazawa. 2004. Evidence from massive siderite beds for a CO_2-rich atmosphere before approximately 1.8 billion years ago. Nature **429**:395–399.

Ohta, K., D. S. Beall, J. P. Mejia, K. T. Shanmugam, and L. O. Ingram. 1991. Genetic improvement of *Escherichia coli* for ethanol production: chromosomal integration of *Zymomonas mobilis* genes encoding pyruvate decarboxylase and alcohol dehydrogenase II. Appl. Environ. Microbiol. **57**:893–900.

Olsen, E. M., M. Heino, G. R. Lilly, M. J. Morgan, J. Brattey, B. Ernande, and U. Dieckmann. 2004. Maturation trends indicative of rapid evolution preceded the collapse of northern cod. Nature **428**:932–935.

Orgel, L. E. 1998. The origin of life–a review of facts and speculations. Trends Biochem. Sci. **23**:491–495.

Orgel, L. E. and F. H. Crick. 1980. Selfish DNA: the ultimate parasite. Nature **284**:604–607.

Orgel, L. E., F. H. Crick, and C. Sapienza. 1980. Selfish DNA. Nature **288**:645–646.

Orgel, L. E. 1968. Evolution of the genetic apparatus. J. Mol. Biol. **38**:381–393.

Orgel, L. E. 2000. Self-organizing biochemical cycles. Proc. Natl. Acad. Sci. U. S. A **97**:12503–12507.

Orr, W. C. and R. S. Sohal. 1994. Extension of life-span by overexpression of superoxide dismutase and catalase in *Drosophila melanogaster*. Science **263**:1128–1130.

Orr, W. C., R. J. Mockett, J. J. Benes, and R. S. Sohal. 2003. Effects of overexpression of copper-zinc and manganese superoxide dismutases, catalase, and thioredoxin reductase genes on longevity in *Drosophila melanogaster*. J. Biol. Chem. **278**:26418–26422.

Ortiz-Garcia, S., E. Ezcurra, B. Schoel, F. Acevedo, J. Soberon, and A. A. Snow. 2005. Absence of detectable transgenes in local landraces of maize in Oaxaca, Mexico (2003–2004). Proc. Natl. Acad. Sci. U. S. A **102**:12338–12343.

Osborne, K. A., A. Robichon, E. Burgess, S. Butland, R. A. Shaw, A. Coulthard, H. S. Pereira, R. J. Greenspan, and M. B. Sokolowski. 1997. Natural behavior polymorphism due to a cGMP-dependent protein kinase of Drosophila. Science **277**:834–836.

Osyczka, A., C. C. Moser, F. Daldal, and P. L. Dutton. 2004. Reversible redox energy coupling in electron transfer chains. Nature **427**:607–612.

Ovchinnikov, I. V., A. Gotherstrom, G. P. Romanova, V. M. Kharitonov, K. Liden, and W. Goodwin. 2000. Molecular analysis of Neanderthal DNA from the northern Caucasus. Nature **404**:490–493.

Owen, J. J., M. D. Cooper, and M. C. Raff. 1974. In vitro generation of B lymphocytes in mouse foetal liver, a mammalian 'bursa equivalent'. Nature **249**:361–363.

Pace, N. R. 1991. Origin of life–facing up to the physical setting. Cell **65**:531–533.

Packer, A. and K. Clay. 2000. Soil pathogens and spatial patterns of seedling mortality in a temperate tree. Nature **404**:278–281.

Packer, C., D. Ikanda, B. Kissui, and H. Kushnir. 2005. Conservation biology: lion attacks on humans in Tanzania. Nature **436**:927–928.

Page, C. C., C. C. Moser, X. Chen, and P. L. Dutton. 1999. Natural engineering principles of electron tunnelling in biological oxidation-reduction. Nature **402**:47–52.

Paine, J. A., C. A. Shipton, S. Chaggar, R. M. Howells, M. J. Kennedy, G. Vernon, S. Y. Wright, E. Hinchliffe, J. L. Adams, A. L. Silverstone, and R. Drake. 2005. Improving the nutritional value of Golden Rice through increased pro-vitamin A content. Nat. Biotechnol. **23**:482–487.

Pal, C., B. Papp, and M. J. Lercher. 2005. Adaptive evolution of bacterial metabolic networks by horizontal gene transfer. Nat. Genet. **37**:1372–1375.

Pal, C., B. Papp, M. J. Lercher, P. Csermely, S. G. Oliver, and L. D. Hurst. 2006. Chance and necessity in the evolution of minimal metabolic networks. Nature **440**:667–670.

Palenik, B., B. Brahamsha, F. W. Larimer, M. Land, L. Hauser, P. Chain, J. Lamerdin, W. Regala, E. E. Allen, J. McCarren, I. Paulsen, A. Dufresne, F. Partensky, E. A. Webb, and J. Waterbury. 2003. The genome of a motile marine Synechococcus. Nature **424**:1037–1042.

Palmer, J. D. 1993. A genetic rainbow of plastids. Nature **364**:762–763.

Palumbi, S. R. 2001. Humans as the world's greatest evolutionary force. Science **293**:1786–1790.

Pandolfi, J. M., R. H. Bradbury, E. Sala, T. P. Hughes, K. A. Bjorndal, R. G. Cooke, D. McArdle, L. McClenachan, M. J. Newman, G. Paredes, R. R. Warner, and J. B. Jackson. 2003. Global trajectories of the long-term decline of coral reef ecosystems. Science **301**:955–958.

Papp, B., C. Pal, and L. D. Hurst. 2004. Metabolic network analysis of the causes and evolution of enzyme dispensability in yeast. Nature **429**:661–664.

Parker, J. D., K. M. Parker, B. H. Sohal, R. S. Sohal, and L. Keller. 2004. Decreased expression of Cu-Zn superoxide dismutase 1 in ants with extreme lifespan. Proc. Natl. Acad. Sci. U. S. A **101**:3486–3489.

Parkes, J. 1999. Cracking anaerobic bacteria. Nature **401**:217–218.

Parkes, R. J., B. A. Cragg, S. J. Bale, J. M. Getlifff, K. Goodman, P. A. Rochelle, J. C. Fry, A. J. Weightman, and S. M. Harvey. 1994. Deep bacterial biosphere in Pacific Ocean sediments. Nature **371**:410–413.

Parkes, R. J., G. Webster, B. A. Cragg, A. J. Weightman, C. J. Newberry, T. G. Ferdelman, J. Kallmeyer, B. B. Jorgensen, I. W. Aiello, and J. C. Fry. 2005. Deep sub-seafloor prokaryotes stimulated at interfaces over geological time. Nature **436**:390–394.

Parmesan, C., N. Ryrholm, C. Stefanescu, J. K. Hill, C. D. Thomas, H. Descimon, B. Huntley, L. Kaila, J. Kullberg, T. Tammaru, and W. John. 1999. Poleward shifts in geographical ranges of butterfly species associated with regional warming. Nature **399**:579–583.

Parsons, I., I, M. R. Lee, and J. V. Smith. 1998. Biochemical evolution II: origin of life in tubular microstructures on weathered feldspar surfaces. Proc. Natl. Acad. Sci. U. S. A **95**:15173–15176.

Partida-Martinez, L. P. and C. Hertweck. 2005. Pathogenic fungus harbours endosymbiotic bacteria for toxin production. Nature **437**:884–888.

Pascual, M., X. Rodo, S. P. Ellner, R. Colwell, and M. J. Bouma. 2000. Cholera dynamics and El Nino-Southern Oscillation. Science **289**:1766–1769.

Pastor, J. and R. A. Moen. 2004. Palaeontology: ecology of ice-age extinctions. Nature **431**:639–640.

Pathela, P., K. Z. Hasan, E. Roy, K. Alam, F. Huq, A. K. Siddique, and R. B. Sack. 2005. Enterotoxigenic *Bacteroides fragilis*-associated diarrhea in children 0–2 years of age in rural Bangladesh. J. Infect. Dis. **191**:1245–1252.

Pauly, D., and V. Christensen. 1995. Primary production required to sustain global fisheries. Nature **374**:255–257.

Pauly, D., V. Christensen, V, J. Dalsgaard, R. Froese, and F. Torres, Jr. 1998. Fishing down marine food webs. Science **279**:860–863.

Pauly, D., V. Christensen, S. Guenette, T. J. Pitcher, U. R. Sumaila, C. J. Walters, R. Watson, and D. Zeller. 2002. Towards sustainability in world fisheries. Nature **418**:689–695.

Pecher, I. A. 2002. Oceanography: Gas hydrates on the brink. Nature **420**:622–623.

Pedulla, M. L., M. E. Ford, J. M. Houtz, T. Karthikeyan, C. Wadsworth, J. A. Lewis, D. Jacobs-Sera, J. Falbo, J. Gross, N. R. Pannunzio, W. Brucker, V. Kumar,

J. Kandasamy, L. Keenan, S. Bardarov, J. Kriakov, J. G. Lawrence, W. R. Jacobs, Jr., R. W. Hendrix, and G. F. Hatfull. 2003. Origins of highly mosaic mycobacteriophage genomes. Cell **113**:171–182.

Peers, G. and N. M. Price. 2006. Copper-containing plastocyanin used for electron transport by an oceanic diatom. Nature **441**:341–344.

Pergams, O. R., W. M. Barnes, and D. Nyberg. 2003. Mammalian microevolution: Rapid change in mouse mitochondrial DNA. Nature **423**:397.

Pernthaler, J., E. Zollner, F. Warnecke, and K. Jurgens. 2004. Bloom of filamentous bacteria in a mesotrophic lake: identity and potential controlling mechanism. Appl. Environ. Microbiol. **70**:6272–6281.

Perry, L., D. H. Sandweiss, D. R. Piperno, K. Rademaker, M. A. Malpass, A. Umire, and L. De, V. 2006. Early maize agriculture and interzonal interaction in southern Peru. Nature **440**:76–79.

Peterson, K. J. and N. J. Butterfield. 2005. Origin of the Eumetazoa: testing ecological predictions of molecular clocks against the Proterozoic fossil record. Proc. Natl. Acad. Sci. U. S. A **102**:9547–9552.

Petit, J. R., J. Jouzel, D. Raynaud, N. I. Barkov, J.-M. Barnola, I. Basile, M. Bender, J. Chappellaz, M. Davis, G. Delaygue, M. Delmotte, V. M. Kotlyakov, and M. Legrand. 1999. Climate and atmospheric history of the past 420,000 years from the Vostok ice core, Antarctica. Nature **399**:429–436.

Pfeiffer, T., S. Schuster, and S. Bonhoeffer. 2001. Cooperation and competition in the evolution of ATP-producing pathways. Science **292**:504–507.

Pfisterer, A. B. and B. Schmid. 2002. Diversity-dependent production can decrease the stability of ecosystem functioning. Nature **416**:84–86.

Phillipson, D. W. 1986. African pre-history: Life in the Lake Victoria basin. Nature **320**:110–111.

Pickett, C. J. H. 1996. The Chatt cycle and the mechanism of enzymic reduction of molecular nitrogen. J. Biol. Inorg. Chem. **1**:601–606.

Piel, J., D. Hui, G. Wen, D. Butzke, M. Platzer, N. Fusetani, and S. Matsunaga. 2004. Antitumor polyketide biosynthesis by an uncultivated bacterial symbiont of the marine sponge *Theonella swinhoei*. Proc. Natl. Acad. Sci. U. S. A **101**:16222–16227.

Piel, J. 2002. A polyketide synthase-peptide synthetase gene cluster from an uncultured bacterial symbiont of Paederus beetles. Proc. Natl. Acad. Sci. U. S. A **99**:14002–14007.

Piffanelli, P., L. Ramsay, R. Waugh, A. Benabdelmouna, A. D'Hont, K. Hollricher, J. H. Jorgensen, P. Schulze-Lefert, and R. Panstruga. 2004. A barley cultivation-associated polymorphism conveys resistance to powdery mildew. Nature **430**:887–891.

Pimentel, D., C. Harvey, P. Resosudarmo, K. Sinclair, D. Kurtz, M. McNair, S. Crist, L. Spritz, L. Fitton, R. Saffouri, and R. Blair. 1995. Environmental and economic costs of soil erosion and conservation benefits. Science **267**:1117–1123.

Piperno, D. R. and K. E. Stothert. 2003. Phytolith evidence for early Holocene Cucurbita domestication in southwest Ecuador. Science **299**:1054–1057.

Piperno, D. R. and H. D. Sues. 2005. Paleontology. Dinosaurs dined on grass. Science **310**:1126–1128.

Piperno, D. R., A. J. Ranere, I. Holst, and P. Hansell. 2000. Starch grains reveal early root crop horticulture in the Panamanian tropical forest. Nature **407**:894–897.

Platt, T ., C. Fuentes-Yaco, and K. T. Frank. 2003. Marine ecology: Spring algal bloom and larval fish survival. Nature **423**:398–399.

Pley, H. W., K. M. Flaherty, and D. B. McKay. 1994. Three-dimensional structure of a hammerhead ribozyme. Nature **372**:68–74.

Poinar, H. N., C. Schwarz, J. Qi, B. Shapiro, R. D. Macphee, B. Buigues, A. Tikhonov, D. H. Huson, L. P. Tomsho, A. Auch, M. Rampp, W. Miller, and S. C. Schuster. 2006. Metagenomics to paleogenomics: large-scale sequencing of mammoth DNA. Science **311**:392–394.

Poinar, H. N., M. Kuch, K. D. Sobolik, I. Barnes, A. B. Stankiewicz, T. Kuder, W. G. Spaulding, V. M. Bryant, A. Cooper, and S. Paabo. 2001. A molecular analysis of dietary diversity for three archaic Native Americans. Proc. Natl. Acad. Sci. U. S. A **98**:4317–4322.

Polacek, N., M. Gaynor, A. Yassin, and A. S. Mankin. 2001. Ribosomal peptidyl transferase can withstand mutations at the putative catalytic nucleotide. Nature **411**:498–501.

Polson, A. G., B. L. Bass, and J. L. Casey. 1996. RNA editing of hepatitis delta virus antigenome by dsRNA-adenosine deaminase. Nature **380**:454–456.

Ponton, F., C. Lebarbenchon, T. Lefevre, D. G. Biron, D. Duneau, D. P. Hughes, and F. Thomas. 2006. Parasitology: parasite survives predation on its host. Nature **440**:756.

Pope, K. O., M. E. Pohl, J. G. Jones, D. L. Lentz, C. von Nagy, F. J. Vega, and I. R. Quitmyer. 2001. Origin and environmental setting of ancient agriculture in the lowlands of Mesoamerica. Science **292**:1370–1373.

Postel, S. L., G. C. Daily, and P. R. Ehrlich. 1996. Human appropriation of renewable fresh water. Science **271**: 785–788.

Poulsen, M. and J. J. Boomsma. 2005. Mutualistic fungi control crop diversity in fungus-growing ants. Science **307**:741–744.

Prasad, B. V., R. Rothnagel, C. Q. Zeng, J. Jakana, J. A. Lawton, W. Chiu, and M. K. Estes. 1996. Visualization of ordered genomic RNA and localization of transcriptional complexes in rotavirus. Nature **382**:471–473.

Prasad, V., C. A. Stromberg, H. Alimohammadian, and A. Sahni. 2005. Dinosaur coprolites and the early evolution of grasses and grazers. Science **310**:1177–1180.

Prust, C., M. Hoffmeister, H. Liesegang, A. Wiezer, W. F. Fricke, A. Ehrenreich, G. Gottschalk, and U. Deppenmeier. 2005. Complete genome sequence of the acetic acid bacterium *Gluconobacter oxydans*. Nat. Biotechnol. **23**:195–200.

Przyborski, J. and M. Lanzer. 2004. Parasitology. The malarial secretome. Science **306**:1897–1898.

Puntis, J. and H. Kirpalani. 1998. Reappraisal of Malthus. Lancet **352**:241–242.

Purcell, J.E. 1980. Influence of siphonophore behavior on their natural diets; evidence for aggressive mimicry. Science **209**:1045–1047.

Qaim, M. and D. Zilberman. 2003. Yield effects of genetically modified crops in developing countries. Science **299**:900–902.

Qiu, X., T. Kumbalasiri, S. M. Carlson, K. Y. Wong, V. Krishna, I. Provencio, and D. M. Berson. 2005. Induction of photosensitivity by heterologous expression of melanopsin. Nature **433**:745–749.

Quist, D. and I. H. Chapela. 2001. Transgenic DNA introgressed into traditional maize landraces in Oaxaca, Mexico. Nature **414**:541–543.

Ragatz, L., Z. Jiang, C. Bauer, and H. Gest. 1994. Phototactic purple bacteria. Nature **370**:104–104.

Raghoebarsing, A. A., A. J. Smolders, M. C. Schmid, W. I. Rijpstra, M. Wolters-Arts, J. Derksen, M. S. Jetten, S. Schouten, J. S. Sinninghe Damste, L. P. Lamers, J. G. Roelofs, H. J. Op den Camp, and M. Strous. 2005. Methanotrophic symbionts provide carbon for photosynthesis in peat bogs. Nature **436**:1153–1156.

Raghoebarsing, A. A., A. Pol, van de Pas-Schoonen KT, A. J. Smolders, K. F. Ettwig, W. I. Rijpstra, S. Schouten, J. S. Damste, H. J. Op den Camp, M. S. Jetten, and

M. Strous. 2006. A microbial consortium couples anaerobic methane oxidation to denitrification. Nature **440**:918–921.

Ram, R. J., N. C. Verberkmoes, M. P. Thelen, G. W. Tyson, B. J. Baker, R. C. Blake, M. Shah, R. L. Hettich, and J. F. Banfield. 2005. Community proteomics of a natural microbial biofilm. Science **308**:1915–1920.

Randau, L., R. Munch, M. J. Hohn, D. Jahn, and D. Soll. 2005. *Nanoarchaeum equitans* creates functional tRNAs from separate genes for their 5′- and 3′-halves. Nature **433**:537–541.

Raoult, D., S. Audic, C. Robert, C. Abergel, P. Renesto, H. Ogata, B. La Scola, M. Suzan, and J. M. Claverie. 2004. The 1.2-megabase genome sequence of Mimivirus. Science **306**:1344–1350.

Rappe, M. S., S. A. Connon, K. L. Vergin, and S. J. Giovannoni. 2002. Cultivation of the ubiquitous SAR11 marine bacterioplankton clade. Nature **418**:630–633.

Rasmann, S., T. G. Kollner, J. Degenhardt, I. Hiltpold, S. Toepfer, U. Kuhlmann, J. Gershenzon, and T. C. Turlings. 2005. Recruitment of entomopathogenic nematodes by insect-damaged maize roots. Nature **434**:732–737.

Rasmussen, B. 2000. Filamentous microfossils in a 3,235-million-year-old volcanogenic massive sulphide deposit. Nature **405**:676–679.

Rattenborg, N. C., S. L. Lima, and C. J. Amlaner. Half-awake to the risk of predation. Nature **397**:397–398.

Ravasz, E., A. L. Somera, D. A. Mongru, Z. N. Oltvai, and A. L. Barabasi. 2002. Hierarchical organization of modularity in metabolic networks. Science **297**: 1551–1555.

Raven, P. H. 2002. Presidential address. Science, sustainability, and the human prospect. Science **297**:954–958.

Rayfield, E. J., D. B. Norman, C. C. Horner, J. R. Horner, P. M. Smith, J. J. Thomason, and P. Upchurch. 2001. Cranial design and function in a large theropod dinosaur. Nature **409**:1033–1037.

Raymond, J., O. Zhaxybayeva, J. P. Gogarten, S. Y. Gerdes, and R. E. Blankenship. 2002. Whole-genome analysis of photosynthetic prokaryotes. Science **298**:1616–1620.

Raymond, J. and D. Segre. 2006. The effect of oxygen on biochemical networks and the evolution of complex life. Science **311**:1764–1767.

Raymond, P. A. and J. E. Bauer. 2001. Riverine export of aged terrestrial organic matter to the North Atlantic Ocean. Nature **409**:497–500.

Redecker, D., R. Kodner, and L. E. Graham. 2000. Glomalean fungi from the Ordovician. Science **289**:1920–1921.

Reid, A. H., T. G. Fanning, J. V. Hultin, and J. K. Taubenberger. 1999. Origin and evolution of the 1918 "Spanish" influenza virus hemagglutinin gene. Proc. Natl. Acad. Sci. U. S. A **96**:1651–1656.

Reid, R. P., P. T. Visscher, A. W. Decho, J. F. Stolz, B. M. Bebout, C. Dupraz, I. G. Macintyre, H. W. Paerl, J. L. Pinckney, L. Prufert-Bebout, T. F. Steppe, and D. J. DesMarais. 2000. The role of microbes in accretion, lamination and early lithification of modern marine stromatolites. Nature **406**:989–992.

Reid, R. T., D. H. Live, D. J. Faulkner, and A. Butler. 1993. A siderophore from a marine bacterium with an exceptional ferric ion affinity constant. Nature **366**:455–458.

Reinfelder, J. R., A. M. Kraepiel, and F. M. Morel. 2000. Unicellular C4 photosynthesis in a marine diatom. Nature **407**:996–999.

Reinisch, K. M., M. L. Nibert, and S. C. Harrison. 2000. Structure of the reovirus core at 3.6 A resolution. Nature **404**:960–967.

Reisz, R. R., D. Scott, H. D. Sues, D. C. Evans, and M. A. Raath. 2005. Embryos of an early Jurassic prosauropod dinosaur and their evolutionary significance. Science **309**:761–764.

Relman, D. A. and S. Falkow. 2001. The meaning and impact of the human genome sequence for microbiology. Trends Microbiol. **9**:206–208.

Rendulic, S., P. Jagtap, A. Rosinus, M. Eppinger, C. Baar, C. Lanz, H. Keller, C. Lambert, K. J. Evans, A. Goesmann, F. Meyer, R. E. Sockett, and S. C. Schuster. 2004. A predator unmasked: life cycle of *Bdellovibrio bacteriovorus* from a genomic perspective. Science **303**:689–692.

Reznick, D. N., M. Mateos, and M. S. Springer. 2002. Independent origins and rapid evolution of the placenta in the fish genus Poeciliopsis. Science **298**:1018–1020.

Rhee, K. H., E. P. Morris, J. Barber, and W. Kuhlbrandt. 1998. Three-dimensional structure of the plant photosystem II reaction centre at 8 A resolution. Nature **396**:283–286.

Ribbins, W. E., J. N. Kaplanis, M. J. Thompson, T. J. Shortino, C. F. Cohen, and S. C. Joyner. 1967. Ecdysones and analogs: effects on development and reproduction of insects. Science **161**:1158–1159.

Ricchetti, M., C. Fairhead, and B. Dujon. 1999. Mitochondrial DNA repairs double-strand breaks in yeast chromosomes. Nature **402**:96–100.

Zhang, M., C. Eddy, K. Deanda, M. Finkelstein, and S. Picataggio. 1995. Metabolic Engineering of a Pentose Metabolism Pathway in Ethanologenic Zymomonas mobilis. Science **267**:240–243.

Rice, G., L. Tang, K. Stedman, F. Roberto, J. Spuhler, E. Gillitzer, J. E. Johnson, T. Douglas, and M. Young. 2004. The structure of a thermophilic archaeal virus shows a double-stranded DNA viral capsid type that spans all domains of life. Proc. Natl. Acad. Sci. U. S. A **101**:7716–7720.

Rice, G., K. Stedman, J. Snyder, B. Wiedenheft, D. Willits, S. Brumfield, T. McDermott, and M. J. Young. 2001. Viruses from extreme thermal environments. Proc. Natl. Acad. Sci. U. S. A **98**:13341–13345.

Richards, M. P., R. J. Schulting, and R. E. Hedges. 2003. Archaeology: sharp shift in diet at onset of Neolithic. Nature **425**:366.

Riebesell, U., D. A. Wolf-Gladrow, and V. Smetacek. 1993. Carbon dioxide limitation of marine phytoplankton growth rates. Nature **361**:249–251.

Riebesell, U., I. Zondervan, B. Rost, P. D. Tortell, R. E. Zeebe, and F. M. Morel. 2000. Reduced calcification of marine plankton in response to increased atmospheric CO_2. Nature **407**:364–367.

Rieger, M. A., M. Lamond, C. Preston, S. B. Powles, and R. T. Roush. 2002. Pollen-mediated movement of herbicide resistance between commercial canola fields. Science **296**:2386–2388.

Righton, D., J. Metcalfe, and P. Connolly. 2001. Fisheries. Different behaviour of North and Irish Sea cod. Nature **411**:156.

Riley, J. R., U. Greggers, A. D. Smith, D. R. Reynolds, and R. Menzel. 2005. The flight paths of honeybees recruited by the waggle dance. Nature **435**:205–207.

Riley, S., C. Fraser, C. A. Donnelly, A. C. Ghani, L. J. Abu-Raddad, A. J. Hedley, G. M. Leung, and R. M. Anderson. 2003. Transmission dynamics of the etiological agent of SARS in Hong Kong: impact of public health interventions. Science **300**:1961–1966.

Rivera, M. C. and J. A. Lake. 2004. The ring of life provides evidence for a genome fusion origin of eukaryotes. Nature **431**:152–155.

Rivera, M. C., R. Jain, J. E. Moore, and J. A. Lake. 1998. Genomic evidence for two functionally distinct gene classes. Proc. Natl. Acad. Sci. U. S. A **95**:6239–6244.

Roberts, C. M., J. A. Bohnsack, F. Gell, J. P. Hawkins, and R. Goodridge. 2001. Effects of marine reserves on adjacent fisheries. Science **294**:1920–1923.

Roberts, L. 2004. Vaccines. Rotavirus vaccines' second chance. Science **305**:1890–1893.

Roberts, M. B., C. B. Stringer, and S. A. Parfitt. 1994. A hominid tibia from Middle Pleistocene sediments at Boxgrove, UK. Nature **369**:311–313.

Roberts, R. G., T. F. Flannery, L. K. Ayliffe, H. Yoshida, J. M. Olley, G. J. Prideaux, G. M. Laslett, A. Baynes, M. A. Smith, R. Jones, and B. L. Smith. 2001. New ages for the last Australian megafauna: continent-wide extinction about 46,000 years ago. Science **292**:1888–1892.

Rocap, G., F. W. Larimer, J. Lamerdin, S. Malfatti, P. Chain, N. A. Ahlgren, A. Arellano, M. Coleman, and S. W. Chisholm. 2003. Genome divergence in two *Prochlorococcus* ecotypes reflects oceanic niche differentiation. Nature **424**:1042–1047.

Roche, H., A. Delagnes, J. P. Brugal, C. Feibel, M. Kibunjia, V. Mourre, and P. J. Texier. 1999. Early hominid stone tool production and technical skill 2.34 Myr ago in West Turkana, Kenya. Nature **399**:57–60.

Roger, A. J. and J. D. Silberman. 2002. Cell evolution: mitochondria in hiding. Nature **418**:827–829.

Rohde, D. L., S. Olson, and J. T. Chang. 2004. Modelling the recent common ancestry of all living humans. Nature **431**:562–566.

Rojstaczer, S., S. M. Sterling, and N. J. Moore. 2001. Human appropriation of photosynthesis products. Science **294**:2549–2552.

Rokas, A., D. Kruger, and S. B. Carroll. 2005. Animal evolution and the molecular signature of radiations compressed in time. Science **310**:1933–1938.

Roman, J. and S. R. Palumbi. 2003. Whales before whaling in the North Atlantic. Science **301**:508–510.

Rondelez, Y., G. Tresset, T. Nakashima, Y. Kato-Yamada, H. Fujita, S. Takeuchi, and H. Noji. 2005. Highly coupled ATP synthesis by F1-ATPase single molecules. Nature **433**:773–777.

Rose, G. A. 1993. Cod spawning on a migration highway in the north-west Atlantic. Nature **366**:458–461.

Rossant, J. and J. C. Cross. 2001. Placental development: lessons from mouse mutants. Nat. Rev. Genet. **2**:538–548.

Roszak, A. W., T. D. Howard, J. Southall, A. T. Gardiner, C. J. Law, N. W. Isaacs, and R. J. Cogdell. 2003. Crystal structure of the RC-LH1 core complex from *Rhodopseudomonas palustris*. Science **302**:1969–1972.

Rota, P. A., M. S. Oberste, S. S. Monroe, W. A. Nix, R. Campagnoli, J. P. Icenogle, and W. J. Bellini. 2003. Characterization of a novel coronavirus associated with severe acute respiratory syndrome. Science **300**:1394–1399.

Rousseau, D. L. 1999. Bioenergetics. Two phases of proton translocation. Nature **400**:412–413.

Rouvinen, J., T. Bergfors, T. Teeri, J. K. Knowles, and T. A. Jones. 1990. Three-dimensional structure of cellobiohydrolase II from *Trichoderma reesei*. Science **249**:380–386.

Rowe, T. 1999. At the roots of the mammalian family tree. Nature **398**:283–284.

Royant, A., K. Edman, T. Ursby, E. Pebay-Peyroula, E. M. Landau, and R. Neutze. 2000. Helix deformation is coupled to vectorial proton transport in the photocycle of bacteriorhodopsin. Nature **406**:645–648.

Ruitenberg, M., A. Kannt, E. Bamberg, K. Fendler, and H. Michel. 2002. Reduction of cytochrome c oxidase by a second electron leads to proton translocation. Nature **417**:99–102.

Ruxton, G. D. and M. P. Speed. 2005. Evolution: a taste for mimicry. Nature **433**:205–207.

Salama, N., K. Guillemin, T. K. McDaniel, G. Sherlock, L. Tompkins, and S. Falkow. 2000. A whole-genome microarray reveals genetic diversity among *Helicobacter pylori* strains. Proc. Natl. Acad. Sci. U. S. A **97**:14668–14673.

Salamini, F., H. Ozkan, A. Brandolini, R. Schafer-Pregl, and W. Martin. 2002. Genetics and geography of wild cereal domestication in the near east. Nat. Rev. Genet. **3**:429–441.

Salehi-Ashtiani, K. and J. W. Szostak. 2001. In vitro evolution suggests multiple origins for the hammerhead ribozyme. Nature **414**:82–84.

Sambongi, Y., Y. Iko, M. Tanabe, H. Omote, A. Iwamoto-Kihara, I. Ueda, T. Yanagida, Y. Wada, and M. Futai. 1999. Mechanical rotation of the c subunit oligomer in ATP synthase (F_0F_1): direct observation. Science **286**:1722–1724.

Sancetta, C. 1993. Green sea, black mud. Nature **362**:108–108.

Sanchez, P. A. 2002. Ecology. Soil fertility and hunger in Africa. Science **295**:2019–2020.

Sanchez, P. A. and M. S. Swaminathan. 2005a. Public health. Cutting world hunger in half. Science **307**:357–359.

Sanchez, P. A. and M. S. Swaminathan. 2005b. Hunger in Africa: the link between unhealthy people and unhealthy soils. Lancet **365**:442–444.

Sanderson, S. L., A. Y. Cheer, J. S. Goodrich, J. D. Graziano, and W. T. Callan. 2001. Crossflow filtration in suspension-feeding fishes. Nature **412**:439–441.

Sanudo-Wilhelmy, S. A., A. B. Kustka, C. J. Gobler, D. A. Hutchins, M. Yang, K. Lwiza, J. Burns, D. G. Capone, J. A. Raven, and E. J. Carpenter. 2001. Phosphorus limitation of nitrogen fixation by Trichodesmium in the central Atlantic Ocean. Nature **411**:66–69.

Sapra, R., K. Bagramyan, and M. W. Adams. 2003. A simple energy-conserving system: proton reduction coupled to proton translocation. Proc. Natl. Acad. Sci. U. S. A **100**:7545–7550.

Sarker, S. A., S. Sultana, G. J. Fuchs, N. H. Alam, T. Azim, H. Brussow, and L. Hammarstrom. 2005. *Lactobacillus paracasei* strain ST11 has no effect on rotavirus but ameliorates the outcome of nonrotavirus diarrhea in children from Bangladesh. Pediatrics **116**:e221–e228.

Sass, H. J., G. Buldt, R. Gessenich, D. Hehn, D. Neff, R. Schlesinger, J. Berendzen, and P. Ormos. 2000. Structural alterations for proton translocation in the M state of wild-type bacteriorhodopsin. Nature **406**:649–653.

Satin, J., J. W. Kyle, M. Chen, P. Bell, L. L. Cribbs, H. A. Fozzard, and R. B. Rogart. 1992. A mutant of TTX-resistant cardiac sodium channels with TTX-sensitive properties. Science **256**:1202–1205.

Sazanov, L. A. and P. Hinchliffe. 2006. Structure of the hydrophilic domain of respiratory complex I from *Thermus thermophilus*. Science **311**:1430–1436.

Scantlebury, M., J. R. Speakman, M. K. Oosthuizen, T. J. Roper, and N. C. Bennett. 2006. Energetics reveals physiologically distinct castes in a eusocial mammal. Nature **440**:795–797.

Schell, M. A., M. Karmirantzou, B. Snel, D. Vilanova, B. Berger, G. Pessi, M. C. Zwahlen, F. Desiere, P. Bork, M. Delley, R. D. Pridmore, and F. Arigoni. 2002. The genome sequence of *Bifidobacterium longum* reflects its adaptation to the human gastrointestinal tract. Proc. Natl. Acad. Sci. U. S. A **99**:14422–14427.

Schenkl, S., F. van Mourik, Z. G. van der, S. Haacke, and M. Chergui. 2005. Probing the ultrafast charge translocation of photoexcited retinal in bacteriorhodopsin. Science **309**:917–920.

Schiermeier, Q. 2003. Europe dithers as Canada cuts cod fishing. Nature **423**:212.

Schimel, J. 2000. Rice, microbes and methane. Nature **403**:375, 377.

Schindelin, H., C. Kisker, J. L. Schlessman, J. B. Howard, and D. C. Rees. 1997. Structure of ADP x AIF$_4$(-)-stabilized nitrogenase complex and its implications for signal transduction. Nature **387**:370–376.

Schippers, A., L. N. Neretin, J. Kallmeyer, T. G. Ferdelman, B. A. Cragg, R. J. Parkes, and B. B. Jorgensen. 2005. Prokaryotic cells of the deep sub-seafloor biosphere identified as living bacteria. Nature **433**:861–864.

Schmidt, B., J. McCracken, and S. Ferguson-Miller. 2003. A discrete water exit pathway in the membrane protein cytochrome c oxidase. Proc. Natl. Acad. Sci. U. S. A **100**: 15539–15542.

Schnitzer, M. J. 2001. Molecular motors. Doing a rotary two-step. Nature **410**:878–9, 881.

Scholin, C. A., F. Gulland, G. J. Doucette, S. Benson, M. Busman, F. P. Chavez, J. Cordaro, and F. M. Van Dolah. 2000. Mortality of sea lions along the central California coast linked to a toxic diatom bloom. Nature **403**:80–84.

Scholte, E. J., K. Ng'habi, J. Kihonda, W. Takken, K. Paaijmans, S. Abdulla, G. F. Killeen, and B. G. Knols. 2005. An entomopathogenic fungus for control of adult African malaria mosquitoes. Science **308**:1641–1642.

Schopf, J. W. 1993. Microfossils of the Early Archean Apex chert: New evidence of the antiquity of life. Science **260**:640–646.

Schopf, J. W. and B. M. Packer. 1987. Early Archean (3.3-billion to 3.5-billion-year-old) microfossils from Warrawoona Group, Australia. Science **237**:70–73.

Schopf, J. W., A. B. Kudryavtsev, D. G. Agresti, T. J. Wdowiak, and A. D. Czaja. 2002. Laser–Raman imagery of Earth's earliest fossils. Nature **416**:73–76.

Schriner, S. E., N. J. Linford, G. M. Martin, P. Treuting, C. E. Ogburn, M. Emond, P. E. Coskun, W. Ladiges, N. Wolf, H. Van Remmen, D. C. Wallace, and P. S. Rabinovitch. 2005. Extension of murine life span by overexpression of catalase targeted to mitochondria. Science **308**:1909–1911.

Schubbert, R., D. Renz, B. Schmitz, and W. Doerfler. 1997. Foreign (M13) DNA ingested by mice reaches peripheral leukocytes, spleen, and liver via the intestinal wall mucosa and can be covalently linked to mouse DNA. Proc. Natl. Acad. Sci. U. S. A **94**:961–966.

Schubert, W. D., O. Klukas, W. Saenger, H. T. Witt, P. Fromme, and N. Krauss. 1998. A common ancestor for oxygenic and anoxygenic photosynthetic systems: a comparison based on the structural model of photosystem I. J. Mol. Biol. **280**:297–314.

Schulte, A. M., S. Lai, A. Kurtz, F. Czubayko, A. T. Riegel, and A. Wellstein. 1996. Human trophoblast and choriocarcinoma expression of the growth factor pleiotrophin attributable to germ-line insertion of an endogenous retrovirus. Proc. Natl. Acad. Sci. U. S. A **93**:14759–14764.

Schultz, T. R. 1999. Ants, plants and antibiotics. Nature **398**:747–748.

Schuster, S., T. Dandekar, and D. A. Fell. 1999. Detection of elementary flux modes in biochemical networks: a promising tool for pathway analysis and metabolic engineering. Trends Biotechnol. **17**:53–60.

Schuster, S., D. A. Fell, and T. Dandekar. 2000. A general definition of metabolic pathways useful for systematic organization and analysis of complex metabolic networks. Nat. Biotechnol. **18**:326–332.

Schwarz, R. and K. Forchhammer. 2005. Acclimation of unicellular cyanobacteria to macronutrient deficiency: emergence of a complex network of cellular responses. Microbiology 151:2503–2514.

Scott, R. S., P. S. Ungar, T. S. Bergstrom, C. A. Brown, F. E. Grine, M. F. Teaford, and A. Walker. 2005. Dental microwear texture analysis shows within-species diet variability in fossil hominins. Nature 436:693–695.

Scott, W. G., J. B. Murray, J. R. Arnold, B. L. Stoddard, and A. Klug. 1996. Capturing the structure of a catalytic RNA intermediate: the hammerhead ribozyme. Science 274:2065–2069.

Searls, D. B. 2002. The language of genes. Nature 420:211–217.

Searls, D. B. 2003. Linguistics: trees of life and of language. Nature 426:391–392.

Seilacher, A., P. K. Bose, and F. Pfluger. 1998. Triploblastic animals more than 1 billion years ago: trace fossil evidence from india. Science 282:80–83.

Selje, N., M. Simon, and T. Brinkhoff. 2004. A newly discovered Roseobacter cluster in temperate and polar oceans. Nature 427:445–448.

Selker, E. U., N. A. Tountas, S. H. Cross, B. S. Margolin, J. G. Murphy, A. P. Bird, and M. Freitag. 2003. The methylated component of the *Neurospora crassa* genome. Nature 422:893–897.

Semaw, S., P. Renne, J. W. Harris, C. S. Feibel, R. L. Bernor, N. Fesseha, and K. Mowbray. 1997. 2.5-million-year-old stone tools from Gona, Ethiopia. Nature 385:333–336.

Semino, O., G. Passarino, P. J. Oefner, A. A. Lin, S. Arbuzova, L. E. Beckman, G. De Benedictis, L. L. Cavalli-Sforza, and P. A. Underhill. 2000. The genetic legacy of Paleolithic *Homo sapiens* sapiens in extant Europeans: a Y chromosome perspective. Science 290:1155–1159.

Seo, J. S., H. Chong, H. S. Park, K. O. Yoon, C. Jung, J. J. Kim, J. H. Hong, H. Kim, and H. S. Kang. 2005. The genome sequence of the ethanologenic bacterium *Zymomonas mobilis* ZM4. Nat. Biotechnol. 23:63–68.

Sesma, A. and A. E. Osbourn. 2004. The rice leaf blast pathogen undergoes developmental processes typical of root-infecting fungi. Nature 431:582–586.

Sharp, P. M. 2002. Origins of human virus diversity. Cell 108:305–312.

Shear, W. A. 1991. The early development of terrestrial ecosystems. Nature 351:283–289.

Sherman, G. and P. K. Visscher. 2002. Honeybee colonies achieve fitness through dancing. Nature 419:920–922.

Shigetani, Y., F. Sugahara, Y. Kawakami, Y. Murakami, S. Hirano, and S. Kuratani. 2002. Heterotopic shift of epithelial-mesenchymal interactions in vertebrate jaw evolution. Science 296:1316–1319.

Shimizu, T., K. Hozumi, S. Horiike, K. Nunomura, S. Ikegami, T. Takao, and Y. Shimonishi. 1996. A covalently crosslinked histone. Nature 380:32.

Shinya, K., M. Ebina, S. Yamada, M. Ono, N. Kasai, and Y. Kawaoka. 2006. Avian flu: influenza virus receptors in the human airway. Nature 440:435–436.

Short, R. V. 1998. Malthus, a prophet without honour. Lancet 351:1676.

Shu, D. G., S. Conway Morris, J. Han, Y. Li, X. L. Zhang, H. Hua, Z. F. Zhang, J. N. Liu, J. F. Guo, Y. Yao, and K. Yasui. 2006. Lower Cambrian vendobionts from China and early diploblast evolution. Science 312:731–734.

Shu, D.-G., S. Conway Morris, and X.-L. Zhang. 1996. A Pikaia-like chordate from the Lower Cambrian of China. Nature 384:157–158.

Shu, D.-G., S. Conway Morris, X.-L. Zhang, L. Chen, Y. Li, and J. Han. 1999. A pipiscid-like fossil from the Lower Cambrian of south China. Nature 400:746–749.

Shu, D-G., H.-L. Luo, S. Conway Morris, X.-L. Zhang, S.-X. Hu, L. Chen, J. Han, M. Zhu, Y. Li, and L.-Z. Chen. 1999. Lower Cambrian vertebrates from south China. Nature 402:42–46.

Shu, D.-G., X. Zhang, and L. Chen. 1996. Reinterpretation of Yunnanozoon as the earliest known hemichordate. Nature 380:428–430.

Shubin, N. H., E. B. Daeschler and F. A. Jenkins. 2006. The pectoral fin of *Tiktaalik roseae* and the origin of the tetrapod limb. Nature 440:764–771.

Siemers, B. M. and H. U. Schnitzler. 2004. Echolocation signals reflect niche differentiation in five sympatric congeneric bat species. Nature 429:657–661.

Silk, J. B. 2006. Behavior. Who are more helpful, humans or chimpanzees? Science 311:1248–1249.

Silk, J. B., S. F. Brosnan, J. Vonk, J. Henrich, D. J. Povinelli, A. S. Richardson, S. P. Lambeth, J. Mascaro, and S. J. Schapiro. 2005. Chimpanzees are indifferent to the welfare of unrelated group members. Nature 437:1357–1359.

Simard, S. W., D. A. Perry, M. D. Jones, D. D. Myrold, D. M. Durall, and R. Molina. 1997. Net transfer of carbon between ectomycorrhizal tree species in the field. Nature 388:579–582.

Simon, L., J. Bousquet, R. C. Lévesque, and M. Lalonde. 1993. Origin and diversification of endomycorrhizal fungi and coincidence with vascular land plants. Nature 363:67–69.

Simpson, S. J., G. A. Sword, P. D. Lorch, and I. D. Couzin. 2006. Cannibal crickets on a forced march for protein and salt. Proc. Natl. Acad. Sci. U. S. A 103:4152–4156.

Sims, D. W., and V. A. Quayle. 1998. Selective foraging behaviour of basking sharks on zooplankton in a small-scale front. Nature 393:460–464.

Sinclair, A. R., S. Mduma, and J. S. Brashares. 2003. Patterns of predation in a diverse predator-prey system. Nature 425:288–290.

Singer, T., B. Seymour, J. P. O'Doherty, K. E. Stephan, R. J. Dolan, and C. D. Frith. 2006. Empathic neural responses are modulated by the perceived fairness of others. Nature 439:466–469.

Sinninghe Damste, J. S., M. Strous, W. I. Rijpstra, E. C. Hopmans, J. A. Geenevasen, A. C. van Duin, L. A. van Niftrik, and M. S. Jetten. 2002. Linearly concatenated cyclobutane lipids form a dense bacterial membrane. Nature 419:708–712.

Sleep, N. H., A. Meibom, T. Fridriksson, R. G. Coleman, and D. K. Bird. 2004. H_2-rich fluids from serpentinization: geochemical and biotic implications. Proc. Natl. Acad. Sci. U. S. A 101:12818–12823.

Sleep, N. H., K. J. Zahnle, J. F. Kasting, and H. J. Morowitz. 1989. Annihilation of ecosystems by large asteroid impacts on the early Earth. Nature 342:139–142.

Smetacek, V. 2001. A watery arms race. Nature 411:745.

Smetacek, V. 2002. Microbial food webs. The ocean's veil. Nature 419:565.

Smith, B. E. 2002. Structure. Nitrogenase reveals its inner secrets. Science 297:1654–1655.

Smith, D. C., M. Simon, A. L. Alldredge, and F. Azam. 1992. Intense hydrolytic enzyme activity on marine aggregates and implications for rapid particle dissolution. Nature 359:139–142.

Smith, M. M. and Z. Johanson. 2003. Separate evolutionary origins of teeth from evidence in fossil jawed vertebrates. Science 299:1235–1236.

Sogin, M. L. 1993. Giants among the prokaryotes. Nature 362:207.

Sokal, R. R., N. L. Oden, and C. Wilson. 1991. Genetic evidence for the spread of agriculture in Europe by demic diffusion. Nature 351:143–145.

Sokolowski, M. B. 2002. Neurobiology: social eating for stress. Nature 419:893–894.

Solan, Z., D. Horn, E. Ruppin, and S. Edelman. 2005. Unsupervised learning of natural languages. Proc. Natl. Acad. Sci. U. S. A **102**:11629–11634.

Sonnenburg, J. L., J. Xu, D. D. Leip, C. H. Chen, B. P. Westover, J. Weatherford, J. D. Buhler, and J. I. Gordon. 2005. Glycan foraging in vivo by an intestine-adapted bacterial symbiont. Science **307**:1955–1959.

Spear, J. R., J. J. Walker, T. M. McCollom, and N. R. Pace . 2005. Hydrogen and bioenergetics in the Yellowstone geothermal ecosystem. Proc. Natl. Acad. Sci. U. S. A **102**:2555–2560.

Sponheimer, M. and J. A. Lee-Thorp. 1999. Isotopic evidence for the diet of an early hominid, *Australopithecus africanus*. Science **283**:368–370.

Spreitzer, R. J., and M. E. Salvucci. 2002. Rubisco: structure, regulatory interaction, and possibilities for a better enzyme. Annu. Rev. Plant Biol. **53**:449–475.

Staal, M., F. J. Meysman, and L. J. Stal. 2003. Temperature excludes N_2-fixing heterocystous cyanobacteria in the tropical oceans. Nature **425**:504–507.

Stanojevic, D., T. Hoey, and M. Levine. 1989. Sequence-specific DNA-binding activities of the gap proteins encoded by *hunchback* and *kruppel* in Drosophila. Nature **341**: 331–335.

Stappenbeck, T. S., L. V. Hooper, and J. I. Gordon. 2002. Developmental regulation of intestinal angiogenesis by indigenous microbes via Paneth cells. Proc. Natl. Acad. Sci. U. S. A **99**:15451–15455.

Stauffer, B. 1999. Climate change: Cornucopia of ice core results. Nature **399**:412–413.

Stein, P. E., A. Boodhoo, G. J. Tyrrell, J. L. Brunton, and R. J. Read. 1992. Crystal structure of the cell-binding B oligomer of verotoxin-1 from *E. coli.* Nature **355**: 748–750.

Stelling, J., S. Klamt, K. Bettenbrock, S. Schuster, and E. D. Gilles. 2002. Metabolic network structure determines key aspects of functionality and regulation. Nature **420**:190–193.

Stevens, T. O. and J. P. McKinley. 1995. Lithoautotrophic Microbia, Ecosystems in Deep Basalt Aquifers. Science **270**:450–454.

Stock, D., A. G. Leslie, and J. E. Walker. 1999. Molecular architecture of the rotary motor in ATP synthase. Science **286**:1700–1705.

Stokstad, E. 2002. Agriculture. Organic farms reap many benefits. Science **296**:1589.

Stokstad, E. 2003. Paleontology. Primitive jawed fishes had teeth of their own design. Science **299**:1164.

Stowell, M. H., T. M. McPhillips, D. C. Rees, S. M. Soltis, E. Abresch, and G. Feher. 1997. Light-induced structural changes in photosynthetic reaction center: implications for mechanism of electron-proton transfer. Science **276**:812–816.

Stringer, C. 2003. Human evolution: Out of Ethiopia. Nature **423**:692–3, 695.

Stroebel, D., Y. Choquet, J. L. Popot, and D. Picot. 2003. An atypical haem in the cytochrome b_6f complex. Nature **426**:413–418.

Strous, M., J. A. Fuerst, E. H. Kramer, S. Logemann, G. Muyzer, van de Pas-Schoonen KT, R. Webb, J. G. Kuenen, and M. S. Jetten. 1999. Missing lithotroph identified as new planctomycete. Nature **400**:446–449.

Strous, M., E. Pelletier, S. Mangenot, T. Rattei, A. Lehner, M. W. Taylor, M. Horn, H. Daims, and D. Le Paslier. 2006. Deciphering the evolution and metabolism of an anammox bacterium from a community genome. Nature **440**:790–794.

Stuart, A. J., P. A. Kosintsev, T. F. Higham, and A. M. Lister. 2004. Pleistocene to Holocene extinction dynamics in giant deer and woolly mammoth. Nature **431**:684–689.

Subramaniam, S. and R. Henderson. 2000. Molecular mechanism of vectorial proton translocation by bacteriorhodopsin. Nature **406**:653–657.

Sugawara, H., H. Yamamoto, N. Shibata, T. Inoue, S. Okada, C. Miyake, A. Yokota, and Y. Kai. 1999. Crystal structure of carboxylase reaction-oriented ribulose 1, 5-bisphosphate carboxylase/oxygenase from a thermophilic red alga, *Galdieria partita*. J. Biol. Chem. **274**:15655–15661.

Sullivan, M. B., M. L. Coleman, P. Weigele, F. Rohwer, and S. W. Chisholm. 2005. Three *Prochlorococcus* cyanophage genomes: signature features and ecological interpretations. PLoS. Biol. **3**:e144.

Sullivan, M. B., J. B. Waterbury, and S. W. Chisholm. 2003. Cyanophages infecting the oceanic cyanobacterium *Prochlorococcus*. Nature **424**:1047–1051.

Summons, R. E., L. L. Jahnke, J. M. Hope, and G. A. Logan. 1999. 2-Methylhopanoids as biomarkers for cyanobacterial oxygenic photosynthesis. Nature **400**:554–557.

Sunda, W., D. J. Kieber, R. P. Kiene, and S. Huntsman. 2002. An antioxidant function for DMSP and DMS in marine algae. Nature **418**:317–320.

Suttle, C. A., A. M. Chan, and M. T. Cottrell. 1990. Infection of phytoplankton by viruses and reduction of primary productivity. Nature **347**:467–469.

Suzuki, K., B. Meek, Y. Doi, M. Muramatsu, T. Chiba, T. Honjo, and S. Fagarasan. 2004. Aberrant expansion of segmented filamentous bacteria in IgA-deficient gut. Proc. Natl. Acad. Sci. U. S. A **101**:1981–1986.

Sword, G. A., P. D. Lorch, and D. T. Gwynne. 2005. Insect behaviour: migratory bands give crickets protection. Nature **433**:703.

Sze, J. Y., M. Victor, C. Loer, Y. Shi, and G. Ruvkun. 2000. Food and metabolic signalling defects in a *Caenorhabditis elegans* serotonin-synthesis mutant. Nature **403**:560–564.

Takai, K., T. Gamo, U. Tsunogai, N. Nakayama, H. Hirayama, K. H. Nealson, and K. Horikoshi. 2004. Geochemical and microbiological evidence for a hydrogen-based, hyperthermophilic subsurface lithoautotrophic microbial ecosystem (HyperSLiME) beneath an active deep-sea hydrothermal field. Extremophiles. **8**:269–282.

Takeda, S. 1998. Influence of iron availability on nutrient consumption ratio of diatoms in oceanic waters. Nature **393**:774–777.

Talbot, N. J. 2003. On the trail of a cereal killer. Annu. Rev. Microbiol. **57**:177–202.

Tanno, K. and G. Willcox. 2006. How fast was wild wheat domesticated? Science **311**:1886.

Tao, H., R. Gonzalez, A. Martinez, M. Rodriguez, L. O. Ingram, J. F. Preston, and K. T. Shanmugam. 2001. Engineering a homo-ethanol pathway in *Escherichia coli*: increased glycolytic flux and levels of expression of glycolytic genes during xylose fermentation. J. Bacteriol. **183**:2979–2988.

Tattersall, D. B., S. Bak, P. R. Jones, C. E. Olsen, J. K. Nielsen, M. L. Hansen, P. B. Hoj, and B. L. Moller. 2001. Resistance to an herbivore through engineered cyanogenic glucoside synthesis. Science **293**:1826–1828.

Taubenberger, J. K., A. H. Reid, A. E. Krafft, K. E. Bijwaard, and T. G. Fanning. 1997. Initial genetic characterization of the 1918 "Spanish" influenza virus. Science **275**:1793–1796.

Taubenberger, J. K., A. H. Reid, R. M. Lourens, R. Wang, G. Jin, and T. G. Fanning. 2005. Characterization of the 1918 influenza virus polymerase genes. Nature **437**:889–893.

Taylor, J. R. and W. M. Kier. 2006. Biomechanics: a pneumo-hydrostatic skeleton in land crabs. Nature **440**:1005.

Taylor, M. R., J. B. Hurley, H. A. Van Epps, and S. E. Brockerhoff. 2004. A zebrafish model for pyruvate dehydrogenase deficiency: rescue of neurological dysfunction and embryonic lethality using a ketogenic diet. Proc. Natl. Acad. Sci. U. S. A **101**: 4584–4589.

Taylor, T. C. and I. Andersson. 1997. The structure of the complex between rubisco and its natural substrate ribulose 1,5-bisphosphate. J. Mol. Biol. **265**:432–444.

Taylor, T. N., W. Remy, and H. Hass. 1992. Parasitism in a 400-million-year-old green alga. Nature **357**:493–494.

Teaford, M. F. and P. S. Ungar. 2000. Diet and the evolution of the earliest human ancestors. Proc. Natl. Acad. Sci. U. S. A **97**:13506–13511.

Templeton, C. N., E. Greene, and K. Davis. 2005. Allometry of alarm calls: black-capped chickadees encode information about predator size. Science **308**:1934–1937.

Tezcan, F. A., J. T. Kaiser, D. Mustafi, M. Y. Walton, J. B. Howard, and D. C. Rees. 2005. Nitrogenase complexes: multiple docking sites for a nucleotide switch protein. Science **309**:1377–1380.

Thacher, T. D., P. R. Fischer, J. M. Pettifor, J. O. Lawson, C. O. Isichei, J. C. Reading, and G. M. Chan. 1999. A comparison of calcium, vitamin D, or both for nutritional rickets in Nigerian children. N. Engl. J. Med. **341**:563–568.

Thaler, J. S. 1999. Jasmonate-inducible plant defences cause increased parasitism of herbivores. Nature **399**:686–688.

Thieme, H. 1997. Lower Palaeolithic hunting spears from Germany. Nature **385**:807–810.

Thomas, G. L. and R. E. Thorne. 2001. Night-time predation by Steller sea lions. Nature **411**:1013.

Thompson, J. R., S. Pacocha, C. Pharino, V. Klepac-Ceraj, D. E. Hunt, J. Benoit, R. Sarma-Rupavtarm, D. L. Distel, and M. F. Polz. 2005. Genotypic diversity within a natural coastal bacterioplankton population. Science **307**:1311–1313.

Tian, F., O. B. Toon, A. A. Pavlov, and H. De Sterck. 2005. A hydrogen-rich early Earth atmosphere. Science **308**:1014–1017.

Tian, J., R. Bryk, M. Itoh, M. Suematsu, and C. Nathan. 2005. Variant tricarboxylic acid cycle in *Mycobacterium tuberculosis*: identification of alpha-ketoglutarate decarboxylase. Proc. Natl. Acad. Sci. U. S. A **102**:10670–10675.

Tiedemann, H. 1997. "Killer" impacts and life's origins. Science **277**:1687–1688.

Tielens, AGM, C. Rotte, C., J.J. van Hellemond, and W. Martin. 2002. Mitochondria as we don't know them. Trends Biochem. Sci. **27**:564–572.

Tilman, D. 1998. The greening of the green revolution. Nature **396**:211–212.

Tilman, D. 2000. Causes, consequences and ethics of biodiversity. Nature **405**:208–211.

Tilman, D., K. G. Cassman, P. A. Matson, R. Naylor, and S. Polasky. 2002. Agricultural sustainability and intensive production practices. Nature **418**:671–677.

Tilman, D., and J. A. Downing. 1994. Biodiversity and stability in grasslands. Nature **367**:363–365.

Tilman, D., J. Fargione, B. Wolff, C. D'Antonio, A. Dobson, R. Howarth, D. Schindler, W. H. Schlesinger, D. Simberloff, and D. Swackhamer. 2001. Forecasting agriculturally driven global environmental change. Science **292**:281–284.

Tilman, D., P. B. Reich, J. M. H. Knops. 2006. Biodiversity and ecosystem stability in a decade-long grassland experiment. Nature **441**:629–632.

Tilman, D., P. B. Reich, J. Knops, D. Wedin, T. Mielke, and C. Lehman. 2001. Diversity and productivity in a long-term grassland experiment. Science **294**:843–845.

Tilman, D., D. Wedin, and J. Knops. 1996. Productivity and sustainability influenced by biodiversity in grassland ecosystems. Nature **379**:718–720.

Timmons, L. and A. Fire. 1998. Specific interference by ingested dsRNA. Nature **395**:854.

Ting, C. N., M. P. Rosenberg, C. M. Snow, L. C. Samuelson, and M. H. Meisler. 1992. Endogenous retroviral sequences are required for tissue-specific expression of a human salivary amylase gene. Genes Dev. **6**:1457–1465.

Tomitani, A., K. Okada, H. Miyashita, H. C. Matthijs, T. Ohno, and A. Tanaka. 1999. Chlorophyll b and phycobilins in the common ancestor of cyanobacteria and chloroplasts. Nature **400**:159–162.

Torsvik, V., J. Goksoyr, and F. L. Daae. 1990. High diversity in DNA of soil bacteria. Appl. Environ. Microbiol. **56**:782–787.

Tortell, P. D., M. T. Maldonado, and N. M. Price. 1996. The role of heterotrophic bacteria in iron-limited ocean ecosystems. Nature **383**:330–332.

Tovar, J., G. Leon-Avila, L. B. Sanchez, R. Sutak, J. Tachezy, G. M. van der, M. Hernandez, M. Muller, and J. M. Lucocq. 2003. Mitochondrial remnant organelles of Giardia function in iron-sulphur protein maturation. Nature **426**:172–176.

Tran, T. H., T. L. Nguyen, T. D. Nguyen, T. S. Luong, P. M. Pham, V. C. Nguyen, T. S. Pham, and J. Farrar. 2004. Avian influenza A (H5N1) in 10 patients in Vietnam. N. Engl. J. Med. **350**:1179–1188.

Tremblay, L. and W. Schultz. 1999. Relative reward preference in primate orbitofrontal cortex. Nature **398**:704–708.

Trewavas, A. 2002. Malthus foiled again and again. Nature **418**:668–670.

Tringe, S. G., C. von Mering, A. Kobayashi, A. A. Salamov, K. Chen, H. W. Chang, M. Podar, J. M. Short, E. J. Mathur, J. C. Detter, P. Bork, P. Hugenholtz, and E. M. Rubin. 2005. Comparative metagenomics of microbial communities. Science **308**: 554–557.

Tromans, A. 2003. Volume control. Nature **423**:815.

Troy, C. S., D. E. MacHugh, J. F. Bailey, D. A. Magee, R. T. Loftus, P. Cunningham, A. T. Chamberlain, B. C. Sykes, and D. G. Bradley. 2001. Genetic evidence for Near-Eastern origins of European cattle. Nature **410**:1088–1091.

Tu, B. P., A. Kudlicki, M. Rowicka, and S. L. McKnight. 2005. Logic of the yeast metabolic cycle: temporal compartmentalization of cellular processes. Science **310**:1152–1158.

Tu, J., G. Zhang, K. Datta, C. Xu, Y. He, Q. Zhang, G. S. Khush, and S. K. Datta. 2000. Field performance of transgenic elite commercial hybrid rice expressing *Bacillus thuringiensis* delta-endotoxin. Nat. Biotechnol. **18**:1101–1104.

Tumpey, T. M., C. F. Basler, P. V. Aguilar, H. Zeng, A. Solorzano, D. E. Swayne, N. J. Cox, J. M. Katz, J. K. Taubenberger, P. Palese, and A. Garcia-Sastre. 2005. Characterization of the reconstructed 1918 Spanish influenza pandemic virus. Science **310**:77–80.

Turlings, T. C. J., J. H. Tumlinson, and W. J. Lewis. 1990. Exploitation of herbivore-induced plant odors by host-seeking parasitic wasps. Science **250**:1251–1253.

Turner, R. B., R. Bauer, K. Woelkart, T. C. Hulsey, and J. D. Gangemi. 2005. An evaluation of *Echinacea angustifolia* in experimental rhinovirus infections. N. Engl. J. Med. **353**:341–348.

Tuttle and Ryan. 1981. Bat predation and the evolution of frog vocalizations in the neotropics. Science **214**:677–678.

Tyler, S. A., and E. S. Barghoorn. 1954. Occurrence of structurally preserved plants in pre-Cambrian rocks of the Canadian Shield. Science **119**:606–608.

Tyson, G. W., J. Chapman, P. Hugenholtz, E. E. Allen, R. J. Ram, P. M. Richardson, V. V. Solovyev, E. M. Rubin, D. S. Rokhsar, and J. F. Banfield. 2004. Community structure and metabolism through reconstruction of microbial genomes from the environment. Nature **428**:37–43.

Uehlein, N., C. Lovisolo, F. Siefritz, and R. Kaldenhoff. 2003. The tobacco aquaporin NtAQP1 is a membrane CO_2 pore with physiological functions. Nature **425**:734–737.

Ueno, Y., K. Yamada, N. Yoshida, S. Maruyama, and Y. Isozaki. 2006. Evidence from fluid inclusions for microbial methanogenesis in the early Archaean era. Nature **440**:516–519.

Ungchusak, K., P. Auewarakul, S. F. Dowell, R. Kitphati, W. Auwanit, P. Puthavathana, M. Uiprasertkul, K. Boonnak, C. Pittayawonganon, N. J. Cox, S. R. Zaki, P. Thawatsupha, M. Chittaganpitch, R. Khontong, J. M. Simmerman, and S. Chunsutthiwat. 2005. Probable person-to-person transmission of avian influenza A (H5N1). N. Engl. J. Med. **352**:333–340.

Vacelet, J., and N. Boury-Esnault. 1995. Carnivorous sponges. Nature **373**:333–335.

Vacelet, J., N. Boury-Esnault, A. Fiala-Medioni, and C. R. Fisher. 1995. A methanotrophic carnivorous sponge. Nature **377**:296.

Valent, B. 2004. Plant disease: Underground life for rice foe. Nature **431**:516–517.

van der Heijden, M. G. A., J. N. Klironomos, M. Ursic, P. Moutoglis, R. Streitwolf-Engel, T. Boller, A. Wiemken, and I. R. Sanders. 1998. Mycorrhizal fungal diversity determines plant biodiversity, ecosystem variability and productivity. Nature **396**:69–72.

Van Dover, C. L., E. Z. Szuts, S. C. Chamberlain, and J. R. Cann. 1989. A novel eye in 'eyeless' shrimp from hydrothermal vents of the Mid-Atlantic Ridge. Nature **337**:458–460.

van Schaik, C. P., M. Ancrenaz, G. Borgen, B. Galdikas, C. D. Knott, I. Singleton, A. Suzuki, S. S. Utami, and M. Merrill. 2003. Orangutan cultures and the evolution of material culture. Science **299**:102–105.

Van Wassenbergh, S., A. Herrel, D. Adriaens, F. Huysentruyt, S. Devaere, and P. Aerts. 2006. Evolution: a catfish that can strike its prey on land. Nature **440**:881.

Vancanneyt, G., C. Sanz, T. Farmaki, M. Paneque, F. Ortego, P. Castanera, and J. J. Sanchez-Serrano. 2001. Hydroperoxide lyase depletion in transgenic potato plants leads to an increase in aphid performance. Proc. Natl. Acad. Sci. U. S. A **98**:8139–8144.

Vandamme, A. M., M. Salemi, and J. Desmyter. 1998. The simian origins of the pathogenic human T-cell lymphotropic virus type I. Trends Microbiol. **6**:477–483.

Vargas, M., K. Kashefi, E. L. Blunt-Harris, and D. R. Lovley. 1998. Microbiological evidence for Fe(III) reduction on early Earth. Nature **395**:65–67.

Varricchio, D. J., F. Jackson, J. J. Borkowski, and J. R. Horner. 1997. Nest and egg clutches of the dinosaur *Troodon formosus* and the evolution of avian reproductive traits. Nature **385**:247–250.

Velicer, G. J. and Y. T. Yu. 2003. Evolution of novel cooperative swarming in the bacterium *Myxococcus xanthus*. Nature **425**:75–78.

Venter, J. C., K. Remington, J. F. Heidelberg, A. L. Halpern, D. Rusch, J. A. Eisen, D. Wu, I. Paulsen, and H. O. Smith. 2004. Environmental genome shotgun sequencing of the Sargasso Sea. Science **304**:66–74.

Verheyen, E., W. Salzburger, J. Snoeks, and A. Meyer. 2003. Origin of the superflock of cichlid fishes from Lake Victoria, East Africa. Science **300**:325–329.

Verkhovsky, M. I., A. Jasaitis, M. L. Verkhovskaya, J. E. Morgan, and M. Wikstrom. 1999. Proton translocation by cytochrome c oxidase. Nature **400**:480–483.

Vezzi, A., S. Campanaro, M. D'Angelo, F. Simonato, N. Vitulo, F. M. Lauro, A. Cestaro, G. Malacrida, B. Simionati, N. Cannata, C. Romualdi, D. H. Bartlett, and G. Valle. 2005. Life at depth: *Photobacterium profundum* genome sequence and expression analysis. Science **307**:1459–1461.

Vila, C., J. A. Leonard, A. Gotherstrom, S. Marklund, K. Sandberg, K. Liden, R. K. Wayne, and H. Ellegren. 2001. Widespread origins of domestic horse lineages. Science **291**:474–477.

Villareal, T. A., M. A. Altabet, and K. Culver-Rymsza. 1993. Nitrogen transport by vertically migrating diatom mats in the North Pacific Ocean. Nature 363:709–712.

Villareal, T. A., C. Pilskaln, M. Brzezinski, F. Lipschultz, M. Dennett, and G. B. Gardner. 1999. Upward transport of oceanic nitrate by migrating diatom mats. Nature 397: 423–425.

Visscher, P. K. 2003. Animal behaviour: How self-organization evolves. Nature 421: 799–800.

Viswanathan, G. M., V. Afanasyev, S. V. Buldyrev, E. J. Murphy, P. A. Prince, and H. E. Stanley. 1996. Lévy flight search patterns of wandering albatrosses. Nature 381:413–415.

Vitousek, P. M., H. A. Mooney, J. Lubchenco, and J. M. Melillo. 1997. Human domination of earth's ecosystems. Science 277:494–499.

Vogelbein, W. K., V. J. Lovko, J. D. Shields, K. S. Reece, P. L. Mason, L. W. Haas, and C. C. Walker. 2002. Pfiesteria shumwayae kills fish by micropredation not exotoxin secretion. Nature 418:967–970.

Vollbrecht, E., P. S. Springer, L. Goh, E. S. Buckler, and R. Martienssen. 2005. Architecture of floral branch systems in maize and related grasses. Nature 436:1119–1126.

von Helversen, D., and O. von Helversen. 1999. Acoustic guide in bat-pollinated flower. Nature 398:759–760.

Vossbrinck, C. R., J. V. Maddox, S. Friedman, B. A. Debrunner-Vossbrinck, and C. R. Woese. 1987. Ribosomal RNA sequence suggests microsporidia are extremelyancient eukaryotes. Nature 326:411–414.

Votier, S. C., R. W. Furness, S. Bearhop, J. E. Crane, R. W. Caldow, P. Catry, K. Ensor, K. C. Hamer, A. V. Hudson, E. Kalmbach, N. I. Klomp, S. Pfeiffer, R. A. Phillips, I. Prieto, and D. R. Thompson. 2004. Changes in fisheries discard rates and seabird communities. Nature 427:727–730.

Wachtershauser, G. 2003. From pre-cells to Eukarya–a tale of two lipids. Mol. Microbiol. 47:13–22.

Wachtershauser, G. 1990. Evolution of the first metabolic cycles. Proc. Natl. Acad. Sci. U. S. A 87:200–204.

Wachtershauser, G. 1994. Life in a ligand sphere. Proc. Natl. Acad. Sci. U. S. A 91: 4283–4287.

Waldor, M. K. and J. J. Mekalanos. 1996. Lysogenic conversion by a filamentous phage encoding cholera toxin. Science 272:1910–1914.

Wang, H. and G. Oster. 1998. Energy transduction in the F_1 motor of ATP synthase. Nature 396:279–282.

Wang, M. B., X. Y. Bian, L. M. Wu, L. X. Liu, N. A. Smith, D. Isenegger, R. M. Wu, C. Masuta, V. B. Vance, J. M. Watson, A. Rezaian, E. S. Dennis, and P. M. Waterhouse. 2004. On the role of RNA silencing in the pathogenicity and evolution of viroids and viral satellites. Proc. Natl. Acad. Sci. U. S. A 101:3275–3280.

Wang, R. L., A. Stec, J. Hey, L. Lukens, and J. Doebley. 1999. The limits of selection during maize domestication. Nature 398:236–239.

Warneken, F. and M. Tomasello. 2006. Altruistic helping in human infants and young chimpanzees. Science 311:1301–1303.

Watson, F. R. and Tabita. 1997. FEMS Microbiol. Lett. 146:13.

Watson, G. M. and D. A. Hessinger. 1989. Cnidocyte mechanoreceptors are tuned to the movements of swimming prey by chemoreceptors. Science 243:1589–1591.

Webby, R. J., D. R. Perez, J. S. Coleman, Y. Guan, J. H. Knight, E. A. Govorkova, L. R. McClain-Moss, J. S. Peiris, J. E. Rehg, E. I. Tuomanen, and R. G. Webster. 2004.

Responsiveness to a pandemic alert: use of reverse genetics for rapid development of influenza vaccines. Lancet **363**:1099–1103.

Webster, R. and D. Hulse. 2005. Controlling avian flu at the source. Nature **435**:415–416.

Weil, A. 2002. Upwards and onwards. Nature **416**:816–822.

Weimerskirch, H. and R. P. Wilson. 2000. Oceanic respite for wandering albatrosses. Nature **406**:955–956.

Weiner, A. M. and N. Maizels. 1987. tRNA-like structures tag the 3' ends of genomic RNA molecules for replication: implications for the origin of protein synthesis. Proc. Natl. Acad. Sci. U. S. A **84**:7383–7387.

Weiner, S., Q. Xu, P. Goldberg, J. Liu, and O. Bar-Yosef. 1998. Evidence for the use of fire at Zhoukoudian, China. Science **281**:251–253.

Weir, A. A. S., J. Chappell, and A. Kacelnik. 2002. Shaping of hooks in New Caledonian crows. Science **297**:981.

Weiss, E., M. E. Kislev, and A. Hartmann. 2006. Autonomous cultivation before domestication. Science **312**:1608–1610.

Weissert, H. 2000. Deciphering methane's fingerprint. Nature **406**:356–357.

Welander, P. V. and W. W. Metcalf. 2005. Loss of the mtr operon in Methanosarcina blocks growth on methanol, but not methanogenesis, and reveals an unknown methanogenic pathway. Proc. Natl. Acad. Sci. U. S. A **102**:10664–10669.

Wellsbury, P., K. Goodman, T. Barth, B. A. Cragg, S. P. Barnes, and R. J. Parke. 1997. Deep marine biosphere fuelled by increasing organic matter availability during burial and heating. Nature **388**:573–576.

Wexler, I. D., S. G. Hemalatha, J. McConnell, N. R. Buist, H. H. Dahl, S. A. Berry, S. D. Cederbaum, M. S. Patel, and D. S. Kerr. 1997. Outcome of pyruvate dehydrogenase deficiency treated with ketogenic diets. Studies in patients with identical mutations. Neurology **49**:1655–1661.

White, H. B., III. 1976. Coenzymes as fossils of an earlier metabolic state. J. Mol. Evol. **7**:101–104.

White, T. D., B. Asfaw, D. DeGusta, H. Gilbert, G. D. Richards, G. Suwa, and F. C. Howell. 2003. Pleistocene *Homo sapiens* from Middle Awash, Ethiopia. Nature **423**:742–747.

White, T. D., G. WoldeGabriel, B. Asfaw, S. Ambrose, Y. Beyene, R. L. Bernor, J. R. Boisserie, and G. Suwa. 2006. Asa Issie, Aramis and the origin of Australopithecus. Nature **440**:883–889.

Whiten, A., J. Goodall, W. C. McGrew, T. Nishida, V. Reynolds, Y. Sugiyama, C. E. Tutin, R. W. Wrangham, and C. Boesch. 1999. Cultures in chimpanzees. Nature **399**:682–685.

Whiten, A., V. Horner, and F. B. de Waal. 2005. Conformity to cultural norms of tool use in chimpanzees. Nature **437**:737–740.

Whitman, W. B., D. C. Coleman, and W. J. Wiebe. 1998. Prokaryotes: the unseen majority. Proc. Natl. Acad. Sci. U. S. A **95**:6578–6583.

Wikelski, M. and C. Thom. 2000. Marine iguanas shrink to survive El Nino. Nature **403**:37–38.

Wikstrom, M. 1989. Identification of the electron transfers in cytochrome oxidase that are coupled to proton-pumping. Nature **338**:776–778.

Wild, C., M. Huettel, A. Klueter, S. G. Kremb, M. Y. Rasheed, and B. B. Jorgensen. 2004. Coral mucus functions as an energy carrier and particle trap in the reef ecosystem. Nature **428**:66–70.

Wildschutte, H., D. M. Wolfe, A. Tamewitz, and J. G. Lawrence. 2004. Protozoan predation, diversifying selection, and the evolution of antigenic diversity in Salmonella. Proc. Natl. Acad. Sci. U. S. A 101:10644–10649.

Wilf, P., C. C. Labandeira, W. J. Kress, C. L. Staines, D. M. Windsor, A. L. Allen, and K. R. Johnson. 2000. Timing the radiations of leaf beetles: hispines on gingers from latest cretaceous to recent. Science 289:291–294.

Williams, B. A., R. P. Hirt, J. M. Lucocq, and T. M. Embley. 2002. A mitochondrial remnant in the microsporidian Trachipleistophora hominis. Nature 418:865–869.

Williams, P. J. 1998.The balance of plankton respiration and photosynthesis in the open oceans. Nature 394:55–57.

Williams, T. M., R. W. Davis, L. A. Fuiman, J. Francis, B. J. Le Boeuf, M. Horning, J. Calambokidis, and D. A. Croll. 2000. Sink or swim: strategies for cost-efficient diving by marine mammals. Science 288:133–136.

Wilson, M. V. H., and Caldwell, M. W. 1993. New Silurian and Devonian fork-tailed 'thelodonts' are jawless vertebrates with stomachs and deep bodies. Nature 361: 442–444.

Winfield, M. D. and E. A. Groisman. 2003. Role of nonhost environments in the lifestyles of Salmonella and Escherichia coli. Appl. Environ. Microbiol. 69:3687–3694.

Winkler, W., A. Nahvi, and R. R. Breaker. 2002. Thiamine derivatives bind messenger RNAs directly to regulate bacterial gene expression. Nature 419:952–956.

Winkler, W. C., A. Nahvi, A. Roth, J. A. Collins, and R. R. Breaker. 2004. Control of gene expression by a natural metabolite-responsive ribozyme. Nature 428:281–286.

Wirth, D. F. 2002. Biological revelations. Nature 419:495–496.

Witte, U., F. Wenzhofer, S. Sommer, A. Boetius, P. Heinz, N. Aberle, M. Sand, A. Cremer, W. R. Abraham, B. B. Jorgensen, and O. Pfannkuche . 2003. In situ experimental evidence of the fate of a phytodetritus pulse at the abyssal sea floor. Nature 424: 763–766.

Woese, C. 1998. The universal ancestor. Proc. Natl. Acad. Sci. U. S. A 95:6854–6859.

Woese, C. R. 2002. On the evolution of cells. Proc. Natl. Acad. Sci. U. S. A 99: 8742–8747.

Wolfe, G. R., F. X. Cunningham, D. Durnfordt, B. R. Green, and E. Gantt. 1994. Evidence for a common origin of chloroplasts with light-harvesting complexes of different pigmentation. Nature 367:566–568.

Wolfe, G. V., M. Steinke, and G. O. Kirst. 1997. Grazing-activated chemical defence in a unicellular marine alga. Nature 387:894–897.

Wolfe, K. H. and D. C. Shields. 1997. Molecular evidence for an ancient duplication of the entire yeast genome. Nature 387:708–713.

Wolfe, N. D., W. M. Switzer, J. K. Carr, V. B. Bhullar, V. Shanmugam, U. Tamoufe, A. T. Prosser, J. N. Torimiro, A. Wright, E. Mpoudi-Ngole, F. E. McCutchan, D. L. Birx, T. M. Folks, D. S. Burke, and W. Heneine. 2004. Naturally acquired simian retrovirus infections in central African hunters. Lancet 363:932–937.

Wolpoff, M. H., B. Senut, M. Pickford, and J. Hawks. 2002. Palaeoanthropology. Sahelan-thropus or 'Sahelpithecus'? Nature 419:581–582.

Wommack, K. E. and R. R. Colwell. 2000. Virioplankton: viruses in aquatic ecosystems. Microbiol. Mol. Biol. Rev. 64:69–114.

Wood, B. 1997. Palaeoanthropology. The oldest whodunnit in the world. Nature 385: 292–293.

Wood, B. and A. Brooks. 1999. Human evolution. We are what we ate. Nature 400: 219–220.

Wood, W. T., J. F. Gettrust, N. R. Chapman, G. D. Spence, and R. D. Hyndman. 2002. Decreased stability of methane hydrates in marine sediments owing to phase-boundary roughness. Nature **420**:656–660.

Wray, G. A., J. S. Levinton, and L. H. Shapiro. 1996. Molecular Evidence for Deep Precambrian Divergences Among Metazoan Phyla. Science **274**:568–573.

Wuarin, J. and P. Nurse. 1996. Regulating S phase: CDKs, licensing and proteolysis. Cell **85**:785–787.

Xia, D., C. A. Yu, H. Kim, J. Z. Xia, A. M. Kachurin, L. Zhang, L. Yu, and J. Deisenhofer. 1997. Crystal structure of the cytochrome bc1 complex from bovine heart mitochondria. Science **277**:60–66.

Xiao, S., Y. Zhang, and A. H. Knoll. 1998. Three-dimensional preservation of algae and animal embryos in a Neoproterozoic phosphorite. Nature **391**:553–558.

Xiong, J., W. M. Fischer, K. Inoue, M. Nakahara, and C. E. Bauer. 2000. Molecular evidence for the early evolution of photosynthesis. Science **289**: 1724–1730.

Xiong, J., K. Inoue, and C. E. Bauer. 1998. Tracking molecular evolution of photosynthesis by characterization of a major photosynthesis gene cluster from *Heliobacillus mobilis*. Proc. Natl. Acad. Sci. U. S. A **95**:14851–14856.

Xiong, Y., F. Li, J. Wang, A. M. Weiner, and T. A. Steitz. 2003. Crystal structures of an archaeal class I CCA-adding enzyme and its nucleotide complexes. Mol. Cell **12**:1165–1172.

Xu, J. and J. I. Gordon. 2003. Inaugural Article: Honor thy symbionts. Proc. Natl. Acad. Sci. U. S. A **100**:10452–10459.

Xu, J., M. K. Bjursell, J. Himrod, S. Deng, L. K. Carmichael, H. C. Chiang, L. V. Hooper, and J. I. Gordon. 2003. A genomic view of the human-*Bacteroides thetaiotaomicron* symbiosis. Science **299**:2074–2076.

Xu, Q., M. Dziejman, and J. J. Mekalanos. 2003. Determination of the transcriptome of *Vibrio cholerae* during intraintestinal growth and midexponential phase in vitro. Proc. Natl. Acad. Sci. U. S. A **100**:1286–1291.

Yack, J. E. and J. H. Fullard. 2000. Ultrasonic hearing in nocturnal butterflies. Nature **403**:265–266.

Yamei, H., R. Potts, G. Baoyin, G. Zhengtang, A. Deino, W. Wei, J. Clark, X. Guangmao, and H. Weiwen. 2000. Mid-Pleistocene Acheulean-like stone technology of the Bose basin, South China. Science **287**:1622–1626.

Yandulov, D. V. and R. R. Schrock. 2003. Catalytic reduction of dinitrogen to ammonia at a single molybdenum center. Science **301**:76–78.

Yankovskaya, V., R. Horsefield, S. Tornroth, C. Luna-Chavez, H. Miyoshi, C. Leger, B. Byrne, G. Cecchini, and S. Iwata. 2003. Architecture of succinate dehydrogenase and reactive oxygen species generation. Science **299**:700–704.

Yasuda, R., H. Noji, M. Yoshida, K. Kinosita, Jr., and H. Itoh. 2001. Resolution of distinct rotational substeps by submillisecond kinetic analysis of F_1-ATPase. Nature **410**:898–904.

Yasuda, R., H. Noji, K. Kinosita, Jr., and M. Yoshida. 1998. F_1-ATPase is a highly efficient molecular motor that rotates with discrete 120 degree steps. Cell **93**: 1117–1124.

Ye, X., S. Al Babili, A. Kloti, J. Zhang, P. Lucca, P. Beyer, and I. Potrykus. 2000. Engineering the provitamin A (beta-carotene) biosynthetic pathway into (carotenoid-free) rice endosperm. Science **287**:303–305.

Yellen, J. E., A. S. Brooks, E. Cornelissen, M. J. Mehlman, and K. Stewart. 1995. A middle stone age worked bone industry from Katanda, Upper Semliki Valley, Zaire. Science **268**:553–556.

Yoder, J. A., S. G. Ackleson, R. T. Barber, P. Flament, and W. M. Balch. 1994. A line in the sea. Nature **371**:689–692.

Yoshida, T., L. E. Jones, S. P. Ellner, G. F. Fussmann, and N. G. Hairston, Jr. 2003. Rapid evolution drives ecological dynamics in a predator-prey system. Nature **424**:303–306.

Yoshikawa, S., K. Shinzawa-Itoh, R. Nakashima, R. Yaono, E. Yamashita, N. Inoue, M. Yao, M. J. Fei, C. P. Libeu, T. Mizushima, H. Yamaguchi, T. Tomizaki, and T. Tsukihara. 1998. Redox-coupled crystal structural changes in bovine heart cytochrome c oxidase. Science **280**:1723–1729.

Yuan, X., S. Xiao, and T. N. Taylor. 2005. Lichen-like symbiosis 600 million years ago. Science **308**:1017–1020.

Yusupov, M. M., G. Z. Yusupova, A. Baucom, K. Lieberman, T. N. Earnest, J. H. Cate, and H. F. Noller. 2001. Crystal structure of the ribosome at 5.5 A resolution. Science **292**:883–896.

Zdobnov, E. M., C. von Mering, I. Letunic, D. Torrents, M. Suyama, R. R. Copley, and P. Bork. 2002. Comparative genome and proteome analysis of *Anopheles gambiae* and *Drosophila melanogaster*. Science **298**:149–159.

Zeder, M. A. and B. Hesse. 2000. The initial domestication of goats (*Capra hircus*) in the Zagros mountains 10,000 years ago. Science **287**:2254–2257.

Zehr, J. P., J. B. Waterbury, P. J. Turner, J. P. Montoya, E. Omoregie, G. F. Steward, A. Hansen, and D. M. Karl. 2001. Unicellular cyanobacteria fix N_2 in the subtropical North Pacific Ocean. Nature **412**:635–638.

Zengler, K., H. H. Richnow, R. Rossello-Mora, W. Michaelis, and F. Widdel. 1999. Methane formation from long-chain alkanes by anaerobic microorganisms. Nature **401**:266–269.

Zhang, B. and T. R. Cech. 1997. Peptide bond formation by in vitro selected ribozymes. Nature **390**:96–100.

Zhang, M., C. Eddy, K. Deanda, M. Finkelstein, and S. Picataggio. 1995. Metabolic Engineering of a Pentose Metabolism Pathway in Ethanologenic *Zymomonas mobilis*. Science 267:240–243.

Zhang, Y., H. Lu, and C. I. Bargmann. 2005. Pathogenic bacteria induce aversive olfactory learning in *Caenorhabditis elegans*. Nature **438**:179–184.

Zhang, Z., L. Huang, V. M. Shulmeister, Y. I. Chi, K. K. Kim, L. W. Hung, A. R. Crofts, E. A. Berry, and S. H. Kim. 1998. Electron transfer by domain movement in cytochrome bc_1. Nature **392**:677–684.

Zhang, Z., B. R. Green, and T. Cavalier-Smith. 1999. Single gene circles in dinoflagellate chloroplast genomes. Nature **400**:155–159.

Zhou, J. M. and P. D. Boyer. 1993. Evidence that energization of the chloroplast ATP synthase favors ATP formation at the tight binding catalytic site and increases the affinity for ADP at another catalytic site. J. Biol. Chem. **268**:1531–1538.

Zhu, G., G. B. Golding, and A. M. Dean. 2005. The selective cause of an ancient adaptation. Science **307**:1279–1282.

Zhu, Y., H. Chen, J. Fan, Y. Wang, Y. Li, J. Chen, J. Fan, S. Yang, L. Hu, H. Leung, T. W. Mew, P. S. Teng, Z. Wang, and C. C. Mundt. 2000. Genetic diversity and disease control in rice. Nature **406**:718–722.

Zilhao, J. 2001. Radiocarbon evidence for maritime pioneer colonization at the origins of farming in west Mediterranean Europe. Proc. Natl. Acad. Sci. U. S. A **98**:14180–14185.

Zollikofer, C. P., M. S. Ponce de Leon, D. E. Lieberman, F. Guy, D. Pilbeam, A. Likius, H. T. Mackaye, P. Vignaud, and M. Brunet. 2005. Virtual cranial reconstruction of *Sahelanthropus tchadensis*. Nature **434**:755–759.

Zou, W. and P. Gambetti. 2005. From Microbes to PrionsThe Final Proof of the Prion Hypothesis. Cell, 121, 155–157.

Zouni, A., H. T. Witt, J. Kern, P. Fromme, N. Krauss, W. Saenger, and P. Orth. 2001. Crystal structure of photosystem II from *Synechococcus elongatus* at 3.8 A resolution. Nature **409**:739–743.

Biochemical Back-ups

Web References

If you have difficulties in following the text of the book because it anticipates too much background knowledge in biochemistry, cell biology, genetics or micro-biology you can consult a number of excellent textbooks from the NCBI (National Center for Biotechnology Information sponsored by the National Library of Medicine and the National Institutes of health) bookshelf. The link will guide you to the search box, you type your unknown term from the text and start the search by "Go". These books will provide you sound information. Below you will find the entry page of selected books:

in biochemistry: http://www.ncbi.nlm.nih.gov/books/bv.fcgi?rid=stryer.TOC&depth=2
in human genetics: http://www.ncbi.nlm.nih.gov/books/bv.fcgi?rid=hmg.TOC&depth=2
in general genetics: http://www.ncbi.nlm.nih.gov/books/bv.fcgi?rid=iga.TOC
in microbiology: http://www.ncbi.nlm.nih.gov/books/bv.fcgi?rid=mmed.TOC&depth=2
in cell biology: http://www.ncbi.nlm.nih.gov/books/bv.fcgi?rid=mboc4.TOC&depth=2

After that general part, the author has listed web sites, which provide a back-up for the specified subjects of each chapter. The subjects are listed in their order of appearance in the book text. The author has looked for web sites from universities, research institutes and research organizations or reliable news agencies. However, many web sites are not the result of a review process and the author cannot guarantee that only sound information is provided. While the author has carefully read all papers quoted in the reference list and reviewed in the text, he has not read through the web sites. The information content has thus to be multiplied by the reputation of the institution or the authors backing it. Some web sites will also cease to exist. The author apologizes for any inconvenience.

Chapter 1

Lactose:
  http://en.wikipedia.org/wiki/Lactose
NADPH:
  http://en.wikipedia.org/wiki/NADPH

CoA:
  http://en.wikipedia.org/wiki/Coenzyme_A

FADH:
  http://en.wikipedia.org/wiki/FADH

ATP:
  http://en.wikipedia.org/wiki/Adenosine_triphosphate

Basic Thermodynamics:
  http://lecturer.ukdw.ac.id/dhira/Metabolism/BasicEnerConcepts.html

Types of Catabolism:
  http://lecturer.ukdw.ac.id/dhira/Metabolism/TypesCatabolism.html

Fermentation:
  http://lecturer.ukdw.ac.id/dhira/Metabolism/Fermentation.html

Chapter 2

Glucose:
  http://en.wikipedia.org/wiki/Glucose

Glycolysis:
  http://en.wikipedia.org/wiki/Glycolysis
  http://www.gwu.edu/~mpb/glycolysis.htm

Glycerate 3 phosphate dehydrogenase:
  http://en.wikipedia.org/wiki/GAPDH
  http://www.ncbi.nlm.nih.gov/books/bv.fcgi?rid=mcb.figgrp.4342

Gluconeogenesis:
  http://en.wikipedia.org/wiki/Gluconeogenesis

Pentose phosphate pathway:
  http://www.gwu.edu/~mpb/pentphos.htm

Entner-Doudoroff Pathway (comparison with glycolysis):
  http://biocyc.org/ECOLI/NEW-IMAGE?type=PATHWAY&object=GLYCOLYSIS/
  E-D&detail-level=3

pyruvate dehydrogenase complex:
   http://www.rpi.edu/dept/bcbp/molbiochem/MBWeb/mb1/part2/krebs.htm
   http://www.med.unibs.it/~marchesi/tca.html

Decarboxylation:
   http://www.ncbi.nlm.nih.gov/books/bv.fcgi?rid=stryer.figgrp.2380

Reactions:
   http://www.ncbi.nlm.nih.gov/books/bv.fcgi?rid=stryer.figgrp.2384
   http://www.ncbi.nlm.nih.gov/entrez/query.fcgi?cmd=Search&db=books&doptcmdl=
   enBookHL&term=Pyruvate+dehydrogenase+complex+AND+stryer%5Bbook%5D+
   AND+216235%5Buid%5D&rid=stryer.section.2375#2376

TCA cycle:
   http://www.gwu.edu/~mpb/citric.htm
   http://www.med.unibs.it/~marchesi/tca2.html

Horseshoe TCA cycle:
   http://biocyc.org/META/NEW-IMAGE?type=PATHWAY&object=REDCITCYC

Reductive TCA cycle and other carbon assimilation pathways:
   http://lecturer.ukdw.ac.id/dhira/Metabolism/CarbonAssim.html

Isocitrate dehydrogenase:
   http://www.chem.uwec.edu/Webpapers2005/mintermm/pages/IDH.html

Glyoxylate cycle:
   http://138.192.68.68/bio/Courses/biochem2/TCA/GlyoxylateCycle.html

Chapter 3

Iron sulfur centers, quinones and electron transport chain:
   http://www.rpi.edu/dept/bcbp/molbiochem/MBWeb/mb1/part2/redox.htm

Electron transport chain:
   http://en.wikipedia.org/wiki/Electron_transport_chain

NADH dehydrogenase/Complex I:
   http://en.wikipedia.org/wiki/NADH_dehydrogenase
   http://www.scripps.edu/mem/biochem/CI/research.html

Cytochrome bc1:
   http://en.wikipedia.org/wiki/Cytochrome_bc1_complex
   http://sb20.lbl.gov/cytbc1/images/bluedimer.jpg

Q Cycle:
   http://metallo.scripps.edu/PROMISE/CYTBC1.html#Q_cycle

Cytochrome b6f:
    http://www.biology.purdue.edu/people/faculty/cramer/Cramer/html/cytbf.html

Cytochrome c oxidase:
    http://www.aecom.yu.edu/home/biophysics/rousseau/cco/cytcox.htm
    http://www.life.uiuc.edu/crofts/bioph354/cyt_ox.html

Bacteriorhodopsin:
    http://nai.arc.nasa.gov/news_stories/news_detail.cfm?article=old/
    powerhouses.htm

ATP synthase:
    http://en.wikipedia.org/wiki/ATP_synthase
    http://www.cnr.berkeley.edu/~hongwang/Project/ATP_synthase/
    http://www.mrc-dunn.cam.ac.uk/research/atpase.html
    http://www.biologie.uni-osnabrueck.de/biophysik/Junge/pics.html

Chapter 4

RNA as a catalyst:
    http://www.mpibpc.gwdg.de/inform/MpiNews/cientif/jahrg5/9.99/scta.html

ribosome:
    http://www.molgen.mpg.de/~ag_ribo/ag_fucini/fucini-history.html

tRNA:
    http://en.wikipedia.org/wiki/Transfer_RNA

ribozymes, riboswitches:
    http://www.yale.edu/breaker/science.htm

viroids:
    http://www-micro.msb.le.ac.uk/3035/Viroids.html

Aquifex:
    http://microbewiki.kenyon.edu/index.php/Aquifex

Thermotoga:
    http://microbewiki.kenyon.edu/index.php/Petrotoga

Methanogenesis:
    http://umbbd.msi.umn.edu/meth/meth_map.html
    http://www.agen.ufl.edu/~chyn/age4660/lect/lect_08x/lect_08.htm

Methanogens:
    http://microbewiki.kenyon.edu/index.php/Methanogens

reverse methanogenesis:
    http://www.mumm-research.de/indexx.php?p=85

Beggiatoa:
   http://microbewiki.kenyon.edu/index.php/Beggiatoa

Thiomargerita:
   http://microbewiki.kenyon.edu/index.php/Thiomargarita

Thiobacillus:
   http://www.spaceship-earth.org/REM/THIOBAC.htm
   http://microbewiki.kenyon.edu/index.php/Thiobacillus

Chlorobium:
   http://microbewiki.kenyon.edu/index.php/Chlorobium

Desulfuromonas:
   http://microbewiki.kenyon.edu/index.php/Desulfuromonas

hydrothermal vents:
   http://www.onr.navy.mil/Focus/ocean/habitats/vents1.htm

Chloroflexus:
   http://microbewiki.kenyon.edu/index.php/Chloroflexus

Cyanobactria:
   http://www.ucmp.berkeley.edu/bacteria/cyanointro.html

Rhodopseudomonas:
   http://microbewiki.kenyon.edu/index.php/Rhodopseudomonas
   http://metallo.scripps.edu/PROMISE/PRCPB.html#PRC_electron_transfer
   http://www.uphs.upenn.edu/biocbiop/local_pages/dutton_lab/etprc.html
   http://nobelprize.org/nobel_prizes/chemistry/laureates/1988/illpres/index.html
   http://gepard.bioinformatik.uni-saarland.de/people/hutter/rcenter.html

Rhodobacter:
   http://microbewiki.kenyon.edu/index.php/Rhodobacter

Photosynthesis:
   http://instruct1.cit.cornell.edu/courses/biomi290/MOVIES/OXYPS.HTML
   http://links.baruch.sc.edu/scael/personals/pjpb/lecture/lecture.html
   http://www.biologie.uni-hamburg.de/b-online/e24/24.htm

water splitting center:
   http://www.reactivereports.com/37/37_1_intr.html

origin of photosynthesis:
   http://www.bio.indiana.edu/~bauerlab/origin.html

Trichodesmium:
   http://microbewiki.kenyon.edu/index.php/Trichodesmium
   http://www.whoi.edu/science/B/people/ewebb/Tricho.html

Calvin cycle:
   http://www.rpi.edu/dept/bcbp/molbiochem/MBWeb/mb2/part1/dark.htm

Rubisco:
   http://www.biologie.uni-hamburg.de/b-online/fo24_1/e1rxoe.htm

Rubisco activase:
   http://4e.plantphys.net/article.php?ch=8&id=81

Carboxysome:
   http://www.jensenlab.caltech.edu/Projects/carboxysome.html

Nitrogenase-cofactors:
   http://metallo.scripps.edu/promise/NITROGENASE_REV.html

Nitrogenase:
   http://www.geocities.com/Vienna/Strasse/6671/n2/nitro.html#nitrogenase
   http://www.microbiologytext.com/index.php?module=Book&func=
   displayfigure&book_id=4&fig_number=17&chap_number=2

Nitrification:
   http://www.abdn.ac.uk/~mbi010/nitrification.htm

Anammox bacteria:
   http://www.mpi-bremen.de/en/Anammox_Bacteria_produce_Nitrogen_Gas_in_
   Oceans_Snackbar.html

Rhizobium:
   http://microbewiki.kenyon.edu/index.php/Bradyrhizobium
   http://www.biologie.uni-hamburg.de/b-online/e34/34b.htm

Nitrogen fixation:
   http://helios.bto.ed.ac.uk/bto/microbes/nitrogen.htm#Top

Stromatolite & microbial mat:
   http://microbes.arc.nasa.gov/movie/small-qt.html (popular)

Winogradsky column:
   http://helios.bto.ed.ac.uk/bto/microbes/winograd.htm

Prochlorococcus:
   http://microbewiki.kenyon.edu/index.php/Prochlorococcus

Synechococcus:
   http://microbewiki.kenyon.edu/index.php/Synechococcus
   http://en.wikipedia.org/wiki/Synechococcus

heterocysts & Anabena:
http://microbewiki.kenyon.edu/index.php/Anabaena

Ocean iron fertilization:
http://www.mbari.org/expeditions/SOFeX2002/

marine snow:
http://www.whoi.edu/oceanus/viewArticle.do?id=2387&archives=true

Roseobacter, Silicibacter:
http://microbewiki.kenyon.edu/index.php/Roseobacter
http://genome.jgi-psf.org/draft_microbes/siltm/siltm.home.html

Bdellovibrio:
http://www.mpg.de/english/illustrationsDocumentation/documentation/pressReleases/
2004/pressRelease20040123/index.html

Myxococcus:
http://microbewiki.kenyon.edu/index.php/Myxococcus

Bacillus & spores:
http://textbookofbacteriology.net/Bacillus.html

Giardia:
http://microbewiki.kenyon.edu/index.php/Giardia

Grypania:
http://www.newarkcampus.org/professional/osu/faculty/jstjohn/Cool%20Fossils/
Grypania%20from%20Negaunee%20Fe-Fm.htm

Microsporidia:
http://microbewiki.kenyon.edu/index.php/Microsporidia

Cyanidioschyzon:
http://merolae.biol.s.u-tokyo.ac.jp/gallery/

Dinoflagellates:
http://www.geo.ucalgary.ca/~macrae/palynology/dinoflagellates/dinoflagellates.html

choanoflagellates:
http://www.ucmp.berkeley.edu/protista/choanos.html

Geological time periods:
http://www.fossilmuseum.net/PaleobiologyVFM.htm

Ediacaran fauna:
http://www.ucmp.berkeley.edu/vendian/critters.html
http://www.peripatus.gen.nz/paleontology/Ediacara.html

Burgess Shale fauna:

  http://www.nmnh.si.edu/paleo/shale/pfoslidx.htm
  http://www.ucmp.berkeley.edu/cambrian/burgess.html
  http://gpc.edu/~pgore/geology/geo102/burgess/burgess.htm

Agnatha:

  http://en.wikipedia.org/wiki/Agnatha
  http://en.wikipedia.org/wiki/Myxini
  http://en.wikipedia.org/wiki/Petromyzontidae

anatomy fishes:

  http://www.auburn.edu/academic/classes/zy/0301/Topic3b/Topic3b.html#Vertebrates
  http://www.ucmp.berkeley.edu/vertebrates/basalfish/placodermi.html
  http://www.palaeos.com/Vertebrates/Units/050Thelodonti/050.100.html

early tetrapodes:
Eusthenopteron:

  http://www.palaeos.com/Vertebrates/Units/140Sarcopterygii/140.860.html
  http://www.palaeos.com/Vertebrates/Units/140Sarcopterygii/
  140.920.html#Panderichthys
  http://www.palaeos.com/Vertebrates/Units/150Tetrapoda/150.150.html#Acanthostega

Ichthyostega:

  http://www.palaeos.com/Vertebrates/Units/150Tetrapoda/150.200.html
  http://en.wikipedia.org/wiki/Tiktaalik Tiktaalik:
  http://research.uchicago.edu/highlights/resources/media/shubin/shubin_package_
  VO.mov?prog=true

Osteolepis:

  http://www.palaeos.com/Vertebrates/Units/140Sarcopterygii/140.830.html

Pederpes:

  http://www.evolutionpages.com/pederpes%20finneyae.htm

Oviraptor:

  http://www.ucmp.berkeley.edu/diapsids/saurischia/oviraptoridae.html

Early mammals:

  http://www.palaeos.com/Vertebrates/Units/Unit420/420.300.html#Hadrocodium
  http://www.palaeos.com/Vertebrates/Units/430Mammalia/430.400.html#Jeholodens
  http://en.wikipedia.org/wiki/Eomaia
  http://www.sinofossa.org/mammal/eomaia.htm
  http://www.mun.ca/biology/scarr/Eomaia_fossil_find.htm
  http://www.carnegiemnh.org/research/sinodelphys/index.htm
  http://www.sinofossa.org/mammal/sinodelphys.htm

Zhangheotherium:
http://www.carnegiemuseums.org/cmag/bk_issue/1998/marapr/feat3.htm
http://www.sinofossa.org/mammal/repenomamus.htm

Castorocauda:
http://www.crystalinks.com/fossilbeavers.html

Fruitafossor:
http://www.carnegiemnh.org/news/05-jan-mar/fossil/

Chapter 5

Emiliania:
http://www.soes.soton.ac.uk/staff/tt/

DMS & climate:
http://saga.pmel.noaa.gov/review/dms_climate.html

Calanus finmarchicus:
http://www.nioz.nl/nioz_nl/a629649f811ee5378d18b46d3ec1367d.php

Basking shark:
http://www.flmnh.ufl.edu/fish/Gallery/Descript/baskingshark/baskingshark.html

Anchovy:
http://en.wikipedia.org/wiki/Anchovy

Sardine:
http://animaldiversity.ummz.umich.edu/site/accounts/pictures/Clupeidae.html

Tuna:
http://www.flmnh.ufl.edu/fish/Gallery/Descript/BluefinTuna/BluefinTuna.html

Mormon cricket:
http://buzz.ifas.ufl.edu/269a.htm

Poison frogs:
http://www.poisonfrogs.nl/e020368.html

Lichens:
http://helios.bto.ed.ac.uk/bto/microbes/lichen.htm#Top
http://helios.bto.ed.ac.uk/bto/microbes/cyano.htm#Top

lignin:
http://en.wikipedia.org/wiki/Lignin

wood decay fungi:
http://helios.bto.ed.ac.uk/bto/microbes/armill.htm

cellulase:

> http://en.wikipedia.org/wiki/Cellulase
> http://www.fao.org/docrep/w7241e/w7241e08.htm
> http://www.wzw.tum.de/mbiotec/cellpage.htm

cellulose:

> http://www.lsbu.ac.uk/water/hycel.html

Chapter 6

Condylura:

> http://animaldiversity.ummz.umich.edu/site/accounts/information/
> Condylura_cristata.html
> http://www.naturalhistorymag.com/master.html?
> http://www.naturalhistorymag.com/features/0600_feature2.html

Impatiens:

> http://www.arkive.org/species/ARK/plants_and_algae/Impatiens_glandulifera/

Corn borer:

> http://creatures.ifas.ufl.edu/field/e_corn_borer.htm

Cryptomys:

> http://www.fun-morph.ugent.be/Projects/Cryptomys/Cryptomys.htm

Bee hive & dance:

> http://www.pbs.org/wgbh/nova/bees/

Fungus growing ants:

> http://www.zi.ku.dk/personal/drnash/atta/Pages/Leafcut.html

crow making tools:

> http://news.nationalgeographic.com/news/2002/08/0808_020808_crow.html

Chimpanzee culture:

> http://news.bbc.co.uk/2/hi/science/nature/4166756.stm

Cultural phylogeny:

> http://www.eva.mpg.de/phylogen/ctp.html

human evolution:

> http://www.mnh.si.edu/anthro/humanorigins/ha/a_tree.html
> http://news.nationalgeographic.com/news/2004/10/1027_041027_homo_
> floresiensis.html

Reinsdorf spear:

> http://www.morieninstitute.org/anomalies.html

Homo heidelbergensis as hunter:
http://www.bbc.co.uk/sn/prehistoric_life/human/human_evolution/
first_europeans1.shtml

Neanderthals and cannibalism:
http://cas.bellarmine.edu/tietjen/Human%20Nature%20S%201999/
neanderthal_cannibalism_at_moula.htm
http://news.bbc.co.uk/1/hi/sci/tech/462048.stm

First fire use:
http://news.bbc.co.uk/2/hi/science/nature/3670017.stm

stone tools:
http://www.amonline.net.au/human_evolution/tools/index.htm

Mousterian:
http://anthro.palomar.edu/homo2/mod_homo_3.htm

Dental microwear:
http://comp.uark.edu/~pungar/images.htm

Australian megafauna:
http://www.museum.vic.gov.au/prehistoric/mammals/australia.html

Noodles in Neolithic China:
http://news.bbc.co.uk/2/hi/science/nature/4335160.stm

Chapter 7

Zoonosis:
http://www.cdc.gov/flu/avian/virus.htm
http://www.who.int/csr/disease/avian_influenza/en/
http://www.who.int/topics/hiv_infections/en/

Avian influenza:
http://www.fao.org/AG/AGAInfo/subjects/en/health/diseasescards/avian.html
http://www.fao.org/AG/AGAInfo/subjects/en/health/diseasescards/avian_videos.html

BSE:
http://www.fao.org/AG/AGAInfo/subjects/en/health/bse/bse_path.html

Prion disease:
http://www.cdc.gov/ncidod/dvrd/prions/

Diverse veterinary diseases:
http://www.fao.org/AG/AGAInfo/subjects/en/health/diseasescards/default.html

Anopheles:
http://www.cdc.gov/malaria/biology/mosquito/index.htm

Malaria:
http://www.cdc.gov/malaria/index.htm

Bacteroides:
http://www.sanger.ac.uk/Projects/B_fragilis/

Cholera:
http://textbookofbacteriology.net/cholera.html
http://gsbs.utmb.edu/microbook/ch024.htm

rotavirus:
http://www.cdc.gov/ncidod/EID/vol4no4/parashar.htm

animal virus pictures:
http://web.uct.ac.za/depts/mmi/stannard/emimages.html

retrovirus:
http://www.accessexcellence.org/RC/VL/GG/ecb/retrovirus_life_cycle.html
http://www.tulane.edu/~dmsander/WWW/335/Retroviruses.html
http://www-micro.msb.le.ac.uk/3035/Retroviruses.html

Chapter 8

Malthus:
http://www.gutenberg.org/dirs/etext03/prppl10.txt
http://en.wikipedia.org/wiki/Thomas_Malthus

organic farming:
http://news.bbc.co.uk/2/hi/science/nature/2017094.stm
http://www.fibl.net/english/research/soilsciences/dok/index.php

Cedar creek plot:
http://www.cedarcreek.umn.edu/

Rice blast fungus:
http://news.bbc.co.uk/2/hi/science/nature/4466783.stm

Index